D0845532

From Finite Sample to Asymptotic Methods in Statistics

Exact statistical inference may be employed in diverse fields of science and technology. As problems become more complex and sample sizes become larger, mathematical and computational difficulties can arise that require the use of approximate statistical methods. Such methods are justified by asymptotic arguments but are still based on the concepts and principles that underlie exact statistical inference. With this in perspective, this book presents a broad view of exact statistical inference and the development of asymptotic statistical inference, providing a justification for the use of asymptotic methods for large samples. Methodological results are developed on a concrete and yet rigorous mathematical level and are applied to a variety of problems that include categorical data, regression, and survival analyses.

This book is designed as a textbook for advanced undergraduate or beginning graduate students in statistics, biostatistics, or applied statistics but may also be used as a reference for academic researchers.

Pranab K. Sen is the Cary C. Boshamer Professor of Biostatistics and Professor of Statistics and Operations Research at the University of North Carolina at Chapel Hill. He is the author or coauthor of numerous textbooks in statistics and biostatistics and editor or coeditor of numerous volumes in the same field. He has more than 600 publications in leading statistics journals and has supervised 83 doctoral students. Sen is a Fellow of both the Institute of Mathematical Statistics and the American Statistical Association. In 2002, he was the Senior Noether Awardee for his lifelong contributions to nonparametrics and received the Commemoration Medal from the Czech Union of Physicists and Mathematicians in 1998.

Julio M. Singer is a Professor in the Department of Statistics, University of São Paulo, Brazil, and is the codirector of the university's Center for Applied Statistics. Professor Singer is the coauthor of books on categorical data and large sample theory and has publications in both methodological and applications-oriented journals. He was the 1993 James E. Grizzle Distinguished Alumnus in Biostatistics from the University of North Carolina at Chapel Hill. He supervised several graduate students and contributed to the development of the doctoral program in statistics at the University of São Paulo, Brazil.

Antonio C. Pedroso de Lima is an Associate Professor at the Department of Statistics, University of São Paulo, Brazil, and is the codirector of the university's Center for Applied Statistics. He received his doctoral degree in biostatistics from the University of North Carolina at Chapel Hill. He is the coauthor of a book on introductory statistics, and his research has been published in theoretical, methodological, and applications-oriented journals. Professor Pedroso de Lima has advised a number of master's degree and doctoral students in the graduate program in statistics at the University of São Paulo.

CAMBRIDGE SERIES IN STATISTICAL AND PROBABILISTIC MATHEMATICS

This series of high-quality upper-division textbooks and expository monographs covers all aspects of stochastic applicable mathematics. The topics range from pure and applied statistics to probability theory, operations research, optimization, and mathematical programming. The books contain clear presentations of new developments in the field and also of the state of the art in classical methods. While emphasizing rigorous treatment of theoretical methods, the books also contain applications and discussions of new techniques made possible by advances in computational practice.

Already Published

1. *Bootstrap Methods and Their Application*, by A. C. Davison and D. V. Hinkley
2. *Markov Chains*, by J. Norris
3. *Asymptotic Statistics*, by A. W. van der Vaart
4. *Wavelet Methods for Time Series Analysis*, by Donald B. Percival and Andrew T. Walden
5. *Bayesian Methods*, by Thomas Leonard and John S. J. Hsu
6. *Empirical Processes in M-Estimation*, by Sara van de Geer
7. *Numerical Methods of Statistics*, by John F. Monahan
8. *A User's Guide to Measure Theoretic Probability*, by David Pollard
9. *The Estimation and Tracking of Frequency*, by B. G. Quinn and E. J. Hannan
10. *Data Analysis and Graphics Using R*, by John Maindonald and John Braun
11. *Statistical Models*, by A. C. Davison
12. *Semiparametric Regression*, by D. Ruppert, M. P. Wand, and R. J. Carroll
13. *Exercise in Probability*, by Loic Chaumont and Marc Yor
14. *Statistical Analysis of Stochastic Processes in Time*, by J. K. Lindsey
15. *Measure Theory and Filtering*, by Lakhdar Aggoun and Robert Elliott
16. *Essentials of Statistical Inference*, by G. A. Young and R. L. Smith
17. *Elements of Distribution Theory*, by Thomas A. Severini
18. *Statistical Mechanics of Disordered Systems*, by Anton Bovier
19. *The Coordinate-Free Approach to Linear Models*, by Michael J. Wichura
20. *Random Graph Dynamics*, by Rick Durrett
21. *Networks*, by Peter Whittle
22. *Saddlepoint Approximations with Applications*, by Ronald W. Butler
23. *Applied Asymptotics*, by A. R. Brazzale, A. C. Davison, and N. Reid
24. *Random Networks for Communication*, by Massimo Franceschetti and Ronald Meester
25. *Design of Comparative Experiments*, by R. A. Bailey
26. *Symmetry Studies*, by Marlos A. G. Viana
27. *Model Selection and Model Averaging*, by Gerda Claeskens and Nils Lid Hjort

From Finite Sample to Asymptotic Methods in Statistics

PRANAB K. SEN
University of North Carolina

JULIO M. SINGER
Universidade de São Paulo

ANTONIO C. PEDROSO DE LIMA
Universidade de São Paulo

CAMBRIDGE UNIVERSITY PRESS
Cambridge, New York, Melbourne, Madrid, Cape Town, Singapore,
São Paulo, Delhi, Dubai, Tokyo

Cambridge University Press
32 Avenue of the Americas, New York, NY 10013-2473, USA

www.cambridge.org
Information on this title: www.cambridge.org/9780521877220

First published 2010

Printed in the United States of America

A catalog record for this publication is available from the British Library.

Library of Congress Cataloging in Publication Data

Sen, Pranab Kumar, 1937–
From finite sample to asymptotic methods in statistics / Pranab K. Sen, Julio M. Singer, Antonio C. Pedroso de Lima.
p. cm. – (Cambridge series in statistical and probabilistic mathematics)
Includes bibliographical references and index.
ISBN 978-0-521-87722-0 (hardback)
1. Mathematical statistics. 2. Probabilities. 3. Estimation theory. 4. Asymptotic expansions.
I. Singer, Julio da Motta, 1950– II. Lima, Antonio C. Pedroso de, 1961– III. Title. IV. Series.
QA276.S358 2009
519.5–dc22 2009034794

ISBN 978-0-521-87722-0 Hardback

To our beloved parents,

Nagendra and Kalyani
José and Edith
Manoel and Neusa

Contents

Preface

Students and investigators working in statistics, biostatistics, or applied statistics, in general, are constantly exposed to problems that involve large quantities of data. This is even more evident today, when massive datasets with an impressive amount of details are produced in novel fields such as genomics or bioinformatics at large. Because, in such a context, exact statistical inference may be computationally out of reach and in many cases not even mathematically tractable, they have to rely on approximate results. Traditionally, the justification for these approximations was based on the convergence of the first four moments of the distributions of the statistics under investigation to those of some normal distribution. Today we know that such an approach is not always theoretically adequate and that a somewhat more sophisticated set of techniques based on asymptotic considerations may provide the appropriate justification. This need for more profound mathematical theory in statistical large-sample theory is patent in areas involving dependent sequences of observations, such as longitudinal and survival data or life tables, in which the use of martingale or related structures has distinct advantages.

Unfortunately, most of the technical background for understanding such methods is dealt with in specific articles or textbooks written for a readership with such a high level of mathematical knowledge that they exclude a great portion of the potential users. We tried to bridge this gap in a previous text (Sen and Singer [1993]: *Large Sample Methods in Statistics: An Introduction with Applications*), on which our new enterprise is based. While teaching courses based on that text in the Department of Biostatistics at the University of North Carolina at Chapel Hill and in the Department of Statistics at the University of São Paulo, during the past few years, we came across a new hiatus originated in the apparent distinction between the theory covered in the exact statistical inference and the approximate approach adopted in the book. While realizing that the foundations of the large-sample statistical methods we proposed to address stem from the basic concepts and principles that underlie finite sample setups, we decided to integrate both approaches in the present text. In summary, our intent is to provide a broad view of finite-sample statistical methods, to examine their merits and caveats, and to judge how far asymptotic results eliminate some of the detected impasses, providing the basis for sound application of approximate statistical inference in large samples.

Chapter 1 describes the type of problems considered in the text along with a brief summary of some basic mathematical and statistical concepts required for a good understanding of the remaining chapters. Chapters 2 and 3 lay out the two basic building blocks of statistical inference, namely, estimation and hypothesis testing. There we address issues relating to the important concepts of likelihood, sufficiency, and invariance, among others.

In Chapter 4 we present a brief review of Decision Theory and Bayesian methods, contrasting them with those outlined in the previous chapters. Chapter 5 is devoted to stochastic processes, given their importance in the development of models for dependent random variables as well as their significance as paradigms for many problems arising in practical applications. Chapters 6 and 7 contain the essential tools needed to prove asymptotic results for independent sequences of random variables as well as an outline of the possible extensions to cover the dependent sequence case. Chapter 8 discusses some general results on the asymptotics of estimators and test statistics; their actual application to categorical data and regression analysis is illustrated in Chapters 9 and 10, respectively. Finally, Chapter 11 deals with an introductory exposition of the technical background required to deal with the asymptotic theory for statistical functionals. The objective here is to provide some motivation and the general flavor of the problems in this area, because a rigorous treatment would require a much higher level of mathematical background than the one we contemplate. The 11 chapters were initially conceived for a two-semester course for second-year students in biostatistics or applied statistics doctoral programs as well as for last-year undergraduate or first-year graduate students in statistics. A more realistic view, however, would restrict the material for such purposes to the first eight chapters, along with a glimpse into Chapter 11. Chapters 9 and 10 could be included as supplementary material in categorical data and linear models courses, respectively. Because the text includes a number of practical examples, it may be useful as a reference text for investigators in many areas requiring the use of statistics.

The authors would like to thank the numerous students who took large-sample theory courses at the Department of Biostatistics, University of North Carolina at Chapel Hill, and Department of Statistics, University of São Paulo, providing important contributions to the design of this text. Finally, we must acknowledge the Cary C. Boshamer Foundation, University of North Carolina at Chapel Hill, as well as Conselho Nacional de Desenvolvimento Científico e Tecnológico, Brazil, and Fundação de Amparo à Pesquisa do Estado de São Paulo, Brazil, for providing financial support during the years of preparation of the text.

Pranab K. Sen
Julio M. Singer
Antonio C. Pedroso de Lima
Chapel Hill and São Paulo, April 2009

1

Motivation and Basic Tools

1.1 Introduction

Statistics is a body of mathematically based methodology designed to organize, describe, model, and analyze data. In this context, *statistical inference* relates to the process of drawing conclusions about the unknown frequency distribution (or some summary measure therefrom) of some characteristic of a population based on a known subset thereof (the *sample data*, or, for short, the *sample*). Drawing statistical conclusions involves the choice of suitable models that allow for random errors, and this, in turn, calls for convenient probability laws. It also involves the ascertainment of how appropriate a postulated probability model is for the genesis of a given dataset, and of how adequate the sample size is to maintain incorrect conclusions within acceptable limits.

Finite statistical inference tools, in use for the last decades, are appealing because, in general, they provide "exact" statistical results. As such, finite methodology has experienced continuous upgrading with annexation of novel concepts and approaches. Bayesian methods are especially noteworthy in this respect. Nevertheless, it has been thoroughly assessed that the scope of exact statistical inference in an optimal or, at least, desirable way, is rather confined to some special classes of probability laws (such as the exponential family of densities). In real-life applications, such optimal statistical inference tools often stumble into impasses, ranging from validity to efficacy and thus, have practical drawbacks. This is particularly the case with large datasets, which are encountered in diverse (and often interdisciplinary) studies, more so now than in the past. Such studies, which include biostatistical, environmetrical, socioeconometrical, and more notably bioinformatical (genomic) problems, in general, cater to statistical reasoning beyond that required by conventional finite-sample laboratory studies. Although this methodology may yet have an appeal in broader interdisciplinary fields, some extensions are needed to capture the underlying stochastics in large datasets and thereby draw statistical conclusions in a valid and efficient manner. The genesis of asymptotic methods lies in this infrastructure of finite-sample statistical inference. Therefore it is convenient to organize our presentation of asymptotic statistical methods with due emphasis on this finite- to large-sample bridge.

To follow this logically integrated approach to statistical inference, it is essential to encompass basic tools in probability theory and stochastic processes. In addition, the scope of applications to a broad domain of biomedical, clinical, and public health disciplines may dictate additional features of the underlying statistical methodology. We keep this dual objective of developing methodology with an application-oriented spirit in mind and thereby provide an overview of finite-sample (exact or small) statistical methods, appraising their scope and integration to asymptotic (approximate or large-sample) inference.

In Section 1.2, we motivate our approach through a set of illustrative examples that range from very simple to more complex application problems. We propose some simple statistical models for such problems, describe their finite-sample analysis perspectives, and, with this in mind, stress the need for asymptotic methods. In Section 1.3, we go further along this line, specifying more realistic models for the problems described previously as well as for those arising from additional practical examples and fortifying the transit from the limited scope of finite sample inference to omnibus asymptotic methods. In Section 1.4, we present a brief description of the basic coverage of the book. We conclude with a summary of some basic tools needed throughout the text.

1.2 Illustrative Examples and Motivation

In general, the strategy employed in statistical inference involves (i) the selection of an appropriate family of *stochastic models* to describe the characteristics under investigation, (ii) an evaluation of the compatibility of chosen model(s) with the data (*goodness of fit*), and (iii) subsequent *estimation* of, or *tests of hypotheses* about, the underlying *parameters*. In this process, we are faced with models that may have different degrees of complexity and, depending on regularity assumptions, different degrees of restrictiveness. They dictate the complexity of the statistical tools required for inference which, in many instances, call for asymptotic perspectives because conventional finite-sample methods may be inappropriate. Let us first appraise this scenario of statistical models through some illustrative examples.

Example* 1.2.1 *(Inspection sampling). Suppose that we are interested in selecting a random sample to estimate the proportion of defective items manufactured in a textile plant. The data consist of the number x of defective items in a random sample of size n. For a random variable X representing the number of defective items in the sample, four possible stochastic models follow.

(a) We assume that the items are packed in lots of N (known) units of which D (unknown) are defective, so that our interest lies in estimating $\pi = D/N$. The probability of obtaining x defective items in the sample of size n may be computed from the probability function of the hypergeometric distribution

$$P(X = x) = \frac{\binom{D}{x}\binom{N-D}{n-x}}{\binom{N}{n}}, \quad x = \max[0, n - (N - D)], \ldots, \min(n, D).$$

(b) If, on the other hand, we assume that the items are manufactured continuously and that the proportion of defective items is $\pi, 0 < \pi < 1$, the probability of obtaining x defective items in the random sample of size n may be computed from the probability function of the binomial distribution,

$$P(X = x) = \binom{n}{x}\pi^x(1 - \pi)^{n-x}, \quad x = 0, \ldots, n.$$

(c) Alternatively, if under the same assumptions as in item (b), we decide to sample until r defective items are selected, the probability of sampling $x + r$ items may be obtained from the probability function of the negative binomial distribution, that is,

$$P(X = x) = \binom{r + x - 1}{x} \pi^r (1 - \pi)^x, \quad x = 0, 1, 2, \ldots.$$

(d) In some cases, it is assumed that X has a Poisson distribution, namely,

$$P(X = x) = \frac{e^{-\lambda}\lambda^r}{r!}, \quad r \geq 0, \ \lambda > 0,$$

where $\lambda > 0$. □

Example* 1.2.2 *(One sample location model). We consider the problem of estimating the average height μ of a (conceptually infinite) population based on a random sample of n individuals. Let Y denote the height of a randomly selected individual and F denote the underlying distribution function. In such a context, three alternative (stochastic) models include the following:

(a) F is assumed to be symmetric and continuous with mean μ and finite variance σ^2, and the observations are the heights $Y_1, \ldots, Y_n$ of the n individuals in the sample.

(b) F is assumed to be normal with mean μ and known variance σ^2, and the observations are like those in item (b).

(c) The assumptions on F are as in either (a) or (b), but the observations correspond to the numbers of individuals falling within each of m height intervals (grouped data). This is, perhaps, a more realistic model, because, in practice, we are only capable of coding the height measurements to a certain degree of accuracy (e.g., to the nearest millimeter). In this case, using a multinomial distribution with ordered categories would be appropriate. □

Example* 1.2.3 *(Paired two-sample location model). Let $X(Y)$ denote the blood pressure of a randomly selected hypertense individual before (after) the administration of an antihypertensive drug and μ_X (μ_Y) represent the corresponding (population) mean. Our objective is to estimate the average reduction in blood pressure, $\Delta = \mu_X - \mu_Y$, based on a sample of n hypertense individuals randomly selected from a conceptually infinite or very large population for which

(i) both X and Y are observed for all n subjects in the sample;

(ii) there are missing observations, that is, X or Y or both are not recorded for all subjects. A common stochastic model for this problem assumes that the vector $(\boldsymbol{X}, \boldsymbol{Y})$ follows a bivariate normal distribution with $\mathbb{V}\text{ar}(X) = \sigma_X^2$, $\mathbb{V}\text{ar}(Y) = \sigma_Y^2$, and $\mathbb{C}\text{ov}(X, Y) = \sigma_{XY}$. □

Example* 1.2.4 *(Multisample location model). A manufacturing company produces electric lamps wherein the cross section of the coil (say, X) may be one of s (≥ 2) possible choices, designated $x_1, \ldots, x_s$, respectively. Our objective is to verify whether the level x_i has some influence on the expected life of the lamps. For each level x_i, we consider a set of n_i lamps taken at random from a (very large) production lot, and let Y_{ij} denote the life length (in hours, say) corresponding to the jth lamp in the ith lot ($i = 1, \ldots, s$; $j = 1, \ldots, n_i$). It may be assumed that the Y_{ij} are independent for different i $(= 1, \ldots, s)$ and j $(= 1, \ldots, n_i)$ and follow distributions characterized by distribution functions F_i, defined on $\mathbb{R}^+ = [0, \infty)$, $i = 1, \ldots, s$. As in Example 1.2.2, we may consider a variety of models for the F_i, among which we pose the following:

(a) F_i is normal with mean μ_i and variance σ_i^2, $i = 1, \ldots, s$, where the σ_i^2 may or may not be the same.

(b) F_i is continuous (and symmetric) with median ν_i and scale parameter τ_i, $i = 1, \ldots, s$, but its form is not specified.

(c) Although F_i may satisfy (a) or (b), because the Y_{ij} are recorded in class intervals (of width one hour, say), we must replace it by some appropriate (ordered) categorical data distribution, such as the multinomial distribution postulated in Example 1.2.2.(c).

Under model (a), if we assume further that $\sigma_1^2 = \cdots = \sigma_s^2 = \sigma^2$, we have the *classical (normal theory) multisample shift in location model* (or simply the *location model*); in this context, if $n_i = n$, $i = 1, \ldots, s$, the data are said to be balanced; in (b), if we let $F_i(y) = F(y - \nu_i)$, $i = 1, \ldots, s$, we have the so-called *nonparametric multisample location model.* Either (a) or (b) may be made more complex when we drop the assumption of homogeneity of the variances σ_i^2 or of the scale parameters τ_i or allow for possible scale perturbations in the location model, that is, if we let $F_i(y) = F[(y - \mu_i)/\sigma_i]$ or $F_i(y) = F[(y - \nu_i)/\tau_i]$, $i = 1, \ldots, s$, where the σ_i^2 (or the τ_i) are not necessarily the same. □

Example* 1.2.5 *(Simple linear regression model). We consider a study designed to evaluate the association between the level of fat ingestion (X) on weight (Y) of children in a certain age group. The data consist of the pairs of measurements of X and Y for n randomly selected subjects (from a large conceptual population) in the appropriate age group, namely, (X_i, Y_i), $i = 1, \ldots, n$. A possible relation between X and Y may be expressed by the simple linear regression model

$$Y_i = \alpha + \beta X_i + e_i, \quad i = 1, \ldots, n,$$

where α and β are unknown parameters and the e_i are random errors with distribution function F. Under this model, X is, in general, assumed fixed and known without error. Again, many assumptions about the stochastic features of the model may be considered.

(a) The error terms e_i are assumed to be independent and to have mean 0 and constant variance σ^2, but the form of F is not specified. This is the *Gauss–Markov setup.*

(b) Additionally to the assumptions in (a), F is required to be normal.

(c) The assumptions in (a) are relaxed to allow for heteroskedasticity, that is, the variances may not be the same for all observations.

(d) Under (a), (b), or (c), X is assumed to be measured with error, that is, $X = W + u$, where W is a fixed (unknown) constant and u is a random term with mean 0 and variance τ^2. This is the *error in variables simple linear regression model.*

(e) In some cases, W in (d) is also a random variable with mean ν and variance σ_W^2. In that case, the regression of Y on W has slope $\gamma = \beta\sigma_W^2/(\sigma_W^2 + \tau^2)$. This is referred to as a *measurement-error model* [Fuller (1986)]. □

Example* 1.2.6 *(Repeated measures). We suppose that in a study similar to that described in Example 1.2.5, the weight of each child (Y) is observed under p fixed, randomly assigned levels (X) of physical activity (e.g., light, medium, or intense). The data consist of $p > 1$ pairs of measurements of X and Y for each of n randomly selected subjects in the appropriate age group, namely, (X_{ij}, Y_{ij}), $i = 1, \ldots, n$, $j = 1, \ldots, p$. Because the same response (Y) is observed two or more times on the same subject, this type of data is generally referred to as *repeated measures*. Under this setup, a commonly used model to express the relationship between Y and X is

$$Y_{ij} = \mu + \alpha_j + e_{ij}, \quad i = 1, \ldots, n, \; j = 1, \ldots, p,$$

where μ and α_j are fixed constants and e_{ij} are random errors with distribution function F. Some possible assumptions for the stochastic components of the model follow:

(a) The error terms e_{ij} are normally distributed with mean 0 and covariance structure given by $\mathbb{V}\mathrm{ar}(e_{ij}) = \sigma^2$, $\mathbb{C}\mathrm{ov}(e_{ij}, e_{ij'}) = \sigma_a^2$, $j \neq j'$, and $\mathbb{C}\mathrm{ov}(e_{i'j}, e_{ij'}) = 0$, $i \neq i'$.
(b) The assumptions are similar to those in (a), but heteroskedasticity is allowed, that is, $\mathbb{V}\mathrm{ar}(e_{ij}) = \sigma_i^2$.
(c) The assumptions are similar to those in (a), but F is allowed to be a member of the exponential family.
(d) The assumptions are as in (a), (b), or (c) but some observations are missing. □

Example* 1.2.7 *(Longitudinal data). In the study described in Example 1.2.5, we suppose that measurements of fat ingestion (X) and weight (Y) are taken on each child at different instants in a one-year period. The data consist of triplets of the form (X_{ij}, Y_{ij}, T_{ij}), $i = 1, \ldots, n$, $j = 1, \ldots, p_i$, where T_{ij} denotes the instants at which the jth measurement on the ith child was taken. This is a special case of repeated measures data where the repeated observations are taken sequentially along an ordered dimension (time). To express the variation of Y with X, taking time into consideration, a useful model is the *random coefficients model*:

$$Y_{ij} = (\alpha + a_i) + (\beta + b_i)X_{ij} + \gamma T_{ij} + e_{ij}, \quad i = 1, \ldots, n, \; j = 1, \ldots, p_i,$$

where α, β and γ are unknown parameters, and a_i, b_i, and e_{ij} are random terms. Again, we may consider different assumptions for the stochastic components of the model.

(a) The pairs (a_i, b_i) follow independent bivariate normal distributions with mean vector $\mathbf{0}$ and positive definite covariance matrix $\mathbf{\Sigma}$ and the e_{ij} follow independent normal distributions with means 0 and variances σ^2. Furthermore, e_{ij} is independent of (a_i, b_i). Because given (a_i, b_i), the Y_{ij} are independent, this model is usually referred to as the *conditional independence model*.
(b) The assumptions are like in (a), but a uniform structure is imposed upon $\mathbf{\Sigma}$ by letting the a_i (b_i) follow independent normal distributions with means 0 and variances $\sigma_a^2 = \sigma_b^2$ and $\mathbb{C}\mathrm{ov}(a_i, b_i) = \sigma_{ab}$. □

Example* 1.2.8 *(Generalized linear model). In a typical ecological study, we are interested in assessing the association between the concentration of a certain atmospheric pollutant like SO_2 (X) and the death count (Y) in a certain region, controlling for temperature (T) and relative humidity (H). The data consist of vectors (Y_i, X_i, T_i, H_i), $i = 1, \ldots, n$, containing the measurements of all variables along a period of n days.

The relation of the response variable (Y) and the explanatory variables (X, T, H) may be postulated as

$$g[\mathbb{E}(Y_i)] = \beta_0 + \beta_1 X_i + \beta_2 T_i + \beta_3 H_i, \quad i = 1, \ldots, n,$$

where β_j, $j = 0, 1, 2, 3$ are unknown regression coefficients and g is a convenient link function, for example, the logarithmic function. To specify the stochastic components of the model we may assume consider the following assumptions:

(a) The Y_i follow independent Poisson distributions, or more generally any distribution in the exponential class, in which case, the model may be classified as a *generalized linear model*.

(b) We do not specify the distribution of Y_i but assume that $\mathbb{V}\text{ar}(Y_i) = v[\mathbb{E}(Y_i)]$, where v is a given function known as the *variance function*. When v is such that $\mathbb{V}\text{ar}(Y_i) > \mathbb{E}(Y_i)$ we say that there is *overdispersion*. □

Example* 1.2.9 *(Dilution bioassay). We consider a bioassay experiment designed to compare the effects of a standard preparation (SP) and an experimental preparation (EP) on the occurrence of some response (Y), such as death or tumor onset. The data consist of pairs (X_{ij}, Y_{ij}), $i = 1, 2$, $j = 1, \ldots, n_i$ where X_{ij} denotes the dose applied to the jth subject submitted to the ith preparation ($1 = SP, 2 = EP$) and Y_{ij} is equal to 1 if the jth subject submitted to the ith preparation presents a positive response (e.g., death) and to 0, otherwise.

Let X_S denote the *dose* of SP above, which the response for a given subject is positive, and let X_T be defined similarly for the EP. The random variables X_S and X_T are called *tolerances*. Assume that X_S follows the (tolerance) distribution F_S defined on $\mathbb{R}^+$ and that X_T follows the distribution F_T, also defined on $\mathbb{R}^+$. Thus, $F_S(0) = 0 = F_T(0)$ and $F_S(\infty) = 1 = F_T(\infty)$. In many assays, it is conceived that the experimental preparation behaves as if it were a *dilution* (or *concentration*) of the standard preparation. In such a case, for $x \in \mathbb{R}^+$ we may set $F_T(x) = F_S(\rho x)$, where ρ (> 0) is called the *relative potency* of the test preparation with respect to the standard one. Note that for $\rho = 1$, the two distribution functions are the same, so that the two preparations are equipotent; for $\rho > 1$, the test preparation produces the same frequency of response with a smaller dose than the standard one and hence is more potent, whereas for $\rho < 1$, the opposite conclusion holds. Possible stochastic models follow:

(a) If we assume that F_S corresponds to a normal distribution with mean μ_S and variance σ_S^2, then F_T is normal with mean $\mu_T = \rho^{-1}\mu_S$ and variance $\sigma_T^2 = \sigma_S^2/\rho^2$. Thus, $\mu_S/\mu_T = \rho$ and $\sigma_S/\sigma_T = \rho$.

(b) Because of the positivity of the dose, a normal tolerance distribution may not be very appropriate, unless μ_S/σ_S and μ_T/σ_T are large. As such, often, it is advocated that instead of the dose, one should work with dose transformations called *dose metameters* or *dosages*. For example, if we take $X_T^* = \log X_T$ and $X_S^* = \log X_S$, the response distributions, denoted by F_T^* and F_S^*, respectively, satisfy

$$F_S^*(x) = F_T^*(x - \log \rho), \quad -\infty < x < \infty, \tag{1.2.1}$$

which brings in the relevance of linear models in such a context. □

Example* 1.2.10 *(Quantal bioassay). We consider a study designed to investigate the mutagenic effect of a certain drug (or a toxic substance). Suppose that the drug is administered at different dose levels, say $0 \le d_1 < \cdots < d_s$, $s \ge 2$ so that at each level, n subjects are tried. We let m_j denote the numbers of subjects having a positive response under dose j. Here the response, Y, assumes the value 1 if there is a mutagenic effect and 0 otherwise.

Thus, the data consist of the pairs (d_j, m_j), $j = 1, \ldots, s$. Possible stochastic models are given below.

(a) At dose level d_j, we denote the probability of a positive response by $\pi_j = \pi(d_j)$. Then $\pi_j = P\{Y = 1|d_j\} = 1 - P\{Y = 0|d_j\}$, $1 \leq j \leq s$, and the joint distribution of $m_1, \ldots, m_s$ is

$$\prod_{j=1}^{s} \binom{n}{m_j} \pi_j^{m_j}(1-\pi_j)^{n-m_j}, \quad 0 \leq m_j \leq n, \ 1 \leq j \leq s.$$

(b) As in the previous example, it is quite conceivable that there is an underlying tolerance distribution, say F, defined on $[0, \infty)$ and a threshold value (the tolerance), say T_0, such that whenever the actual dose level exceeds T_0, one has $Y = 1$ (i.e., a positive response); otherwise, $Y = 0$. Moreover, we may also quantify the effect of the dose level d_j by means of a suitable function, say $\beta(d_j)$, $1 \leq j \leq s$, so that

$$\pi_j = \pi(d_j) = 1 - F[T_0 - \beta(d_j)], \quad 1 \leq j \leq s.$$

With this formulation, we are now in a more flexible situation wherein we may assume suitable regularity conditions on F and $\beta(d_j)$, leading to appropriate statistical models that can be more convenient for analysis. For example, taking $x_j = \log d_j$, we may put $\beta(d_j) = \beta^*(x_j) = \beta_0^* + \beta_1^* x_j$, $1 \leq j \leq s$, where β_0^* and β_1^* are unknown and, as such,

$$1 - \pi_j = F(T_0 - \beta_0^* - \beta_1^* x_j), \quad 1 \leq j \leq s.$$

It is common to assume that the tolerance follows a logistic or a normal distribution. In the first case, we have $F(y) = [1 + \exp(-y/\sigma)]^{-1}$, $-\infty < y < \infty$, with σ (> 0) denoting a scale factor. Then, $F(y)/[1 - F(y)] = \exp(y/\sigma)$ or $y = \sigma \log\{F(y)/[1 - F(y)]\}$, implying that for $1 \leq j \leq s$,

$$\log \frac{\pi_j}{1 - \pi_j} = \log \frac{1 - F(T_0 - \beta_0^* - \beta_1^* x_j)}{F(T_0 - \beta_0^* - \beta_1^* x_j)} = \alpha + \beta x_j, \qquad (1.2.2)$$

where $\alpha = (\beta_0^* - T_0)/\sigma$ and $\beta = \beta_1^*/\sigma$. The quantity $\log[\pi_j/(1 - \pi_j)]$ is called a *logit* (or *log-odds*) at dose level d_j. Similarly, when F is normal, we have

$$\pi_j = 1 - F(T_0 - \beta_0^* - \beta_1^* x_j) = \Phi(\alpha + \beta x_j), \quad 1 \leq j \leq s,$$

where $\Phi(y) = (\sqrt{2\pi})^{-1} \int_{-\infty}^{y} \exp(-x^2/2)dx$. Therefore,

$$\Phi^{-1}(\pi_j) = \alpha + \beta x_j, \quad 1 \leq j \leq s, \qquad (1.2.3)$$

and the quantity $\Phi^{-1}(\pi_j)$ is called a *probit* or a *normit* at the dose level d_j. Both logit and probit models may be employed in more general situations where we intend to study the relationship between a dichotomous response and a set of explanatory variables along the lines of linear models. For the logit case, such extensions are known as *logistic regression models*. In a broader sense, the models discussed above may be classified as *generalized linear models*. □

Example* 1.2.11 *(Hardy–Weinberg equilibrium). We consider the OAB blood classification system (where the O allele is recessive and the A and B alleles are codominant). Let p_O, p_A, p_B and p_{AB}, respectively, denote the probabilities of occurrence of the phenotypes

OO, (AA, AO), (BB, BO), and AB in a given (conceptually infinite) population; also, we let q_O, q_A, and q_B, respectively, denote the probabilities of occurrence of the O, A, and B alleles in that population. This genetic system is said to be in *Hardy–Weinberg equilibrium* if the following relations hold: $p_O = q_O^2$, $p_A = q_A^2 + 2q_O q_A$, $p_B = q_B^2 + 2q_O q_B$, and $p_{AB} = 2q_A q_B$. A problem of general concern to geneticists is to test whether a given population satisfies the Hardy–Weinberg conditions based on the evidence provided by a sample of n observational units for which the observed phenotype frequencies are n_O, n_A, n_B, and n_{AB}.

Under the assumption of random sampling, an appropriate stochastic model for such a setup corresponds to the multinomial model specified by the probability function

$$p(n_O, n_A, n_B, n_{AB}) = \frac{n!}{n_O!n_A!n_B!n_{AB}!} p_O^{n_O} p_A^{n_A} p_B^{n_B} p_{AB}^{n_{AB}},$$

with $p_O + p_A + p_B + p_{AB} = 1$. □

1.3 Synthesis of Finite to Asymptotic Statistical Methods

The illustrative examples in the preceding section are useful to provide motivation for some statistical models required for inference; even though exact inferential methods are available under many of such models, alternative approximate solutions are required when more realistic assumptions are incorporated. In this section, we build on the examples discussed earlier, as well as on additional ones, by considering more general statistical models that seem more acceptable in view of our limited knowledge about the data generation process. In this context, we outline the difficulties associated with the development of exact inferential procedures and provide an overview of the genesis of approximate large-sample methods along with the transit from their small-sample counterparts.

We start with Example 1.2.2(b). When F, the distribution of Y is assumed to be normal, with mean μ and variance σ^2, the sample mean $\overline{Y}_n = n^{-1}\sum_{i=1}^n Y_i$ has also a normal distribution with mean μ and variance $n^{-1}\sigma^2$, so that the *pivotal quantity*

$$Z_n = \sqrt{n}(\overline{Y}_n - \mu)/\sigma \tag{1.3.1}$$

has the standard normal distribution, that is, with null mean and variance equal to 1. If μ, the unknown parameter, is of prime interest and the variance σ^2 is known, then Z_n in (1.3.1) can be used for drawing statistical conclusions on μ. In the more likely case of σ^2 being unknown too, for $n \geq 2$, we may take the sample variance

$$s_n^2 = \frac{1}{n-1}\sum_{i=1}^n (Y_i - \overline{Y}_n)^2,$$

and consider

$$t_n = \sqrt{n}(\overline{Y}_n - \mu)/s_n, \tag{1.3.2}$$

which follows the Student t-distribution with $n-1$ degrees of freedom, for drawing statistical conclusions on μ. Both Z_n and t_n have distributions that are symmetric about 0, but t_n is more spread out than Z_n in the sense that

$$P(|t_n| > c) \geq P(|Z_n| > c), \quad \forall c > 0, \tag{1.3.3}$$

more dominantly when n is small. Thus, assuming σ^2 to be unknown, may result in less precise statistical conclusions when n is not large, the basic assumption of normality of F is crucial in this respect.

We consider now the dilution bioassay problem in Example 1.2.9, and suppose that we are interested in estimating the mean response Y of one of the preparations. Being a nonnegative random variable, Y has, typically, a positively skewed distribution on $\mathbb{R}^+ = [0, \infty)$. Thus, often, a logarithmic transformation ($Y^* = \log Y$) is advocated to achieve more symmetry for the distribution F^* (of Y^*), albeit there is still no assurance that F^* be normal. Indeed, in many applications, F^* is assumed to be logistic instead of normal. For a logistic distribution, the scale parameter and the standard deviation are not the same as in the case of the normal distribution, and the pivotal quantity Z_n (or t_n) may not be (even approximately) normal, more noticeably when n is small. Thus, assuming normality for F, when truly it is not normal, may induce a loss of precision of the statistical conclusions. For some simple specifications of the distribution function F, the *exact* sample distribution of $\overline{Y}_n$ may be obtained in closed form and that can be used to draw conclusions on $\mu = \mathbb{E}(\overline{Y}_n)$. This occurs, for example, when F belongs to the regular exponential family where optimal estimators [often characterized as *maximum likelihood estimators* (MLE)] may have a mean-like structure. However, for a general nonnormal F, the MLE is not $\overline{Y}_n$, and thereby the use of $\overline{Y}_n$ may entail further loss of statistical information about the parameter of interest. As an example, we consider the case of F as a Laplace distribution with location parameter μ and scale parameter $\lambda > 0$. The MLE of μ is the sample median, not the sample mean, and hence, inference based on $\overline{Y}_n$ may be less efficient. Recall that the logistic and the Laplace distributions are not members of the exponential family.

Two issues transpire from the above discussion. First, if some specific distributional assumption is made, how easy is it to extract full statistical information from the sample observations to draw optimal statistical conclusions? Computational complexity or mathematical intractability generally dominate the scenario, specially for nonnormal F. Second, how sensitive is the derived statistical conclusion to plausible departures from the assumed model? In many instances, they may be severely affected even by a single *outlier*. The first issue dominated the statistical literature from the 1930s to 1960s with some hope that advances in computer technology would make the situation more tractable. Nevertheless, the second issue has raised some basic questions regarding *robustness* of classical (finite-sample) parametric inference procedures to plausible model departures, either in a local sense, or more typically in a global sense; developments in nonparametric and robust statistical procedures have their genesis in this complex. We illustrate these points with some additional examples.

***Example* 1.3.1.** In Example 1.2.2.(a), we assumed that F was symmetric and continuous with mean μ and variance $\sigma^2 < \infty$. We consider several alternative models that incorporate different levels of knowledge about the nature of the data:

(a) Let $F(x) = F_o[(x - \mu)/\lambda], x \in \mathbb{R}$, with μ and $\lambda > 0$ being, respectively, the *location* and *scale parameter* and let the distribution F_o be free from (μ, λ). The normal, logistic, Laplace, and Cauchy distributions are all members of this *location-scale family*, denoted by $\mathcal{F}_{LS}$.

(b) Let $\mathcal{F}$ be the class of all distributions defined on $\mathbb{R}$ with finite first- and second-order moments. Note that $\mathcal{F}$ includes $\mathcal{F}_o$, the class of all symmetric distributions having finite second-order moments, as a subclass. In this setup, the parameter μ (or σ^2) may be viewed as a *functional* $\theta(F)$ of the distribution $F \in \mathcal{F}$. Many other parameters may be expressed in this form.

(c) Consider the subclass $\mathcal{F}_{oS}$ of all symmetric distributions defined on $\mathbb{R}$ with the origin of symmetry regarded as the location parameter. In this setup, the location parameter is the *median* of F, and will be its mean whenever F admits a finite first-order moment.

In (a) exact statistical inference pertaining to (μ, λ) may be obtained in some cases when the functional form of F_o (in a parametric setup) is specified; however, when F_o is treated nonparametrically (i.e., without specification of its functional form), approximate methods are called for. In (b) or (c), the generality of the assumptions clearly point toward the need for nonparametric methods. In this regard, the *nonparametric estimation* of statistical functionals developed by Halmos (1946) and Hoeffding (1948) constitutes a paradigm. □

Example* 1.3.2 *(Quantile function). For a continuous distribution F defined on $\mathbb{R}$, and for every p $(0 < p < 1)$, we let

$$Q_F(p) = \inf\{x : F(x) \geq p\}, \tag{1.3.4}$$

be the p-quantile of F. The function $Q_F = \{Q_F(p) : 0 < p < 1\}$ is called the quantile function. Nonparametric measures of location and dispersion are often based on Q_F. For example, $Q_F(0.5)$, the *median* is a measure of location, while the *interquartile range* $Q_F(0.75)$–$Q_F(0.25)$ is a measure of *dispersion*.

Even in a parametric setup [e.g., (c) of Example 1.3.1)], statistical inference under this model must rely on linear functionals of Q_F (called linear combinations of *order statistics*) for which only approximate (large-sample) methods are available. □

Example* 1.3.3 *(Contamination model). Instead of letting $F \in \mathcal{F}$, we consider a model that allows for some local departures from a stipulated distribution $F_o \in \mathcal{F}$ (usually normal). For example, we take for some (small) $\varepsilon > 0$ and let

$$F(x) = (1 - \varepsilon)F_o(x) + \varepsilon H(x), \quad x \in \mathbb{R}, \tag{1.3.5}$$

where H has a heavier tail than F_o in the sense of (1.3.3).

Depending on the divergence between F_o and H, even a small $\varepsilon(> 0)$ can create a great damage to the optimality properties of standard (exact) statistical inference procedures and, again, we must rely on approximate (large-sample) statistical methods. This is the genesis of *robustness* in a local sense as discussed in Huber (1981), for example. □

Example* 1.3.4 *(Nonparametric regression). In Example 1.2.8 we consider a generalized linear model where the regression function is specified by means of a finite dimensional vector $\boldsymbol{\beta}$ of regression parameters and known explanatory variables. In a more general setup, corresponding to a (possibly vector valued) regressor $\mathbf{Z}$ ($\in \mathbb{R}^q$, for some $q \geq 1$), we assume that the dependence of the response Y on $\mathbf{Z}$ is described by

$$m(z) = \mathbb{E}(Y|\mathbf{Z} = z) = \int y f(y|z) \mathrm{d}y, \quad \mathbf{Z} \in \mathbb{R}^q, \tag{1.3.6}$$

where f, the conditional density (or probability function) of Y given $\mathbf{Z}$ does not have a specified functional form. The objective is to draw statistical conclusions about $m(z)$, without further assumptions about f. Apart from some smoothness conditions on m, it is not assumed that $m(z)$ can be expressed as a function of $\mathbf{Z}$ involving a finite number of parameters (either linearly or even in a more general nonlinear form).

Here, not only the unspecified form of the underlying distribution but also the functional relation between the expected response and the explanatory variables are clear signs that no exact statistical methods may be considered for inferential purposes. □

Example* 1.3.5 *(Survival analysis). For a nonnegative random variable Y with a distribution function F, defined on $\mathbb{R}^+$, the *survival function* S is

$$S(y) = 1 - F(y) = P(Y \geq y), \quad y \in \mathbb{R}^+. \tag{1.3.7}$$

Often, there is a vector of concomitant variables, $\mathbf{Z}$, associated with Y, so that the *conditional survival function*

$$P(Y \geq y|\mathbf{Z}) = S(y|\mathbf{Z}), \quad \mathbf{Z} \in \mathbb{R}^q, \; y \in \mathbb{R}^+, \tag{1.3.8}$$

is of prime statistical interest. Equivalently, we may write

$$S(Y|\mathbf{Z}) = \exp[-H(y|\mathbf{Z})], \tag{1.3.9}$$

where $H(y|\mathbf{Z}) = \int_0^y h(u|\mathbf{Z})du$, $h(u|\mathbf{Z})$ is the conditional *hazard function*, given $\mathbf{Z}$. The (*semiparametric*) Cox (1972) *proportional hazards model* stipulates that

$$h(y|\mathbf{Z}) = h_o(y)\exp\{\boldsymbol{\beta}'\mathbf{Z}\}, \quad \boldsymbol{\beta} \in \mathbb{R}^q, \tag{1.3.10}$$

where the functional form of h_o is not specified. Statistical inference is often directed at $\boldsymbol{\beta}$, treating h_o as a nuisance functional.

Once more, the semiparametric nature of the model points toward the inexistence of exact statistical methods for inferential purposes. □

Let us try to contrast the examples described above in terms of model complexities as well as of statistical perspectives. For the examples in Section 1.2, exact statistical inference is generally possible for some specific distributions. When such distributions do not belong to the exponential family, or even when they belong to that class, but there are restraints on the parameter/sample spaces, optimality, or desirable properties of exact statistical procedures may not transpire. Further, obtaining the sampling distribution of an (optimal) estimator or test statistic (whenever they exist) in a closed and amenable form for statistical analysis, may be a challenging computational task, often prohibitively laborious if the sample size is not small. As an example, consider the OAB blood group data model in Example 1.2.11. The maximum likelihood estimator (MLE) of the gene frequencies q_O, q_A, and q_B (subject to the restraints that $q_O \geq 0$, $q_A \geq 0$, $q_B \geq 0$, $q_O + q_A + q_B = 1$) are nonlinear functions of the cell frequencies n_O, n_A, n_B, and n_{AB}, and no closed expression for its (exact) joint distribution is available; the complexities intensify with large n, even though the multinomial law is a bonafide member of the exponential family. Another classical example relates to the Cauchy distribution with location parameter μ and scale parameter $\lambda > 0$. Because the distribution does not admit finite moments of any order, the *method of moments* is not appropriate, while the MLE is a highly nonlinear estimator. Obtaining their distribution in a closed form is again a challenging task. The scenario is evidently more complex with all the examples in this section, creating impasses for establishing exact statistical properties by direct analytical manipulations.

In summary, with the exception of some very specific simple cases, the exact statistical properties of many useful estimators and test statistics are not directly obtainable for many models of practical interest. Although the use of exact statistical methods, whenever manageable, should be encouraged, methods for approximating the relevant probability distributions (or some appropriate summary measures) are of special concern in the field of statistical inference. The main tools for such approximations are based on statistical properties of large samples and are investigated under the general denomination of *asymptotic statistical theory* (or *large-sample methods*). Essentially, such large-sample methods are derived under the assumption that the sample size increases indefinitely, that is, that $n \to \infty$, which justifies both their denomination as asymptotic methods as well the denomination of finite-sample methods for their small-sample counterparts. Although such large sample procedures are only directed at the development of approximate results, they are associated with the following advantages:

(i) They allow for some flexibility in the selection of models; asymptotic properties of estimators and test statistics are usually less dependent on the particular functional form of the underlying distribution, and thus, are relatively robust against departures from some assumed model.
(ii) In many cases, they allow for some relaxation of the assumptions of independence and identical distribution of the sample elements, usually required in exact statistical procedures.
(iii) They are generally associated to simple and well-studied limiting distributions, such as the normal, chi-squared, Weibull, and soforth.
(iv) They provide ample flexibility in the selection of estimators and test statistics and thus allow for the choice of those with more appealing interpretation.

The evaluation of the rates of convergence of the distribution of interest to some limiting distribution is an important topic in this context. Although such an issue is usually associated with some rather complicated technical problems, it should not be underemphasized, because the applied statistician must deal with finite albeit large sample sizes in practice.

The recent advances in the use of computers for data management purposes have triggered a general interest in the development of statistical methods designed to deal with large data sets. As a consequence, many new approximate techniques have been proposed, which are rather more complicated (and realistic) than those that can be handled by exact methods. The advent of genomics and bioinformatics, at large, has added new dimensions to the need for statistical methodology to accommodate high-dimensionality problems that mar the use of conventional methods. Yet, such traditional statistical reasoning still holds the keys to motivation as well as to intuition for the highly nonstandard and nonregular methodology that needs to be developed for this evolving interdisciplinary field. Therefore, asymptotic methods in Statistics have become increasingly necessary tools for both theoretically and practically oriented statisticians. However, the concepts that underlie standard exact statistical methods are crucial for the development of the approximate methods that we have been discussing. With this basic consideration, we proceed to present an outline of the material covered in this text, emphasizing potential applications and the interrelationships among the several probabilistic concepts required for such purposes.

1.4 The Organization of the Book

This intermediate level book is primarily meant for graduate students in (applied or theoretical) Statistics and Biostatistics, as well as for statisticians who intend to make use of sound methodology in their professional activities. Although we try to avoid excessive mathematical technicalities, some basic concepts in the fields of Real Analysis, Matrix Algebra, Probability, and Statistics are required for a fair understanding of the material covered in the text. For the sake of self-containment, a summary of such results is presented (without proofs) in Section 1.5.

Chapter 2 deals with the basics of statistical estimation theory in finite sample setups, along with the motivation for the asymptotics. There are several perspectives in statistical estimation theory that deserve our attention. They are essentially finite-sample properties, but there is a genuine need to hook them up with general asymptotics prevailing only in large samples. *Unbiasedness*, *minimum variance*, *efficiency*, *information bounds*, and *sufficiency* of statistical estimators are among those deserving special attention. Sufficiency along with the *factorization theorem* and the Rao–Blackwell dominance provide the basic finite-sample optimality properties, and the exponential family of densities plays a basic role in this respect. Also, the concepts of *invariance* and *equivariance* along with *ancilarity* have important impact in the estimation theory. In many nonregular cases, including those relating to distributions not belonging to the exponential family, finite-sample optimality properties may not be attainable; nevertheless, extensions of such concepts to the asymptotic cases are useful to show that optimality is (approximately) tenable under fairly general regularity conditions when large samples are available. Given the good properties of the estimators derived therefrom, there is a dominant emphasis on *likelihood functions* in the finite-sample case, and again, in the asymptotic setup, likelihood, though very basic, is not the only superstar, and there are ramifications as well as other approaches that share the limelight to a greater extend. Along with *maximum likelihood* for which the good properties are even more evident in large samples, other methods of estimation, such as the *method of moments*, *least squares*, *weighted least squares*, and *generalized estimating equations*, along with some robust and nonparametric alternatives are appraised in this light. *Bayes* and *empirical Bayes* estimators also constitute important options; given their peculiar nature, their treatment is delayed to Chapter 4. For better understanding, computational algorithms, and associated resampling plans are relegated to Chapters 8 and 11.

Chapter 3 is devoted to the basic principles of statistical hypothesis testing. The classical Neyman–Pearson concepts of *critical region/similar region*, *type I* and *type II errors*, *significance level* (*size*), and *power*, covering both *simple* and *composite* hypotheses are approached through the *(generalized) Neyman–Pearson Lemma*. Optimality properties in the global as well as in the local sense are discussed, and various methods of deriving tests (e.g., *likelihood ratio*, *union-intersection*, *score statistics*) are considered in this context. *Bayes tests* and the dual problem of confidence intervals (sets) are discussed partly here, and partly in Chapter 4. Asymptotics for hypothesis testing are discussed in Chapter 8.

Elements of statistical *decision theory* constitute the kernel of Chapter 4. The notions of *loss* and *risk* function lead us to the formulation of *minimum risk* and *minimax* estimation theory where *Bayes methods* are appraised along with Empirical Bayes procedures, which are also briefly considered.

The elements of stochastic processes are outlined in Chapter 5 because much of the reasoning that underlie statistical inference rests on appropriate probabilistic tools and allied methodology. *Random walks* and *Markov processes*, including *Poisson processes* and *Markov chains*, are useful for statistical inference. The passage from random walks to *diffusion processes* and *Brownian motion* (or Wiener) and *Brownian bridge* (or tied-down Wiener) processes has an important bearing on asymptotic methods relying on weak convergence results; *martingales* and *submartingales* also play a basic role in this context.

Chapter 6 deals with the basic notion of *stochastic order*, *stochastic convergence* in various modes, and some of the basic *probability inequalities* as well as *laws of large numbers*, encompassing some dependent sequences as well. These probability tools are then incorporated in the study of *weak* as well as *strong convergence* (or *consistency*) results of general classes of statistics, which are abundant in statistical inference. Along with a *reversed martingale* characterization of *U-statistics*, such convergence results are extended to more complex statistics.

Chapter 7 is devoted to *central limit theorems* (CLT) in various environments, including some with dependent variables. The study of sampling distributions of estimators and test statistics occupies a focal stand in statistical inference. Many of these topics rely heavily on probability tools, and their finite-sample perspectives are not usually strong enough to percolate to large samples. Here, *Weak convergence* (or *convergence in law*) play an important role that paves the way for asymptotics of statistical inference far beyond the simple scenario of sums of independent random variables. Weak convergence often relates to asymptotically normal distributions, and in such a setting, it emerges in the form of central limit theorems. *Characteristic functions* have an important role in CLT and are reviewed briefly. *Projection results* and *variance-stabilizing* transformations are outlined in this context, too. Asymptotics for quadratic forms are presented and convergence rates for asymptotic distributions are summarized. Asymptotics for empirical distributions and order statistics are also discussed.

Chapter 8 comprises the asymptotic behavior of a general class of estimators and test statistics. On the estimation frontier, asymptotics for the MLE form its nucleus. Statistical functionals rely heavily on the asymptotic properties of empirical distribution functions, and these are appraised as well. The approximate behavior of the distribution of U-statistics and related functionals are also covered. The concept of *asymptotic efficiency* of estimators and its relation to finite-sample efficiency properties are elaborated. The second part of the chapter is devoted to the study of asymptotic properties of the likelihood ratio – Wald-type and Raos score tests along with their ramifications. Asymptotic efficiency of statistical tests is studied. Asymptotics under nonstandard conditions are discussed along the same vein. Finally, resampling plans are appraised in these asymptotic setups.

Chapters 9 and 10 are directed at the specific application of asymptotic methods in some of the most commonly used families of (statistical) models: categorical data models and linear models. The corresponding large-sample results are related to those discussed in Chapter 8 under a more general setting; however, some particularities associated with such models justify a somewhat more specific analysis. The inclusion of a detailed treatment of large-sample methods for categorical data models in this text may be justified not only on the grounds of their importance in statistical methodology but also because they provide good examples for the application of almost all asymptotic concepts discussed in earlier chapters. In particular, they are useful in emphasizing the asymptotic equivalence of statistical procedures, such as maximum likelihood estimation for *product multinomial distributions* and the corresponding *weighted least-squares methods*. Linear regression models pose special problems, because the observed random variables are not identically distributed. Also, in many cases, the exact functional form of the underlying distribution is not completely specified and, in such situations, asymptotic results are of major interest. Least-squares methods are attractive under these conditions, because they may be employed in a rather general setup. However, other alternatives, including generalized and weighted least-squares procedures as well as robust M-estimation procedures have also been given due attention. In this context, generalized linear models (in the spirit referred to in Example 1.2.8) deserve special mention. The study of their asymptotic properties involves methodology having roots in the general theory of maximum likelihood estimation discussed in Chapter 8 but also bear similarity with that of generalized least squares for linear models. The last topic covered in Chapter 10 is nonparametric regression; we essentially describe the difficulties in establishing asymptotic results for some of the usual techniques employed in this field.

Chapter 11 is devoted to some basic, but more complicated technical problems that have been deferred from earlier discussions. In particular, the concept of stochastic convergence and convergence in law, as developed in Chapters 6 and 7 for the finite dimensional cases, have gone through some extensive generalizations in the recent past. This involves *weak convergence in metric spaces* for which a detailed treatment is undoubtedly beyond the intended level of presentation of this text. Nevertheless, a unified introduction to such weak invariance principles with due emphasis on the partial sum and empirical distributional processes is outlined along with some applications. Weak convergence of *statistical functionals* (which are differentiable in some sense) and the *Bahadur–Kiefer representation* for sample quantiles are considered in the same vein. Finally, some insight into the role of weak convergence in nonparametrics is given. Although less technical than required for a rigorous development, this chapter may provide good motivation for the use of these novel techniques in various applied problems.

1.5 Basic Tools and Concepts

In general, boldface lowercase (uppercase) characters denote vectors (matrices). Given $a \in \mathbb{R}^+$, we denote the largest integer less then or equal to a as $[a]$.

Some results on matrix algebra

The *transpose* of $\boldsymbol{a}$ ($\boldsymbol{A}$) is denoted $\boldsymbol{a}'$ ($\boldsymbol{A}'$). Also, $\text{tr}(\boldsymbol{A})$ and $|\boldsymbol{A}|$, respectively, denote the *trace* and the *determinant* of a matrix $\boldsymbol{A}$. Let $\boldsymbol{a} = (a_1, \ldots, a_n)'$ and $\boldsymbol{b} = (b_1, \ldots, b_n)'$; then $\boldsymbol{a} \leq \boldsymbol{b}$ ($\boldsymbol{a} \geq \boldsymbol{b}$) stands for the simultaneous inequalities $a_i \leq b_i$ ($a_i \geq b_i$), $i = 1, \ldots, n$. In the text, $\boldsymbol{I}_n$ denotes the identity matrix of order n and $\boldsymbol{1}_n$ ($\boldsymbol{0}_n$) denotes a column vector of dimension n with all elements equal to 1 (0). The subscript might be omitted when the dimension is clear from the discussion. If $\boldsymbol{x}$ is a p-vector and $\boldsymbol{A}$ is a $p \times p$ matrix, the function $Q(\boldsymbol{x}) = \boldsymbol{x}'\boldsymbol{A}\boldsymbol{x}$ is called a *quadratic form*. If $Q(\boldsymbol{x}) \geq 0$ for all $\boldsymbol{x} \in \mathbb{R}^p$, $\boldsymbol{x} \neq \boldsymbol{0}$, the matrix $\boldsymbol{A}$ is *positive semidefinite* (p.s.d.); if $Q(\boldsymbol{x}) > 0$ for all $\boldsymbol{x} \in \mathbb{R}^p$, $\boldsymbol{x} \neq \boldsymbol{0}$, the matrix $\boldsymbol{A}$ is *positive definite* (p.d.). If $Q(\boldsymbol{x}) \leq 0$ for all $\boldsymbol{x} \in \mathbb{R}^p$, $\boldsymbol{x} \neq \boldsymbol{0}$, the matrix A is negative semidefinite (n.s.d).

The *generalized inverse* of a matrix $\boldsymbol{A}$ is any matrix $\boldsymbol{A}^-$, such that

$$\boldsymbol{A}\boldsymbol{A}^-\boldsymbol{A} = \boldsymbol{A}. \tag{1.5.1}$$

Note that if we let $\boldsymbol{H} = \boldsymbol{A}^-\boldsymbol{A}$, then $\boldsymbol{H}^2 = \boldsymbol{H}\boldsymbol{H} = \boldsymbol{A}^-\boldsymbol{A}\boldsymbol{A}^-\boldsymbol{A} = \boldsymbol{A}^-\boldsymbol{A} = \boldsymbol{H}$, so that $\boldsymbol{H}$ is *idempotent*, and, hence,

$$\text{rank}(\boldsymbol{A}) = \text{rank}(\boldsymbol{H}) = \text{trace}(\boldsymbol{H}). \tag{1.5.2}$$

In various statistical applications, we may need to partition a symmetric matrix $\boldsymbol{A}$ as

$$\begin{pmatrix} \boldsymbol{A}_{11} & \boldsymbol{A}_{12} \\ \boldsymbol{A}_{21} & \boldsymbol{A}_{22} \end{pmatrix},$$

If $\boldsymbol{A}$ and $\boldsymbol{A}_{11}$ are p.d., then we may verify that

$$\boldsymbol{A}^{-1} = \begin{pmatrix} \boldsymbol{A}_{11}^{-1} + \boldsymbol{A}_{11}^{-1}\boldsymbol{A}_{12}\boldsymbol{A}_{22.1}^{-1}\boldsymbol{A}_{21}\boldsymbol{A}_{11}^{-1} & -\boldsymbol{A}_{11}\boldsymbol{A}_{12}\boldsymbol{A}_{22.1}^{-1} \\ -\boldsymbol{A}_{22.1}^{-1}\boldsymbol{A}_{21}\boldsymbol{A}_{11}^{-1} & \boldsymbol{A}_{22.1}^{-1} \end{pmatrix}, \tag{1.5.3}$$

where

$$\boldsymbol{A}_{22.1} = \boldsymbol{A}_{22} - \boldsymbol{A}_{21}\boldsymbol{A}_{11}^{-1}\boldsymbol{A}_{12}. \tag{1.5.4}$$

Whenever $\boldsymbol{A}$ is a $p \times p$ matrix, the determinantal equation $|\boldsymbol{A} - \lambda\boldsymbol{I}| = 0$ has p roots, say, $\lambda_1, \ldots, \lambda_p$, which are called the *characteristic roots* or *eigenvalues* of $\boldsymbol{A}$. If $\boldsymbol{A}$ is p.s.d., then all of the λ_i are nonnegative, and we take them in the order $\lambda_1 \geq \lambda_2 \geq \cdots \geq \lambda_p \geq 0$. Then, corresponding to each λ_i, there is a vector $\boldsymbol{x}_i$ (called the *characteristic vector* or *eigenvector*), such that

$$\boldsymbol{A}\boldsymbol{x}_i = \lambda_i\boldsymbol{x}_i, \quad i = 1, \ldots, p. \tag{1.5.5}$$

Hence, there exists an orthogonal matrix $\boldsymbol{B}$, such that

$$\boldsymbol{B}'\boldsymbol{A}\boldsymbol{B} = \boldsymbol{D}_\lambda = \text{diag}(\lambda_1, \ldots, \lambda_p), \tag{1.5.6}$$

where diag($\boldsymbol{x}$) denotes a diagonal matrix with the elements of $\boldsymbol{x}$ along the main diagonal. This decomposition is useful for studying quadratic forms.

Writing $\boldsymbol{x} = \boldsymbol{B}\boldsymbol{y}$, we obtain from (1.5.6) that

$$Q(\boldsymbol{y}) = \boldsymbol{y}'\boldsymbol{B}'\boldsymbol{A}\boldsymbol{B}\boldsymbol{y} = \boldsymbol{y}'\boldsymbol{D}_\lambda\boldsymbol{y} = \sum_{i=1}^{p} \lambda_i y_i^2. \tag{1.5.7}$$

From (1.5.7) and using the ordering of the λ_i, it follows that

$$\lambda_p\boldsymbol{y}'\boldsymbol{y} \leq Q(\boldsymbol{y}) \leq \lambda_1\boldsymbol{y}'\boldsymbol{y}. \tag{1.5.8}$$

In fact, (1.5.8) extends to the following more general result:

Theorem* 1.5.1 *(Courant). *Let* $\boldsymbol{A}$ *be a symmetric matrix and* $\boldsymbol{B}$ *be a p.d. matrix. Then if* λ_i *denotes the* ith *characteristic root of* $\boldsymbol{A}\boldsymbol{B}^{-1}$ *for* $i = 1, \ldots, p$, *we have*

$$\text{ch}_p(\boldsymbol{A}\boldsymbol{B}^{-1}) = \lambda_p = \inf_x \frac{\boldsymbol{x}'\boldsymbol{A}\boldsymbol{x}}{\boldsymbol{x}'\boldsymbol{B}\boldsymbol{x}} \leq \sup_x \frac{\boldsymbol{x}'\boldsymbol{A}\boldsymbol{x}}{\boldsymbol{x}'\boldsymbol{B}\boldsymbol{x}} = \lambda_1 = \text{ch}_1(\boldsymbol{A}\boldsymbol{B}^{-1}). \tag{1.5.9}$$

Some results on real analysis

The notation $y = f(x)$ is used to mean that y is a function of x for a certain set of values of x, that is, for each x of the *domain* A, $f(x)$ associates a corresponding value of y of the *range* B; this may be also represented by $f\colon A \to B$. If both A and B are collections of real numbers, f is called a *real-valued function* of a *real variable*. On the other hand, f is said to be a *vector-valued function* of dimension m of p *real variables* if A and B are collections of p and m real vectors, respectively; the value of the function f of dimension m at the point $\boldsymbol{x} = (x_1, \ldots, x_p)' \in A \subset \mathbb{R}^p$ is denoted by $\boldsymbol{f}(\boldsymbol{x}) = [f_1(\boldsymbol{x}), \ldots, f_m(\boldsymbol{x})]'$.

Let $f(x)$, $x \in [a, b)$, be a real-valued function. If for every $\varepsilon > 0$, there exists an $\eta > 0$, such that for some c^-

$$|f(x) - c^-| < \varepsilon, \quad \text{whenever} \quad b - \eta < x < b, \tag{1.5.10}$$

then c^- is the *left-hand limit* of f at b. The *right-hand limit* c^+ is defined similarly. If both c^- and c^+ exist and $c^- = c^+ = c$, then $f(x)$ is said to be *continuous* at b. Also, $f(x)$ is said to be continuous over the interval (a, b) if it is continuous at each $x \in (a, b)$; if the interval is $[a, b]$, we also need f to be continuous to the left (right) at b (a). Some care is needed to have this continuity over $[a, b]$ when a is $-\infty$ or b is $+\infty$. In other words, we say that f is continuous at x_0 if, given $\varepsilon > 0$, there exists $\delta = \delta(x_0, \varepsilon)$, such that $|x - x_0| < \delta$ implies $|f(x) - f(x_0)| < \varepsilon$. If, for a given $\mathfrak{I}$ interval (open or closed), δ does not depend on x_0, we say that f is *uniformly continuous* over $\mathfrak{I}$. Note that if f is continuous over a closed interval, it is uniformly continuous over that interval.

Let the function $f(x)$ be defined on (a, b). If for every $x_0 \in (a, b)$ the limit

$$g(x_0) = \lim_{\Delta x \to 0} \left\{ \frac{1}{\Delta x}[f(x_0 + \Delta x) - f(x_0)] \right\} \tag{1.5.11}$$

exists, then $g(x)$ is the *derivative* of $f(x)$ and is denoted by $f'(x)$. A function $f(x)$ is said to be *differentiable* if $f'(x)$ exists. Note that continuity of f may not ensure its differentiability. We may rewrite (1.5.11) as

$$f(x) - f(x_0) = (x - x_0)f'(x_0) + o(x - x_0) \quad \text{as} \quad x \to x_0, \tag{1.5.12}$$

where $o(x - x_0)$ denotes a quantity that is negligible when compared to $(x - x_0)$ as $x \to x_0$ [the notation $o(\cdot)$ and $O(\cdot)$ will be elaborated on in Section 6.2]. This, in turn, provides two important results:

(i) The *first mean value theorem* in Differential Calculus: If $f(x)$ is continuous over $[a, b]$ and differentiable for $a < x < b$, then

$$f(b) - f(a) = (b - a)f'(x^*), \quad \text{for some} \quad x^* \in (a, b). \tag{1.5.13}$$

(ii) The *fundamental theorem* of Integral Calculus: For $b > a$

$$f(b) - f(a) = \int_a^b g(x)\mathrm{d}x, \quad \text{where} \quad g(x) = f'(x). \tag{1.5.14}$$

The derivative of $f'(x)$, denoted by $f^{(2)}(x)$, may be defined as in (1.5.11), wherein we replace f by f' and the chain rule applies to define the nth derivative $f^{(n)}(x)$, for every $n \geq 1$ (whenever it exits). In some cases, we also use the common notation f'' and f''' for $f^{(2)}$ and $f^{(3)}$, respectively. Let $f(x)$, $a \leq x \leq b$, be a continuous function and have continuous derivatives up to the kth order for some $k \geq 0$. Then, for every $x \in [a, b]$, $x_0 \in (a, b)$,

$$f(x) = f(x_0) + \sum_{j=1}^{k} \frac{(x - x_0)^j}{j!} f^{(j)}(x_0) + R_k(x, x_0), \tag{1.5.15}$$

where

$$R_k(x, x_0) = \frac{(x - x_0)^k}{k!}\{f^{(k)}[hx_0 + (1 - h)x] - f^{(k)}(x_0)\}, \tag{1.5.16}$$

for some $0 < h < 1$. This is known as the *Taylor expansion* (up to the kth order) with a remainder term. There is no need to take $x = x_0$ in (1.5.15), while the case $k = 0$ is treated in (1.5.13); also, $k = +\infty$ leads to the *Taylor series expansion*.

Now let f be defined on (a, b) and suppose that there exists an interior point c, such that $f(c) = \sup\{f(x) : a \le x \le b\}$ or $f(c) = \inf\{f(x): a \le x \le b\}$, *i.e.*, f has a *relative (local) maximum* or *minimum* at c. If the derivative $f'(c)$ exists, then $f'(c) = 0$. Moreover, if $f^{(2)}(c)$ also exists and $f^{(2)}(c) \neq 0$, then,

(i) $f^{(2)}(c) < 0$ implies that $f(c)$ is a relative maximum;
(ii) $f^{(2)}(c) > 0$ implies that $f(c)$ is a relative minimum.

Consider now the computation of the limit (1.5.11) of $f(x) = f_1(x)/f_2(x)$. If, at any point x_0, $f_1(x)$ and $f_2(x)$ both have limits, say c_1 and c_2, where $c_2 \neq 0$, then the limit of $f(x)$ at x_0 is c_1/c_2. If $c_2 = 0$, but $c_1 \neq 0$, by working with $1/f(x) = f_2(x)/f_1(x)$, we may claim that $f(x) \to \text{sign}(c_1)\infty$ as $x \to x_0$. However, there is an indeterminate form when both c_1 and c_2 are zero and in such a case we may apply the *l'Hôpital rule*: assume that f_1 and f_2 have derivatives f_1' and f_2', respectively [at each point of an open interval (a, b)], and that as $x \to a^+$, $f_1(x) \to 0$, $f_2 \to 0$, but $f_2'(x) \neq 0$ for each $x \in (a, b)$ and $f_1'(x)/f_2'(x)$ has a limit as $x \to a^+$. Then,

$$\lim_{x\to a^+}[f_1(x)/f_2(x)] = \lim_{x\to a^+}[f_1'(x)/f_2'(x)]. \tag{1.5.17}$$

Consider now a vector-valued function $f(\boldsymbol{x})$ of p real variables. The (first-order) *partial derivatives* of f with respect to x_i, $i = 1, \ldots, p$ at the point $\boldsymbol{x} = (x_1, \ldots, x_p)'$ is defined as

$$\begin{aligned} f_i'(\boldsymbol{x}) &= \frac{\partial}{\partial x_i} f(\boldsymbol{x}) \\ &= \lim_{\Delta x_i \to 0} \frac{f(x_1, \ldots, x_{i-1}, x_i + \Delta x_i, x_{i+1}, \ldots, x_p) - f(x_1, \ldots, x_p)}{\Delta x_i} \end{aligned} \tag{1.5.18}$$

when the limit exists. The *gradient* of f at the point $\boldsymbol{x}$ is the p-vector of (first-order) partial derivatives:

$$\overset{\bullet}{f}(\boldsymbol{x}) = \frac{\partial}{\partial \boldsymbol{x}} f(\boldsymbol{x}) = \left[\frac{\partial}{\partial x_1} f(\boldsymbol{x}), \frac{\partial}{\partial x_2} f(\boldsymbol{x}), \ldots, \frac{\partial}{\partial x_p} f(\boldsymbol{x})\right]'. \tag{1.5.19}$$

Along the same lines, we may define the second-order partial derivatives of f with respect to x_i, x_j, $i, j = 1, \ldots, p$ at the point $\boldsymbol{x}$ [denoted by $(\partial^2/\partial x_i \partial x_j) f(\boldsymbol{x})$] by replacing $f'(\boldsymbol{x})$ for $f_i(\boldsymbol{x})$ in (1.5.18). The extension to higher-order partial derivatives is straightforward. Within this context, the *Hessian* of f at the point $\boldsymbol{x}$ is defined as the $(p \times p)$ matrix of second-order partial derivatives:

$$\begin{aligned}\boldsymbol{H}(\boldsymbol{x}) =& \frac{\partial^2}{\partial \boldsymbol{x}\partial \boldsymbol{x}'} f(\boldsymbol{x}) = \frac{\partial}{\partial \boldsymbol{x}}\left\{\left[\frac{\partial}{\partial \boldsymbol{x}} f(\boldsymbol{x})\right]'\right\} \\ =& \begin{pmatrix} \frac{\partial^2}{\partial x_1^2} f(\boldsymbol{x}) & \frac{\partial^2}{\partial x_1 \partial x_2} f(\boldsymbol{x}) & \cdots & \frac{\partial^2}{\partial x_1 \partial x_p} f(\boldsymbol{x}) \\ \frac{\partial^2}{\partial x_2 \partial x_1} f(\boldsymbol{x}) & \frac{\partial^2}{\partial x_2^2} f(\boldsymbol{x}) & \cdots & \frac{\partial^2}{\partial x_2 \partial x_p} f(\boldsymbol{x}) \\ \vdots & \vdots & \cdots & \vdots \\ \frac{\partial^2}{\partial x_p \partial x_1} f(\boldsymbol{x}) & \frac{\partial^2}{\partial x_p \partial x_2} f(\boldsymbol{x}) & \cdots & \frac{\partial^2}{\partial x_p^2} f(\boldsymbol{x}) \end{pmatrix}. \end{aligned} \quad (1.5.20)$$

Now let $f(\boldsymbol{x}), \boldsymbol{x} \in A \subseteq \mathbb{R}^p$, be a continuous function and have continuous partial derivatives up to $(k+1)$th order for some $k \geq 1$. Then, for every $\boldsymbol{x} \in A, \boldsymbol{x}_0 \in A$, we have the following *multivariate Taylor expansion*:

$$f(\boldsymbol{x}) = f(\boldsymbol{x}_0) + \sum_{j=1}^{k} \frac{1}{j!} \sum_{i_1=1}^{p} \cdots \sum_{i_j=1}^{p} \frac{\partial^j}{\partial x_{i_1} \cdots \partial x_{i_j}} f(\boldsymbol{x})\Big|_{\boldsymbol{x}_0} \prod_{l=1}^{j} (x_{i_l} - x_{0_{i_l}}) + R_k(\boldsymbol{x}, \boldsymbol{x}_0), \quad (1.5.21)$$

where $R_k(\boldsymbol{x}, \boldsymbol{x}_0)$ is given by

$$\frac{1}{(k+1)!} \sum_{i_1=1}^{p} \cdots \sum_{i_{k+1}=1}^{p} \frac{\partial^{k+1}}{\partial x_{i_1} \cdots \partial x_{i_{k+1}}} f[h\boldsymbol{x}_0 + (1-h)\boldsymbol{x}] \prod_{l=1}^{k+1} (x_{i_l} - x_{0_{i_l}}), \quad (1.5.22)$$

for some $0 < h < 1$. In matrix notation, it may be expressed as

$$\begin{aligned} f(\boldsymbol{x}) =& f(\boldsymbol{x}_0) + (\boldsymbol{x} - \boldsymbol{x}_0)' \frac{\partial}{\partial \boldsymbol{x}} f(\boldsymbol{x})\Big|_{\boldsymbol{x}_0} \\ &+ \frac{1}{2} (\boldsymbol{x} - \boldsymbol{x}_0)' \frac{\partial^2}{\partial \boldsymbol{x} \partial \boldsymbol{x}'} f(\boldsymbol{x})\Big|_{\boldsymbol{x}_0} (\boldsymbol{x} - \boldsymbol{x}_0) + o(||\boldsymbol{x} - \boldsymbol{x}_0||^2). \end{aligned} \quad (1.5.23)$$

The notation $f(\boldsymbol{x})|_{\boldsymbol{x}_0}$ is used to indicate that the function $f(\boldsymbol{x})$ is evaluated at $\boldsymbol{x} = \boldsymbol{x}_0$.

In many applications, we have integrals of the form:

$$F(t) = \int_a^\infty f(x, t)\mathrm{d}x; \quad (1.5.24)$$

in such cases t is usually called a *parameter*. Suppose that $f(x, t)$ is continuous in x and t and has a (partial) derivative $\partial f/\partial t$, which is also continuous in x and t (for $x \in [a, \infty)$ and $t \in [c, d]$), such that

$$\int_a^\infty f(x, t)\mathrm{d}x \quad \text{and} \quad \int_a^\infty \frac{\partial}{\partial t} f(x, t)\mathrm{d}x \quad (1.5.25)$$

converge (the second one, uniformly in t). Then F has a continuous derivative for $t \in [c, d]$, and

$$F'(t) = \int_a^\infty \frac{\partial}{\partial t} f(x, t)\mathrm{d}x. \quad (1.5.26)$$

In regard to the uniformity condition in (1.5.25), we may introduce the notion of *uniform integrability*; we say that $h(t) = \{h(x,t),\ x \in \mathbb{R}\}$ is uniformly (in t) integrable, if

$$\int_{|x|>c} |h(x,t)|\mathrm{d}x \to 0 \quad \text{as} \quad c \to \infty, \quad \text{uniformly in } t. \tag{1.5.27}$$

In some cases, the limits of integration in (1.5.24) are themselves functions of the "parameter t"; suppose, for example, that $a(t)$ and $b(t)$ have continuous derivatives for $t \in [c,d]$ so that $F(t) = \int_{a(t)}^{b(t)} f(x,t)\mathrm{d}x$. Then, under the same conditions above, we have

$$F'(t) = f[b(t),t]b'(t) - f[a(t),t]a'(t) + \int_{a(t)}^{b(t)} \frac{\partial}{\partial t} f(x,t)\mathrm{d}x. \tag{1.5.28}$$

There are some other properties of functions that will be introduced in later chapters in the appropriate context. We may, however, mention here a few functions of special importance to our subsequent analysis.

(i) *Exponential function*:

$$f(x) = \exp(x) = 1 + x + \frac{x^2}{2!} + \frac{x^3}{3!} + \cdots; \tag{1.5.29}$$

(ii) *Logarithmic function*:

$$f(x) = \log(x), \quad \text{that is,} \quad x = \exp[f(x)] \tag{1.5.30}$$

and also

$$f(x) = \log(1+x) = x - \frac{1}{2}x^2 + \frac{1}{3}x^3 - \frac{1}{4}x^4 + \cdots; \tag{1.5.31}$$

(iii) *Gamma function*:

$$f(x) = \Gamma(x) = \int_0^\infty e^{-y}y^{x-1}\mathrm{d}y, \quad x \geq 0. \tag{1.5.32}$$

Note that if $x = n+1$ is a positive integer, then $\Gamma(x) = \Gamma(n+1) = n!$ $(= 1 \times 2 \times \cdots \times n)$. Also, for $n \geq 5$, we have the *Stirling approximation* to the factorials:

$$n! \approx (2\pi)^{1/2} n^{n+1/2} \exp\left[-n - \frac{1}{12n} - \cdots\right]. \tag{1.5.33}$$

Let us now consider certain inequalities that are very useful in statistical applications.

Cauchy–Schwarz Inequality: Let $\boldsymbol{a} = (a_1, \ldots, a_n)'$ and $\boldsymbol{b} = (b_1, \ldots, b_n)'$ be two n-vectors. Then,

$$(\boldsymbol{a}'\boldsymbol{b})^2 = \left(\sum_{i=1}^n a_i b_i\right)^2 \leq \left(\sum_{i=1}^n a_i^2\right)\left(\sum_{i=1}^n b_i^2\right) = (\boldsymbol{a}'\boldsymbol{a})(\boldsymbol{b}'\boldsymbol{b}), \tag{1.5.34}$$

where the equality holds when $\boldsymbol{a} \propto \boldsymbol{b}$ (i.e., $\boldsymbol{a} = k\boldsymbol{b}$ for some constant k). The integral version of (1.5.34) is given by

$$\begin{aligned}\langle f_1, f_2\rangle^2 &= \left[\int_A f_1(x)f_2(x)\mathrm{d}\mu(x)\right]^2 \\ &\leq \left[\int_A f_1^2(x)\mathrm{d}\mu(x)\right]\left[\int_A f_2^2(x)\mathrm{d}\mu(x)\right] \\ &= \langle f_1, f_1\rangle\langle f_2, f_2\rangle, \qquad (1.5.35)\end{aligned}$$

where f_1 and f_2 are real functions defined on a set A and are square integrable with respect to a measure μ (on A).[1]

Hölder Inequality: Let $\boldsymbol{a} = (a_1, \ldots, a_n)'$ and $\boldsymbol{b} = (b_1, \ldots, b_n)'$ be n-vectors and let r and s be positive numbers such that $r^{-1} + s^{-1} = 1$. Then,

$$|\boldsymbol{a}'\boldsymbol{b}| = \left|\sum_{i=1}^n a_i b_i\right| \leq \left(\sum_{i=1}^n |a_i|^r\right)^{1/r}\left(\sum_{i=1}^n |b_i|^s\right)^{1/s}, \qquad (1.5.36)$$

where the equality holds only if a_i, b_i are of the same sign and $|b_i| \propto |a_i|^{r-1}$. As in (1.5.35), an integral version of (1.5.36) is easy to write. Note that if $r = s = 2$, (1.5.36) reduces to (1.5.34). A direct corollary to (1.5.36) is the Minkowski Inequality.

Minkowski Inequality: For $k \geq 1$ and $\boldsymbol{a}, \boldsymbol{b}$ as in (1.5.36),

$$\left[\sum_{i=1}^n |a_i + b_i|^k\right]^{1/k} \leq \left(\sum_{i=1}^n |a_i|^k\right)^{1/k} + \left(\sum_{i=1}^n |b_i|^k\right)^{1/k}. \qquad (1.5.37)$$

A closely related inequality is the C_r Inequality.

C_r Inequality: For every $r > 0$, a and b real,

$$|a + b|^r \leq C_r[|a|^r + |b|^r], \qquad (1.5.38)$$

where

$$C_r = \begin{cases} 1, & 0 \leq r \leq 1 \\ 2^{r-1}, & r > 1. \end{cases} \qquad (1.5.39)$$

The inequality extends to any m (≥ 1) numbers $a_1, \ldots, a_m$, where C_r is then equal to m^{r-1}, $r > 1$, and 1 for $0 \leq r \leq 1$. For $r = 1$, (1.5.38) reduces to the *triangular inequality*.

Arithmetic/Geometric/Harmonic Mean Inequality: Let $a_1, a_2, \ldots, a_n$ be nonnegative numbers. Then, for every n (≥ 1),

$$AM = \frac{1}{n}\sum_{i=1}^n a_i \geq GM = \left(\prod_{i=1}^n a_i\right)^{1/n} \geq HM = \left[\frac{1}{n}\sum_{i=1}^n \frac{1}{a_i}\right]^{-1}, \qquad (1.5.40)$$

where the equality sign holds only when $a_1 = a_2 = \cdots = a_n$. An integral version of (1.5.40) can be worked on as in (1.5.35), provided the arithmetic mean exists.

[1] The measure μ, commonly taken as the Lebesgue measure in the continuous case and the sign measure in the discrete case, will be explained later on.

Entropy Inequality: Let $\sum_{i=1}^{n} a_i$ and $\sum_{i=1}^{n} b_i$ be nonnegative numbers, such that $\sum_{i=1}^{n} a_i \geq \sum_{i=1}^{n} b_i$. Then,

$$\sum_{i=1}^{n} a_i \log(b_i/a_i) \leq 0, \tag{1.5.41}$$

where the equality holds when and only when $a_i = b_i$, for all $i \geq 1$.

Some results on probability distributions

A real-valued random variable X is characterized by its *distribution function*

$$F_X(x) = P(X \leq x), \quad x \in \mathbb{R}, \tag{1.5.42}$$

which may be continuous everywhere or may have jump discontinuities. Two important classes of distribution functions are the *absolutely continuous type* and the *discrete type*. In the former case, $F_X(x)$ has a derivative $f_X(x) = (\mathrm{d}/\mathrm{d}x)F_X(x)$ almost everywhere (a.e.),[2] which is also called the *probability density function* (p.d.f.) of X. In the latter case, $F_X(x)$ is a step function and its jump discontinuities represent the probability masses $f_X(x)$ associated with the discrete points $\{x\}$; in this setup, $f_X(x)$ is often called the *probability function* (p.f.). The pdf and pf are also referred to as the density with respect to the Lebesgue and sign measures, respectively. Given a function g, we use the notation

$$\int_{x\in\mathbb{R}} g(x)\mathrm{d}F_X(x)$$

to represent either

$$\int_{x\in\mathbb{R}} g(x)f_X(x)\mathrm{d}x \quad \text{or} \quad \sum_{x_i\in\mathbb{N}} g(x_i)f_X(x_i)$$

according as the density exists or otherwise. In particular, for $g(x) = x$, this is the expectation $\mathbb{E}(X)$ and for $g(x) = [x - \mathbb{E}(X)]^2$, the variance, $\mathbb{V}\mathrm{ar}(X)$.

Jensen Inequality: Let X be a random variable and $g(x)$, $x \in \mathbb{R}$, be a convex function such that $\mathbb{E}[g(X)]$ exists. Then,

$$g[\mathbb{E}(X)] \leq \mathbb{E}[g(X)], \tag{1.5.43}$$

with the equality sign holding only when g is linear a.e.

Proof. Let $a < b < c$ be three arbitrary points in $\mathbb{R}$, and define

$$\alpha = \frac{c-b}{c-a} \quad \text{and} \quad \beta = 1 - \alpha = \frac{b-a}{c-a}.$$

Note that $b = \alpha a + \beta c$. Also, by the definition of a convex function, as displayed in Figure 1.5.1, we have

$$\begin{aligned} g(b) = g(\alpha a + \beta b) &\leq \alpha g(a) + \beta g(c) \\ &= (c-a)^{-1}[(c-b)g(a) + (b-a)g(c)], \end{aligned}$$

[2] The term *almost everywhere* (a.e.) following a statement means that it holds for all sets excepting those with measure zero. A similar comment holds for the term *almost all*. For details and other measure theoretical concepts we may refer to Billingsley (1995).

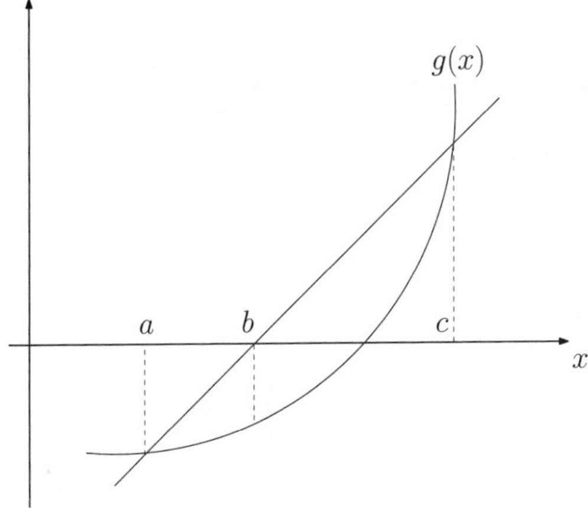

Figure 1.5.1. Representation of the convex function $g(x)$.

leading to

$$\frac{g(b) - g(a)}{(b - a)} \leq \frac{g(c) - g(a)}{(c - a)}, \quad a < b < c,$$

and this implies that for $h > 0$,

$$h^{-1}[g(x + h) - g(x)] \quad \text{is nondecreasing in } x. \tag{1.5.44}$$

Therefore, letting $\mathbb{E}(X) = \theta$, we obtain, by using (1.5.44), that

$$g(x) - g(\theta) \geq g'_+(\theta)(x - \theta), \quad x \geq \theta, \tag{1.5.45}$$

$$g(x) - g(\theta) \geq g'_-(\theta)(x - \theta), \quad x < \theta, \tag{1.5.46}$$

where

$$g'_+(\theta) = \lim_{h \downarrow 0} h^{-1}[g(\theta + h) - g(\theta)],$$
$$g'_-(\theta) = \lim_{h \downarrow 0} h^{-1}[g(\theta) - g(\theta - h)],$$

and, by (1.5.44), it follows that $g'_+(\theta) \geq g'_-(\theta)$. Letting $g'(\theta) = g'_+(\theta) = g'_-(\theta)$ whenever $g'_+(\theta) = g'_-(\theta)$ and $g'(\theta)$ be some value in the interval $[g'_-(\theta),\ g'_+(\theta)]$ whenever $g'_-(\theta) \neq g'_+(\theta)$, we obtain from (1.5.45) and (1.5.46) that

$$g(x) - g(\theta) \geq g'(\theta)(x - \theta), \quad x \in \mathbb{R}, \tag{1.5.47}$$

and this implies that $\mathbb{E}[g(X)] - g(\theta) \geq g'(\theta)\mathbb{E}(X - \theta) = 0$, that is, $\mathbb{E}[g(X)] \geq g(\theta)$. Furthermore, in (1.5.47), a strict equality sign holds for all x only when $g(x)$ is linear in x, and hence, for $g(\theta) = \mathbb{E}[g(X)]$, we need that $g(x)$ be linear a.e. ■

Next, we consider some of the distribution functions more commonly adopted in statistical inference.

(i) *Binomial distribution.* The random variable X has the $\text{Bin}(n, \pi)$ distribution, $0 < \pi < 1$, if its probability function is

$$P(X = x) = f_X(x) = \binom{n}{x}\pi^x(1-\pi)^{n-x}, \quad x = 0, 1, \ldots, n. \tag{1.5.48}$$

The corresponding distribution function is

$$F_X(y) = \sum_{x=0}^{\min([y],n)} \binom{n}{x}\pi^x(1-\pi)^{n-x}, \quad y \geq 0.$$

Also,

$$\mathbb{E}(X) = \sum_{x=0}^{n} x f_X(x) = n\pi \quad \text{and} \quad \mathbb{V}\text{ar}(X) = n\pi(1-\pi). \tag{1.5.49}$$

(ii) *Negative binomial distribution.* The random variable X has the $\text{NBin}(n, \pi)$ distribution, with parameters $r > 0$ and $0 < \pi < 1$ if its probability function is

$$P(X = x) = f_X(x) = \binom{r+x-1}{x}\pi^r(1-\pi)^x, \quad x = 0, 1, \ldots. \tag{1.5.50}$$

Here we have

$$\mathbb{E}(X) = r(1-\pi)/\pi \quad \text{and} \quad \mathbb{V}\text{ar}(X) = r(1-\pi)/\pi^2. \tag{1.5.51}$$

(iii) *Poisson distribution.* The random variable X follows a $\text{Poisson}(\lambda)$ distribution if the probability function is

$$P(X = x) = f_X(x) = e^{-\lambda}\lambda^x/x!, \quad x \geq 0, \ \lambda > 0. \tag{1.5.52}$$

For this discrete distribution,

$$\mathbb{E}(X) = \mathbb{V}\text{ar}(X) = \lambda. \tag{1.5.53}$$

(iv) *Uniform*$[0, 1]$ *distribution.* If the random variable X follows the $\text{Uniform}[0, 1]$ distribution, the density function is defined as

$$f_X(x) = \begin{cases} 1, & x \in [0, 1] \\ 0, & \text{otherwise;} \end{cases} \tag{1.5.54}$$

so that $F_X(x) = x, 0 \leq x \leq 1$, $F_X(x) = 0$, $x \leq 0$ and $F_X(x) = 1$, $x \geq 1$. Here,

$$\mathbb{E}(X) = \frac{1}{2} \quad \text{and} \quad \mathbb{V}\text{ar}(X) = \frac{1}{12}. \tag{1.5.55}$$

(v) *Gamma distribution.* A random variable X has the $\text{Gamma}(c, p)$ distribution if the corresponding density function is specified by

$$f_X(x) = c^p[\Gamma(p)]^{-1}e^{-cx}x^{p-1}, \quad x \geq 0, \tag{1.5.56}$$

where $c > 0$ and $p > 0$ are associated parameters and $\Gamma(p) = \int_0^\infty \exp(-y)y^{p-1}\mathrm{d}y$ is the *gamma function*. The distribution function F_X is

$$F_X(x) = \int_0^x f_X(y)\mathrm{d}y = \int_0^{cx} e^{-y}y^{p-1}\mathrm{d}y \Big/ \int_0^{+\infty} e^{-y}y^{p-1}\mathrm{d}y$$
$$= I_p(cx), \quad x \geq 0. \tag{1.5.57}$$

The function $I_p(x)$ is known as the *incomplete gamma function*. It is easy to verify that

$$\mathbb{E}(X) = pc^{-1} \quad \text{and} \quad \mathbb{V}\text{ar}(X) = pc^{-2}. \tag{1.5.58}$$

The gamma distribution with parameters $p = k/2$, $k \geq 1$, integer and $c = 1/2$ is known as the (central) *chi-squared* distribution with k degrees of freedom.

(vi) *(Negative) exponential distribution.* A random variable X has the Exp(c), $c > 0$ distribution if its density function is given by

$$f_X(x) = ce^{-cx}, \quad x \geq 0, \quad c > 0. \tag{1.5.59}$$

Hence, the exponential distribution is a special case of the gamma distribution, where $p = 1$, and

$$\mathbb{E}(X) = c^{-1} \quad \text{and} \quad \mathbb{V}\text{ar}(X) = c^{-2}. \tag{1.5.60}$$

(vii) *Beta distribution.* If X is a random variable with a Beta(p, q) distribution, $p > 0$ and $q > 0$, then its density function is

$$f_X(x) = \frac{\Gamma(p+q)}{\Gamma(p)\Gamma(q)} x^{p-1}(1-x)^{q-1}, \quad 0 \leq x \leq 1. \tag{1.5.61}$$

The distribution function $F_X(x) = \int_0^x f_X(y)\mathrm{d}y$ is, therefore, the *incomplete beta function.* We may verify that

$$\mathbb{E}(X) = \frac{p}{p+q} \quad \text{and} \quad \mathbb{V}\text{ar}(X) = \frac{pq}{(p+q)^2(p+q+1)}. \tag{1.5.62}$$

(viii) *Double exponential distribution.* The random variable X has DExp(θ, λ) distribution, $\theta \in \mathbb{R}$, $\lambda > 0$, if its density function is given by

$$f_X(x) = \frac{\lambda}{2} e^{-\lambda|x-\theta|}, \quad -\infty < x < \infty. \tag{1.5.63}$$

We may verify that

$$\mathbb{E}(X) = \theta \quad \text{and} \quad \mathbb{V}\text{ar}(X) = 2\lambda^{-2}. \tag{1.5.64}$$

(ix) *Cauchy distribution.* The density function for a random variable with the Cauchy(θ, λ) distribution is

$$f_X(x) = (\lambda/\pi)\{\lambda^2 + (x-\theta)^2\}^{-1}, \quad -\infty < x < \infty. \tag{1.5.65}$$

It is easy to verify that $\mathbb{E}(X)$ and $\mathbb{V}\text{ar}(X)$ do not exist.

(x) *Normal distribution.* A random variable X has the $\mathcal{N}(\mu, \sigma^2)$ distribution function with parameters $\mu \in \mathbb{R}$ and $\sigma^2 > 0$ if its density function is given by

$$f(x; \mu, \sigma) = \frac{1}{\sqrt{2\pi}\sigma} e^{-(x-\mu)^2/2\sigma^2}, \quad -\infty < x < \infty. \tag{1.5.66}$$

It is easily seen that

$$\mathbb{E}(X) = \mu \quad \text{and} \quad \mathbb{V}\text{ar}(X) = \sigma^2. \tag{1.5.67}$$

When $\mu = 0$ and $\sigma = 1$, (1.5.66) reduces to $\varphi(x) = (1/\sqrt{2\pi})\exp(-x^2/2)$, the *standard normal density function*. Note that

$$f(x; \mu, \sigma) = \frac{1}{\sigma}\varphi\left(\frac{x-\mu}{\sigma}\right).$$

The *standard normal distribution function* is

$$\Phi(x) = \int_{-\infty}^{x} \varphi(y)\mathrm{d}y, \qquad -\infty < x < \infty.$$

We denote the α-quantile by z_α, that is,

$$z_\alpha = \Phi^{-1}(\alpha).$$

If the independent random variables $X_1, \ldots, X_n$ follow normal distributions with $\mathbb{E}(X_i) = 0$ and $\mathbb{V}\text{ar}(X_i) = \sigma^2$ $i = 1, \ldots, n$ then

$$Q_n = \sum_{i=1}^{n} X_i^2/\sigma^2$$

follows a chi-squared distribution with n degrees of freedom, denoted by χ_n^2. The corresponding α-quantile is denoted by $\chi_{n,\alpha}^2$.

If $\mathbb{E}(X_i) = \mu_i$, $i = 1, \ldots, n$, then Q_n follows the *noncentral chi-squared distribution* with n degrees of freedom and noncentrality parameter $\sum_{i=1}^{n} \mu_i^2/\sigma^2$. Consider now the case where $\mathbb{E}(X_i) = 0$, $i = 1, \ldots, n$, and let X_0 be a random variable following a normal distribution with $\mathbb{E}(X_0) = \delta$ and $\mathbb{V}\text{ar}(X_0) = 1$; then,

$$t_n = \sqrt{n}X_o/\sqrt{Q_n}$$

follows a central (noncentral) *Student t distribution* with n degrees of freedom whenever $\delta = 0$ ($\delta \neq 0$); δ is the noncentrality parameter. The α-quantile for a Student t distribution with n degrees of freedom is denoted by $t_{n,\alpha}$. Finally, if we let $X_1, \ldots, X_{n_1}$ and $Y_1, \ldots, Y_{n_2}$ denote two sets of independent normal random variables with $\mathbb{E}(X_i) = \mu$, $\mathbb{E}(Y_j) = 0$, $\mathbb{V}\text{ar}(X_i) = \mathbb{V}\text{ar}(Y_j) = \sigma^2$, $i = 1, \ldots, n_1$, $j = 1, \ldots, n_2$ and let $Q_{n_1} = \sum_{i=1}^{n_1} X_i^2/\sigma^2$ and $Q_{n_2} = \sum_{j=1}^{n_2} Y_j^2/\sigma^2$, then,

$$F = n_2 Q_{n_1}/n_1 Q_{n_2}$$

follows a central (noncentral) *F distribution* with n_1 degrees of freedom in the numerator and n_2 degrees of freedom in the denominator whenever $\mu = 0$ ($\mu \neq 0$); μ^2/σ^2 is the noncentrality parameter.

(xi) *Logistic distribution.* The random variable X follows a Logistic(α, β) distribution, $\alpha \in \mathbb{R}$, $\beta > 0$ if the corresponding density function is

$$f_X(x) = \exp[-(x-\alpha)/\beta]/\{\beta(1 + \exp[-(x-\alpha)/\beta]^2\}, \qquad -\infty < x < \infty. \tag{1.5.68}$$

It is easy to see that

$$\mathbb{E}(X) = \alpha \quad \text{and} \quad \mathbb{V}\text{ar}(X) = \beta^2\pi^2/3. \tag{1.5.69}$$

There are several functions that are associated with a distribution function F. Among these, the *probability generating function*, *moment generating function*, and *characteristic function* deserve special mention. The probability generating function $g(s)$ corresponding to the probability law $P = (p_o, p_1, p_2, \ldots)$ is defined as

$$g(s) = E(s^X) = \sum_{k \geq 0} s^k p_k = p_o + \sum_{k \geq 1} s^k p_k. \tag{1.5.70}$$

Note that $g(0) = p_o$, $g(1) = \sum_{k \geq 0} p_k = 1$, and $g'(s) = \sum_{k \geq 1} k s^{k-1} p_k$ is nonnegative, so that $g(s)$ is increasing in s $(0 \leq s \leq 1)$. Furthermore, note that $g'(0) = p_1$, $g'(1) = \sum_{k \geq 1} k p_k = \mathbb{E}(X)$. Similarly, $g^{(2)}(s) = \sum_{k \geq 1} k(k-1)s^{k-2} p_k$ so that $g^{(2)}(0) = 2!p_2$ and $g^{(2)}(1) = \mathbb{E}[X(X-1)] = \mathbb{E}(X^{[2]})$, the second factorial moment. Then, for (Exercise 1.5.2), it can be shown that

$$\frac{1}{k!} g^{(k)}(0) = p_k, \ k \geq 0 \quad \text{and} \quad g^{(k)}(1) = \mathbb{E}(X^{[k]}), \ k \geq 0.$$

Thus, the successive derivatives of $g(s)$ at $s = 0$, generate the probabilities $p_k, k \geq 0$, while the $g^{(k)}(1)$ provide the factorial moment.

Let X and Y be two independent random variables with probability Laws P_1 and P_2 on the common support $\{0, 1, 2, \ldots\}$. Then,

$$g_{X+Y}(s) = \mathbb{E}(s^{X+Y}) = \mathbb{E}(s^X)\mathbb{E}(s^Y) = g_X(s) g_Y(s),$$

for all $s \in (0, 1)$. Exercise 1.5.3 is set to verify that for independent $X_1, \ldots, X_n$,

$$g_{X_1 + \cdots + X_n}(s) = \prod_{i=1}^{n} g_{X_i}(s);$$

moreover, if the X_i are identically distributed, then

$$g_{X_1 + \cdots + X_n}(s) = [g_{X_1}(s)]^n. \tag{1.5.71}$$

The moment generating function of a distribution function F is defined by

$$M_F(t) = \mathbb{E}_F[\exp(tX)] = \int e^{tx} \mathrm{d}F_X(x), \quad t \in \mathbb{R} \tag{1.5.72}$$

whenever the integral on the right-hand side exists. The function $M_F(t)$ provides the moments of F by successive differentiation, that is, for every positive integer k,

$$\mathbb{E}_F(X^k) = \mu^{(k)} = (\mathrm{d}^k/\mathrm{d}t^k) M_F(t)\big|_{t=0}$$

and, hence, it bears its name. Unfortunately, $M_F(t)$ may not always exist [e.g., for the Cauchy distribution in (1.5.65)]. For this reason, often, it is more convenient to work with the characteristic function, which is defined as follows. For t real, the characteristic function corresponding to the distribution function F is

$$\begin{aligned} \phi_F(t) &= \int e^{it} \mathrm{d}F_X(x) = \int \cos(tx) \mathrm{d}F_X(x) + i \int \sin(tx) \mathrm{d}F_X(x) \\ &= \phi_F^{(1)}(t) + i\phi_F^{(2)}(t), \end{aligned} \tag{1.5.73}$$

where $i = \sqrt{-1}$. Since $|\exp(itx)| = 1$, for all $x \in \mathbb{R}$, $\phi_F(t)$ exists for all F. Note that $\phi_F(0) = 1$, $|\phi_F(t)| \leq 1$ and $\phi_F(t)$ is uniformly continuous on the entire real line. Furthermore, note that $\phi_F(-t)$ is the characteristic function of the random variable $(-1)X$ [whose distribution function is $1 - F(-x)$], so that if X has a distribution function F_X symmetric about 0 [i.e., $F_X(-x) + F_X(x) = 1$, for all x], then $\phi_F(-t) = \phi_F(t)$, and $\phi_F(t)$ is real. In general, by (1.5.72), $\phi_F(t)$ is a complex-valued function. Exercises (1.5.1)–(1.5.12) are set to compute the characteristic functions of the distribution functions presented above. As we see, they all have different characteristic functions. In fact, there is a one-to-one correspondence between distribution functions and their characteristic functions. Two different distribution functions F and G cannot have the same characteristic function, and, conversely, two different characteristic functions cannot relate to a common distribution function. Characteristic functions have nice convergence properties: we will present such results in Chapter 6. If moments of F up to a certain order exist, then the characteristic function $\phi_F(t)$ can be expanded in a form, which provides a very handy tool for analysis. We present this basic result in the following theorem.

***Theorem* 1.5.2.** *Let Y be a random variable with distribution function F, such that $\nu^{(k+\delta)} = \mathbb{E}(|Y|^{k+\delta}) < \infty$ for some integer $k \geq 1$ and $0 < \delta < 1$. Also let $\mu^{(r)} = \mathbb{E}(Y^r)$, $r = 1, \ldots, k$. Then*

$$\phi_F(t) = \mathbb{E}[\exp(itY)] = 1 + it\mu^{(1)} + \cdots + (it)^k\mu^{(k)}/k! + R_k(t),$$

where $|R_k(t)| \leq c|t|^{k+\delta}\nu^{(k+\delta)}$ with $c < \infty$ independent of t.

Proof. First consider the expansion

$$e^{iu} = 1 + iu + \cdots + (iu)^k/k! + (iu)^k(e^{iu\xi} - 1)/k!,$$

where $0 < \xi < 1$. For $|u| < 2$, let $k = 1$ and note that

$$|e^{iu} - 1| = |u(i - ue^{i\xi u}/2)| \leq 2|u| = 4|u/2| \leq 4|u/2|^\delta. \tag{1.5.74}$$

For $|u| \geq 2$, observe that

$$|e^{iu} - 1| \leq 2 \leq 2|u/2|^\delta \leq 4|u/2|^\delta. \tag{1.5.75}$$

From (1.5.74) and (1.5.75), we may conclude that

$$|e^{iu} - 1| \leq 4|u/2|^\delta = 2^{2-\delta}|u|^\delta. \tag{1.5.76}$$

Using the same expansion as above, we may write

$$\phi_F(t) = \mathbb{E}(e^{itY}) = 1 + it\mathbb{E}(Y) + \cdots + (it)^k\mathbb{E}(Y^k)/k! + R_k(t),$$

where

$$R_k(t) = \frac{(it)^k}{k!}\int_{-\infty}^{+\infty} y^k(e^{it\xi y} - 1)\mathrm{d}F(y),$$

and, using (1.5.76), we have

$$\begin{aligned}
|R_k(t)| &\le \frac{|t|^k}{k!}\int_{-\infty}^{+\infty} |y|^k|e^{it\xi y} - 1|\mathrm{d}F(y) \\
&\le \frac{|t|^k}{k!}\int_{-\infty}^{+\infty} |y|^k 2^{2-\delta}|ty|^\delta \mathrm{d}F(y) \\
&= \frac{2^{2-\delta}|t|^{k+\delta}}{k!}\int_{-\infty}^{+\infty} |y|^{k+\delta}\mathrm{d}F(y) \\
&= C|t|^{k+\delta}\nu^{(k+\delta)}. \qquad (1.5.77)
\end{aligned}$$

■

Let us also present some parallel results for the multivariate case, which will be considered in subsequent chapters. The distribution function of a random vector $\boldsymbol{X} = (X_1, \ldots, X_p)'$, $p \ge 1$, is defined as

$$F(\boldsymbol{x}) = P(\boldsymbol{X} \le \boldsymbol{x}), \quad \boldsymbol{x} \in \mathbb{R}^p. \qquad (1.5.78)$$

The characteristic function of $\boldsymbol{X}$ (or F), a function of $\boldsymbol{t} = (t_1, \ldots, t_p)'$, is defined as

$$\phi_F(\boldsymbol{t}) = \int \exp(i\boldsymbol{t}'\boldsymbol{x})\mathrm{d}F(\boldsymbol{x}). \qquad (1.5.79)$$

The one-to-one correspondence between F and ϕ_F remains intact. Of particular importance is the class of *multivariate normal* distribution functions that can be uniquely represented by the characteristic functions

$$\phi_\Phi(\boldsymbol{t}) = \exp\left(i\boldsymbol{t}'\boldsymbol{\mu} - \frac{1}{2}\boldsymbol{t}'\boldsymbol{\Sigma}\boldsymbol{t}\right), \quad \boldsymbol{t} \in \mathbb{R}^p, \qquad (1.5.80)$$

where $\boldsymbol{\mu} = (\mu_1, \ldots, \mu_p)'$ is the mean vector and $\boldsymbol{\Sigma} = ((\sigma_{ij}))$ is the dispersion or covariance matrix corresponding to the distribution function Φ. In general, $\boldsymbol{\Sigma}$ is positive semidefinite (p.s.d.); if $\boldsymbol{\Sigma}$ is positive definite (p.d.), then the distribution function Φ is of full rank and has a density function given by

$$\varphi(\boldsymbol{x}) = (2\pi)^{-p/2}|\boldsymbol{\Sigma}|^{-1/2}\exp\left[-\frac{1}{2}(\boldsymbol{x}-\boldsymbol{\mu})'\boldsymbol{\Sigma}^{-1}(\boldsymbol{x}-\boldsymbol{\mu})\right], \quad \boldsymbol{x} \in \mathbb{R}^p. \quad (1.5.81)$$

Note that the density function does not exist if $|\boldsymbol{\Sigma}| = 0$, but through the characteristic function in (1.5.79) still characterize a (possibly degenerate) multinormal distribution.

Some results on order statistics and sample quantiles

In statistical inference the sample order statistics and empirical distributions play an important role. We present some related concepts here.

Let X be a real-valued random variable having a distribution function F. Consider a sample of n independent and identically distributed (i.i.d.) random variables $\{X_1, \ldots, X_n\}$ drawn from the distribution function F. We may define the *sample* or *empirical distribution* by

$$F_n(x) = n^{-1}\sum_{i=1}^{n} I(X_i \le x), \quad x \in \mathbb{R}. \qquad (1.5.82)$$

For each fixed sample, F_n is a distribution function when considered as a function of x. For every fixed $x \in \mathbb{R}$, when considered as a function of $X_1, \ldots, X_n$, $F_n(x)$ is a random variable; in this context, since the $I(X_i \leq x)$, $i = 1, \ldots, n$, are i.i.d. zero-one–valued random variables, we have

$$\mathbb{E}[F_n(x)] = \frac{1}{n}\sum_{i=1}^{n}\mathbb{E}[I(X_i \leq x)] = \frac{1}{n}\sum_{i=1}^{n}P(X_i \leq x) = F(x) \tag{1.5.83}$$

and

$$\mathbb{V}\text{ar}[F_n(x)] = \frac{1}{n^2}\sum_{i=1}^{n}\mathbb{V}\text{ar}[I(X_i \leq x)] = \frac{F(x)[1-F(x)]}{n} \leq \frac{1}{4n}. \tag{1.5.84}$$

For the time being, in order to simplify the analysis, let us assume that F is continuous a.e., so that

$$P(X_i = X_j, \text{ for some } i \neq j, i, j = 1, \ldots, n) = 0. \tag{1.5.85}$$

Let us arrange the observations $X_1, \ldots, X_n$ in ascending order of magnitude and denote these ordered random variables by

$$X_{n:1} \leq X_{n:2} \leq \cdots \leq X_{n:k} \leq \cdots \leq X_{n:n}. \tag{1.5.86}$$

By virtue of (1.5.85), ties among the X_i and (hence, the $X_{n:k}$) can be neglected with probability 1, so that in (1.5.86) we may as well replace all the "$\leq$" signs by strict "$<$" signs. Note that $X_{n:k}$ is the kth smallest observation in the sample and $X_{n:n-k+1}$, is the kth largest observation, for $k = 1, \ldots, n$. Also,

$$\boldsymbol{t}_n = (X_{n:1}, \ldots, X_{n:n})' \tag{1.5.87}$$

is the vector of *sample order statistics*. Depending on whether k/n is close to 0 (or 1) or converging to some p, $0 < p < 1$, $X_{n:k}$ will be called an *extreme value* or a *sample quantile*, and we will make this point clear in Chapter 6. Let

$$R_i = \sum_{j=i}^{n} I(X_j \leq X_i) \tag{1.5.88}$$

be the *rank* of X_i among $X_1, \ldots, X_n$, for $i = 1, \ldots, n$, so that $R_1, \ldots, R_n$ represent the natural integers, permuted in some (random) manner. We let

$$\boldsymbol{R} = (R_1, \ldots, R_n)', \tag{1.5.89}$$

so that it takes on the permutations of $(1, \ldots, n)$, there being $n!$ possible realizations. Then, note that by (1.5.86) and (1.5.88),

$$X_i = X_{n:R_i}, \quad i = 1, \ldots, n. \tag{1.5.90}$$

By reversing the order of the index, it is also possible to write

$$X_{n:i} = X_{S_i}, \quad i = 1, \ldots, n, \tag{1.5.91}$$

where S_i is termed the *antirank*, $i = 1, \ldots, n$. For the last two equations, we have, therefore,

$$R_{S_i} = i = S_{R_i}, \quad i = 1, \ldots, n, \tag{1.5.92}$$

so that $\boldsymbol{S} = (S_1, \ldots, S_n)^t$ also takes on the permutations of $(1, \ldots, n)$. Although the X_i are assumed to be i.i.d. random variables, the $X_{n:k}$ are neither independent nor identically distributed. Moreover, by (1.5.82) and (1.5.86), we have, for $1 \le k \le n-1$,

$$F_n(x) = \begin{cases} 0, & x < X_{n:1} \\ k/n, & X_{n:k} \le x \le X_{n:k+1} \\ 1, & x \ge X_{n:n}. \end{cases} \tag{1.5.93}$$

Thus, F_n and $\boldsymbol{t}_n$ are very much interrelated, and this will have consequences on the study of the properties of the empirical distribution functions and order statistics.

The empirical distribution function F_n may generally be defined in a multivariate setup as follows. Let $\boldsymbol{X}_1, \ldots, \boldsymbol{X}_n$ be p-vectors, for some $p \ge 1$. Then, (1.5.82) extends to the p-variate case as

$$F_n(\boldsymbol{x}) = n^{-1} \sum_{i=1}^{n} I(\boldsymbol{X}_i \le \boldsymbol{x}), \quad \boldsymbol{x} \in \mathbb{R}^p. \tag{1.5.94}$$

An empirical distributional process is defined as $W_n = \{W_n(\boldsymbol{x}); \boldsymbol{x} \in \mathbb{R}^p\}$, where

$$W_n(\boldsymbol{x}) = \sqrt{n}[F_n(\boldsymbol{x}) - F(\boldsymbol{x})], \quad \boldsymbol{x} \in \mathbb{R}^p. \tag{1.5.95}$$

The univariate case (i.e., $p = 1$) is the most used one and will be considered in detail; multivariate cases will be treated only briefly.

In the univariate case, to be more general, we drop the assumption that F is continuous a.e. In that case, (1.5.85) may not hold, so that the *ties* among the X_i may not be negligible. Suppose that in the sample of size n there are m $(= m_n)$ distinct values (arranged in ascending order)

$$X^*_{n:1} < \cdots < X^*_{n:m}, \tag{1.5.96}$$

and that $X^*_{n:k}$ occurs with the frequency $f^*_k \ge 1, k = 1, \ldots, m$, so that $f^*_1 + \cdots + f^*_m = n$. In this setup, the vectors $\boldsymbol{t}_n$ of sample order statistics can be equivalently expressed as

$$\boldsymbol{t}_n = (X^*_{n:1}, f^*_1; \ldots, X^*_{n:k}, f^*_k; \ldots; X_{n:m}, f^*_m), \tag{1.5.97}$$

where m $(= m_n)$ is itself a positive integer-valued random variable, such that $P(1 \le m \le n) = 1$. Let $F^*_k = f^*_1 + \cdots + f^*_k$, $1 \le k \le m$, stand for the cumulative frequencies. With these notations, for $1 \le k \le m-1$, (1.5.93) extends to

$$F_n(x) = \begin{cases} 0, & x < X^*_{n:1} \\ n^{-1}F^*_k, & X^*_{n:k} \le x < X^*_{n:k+1} \\ 1 & x \ge X^*_{n:m}. \end{cases} \tag{1.5.98}$$

Thus, F_n is a *step function* having m *jump-points* $X^*_{n:1} < \cdots < X^*_{n:m}$, and the jump at $X^*_{n:k}$ has magnitude $n^{-1}f^*_k$, $k = 1, \ldots, m$. In the particular case of F being continuous a.e., we have $m = n$, $X^*_{n:k} = X_{n:k}$ and $f^*_k = 1$, $1 \le k \le n$, with probability 1, so that F_n has n jumps of equal magnitude (i.e., n^{-1}) at each of the sample order statistics $X_{n:1}, \ldots, X_{n:n}$. In any case, there is a one-to-one correspondence between F_n and $\boldsymbol{t}_n$, as defined above.

Consider next a fixed positive number $0 < p < 1$. If for an order statistic $X_{n:k}$, k/n converges to p in a suitable manner, then $X_{n:k}$ is called a *sample p-quantile*. Although

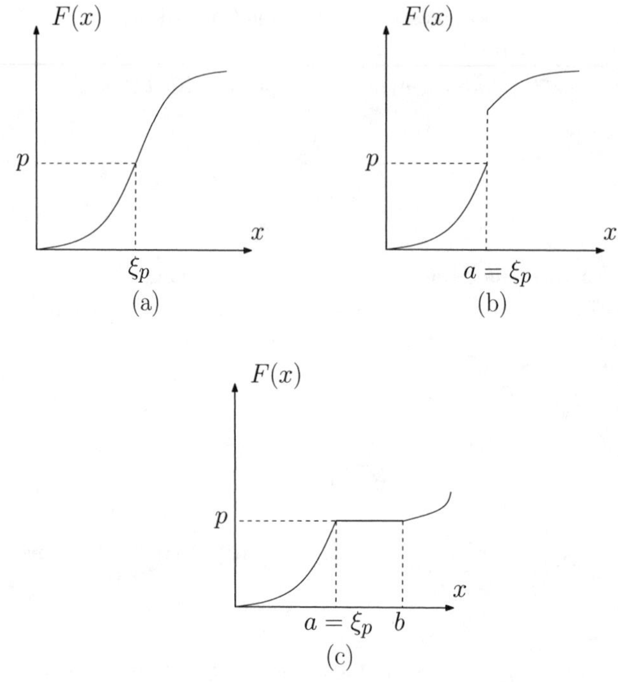

Figure 1.5.2. Population Quantiles.

there are various ways of defining such a k, the following are the most adopted ones:

$$k = k_p = [np] + 1, \quad \text{and} \tag{1.5.99}$$

$$k = k_p = [(n+1)p]. \tag{1.5.100}$$

Perhaps, it will be in order to define the population counterpart in an unambiguous manner. If the equation

$$F(x) = p \tag{1.5.101}$$

admits a unique solution ξ_p then it is called the *population p-quantile*. Such a definition entails that F is strictly monotone at ξ_p, so that no multiple roots exist [see plot (a) in Figure 1.5.2]. In plot (b), a is a jump point for F, so that $F(a-) < p < F(a)$, and, hence, (1.5.101) is not properly defined. On the other hand, in plot (c), $F(x) = p$, for $a \leq x \leq b$, so that there are multiple roots for (1.5.101).

To avoid such technical difficulties, often we define

$$\xi_p = F^{-1}(p) = \inf\{x\colon F(x) \geq p\}. \tag{1.5.102}$$

In the same manner, we may define the sample p-quantile as

$$\widehat{\xi}_{p,n} = F_n^{-1}(p) = \inf\{x\colon F_n(x) \geq p\}. \tag{1.5.103}$$

If F is continuous and strictly monotone at ξ_p, then, of course, ξ_p as defined by (1.5.101) and (1.5.102) are the same. But, (1.5.103) may not be in agreement with some other conventional definitions of sample quantiles. To clarify this point, let us consider the case of $p = 1/2$

for which $\xi_{0.5}$ is also called the *median*. If n is an odd number ($= 2m + 1$, say), then both (1.5.99) and (1.5.100) lead to $k_p = m + 1$. In the case of all n distinct order statistics, (1.5.103) also leads to $\widehat{\xi}_{p,n} = X_{n:m+1}$. On the other hand, if n is even ($= 2m$, say), (1.5.99) yields $k_p = m + 1$, (1.5.100) leads to $k_p = m$ and (1.5.103) leads to $\widehat{\xi}_{p,n} = X_{n:m}$. In this case, a more conventional definition of the sample median is $\widetilde{X}_n = \frac{1}{2}(X_{n:m} + X_{n:m+1})$. Since F_n is a step-function, in general, $\widehat{\xi}^{(1)}_{p,n} = \sup\{x\colon F_n(x) < p\}$ and $\widehat{\xi}^{(2)}_{p,n} = \inf\{x\colon F_n(x) > p\}$ define an open interval. For a given k, $(k-1)/n < p < k/n$, so that by linear interpolation, we may let

$$\widehat{\xi}_{p,n} = (k - np)\widehat{\xi}^{(1)}_{p,n} + (np - k + 1)\widehat{\xi}^{(2)}_{p,n}. \tag{1.5.104}$$

For simplicity, we let

$$\widehat{\xi}_{p,n} = \frac{1}{2}(\widehat{\xi}^{(1)}_{p,n} + \widehat{\xi}^{(2)}_{p,n}). \tag{1.5.105}$$

However, as we will see later that for large n, this modification is of minor importance, and any one of (1.5.99), (1.5.100), (1.5.103), or (1.5.104) will work out well.

Note that $X_{n:1}$ ($X_{n:n}$) is the sample smallest (largest) observation, and, in general, for any k (≥ 1), $X_{n:k}$ ($X_{n:n-k+1}$) is the kth smallest (largest) observation. In an asymptotic setup (where n is taken to be large), whenever $k/n \to 0$, these order statistics are called *extreme values*. Their populations counterparts may be introduced as follows. Suppose that there exists ξ_0 ($> -\infty$), such that

$$F(x) > 0, \quad x > \xi_0 \qquad \text{and} \qquad F(x) = 0, \quad x \leq \xi_0. \tag{1.5.106}$$

Then ξ_0 is called a *lower end point* of the distribution function F. Similarly, if there exists $\xi_1 < \infty$, such that

$$F(x) < 1, \quad x < \xi_1 \qquad \text{and} \qquad F(x) = 1, \quad x \geq \xi_1, \tag{1.5.107}$$

then ξ_1 is an *upper end point* of the distribution function F. If $\xi_0 = -\infty$ ($\xi_1 = +\infty$), the distribution function F is said to have an *infinite lower (upper) end point*. This can be further clarified through the following example.

***Example* 1.5.1.** If X is a random variable following an Exp(θ) distribution, then $\xi_0 = 0$ and $\xi_1 = \infty$ as depicted in plot (a) of Figure 1.5.3. On the other hand, if X is a random variable following a Uniform$[\mu - \theta/2, \mu + \theta/2]$ distribution, then $\xi_0 = \mu - \theta/2$ and $\xi_1 = \mu + \theta/2$ as shown in plot (b). $\square$

The behavior of the sample extreme values depends heavily on whether the population end points are finite or not, and also on how the distribution function F behaves in its tails. These results will be considered in Chapter 7.

Next we consider some results on the distribution function of sample order statistics. First, consider the kth order statistic $X_{n:k}$ and let us obtain the distribution function $Q_{n,k}(x) = P(X_{n:k} \leq x)$ assuming that the original distribution function (of the X_i) is F.

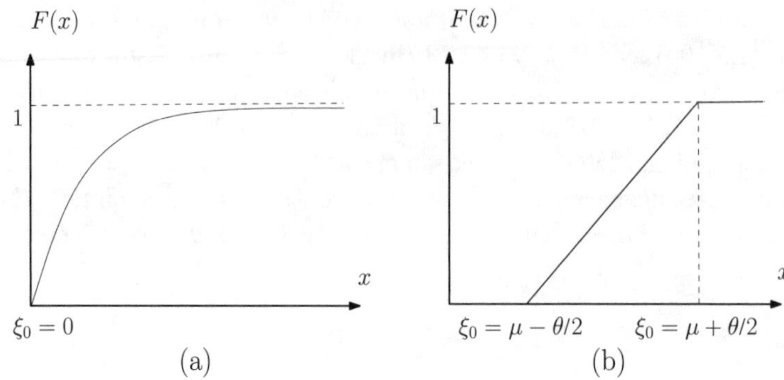

Figure 1.5.3. End points of Exponential and Uniform distributions.

Note that by definition,

$$\begin{aligned}
Q_{n,k}(x) &= P(X_{n:k} \leq x) = P(k \text{ or more of } (X_1, \ldots, X_n) \text{ are } \leq x) \\
&= P\left[\sum_{i=1}^{n} I(X_i \leq x) \geq k\right] = P\left[\frac{1}{n}\sum_{i=1}^{n} I(X_i \leq x) \geq \frac{k}{n}\right] \\
&= P\left[F_n(x) \geq \frac{k}{n}\right] = P[nF_n(x) \geq k].
\end{aligned}$$

Since $nF_n(x) \sim \text{Bin}[n, F(x)]$, it follows that

$$Q_{n,k(x)}(x) = \sum_{r=k}^{n} \binom{n}{r} [F(x)]^r [1 - F(x)]^{n-r}. \tag{1.5.108}$$

Now, recalling the incomplete beta function (1.5.61), we have

$$\begin{aligned}
B_{n,k}(u) &= \frac{\Gamma(n+1)}{\Gamma(k)\Gamma(n-k+1)} \int_0^u t^{k-1}(1-t)^{n-k} \mathrm{d}t \\
&= \frac{n!}{(k-1)!(n-k)!} \int_0^u t^{k-1}(1-t)^{n-k} \mathrm{d}t \\
&= \frac{n!}{k!(n-k)!} \int_0^u (1-t)^{n-k} \mathrm{d}t^k, \quad 0 < u < 1.
\end{aligned} \tag{1.5.109}$$

Integrating by parts, we get

$$\begin{aligned}
B_{n,k}(u) &= \binom{n}{k} \left[t^k(1-t)^{n-k}\right]_0^u + \binom{n}{k} \int_0^u t^k (n-k)(1-t)^{n-k-1} \mathrm{d}t \\
&= \binom{n}{k} u^k (1-u)^{n-k} + \frac{n!}{k!(n-k-1)!} \int_0^u t^k (1-t)^{n-k-1} \mathrm{d}t.
\end{aligned}$$

Repeating this process $(n - k - 1)$ times, we obtain

$$B_{n,k}(u) = \sum_{r=k}^{n} \binom{n}{r} u^r (1-u)^{n-r}, \quad 0 < u < 1. \tag{1.5.110}$$

From (1.5.108)–(1.5.110), it follows that

$$Q_{n,k}(x) = \frac{\Gamma(n+1)}{\Gamma(k)\Gamma(n-k+1)} \int_0^{F(x)} t^{k-1}(1-t)^{n-k}\mathrm{d}t, \quad x \in \mathbb{R}. \tag{1.5.111}$$

Now we suppose that F is absolutely continuous with density f; then the density of $X_{n:k}$ is given by

$$\begin{aligned} q_{n,k}(x) &= \frac{\mathrm{d}}{\mathrm{d}x} Q_{n,k}(x) = \frac{\mathrm{d}}{\mathrm{d}F(x)} Q_{n,k}(x) \frac{\mathrm{d}F(x)}{\mathrm{d}x} \\ &= \frac{\Gamma(n+1)}{\Gamma(k)\Gamma(n-k+1)} [F(x)]^{k-1}[1-F(x)]^{n-k} f(x) \\ &= k\binom{n}{k}[F(x)]^{k-1}[1-F(x)]^{n-k} f(x). \end{aligned} \tag{1.5.112}$$

Let us consider the joint distribution of a pair $(X_{n:k}, X_{n:q})$, for $k < q$. Note that

$$\begin{aligned} Q_{n,kq}(x,y) &= P(X_{n:k} \le x, X_{n:q} \le y) \\ &= P\left[\sum_{i=1}^{n} I(X_i \le x) \ge k, \sum_{i=1}^{n} I(X_i \le y) \ge q\right] \\ &= P[F_n(x) \ge k/n, F_n(y) \ge q/n] \\ &= \sum_{j=0}^{n-q} P[F_n(x) \ge k/n, F_n(y) = (q+j)/n] \\ &= \sum_{j=0}^{n-q} P[F_n(x) = (q+j)/n) P[F_n(x) \ge k/n | F_n(y) = (q+j)/n] \\ &= \sum_{j=0}^{n-q} \binom{n}{q+j} [F(y)]^{q+j}[1-F(y)]^{n-q+j} \sum_{r=k}^{q+j} \binom{q+j}{r} \\ &\quad \times \frac{[F(x)]^r[F(y)-F(x)]^{q+j-r}}{[F(y)]^{q+j}} \\ &= \sum_{j=0}^{n-q}\sum_{r=k}^{q+j} \binom{n}{q+j}\binom{q+j}{r} [F(x)]^r[F(y)-F(x)]^{q+j-r}[1-F(y)]^{n-q+j} \\ &= \sum_{j=q}^{n}\sum_{r=k}^{s} \binom{n}{s}\binom{s}{r} [F(x)]^r[F(y)-F(x)]^{s-r}[1-F(y)]^{n-s}. \end{aligned} \tag{1.5.113}$$

If F is absolutely continuous, then we have

$$\begin{aligned} \frac{\partial^2 Q_{n,kq}(x,y)}{\partial x \partial y} &= q_{n,kq}(x,y) \\ &= \frac{n!}{(k-1)!(q-k-1)!(n-q)!} \\ &\quad \times [F(x)]^{k-1}[F(y)-F(x)]^{q-k-1}[1-F(y)]^{n-q} f(x)f(y). \end{aligned} \tag{1.5.114}$$

We leave the proof of (1.5.114) as an exercise (Exercise 1.5.16).

Obviously, the use of (1.5.111)–(1.5.114) is restricted to situations where the form of the underlying distribution is known. In general, this is not the case and we must rely on the asymptotic properties of the sample quantiles, which are discussed in later chapters.

Next we consider the problem of existence of moments of order statistics; this issue is important because linear functions of order statistics, like *trimmed* or *Winsorized* means, constitute appealing alternatives to estimating parameters. Let $X_1, \dots, X_n$ be a random sample corresponding to a random variable X with distribution function F and suppose that $\mathbb{E}(|X|^a) < \infty$ for some $a > 0$. Let us show that if n, k and r satisfy $r \le a \min(k, n-k+1)$, then $\mathbb{E}(|X_{n:k}|^r) < \infty$. First, we note that $\mathbb{E}(|X|^a) < \infty$ implies $\mathbb{E}(|X|^r) < \infty$ for all $r \le a$. Next we observe that as $x \to \infty$,

$$\infty > \mathbb{E}(|X|^r) \ge \overline{c}_r(x) = \int_x^\infty |t|^r \mathrm{d}P(X \le t) \to 0, \tag{1.5.115}$$

and also that

$$\overline{c}_r(x) \ge |x|^r P(X \ge x) = |x|^r [1 - F(x)],$$

which implies

$$\begin{aligned} 0 = \lim_{x\to\infty} \overline{c}_r(x) &\ge \lim_{x\to\infty} |x|^r [1 - F(x)] \\ &\ge \lim_{x\to\infty} |x|^r [1 - F(x)] F(x). \end{aligned}$$

Since $|x|^r F(x)[1 - F(x)] \ge 0$, it follows that

$$\lim_{x\to\infty} |x|^r [1 - F(x)] = \lim_{x\to\infty} |x|^r [1 - F(x)] F(x) = 0. \tag{1.5.116}$$

Furthermore, we note that

$$\begin{aligned} \infty > \mathbb{E}(|X|^a) &\ge \int_0^\infty x^a \mathrm{d}P(X \le x) \\ &= \int_0^\infty x^a \mathrm{d}[1 - P(X \ge x)] \\ &= -\int_0^\infty x^a \mathrm{d}P(X \ge x). \end{aligned}$$

Integrating by parts, we obtain

$$\begin{aligned} &-\int_0^\infty x^a \mathrm{d}P(X \ge x) \\ &= -x^{a-1} P(X \ge x)\Big|_0^\infty + a \int_0^\infty x^{a-1} P(X \ge x) \mathrm{d}x \\ &= a \int_0^\infty x^{a-1} [1 - F(x)] \mathrm{d}x < \infty \end{aligned} \tag{1.5.117}$$

in view of (1.5.116). Then putting $P(X_{n:k} \le x) = Q_{n,k}(x)$, we may write

$$\mathbb{E}(|X_{n:k}|^r) = \int_{-\infty}^0 |x|^r \mathrm{d}Q_{n,k}(x) + \int_0^\infty x^r \mathrm{d}Q_{n,k}(x). \tag{1.5.118}$$

Using an argument similar to that considered in (1.5.117) and recalling (1.5.108), we get

$$\begin{aligned}
\int_0^\infty x^r \mathrm{d}Q_{n,k}(x) &= r\int_0^\infty x^{r-1}P(X_{n:k}\geq x)\mathrm{d}x \\
&= r\sum_{i=0}^{k-1}\binom{n}{i}\int_0^\infty x^{r-1}[F(x)]^i[1-F(x)]^{n-i}\mathrm{d}x \\
&= r\sum_{i=0}^{k-1}\binom{n}{i}\int_0^\infty x^{r-a}[1-F(x)]^{n-i-1}[F(x)]^i x^{a-1}[1-F(x)]\mathrm{d}x \\
&= r\sum_{i=0}^{k-1}\binom{n}{i}\int_0^\infty \{x^a[1-F(x)]\}^{(r-a)/a}[1-F(x)]^{n-i-r/a} \\
&\quad\times [F(x)]^i x^{a-1}[1-F(x)]\mathrm{d}x.
\end{aligned} \tag{1.5.119}$$

From (1.5.115) it follows that $\{x^a[1-F(x)]\}^{(r-a)/a}$ is finite; we note also that $[F(x)]^i$ is bounded and that $[1-F(x)]^{n-i-r/a}$ is bounded if $n-i-r/a\geq 0$ for $i=0,\ldots,k-1$ or equivalently if $r\leq a(n-k+1)$. Thus, if $r\leq a(n-k-1)$ it follows from (1.5.117) and (1.5.119) that

$$\int_0^\infty x^r\mathrm{d}Q_{n,k}(x)\leq M\int_0^\infty x^{a-1}[1-F(x)]\mathrm{d}x<\infty \tag{1.5.120}$$

Using similar arguments we may show that

$$\int_{-\infty}^0 |x|^r\mathrm{d}Q_{n,k}(x)<\infty \quad \text{if} \quad r\leq ak. \tag{1.5.121}$$

Then, from (1.5.120) and (1.5.121) if $r\leq a\min(k,n-k+1)$, it follows that

$$\mathbb{E}(|X_{n:k}|^r)=\int_{-\infty}^0|x|^r\mathrm{d}Q_{n,k}(x)+\int_0^\infty x^r\mathrm{d}Q_{n,k}(x)<\infty. \tag{1.5.122}$$

***Example* 1.5.2.** Let X be a random variable with the Cauchy distribution in (1.5.65). First we observe that, since

$$\mathbb{E}(|X|)=\frac{2}{\pi}\int_0^\infty\frac{x\mathrm{d}x}{1+x^2}=\frac{1}{\pi}\int_0^\infty\frac{\mathrm{d}(1+x^2)}{1+x^2}=\frac{1}{\pi}\log x\Big|_1^\infty=\infty,$$

the mean $\overline{X}_n$ of a random sample $X_1,\ldots,X_n$ from that distribution is such that $\mathbb{E}(|\overline{X}_n|^r)=\infty$ for all $r\geq 1$. However, we note that, for all $\varepsilon>0$,

$$\mathbb{E}(|X|^{1-\varepsilon})=\frac{1}{\pi}\int_0^\infty x^{-\varepsilon}\frac{2x}{1+x^2}\mathrm{d}x,$$

which upon integration by parts yields

$$\begin{aligned}
\mathbb{E}(|X|^{1-\varepsilon}) &= \frac{1}{\pi}\left[\frac{\log(1+x^2)}{x^\varepsilon}\Big|_0^\infty+\varepsilon\int_0^\infty x^{-1-\varepsilon}\log(1+x^2)\mathrm{d}x\right] \\
&= \frac{\varepsilon}{\pi}\int_0^\infty x^{-1-\varepsilon}\log(1+x^2)\mathrm{d}x \\
&= \frac{\varepsilon}{\pi}\int_0^\infty x^{-1-\varepsilon/2}\frac{\log(1+x^2)}{x^{\varepsilon/2}}\mathrm{d}x.
\end{aligned} \tag{1.5.123}$$

Observing that $\lim_{x\to\infty} x^{-\varepsilon/2}\log(1+x^2)=0$, it follows that there exist $x_0>0$ and $M>0$ such that for all $x>x_0$ we have $x^{-\varepsilon/2}\log(1+x^2)<M$. Therefore, from (1.5.123) we may write

$$\mathbb{E}(|X|^{1-\varepsilon}) = \frac{\varepsilon}{\pi}\left[\int_0^{x_0} x^{-1-\varepsilon}\log(1+x^2)\mathrm{d}x + M\int_{x_0}^{\infty} x^{-1-\varepsilon/2}\mathrm{d}x\right] < \infty.$$

Suppose that $n=3$ and consider the sample median (i.e., take $k=2$); using (1.5.122), it follows that $\mathbb{E}(|X_{3:2}|^r)$ exists for $r\le(1-\varepsilon)\min(2,2)=2-2\varepsilon$, which implies that the sample median from a sample of size 3 from a Cauchy distribution will have finite mean but infinite variance.

Now we consider the sample quantile $X_{n:k}$ where $k=[np]+1$, $0<p<1$. Then,

$$\begin{aligned} r &\le (1-\varepsilon)\min(k, n-k+1) \\ &= (1-\varepsilon)\min\{[np]+1, n-[np]\} = (1-\varepsilon)O(np). \end{aligned}$$

If, for example, $n=50$ and $p=1/2$, we may choose ε conveniently so that $r\le 24$; in other words, the sample median from a sample of size 50 from a Cauchy distribution has finite moments of order up to 24. □

1.6 Exercises

1.2.1 A sample of n portions of lemon juice was obtained from a large batch. Each sample was subdivided into m portions and for each portion the content of essential oils (expressed in *kg* of oil/*ton* of juice) was obtained. Specify different statistical models that allow the estimation of the mean content of essential oils and identify the appropriate methods to analyze the data, indicating whether they are exact or approximate. In each case, state the assumptions clearly and discuss whether they seem reasonable in the light of the practical aspects of the study. [Suggestions may be obtained in Singer, Pedroso de Lima, Tanaka, and González-López (2007).]

1.2.2 A sample of n three-child families was selected from a certain community and each of them was classified according to the gender of the children (three males and no females, two males and one female, etc.). The objective was to verify whether the probability of a male birth is one-half. Specify statistical models to answer the study question, stating the assumptions and discussing whether they are reasonable or not from a practical point of view. Identify the appropriate methods required to analyze the data, indicating whether they are exact or approximate.

1.2.3 The data in Table 1.6.1 were extracted from a study designed to evaluate the effect of salt ingestion on blood pressure for hypertense patients. Specify statistical models to answer the study question, stating the assumptions and indicate the appropriate methods for analysis of the data. Suppose now that the high sodium values for subjects 2 and 9 were erroneously recorded and thus that the actual values must be considered as missing. What changes must be introduced in the statistical models and analyses to account for the missing data?

1.2.4 The data in Table 1.6.2 were extracted from a study conducted at the Dental School of the University of São Paulo. They correspond to the measurements of a dental plaque index (on a scale that ranges from zero to four) obtained for each of fourteen male and twelve female preschoolers, before and after the use of a conventional or an experimental toothbrush. One of the objectives of the study was to compare the two toothbrushes with respect to the reduction of dental plaque controlling for the motor ability of the female and male children

Table 1.6.1. *Systolic blood pressure (mm Hg).*

Subject	Type of Diet: Low sodium	High sodium
1	138	143
2	147	154
3	146	147
4	154	147
5	142	157
6	156	158
7	134	164
8	146	156
9	143	151
10	175	182
11	117	116
12	128	125

Table 1.6.2. *Dental plaque index.*

		Type of toothbrush			
		Experimental		Conventional	
Subject	Sex	Before brushing	After brushing	Before brushing	After brushing
1	F	2.18	0.43	1.20	0.75
2	F	2.05	0.08	1.43	0.55
3	F	1.05	0.18	0.68	0.08
4	F	1.95	0.78	1.45	0.75
5	F	0.28	0.03	0.50	0.05
6	F	2.63	0.23	2.75	1.60
7	F	1.50	0.20	1.25	0.65
8	F	0.45	0.00	0.40	0.13
9	F	0.70	0.05	1.18	0.83
10	F	1.30	0.30	1.43	0.58
11	F	1.25	0.33	0.45	0.38
12	F	0.18	0.00	1.60	0.63
13	F	3.30	0.90	0.25	0.25
14	F	1.40	0.24	2.98	1.03
15	M	0.90	0.15	3.35	1.58
16	M	0.58	0.10	1.50	0.20
17	M	2.50	0.33	4.08	1.88
18	M	2.25	0.33	3.15	2.00
19	M	1.53	0.53	0.90	0.25
20	M	1.43	0.43	1.78	0.18
21	M	3.48	0.65	3.50	0.85
22	M	1.80	0.20	2.50	1.15
23	M	1.50	0.25	2.18	0.93
24	M	2.55	0.15	2.68	1.05
25	M	1.30	0.05	2.73	0.85
26	M	2.65	0.25	3.43	0.88

and to estimate the corresponding average reduction in dental plaque after toothbrushing. Specify different statistical models for the the problem stating the assumptions and indicating whether the corresponding inferential methods are exact or approximate. [Suggestions may be obtained in Singer and Andrade (1997).]

1.2.5 Specify three situations (different from the ones contained in the text) for which exact statistical methods are available. In this context, you should describe the problem clearly, specify the statistical model considered for inference, discuss the plausibility of the assumptions, and detail the analytical methods suggested.

1.5.1 For the binomial law (1.5.48), show that the characteristic function is $\phi(t) = (\pi \exp(it) + 1 - \pi)^n$, $t \in \mathbb{R}$. Use this to verify (1.5.49).

1.5.2 Show that if g is a probability generating function associated to the probability distribution $\boldsymbol{P} = (p_0, p_1, \ldots)$, we have

$$\frac{1}{k!} g^{(k)}(0) = p_k,\ k \geq 0 \quad \text{and} \quad g^{(k)}(1) = \mathbb{E}\big(X^{[k]}\big),\ k \geq 0.$$

1.5.3 Let $X_1, \ldots, X_n$ be independent random variables, g_{X_i} denote the probability generating function of X_i, $i = 1, \ldots, n$, and let $g_{X_1+\cdots+X_n}$ denote the probability generating function of the sum $X_1 + \cdots + X_n$. Show that

$$g_{X_1+\cdots+X_n}(s) = \prod_{i=1}^{n} g_{X_i}(s);$$

furthermore, if the X_i are identically distributed, show that

$$g_{X_1+\cdots+X_n}(s) = [g_{X_1}(s)]^n.$$

1.5.4 In the context of Exercise 1.5.3, let $Z_n = X_1 + \cdots + X_n$. Show that for every $n \geq 1$,

$$g_n(s) = \mathbb{E}(s_n^{Z_n}) = g_{n-1}[g(s)] = g[g_{n-1}(s)],\ 0 \leq s \leq 1,$$

1.5.5 Obtain the characteristic function for the Poisson law (1.5.52) and use it to verify (1.5.53).

1.5.6 For the Uniform[0, 1] distribution, obtain the characteristic function and verify (1.5.55).

1.5.7 For the gamma distribution (1.5.56), derive the characteristic function. Hence or otherwise, derive the characteristic function for the chi-squared distribution with k degrees of freedom.

1.5.8 Derive the characteristic function for the beta distribution (1.5.61) and hence, or otherwise, verify (1.5.62).

1.5.9 For the double-exponential distribution (1.5.63) derive the characteristic function and verify (1.5.64).

1.5.10 Show that the characteristic function of the Cauchy distribution does not satisfy the condition for the existence of its first two moments.

1.5.11 For the normal distribution (1.5.66), show that the characteristic function is $\phi(t) = \exp(it\mu - t^2\sigma^2/2)$, $t \in \mathbb{R}$.

1.5.12 Using the results from Exercise 1.5.8, derive the characteristic function for the $F(p, n-p)$ distribution.

1.5.13 Use (1.5.111) to show that for $n = 2m + 1$ and $k = m + 1$, when F is symmetric about its median θ, then $Q_{n,k}(x)$ is also symmetric about θ.

1.5.14 Define $q_{n,k}$ as in (1.5.111) and assume that $f(x) = F'(x)$ is symmetric about θ. Then show that

$$q_{n,k}(\theta + x) = q_{n,n-k+1}(\theta - x), \quad x \in \mathbb{R}.$$

Extend this symmetry result to the bivariate case, i.e., to the density of $(X_{n:k}, X_{n:n-k+1})$. Hence, or otherwise, show that for $n = 2m$ and $k = m$ when the median is defined as in (1.5.104), under the symmetry of F, it has a symmetric distribution too.

1.5.15 Let θ be the median of F [i.e., $F(\theta) = 1/2$], and use (1.5.108) to verify that for every $n > 1$ and $0 \leq r < s \leq n + 1$ (where $X_{n:0} = -\infty$ and $X_{n:n+1} = +\infty$), we have

$$P(X_{n:r} \leq \theta \leq X_{n:s}) = 2^{-n} \sum_{i=r}^{s-1} \binom{n}{i}.$$

Choose $s = n - r + 1, r \leq n/2$ and show that the above formula provides a distribution-free confidence interval for θ. What can be said about the confidence coefficient (i.e., coverage probability) when n is not large?

1.5.16 Obtain expression (1.5.114).

2

Estimation Theory

2.1 Introduction

We consider a set of independent and identically distributed (i.i.d.) random variables $\boldsymbol{X}_1, \ldots, \boldsymbol{X}_n$ following a probability law $P_{\boldsymbol{\theta}}$, the form of which is known up to some associated parameter $\boldsymbol{\theta}$, appearing as a constant. Both the $\boldsymbol{X}_i$ and $\boldsymbol{\theta}$ can be vector-valued. The set $\mathcal{X}$ of possible values of the sample $\boldsymbol{X}_1, \ldots, \boldsymbol{X}_n$ is called the *sample space*. We assume that $\boldsymbol{\theta} \in \boldsymbol{\Theta}$, the *parameter space*, so that a parametric family of probability laws may be represented as $\mathcal{P}_{\boldsymbol{\theta}} = \{P_{\boldsymbol{\theta}}; \boldsymbol{\theta} \in \boldsymbol{\Theta}\}$. One of the objectives of statistical estimation theory is to develop methods of choosing appropriate *statistics* $\boldsymbol{T}_n = \boldsymbol{T}(\boldsymbol{X}_1, \ldots, \boldsymbol{X}_n)$, that is, functions of the sample observations, to estimate $\boldsymbol{\theta}$ (i.e., to guess the true value of $\boldsymbol{\theta}$) in a reproducible way. In this context, $\boldsymbol{T}_n$ is an *estimator* of $\boldsymbol{\theta}$.

In an alternative (*nonparametric*) setup, we may allow the probability law P to be a member of a general class $\mathcal{P}$, not necessarily indexed by some parameter, and then our interest lies in estimating the probability law P itself or some *functional* $\boldsymbol{\theta}(P)$ thereof, without specifying the form of P.

There may be many variations of this setup wherein the $\boldsymbol{X}_i$ may not be independent nor identically distributed as in multisample models, linear models, time-series models, stochastic processes, or unequal probability sampling models. However, to motivate the basic statistical problems, we start with the most simple set of i.i.d. real-valued random variables $X_1, \ldots, X_n$ and a real-valued parameter θ, so that the parameter space is $\Theta \subseteq \mathbb{R} = (-\infty, \infty)$.

The choice of estimators is governed by some principles and criteria that provide some objective basis for comparing the existing alternatives. Section 2.2 deals with some of these basic tools. Section 2.3 is devoted to the foundations of *likelihood* and *information bounds* and their role in estimation theory. *Sufficiency*, *completeness*, and *ancillarity* concepts are also illustrated in Sections 2.2 and 2.3. Various methods of estimation are presented in Section 2.4. The following section deals with finite sample *optimality* properties of estimators. The transit from finite (small) to asymptotic (large) sample perspectives is briefly outlined in the concluding section, with cross reference to Chapter 8 for full exploitation. *Confidence intervals* (dual to appropriate *hypothesis testing* problems) are considered in Chapter 4.

2.2 Basic Concepts

An estimator $T_n = T(X_1, \ldots, X_n)$, being a statistic, is itself a random variable; its distribution function, $G_{T_n}(\cdot, \theta)$, is determined by the probability law P_θ, the sample size n, and its functional form. Thus, it may not be possible to have $T_n = \theta$ (as desired) all the time, unless $G_{T_n}(\cdot, \theta)$ is degenerate at θ. In fact, if $G_{T_n}(\cdot, \theta)$ is continuous and nondegenerate, $P_\theta(T_n = \theta) = 0$. However, it is quite natural to expect that on the average, $T_n = \theta$, that is,

$$\mathbb{E}_\theta(T_n) = \theta, \quad \forall \theta \in \Theta. \tag{2.2.1}$$

The estimator T_n is said to be *unbiased* for θ if (2.2.1) holds. The *bias* in estimating θ by T_n is defined in the same vein as

$$b_n(T_n, \theta) = \mathbb{E}_\theta(T_n) - \theta, \quad \theta \in \Theta, \tag{2.2.2}$$

so that for unbiased estimators, the bias is zero. This feature is tied down to estimators for whose the first moment is finite. In some problems, an estimator T_n may not have a finite expectation and yet it can be *very good* in terms of other criteria that will be discussed here. For example, let X_i be a binary random variable with

$$P(X_i = 1) = 1 - P(X_i = 0) = \pi, \quad 0 < \pi < 1.$$

In some situations (see Example 6.3.5), we may be interested in estimating $\theta = \pi^{-1}$. A natural estimator of θ is $T_n = n / \sum_{i=1}^n X_i$. Since

$$P_\pi\left(\sum_{i=1}^n X_i = 0\right) = (1 - \pi)^n > 0,$$

$\mathbb{E}_\theta(T_n) = +\infty$, $\forall \theta \in \Theta$, $n \geq 1$. The modified estimator

$$T_n^* = (n + 1) \bigg/ \left(\sum_{i=1}^n X_i + 1\right)$$

has a finite first moment but is biased, since $\mathbb{E}_\theta(T_n^*) \neq \theta$.

Unbiasedness, though desirable, is not essential. In Chapter 4 while studying *risk* in a decision theoretic setup, we will see that often biased estimators may have better performance than unbiased ones. An alternative criterion in this respect is *median unbiasedness*, which is defined as follows. If for an estimator T_n of θ,

$$P_\theta(T_n \leq \theta) = P_\theta(T_n \geq \theta), \quad \forall \theta \in \Theta, \tag{2.2.3}$$

then T_n is median unbiased for θ. This criterion does not require the existence of the first moment of T_n.

On the other hand, a desirable characteristic for an estimator is that its sampling distribution should become more concentrated around θ as the sample size n increases, that is, as more information on θ becomes available from the sample observations. Thus, with $n \to \infty$, the estimator T_n should converge to θ in some well-defined sense. Although this topic will be elaborated in Chapter 6, the notion of *consistency* is presented here in a more intuitive way. An estimator T_n is consistent for θ if, for all $\varepsilon > 0$,

$$P(|T_n - \theta| > \varepsilon) \to 0, \quad \text{as } n \to \infty. \tag{2.2.4}$$

Note that consistency is essentially a large sample property, though even in finite samples, inconsistent estimators are rarely advocated.

Generally, there are many estimators for some parameter θ, even if we confine ourselves to the class of consistent ones. Hence, further scrutiny is necessary to choose some appropriate one(s) from among them. Intuitively speaking, given a set of competing estimators

$\{T_n\}$, it is appealing to choose the estimator T_n^* for which the sampling distribution, $G_{T_n^*}$, is most concentrated around θ. A possible way to characterize this feature is to write

$$P(|T_n^* - \theta| > \varepsilon) = \inf_{T_n} P_\theta(|T_n - \theta| > \varepsilon), \quad \forall \varepsilon > 0. \tag{2.2.5}$$

However intuitive it might be, it is generally hard to implement (2.2.5) as a discriminating criterion in finite samples. For that reason, moment-based measures are generally used. In this context, we define the *mean squared error* (MSE) of T_n as

$$\text{MSE}(T_n, \theta) = \mathbb{E}_\theta(T_n - \theta)^2, \quad \theta \in \Theta, \ T_n \in \Im, \tag{2.2.6}$$

where $\Im$ is the class of all estimators having finite MSE. Note that if T_n were unbiased for θ, then its MSE would reduce to its variance; otherwise,

$$\text{MSE}(T_n, \theta) = \mathbb{V}\text{ar}_\theta(T_n) + b_n^2(T_n, \theta), \quad \theta \in \Theta, \tag{2.2.7}$$

where the bias $b_n(T_n, \theta)$ is defined in (2.2.2). An estimator T_n^* of θ is called the *uniformly minimum MSE* (UMMSE) estimator within a class $\Im$ of estimators if

$$\text{MSE}(T_n^*, \theta) = \inf_{T_n \in \Im} \nu_{T_n}^2(\theta), \quad \forall \theta \in \Theta. \tag{2.2.8}$$

If $\Im$ relates to the class of unbiased estimators with finite second moments, then (2.2.8) yields the *uniformly minimum variance unbiased estimator* (UMVUE) of θ. Having said this, either way, it may still be hard to obtain a UMMSE estimator (or UMVUE) of θ in finite samples, except under certain regularity conditions. As such, for further discussion of *efficient* estimators in finite samples, we introduce first the notions of *sufficiency, likelihood,* and *information* (on θ).

A statistic $T_n = T(X_1, \ldots, X_n)$, is called a *sufficient statistic* for some parameter θ if it accounts for all information on θ contained in the entire sample $X_1, \ldots, X_n$. Let $L_n(X_1, \ldots, X_n; \theta)$ be the joint density of $X_1, \ldots, X_n$, and $g_n(T_n|\theta)$ be the density of T_n; both of these densities generally depend on $\theta \in \Theta$. Consider then the conditional density of $(X_1, \ldots, X_n)$, given T_n, which is

$$p(X_1, \ldots, X_n|T_n; \theta) = L_n(X_1, \ldots, X_n; \theta)/g_n(T_n|\theta). \tag{2.2.9}$$

If this conditional density does not depend on θ (but on T_n), then T_n is sufficient for θ. Since $g_n(T_n|\theta)$ depends only on θ (and T_n) while $p(X_1, \ldots, X_n|T_n; \theta)$ does not depend on θ ($\in \Theta$), when T_n is sufficient for θ, any one-to-one function $h(T_n) = T_n^*$ is also sufficient for θ in the sense of (2.2.9). This is summarized in the following theorem.

Theorem 2.2.1 (Factorization Theorem). *A necessary and sufficient condition for T_n to be sufficient for θ with respect to $\mathcal{P}_\theta = \{P_\theta;\ \theta \in \Theta\}$ is that the joint density $L_n(X_1, \ldots, X_n; \theta)$ can be factorized as*

$$L_n(X_1, \ldots, X_n; \theta) = g_1(T_n; \theta) g_2(X_1, \ldots, X_n), \quad \theta \in \Theta, \tag{2.2.10}$$

where g_2 is free from θ.

In fact, the *sufficiency principle* is an immediate byproduct of (2.2.9) and (2.2.10); it stipulates that whenever a sufficient statistic T_n exists, to make inferences about θ, it suffices to formulate statistical tools that depend only on T_n. The Factorization Theorem 2.2.1 and the sufficiency principle play a focal role in statistical estimation theory and to appraise this scenario we need to explore further developments in the domain of sufficiency; this will be done in the next section. Also, note that there is room for flexibility in the choice of g_1 and g_2.

A special class of densities deserves some attention in respect to its prominent role in statistical inference under sufficiency. A density (or probability function) $f(x;\boldsymbol{\theta})$, of a random variable X is said to be a member of the *exponential family of densities* if it can be expressed as

$$f(x;\boldsymbol{\theta}) = c(x)\exp\left[\sum_{j=1}^{q} t_j(x)a_j(\boldsymbol{\theta}) - b(\boldsymbol{\theta})\right], \quad \boldsymbol{\theta} \in \Theta, \tag{2.2.11}$$

where $c(x)$ and the $t_j(x)$ do not depend on $\boldsymbol{\theta}$, and the $a_j(\boldsymbol{\theta})$ and $b(\boldsymbol{\theta})$ are functions of $\boldsymbol{\theta}$ alone. If $X_1, \ldots, X_n$ are i.i.d. random variables, each having the density $f(x;\boldsymbol{\theta})$ in (2.2.11), then their joint density is

$$L_n(X_1, \ldots, X_n;\boldsymbol{\theta}) = \left[\prod_{i=1}^{n} c(X_i)\right]\exp\left[\sum_{j=1}^{q} a_j(\boldsymbol{\theta})T_{nj} - nb(\boldsymbol{\theta})\right], \tag{2.2.12}$$

where $T_{nj} = \sum_{i=1}^{n} t_j(X_i)$, $1 \le j \le q$, are statistics free from $\boldsymbol{\theta}$. As such, by (2.2.10) and (2.2.12), we conclude that $\boldsymbol{T}_n = (T_{n1}, \ldots, T_{nq})'$ is a sufficient statistic for $\boldsymbol{\theta}$. The normal, negative exponential, gamma, beta, binomial, Poisson, and multinomial distributions all belong to this exponential family; the uniform, Laplace, logistic, Cauchy, and some other distributions do not.

2.3 Likelihood, Information, and Sufficiency

When the joint density of the sample observations, $L_n(X_1, \ldots, X_n;\theta)$ is viewed as a function of θ given $X_1, \ldots, X_n$, it is called the *likelihood function*. For a given θ ($\in \Theta$), L_n is a function of the X_i. On the other hand, given the sample $X_1, \ldots, X_n$, L_n is a function of θ ($\in \Theta$), and hence, is sometimes expressed as $L_n(\theta; X_1, \ldots, X_n)$, $\theta \in \Theta$. For drawing inference on θ, based on the sample observations, this concept has long been recognized in the statistical literature; both *fiducial inference* and *Bayesian methods* exploit this principle. We will make more comments on that in Chapter 4.

It is often more convenient to work with the *log-likelihood* function, namely, $l_n(\theta; X_1, \ldots, X_n) = \log L_n(\theta; X_1, \ldots, X_n)$ and an important statistic derived therefrom is the *likelihood score statistic* or simply *score statistic*, defined as

$$U_n(\theta) = \frac{\partial}{\partial\theta} l_n(\theta; X_1, \ldots, X_n). \tag{2.3.1}$$

In the case of i.i.d. random variables, $X_1, \ldots, X_n$ we have

$$U_n(\theta) = \sum_{i=1}^{n} \frac{\partial}{\partial\theta}\log f(X_i;\theta) = \sum_{i=1}^{n} \frac{f'_\theta(X_i;\theta)}{f(X_i;\theta)}, \tag{2.3.2}$$

where $f'_\theta(x;\theta) = (\partial/\partial\theta)f(x;\theta)$. Note that

$$\mathbb{E}_\theta[U_n(\theta)] = n\mathbb{E}_\theta\left[\frac{f'_\theta(X_1;\theta)}{f(X_1;\theta)}\right] = n\int \frac{f'_\theta(x;\theta)}{f(x;\theta)}\mathrm{d}F(x;\theta), \tag{2.3.3}$$

where the integration extends over the support of f (i.e., the range of x for which $f(x;\theta) > 0$); for simplicity, we consider the case of continuous F only. Then we note that under the standard regularity assumptions permitting the interchange of the order of integration and differentiation, namely,

$$\{x : f(x;\theta) > 0\} \text{ does not depend on } \theta; \tag{2.3.4}$$

and

$$\mathbb{E}_\theta\left[\left|\frac{\partial}{\partial\theta}\log f(X;\theta)\right|\right] = \int |f'_\theta(x;\theta)|\mathrm{d}x < \infty, \quad \forall\theta \in \Theta, \tag{2.3.5}$$

(2.3.3) implies that

$$\mathbb{E}_\theta[U_n(\theta)] = n\int_{\mathbb{R}} \frac{\partial}{\partial\theta}f(x;\theta)\mathrm{d}x = n\left(\frac{\partial}{\partial\theta}\right)\int_{\mathbb{R}} f(x;\theta)\mathrm{d}x = 0, \tag{2.3.6}$$

since the integral is equal to 1. Next we define *Fisher's information* on θ *per sample unit* as

$$I(\theta) = \mathbb{E}_\theta\left[\frac{\partial}{\partial\theta}\log f(X;\theta)\right]^2 < \infty, \quad \theta \in \Theta \tag{2.3.7}$$

so that

$$\mathbb{E}_\theta[U_n^2(\theta)] = nI(\theta), \tag{2.3.8}$$

is the information on θ in the sample of size n.

Here, using the Jensen inequality (1.5.43) we note that

$$I(\theta) \ge \left[\mathbb{E}\left(\left|\frac{\partial}{\partial\theta}\log f(X;\theta)\right|\right)\right]^2, \quad \theta \in \Theta, \tag{2.3.9}$$

so that (2.3.6) holds under a less restrictive condition than (2.3.7), since $I(\theta) < \infty$ implies $\{\mathbb{E}[|(\partial/\partial\theta)\log f(X;\theta)|]\}^2 < \infty$, but the converse may not be true.

Let us next elaborate on the role of sufficiency in this context. The information (on θ) contained in a statistic T_n with density $g_n(T_n;\theta)$, is defined analogously as

$$I_{T_n}(\theta) = \mathbb{E}_\theta\left[\frac{\partial}{\partial\theta}\log g_n(T_n;\theta)\right]^2. \tag{2.3.10}$$

If T_n is a sufficient statistic, the conditional density $p(X_1, \ldots, X_n|T_n;\theta)$ is free from θ, and then, from (2.2.9) it follows that for almost all $X_1, \ldots, X_n$,

$$\frac{\partial}{\partial\theta}\log L_n(X_1, \ldots, X_n;\theta) = \frac{\partial}{\partial\theta}\log g_n(T_n;\theta) + 0, \quad \theta \in \Theta. \tag{2.3.11}$$

Consequently,

$$\begin{aligned} nI(\theta) &= \mathbb{E}\left[\frac{\partial}{\partial\theta}\log L_n(X_1, \ldots, X_n;\theta)\right]^2 \\ &= \mathbb{E}\left[\frac{\partial}{\partial\theta}\log g_n(T_n;\theta)\right]^2 = I_{T_n}(\theta), \quad \forall\theta \in \Theta. \end{aligned} \tag{2.3.12}$$

It is in this sense that a sufficient statistic preserves all the information on θ contained in the sample, and thereby allows a possible data reduction by ignoring the part that is irrelevant for drawing inference on θ.

Suppose now that there are two sample points $\boldsymbol{X} = (X_1, \ldots, X_n)'$ and $\boldsymbol{X}^* = (X_1^*, \ldots, X_n^*)'$, such that $L_n(\boldsymbol{X};\theta) = L_n(\boldsymbol{X}^*;\theta)$, for all $\theta \in \Theta$. Then, if there is a sufficient statistic T_n taking on values $T(\boldsymbol{X})$ and $T(\boldsymbol{X}^*)$ at $\boldsymbol{X}$ and $\boldsymbol{X}^*$, respectively, we must have $T(\boldsymbol{X}) = T(\boldsymbol{X}^*)$; conversely, $T(\boldsymbol{X}) = T(\boldsymbol{X}^*)$ implies that $L_n(\boldsymbol{X};\theta) = L_n(\boldsymbol{X}^*;\theta)$, $\forall \theta \in \Theta$. On the sample space $\mathcal{X}$ we define the *orbit*

$$\mathcal{X}_T = \{\boldsymbol{X} \in \mathcal{X} : T(\boldsymbol{X}) = T\}, \quad T \in \mathbb{R}. \tag{2.3.13}$$

Then on $\mathcal{X}_T$, the likelihood function has a constant value, possibly different for different T, and hence, for statistical inference we may rely on sufficient statistics alone. This theme will be further elaborated in the light of the likelihood and sufficiency principles presented in Chapter 4.

In addition to (2.3.5), let us assume that

$$\int |f_\theta''(x;\theta)| \mathrm{d}x < \infty, \quad \forall \theta \in \Theta, \tag{2.3.14}$$

where, $f_\theta''(x;\theta) = (\partial^2/\partial\theta^2) f(x;\theta)$. Then, since

$$\begin{aligned}\frac{\partial^2}{\partial\theta^2} \log f(x;\theta) &= \frac{\partial}{\partial\theta}\left[\frac{f_\theta'(x;\theta)}{f(x;\theta}\right] \\ &= \frac{f_\theta''(x;\theta)}{f(x;\theta)} - \left[\frac{f_\theta'(x;\theta)}{f(x;\theta}\right]^2, \quad \theta \in \Theta,\end{aligned} \tag{2.3.15}$$

we obtain

$$\begin{aligned}\mathbb{E}_\theta\left[-\frac{\partial^2}{\partial\theta^2} \log f(X;\theta)\right] &= \mathbb{E}_\theta\left\{\left[\frac{f_\theta'(X;\theta)}{f(X;\theta)}\right]^2\right\} - \mathbb{E}_\theta\left[\frac{f_\theta''(X;\theta)}{f(X;\theta)}\right] \\ &= I(\theta) - \int f_\theta''(x;\theta)\mathrm{d}x \\ &= I(\theta) - \frac{\partial^2}{\partial\theta^2}\int f(x;\theta)\mathrm{d}x = I(\theta), \quad \forall\theta \in \Theta.\end{aligned} \tag{2.3.16}$$

Thus, defining the *curvature of the log-likelihood* as

$$\begin{aligned}V_n(\theta) &= -\frac{\partial^2}{\partial\theta^2} \log L_n(X_1, \ldots, X_n;\theta) \\ &= \sum_{i=1}^n \left[-\frac{\partial^2}{\partial\theta^2} \log f(X_i;\theta)\right], \quad \theta \in \Theta,\end{aligned} \tag{2.3.17}$$

and assuming the regularity condition above, it follows that

$$\mathbb{E}[V_n(\theta)] = nI(\theta), \quad \forall\theta \in \Theta. \tag{2.3.18}$$

The next theorem shows that the results on the Fisher information may be used to provide a lower bound to the MSE of θ.

Theorem 2.3.1 (Cramér–Rao–Fréchet). *Let T_n denote an estimator of some parameter θ. Assume that the regularity conditions (2.3.4), (2.3.5), and (2.3.7) hold and let*

$\psi_n(\theta) = \mathbb{E}_\theta(T_n)$, $\theta \in \Theta$. *Then,*

$$\mathbb{E}_\theta[T_n - \psi_n(\theta)]^2 \geq [\psi_n'(\theta)]^2/[nI(\theta)], \quad \theta \in \Theta, \tag{2.3.19}$$

with equality holding only when

$$T_n - \psi_n(\theta) = K_n U_n(\theta) \quad \text{a.e.}, \tag{2.3.20}$$

where K_n *is a constant that does not depend on* $\boldsymbol{X} = (X_1, \ldots, X_n)'$.

Proof. First we note that

$$\mathbb{E}_\theta(T_n) = \psi_n(\theta) = \int \cdots \int T_n L_n(\boldsymbol{x};\theta)\mathrm{d}\boldsymbol{x},$$

where $\boldsymbol{x} = (x_1, \ldots, x_n)'$. Then, we use (2.3.4) and (2.3.5) to see that

$$\begin{aligned}
\psi_n'(\theta) &= \frac{\partial}{\partial\theta}\psi_n(\theta) = \frac{\partial}{\partial\theta}\int \cdots \int T_n L_n(\boldsymbol{x};\theta)\mathrm{d}\boldsymbol{x} \\
&= \int \cdots \int T_n \frac{\partial}{\partial\theta} L_n(\boldsymbol{x};\theta)\mathrm{d}\boldsymbol{x} \\
&= \int \cdots \int T_n U_n(\theta) L_n(\boldsymbol{x};\theta)\mathrm{d}\boldsymbol{x} \\
&= \int \cdots \int [T_n - \psi_n(\theta)] U_n(\theta) L_n(\boldsymbol{x};\theta)\mathrm{d}\boldsymbol{x} \\
&\quad + \psi_n(\theta) \int \cdots \int U_n(\theta) L_n(\boldsymbol{x};\theta)\mathrm{d}\boldsymbol{x} \\
&= \mathbb{E}_\theta\{[T_n - \psi_n(\theta)]U_n(\theta)\} + \psi_n(\theta)\mathbb{E}_\theta[U_n(\theta)] \\
&= \mathbb{E}_\theta\{[T_n - \psi_n(\theta)]U_n(\theta)\}.
\end{aligned} \tag{2.3.21}$$

Therefore, by the Cauchy–Schwarz inequality (1.5.34),

$$[\psi_n'(\theta)]^2 = \{\mathbb{E}_\theta[(T_n - \psi_n(\theta)]U_n(\theta)\}^2 \leq \mathbb{E}_\theta[T_n - \psi_n(\theta)]^2 \mathbb{E}_\theta[U_n^2(\theta)], \tag{2.3.22}$$

so that

$$\mathbb{E}_\theta[T_n - \psi_n(\theta)]^2 \geq [\psi_n'(\theta)]^2/\mathbb{E}_\theta[U_n^2(\theta)] = [\psi_n'(\theta)]^2/[nI(\theta)]. \tag{2.3.23}$$

Equality in (2.3.22) holds only when

$$T_n - \psi_n(\theta) = K_n U_n(\theta) \quad \text{a.e.}, \tag{2.3.24}$$

where K_n is a constant, possibly dependent on θ but not on $\boldsymbol{X}$. ■

Let us now illustrate the utility of the Cramér–Rao–Fréchet Theorem 2.3.1 in estimation theory. First, we note that if T_n is unbiased for θ, that is, if $\psi_n(\theta) = \theta$, then $\psi_n'(\theta) = 1$, and it follows that

$$\mathbb{E}_\theta(T_n - \theta)^2 = \mathbb{V}\mathrm{ar}(T_n) \geq [nI(\theta)]^{-1}, \quad \forall \theta \in \Theta, \tag{2.3.25}$$

where equality holds only when

$$T_n - \theta = K_n U_n(\theta) \quad \text{a.e.} \tag{2.3.26}$$

Note that (2.3.26) implies

$$\log L_n(\boldsymbol{X}, \theta) = \int K_n^{-1}(T_n - \theta)\mathrm{d}\theta + \log C_n,$$

where $\log C_n$ is an integration constant that does not depend on θ. We may then express the above equation as

$$L_n(\boldsymbol{x}; \theta) = C_n \exp[T_n a_n(\theta) - b_n(\theta)], \tag{2.3.27}$$

where $a_n(\theta) = \int K_n^{-1}\mathrm{d}\theta$ and $b_n(\theta) = \int \theta K_n^{-1}\mathrm{d}\theta$. Now, since (2.3.27) relates to a one-parameter exponential density of the type (2.2.11) with $q = 1$, it follows that T_n is a sufficient statistic for θ.

Conversely, if $f(x; \theta)$ belongs to the exponential family (2.2.11) [with $q = 1$, and $a_1(\theta) = a(\theta)$], then

$$L_n(\boldsymbol{X}; \theta) = \left[\prod_{i=1}^{n} c(X_i)\right] \exp[T_n a(\theta) - nb(\theta)],$$

where $\mathbb{E}_\theta(T_n) = nb'(\theta)/a'(\theta)$, so that

$$U_n(\theta) = \frac{\partial}{\partial\theta}\log L_n(\boldsymbol{X}; \theta) = T_n a'(\theta) - nb'(\theta) = a'(\theta)[T_n - \mathbb{E}_\theta(T_n)],$$

and hence (2.3.24) holds. The proof is the object of Exercise 2.3.1. As such, if we define $\overline{T}_n = T_n/n$ and note that $\mathbb{E}_\theta(\overline{T}_n) = b'(\theta)/a'(\theta) = \psi(\theta)$, say, we may rewrite

$$U_n(\theta) = [na'(\theta)][\overline{T}_n - \psi(\theta)].$$

When $\psi(\theta)$ is a one-to-one transformation of θ, we let $\lambda = \psi(\theta)$ so that $\partial\lambda/\partial\theta = \psi'(\theta)$. Thus,

$$\frac{\partial}{\partial\lambda}\log L_n(\boldsymbol{X}; \theta) = \left[\frac{\partial}{\partial\theta}\log L_n(\boldsymbol{X}; \theta)\right] \Big/ \left(\frac{\partial\lambda}{\partial\theta}\right) = U_n(\theta)/\psi'(\theta),$$

and consequently,

$$nI(\lambda) = \mathbb{E}_\lambda\left[\frac{\partial}{\partial\lambda}\log L_n(\boldsymbol{X}; \theta)\right]^2 = nI(\theta)/[\psi'(\theta)]^2.$$

As a result, from the Cramér–Rao–Fréchet Theorem 2.3.1, we conclude that for one-parameter exponential models with densities of the form

$$L_n(\boldsymbol{X}; \theta) = \left[\prod_{i=1}^{n} c(X_i)\right] \exp\{na'(\theta)[\overline{T}_n - \lambda]\},$$

it follows that

$$\mathbb{E}_\theta[\overline{T}_n - \mathbb{E}(\overline{T}_n)]^2 = \frac{[\psi'(\theta)]^2}{nI(\theta)} = \frac{1}{nI(\lambda)}, \tag{2.3.28}$$

and as such, $\overline{T}_n$ is the UMVUE of $\lambda = \psi(\theta) = b'(\theta)/a'(\theta)$. Exercises 2.3.2–2.3.6 are set to work out the specific cases of the one-parameter binomial, Poisson, and normal distributions, the latter with $(\mu = \theta, \sigma^2 = 1)$ or $(\mu = 0, \sigma^2 = \theta)$.

There are many distributions that do not belong to the exponential family or do not satisfy the regularity conditions underlying the Cramér–Rao–Fréchet Theorem 2.3.1, some notable examples of which are presented in the sequel.

***Example* 2.3.1.** Let $X_1, \ldots, X_n$ be i.i.d. random variables following the Laplace distribution, for which the density is $f(x;\theta) = (1/2)\exp(-|x-\theta|)$, $-\infty < x < \infty$, $\theta \in \mathbb{R}$. Here, $\mathbb{E}_\theta(X) = \theta$, $\mathbb{V}\text{ar}(X) = \Gamma(2) = 1$, $I(\theta) = 1$, for all $\theta \in \Theta$ so that $U_n(\theta) = \sum_{i=1}^n \text{sign}(X_i - \theta)$. Thus, the sample mean $\overline{X}_n$ is unbiased for θ with MSE $= 2/n > 1/n$, which shows that it does not attain the Cramér–Rao–Fréchet lower bound. Furthermore, (2.3.20) does not hold for finite n, even for the sample median (which has a smaller MSE than $\overline{X}_n$). For large samples, however, as we will see in Chapter 8, the sample median has some optimal properties in this setup. □

***Example* 2.3.2.** Let $X_1, \ldots, X_n$ be i.i.d. random variables following the logistic distribution, for which the distribution function is $F_\theta(x) = \{1 + \exp[-(x-\theta)]\}^{-1}$, $-\infty < x < \infty$, $\theta \in \mathbb{R}$. Here,

$$U_n(\theta) = \sum_{i=1}^n \left[\frac{2}{1+\exp(X_i - \theta)} - 1\right], \quad \theta \in \mathbb{R},$$

and the bound in (2.3.19) is not attainable for finite n. □

***Example* 2.3.3.** The Cauchy distribution, with density $f(x;\theta) = \pi^{-1}[1 + (x-\theta)^2]^{-1}$, $-\infty < x < \infty, \theta \in \mathbb{R}$, does not have moments of any order; the sample mean is inconsistent and does not have a finite mean or variance. Here, for a sample $X_1, \ldots, X_n$, we have

$$U_n(\theta) = 2\sum_{i=1}^n \left[\frac{(X_i - \theta)}{1 + (X_i - \theta)^2}\right],$$

and, hence, for finite n, there is no estimator T_n of θ for which (2.3.20) can be worked out. □

***Example* 2.3.4.** The Uniform$(0, \theta)$ distribution with density $f(x;\theta) = \theta^{-1}I(0 \le x \le \theta)$, $\theta > 0$, has a range $[0, \theta]$ that depends on θ, and, hence, the Cramér–Rao–Fréchet bound is not applicable, too. □

***Example* 2.3.5.** The shifted exponential distribution has density given by $f(x;\theta) = \exp[-(x-\theta)]I(x > \theta)$, $\theta \in \mathbb{R}$. Here also the range (θ, ∞) depends on θ, and hence the Cramér–Rao–Fréchet bound is not applicable. □

***Example* 2.3.6.** Consider a random variable with density $f(x;\theta) = [g(x)/h(\theta)]I(a_\theta \le x \le b_\theta)$, where $a_\theta < b_\theta$ with at least one of them depends on θ, g is a function depending only on x, and h is a function depending only on θ. Again, the regularity conditions of the Cramér–Rao–Fréchet Theorem 2.3.1 do not all hold, and hence the Cramér–Rao–Fréchet bound may not be applicable. Examples 2.3.4 and 2.3.5 are special cases of the density function f. □

***Example* 2.3.7.** The contaminated normal has density $f(x;\theta) = (1-\varepsilon)\varphi(x-\theta) + (\varepsilon/K)\varphi[(x-\theta)/K]$, $-\infty < x < \infty$, where $\varepsilon > 0$ is a small contamination factor, $K > 1$ and φ is the standard normal density. Even if ε and K are specified, the score function $U_n(\theta) = f'_\theta(x-\theta)/f(x-\theta)$ is highly nonlinear and is not separable into a component depending on a statistic and another depending on a parameter. As as result, (2.3.20) does not hold. □

In all these examples, although the Cramér–Rao–Fréchet bound is not attainable, the asymptotic perspectives are quite pertinent, as we will see in Chapter 8.

Two issues that transpire from Cramér–Rao–Fréchet Theorem 2.3.1 are (i) the attainability of the Cramér–Rao–Fréchet bound in finite samples, and (ii) the role of sufficiency and unbiasedness in this context. The first issue is particularly important when $\boldsymbol{\theta}$ is a vector-valued parameter, or when there are possible nuisance parameters, that is, parameters in which we are not interested. As an example, consider the normal density

$$f(x;\theta,\sigma^2) = \frac{1}{\sqrt{\pi}\sigma}\exp\left[-\frac{(x-\theta)^2}{2\sigma^2}\right], \qquad -\infty < x < \infty, \theta \in \mathbb{R}, \sigma^2 > 0.$$

An optimal unbiased estimator of σ^2 (when μ is treated as a nuisance parameter) is the sample variance $s_n^2 = (n-1)^{-1}\sum_{i=1}^n (X_i - \overline{X}_n)^2$, and it is easy to verify that $\mathbb{E}_{\sigma^2}(s_n^2 - \sigma^2)^2 = 2\sigma^4/(n-1) > 2\sigma^4/n = [nI(\sigma^2)]^{-1}$ (see Exercise 2.3.7). A similar problem crops up when we consider the simultaneous estimation of (θ, σ^2). The second issue has led to a stream of highly skilled and yet simple results on unbiased estimators in the presence of sufficient statistics.

First, we provide a natural extension of the Cramér–Rao–Fréchet Theorem 2.3.1 according to Bhattacharya (1946) where higher-order derivatives of the likelihood have is exploited in achieving better bounds. Let us define a vector $\boldsymbol{U}_n(\theta) = (U_{n1}(\theta), \ldots, U_{nr}(\theta))'$, for some $r \geq 1$, where

$$U_{nj}(\theta) = \frac{1}{L_n(\boldsymbol{X};\theta)}\frac{\partial^j}{\partial\theta^j}L_n(\boldsymbol{X};\theta), \quad 1 \leq j \leq r,$$

assuming that, of course, these derivatives exist and that condition (2.3.5) holds for such higher-order derivatives too. Note that $U_{n1}(\theta)$ is the score statistic defined in (2.3.2). For an estimator T_n of θ, we define $\boldsymbol{W}_n = \{T_n - \theta, [\boldsymbol{U}_n(\theta)]'\}'$ and let $\boldsymbol{\Gamma}_\theta = \mathbb{E}_\theta(\boldsymbol{W}_n\boldsymbol{W}_n')$ be partitioned as

$$\boldsymbol{\Gamma}_\theta = \begin{pmatrix} \mathbb{E}(T_n-\theta)^2 & \boldsymbol{\gamma}'_\theta \\ \boldsymbol{\gamma}_\theta & \boldsymbol{\Delta}_\theta \end{pmatrix}, \tag{2.3.29}$$

where $\boldsymbol{\gamma}'_\theta = \{\mathbb{E}_\theta[T_nU_{n1}(\theta)], \ldots, \mathbb{E}_\theta[T_nU_{nr}(\theta)]\}$ and $\boldsymbol{\Delta}_\theta$ is the dispersion matrix of $\boldsymbol{U}_n(\theta)$. Then,

$$\mathbb{E}(T_n-\theta)^2 - \boldsymbol{\gamma}'_\theta\boldsymbol{\Delta}_\theta^{-1}\boldsymbol{\gamma}_\theta \geq 0, \tag{2.3.30}$$

where the equality sign holds only when

$$(T_n - \theta) = \boldsymbol{\lambda}'\boldsymbol{U}_n(\theta) \quad \text{a.e., for some } \boldsymbol{\lambda} \in \mathbb{R}^r. \tag{2.3.31}$$

If $\psi(\theta) = \mathbb{E}_\theta(T_n)$ and we let

$$\boldsymbol{\psi}^{(*)}(\theta) = [\psi^{(1)}(\theta), \ldots, \psi^{(r)}(\theta)]' = \left[\frac{\mathrm{d}}{\mathrm{d}\theta}\psi(\theta), \ldots, \frac{\mathrm{d}^r}{\mathrm{d}\theta^r}\psi(\theta)\right]',$$

then $\boldsymbol{\gamma}'_\theta = \mathbb{E}_\theta[T_n \boldsymbol{U}'_n(\theta)] = [\boldsymbol{\psi}^{(*)}(\theta)]'$, $\theta \in \Theta$. The elements of $\boldsymbol{\Delta}_\theta = n((\delta_{kq\theta}))$ are given by

$$\delta_{kq\theta} = \int \left[\frac{\partial^k}{\partial\theta^k} f(x;\theta)\right]\left[\frac{\partial^q}{\partial\theta^q} f(x;\theta)\right] f(x;\theta)^{-1}\mathrm{d}x,$$

$k, q = 1, \ldots, r$, and $\delta_{11\theta} = I(\theta)$. We also note that for any p.d. matrix $\boldsymbol{A} = ((a_{ij}))$, $i, j = 1, \ldots, r$, $\boldsymbol{x}'\boldsymbol{A}^{-1}\boldsymbol{x} = x_1^2/a_{11} + \boldsymbol{x}'_{2:1}\boldsymbol{A}_{22:1}^{-1}\boldsymbol{x}_{2:1} \geq x_1^2/a_{11}$ where $\boldsymbol{x}'_{2:1} = (x_2, \ldots, x_r)' - \boldsymbol{a}_1 x_1/a_{11}$, $\boldsymbol{a}_1 = (a_{21}, \ldots, a_{r1})'$, $\boldsymbol{A}_{22:1} = \boldsymbol{A}_{22} - \boldsymbol{a}_{11}a_{11}^{-1}\boldsymbol{a}'_{11}$ and the equality sign holds only when $\boldsymbol{x}_{2:1} = \boldsymbol{0}$ (see Exercise 2.3.8). Using (2.3.30) and (2.3.31), for an estimator T_n of θ with $\mathbb{E}_\theta(T_n) = \psi(\theta)$, we arrive at the *Bhattacharya information bound* given by

$$\mathbb{E}_\theta(T_n - \theta)^2 \geq (\boldsymbol{\psi}^{(*)})'\boldsymbol{\Delta}_\theta^{-1}(\boldsymbol{\psi}^{(*)}) \geq [\psi^{(1)}(\theta)]^2/[nI(\theta)], \quad \theta \in \Theta. \tag{2.3.32}$$

Exercise 2.3.9 is set to verify that for the normal variance estimation problem, (2.3.32) provides a better (larger) lower bound than (2.3.23).

Some of the regularity conditions in the Cramér–Rao–Fréchet Theorem 2.3.1 have been relaxed to some extent, and, in such nonregular cases, lower bounds for the MSE of an estimator have been studied by Chapman and Robbins (1951), Kiefer (1952), and others; see also Sen and Ghosh (1976) for a general appraisal of such nonregular cases.

Let us now outline such information bounds in the multiparameter case. For $1 \leq i \leq n$, let $f(x;\boldsymbol{\theta})$ be the density of X_i, indexed by the parameter $\boldsymbol{\theta} = (\theta_1, \ldots, \theta_r)'$, for some $r \geq 1$. Also, let $\boldsymbol{\psi}(\boldsymbol{\theta}) = (\psi_1(\boldsymbol{\theta}), \ldots, \psi_t(\boldsymbol{\theta}))'$, $t \leq r$, be a parametric function vector and let $\boldsymbol{T}_n = (T_{n1}, \ldots, T_{nt})'$ be an unbiased estimator of $\boldsymbol{\psi}(\boldsymbol{\theta})$. Furthermore, let

$$\boldsymbol{V}_{\boldsymbol{T}} = \mathbb{E}_{\boldsymbol{\theta}}[\boldsymbol{T}_n - \boldsymbol{\psi}(\boldsymbol{\theta})][\boldsymbol{T}_n - \boldsymbol{\psi}(\boldsymbol{\theta})]', \tag{2.3.33}$$

$$\boldsymbol{I}(\boldsymbol{\theta}) = \mathbb{E}_{\boldsymbol{\theta}}\left\{\left[\frac{\partial}{\partial\boldsymbol{\theta}}\log f(X;\boldsymbol{\theta})\right]\left[\frac{\partial}{\partial\boldsymbol{\theta}'}\log f(X;\boldsymbol{\theta})\right]\right\} \tag{2.3.34}$$

$$\boldsymbol{\Gamma}_{\boldsymbol{\theta}} = \left(\left(\frac{\partial}{\partial\theta_j}\psi_i(\boldsymbol{\theta})\right)\right)_{i=1,\ldots,t;j=1,\ldots,r}. \tag{2.3.35}$$

Then, under regularity conditions similar to that of the Cramér–Rao–Fréchet Theorem 2.3.1, it follows that

$$\boldsymbol{V}_{\boldsymbol{T}} - \frac{1}{n}\boldsymbol{\Gamma}_{\boldsymbol{\theta}}[\boldsymbol{I}(\boldsymbol{\theta})]^{-1}\boldsymbol{\Gamma}'_{\boldsymbol{\theta}} \quad \text{is p.s.d.}, \boldsymbol{\theta} \in \Theta. \tag{2.3.36}$$

The proof follows as in the Cramér–Rao–Fréchet Theorem 2.3.1 and is left as Exercise 2.3.10.

At this time, we introduce the notion of *minimal sufficiency*. Recall the definition of sufficient statistics in (2.2.10); there, T_n need not be a real-valued statistic. As an extreme case, we consider a nonparametric model based on i.i.d. random variables $X_1, \ldots, X_n$ with density f belonging to the class of continuous densities on $\mathbb{R}$. Let $X_{n:1} < \cdots < X_{n:n}$ be the corresponding order statistics (1.5.86), ties being neglected with probability one. Their joint density function is

$$n!f(X_{n:1})\cdots f(X_{n:n}) \quad \text{for} \quad X_{n:1} < \cdots < X_{n:n}. \tag{2.3.37}$$

Given the order statistics $\{X_{n:1}, \ldots, X_{n:n}\}$, the possible values of $(X_1, \ldots, X_n)$ correspond to the $n!$ possible permutations thereof; hence, if follows that the conditional density of the X_i, given the order statistics, is discrete uniform, with the common probability $1/n!$ attached to each permutation. Thus $\{X_{n:1}, \ldots, X_{n:n}\}$ are sufficient statistics for the density f in this nonparametric mode. On the other hand, in a parametric setup, for example, with the class of exponential densities in (2.2.11), if we take the statistics $T_{n1}, \ldots, T_{nq}$, where

$$T_{nj} = \sum_{j=1}^{n} t_j(X_i) = \sum_{j=1}^{n} t_j(X_{n:i}), \quad j = 1, \ldots, q, \tag{2.3.38}$$

we may have a smaller set (when $q < n$) of statistics, which are jointly sufficient for the parameter (vector) $\boldsymbol{\theta}$. If in addition to $\boldsymbol{T}_n = (T_{n1}, \ldots, T_{nq})$, we take an extra component $T_{n(q+1)}$, defined arbitrarily as a function of the order statistics, then $(\boldsymbol{T}'_n, T_{n(q+1)})'$ will still be jointly sufficient, although sufficiency holds without $T_{n(q+1)}$ as well. Thus, we prefer to use the smaller set which still preserves sufficiency. If no further subset does so, then this set is called a *minimal sufficient statistic*. Any other set of sufficient statistics can be expressed as a function of the latter. Thus, for statistical inference, it suffices to base our analysis on minimal sufficient statistics whenever they exist.

Another property of (minimal) sufficient statistics is *completeness*. Consider a statistic $T_n = T(\boldsymbol{X})$, with density $g_n(T_n, \theta), \theta \in \Theta$. The family of densities $\{g_n(\cdot, \theta),\ \theta \in \Theta\}$ is said to be *complete* if, for all $\theta \in \Theta$,

$$\mathbb{E}_\theta[h(T_n)] = 0 \Rightarrow h(T_n) = 0 \ \text{ a.e.}, \tag{2.3.39}$$

and then T_n is also called a *complete statistic*. If we confine our attention only to bounded h, then it is called a *bounded complete statistic*. Statistics for which the distribution does not depend on $\theta \in \Theta$ are called *ancillary* statistics [see, e.g., Cox and Hinkley (1974)]. The well known Basu Theorem [see, e.g., Lehmann (1983)], asserts that if T_n is complete and minimal sufficient, then it is independent of every ancillary statistic (see Exercise 2.3.12).

We explore now the role of sufficiency in the context of uniformly minimum variance unbiased estimation. If we have a sufficient statistic T_n, which is itself unbiased for θ, then it is an UMVUE of θ. Even if T_n is not unbiased but an unbiased estimator W_n of θ exists, we can arrive at a better estimator by using the following result.

Theorem 2.3.2 (Rao–Blackwell). *If there exists an unbiased estimator W_n for a parametric function $\psi(\theta)$, and if T_n is a sufficient statistic for θ, then $T_n^* = \mathbb{E}(W_n|T_n)$, a sufficient statistic too, and dominates W_n in variance (or any convex risk function). If T_n is a complete sufficient statistic, then T_n^* is the unique UMVUE of $\psi(\theta)$.*

Proof. Since $\mathbb{E}_\theta(T_n^*) = \mathbb{E}_\theta[\mathbb{E}(W_n|T_n)] = \mathbb{E}_\theta(W_n) = \psi(\theta),\ \theta \in \Theta$, we immediately get that

$$\begin{aligned}\mathbb{E}_\theta[W_n - \psi(\theta)]^2 &= \mathbb{E}_\theta(W_n - T_n^*)^2 + \mathbb{E}_\theta[T_n^* - \psi(\theta)]^2 \\ &\geq \mathbb{E}_\theta[T_n^* - \psi(\theta)]^2,\end{aligned} \tag{2.3.40}$$

where the equality sign holds only when $W_n = T_n^*$ with probability one. The case of convex loss, to be introduced in Chapter 4, follows along parallel lines. If there are two or more such T_n^* having the same variance, by appealing to the completeness criterion in (2.3.39), we claim that their difference is 0 a.e. and $\theta \in \Theta$. ■

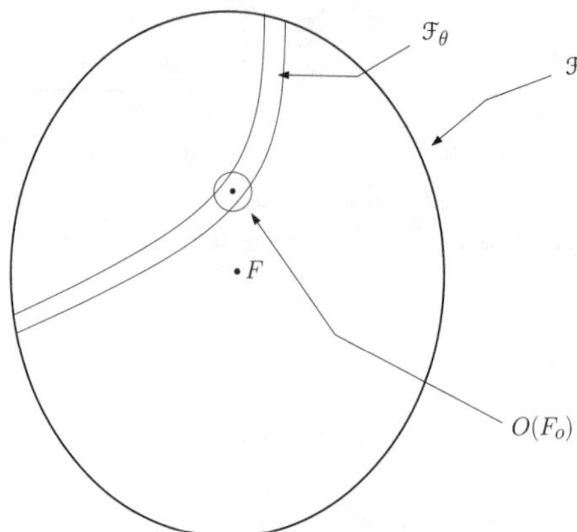

Figure 2.3.1. Families of distributions and their neighborhoods.

It is not an easy task to verify completeness, in general (even under sufficiency). For the exponential family of densities, this has been treated in detail in the literature. However, missing data or other complications may generally distort the completeness property. Nevertheless, the dominance result in (2.3.40) goes through in a much broader setup as indicated in Exercise 2.3.11.

Often, the class of *good* estimators, in some sense, is too large to allow the search for a suitable one (especially when minimal sufficiency may not hold). In such a situation, usually a subclass of estimators is chosen by incorporating some natural constraints that the parameter(s) satisfy. Consider, for example, the problem of estimating the difference between the location parameters θ_1 and θ_2 of two distributions with distribution functions F_1 and F_2, such that $F_1(x) = F(x - \theta_1)$ and $F_2(x) = F(x - \theta_2)$ for a common F. Since the estimation problem is invariant under the transformation $\theta_1 \to \theta_1 + a$, $\theta_2 \to \theta_2 + a$, $a \in \mathbb{R}$, we may impose that the estimators of $\theta_1 - \theta_2$ be *invariant* under transformations $X \to X + a$, $Y \to Y + a$, a arbitrary. *Invariance* (and related topics) will be dealt with in more detail in Chapters 3 and 4.

We conclude this section with some remarks on *robustness*. In essence, we generally assume that the sample observations $\boldsymbol{X} = (X_1, \ldots, X_n)$ come from a distribution F, which is a member of some family $\mathcal{F}$. In a parametric setup, we usually take $F \equiv F_0(x; \boldsymbol{\theta})$, where F_0 has a specified form defining an arch $\mathcal{F}_{\boldsymbol{\theta}} = \{F_0(x; \boldsymbol{\theta}) : \boldsymbol{\theta} \in \boldsymbol{\Theta}\}$ in $\mathcal{F}$. The true F, however, may not belong to this arch. It may or may not lie in a small neighborhood of the specified F_0 (e.g., when F_0 is normal and F is the Laplace distribution); the neighborhood $O(F_0)$ being defined in accordance with a suitable probability law distance measure. This is schematically displayed in Figure 2.3.1.

The impact of local departures can be much different than global ones, when inference is based on an assumed distribution. Robust estimation primarily addresses the damage in optimality properties of conventional estimators caused by local model departures, and formulates alternative estimators, which are much less sensitive to small model distortions but still retain good performance characteristics over such neighborhoods. Global robustness stresses more on validity aspects even for nonlocal distortions and examine the superiority of some alternative estimators to the assumed model-based ones. In either case, the developments are not so amenable for small sample situations and hence large-sample perspectives are generally appraised. We will provide a coverage of these robust estimators in Chapter 8. For a detailed presentation, see Huber (1981).

2.4 Methods of Estimation

Initially estimation procedures were tailored to specific problems, largely guided by intuition. It did not take long to classify these ad hoc methods into some broad categories. Among these, commonly used are the *method of moments* (MM), (*weighted*) *least squares*[(W)LS], *maximum likelihood* (ML), *best linear unbiased* (BLU) estimation, and *generalized estimating equations*-based (GEE) methods. In a sense, all these methods rely on appropriate estimating equations, but the basic motivation for specifying them varies from *alignment* to appropriate *minimization/maximization* principles.

We consider first the method of moments. Let $X_1, \ldots, X_n$ be i.i.d. random variables with a distribution function $F(x; \boldsymbol{\theta}), x \in \mathbb{R}, \boldsymbol{\theta} \in \boldsymbol{\Theta} \subset \mathbb{R}^p$, for some $p \geq 1$. Assume that the distribution admits finite moments at least up to the order p. Then let

$$\mu^{(r)}(\boldsymbol{\theta}) = \mathbb{E}_{\boldsymbol{\theta}}(X^r), \quad r = 1, \ldots, p, \tag{2.4.1}$$

and assume that the $\mu^{(r)}(\boldsymbol{\theta})$ are of known functional form. The corresponding sample moments are defined as

$$m_n^{(r)} = \frac{1}{n} \sum_{i=1}^{n} X_i^r, \quad r = 1, \ldots, p. \tag{2.4.2}$$

Note that $\mathbb{E}_{\boldsymbol{\theta}}(m_n^{(r)}) = \mu^{(r)}(\boldsymbol{\theta}), r = 1, \ldots, p$, so that when the functions $\mu^{(1)}(\boldsymbol{\theta}), \ldots, \mu^{(p)}(\boldsymbol{\theta})$ are linearly independent, the method of moments estimator is defined as a solution $\widehat{\boldsymbol{\theta}}_n$ to the estimating equations

$$m_n^{(r)} = \mu^{(r)}(\widehat{\boldsymbol{\theta}}_n), \quad r = 1, \ldots, p. \tag{2.4.3}$$

In other words, $\widehat{\boldsymbol{\theta}}_n = \boldsymbol{h}(\boldsymbol{m}_n)$, where $\boldsymbol{m}_n = (m_n^{(1)}, \ldots, m_n^{(p)})'$, and $\boldsymbol{h}$ is a vector of inverse transformations. If the $\mu^{(r)}(\boldsymbol{\theta})$ are well behaved functions, we may solve for (2.4.3) directly. Here, unbiasedness and consistency of the $m_n^{(r)}$ provide the basic motivation to the estimating equations (2.4.3). In this setup, Exercises 2.4.1 and 2.4.2 provide examples in the context of the exponential family of densities and the normal distribution in particular. Note that the finiteness of first p moments is a prerequisite for (2.4.3). To clarify this, consider the Cauchy density $f(x; \theta_1, \theta_2) = (\theta_2/\pi)\{\theta_2 + (x - \theta_1)^2\}^{-1}, -\infty < x < \infty, \theta_1 \in \mathbb{R}, \theta_2 > 0$. Since, for this density, moments of order one or higher do not exist, (2.4.3) is not operable.

Next, we consider the least-squares estimation method. Let $X_1, \ldots, X_n$ be random variables, not necessarily identically distributed and set $\mathbb{E}_{\boldsymbol{\theta}}(X_i) = g_i(\boldsymbol{\theta}), i = 1, \ldots, n$, where the $g_i(\boldsymbol{\theta})$ depend on a parameter $\boldsymbol{\theta} = (\alpha, \boldsymbol{\beta})$, and possibly involve some other known (design/concomitant) variables. For example, $g_i(\boldsymbol{\theta}) = \alpha + \boldsymbol{\beta}'\boldsymbol{c}_i$, $1 \leq i \leq n$, where the $\boldsymbol{c}_i$ are known vectors. Then, $X_i - g_i(\boldsymbol{\theta})$ accounts for the deviation of the observed X_i from its expected value. For given $\boldsymbol{X} = (X_1, \ldots, X_n)'$, the sum of squares of such deviations, namely,

$$Q(\boldsymbol{X}, \boldsymbol{\theta}) = \sum_{i=1}^{n} [X_i - g_i(\boldsymbol{\theta})]^2, \quad \boldsymbol{\theta} \in \Theta, \tag{2.4.4}$$

is a measure of the distance between the observed sample point and its expectation; in fact, in this simple form, it is the square of the Euclidean distance $||\boldsymbol{X} - \boldsymbol{g}(\boldsymbol{\theta})||$. The *least squares estimator* (LSE) of $\boldsymbol{\theta}$ is obtained by minimizing $Q(\boldsymbol{X}, \boldsymbol{\theta})$ over $\boldsymbol{\theta} \in \boldsymbol{\Theta}$. Whenever the g_i are twice differentiable with respect to $\boldsymbol{\theta}$, and

$$\left(\left(\frac{\partial^2 Q(\boldsymbol{X}, \boldsymbol{\theta})}{\partial\boldsymbol{\theta}\partial\boldsymbol{\theta}'}\right)\right)\bigg|_{\widehat{\boldsymbol{\theta}}_n}$$

is p.s.d., the LSE, $\widehat{\boldsymbol{\theta}}_n$, is the solution to the estimating equation

$$\frac{\partial Q(\boldsymbol{X}, \boldsymbol{\theta})}{\partial\boldsymbol{\theta}}\bigg|_{\widehat{\boldsymbol{\theta}}_n} = \boldsymbol{0}. \tag{2.4.5}$$

The rationale behind (2.4.5) is the implicit assumption that $\mathbb{E}_{\boldsymbol{\theta}}[X_i - \boldsymbol{g}_i(\boldsymbol{\theta})]^2 = \sigma_{\boldsymbol{\theta}}^2$ exists, $i = 1, \ldots, n$. For some models, we may have a more general situation, where

$$\mathbb{E}_{\boldsymbol{\theta}}[X_i - \boldsymbol{g}_i(\boldsymbol{\theta})]^2 = \sigma_{\boldsymbol{\theta}}^2 w_i, \quad i = 1, \ldots, n, \tag{2.4.6}$$

with w_i denoting known positive constants, not all equal. In that case, we extend (2.4.4) to

$$Q_{\boldsymbol{w}}(\boldsymbol{X}, \boldsymbol{\theta}) = \sum_{i=1}^{n} \frac{1}{w_i}[X_i - g_i(\boldsymbol{\theta})]^2 \bigg/ \sum_{i=1}^{n} \frac{1}{w_i}. \tag{2.4.7}$$

Minimization of (2.4.7) with respect to $\boldsymbol{\theta}$, yields the *weighted least squares estimator* (WLSE). Examples are given in Exercises (2.4.3)–(2.4.5). There are situations like with generalized linear models, where the weights w_i depend on $\boldsymbol{\theta}$, and this will be discussed in Chapter 10.

We may now go one step further. Suppose that $\boldsymbol{X} = (X_1, \ldots, X_n)'$ has expectation $\boldsymbol{g}(\boldsymbol{\theta})$ and covariance matrix $\boldsymbol{\Sigma} = \sigma^2 \boldsymbol{V}_n$, where $\boldsymbol{V}_n$ is a specified p.d. matrix. Then, in lieu of (2.4.4), consider the *quadratic form*

$$Q_{\boldsymbol{V}_n}(\boldsymbol{X}, \boldsymbol{\theta}) = [\boldsymbol{X} - \boldsymbol{g}(\boldsymbol{\theta})]'\boldsymbol{V}_n^{-1}[\boldsymbol{X} - \boldsymbol{g}(\boldsymbol{\theta})]. \tag{2.4.8}$$

Assuming that $\boldsymbol{g}$ is differentiable with respect to $\boldsymbol{\theta}$ and denoting the $n \times p$ matrix $(\partial/\partial\boldsymbol{\theta}')\boldsymbol{g}(\boldsymbol{\theta})$ by $\dot{\boldsymbol{G}}_{\boldsymbol{\theta}}$, minimization of $Q_{\boldsymbol{V}_n}(\boldsymbol{X}, \boldsymbol{\theta})$ leads to the estimating equations

$$\dot{\boldsymbol{G}}_{\boldsymbol{\theta}}\boldsymbol{V}_n^{-1}[\boldsymbol{X} - \boldsymbol{g}(\boldsymbol{\theta})]|_{\widehat{\boldsymbol{\theta}}_n} = \boldsymbol{0}. \tag{2.4.9}$$

The solution to (2.4.9) is the *generalized least squares estimator* (GLSE) of $\boldsymbol{\theta}$. In particular, if $\boldsymbol{V}_n$ is a diagonal matrix, we have the WLSE, and if $\boldsymbol{V}_n \equiv \boldsymbol{I}_n$, we have the LSE (see Exercise 2.4.6). If $\boldsymbol{g}(\boldsymbol{\theta})$ is linear in $\boldsymbol{\theta}$, that is, $\boldsymbol{g}(\boldsymbol{\theta}) = \boldsymbol{L}\boldsymbol{\theta}$, for some matrix $\boldsymbol{L}$, then $\dot{\boldsymbol{G}}_{\boldsymbol{\theta}}$ does not depend on $\boldsymbol{\theta}$, so that (2.4.9) yields

$$\dot{\boldsymbol{G}}_{\boldsymbol{\theta}}'\boldsymbol{V}_n^{-1}\boldsymbol{X} = \dot{\boldsymbol{G}}_{\boldsymbol{\theta}}'\boldsymbol{V}_n^{-1}\boldsymbol{g}(\widehat{\boldsymbol{\theta}}_n) = \dot{\boldsymbol{G}}_{\boldsymbol{\theta}}'\boldsymbol{V}_n^{-1}\boldsymbol{L}\widehat{\boldsymbol{\theta}}_n. \tag{2.4.10}$$

Thus,

$$\widehat{\boldsymbol{\theta}}_n = (\dot{\boldsymbol{G}}_{\boldsymbol{\theta}}'\boldsymbol{V}_n^{-1}\boldsymbol{L})^{-1}\dot{\boldsymbol{G}}_{\boldsymbol{\theta}}'\boldsymbol{V}_n^{-1}\boldsymbol{X},$$

so that a closed expression is available. On the other hand, for nonlinear $\boldsymbol{g}(\boldsymbol{\theta})$, $\dot{\boldsymbol{G}}$ will depend on $\boldsymbol{\theta}$, and hence, an iterative solution may be needed to obtain the GLSE.

As in the method of moments, the (G/W)LS methods rest on the assumption of finite variance, and can be quite sensitive to heavy tails (i.e., outliers/error contamination). Therefore, it is not appropriate for the Cauchy distribution, for example. For Gaussian distributions, the LS method has some optimality properties shared with ML methods, and this will be elaborated later on.

The method of maximum likelihood is perhaps the most widely used one in parametric inference; along with its ramifications, the likelihood methodology umbrella extends beyond parametrics as well. As in Section 2.3, given $\boldsymbol{X}$, in a parametric setup, the likelihood function $L_n(\boldsymbol{\theta}; \boldsymbol{X})$ is regarded as a function of $\boldsymbol{\theta} \in \Theta$. Therefore, intuitively, at least, it might be appealing to choose an estimator $\widehat{\boldsymbol{\theta}}_n$ of $\boldsymbol{\theta}$ that maximizes $L_n(\boldsymbol{\theta}; \boldsymbol{X})$ over Θ, that is,

$$L_n(\widehat{\boldsymbol{\theta}}; \boldsymbol{X}) = \sup_{\boldsymbol{\theta} \in \Theta}[L_n(\boldsymbol{X}, \boldsymbol{\theta})], \tag{2.4.11}$$

or, equivalently,

$$\widehat{\boldsymbol{\theta}}_n = \arg\max_{\boldsymbol{\theta} \in \Theta}[L_n(\boldsymbol{\theta}; \boldsymbol{X})], \tag{2.4.12}$$

where it is tacitly assumed that such a maximizer exists and is unique. For densities not belonging to the exponential family, in finite samples, a MLE, as defined by (2.4.12) may not be unique, even if it exists. The Cauchy density (1.5.65) is a good example in this respect: not only there are multiple roots to the ML estimating equations but also a closed expression for $\widehat{\theta}_n$ does not exist. Exercise 2.4.7 is set to verify the existence/uniqueness of the MLE for Examples 2.3.1 to 2.3.7.

In a majority of cases, whenever the density $f(x; \theta)$ is well-behaved (in the sense of Section 2.3), the maximization process may simplify if we work with the logarithmic transformation $l_n(\boldsymbol{X}; \boldsymbol{\theta}) = \log L_n(\boldsymbol{X}; \boldsymbol{\theta})$. Thus, whenever $l_n(\boldsymbol{X}; \boldsymbol{\theta})$ is twice differentiable with respect to $\boldsymbol{\theta}$, we can set the ML estimating equations as

$$\left.\frac{\partial l_n(\boldsymbol{\theta}; \boldsymbol{X})}{\partial \boldsymbol{\theta}}\right|_{\widehat{\boldsymbol{\theta}}_n} = \mathbf{0}. \tag{2.4.13}$$

The maximization process requires that the Hessian (or *curvature matrix*),

$$\left.\frac{\partial^2}{\partial \boldsymbol{\theta} \partial \boldsymbol{\theta}'} l_n(\boldsymbol{\theta}; \boldsymbol{X})\right|_{\widehat{\boldsymbol{\theta}}_n}$$

be n.s.d. Even so, there might not be a guarantee for a unique solution, especially for densities not belonging to the exponential family. For large samples, the prospects are better, and we will discuss those in Chapter 8.

Note that $l_n(\boldsymbol{\theta}; \boldsymbol{X}) = \sum_{i=1}^n \log f(X_i; \boldsymbol{\theta})$ so that the ML estimating equations in (2.4.13) can be written as

$$\sum_{i=1}^n \left.\frac{\partial}{\partial \boldsymbol{\theta}} \log f(X_i; \boldsymbol{\theta})\right|_{\widehat{\boldsymbol{\theta}}_n} = \mathbf{0}. \tag{2.4.14}$$

This enables us to use the score statistic $\boldsymbol{U}_n(\boldsymbol{\theta})$ and rewrite such estimating equations as

$$\boldsymbol{U}_n(\widehat{\boldsymbol{\theta}}_n) = \mathbf{0}. \tag{2.4.15}$$

Similarly to the scalar parameter case detailed in Section 2.3, $\boldsymbol{U}_n(\boldsymbol{\theta})$ involves independent summands, $\mathbb{E}_{\boldsymbol{\theta}}[\boldsymbol{U}_n(\boldsymbol{\theta})] = \mathbf{0}$, and $\mathbb{E}\{[\boldsymbol{U}_n(\boldsymbol{\theta})][\boldsymbol{U}_n(\boldsymbol{\theta})]'\} = n\boldsymbol{I}(\boldsymbol{\theta})$. Thus, if a Taylor expansion may be applied to $\boldsymbol{U}_n(\boldsymbol{\theta})$, a computational scheme can be adapted to solve for $\widehat{\boldsymbol{\theta}}_n$, whenever

(2.4.15) does not provide a closed-form algebraic solution. Such iterative solutions, that include the well-known *Newton–Raphson*, *Fisher scoring*, and *EM* algorithms will be explored in Chapter 8.

For the exponential family (2.3.11) we noted that if there is a sufficient statistic $\boldsymbol{T}_n$ for the parameter $\boldsymbol{\theta}$, then,

$$\boldsymbol{U}_n(\boldsymbol{\theta}) = \frac{\partial}{\partial \boldsymbol{\theta}} l_n(\boldsymbol{\theta}; \boldsymbol{X}) = \frac{\partial}{\partial \boldsymbol{\theta}} \log g_n(\boldsymbol{T}_n; \boldsymbol{\theta}) = \boldsymbol{U}^*_{nT}(\boldsymbol{\theta}), \tag{2.4.16}$$

say, where $\boldsymbol{U}^*_{nT}(\boldsymbol{\theta})$ depends on $\boldsymbol{\theta}$ and on $\boldsymbol{X}$ only through $\boldsymbol{T}_n$. This implies that the MLE are functions of sufficient statistics whenever they exist. As such, in the presence of a sufficient statistic, the MLE can often be modified as in the Rao–Blackwell Theorem 2.3.2 to yield a better estimator; completeness and minimal sufficiency add more incentives to this process.

We now show that these concepts might be easily extended to the multivariate case via the following example.

Example* 2.4.1 *(Multinomial distribution). Let $X_1, \ldots, X_n$ be independent and identically distributed random c-vectors, which may assume the values $(1, 0, \ldots, 0)'$, $(0, 1, \ldots, 0)'$ or $(0, 0, \ldots, 1)'$ with probabilities $\pi_1, \ldots, \pi_c$, respectively, such that $\sum_{j=1}^{c} \pi_j = 1$. In other words, the X_i follow a c-variate generalization of the Bernoulli distribution, with parameter vector $\boldsymbol{\pi} = (\pi_1, \ldots, \pi_c)'$. Then, $\boldsymbol{n} = (n_1, \ldots, n_c)' = \sum_{i=1}^{n} \boldsymbol{X}_i$ follows a *multinomial distribution* having probability function

$$f(\boldsymbol{n}|n; \boldsymbol{\pi}) = \frac{n!}{n_1! \cdots n_c!} \pi_1^{n_1} \cdots \pi_c^{n_c}, \tag{2.4.17}$$

where $n = \sum_{j=1}^{c} n_j$. The corresponding log-likelihood function is

$$l_n(\boldsymbol{\pi}|\boldsymbol{n}) = K_n + \sum_{j=1}^{c} n_j \log \pi_j, \tag{2.4.18}$$

where K_n is a constant not depending on $\boldsymbol{\pi}$. Maximization of (2.4.18) subject to the natural restriction $\sum_{j=1}^{c} \pi_j = 1$ may be easily conducted via *Lagrangean multiplier methods* leading to the MLE

$$\widehat{\boldsymbol{\pi}} = \frac{1}{n} \boldsymbol{n} = \frac{1}{n} \sum_{j=1}^{n} \boldsymbol{X}_j. \tag{2.4.19}$$

Given the nature of the underlying random vectors $\boldsymbol{X}_i$, it is not difficult to show that

$$\mathbb{E}(\widehat{\boldsymbol{\pi}}) = \boldsymbol{\pi} \quad \text{and} \quad \mathbb{V}\text{ar}(\widehat{\boldsymbol{\pi}}) = \frac{1}{n}(\boldsymbol{D}_\pi - \boldsymbol{\pi}\boldsymbol{\pi}'), \tag{2.4.20}$$

where $\boldsymbol{D}_\pi = \text{diag}(\pi_1, \ldots, \pi_c)$. Also, because (2.4.17) belongs to the exponential family, it follows that $\boldsymbol{n}$ (and consequently, $\widehat{\boldsymbol{\pi}}$) is sufficient and complete for $\boldsymbol{\pi}$. Therefore, using a direct generalization of the Rao–Blackwell Theorem 2.3.2, we may claim that $\widehat{\boldsymbol{\pi}}$ is an optimal estimator of $\boldsymbol{\pi}$. □

Now we extend the idea underlying the ML estimation a step further. In this context, we recall that $-l_n(\boldsymbol{\theta}; \boldsymbol{X}) = \sum_{i=1}^{n} \{-\log f(X_i; \boldsymbol{\theta})\}$, and write $\rho(X_i; \boldsymbol{\theta}) = -\log f(X_i; \boldsymbol{\theta})$; then, it follows that the maximizer of $l_n(\boldsymbol{\theta})$, is also the minimizer of $\sum_{i=1}^{n} \rho(X_i; \boldsymbol{\theta})$, $\boldsymbol{\theta} \in \Theta$.

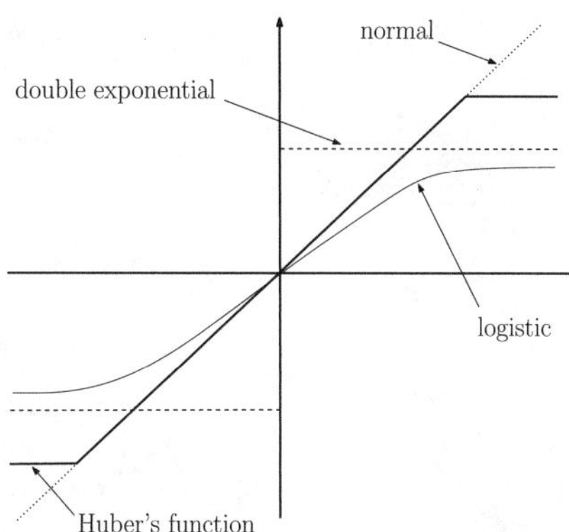

Figure 2.4.1. Score functions for the normal, double exponential, logistic distributions, and the Huber score function.

In fact, for a general class of location scale (regression) family of densities, $\rho(X_i;\boldsymbol{\theta})$ is convex in $\boldsymbol{\theta}$ (Exercise 2.4.8). Let $\rho(a,b)=\rho(|a-b|)$, so that for the location model, we have $\sum_{i=1}^n \rho(X_i,\theta)=\sum_{i=1}^n \rho(|X_i-\theta|)$. Assume that ρ is a nonnegative function and that the (generalized) *score function* $\psi(y)=(\partial/\partial y)\rho(y)$ exists a.e. Note that for the normal distribution, $\rho(y)=y^2/2$ while for the Laplace density, $\rho(y)=+|y|$, and that $\psi(y)=y$ and $\psi(y)=\text{sign}(y)$, respectively.

Estimators obtained as the solutions to implicit equations, such as

$$M_n(\widehat{\theta}_n)=\sum_{i=1}^n \psi(X_i;\widehat{\theta}_n)=0, \tag{2.4.21}$$

are generally called *M-estimators* and are of special interest for location problems, i.e., for $\psi(x;\theta)=\psi(x-\theta)$, because they may be easily defined in a way to produce *robust* estimates of θ, i.e., estimates that have stable statistical properties under fairly general specifications for the underlying probability model. To clarify this issue we bring into perspective the *influence function*, which loosely speaking, measures the effect of a single observation on the performance of an estimator. For M-estimators of location, the influence function is proportional to the *score function* ψ; thus, a single outlier (i.e., an observation that is distant from the "bulk" of the data) may severely affect M-estimators obtained via the use of unbounded ψ.

In Figure 2.4.1 we depict the score functions corresponding to $\rho(x)=-\log f(x)$ for:

(i) a normal distribution, where, $\psi_1(x)=x$, $x\in\mathbb{R}$,
(ii) a double-exponential distribution, where, $\psi_2(x)=\text{sign}(x)$, $x\in\mathbb{R}$,
(iii) a logistic distribution, where $\psi_3(x)=[\exp(x)-1]/[\exp(x)+1]$, $x\in\mathbb{R}$

along with the Huber score function (2.4.22).

The fact that ψ_1 is unbounded implies that a single outlier may have a disastrous effect on the sample mean (the M-estimator based on ψ_1); on the other hand, M-estimators based on either ψ_2 or ψ_3 are not affected by outliers although they do not preserve the nice linear influence provided by ψ_1. With this picture in mind, Huber (1964) proposed the following score function, which incorporates the desirable features of ψ_1, ψ_2, and ψ_3:

$$\psi(t) = \begin{cases} t, & |t| \le k \\ k \operatorname{sign}(t), & |t| > k, \end{cases} \tag{2.4.22}$$

where k is a positive constant; this function corresponds to

$$\rho(t) = \begin{cases} \frac{1}{2}t^2, & |t| \le k \\ k|t| - \frac{1}{2}k^2, & |t| > k. \end{cases}$$

More specifically, the Huber score function (2.4.22) assigns a limited influence for observations in the tails of the distribution (i.e., for $|x| > k$) like ψ_2, and a linear influence for the central portion (i.e., for $|x| \le k$), like ψ_1, while maintaining some similarity to ψ_3.

In general, equations like (2.4.21) may have many roots, and they have to be examined in order to verify which of them corresponds to the required minimum of $\sum_{i=1}^n \rho(X_i, \theta)$. In many location problems, however, the estimating function $M_n(\theta) = \sum_{i=1}^n \psi(X_i - \theta)$ is nonincreasing in θ and even in the presence of multiple roots, the corresponding M-estimator may be easily defined, i.e., as any value $\widehat{\theta}_n \in [\widehat{\theta}_{n_1}, \widehat{\theta}_{n_2}]$ where $\widehat{\theta}_{n_1} = \sup\{\theta : M_n(\theta) > 0\}$ and $\widehat{\theta}_{n_2} = \inf\{\theta : M_n(\theta) < 0\}$. Finite sample properties of such interesting estimators are practically impossible to obtain analytically. This is another area where asymptotic methods play an important role. See Huber (1981), Serfling (1980), or Jurečková and Sen (1996) for details.

A generalization of the robust techniques presented above along with the method of GEE will be considered in Chapter 10, in the context of linear regression problems.

In a nonparametric setup, a parameter $\theta = \theta(F)$ is defined and interpreted as a functional of the distribution function F. For example, if we let

$$\theta(F) = \int x \mathrm{d}F(x), \tag{2.4.23}$$

then $\theta(F)$ is the mean of the distribution function F. Likewise, the variance corresponds to

$$\mathbb{V}\mathrm{ar}_F(X) = \int\int \frac{1}{2}(x - y)^2 \mathrm{d}F(x)\mathrm{d}F(y).$$

In the same vein,

$$\theta_p(F) = F^{-1}(p) = \inf\{x : F(x) \ge p\}, \quad 0 < p < 1 \tag{2.4.24}$$

is the *quantile function* defined in (1.3.4).

If $\theta(F)$, $F \in \mathcal{F}$ where $\mathcal{F}$ is the class of functions F for which $\theta(F)$ is well-behaved, is sufficiently smooth (to be more precisely defined in Chapter 8), then it might be intuitive to consider the *plug-in estimator* of $\theta(F)$ given by

$$V_n = \theta(F_n). \tag{2.4.25}$$

Such functionals, according to von Mises (1947), are known as *V-statistics* or *von Mises functionals*.

For example, if $\theta(F)$ is defined as (2.4.23), we have $\theta(F_n) = \int x \mathrm{d}F_n(x) = \overline{X}_n$, the sample mean. Similarly, for the variance functional,

$$\theta(F_n) = \frac{1}{2n^2}\sum_{i=1}^n\sum_{j=1}^n (X_i - X_j)^2 = \frac{1}{n}\sum_{i=1}^n (X_i - \overline{X}_n)^2,$$

is the (biased) sample variance. For the quantile function, we have,

$$\theta_p(F_n) = F_n^{-1}(p) = \min\{x : F_n(x) \geq p\} = X_{n:[np]+1}, \tag{2.4.26}$$

If the functional $\theta(F)$ is linear, then $\mathbb{E}_F[\theta(F_n)] = \theta[\mathbb{E}_F(F_n)] = \theta(F)$, but nonlinear functionals $\theta(F_n)$ may not be generally unbiased for $\theta(F)$. Exercise 2.4.9 is set to verify this for the variance and quantile functions.

More generally, we consider a functional

$$\theta(F) = \int \cdots \int g(x_1, \ldots, x_m) \mathrm{d}F(x_1) \cdots \mathrm{d}F(x_m), \quad F \in \mathcal{F}, \tag{2.4.27}$$

where $g(x_1, \ldots, x_m)$ is a symmetric *kernel* of degree $m \geq 1$. Then, when $n \geq m$, we can rewrite $\theta(F)$ as $\mathbb{E}_F[g(X_1, \ldots, X_n)]$. As such, we consider all possible $\binom{n}{m}$ subsamples of size m, and for each subsample, consider the statistic $g(X_{i_1}, \ldots, X_{i_m})$. The average of such statistics over all subsamples will then provide an unbiased estimator of $\theta(F)$, given by

$$U_n = \binom{n}{m}^{-1} \sum_{1 \leq i_1 \leq \cdots \leq i_m \leq n} g(X_{i_1}, \ldots, X_{i_m}). \tag{2.4.28}$$

Hoeffding (1948) called statistics defined as in (2.4.28) *U-statistics* (U for unbiased).

In fact, the von Mises functionals (2.4.25), which in this case are given by

$$\begin{aligned} V_n = \theta(F_n) &= \int \cdots \int g(x_1, \ldots, x_m) \mathrm{d}F_n(x_1) \cdots \mathrm{d}F_n(x_m) \\ &= n^{-m} \sum_{i_1=1}^{n} \cdots \sum_{i_m=1}^{n} g(X_{i_1}, \ldots, X_{i_m}), \end{aligned} \tag{2.4.29}$$

are intimately related to such U-statistics. For $m = 1$, $U_n = V_n$, so that they are the same and estimate $\theta(F)$ unbiasedly. For $m \geq 2$, this equivalence is not generally true, and V_n may not be unbiased for $\theta(F)$. For example, when $m = 2$, by (2.4.25) and (2.4.28), we have

$$\begin{aligned} V_n &= n^{-2} \left[n(n-1)U_n + \sum_{i=1}^{n} g(X_i, X_i) \right] \\ &= U_n + n^{-1} \left\{ \frac{1}{n} \sum_{i=1}^{n} [g(X_i, X_i) - U_n] \right\}, \end{aligned}$$

so that whenever $g(X_1, X_1)$ is not unbiased for $\theta(F)$, V_n fails to be so. However, for large values of n, this bias may be small, so that both V_n and U_n share the same asymptotic properties, as we will see in Chapter 7.

As U_n is a symmetric function of all $X_1, \ldots, X_n$, it is also a symmetric function of the order statistics, that is, (Exercise 2.4.10),

$$U_n = U(X_1, \ldots, X_n) = U(X_{n:1}, \ldots, X_{n:n}), \tag{2.4.30}$$

so that invoking the sufficiency of order statistics in a nonparametric sense [see (2.3.37)], we can claim that U_n is an optimal nonparametric estimator of $\theta(F)$, for $F \in \mathcal{F}$. This result extends readily to generalized U-statistics, covering multisample models and to vector valued $\theta(F)$ as well as to more complex sample spaces. For details, see Sen (1981).

Bayes and empirical Bayes estimation methods have recently been extensively advocated in a vast area of applications. We find it more convenient to introduce them in Chapter 4 in a decision theoretic setup, where a brief discussion of their relation to the topics considered in this text is outlined; see Berger (1993) for a full account.

2.5 Finite Sample Optimality Perspectives

With the basic concepts introduced in the preceding sections, we like to gather some general consensus on optimality perspectives in finite sample setups. Although unbiasedness of estimators is a desirable property, it is not to be emphasized strongly; indeed, there are some biased estimators that dominate unbiased ones in finite-sample setups, even when (minimal) sufficiency and completeness hold for the model. A classical example is the *James–Stein estimator* of a multivariate normal mean vector [Stein (1956)] that we will briefly revisit in Chapter 4. Consistency is also a property shared by many estimators, even in a finite setup. This leaves us to appraise estimation theory, in the light of information bounds and sufficiency. Yet, there are some points that merit some further discussion.

Attainment of the (Cramér–Rao–Fréchet, Bhattacharya, or Chapman–Robbins) information bounds is basically related to some *regularity conditions*, which may not be satisfied by densities not belonging to the regular exponential family. Thus, the Laplace, logistic, Cauchy, and many other densities, do not allow finite-sample efficient estimators in accordance to the information bound. On the other hand, for some *nonregular* densities, as the uniform density, sufficiency holds, and the MLE has a MSE smaller than the information bound! In such a case, if we define the efficiency of an estimator T_n with respect to the information bound, that is,

$$\begin{aligned} e(T_n;\theta) &= [nI(\theta)]^{-1}/(\text{MSE of } T_n) \\ &= \{[nI(\theta)]\mathbb{E}_\theta(T_n-\theta)^2\}^{-1}, \end{aligned} \tag{2.5.1}$$

then $e(T_n;\theta)$ will be greater than 1, and thus leads to misinterpretation. The use of the Rao–Blackwell Theorem 2.3.2, in the presence of sufficiency, conveys a better sense, though confining the attention to the class of unbiased estimators. In that way too, there could be some biased estimators that dominate the unbiased sufficient estimator; the *Stein rule estimator* [Berger (1993)] is a classical example.

In either setup, there is a strong emphasis on MSE (or variance) of estimators, and those that do not have finite second moment are ruled out as possible competitors. In this context, the criterion introduced in (2.2.5) may make more sense, especially when the distribution of T_n is symmetric about θ. Again, for asymmetric distributions, such a criterion needs some modifications to adjust for unequal tails. The situation becomes much more complex when $\boldsymbol{\theta}$ is a vector, so that $||\boldsymbol{T}_n-\boldsymbol{\theta}||$ needs to be defined in a convenient way so as to reflect the interdependence of the coordinate elements of $\boldsymbol{T}_n$ as well as the possible heterogeneity of their scale factors.

In multiparametric models some of the parameters are of primary interest while others may be nuisance parameters. Yet, the estimators of the parameters of interest generally have distributions that depend on the latter. For different estimators, this dependence may show up in possibly different ways, thus creating an impasse for direct comparison or for drawing conclusions on their finite-sample optimality. The basic difficulty in selecting the best among competing estimators in finite samples stems from the fact that excepting in some very simple models, *exact sampling distributions* of the estimators may not be available in closed forms, which permit analytic comparisons based on any criterion. This is particularly noteworthy in multiparameter estimation theory.

In single parameter estimation problems, often some other criteria are advocated (in lieu of the MSE). Among these, special mention may be made to the *mean absolute deviation* (MAD) criterion; for an estimator T_n of θ, the MAD is defined by

$$\mathbb{E}_\theta(|T_n - \theta|), \quad \theta \in \Theta, \; T_n \in \mathcal{T}, \tag{2.5.2}$$

where $\mathcal{T}$ is the class of all estimators of θ having finite first moment. Note that the MAD is a minimum about the median [Exercise 2.5.1]. Thus, if θ_T is the median (of the distribution) of T_n, then

$$\mathbb{E}_\theta(|T_n - \theta|) \geq \mathbb{E}_\theta(|T_n - \theta_T|). \tag{2.5.3}$$

Here, however, the nice additivity property encountered in the case of the MSE may not hold. From this perspective, it may be better to consider the class of median-unbiased estimators [see (2.2.3)], so that $\theta_T = \theta$, and then the comparison in (2.5.3) would make more sense. Except for some simple models, the computation of $\mathbb{E}_\theta(T_n - \theta)^2$ can be done more routinely than that of $\mathbb{E}_\theta(|T_n - \theta|)$, especially when T_n involves an average of independent terms (as is the case with the exponential family of densities). Further extensions of (2.5.2) to the multiparameter case are more complex than that of the MSE. For example, if we take $||\boldsymbol{T}_n - \boldsymbol{\theta}||$ as the Euclidean distance, then $E_\theta(||\boldsymbol{T}_n - \boldsymbol{\theta}||)$ would depend on the distribution of $||\boldsymbol{T}_n - \boldsymbol{\theta}||^2$, which is generally difficult to obtain.

In decision theoretic setups, as we will see in Chapter 4, there could be general consideration of loss risk functions, in the light of which, optimality of estimators can be appraised for finite-sample setups. This problem is more acute in multiparameter estimation problems. Beyond parametric scenarios often offer greater challenges for finite-sample (exact) inference. There are, however, good prospects when optimization is confined to some specific subclasses of estimators.

We illustrate this with the two-parameter (location-scale) logistic distribution that is, for which the distribution function is $F(x;\theta_1,\theta_2) = \{1 + \exp[-(x - \theta_1)/\theta_2]\}^{-1}$, $-\infty < x < \infty$, $\theta_1 \in \mathbb{R}$, $\theta_2 > 0$. Note that though θ_1 is the mean of F, its variance is different from θ_2^2, and hence, we regard θ_1 and θ_2 as location and scale parameters, respectively. The corresponding density function is

$$f(x;\theta_1,\theta_2) = \frac{\theta_2^{-1}\exp[-(x - \theta_1)/\theta_2]}{\{1 + \exp[-(x - \theta_1)/\theta_2]\}^2}, \tag{2.5.4}$$

so that the log-likelihood is

$$l_n(\boldsymbol{\theta}, \boldsymbol{X}) = -n\log\theta_2 - \frac{1}{\theta_2}\sum_{i=1}^{n}(X_i - \theta_1) - 2\sum_{i=1}^{n}\log[1 + e^{-(X_i - \theta_1)/\theta_2}]. \tag{2.5.5}$$

Because of the last term in (2.5.5), the Factorization Theorem 2.2.1 does not hold, ruling out the existence of (minimal) sufficient statistics. Thus, neither the information bounds nor the Rao–Blackwellization process work here. However, we may note that $Y_i = (X_i - \theta_1)/\theta_2$ has the standard logistic density $f(y) = \exp(-y)/[1 + \exp(-y)]^2$, completely free from $\boldsymbol{\theta}$. Then, let $X_{n:1} < \cdots < X_{n:n}$ be the order statistics corresponding to $X_1, \ldots, X_n$, and let $Y_{n:1} < \cdots < Y_{n:n}$ be the order statistics for $Y_1, \ldots, Y_n$, so that

$$X_{n:i} = \theta_1 + \theta_2 Y_{n:i}, \quad i = 1, \ldots, n. \tag{2.5.6}$$

Furthermore, let $\boldsymbol{a}_n = (a_{n1}, \ldots, a_{nn})'$, where $a_{ni} = \mathbb{E}(Y_{n:i})$, $1 \leq i \leq n$. Also, let $\boldsymbol{B}_n = ((b_{nij}))$ with $b_{nij} = \mathbb{C}\text{ov}(Y_{n:i}, Y_{n:j})$, $1 \leq i, j, \leq n$. Neither the vector $\boldsymbol{a}_n$, nor the matrix $\boldsymbol{B}_n$ depend on $\boldsymbol{\theta}$ and have been extensively tabulated for finite n [see Sarhan and Greenberg (1962)]; hence, we assume that they are known. Let $\boldsymbol{X}^* = (X_{n:1}, \ldots, X_{n:n})'$, so that

$$\mathbb{E}(\boldsymbol{X}^*) = \theta_1 \mathbf{1}_n + \theta_2 \boldsymbol{a}_n, \quad \text{and} \quad \mathbb{V}\text{ar}(\boldsymbol{X}^*) = \theta_2^2 \boldsymbol{B}_n. \tag{2.5.7}$$

Consider then the GLSE in (2.4.8)–(2.4.10). Here, $\boldsymbol{g}(\boldsymbol{\theta}) = \theta_1 \mathbf{1}_n + \theta_2 \boldsymbol{a}_n$ so that $\dot{\boldsymbol{G}}_{\boldsymbol{\theta}} = (\mathbf{1}_n, \boldsymbol{a}_n)$, does not depend on $\boldsymbol{\theta}$. Hence,

$$\widehat{\boldsymbol{\theta}}_n = \left[\begin{pmatrix} \mathbf{1}'_n \\ \boldsymbol{a}'_n \end{pmatrix} \boldsymbol{B}_n^{-1} (\mathbf{1}_n, \boldsymbol{a}_n) \right]^{-1} \begin{pmatrix} \mathbf{1}'_n \\ \boldsymbol{a}'_n \end{pmatrix} \boldsymbol{B}_n^{-1} \boldsymbol{X}_n^*. \tag{2.5.8}$$

Because it is based on linear functions of ordered observations, the estimator $\widehat{\boldsymbol{\theta}}_n$ in (2.5.8) is the BLUE of $\boldsymbol{\theta}$. Of course, such optimality is confined to the class of linear functions of order statistics, and for this estimator, (2.3.36) is not a null matrix, so that $\widehat{\boldsymbol{\theta}}_n$ is not optimal in the class of unbiased estimators of $\boldsymbol{\theta}$.

The BLU estimation theory applies to a larger class of location-scale family of distributions that includes the normal, Laplace, Cauchy, and others. Thus, in finite samples, suboptimal estimators exist within some specific classes, but may not be globally optimal, in general. For example, in the normal case, minimal sufficiency holds and an optimal unbiased sufficient estimator exists according to the Rao–Blackwell prescription. The BLUE of σ is not a minimal sufficient statistic and, hence, is not optimal. However, it is highly efficient. In this normal case, the optimal estimator of θ_1 is the sample mean $\overline{X} = n^{-1} \sum_{i=1}^n X_{n:i}$, which is also equal to the BLUE of θ_1. On the other hand, the optimal estimator of θ_2^2 is the sample variance $s_n^2 = (n-1)^{-1} \sum_{i=1}^n (X_i - \overline{X})^2$. The Rao–Blackwell Theorem 2.3.2 then yields an optimal estimator of θ_2, which is not linear in the $X_{n:i}$. As such, the BLUE of θ_2 may not have full efficiency unless $n = 2$. Such BLUE are quite amenable for censoring/truncation, and also are better from robustness perspectives and preferred to the MLE/LSE, even if a small sacrifice in efficiency is needed as a compromise.

Linear functions of order statistics (not necessarily BLUE) have been considered as estimators on robustness considerations. Among them, the trimmed mean

$$T_{n,k} = (n - 2k)^{-1} \sum_{i=k+1}^{n-k} X_{n:i}$$

and of the Winsorized mean

$$W_{n,k} = n^{-1} \left[(k+1) X_{n:k+1} + \sum_{j=k+2}^{n-k+1} X_{n:j} + (k+1) X_{n:n-k} \right]$$

have especial interest. In particular, the requirements of first and second moments is less stringent for such estimators than those in (2.5.7). If we take $k = 2$, both the trimmed and the Winsorized means from a sample of size $n \geq 5$ from a Cauchy distribution will have finite variance.

2.6 Concluding Notes

Likelihood, sufficiency, and invariance play a key role in finite-sample estimation theory, providing (at least, sub-) optimal estimators for a broad class of probability distributions. The regular exponential family of densities, admitting minimal sufficiency and completeness, has optimal estimators in the light of information bounds as well as of the uniformly minimum variance unbiased property, but for densities not belonging to this mathematically tractable class, optimality may generally have to be compromised to suboptimality within specific subclasses of estimators (such as the BLUE). Even for the regular exponential family, when the MSE is used as a discriminating criterion, optimality depends heavily on moment conditions, and hence, from robustness considerations, it might be better to consider estimators that are less sensitive to gross error contamination/outliers or some model distortion, albeit at the cost of some compromise on efficiency.

Also, a basic question relates to the adequacy of sample sizes in order to have satisfactory levels of performance of estimators. This again depends on the characteristics of there sampling distributions; the most adopted measure being the *standard error*. However, the standard error is, by itself alone, a meaningful measure, only when the sampling distribution of $T_n - \theta$ is completely determined by it, a case that holds essentially only when this sampling distribution is exactly normal. In finite-sample estimation problems, (exact) normal sampling distribution of estimators is practically confined to the normal density with known variance; even for the binomial or Poisson distributions, this is not the case in finite samples. For some other members of the exponential family, the exact form of the distribution of T_n may be known (up to some unknown parameters), but their standard errors may not completely characterize their distributions; estimation of other parameters of the sampling distribution might provide additional information, and yet, may not be of full utility in finite samples.

We now summarize the basic difficulties with finite-sample approaches to statistical estimation theory. Excepting in some special (and simplified) problems, the exact sampling distribution of estimators may not be fully known. This is particularly true for nonlinear estimators, and finding a closed form for their distribution could be a problem.

The situation with nonparametric estimation may even be worse as the true underlying distribution F belongs to a general class, say $\mathcal{F}$, and the specific sampling distribution may vary over F in $\mathcal{F}$.

As the sample size increases, analytical expressions for exact sampling distributions of estimators generally become more difficult to obtain, creating impasses for inference.

Excepting for the exponential family of densities, optimal or desirable estimators are nonlinear and often not expressible in closed form. Thus, computationally, such (sub-) optimal estimators are less appealing, while simpler estimators are generally less efficient.

For ML/M-estimators and most GEE based estimators, iterative solutions are generally needed. Although various computational algorithms are available evaluation of the convergence properties may often be difficult for finite samples.

Optimality criteria of diverse types, as discussed earlier, may not lead to a general consensus in a finite-sample setup, more so in a multiparameter situation. Exact evaluation of such measures may not always be easy to implement.

Bayes estimators, as we will see in Chapter 4, though geared for finite samples, have *prior distributions* attached on the parameters, and they may not always be interpretable (though often posed as conjugate to the density of observed random variables, for mathematical convenience).

In a sense, a unified, robust, optimal estimation theory is generally lacking in finite-sample setups (normal densities being exceptions).

Asymptotic approaches attempt to minimize, if not eliminate, most of these shortcomings. The basic idea is to provide suitable approximations to the exact sampling distributions by simpler (e.g., normal, chi-squared, gamma, Poisson) distributions, that may be used to justify robustness and optimality criteria in a unified way. In this context, computational algorithms generally have better convergence prospects, and the scope of statistical inference extends to a much broader setup covering many complex cases and matching needs for fruitful applications in modern interdisciplinary studies. This is considered in Chapter 8.

2.7 Exercises

2.3.1 For the density function $f(x;\theta) = c(x)\exp[a(\theta)t(x) - b(\theta)]$ belonging to the exponential family, use the identity

$$\int f(x;\theta)\mathrm{d}x = 1 \text{ or } \int c(x)e^{a(\theta)t(x)}\mathrm{d}x = e^{b(\theta)},$$

and by differentiation with respect to θ, show that

$$E_\theta[t(X)] = b'(\theta)/a'(\theta).$$

2.3.2 Show that the binomial law

$$P(X = x) = \binom{n}{x}\theta^x(1-\theta)^{n-x},\ 0 \le \theta \le 1,\ x = 0, \ldots, n$$

belongs to the one-parameter exponential family. Hence, or otherwise, verify (2.3.20) and (2.3.28).

2.3.3 Show that the Poisson law

$$P\{X = x\} = e^{-\theta}\theta^x/x!, \quad x = 0, 1, 2, \ldots \ge 0,$$

belongs to the exponential family, and hence, or otherwise, show that (2.3.20) and (2.3.28) hold.

2.3.4 Show that the exponential law

$$f(x;\theta) = \theta e^{-\theta x}I(x > 0),\ \theta > 0,\ x \in \mathbb{R}^+$$

belongs to the exponential family. Hence, or otherwise, verify whether (2.3.19) or (2.3.28) holds or not.

2.3.5 Consider the normal density function

$$f(x;\theta) = (2\pi)^{-1/2}\exp[-(x-\theta)^2/2], \quad x \in \mathbb{R},$$

and verify (2.3.19).

2.3.6 Consider the normal density

$$f(x;\theta) = (2\pi)^{-1/2}\theta^{-1}\exp[-x^2/(2\theta^2)], \quad \theta > 0,\ x \in \mathbb{R},$$

and verify whether or not (2.3.19) holds for an estimator of (a) θ and (b) θ^2.

2.3.7 Consider the two-parameter normal density function

$$f(x;\theta,\sigma^2) = \frac{1}{\sqrt{2\pi}\sigma}\exp\left[-\frac{(x-\theta)^2}{2\sigma^2}\right], \quad \theta \in \mathbb{R},\ \sigma \in \mathbb{R}^+,\ x \in \mathbb{R}.$$

Show that the unbiased estimator $s_n^2 = (n-1)^{-1}\sum_{i=1}^n (X_i - \overline{X}_n)^2$ does not attain the information bound for its variance.

2.3.8 Show that for $\boldsymbol{x}' = (\boldsymbol{x}_1', \boldsymbol{x}_2')$ and $\boldsymbol{A} = \begin{pmatrix} \boldsymbol{A}_{11} & \boldsymbol{A}_{12} \\ \boldsymbol{A}_{21} & \boldsymbol{A}_{22} \end{pmatrix}$, we have

$$\boldsymbol{x}'\boldsymbol{A}^{-1}\boldsymbol{x} = \boldsymbol{x}_1'\boldsymbol{A}_{11}^{-1}\boldsymbol{x} + \boldsymbol{x}_{2:1}'\boldsymbol{A}_{22}^{-1}\boldsymbol{x}_{2:1},$$

where $\boldsymbol{A}_{22\cdot 1} = \boldsymbol{A}_{22} - \boldsymbol{A}_{21}\boldsymbol{A}_{11}^{-1}\boldsymbol{A}_{12}$ and $\boldsymbol{x}_{2:1} = \boldsymbol{x}_2 - \boldsymbol{A}_{21}\boldsymbol{A}_{11}^{-1}\boldsymbol{x}_1$.

2.3.9 Use the result in Exercise 2.3.8 and show that for s_n^2 in Exercise 2.3.7, the Bhattacharya bound is attained.

2.3.10 Verify (2.3.36).

2.3.11 Let $f(x;\theta) = I(\theta - 1/2 \le x \le \theta + 1/2)$. Show that $\mathbb{E}_\theta(\overline{X}_n) = \theta$ and $\mathbb{V}\text{ar}_\theta(\overline{X}_n) = 1/(12n)$. Also, show that $(X_{n:1}, X_{n:n})$ are jointly sufficient for θ. Hence, show that the midrange $(X_{n:1} + X_{n:n})/2$ is unbiased for θ and has a variance smaller than $1/(12n)$ for all n and θ.

2.3.12 Prove the Basu Theorem.

2.4.1 Let $X_1, \dots, X_n$ be n i.i.d. random variables having normal distribution with location parameter μ and scale parameter $\sigma > 0$. Express $\mathbb{E}(X)$ and $\mathbb{V}\text{ar}(X)$ in terms of $\boldsymbol{\theta} = (\mu, \sigma)'$ and, hence, use the method of moment to estimate $\boldsymbol{\theta}$. Also, work out the case when X has a Laplace/logistic distribution with location and scale parameters θ_0 and θ_1, respectively. Comment on their optimality properties.

2.4.2 Let $X_1, \dots, X_n$ be n i.i.d. random variables having p.d.f. $f(x;\boldsymbol{\theta}) = (1/\sqrt{\theta_2})\theta_1^{\theta_2}$ $[\exp(-\theta_1 x)]x^{\theta_1 - 1}$, $\theta_1 > 0$ and $\theta_2 > 0$. Obtain the method of moments estimators of θ_1 and θ_2 and examine whether they are efficient and sufficient statistics.

2.4.3 Let $X_1, \dots, X_n$ be i.i.d. random variables with mean μ and variance $\sigma^2 < \infty$. Show that the LSE of μ is $\overline{X} = n^{-1}\sum_{i=1}^n X_i$. What is the WLSE of μ when $\mathbb{V}\text{ar}(X_i) = w_i\sigma^2$, $i \ge 1$, w_i known?

2.4.4 Let $X_i = \theta_o + \theta c_i + e_i$, $1 \le i \le n$, where the e_i are i.i.d. random variables with mean 0 and variance $\sigma^2 < \infty$, while the c_i, known as regression constants, are not all equal. Obtain the LSE of the parameters θ_o and θ as well as an expression for their covariance matrix.

2.4.5 Let $X_i = \theta c_i + e_i$, where $\mathbb{V}\text{ar}(e_i) = \sigma^2/c_i^\beta$, with c_i denoting regression constants, $i = 1, \dots, n$, and $\beta \ge 0$ is known. Obtain the WLSE of the parameter θ when $\beta = 1, 2, 0$. The result is known as the *ratio estimator*.

2.4.6 Show that when the covariance matrix $\boldsymbol{V}_n$ is diagonal, (2.4.8) may be expressed as (2.4.7). Furthermore, show that if all its nonnull elements are equal, we have (2.4.4).

2.4.7 Given i.i.d. random variables $X_1, \dots, X_n$ following each of the distributions considered in Examples 2.3.1–2.3.7, obtain the MLE of the corresponding parameters.

2.4.8 Consider a location model with density function $f(x;\theta) = f_0(x - \theta)$, where $\log f_0(y)$ is convex in $y \in \mathbb{R}$. Show that the normal density is a special case. Let $\psi = -(\partial/\partial y)\log f_0(y) = -f_0'(y)/f_0(y)$. Then, $\psi(y)$ is increasing in y. Therefore, $-\log f_0(y)$ is convex. Show that the convexity holds for the location-scale family also, where $f(x;\theta,\lambda) = (1/\lambda)f_0[(x - \theta)/\lambda]$, $\lambda > 0$, and $-\log f_0(y)$ is convex.

2.4.9 Let $\theta(F) = F^{-1}(p)$, for some $p \in (0, 1)$. What is $\theta(F_n)$? Is it unbiased for $\theta(F)$? Also, let $\theta(F) = \mathbb{E}_\theta[X - E(X)]^2$. What is $\theta(F_n)$ in this case? Is it unbiased for $\theta(F)$?

2.4.10 Show that U_n in (1.5.108) is a symmetric function of $X_1, \dots, X_n$ (i.e., it remains invariant under any permutation of $(X_1, \dots, X_n)$; hence, show that $U(X_1, \dots, X_n) = U(X_{n:1}, \dots, X_{n:n})$.

2.5.1 Show that for a distribution with median θ, $E(|X - a|) \ge E(|X - \theta|)$ for all real a, where the equality sign holds when $\theta \equiv a$.

3

Hypothesis Testing

3.1 Introduction

Testing statistical hypotheses, a dual problem to estimation, has the prime objective of making decisions about some population characteristic(s) with information obtained from sample data. A statistical hypothesis is a statement regarding a target distribution or some parameters associated with it, the tenacity of which is to be ascertained via statistical reasoning. In this context, the decision based on random samples may not always be correct, so appropriate strategies are needed to control the frequency of such errors. In this respect, the genesis of finite-sample principles of hypotheses testing stemmed primarily from the pioneering work of J. Neyman and E. S. Pearson in the 1930s. The Neyman–Pearsonian foundation for parametric as well as nonparametric setups in conjunction with other tributaries are appraised here under a finite-sample (exact) methodological framework, along with its transit to asymptotic reasoning.

Section 3.2 deals primarily with the basic concepts and the formulation of *simple hypotheses* testing problems. The more likely situation of *composite hypotheses* testing is considered with more detail in Section 3.3. There, diverse statistical approaches yielding different testing procedures are considered. In particular, *invariant tests* are highlighted. The interplay of *invariance* and *sufficiency* in parametric as well as nonparametric setups is analyzed in Section 3.4. Bayes procedures are to be discussed in Chapter 4.

3.2 The Neyman–Pearson Paradigm

Consider a sample $\boldsymbol{X} = (X_1, \ldots, X_n)'$ defined by i.i.d. random variables, and denote the corresponding sample space by $\mathcal{X}$. Associated with each X_i is a probability law P; in a parametric setup, P is indexed by a parameter θ, and the parameter space is denoted by Θ. For simplicity, we first consider the case where θ is a scalar; the multiparametric case will be considered later. In a nonparametric setup, on the other hand, P is allowed to be a member of a general class $\mathcal{P}$. Statistical hypotheses are framed in terms of θ (assuming P to be of given form) in the parametric case, while they are defined in terms of P belonging to some subset of $\mathcal{P}$ in the nonparametric case, that is also related to the *goodness-of-fit* testing problem. For simplicity of presentation, we start here with the parametric case.

In the Neyman–Pearson prescription, customarily, a *null* (H_0) and an *alternative* (H_1) hypothesis are simultaneously specified and serve as the basis for test construction. For example, consider the null hypothesis

$$H_0 : \theta \in \Theta_0 \subseteq \Theta, \tag{3.2.1}$$

Table 3.2.1. *Possible situations in a hypothesis testing setup.*

		True Hypothesis	
		H_0	H_1
Sample based decision	H_0	No error	Incorrect decision
	H_1	Incorrect decision	No error

the validity of which is to be contrasted with that of the alternative hypothesis,

$$H_1 : \theta \in \Theta_1 \subseteq \Theta \setminus \Theta_0. \tag{3.2.2}$$

The null (alternative) hypothesis H_0 (H_1) is said to be *simple*, if P_θ is completely specified when $\theta \in \Theta_0$ (Θ_1). If, otherwise, possibly due to some nuisance parameter(s), P_θ is not completely specified under H_0 (H_1), then H_0 (H_1) is termed a *composite* hypothesis. Thus, there are four possible pairing of hypotheses, namely,

$$H_0 \text{ simple/composite versus } H_1 \text{ simple/composite.}$$

The hypotheses testing paradigm is to use proper statistical methodology to decide (on the basis of the sample $\boldsymbol{X}$), which of H_0 or H_1 is the right statement. Thus, the hypothesis testing problem involves a partitioning of the sample space $\mathcal{X}$ into two disjoint subsets $\mathcal{X}_0$ and $\mathcal{X}_1$, with $\mathcal{X}_0 \cup \mathcal{X}_1 = \mathcal{X}$, such that H_0 is rejected (in favor of H_1) or not, according to $\boldsymbol{X} \in \mathcal{X}_1$ or $\boldsymbol{X} \in \mathcal{X}_0$. In this setup, $\mathcal{X}_0$ and $\mathcal{X}_1$ are called the *acceptance* and *rejection (critical) regions* for H_0. As $\boldsymbol{X}$ is random with density $L_n(\boldsymbol{x};\theta) = \prod_{i=1}^n f(x_i;\theta)$ (when it exists), the probability that $\boldsymbol{X}$ lies in $\mathcal{X}_0$ (or $\mathcal{X}_1$) not only depends on $\mathcal{X}_0$ (or $\mathcal{X}_1$) but also on $\theta \in \Theta$. The possible incorrect decision in favor of H_1 when H_0 is true is called a *Type I error*; otherwise, when the possible incorrect decision in favor of H_0 when H_1 is true, is called a *Type II error*. The outcomes of this decision process are schematically displayed in Table 3.2.1. For $\theta \in \Theta$, let $\beta(\theta) = P_\theta(\boldsymbol{X} \in \mathcal{X}_1)$. For the special case where H_0 corresponds to a simple hypothesis, that is, $H_o : \theta = \theta_0$, the *probability of Type I error* is

$$\beta(\theta_0) = P_{\theta_0}(\boldsymbol{X} \in \mathcal{X}_1) = P(H_0 \text{ is rejected} | H_0 \text{ is true}). \tag{3.2.3}$$

Similarly, when $H_1 : \theta = \theta_1$, the *probability of Type II error* is

$$P_{\theta_1}(\boldsymbol{X} \in \mathcal{X}_0) = 1 - \beta(\theta_1) = P(H_1 \text{ is rejected} | H_1 \text{ is true}). \tag{3.2.4}$$

In the more general case when H_1 in (3.2.2) is composite, (3.2.4) depends on the particular choice of $\theta \in \Theta_1$. Motivated by this, we may consider the *power function*

$$\beta(\theta) = P(H_0 \text{ is rejected} | \theta), \quad \theta \in \Theta.$$

By an abuse of language, the power at θ_0 is the Type I error probability.

Ideally, we would like to make both Type I and II error probabilities as small as possible; however, this is not feasible. Reducing the dimension of $\mathcal{X}_1$ (i.e., increasing that of $\mathcal{X}_0$) will make the Type II error probability larger, while increasing the dimension of $\mathcal{X}_1$ results in greater Type I error probability. Hence, the Neyman–Pearson strategy is to hold $\beta(\theta_0)$ at a preassigned level $0 < \alpha < 1$, called the *significance level*, and to maximize $\beta(\theta_1)$, $\theta_1 \in \Theta_1$, by skillful choice of $\mathcal{X}_1$. This can be done in a slightly more abstract way.

Let $\tau(\boldsymbol{X})$ be a *test function* (or simply *test*) taking values on [0, 1], and such that

$$\mathbb{E}_{\theta_0}[\tau(\boldsymbol{X})] = \alpha, \tag{3.2.5}$$

and

$$\mathbb{E}_{\theta_1}[\tau(\boldsymbol{X})] = \beta(\theta_1) \text{ is a maximum,} \tag{3.2.6}$$

where the maximization is with respect to the class of all τ. Whenever such a test function exists, it is called a *most powerful* (MP) test. If $\mathcal{X}_1$ has zero probability on its boundary, then $\tau(\boldsymbol{X})$ takes on the value 0 and 1 on $\mathcal{X}_0$ and $\mathcal{X}_1$, respectively, with probability $1 - \beta(\theta_1)$ and $\beta(\theta_1)$, so that $\mathcal{X}_1$ is called a MP *critical region* (of size α) at $\theta = \theta_0$ against $\theta = \theta_1$. If the boundary of $\mathcal{X}_1$ has nonzero probability under θ_0, then a *randomization test* retains the equality in (3.2.5) [see Exercise 3.2.1].

For testing a simple hypothesis $H_0 : \theta = \theta_0$ versus a simple hypothesis $H_1 : \theta = \theta_1$ ($\neq \theta_0$) hypothesis, we may consider the *likelihood ratio statistic*

$$\lambda_n = \lambda_n(\theta_0, \theta_1) = \frac{L_n(\theta_1; \boldsymbol{X})}{L_n(\theta_0; \boldsymbol{X})}. \tag{3.2.7}$$

When λ_n has a continuous distribution under θ_0, that is, $P_{\theta_0}(\lambda_n = t) = 0$, $\forall t \in \mathbb{R}^+$, we define

$$\mathcal{X}_0 = \{\boldsymbol{x} \in \mathcal{X} : \lambda_n < k_\alpha\}, \quad \mathcal{X}_1 = \mathcal{X} \setminus \mathcal{X}_0, \tag{3.2.8}$$

where k_α is chosen so that $P_{\theta_0}(\lambda_n \geq k_\alpha) = \alpha \in (0, 1)$. Let $\mathcal{X}_1^*$ be any other critical with significance level α so that $\mathcal{X}_0^* = \mathcal{X} \setminus \mathcal{X}_1^*$, and let $\beta^*(\theta_1)$ be the power of the test based on $\mathcal{X}_1^*$. Let $\mathcal{X}_1 \cap \mathcal{X}_1^* = \mathcal{X}_1^o$ and $\mathcal{X}_0 \cap \mathcal{X}_1^* = \mathcal{X}_0^o$. Then $\mathcal{X}_1^* = \mathcal{X}_1^o \cup \mathcal{X}_0^o$, where $\mathcal{X}_1^o \cap \mathcal{X}_0^o = \emptyset$. Thus,

$$\begin{aligned}
\beta^*(\theta_1) &= \int_{\mathcal{X}_1^*} L_n(\boldsymbol{x}; \theta_1) d\boldsymbol{x} = \int_{\mathcal{X}_1^o} L_n(\boldsymbol{x}; \theta_1) d\boldsymbol{x} + \int_{\mathcal{X}_0^o} L_n(\boldsymbol{x}; \theta_1) d\boldsymbol{x} \\
&= \int_{\mathcal{X}_1} L_n(\boldsymbol{x}; \theta_1) d\boldsymbol{x} + \int_{\mathcal{X}_0^o} L_n(\boldsymbol{x}; \theta_1) d\boldsymbol{x} - \int_{\mathcal{X}_1 \cap \mathcal{X}_0^*} L_n(\boldsymbol{x}; \theta_1) d\boldsymbol{x} \\
&= \int_{\mathcal{X}_1} L_n(\boldsymbol{x}; \theta_0) d\boldsymbol{x} + \int_{\mathcal{X}_0^o} \lambda_n L_n(\boldsymbol{x}; \theta_0) d\boldsymbol{x} - \int_{\mathcal{X}_1 \cap \mathcal{X}_0^*} \lambda_n L_n(\boldsymbol{x}; \theta_0) d\boldsymbol{x} \\
&\leq \int_{\mathcal{X}_1} \lambda_n L_n(\boldsymbol{x}; \theta_0) d\boldsymbol{x} = \beta(\theta_1),
\end{aligned} \tag{3.2.9}$$

as on $\mathcal{X}_1 \cap \mathcal{X}_0^*$, we have $\lambda_n \geq k_\alpha$ while on $\mathcal{X}_0^o$, we get $\lambda_n < k_\alpha$. Therefore, the test function $\tau(\boldsymbol{X})$ based on the critical region $\mathcal{X}_1$ in (3.2.8) is MP for testing H_0 versus H_1.

If H_0 is simple but H_1 is composite ($\theta \in \Theta_1 \subseteq \Theta \setminus \{\theta_0\}$), then for each $\theta_1 \in \Theta_1$, we can construct a MP test function, given by (3.2.8). If for every $\theta_1 \in \Theta_1$, the same MP test function reigns, the test is said to be *uniformly most powerful* (UMP) for H_0 versus H_1.

***Example* 3.2.1.** Let X be a random variable with density function $f(x; \theta) = (2\pi)^{-1/2} \exp[-(x - \theta)^2/2]$, $-\infty < x < \infty$, $\theta \in \mathbb{R}$ and suppose that we wish obtain a test function τ to decide between $H_0 : \theta = 0$ and $H_1 : \theta > 0$. For a given pair $(0, \theta_1)$, we have $\log \lambda_n(0, \theta_1) = n\theta_1\{\overline{X}_n - \theta_1/2\}$. So, for any $\theta_1 > 0$, $\mathcal{X}_1$ in (3.2.8) is characterized as $\{\boldsymbol{x} \in \mathcal{X} : \sqrt{n}\overline{X}_n >_\alpha\}$. Since this does not depend on the specific $\theta_1 > 0$, it is UMP for $\theta_1 > 0$.

If we let $H_0 : \theta = 0$ and $H_1 : \theta \neq 0 = H_1^+ \cup H_1^-$ where $H_1^+ : \theta > 0$ and $H_1^- : \theta < 0$, then we get UMP tests for H_1^+ and for H_1^-, but none would be UMP for H_0 versus $H_1 : \theta \neq 0$. This example shows that an UMP test may not exist for two-sided alternatives, or more generally, for testing $H_0 : \theta = \theta_0$ versus $H_1 : \theta \in \Theta \setminus \{\theta_0\}$. □

A *test function* τ is said to be *unbiased* for testing $H_0 : \theta = \theta_0 \in \Theta$ against $H_1 : \theta \in \Theta \setminus \{\theta_0\}$, if

$$\mathbb{E}_\theta[\tau(\boldsymbol{X})] \geq \alpha, \ \forall \theta \in \Theta \setminus \{\theta_0\}. \tag{3.2.10}$$

If we restrict ourselves to the class of unbiased tests, and if within that class, a UMP test exists, it is called a *uniformly most powerful unbiased* (UMPU) test. When $\Theta \subseteq \mathbb{R}$ and θ_0 is an inner point of Θ, an unbiased test requires that

$$\beta(\theta) \geq \beta(\theta_0) = \alpha, \quad \text{for all } \theta \in \Theta, \tag{3.2.11}$$

so that the power function attains a minimum at θ_0. If $\beta(\theta)$ is differentiable with respect to θ, we have

$$\beta'(\theta)|_{\theta_0} = 0. \tag{3.2.12}$$

If we intend to confine ourselves only to the class of test functions satisfying (3.2.12), whenever differentiation with respect to θ is permissible under integration with respect to $\boldsymbol{x}$ [see (2.3.4)–(2.3.5)], we may observe that

$$\begin{aligned}
\beta'(\theta) &= \frac{\partial}{\partial\theta}\mathbb{E}_\theta[\tau(\boldsymbol{X})] = \frac{\partial}{\partial\theta}\int_{\mathcal{X}} \tau(\boldsymbol{x})L_n(\boldsymbol{x};\theta)\mathrm{d}\boldsymbol{x} \\
&= \int_{\mathcal{X}} \tau(\boldsymbol{x})\frac{\partial}{\partial\theta}L_n(\boldsymbol{x};\theta)\mathrm{d}\boldsymbol{x} \\
&= \int_{\mathcal{X}} \tau(\boldsymbol{x})\frac{\partial}{\partial\theta}\log L_n(\boldsymbol{x};\theta)L_n(\boldsymbol{x};\theta)\mathrm{d}\boldsymbol{x} \\
&= \int_{\mathcal{X}} \tau(\boldsymbol{x})U_n(\theta)L_n(\boldsymbol{x};\theta)\mathrm{d}\boldsymbol{x} \\
&= \mathbb{E}_\theta[\tau(\boldsymbol{X})U_n(\theta)],
\end{aligned} \tag{3.2.13}$$

where $U_n(\theta)$ is the score statistic defined by (2.3.1). The unbiasedness condition reduces to

$$\mathbb{E}_{\theta_0}[\tau(\boldsymbol{X})U_n(\theta_0)] = 0. \tag{3.2.14}$$

If $\tau(\boldsymbol{x})$ is equal to 0 or 1, according as $\boldsymbol{x} \in \mathcal{X}_0$ or $\mathcal{X}_1$, then (3.2.14) reduces to

$$\int_{\mathcal{X}_1} U_n(\theta_0)L_n(\theta_0;\boldsymbol{x})\mathrm{d}\boldsymbol{x} = 0,$$

which is equivalent to

$$\frac{\partial}{\partial\theta}P_\theta(\boldsymbol{X} \in \mathcal{X}_1)\bigg|_{\theta_0} = 0. \tag{3.2.15}$$

Looking back at Example 3.2.1, we observe that (3.2.15) limits unbiased critical regions to the form $h(|\overline{X}_n|)$ for suitable h, and within this class, the UMPU critical region is given by

$$\mathcal{X}_1 = \{\boldsymbol{x} \in \mathcal{X} : \sqrt{n}|\overline{X}_n| > z_{\alpha/2}\}. \tag{3.2.16}$$

This also brings us to appraise the role of *sufficiency* in hypothesis testing contexts. If a (minimal) sufficient statistics T_n exists, then using the factorization (2.2.10) in Theorem 2.2.1 it follows that for every $\theta_0, \theta_1 \in \Theta$,

$$\lambda_n = \lambda_n(\theta_0, \theta_1) = h_n(T_n; \theta_0, \theta_1), \tag{3.2.17}$$

where $h_n(T_n; \theta_0, \theta_1) = g_1(T_n; \theta_1)/g_2(T_n; \theta_0)$ depends only on $\boldsymbol{X}$ through T_n. Thus, not only MP tests depend only on T_n alone, but also UMP and UMPU tests, whenever they exist, will be based on T_n alone. In this context, it is worthwhile to introduce the concept of *monotone likelihood ratio* (MLR). If $\theta \in \Theta \subseteq \mathbb{R}$, the density $f(x, \theta)$ has the MLR property if for $\theta < \theta'$, $f(x, \theta)$ and $f(x, \theta')$ are not identical everywhere, and $f(x, \theta')/f(x; \theta)$ is an increasing function of some statistic $T(x)$. For densities with the MLR property (in T_n), $\beta(\theta)$ is increasing in θ and one-sided UMP tests exist. More details may be found in Lehmann (1986), for example.

Example* 3.2.2 *(One-parameter exponential family). Consider a random variable X with density function belonging to the exponential family (2.2.11), where $T_n = \sum_{i=1}^n t(X_i)$. Thus, the MLR property holds. Exercise 3.2.2 is set to verify the UMPU test property when either (i) $a(\theta) = \theta, t(x) = x$ and $b(\theta) = -(1/2)\theta^2$, or (ii) $a(\theta) = \theta, t(x) = x, b(\theta) = \log\theta$, that is, when f corresponds to the normal and simple exponential densities. □

Let us now briefly present the multiparameter case where $\boldsymbol{\theta} = (\theta_1, \ldots, \theta_q)'$, $q > 1$. Suppose that $H_0 : \boldsymbol{\theta} = \boldsymbol{\theta}_0$ is a simple hypothesis while $H_1 : \boldsymbol{\theta} = \boldsymbol{\theta}_1$ may not be. For a specific $\boldsymbol{\theta}_1$, the MP test can be constructed as before. However, characterizing that as a UMP test may not be generally possible. In this case sufficiency by itself may not be of much help, and we may need to restrict ourselves to more specific subclasses of test functions (e.g., *invariant tests*) where an UMP property can be established. Again, such invariance may not pertain to *one-sided alternatives* as in a multiparameter setup [Silvapulle and Sen (2004)]. To this end, we consider an illustrative example.

***Example* 3.2.3.** Let $\boldsymbol{X}_i = (X_{i1}, X_{i2})$, $1 \le i \le n$, be i.i.d. random vectors with density

$$f(\boldsymbol{x}; \boldsymbol{\theta}) = (2\pi)^{-1} \exp\left[-\frac{1}{2}||\boldsymbol{x} - \boldsymbol{\theta}||^2\right], \quad \boldsymbol{x} \in \mathbb{R}^2,\ \boldsymbol{\theta} \in \mathbb{R}^2. \tag{3.2.18}$$

The MLE of $\boldsymbol{\theta}$ is the sample mean vector $\overline{\boldsymbol{X}}_n$, a sufficient statistic; also, the density is complete. Suppose that we want to test for $H_0 : \boldsymbol{\theta} = \boldsymbol{0}$ against $H_1 : \boldsymbol{\theta} \ge \boldsymbol{0}$, $||\boldsymbol{\theta}|| > 0$. Here, for a given $\boldsymbol{\theta}^* \ge 0$, the logarithm of (3.2.7) reduces to

$$\log\lambda_n(\boldsymbol{0}, \boldsymbol{\theta}^*) = n\left[\boldsymbol{\theta}^{*\prime}\left(\overline{\boldsymbol{X}}_n - \frac{1}{2}\boldsymbol{\theta}\right)\right], \tag{3.2.19}$$

and $\sqrt{n}\boldsymbol{\theta}^{*\prime}\overline{\boldsymbol{X}}_n \sim \mathcal{N}(\sqrt{n}||\boldsymbol{\theta}^*||^2, ||\boldsymbol{\theta}^*||^2)$. Thus, the MP test function for testing $H_0 : \boldsymbol{\theta} = \boldsymbol{0}$ versus $H_1 : \boldsymbol{\theta} = \boldsymbol{\theta}^* \ge \boldsymbol{0}$ can be equivalently written as

$$\tau(\boldsymbol{X}) = \begin{cases} 1, & \text{if } ||\boldsymbol{\theta}^*||^{-1}\sqrt{n}\boldsymbol{\theta}^{*\prime}\overline{\boldsymbol{X}}_n > z_{1-\alpha}, \\ 0, & \text{otherwise.} \end{cases} \tag{3.2.20}$$

As such, without loss of generality, we may let $\boldsymbol{\theta}^*$ be such that $||\boldsymbol{\theta}^*|| = 1$. Even so, (3.2.20) depends on $\boldsymbol{\theta}^*$ (given $\overline{X}_n$) through $\overline{X}_n^{'}\boldsymbol{\theta}^*$ and would not be the same for all $\boldsymbol{\theta}^*$ on the segment $\boldsymbol{\theta}^* \geq \mathbf{0}$, $||\boldsymbol{\theta}^*|| = 1$. Therefore, $\tau(\boldsymbol{X})$ is not UMP for $H_1 : \boldsymbol{\theta} \geq \mathbf{0}$, $||\boldsymbol{\theta}|| = 1$. Thus, an UMP test for $H_0 : \boldsymbol{\theta} = \mathbf{0}$ against $H_1 : \boldsymbol{\theta} \geq \mathbf{0}$ does not exist. □

We now introduce another tool that is commonly used in hypothesis testing problems. Obtaining exact and optimal tests in finite samples depends on the existence of the exact sampling distribution of the test statistic under H_0, in a form that leads to exact demarcation of the critical region. This feature holds for some specific densities but for others, specially those not belonging to the exponential family may this might not be true. Now we suppose that an optimal (or desirable) test for H_0 versus H_1 is based on a test statistic T_n, the distribution of which, under H_0, is denoted by $F^o_{T_n}$, and that the critical region is composed by its upper tail. We also assume that $F^o_{T_n}$ is continuous. For an observed value of T_n $(= t$, say), we define the *observed significance level* (OSL) or *p-value* by

$$L_{T_n}(t) = 1 - F^o_{T_n}(t) = P(T_n \geq t | H_0). \tag{3.2.21}$$

If the null hypothesis H_0 is true, $L_{T_n}(t)$ has the Uniform[0, 1] distribution. On the other hand, if T_n is stochastically larger[1] under the alternative hypothesis than under the null hypothesis so that L_{T_n} will be stochastically tilted toward the lower end point 0. As a result, $-\log L_T(t)$ will be stochastically larger under the alternative hypothesis. This underlies R. A. Fisher's fiducial inference principle, so the smaller is $L_{T_n}(T_n)$, the greater is the evidence that H_0 is not likely to be the right statement concerning θ. In the Neyman–Pearson setup, whenever $L_{T_n}(T_n) \leq \alpha$, H_0 is rejected, and it does not matter how small it is. In that sense, it is a two-alternative decision problem with the decisions to be made on the basis of $L_{T_n}(T_n)$ being $\leq \alpha$ or $> \alpha$, for a specified significance level $0 < \alpha < 1$. The p-values have greater appeal when evidence from multiple samples on a common hypothesis need to be pooled toward a more consensual decision. This is indeed the case in *meta analysis* [Hedges and Olkin (1985)] and *combination of independent tests*, where the detailed information on the individual p-values generally provides better statistical resolutions. On the other hand, the Neyman–Pearson dichotomy of p-values ($<$ or $\geq \alpha$) of the individual tests may convey some loss of information, resulting in less precise decisions. However, such analyses depend on the exact distribution of L_{T_n}, under H_0, especially for very small p-values, and in many testing problems, this is a prohibitively laborious task, moreover if the density does not belong to the regular exponential family. In this setup, the large-sample prospects are somewhat mixed. We will discuss that situation in a more general setup in Chapter 8.

3.3 Composite Hypotheses: Beyond the Neyman–Pearson Paradigm

When the hypotheses H_0 and H_1 in (3.2.1) and (3.2.2) are composite, under $H_0 : \boldsymbol{\theta} \in \boldsymbol{\Theta}_0$ $(H_1 : \boldsymbol{\theta} \in \boldsymbol{\Theta}_1)$ the density $f(x; \boldsymbol{\theta})$ is not completely specified, so that the likelihood ratio

[1] If $P(X > Y) \geq 1/2$ then X is said to be stochastically larger than Y. This is equivalent to $\int f_Y(x) \mathrm{d}F(x) \geq 1/2$ and is implied by $F_X(x) \leq F_Y(x)$, for all x. Sometimes stochastic ordering is also defined by the latter ordering.

statistic in (3.2.7) needs to be defined in a more general way, namely,

$$\lambda_n = \frac{\sup_{\boldsymbol{\theta}\in\Theta_1} L_n(\boldsymbol{\theta}; \boldsymbol{X})}{\sup_{\boldsymbol{\theta}\in\Theta_0} L_n(\boldsymbol{\theta}; \boldsymbol{X})}, \tag{3.3.1}$$

and in that process, the MLE of $\boldsymbol{\theta}$ restricted to Θ_0 and Θ_1 are needed to evaluate the right-hand side of (3.3.1). Thus, unlike in the previous section, $\log \lambda_n$ may no longer involve independent summands, thus possibly vitiating the direct Neyman–Pearson approach considered in Section 3.2.

The first problem that arises in this setup is how to construct a rejection (critical) region that has a significance level $0 < \alpha < 1$, when the density is not fully specified under the null hypothesis. Thus, we require that

$$P_{\boldsymbol{\theta}}(\boldsymbol{X} \in \mathcal{X}_1) = \alpha, \quad \text{for all } \boldsymbol{\theta} \in \Theta_0, \tag{3.3.2}$$

although $f(x, \boldsymbol{\theta})$ is not fully known. If (3.3.2) holds, the rejection region is called a *similar region*. If, instead, we have

$$\sup_{\boldsymbol{\theta}\in\Theta_0} P_{\boldsymbol{\theta}}(\boldsymbol{X} \in \mathcal{X}_1) \leq \alpha, \tag{3.3.3}$$

α is called the *size* of the test, albeit the rejection region might not be a similar region in a strict sense. In general, a similar region may not exist. As an illustration, consider the following example.

***Example* 3.3.1.** Let X be a random variable with density function $f(x; \boldsymbol{\theta}) = (2\theta_2)^{-1}\exp[-|x - \theta_1|/\theta_2]$, $-\infty < x < \infty$, $\theta_1 \in \mathbb{R}$, $\theta_2 > 0$. Thus, $\Theta = \mathbb{R} \times \mathbb{R}^+$. Also, assume that we are interested in testing $H_0 : \theta_1 = 0$ versus $H_1 : \theta_1 > 0$, treating θ_2 as nuisance parameter. Here

$$\log L_n(\boldsymbol{\theta}; \boldsymbol{X}) = -n \log 2 - n \log \theta_2 - \frac{1}{\theta_2}\sum_{i=1}^{n} |X_i - \theta_1|. \tag{3.3.4}$$

Thus, under H_0, (3.3.4) depends on θ_2 and on $\sum_{i=1}^n |X_i|$, while under H_1, it depends on (θ_1, θ_2) as well as on $\sum_{i=1}^n |X_i - \theta_1|$. The MLE of θ_2 under H_0 is $\widehat{\theta}_{20} = n^{-1}\sum_{i=1}^n |X_i|$. Also, for a specified $\theta_1 > 0$, under H_1, the MLE of θ_2 is $\widehat{\theta}_{21} = n^{-1}\sum_{i=1}^n |X_i - \theta_1|$. As a result, for θ_1 specified, with θ_2 treated as a nuisance parameter, we have

$$\begin{aligned} log\lambda_n &= -n \log\left(\frac{\sum_{i=1}^n |X_i - \theta_1|}{\sum_{i=1}^n |X_i|}\right) \\ &= -n \log\left(\frac{\sum_{i=1}^n \theta_2^{-1}|X_i - \theta_1|}{\sum_{i=1}^n \theta_2^{-1}|X_i|}\right), \end{aligned} \tag{3.3.5}$$

where under H_0, $\theta_2^{-1}\sum_{i=1}^n |X_i - \theta_1|/[\theta_2^{-1}\sum_{i=1}^n |X_i|]$ has a distribution depending on $\boldsymbol{\theta}$ only through $\theta_1/\theta_2 = \delta$, say. However, $\sum_{i=1}^n |X_i/\theta_2 - \delta|/\sum_{i=1}^n |X_i/\theta_2|$ cannot be additively decomposed in terms of a statistic based on the scale free X_i/θ alone and δ, and hence, (3.3.2) does not hold when λ_n in (3.3.5) is used. We may avoid this impasse by restricting ourselves to the class scale-invariant tests, for which a similar region may be obtained. Note that for testing $H_0 : \theta_1 = 0$ versus $H_1 : \theta_1 > 0$, with θ_2 treated as a nuisance parameter, the *maximal invariants* are $Y_i = \text{sign}(X_i), 1 \leq i \leq n$. Thus, if $\tau(\boldsymbol{X}) = \tau(\boldsymbol{Y})$, then $\tau(c\boldsymbol{X}) = \tau(\boldsymbol{Y}) = \tau(\boldsymbol{X}), \forall c > 0$. Furthermore $S_n = (n + \sum_{i=1}^n Y_i)/2$ is Bin(n, 1/2), under

H_0, and Bin(n, p), $p > 1/2$ under H_1, so that a one-sided UMP test based on S_n provides an exact size α-test for H_0 versus H_1 in this composite hypothesis case. Thus finite-sample minimal sufficiency is not attained, but invariance provides us with the solution. □

***Example* 3.3.2.** Let X be a random variable with density function $f(x, \boldsymbol{\theta}) = (2\pi\theta_2)^{-1/2} \exp[-(x - \theta_1)^2/(2\theta_2)], -\infty < x < \infty, \boldsymbol{\theta} \in \boldsymbol{\Theta} = \mathbb{R} \times \mathbb{R}^+, \boldsymbol{\Theta}_0 = \{0\} \times \mathbb{R}^+$ and $\boldsymbol{\Theta}_1 = \mathbb{R}^+ \times \mathbb{R}^+$. Here the sample mean $\overline{X}_n$ and the variance s_n^2 are (jointly) sufficient for $\boldsymbol{\theta}$. As such, the likelihood ratio statistic in (3.3.1) becomes a function of $(\overline{X}_n, s_n^2)$. Note that the MLE of θ_1 is $\overline{X}_n$ if $\overline{X}_n > 0$ and 0 if $\overline{X}_n < 0$; the latter is due to the fact that $\theta_1 \in \mathbb{R}^+$. Exercise 3.3.1 is set to show that, for $\overline{X}_n \geq 0$, (3.3.1) reduces to

$$U_n = \left[\sum_{i=1}^{n} X_i^2 \Big/ \sum_{i=1}^{n} (X_i - \overline{X}_n)^2 \right]^{n/2},$$

which is a one-to-one function of

$$t_n = \sqrt{n}\overline{X}_n \Big/ \left[\sum_{i=1}^{n} (X_i - \overline{X}_n)^2 \Big/ (n-1) \right]^{1/2},$$

while it is equal to 1 for $\overline{X}_n < 0$. Furthermore, under $H_0 : \theta_1 = 0$, θ_2 being a nuisance parameter, t_n has the Student t-distribution with $n - 1$ degrees of freedom, so that an exact size α test can be based on that distribution. In this example, sufficiency along with the independence of $\overline{X}_n$ and s_n^2 have been exploited for a finite-sample solution. □

As we have seen, the Neyman–Pearson paradigm places emphasis on the likelihood ratio statistic, and thereby for finite-sample solutions it relies heavily on sufficiency. Alternative approaches exploit sufficiency along with conditionality in some cases, and invariance in some other cases. We illustrate this with some additional examples.

***Example* 3.3.3.** Let $X_1 \sim$ Poisson(λ_1) and $X_2 \sim$ Poisson(λ_2) be independent random variables. Consider the null hypothesis $H_0 : \lambda_1 = \lambda_2$ and the alternative $H_1 : \lambda_1 > \lambda_2$, both composite. Note that $Y = X_1 + X_2 \sim$ Poisson$(\lambda_1 + \lambda_2)$, and that conditionally on $Y = y$, $X_1 \sim$ Bin$[y, \lambda_1/(\lambda_1 + \lambda_2)]$. Thus, under H_0, given $Y = y$, $X_1 \sim$ Bin$(y, 1/2)$ while under H_1, given $Y = y$, $X_1 \sim$ Bin(y, p), $p > 1/2$. Thus, a one-sided (conditional) similar region can be obtained by using a (possibly randomized) test function based on X_1 given Y, and this will have unconditionally the exact size α. □

The conditionality principle works out well under sufficiency (which need not be minimal), and we elaborate on that in the next section. Similarly, the invariance principle also works out better under sufficiency, though we explain through Example 3.3.1 that minimal sufficiency is not that crucial in some cases. The next example illustrates this interplay.

***Example* 3.3.4.** Let $\boldsymbol{X}_i, i = 1, \ldots, n$ be random variables with density function $f(\boldsymbol{x}; \boldsymbol{\theta}) = (2\pi)^{-p/2} |\boldsymbol{\Sigma}|^{-1/2} \exp[(\boldsymbol{x} - \boldsymbol{\mu})' \boldsymbol{\Sigma}^{-1} (\boldsymbol{x} - \boldsymbol{\mu})]$, $\boldsymbol{x} \in \mathbb{R}^p$, $\boldsymbol{\mu} \in \mathbb{R}^p$, $\boldsymbol{\Sigma}$ p.d. (arbitrary), and set $\boldsymbol{\theta} = (\boldsymbol{\mu}, \boldsymbol{\Sigma})$. Suppose that we want to test for $H_0 : \boldsymbol{\mu} = \boldsymbol{0}$ versus $H_1 : \boldsymbol{\mu} \neq \boldsymbol{0}$, treating $\boldsymbol{\Sigma}$ as a nuisance parameter. Consider the affine group of transformations $\mathcal{G} = \{g\}$, where $g\boldsymbol{x} = \boldsymbol{a} + \boldsymbol{B}\boldsymbol{x}$, $\boldsymbol{a} \in \mathbb{R}^p$ and $\boldsymbol{B}$ nonsingular. Then, $g\boldsymbol{X}_i \sim \mathcal{N}_p(\boldsymbol{a} + \boldsymbol{B}\boldsymbol{\mu}, \boldsymbol{B}\boldsymbol{\Sigma}\boldsymbol{B}')$ and the hypothesis testing problem remains invariant under the transformation $\boldsymbol{\mu} \rightarrow \boldsymbol{a} + \boldsymbol{B}\boldsymbol{\mu}$. Letting

$\boldsymbol{X} = (\boldsymbol{X}_1, \ldots, \boldsymbol{X}_n)$ denote the sample, we would like to have a test function $\tau(\boldsymbol{X})$ that has the same invariance property, that is, that for testing H_0 versus H_1, the decision conveyed by $\tau(\boldsymbol{X})$ be the same as that conveyed by $\tau(g\boldsymbol{X})$ for testing the induced hypotheses H_{g0} versus H_{g1}, $g \in \mathcal{G}$. Without any loss of generality, we let $\boldsymbol{a} = \boldsymbol{0}$. Then, we need to show that there exists a test function $\tau(\boldsymbol{X})$, such that

$$\tau(\boldsymbol{X}) = \tau(\boldsymbol{B}\boldsymbol{X}), \quad \text{for all } \boldsymbol{B} \text{ nonsingular.} \tag{3.3.6}$$

For testing $H_0 : \boldsymbol{\mu} = \boldsymbol{0}$ versus $H_1 : \boldsymbol{\mu} = \boldsymbol{\mu}_1 (\neq \boldsymbol{0})$, we apply the likelihood ratio test in (3.3.1) where we allow $\boldsymbol{\mu}$ to vary over $\mathbb{R}^p \setminus \{\boldsymbol{0}\}$. Then (see Exercise 3.3.2) λ_n in (3.3.1) as a one-to-one function of the Hotelling T^2-statistic

$$T_n^2 = n(\overline{\boldsymbol{X}}_n - \boldsymbol{0})' \boldsymbol{S}_n^{-1} (\overline{\boldsymbol{X}}_n - \boldsymbol{0}), \tag{3.3.7}$$

where

$$\boldsymbol{S}_n = (n-1)^{-1} \sum_{i=1}^{n} (\boldsymbol{X}_i - \overline{\boldsymbol{X}})(\boldsymbol{X}_i - \overline{\boldsymbol{X}})'$$

is the sample covariance matrix. That this test is not uniformly most powerful over $\boldsymbol{\mu} \in \mathbb{R}^p$ can be easily verified by considering some specific directions of $\boldsymbol{\mu}$, e.g., $\boldsymbol{\mu} = (\theta_1, \boldsymbol{0}')'$ and showing that the univariate t-test is more powerful than the T_n^2-test if $\boldsymbol{\mu}$ lies on such an edge. However, the competing test could be worthless if $\boldsymbol{\mu}$ lies on some other edge. Thus, it seems natural to seek a test that is good overall in some meaningful way.

Note that $\boldsymbol{\mu} = \boldsymbol{0} \Leftrightarrow \boldsymbol{B}\boldsymbol{\mu} = \boldsymbol{0}$ for any nonsingular $\boldsymbol{B}$. Based on the transformation $\boldsymbol{B}\boldsymbol{X}$, the corresponding transformed sample mean vector is $\boldsymbol{B}\overline{\boldsymbol{X}}_n$ and the transformed sample covariance matrix is $\boldsymbol{B}\boldsymbol{S}_n\boldsymbol{B}'$, so that $\tau(\boldsymbol{B}\boldsymbol{X}) = n(\boldsymbol{B}\overline{\boldsymbol{X}}_n)'(\boldsymbol{B}\boldsymbol{S}_n\boldsymbol{B}')^{-1}\boldsymbol{B}\overline{\boldsymbol{X}}_n = n\overline{\boldsymbol{X}}_n'\boldsymbol{S}_n^{-1}\overline{\boldsymbol{X}}_n = \tau(\boldsymbol{X})$. Therefore, the test based on T_n^2 is invariant under the group $\mathcal{G}$ of transformations $g : \boldsymbol{X} \to \boldsymbol{B}\boldsymbol{X}$, $\boldsymbol{B}$ p.d. Invoking this invariance property, we may choose $\boldsymbol{B}$ such that

$$\boldsymbol{\mu} \to \boldsymbol{B}\boldsymbol{\mu} = (\theta, \boldsymbol{0}')' \quad \text{and} \quad \boldsymbol{B}\boldsymbol{\Sigma}\boldsymbol{B}' = \boldsymbol{I}_p \tag{3.3.8}$$

(see Exercise 3.3.3), and, as such, we replace $H_0 : \boldsymbol{\mu} = \boldsymbol{0}$ versus $H_1 : \boldsymbol{\mu} \neq \boldsymbol{0}$ with $H_0^* : \theta = 0$ versus $H_1^* : \theta > 0$, for which a uniformly most powerful invariant (UMPI) test is based on $\tau(\boldsymbol{B}\boldsymbol{X})$ when we confine ourselves to invariant tests. Therefore, this test based on T_n^2 is an UMPI test with respect to the group $\mathcal{G}$ of affine transformations. For this problem, minimal sufficiency of $(\overline{\boldsymbol{X}}_n, \boldsymbol{S}_n)$ and affine invariance of T_n^2 provide the resolution. □

The original Neyman and Pearson prescription was to exploit sufficiency under additional structural setups (known as Neyman structures) which work out for the exponential family of densities. Invariance has the advantage of going beyond the exponential family in finite samples achieving suboptimality within the class of invariant tests. This will be exploited more in the next two sections. Thus, the role of likelihood, sufficiency and invariance must be properly assessed in any specific testing problem, and one should choose the largest group $\mathcal{G}$ of transformations if invariant tests are to be explored.

There are some alternative approaches to hypotheses testing problems having their genesis in the likelihood principle but with somewhat different motivations and scope of applications. We introduce some of these here, although an in-depth discussion needs the asymptotic perspectives to be considered in Chapter 8.

Likelihood ratio tests (LRT). The primary motivation of the LRT stems from the Neyman–Pearson lemma, from which MP tests and often going beyond that, UMP or UMPU tests in simpler models are derived. In general, we may, however, note that under neither $\boldsymbol{\Theta}_0$ nor $\boldsymbol{\Theta}_1$, $L_n(\boldsymbol{X}, \theta)$ is completely specified, so that instead of the probability ratio we need to get a likelihood ratio wherein we substitute the MLE for the unspecified (nuisance) parameters in (3.3.1). Thus, we need two sets of MLE (UMLE and RMLE), the first under H_0 and the second under H_1. Such MLE under restrictions on $\boldsymbol{\Theta}$ are called *restricted maximum likelihood estimators* (RMLE).[2] For distributions admitting (minimal) sufficient statistics, such RMLE are functions of the original sufficient statistics, so that the LRT can be expressed in terms of these. Even so, for general composite hypotheses, the search for the exact sampling distribution of the RMLE and the LRT statistic may encounter considerable roadblocks. The situation is generally much more complex for distributions not admitting minimal sufficient statistics or not belonging to the exponential family. For this reason, if there is an underlying invariance structure, we may restrict the attention to invariant tests, and within that class, explore the advantages of the (restricted) LRT. This prescription works out well for the exponential family of densities when the hypotheses, though composite, are linked by simpler functional interrelations. However, the exact distribution theory, as is needed for finite-sample inference, may not always transpire in closed simple pathways. As an illustration, we revisit Example 3.3.4 but take $H_0 : \boldsymbol{\mu} = \mathbf{0}$ and $H_1 : \boldsymbol{\mu} \geq \mathbf{0}$ with $||\boldsymbol{\mu}|| > 0$ as the hypotheses of interest, treating $\boldsymbol{\Sigma}$ as a nuisance (p.d.) matrix. The group g of affine transformations $\boldsymbol{X} \rightarrow \boldsymbol{B}\boldsymbol{X}$ with nonsingular $\boldsymbol{B}$ yielding the maximal invariant Hotelling T^2-statistic may not generally provide the best invariant test in the present context (as the affine group of transformations is not range-preserving where invariance is restricted to a smaller group g_0 of diagonal p.d. $\boldsymbol{B}$. As such, the Hotelling T^2-test is no longer the best invariant (i.e., UMPI) test for the one-sided restricted alternative problem. In fact, it may not even be admissible – a feature that will be discussed in Chapter 4. For general restricted alternative hypotheses, the restricted LRT may not possess finite-sample optimality properties even within subclasses. See Silvapulle and Sen (2004) for a broad coverage of this topic. Nevertheless, under fairly general regularity conditions, at least for large samples, the LRT may have some optimality properties, although it may run into computational complexities. For this reason, some alternative procedures are often advocated.

Wald tests. For location or general linear hypotheses in normal distributional setups, MLE and LSE are isomorphic, and as a result, LRT statistics are expressible as suitable quadratic forms in the MLE. Wald (1943) capitalized this idea in the formulation of a general class of tests based explicitly on the UMLE and thereby avoided the complexities associated with the LRT brought in by requiring the computation of the UMLE and the RMLE. Suppose that the null and alternative hypotheses can be formulated as

$$H_0 : \boldsymbol{h}(\boldsymbol{\theta}) = \mathbf{0} \quad \text{versus} \quad H_1 : \boldsymbol{h}(\boldsymbol{\theta}) \neq \mathbf{0}, \tag{3.3.9}$$

where $\boldsymbol{h}$ is possibly a vector-valued function. Exercise 3.3.4 is set to verify this for the multisample location model in normal distributional setups. Let $\widehat{\boldsymbol{\theta}}$ be the UMLE of $\boldsymbol{\theta}$ and consider the vector statistic $\boldsymbol{h}(\widehat{\boldsymbol{\theta}})$. In large samples, incorporation of $\boldsymbol{h}(\widehat{\boldsymbol{\theta}})$ directly into a

[2] This terminology is adapted from Silvapulle and Sen (2004), which differs somewhat from the allied term REML [see Diggle, Heagerty, Liang, and Zeger (2002)].

test statistic can be done by invoking suitable asymptotic results on the MLE and LRT, resulting in a quadratic form that is asymptotically equivalent to the LRT. However, in finite samples, construction of a test statistic having an exact (null) distribution may depend on the particular nature of $\boldsymbol{h}$ as well as on the underlying probability model. On the other hand, the Wald proposal goes far beyond the use of MLE, and has turned out to be a major advantage in nonparametric and semiparametric tests. Exercise 3.3.5 is set to verify that for the normal distribution model in a general linear hypothesis setup, the LRT statistic is isomorphic to the Wald test statistic. If $h(\boldsymbol{\theta})$ is real valued, and $\mathbb{V}\text{ar}[h(\widehat{\boldsymbol{\theta}})]$ is known under H_0, then a simple type of Wald test statistic is based on the standardized form $h(\widehat{\boldsymbol{\theta}})/\{\mathbb{V}\text{ar}[h(\widehat{\boldsymbol{\theta}})]\}^{1/2}$ or on the distribution of $h(\widehat{\boldsymbol{\theta}})$ itself if it is completely known. In a class of models where the variance of $h(\widehat{\boldsymbol{\theta}})$ under H_0 is known up to a scalar constant σ^2 for which an estimator is also available along with a simple sampling distribution, a studentized form of the Wald statistic may be used. Exercise 3.3.6 is set to verify this for the two-sample normal mean problem. The general form of Wald statistic, to be considered in detail in Chapter 8, incorporates an estimator of the covariance matrix of $\boldsymbol{h}(\widehat{\boldsymbol{\theta}})$ in a quadratic form, namely,

$$Q_W = [\boldsymbol{h}(\widehat{\boldsymbol{\theta}})]'\{\widehat{\mathbb{V}\text{ar}}[\boldsymbol{h}(\widehat{\boldsymbol{\theta}})]\}^{-1}\boldsymbol{h}(\widehat{\boldsymbol{\theta}}). \tag{3.3.10}$$

This reduces to the simpler form of Exercise 3.3.5 for normal linear models even in the finite sample setup. Exercise 3.3.7 is set to test for the identity of two binomial probabilities, where both the LRT and Wald test statistics run into finite sample dependence of the sampling distribution on nuisance parameter; as a consequence, they result in inexact tests (see Sen 2007).

Rao score tests (RST). The motivation for this type of tests also stems from the linear relation between the MLE and the score statistic in the exponential family of densities. Here, instead of the Wald method of employing the MLE, one may as well use the score statistic, which has the advantage of having independent summands. Also, score tests are directly based on the RMLE and do not require the UMLE. We will explain this with a simple model. Suppose that $\boldsymbol{\theta} \in \boldsymbol{\Theta}$ and consider a simple null hypothesis $H_0 : \boldsymbol{\theta} = \boldsymbol{\theta}_0$ (specified) versus $H_1 : \boldsymbol{\theta} \neq \boldsymbol{\theta}_0$ (a composite alternative). In that case, the score statistic is

$$\boldsymbol{U}_n^o = \frac{\partial}{\partial \boldsymbol{\theta}} \log L_n(\boldsymbol{\theta}; \boldsymbol{X})\bigg|_{\boldsymbol{\theta}_0}.$$

If $\boldsymbol{U}_n^o$ has a known distribution under H_0, it can be directly incorporated to formulate a testing procedure by taking $\boldsymbol{I}(\boldsymbol{\theta}_0) = \mathbb{E}_0[(\boldsymbol{U}_n^o)(\boldsymbol{U}_n^o)']$ and constructing the (score) test statistic

$$Q_R = n(\boldsymbol{U}_n^o)'[\boldsymbol{I}(\boldsymbol{\theta}_0)]^{-1}\boldsymbol{U}_n^o. \tag{3.3.11}$$

This does not require the UMLE of $\boldsymbol{\theta}$. The statistic in (3.3.11) is also used when the finite-sample distribution (under H_0) is not available in closed form, but then it requires some large sample approximations. In any case, if there are nuisance parameters, we need to estimate them only under the null hypothesis. In finite-sample setups, score test statistics are isomorphic to LRT statistics for linear models both in normal distributional setups and for some other densities belonging to the exponential family. However, in general, finite-sample distributions of score statistics may not be available in closed form. Yet, in large samples, they are often computationally simpler and share the same asymptotic optimality properties with LRT and Wald tests. Exercise 3.3.8 is set to verify that for the

Laplace density, the score statistic is simple and its distribution is easily obtainable in finite samples, whereas this is not true for the LRT or Wald tests.

Union-Intersection tests (UIT). Although the genesis of UIT [Roy (1953)] is in LRT, the *union-intersection principle* (UIP) governing it is flexible enough to amend in many non-standard problems where LRT may have serious computational or distributional impasses. Consider, for example, a test of a null hypothesis H_0 against an alternative H_1, both composite, for which an optimal test may not exist (or may be difficult to formulate). Assume that it is possible to express H_0 and H_1 as

$$H_0 = \cap_{a \in \mathcal{A}} H_{0a} \quad \text{and} \quad H_1 = \cup_{a \in \mathcal{A}} H_{1a}, \tag{3.3.12}$$

where the set $\mathcal{A}$ may be discrete or may even be a subset of $\mathbb{R}^p$, for some $p \geq 1$. Furthermore, for every $a \in \mathcal{A}$, assume that it is possible to construct an optimal test (possibly based on the Neyman–Pearson Lemma) for H_{0a} versus H_{1a}. Then, the UIP advocates that H_0 be accepted only when all the H_{0a}, $a \in \mathcal{A}$ are accepted (while H_1 is accepted when at least for some $a \in \mathcal{A}$, H_{1a} is accepted). The crux of the problem is therefore to set the size of the tests for the component hypotheses (H_{0a} versus H_{1a}), say α^*, in such a way that the overall test has size α. For clarity of ideas, we revisit Example 3.3.4 in light of the UIP. Let $\mathcal{A} = \mathbb{R}^p$, and for each $\boldsymbol{a} \in \mathcal{A}$, let $H_{0\boldsymbol{a}} : \boldsymbol{a}'\boldsymbol{\mu} = 0$ and $H_{1\boldsymbol{a}} : \boldsymbol{a}'\boldsymbol{\mu} \neq 0$. For the corresponding (univariate) hypothesis testing problem, an UMPU test is based on the test statistic

$$t_n(\boldsymbol{a}) = (\sqrt{n}\boldsymbol{a}'\overline{\boldsymbol{X}}_n)/(\boldsymbol{a}'\boldsymbol{S}_n\boldsymbol{a})^{1/2}, \quad \boldsymbol{a} \in \mathbb{R}^p. \tag{3.3.13}$$

Under the null hypotheses H_0 ($\Rightarrow H_{0\boldsymbol{a}}$), $t_n(\boldsymbol{a})$ has the Student t-distribution with $(n-1)$ degrees of freedom. Thus, $H_{0\boldsymbol{a}}$ is rejected in favor of $H_{1\boldsymbol{a}}$ at a significance level α^* if $|t_n(\boldsymbol{a})| > t_{n-1,\alpha^*/2}$ or, equivalently, if

$$t_n^2(\boldsymbol{a}) \geq t_{n-1,\alpha^*/2}^2.$$

Therefore, H_0 is accepted only when $t_n^2(\boldsymbol{a}) < t_{n-1,\alpha^*/2}^2$, for every $\boldsymbol{a} \in \mathbb{R}^p$ while it is rejected if $t_n^2(\boldsymbol{a}) > t_{n-1,\alpha^*/2}^2$, for some $\boldsymbol{a} \in \mathbb{R}^p$. Thus, the UIT statistic is

$$t_n^* = \sup_{\boldsymbol{a}}\{|t_n(\boldsymbol{a})| : \boldsymbol{a} \in \mathbb{R}^p\}, \tag{3.3.14}$$

where α^* has to be so determined that

$$P_0(t_n^* \geq t_{n-1,\alpha^*/2}) = \alpha. \tag{3.3.15}$$

Using the Courant Theorem 1.5.1, it follows that

$$t_n^{*2} = \sup_{\boldsymbol{a} \in \mathbb{R}^p} \frac{n\boldsymbol{a}'\overline{\boldsymbol{X}}_n\overline{\boldsymbol{X}}_n'\boldsymbol{a}}{\boldsymbol{a}'\boldsymbol{S}_n\boldsymbol{a}} = n\overline{\boldsymbol{X}}_n'\boldsymbol{S}_n^{-1}\overline{\boldsymbol{X}}_n = T_n^2, \tag{3.3.16}$$

namely, *Hotelling T^2-statistic* (3.3.7). This provides an easy way to determine α^* through the distribution of T_n^2. In this hypothesis testing problem (against global alternatives), the UIT and LRT are isomorphic. This isomorphism, though holding in some other problems, may not generally hold for restricted alternatives and many other nonstandard problems (where generally the UIP has greater flexibility and amenability). We illustrate this again with Example 3.3.4, but with the alternative hypothesis $H_1 : \boldsymbol{\mu} \geq \boldsymbol{0}$ with at least one component > 0. Let $\mathbb{R}^{+p} = \{\boldsymbol{x} \in \mathbb{R}^p : \boldsymbol{x} \geq \boldsymbol{0}\}$ be the positive orthant of $\mathbb{R}^p$. Then, we may set the problem as one of testing $H_0 = \cap_{\boldsymbol{a} \in \mathbb{R}^{+p}} H_{0\boldsymbol{a}}$ versus $H_1 = \cup_{\boldsymbol{a} \in \mathbb{R}^{+p}} H_{1\boldsymbol{a}}$, where $H_{0\boldsymbol{a}} : \boldsymbol{a}'\boldsymbol{\mu} = 0$, for all $\boldsymbol{a} \in \mathbb{R}^{+p}$ and $H_{1\boldsymbol{a}} : \boldsymbol{a}'\boldsymbol{\mu} \geq 0, \boldsymbol{a} \in \mathbb{R}^{+p}$, with a strict inequality for at least one $\boldsymbol{a}$. For testing $H_{0\boldsymbol{a}}$ versus $H_{1\boldsymbol{a}}$, a one-sided UMP test is based on the test statistic $t_n(\boldsymbol{a})$ in (3.3.13) but using only the right-hand tail as its critical region. Therefore, the UIP leads to the UIT

$$t_n^{+*} = \sup_{\boldsymbol{a}}\{t_n(\boldsymbol{a}) : \boldsymbol{a} \in \mathbb{R}^{+p}\}. \tag{3.3.17}$$

The LRT for this (restricted alternative) hypothesis testing problem can be constructed by using the RMLE of $\boldsymbol{\mu}$ and $\boldsymbol{\Sigma}$ (under H_1) and is not isomorphic to the UIT. Computationally, it is clear that $t_n^{+*} \leq t_n^*$ with probability 1, so that t_n^* is stochastically larger than t_n^{+*}. This leads to a smaller critical value for t_n^{+*} (than t_n^*), resulting in better power properties. We refer to Silvapulle and Sen (2004, Chapters 4 and 5), where the UIP has been illustrated in greater generality. In (3.3.16) and (3.3.17), the UIP has been directly related to the likelihood function. In a similar way, it may be related to the Rao score tests and the Wald tests. The latter approach is extensively incorporated in *beyond parametric* setups [Sen (2007)].

3.4 Invariant Tests

In the preceding section, while considering the multivariate normal mean vector problem, we provided the optimal test within the class of *affine invariant* tests. The idea of invoking invariance and equivariance in statistical inference, however, has been capitalized in a far more abstract setup [see Eaton (2007), for example] encompassing both parametrics and nonparametrics. We provide here only an outline of invariant tests. Let the observable random variable X have a sample space $\mathcal{X}$, equipped with a family of probability laws indexed by $P_\theta, \theta \in \Theta$. Let g be a transformation of $\mathcal{X}$ onto itself and suppose that $x \to gx$ leads to a mapping of $P_\theta \to P_{\theta^*}$, where θ^* is also in Θ so that $\overline{g}\Theta = \Theta$, with $\overline{g}$ denoting the conjugate transformation under g, that is, $\mathcal{X} \to g\mathcal{X}$ and $\overline{g} : \theta \to \overline{g}\Theta$. A test satisfying this invariance property is called an *invariant test*. If this invariance holds for all g belonging to a class $\mathcal{G}$, then we say that a function $T(x)$ is *maximal invariant* if for two elements x and y, belonging to $\mathcal{X}$,

$$T(x) = T(y) \Rightarrow y = gx, \quad \text{for some } g \in \mathcal{G}. \tag{3.4.1}$$

This will be illustrated in Example 3.4.1.

A test is invariant (under $\mathcal{G}$) if it depends on x only through a maximal invariant. The allied concept of *equivariance* arises in estimation theory; $T(x)$ is said to be an equivariant estimator if

$$T(gx) = gT(x), \quad \forall x \in \mathcal{X},\ g \in \mathcal{G}. \tag{3.4.2}$$

Thus, invariance and equivariance are complementary. It is quite natural to have tests for location which are *scale-invariant* (and the t-test is a good example thereof) and also tests for scale parameters that are *location-invariant*. The scope for invariant tests goes far beyond normal models, but we refrain ourselves from such details to avoid more sophisticated technical incursions.

Let us outline the role of invariance in nonparametrics through a couple of examples.

***Example* 3.4.1.** We consider continuous distributions F and G, defined on $\mathbb{R}$, and set $F(x) = G(x + \Delta)$ for some real Δ; the null hypothesis is $H_0 : \Delta = 0$, and the alternative hypothesis is $H_1 : \Delta > 0$ (or $<$ or $\neq 0$), but G is arbitrary. As such, it is natural to seek a transformation g such that $g(x)$ is strictly monotone function of x (but need not be linear). If $X_1, \ldots, X_{n_1}$ come from the distribution F while $Y_1, \ldots, Y_{n_2}$ are obtained from G, under H_0, all the n $(= n_1 + n_2)$ observations have a common distribution ($F \equiv G$). Even if we let $X_i^* = g(X_i)$, $1 \leq i \leq n_1$ and $Y_j^* = g(Y_j)$, $1 \leq j \leq n_2$, the X_i^* and Y_j^* have a common distribution. The maximal invariants in this problem are the ranks $R_1, \ldots, R_{n_1}, R_{n_1+1}, \ldots, R_n$ of X and Y in the combined sample. Clearly the ranks remain invariant under $X \to X^*$, $Y \to Y^*$ for any strictly monotone g. Furthermore, under H_0, the vector $(R_1, \ldots, R_n)$ assumes all possible $n!$ permutations of $(1, \ldots, n)$ with the common probability $1/n!$, whatever the form of G. Thus, letting $X^* = G(X)$ and $Y^* = G(Y)$, we may note that under H_0, X^* and Y^* are i.i.d. random variables with the Uniform[0, 1] distribution. Hence, a test based on the R_i is distribution-free under H_0. We leave Exercises 3.4.1 and 3.4.2 to illustrate this permutation-invariance concepts. □

***Example* 3.4.2.** We consider next the *sign-invariance* problem. Let $X_1, \ldots, X_n$ be i.i.d. random variables with a continuous distribution F, defined on $\mathbb{R}$. The null hypothesis H_0 relates to the symmetry of F around a specified location which, without loss of generality, we take as 0. The alternative hypothesis may relate to asymmetry of F around 0, and more often to the symmetry of F around some Δ ($\neq 0$, or $<$, or > 0). Note that under H_0, both X and $-X$ have the common distribution F, which remains symmetric under a strictly monotone transformation $g(x)$ satisfying $g(-x) = -g(x)$, for all $x \geq 0$, that is, under odd functions g. Let then $\boldsymbol{S}_n = (S_1, \ldots, S_n)'$ and $\boldsymbol{R}_n^+ = (R_1^+, \ldots, R_n^+)'$, where $S_i = \text{sign}(X_i)$ and R_i^+ = rank of $|X_i|$ among $|X_1|, \ldots, |X_n|$, $1 \leq i \leq n$. Then $(\boldsymbol{S}_n', \boldsymbol{R}_n^{+\prime})$ form the *maximal invariants*. Also, under H_0, $\boldsymbol{S}_n$ and $\boldsymbol{R}_n^+$ are stochastically independent (see Exercise 3.4.3), and under H_0, $\boldsymbol{S}_n$ takes on all possible 2^n values $(\pm 1, \ldots, \pm 1)$ with the common probability 2^{-n}, while $\boldsymbol{R}_n^+$ takes on each permutation of $(1, \ldots, n)$ with the common probability $1/n!$. Therefore, a test based on $(\boldsymbol{S}_n, \boldsymbol{R}_n^+)$ is distribution-free under H_0. Exercise 3.4.5 refers to this basic invariance. □

3.5 Concluding Notes

The emphasis on tests in this chapter has been laid down from small sample perspectives. In large samples, a somewhat different picture emerges and that will be presented in greater detail in Chapter 8. There are certain robustness concerns regarding LRT and their ramifications and in that respect, nonparametric tests fare better. We will illustrate those perspectives in Chapter 8. The intricate relation between hypothesis testing theory in the Neyman–Pearson setup and Bayes methods will be appraised in Chapter 4. There is a lot of development on nonparametric and semiparametric tests in survival analysis, clinical trials, and environment problems; The Neyman–Pearson approach based on the classical likelihood function stumbles into impasses and other variants of the likelihood are often advocated. We will present a brief outline of some of these methods in Chapters 8–11.

3.6 Exercises

3.2.1 Let $X \sim \text{Bin}(6, \theta)$, $0 < \theta < 1$ and consider the problem of testing $H_0 : \theta = 1/2$ and $H_1 : \theta > 1/2$. Show that if we choose the critical region as (i) $X \geq 6$, or (ii) $X \geq 5$, then the level of significance of the test is 0.015 and 0.109, respectively. Consider a randomization test where $\tau(X)$ is 1, γ or 0, according as X is ≥ 6, $= 5$, and < 5, respectively. Determine γ in such a way that the size of the test is exactly equal to 0.05.

3.2.2 In the context of Example 3.2.2, verify whether UMPU tests exist for the normal and simple exponential distribution models.

3.3.1 In the context of Example 3.3.2, show that the LRT in (3.3.1) is based on $t_n = \sqrt{n}\overline{X}_n/\sqrt{\sum_{i=1}^{n}(X_i - \overline{X}_n)^2/(n-1)}$ when $\overline{X} > 0$ and it is taken as 0 if $\overline{X}_n < 0$.

3.3.2 In the context of Example 3.3.4, show that for testing $H_o : \boldsymbol{\mu} = \mathbf{0}$ versus $H_1 : \boldsymbol{\mu} \neq \mathbf{0}$, the LRT is a random monotone function of T_n^2 in (3.3.7).

3.3.3 In Example 3.3.4, show that affine invariance implies (3.3.8), and as such, a best invariant test exists and is based on T_n^2 in (3.3.7).

3.3.4 For the multisample location model in Example 1.2.4, formulate the hypotheses (3.3.9) in an appropriate way and show that the Wald test is also a function of the classical ANOVA test.

3.3.5 For a general linear model $\boldsymbol{Y} = \boldsymbol{X}\boldsymbol{\beta} + \boldsymbol{\varepsilon}$ with normally distributed errors, show that the Wald test is a monotone function of the classical variance-ratio test.

3.3.6 For the two-sample location problem, show that a studentized Wald test is equivalent to the t-test.

3.3.7 Let $X \sim \text{Bin}(n_1, \theta_1)$, $Y \sim \text{Bin}(n_2, \theta_2)$, and consider the problem of testing $H_0 : \theta_1 = \theta_2$ versus $H_1 : \theta_1 \neq \theta_2$. Show that both the LRT and the Wald tests involve test statistics for which the exact distributions depend on nuisance parameters.

3.3.8 Obtain the score test statistic for the Laplace distribution in Example 3.3.1 and show that its null distribution (i.e., under $H_0 : \theta = 0$) can be obtained from the binomial distribution.

3.4.1 In Example 3.4.1, let $W_n = \sum_{i=1}^{n_1}[R_i - (n+1)/2]$ be the two-sample, Wilcoxon–Mann–Whitney rank sum test statistic. Show that it is invariant under monotone transformation on the observations.

3.4.2 In Example 3.4.1, let $a_n(i) = \sum_{j=1}^{i}(n - j + 1)^{-1}$, $1 \leq i \leq n$. The two-sample log-rank test statistic is $L_n = \sum_{i=1}^{n_1} a_n(R_i)$. Verify the invariance structure for L_n.

3.4.3 In Example 3.4.2 define the Wilcoxon signed-rank test as $\sum_{i=1}^{n} \text{sign}(X_i)R_i^+$. Show that it is invariant under sign inversions and arbitrary monotone transformations.

3.4.4 In Example 3.4.2 set $S_n = \sum_{i=1}^{n} \text{sign}(X_i)$; show that it is invariant under arbitrary monotone odd function.

3.4.5 Show that in Example 3.4.2, under H_0, $\boldsymbol{S}$ and $\boldsymbol{R}^+$ are stochastically independent. What is their joint distribution under H_0?

4

Elements of Statistical Decision Theory

4.1 Introduction

Statistical estimation as well as hypothesis testing may be viewed as important topics of a more general (and admittedly, more abstract) *statistical decision theory* (SDT). Having genesis in the *theory of games* and affinity with *Bayes methods*, SDT has been continuously fortified with sophisticated mathematical tools as well as with philosophical justifications. In conformity with the general objectives and contended intermediate level of this monograph, we intend to provide an overall introduction to the general principles of SDT with some emphasis on Bayes methodology (as well as some of its variants), avoiding the usual philosophical deliberations and mathematical sophistication, to the extent possible. See Berger (1993) for a detailed exposition.

The connection between estimation and hypothesis testing theories treated in the preceding chapters and SDT relates to the uncertainty of statistical conclusions or decisions based on observed data sets and to the adequate provision for quantifying the frequency of incorrect ones. This has generated the notion of *loss* and *risk* functions that form the foundation of SDT. This notion serves as a building block for the formulation of *minimum risk* and *minimax (risk)* estimation theory, where *Bayes estimates* have a focal stand. In the same vein, *Bayes tests*, which are not necessarily isomorphic to the Neyman–Pearson–Wald likelihood-based tests have cropped up in SDT. In either case, the basic difference comes from the concepts of *prior* and *posterior* distributions that bring in more room for subjective judgement in the inferential process. Without much digression, *empirical Bayes* procedures are also outlined here.

The concepts of loss and risk are discussed in Section 4.2 under a statistical perspective. Bayes estimation and hypothesis testing methodology are summarized in Sections 4.3 and 4.4, respectively. The interplay between these methods and those governing the construction of confidence sets is discussed in Section 4.5.

4.2 Basic Concepts

To start with, consider a real-, vector- or even matrix-valued random sample $\boldsymbol{X}$. The sample belongs to a space defined by the set of all possible realizations of $\boldsymbol{X}$ (in the discrete case) or, more generally, to the domain of $\boldsymbol{X}$; as in the previous chapters, we denote this *sample space* by $\mathcal{X}$. Associated with $\boldsymbol{X}$ and $\mathcal{X}$ is the *probability law* $p(\boldsymbol{X};\boldsymbol{\theta})$, where $\boldsymbol{\theta}$ stands for the vector of corresponding parameters. The set of all possible values of $\boldsymbol{\theta}$, denoted by $\boldsymbol{\Theta}$, is the *parameter space*. In a comparatively more general setup, we could imagine a probability law $P(\boldsymbol{X})$ for $\boldsymbol{X}$, and a class $\mathcal{P}$ of probability laws, such that $P \in \mathcal{P}$. For example, in a nonparametric setup, $\mathcal{P}$ may be the class of all continuous distributions on $\mathbb{R}^p$ for some $p \geq 1$.

A *decision rule* $d = d(\boldsymbol{X})$ is a function with domain $\mathcal{X}$ and range $\mathcal{A}$, the set of all possible decisions or actions; we call $\mathcal{A}$ the *action space*. Let $\mathcal{M}$ be the set of all states of Nature. The action space may often be taken as the *decision space*, though they may not be generally the same. In an abuse of a game theoretic terminology, the *loss function* $L(a, b)$, $a \in \mathcal{M}$, $b \in \mathcal{A}$, associates the loss due to an action (decision) b when the true state is a. Naturally, when a and b are the same, there is no loss. Although, typically the loss function is taken to be nonnegative, in general, it need not be so. It is convenient now to use this loss function in the context of stochastic decisions, that is, decisions based on the outcome $\boldsymbol{X}$. We consider a loss function as $L[\delta, d(\boldsymbol{X})]$, $\delta \in \mathcal{M}$, $d(\boldsymbol{X}) \in \mathcal{A}$, $\boldsymbol{X} \in \mathcal{X}$, so that it is governed by the probability law of $\boldsymbol{X}$. As such, it might be more natural to talk of the *expected loss* or *risk function*, namely,

$$R(\delta, d) = E_P\{L[\delta, d(\boldsymbol{X})]\} = \int L[\delta, d(\boldsymbol{X})]\mathrm{d}P(\boldsymbol{X}). \tag{4.2.1}$$

Let us illustrate some loss functions and their risk function counterparts in specific SDT contexts. First, consider the estimation problem, and for simplicity, assume that $\theta \in \mathbb{R}$. Here δ refers to the true parameter value and $d(\boldsymbol{X})$ to its estimator based on the sample point $\boldsymbol{X}$. It is customary to let loss functions depend on the distance between δ and $d(\boldsymbol{X})$, although, in many cases, it may not be a symmetric function of such a distance. For example, in the context of producer and consumer risks, the loss function is typically one sided. In this estimation problem setup, $\mathcal{A} = \mathcal{M} = \Theta$. Some conventional loss functions are:

(i) Absolute error (deviation):

$$L(a, b) = |a - b|, \ a, b \in \mathbb{R};$$

(ii) Squared error (Euclidean norm):

$$L(a, b) = (a - b)^2, \ a, b \in \mathbb{R};$$

(iii) Threshold (large) deviation:

$$L(a, b) = I(|a - b| > c), \ \text{for some } c > 0;$$

(iv) Bowl-shaped (or bath-tub) loss:

$$L(a, b) = (a - b)^2 I(|a - b| \leq c) + c^2 I(|a - b| > c), \ \text{for some } c > 0.$$

A general case of (iv) is represented in Figure 4.2.1.

If $\boldsymbol{\theta}$ is vector-valued, so is its estimator and a quadratic loss function may be defined as

$$L(\boldsymbol{a}, \boldsymbol{b}) = ||\boldsymbol{a} - \boldsymbol{b}||^2_{\boldsymbol{W}} = (\boldsymbol{a} - \boldsymbol{b})' \boldsymbol{W}^{-1} (\boldsymbol{a} - \boldsymbol{b}), \tag{4.2.2}$$

for a suitable positive definite (p.d.) matrix $\boldsymbol{W}$. In a similar way, large deviation and bowl-shaped loss functions may be extended to the vector case.

The risk function corresponding to (i) is the mean absolute deviation (MAD) or L_1-norm risk; for (ii) it is the mean squared error (MSE), and for (iii) it is the probability of large deviations. In the vector case, they may be termed similarly. As the very definition of the risk involves the expectation over the values of $d(\boldsymbol{X})$ with respect to its probability law, it

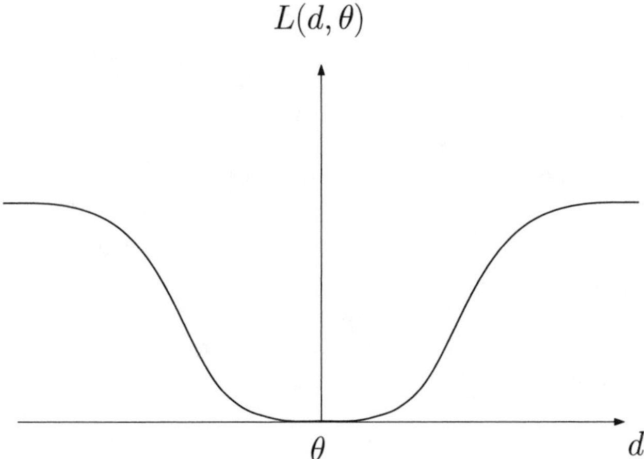

Figure 4.2.1. A bowl-shaped loss function.

also depends on the estimator d itself. In all these cases, the risk function is nonnegative, and the smaller the risk, intuitively at least, the better the decision rule. It is therefore customary to consider a class $\mathcal{D}$ of estimators, and to choose a specific member from this class for which the risk is a minimum, provided such an optimal estimator exits.

The notion of loss and risk function is not limited to point estimation of a scalar or vector parameter. As an illustration, consider the multinomial distribution (Example 2.4.1) relating to $\boldsymbol{\pi} \in \mathcal{S}_{c-1} = \{\boldsymbol{x} \geq \boldsymbol{0} : \boldsymbol{x}'\boldsymbol{1} = 1\}$, the $c-1$ Simplex, with the observed random vector $\boldsymbol{n} = (n_1, \ldots, n_c)'$, for some $c \geq 2$. Let $\omega = (n_1, \ldots, n_c)$ be the observed event while Ω stands for the set of all possible $\boldsymbol{n}$ such that $\boldsymbol{n}'\boldsymbol{1} = n$. For every $\omega \in \Omega$, $\pi(\omega)$ stands for the multinomial law while let $p_n(\omega)$ be some estimator, for $\omega \in \Omega$. Consider the loss function $L(\boldsymbol{p}_n, \boldsymbol{\pi}) = \log[p_n(\omega)/\pi(\omega)]$, $\omega \in \Omega$. The corresponding risk function is

$$\mathbb{E}_\pi[L(\boldsymbol{p}_n, \boldsymbol{\pi})] = \sum_{\omega\in\Omega} \pi(\omega)\log[\pi(\omega)] - \sum_{\omega\in\Omega} \pi(\omega)\log[p_n(\omega)] = I(\boldsymbol{p}_n, \boldsymbol{\pi}),$$

say. This is called the *information discriminant function* or the *entropy risk function.* Now $I(\boldsymbol{p}_n, \boldsymbol{\pi}) \geq 0$, where the equality sign holds only when $\boldsymbol{p}_n(\omega) = \boldsymbol{\pi}(\omega)$, $\forall\omega \in \Omega$. As such, it seems intuitive to minimize $I(\boldsymbol{p}_n, \boldsymbol{\pi})$ over the class of all $\boldsymbol{p}_n \geq \boldsymbol{0}$, which is equivalent to maximizing $\sum_{\omega\in\Omega} \pi(\omega)\log[p_n(\omega)]$. Now writing

$$p_n(\omega) = \pi(\omega) + [p_n(\omega) - \pi(\omega)] = \pi(\omega)\{1 - [1 - p_n(\omega)/\pi(\omega)]\},$$

we have

$$\begin{aligned} I(\boldsymbol{p}_n, \boldsymbol{\pi}) &= \sum_{r\geq 1} \frac{1}{r} \sum_{\omega\in\Omega} \frac{[p_n(\omega) - \pi(\omega)]^r}{[\pi(\omega)]^{r-1}} \\ &= \frac{1}{2} \sum_{\omega\in\Omega} \frac{[p_n(\omega) - \pi(\omega)]^2}{\pi(\omega)} + \sum_{r\geq 3} \frac{1}{r} \frac{[p_n(\omega) - \pi(\omega)]^r}{[\pi(\omega)]^{r-1}}, \end{aligned}$$

where the first term on the right-hand side represents the classical Pearson goodness-of-fit measure [Pearson (1900)].

In the continuous case, for a sample point $\boldsymbol{x}$ with true density f and an estimator $p(\boldsymbol{x})$ of $f(\boldsymbol{x})$, the analogue of $I(\boldsymbol{p}_n, \boldsymbol{\pi})$ is the *entropy* loss $\log[f(\boldsymbol{x})/p_n(\boldsymbol{x})]$ with risk function

$$I(f, p) = \int f(x)\log[f(x)]\mathrm{d}x - \int f(x)\log[p(x)]\mathrm{d}x.$$

There are other variants that will not be discussed here.

A similar formulation of loss and risk functions can be made for the hypothesis testing problem, and this will be detailed in Section 4.4. Motivated by such specific cases, we now consider a more general SDT setup. We set $\mathcal{D}$ as a class of possible decision rules d. For any specific d the associated probability law for $d(\boldsymbol{X})$ would depend not only on $P(\boldsymbol{X}, \boldsymbol{\theta})$ but also on the choice of the specific decision rule itself. As a result, for a given probability law, the risk function $R[\delta, d]$ is defined on $\mathcal{M} \times \mathcal{D}$. If for some $d_0 \in \mathcal{D}$,

$$R[\delta, d_0] = \min\{R[\delta, d] : d \in \mathcal{D}\}, \ \forall\delta \in \mathcal{M}, \tag{4.2.3}$$

then $d_0(\boldsymbol{X})$ is called the *uniformly minimum risk* (UMR) decision rule. Of course, this minimization is based on the tacit assumption that the risk is nonnegative a.e. In particular, for the estimation problem, such an estimator $d_0(\boldsymbol{X})$, if it exists, is called the *UMR estimator*. Unfortunately, UMR decision rules may not universally exist or may be hard to formulate, and hence, we may need to look for alternative optimality criteria.

Typically, $R[\delta, d]$ may not be the same for all $\delta \in \mathcal{M}$. As such, keeping in mind the possible nonconstancy of the risk due to the possible lack of *invariance/equivariance*, we define the *maximum risk* as

$$R^*[d] = \sup_{\delta}\{R[\delta, d] : \delta \in \mathcal{M}\}, \quad d \in \mathcal{D}. \tag{4.2.4}$$

If within the class $\mathcal{D}$, there exists a decision rule d_0 such that

$$R^*[d_0] = \min_{d}\{R^*(d : d \in \mathcal{D}\}, \tag{4.2.5}$$

then d_0 is called the *minimax decision rule*. Generally, such minimax decision rules are easier to formulate than the UMR rules.

If, for a pair of decision rules d_1, d_2, we have

$$R[\delta, d_1] \leq R[\delta, d_2], \quad \forall\delta \in \mathcal{D}, \tag{4.2.6}$$

with the strict inequality sign holding for some δ, then d_1 is said to be *R-better* than d_2, or *d_1 dominates d_2 in risk*. If $R[\delta, d_1] = R[\delta, d_2]$, $\forall\delta \in \mathcal{D}$, then d_1 and d_2 are *R-equivalent*. Based on this perception of risk dominance, a decision rule d_0 is said to be *admissible* if there is no $d_1 \in \mathcal{D}$, which is R-better than d_0. To the contrary, if there is some R-better estimator then d_0, it is said to be *inadmissible*. Essentially, we can confine ourselves to a class of admissible decision rules, whenever such a class can be identified – no member within this class is risk-dominated by other members. We will call such a class a *complete* class of decision rules. UMR and minimax decision rules are members of such a class. Note that the concept of risk dominance essentially rests on a suitable formulation of risks via a specific choice of a loss function.

In the context of estimation problems, much of the discussion made in Chapter 2 can be streamlined in such a SDT setup. However, the manipulations needed to establish optimality properties of specific estimators may not always be manageable. This awkward feature is particularly noticeable in multiparameter estimation problems as well as in multiple decision problems. The *Stein paradox* [Stein (1956)] is a classical illustration of the simple fact that the usual risk-dominance results for the MLE in the single parameter case may not transpire to their multiparameter counterparts. If $t(\boldsymbol{X})$ is a *translation-equivariant* estimator of a location parameter θ, that is, for every $a \in \mathbb{R}$,

$$t(\boldsymbol{X} + a\mathbf{1}) = t(\boldsymbol{X}) + a, \tag{4.2.7}$$

then either the squared error or absolute risk depend only on $|t(\boldsymbol{X}) - \theta|$, that is, the risk is translation-invariant. Within the class of translation-equivariant estimators one may get a *minimum risk estimator* (MRE) under some conditions. A similar feature holds under *scale-equivariance*. However, in general, multiparameter cases, there exist some estimators that are not translation-equivariant but may dominate (in risk) translation-equivariant estimators; this is the *Stein paradox*. An estimator having this risk-dominance property is known as *Stein rule* or *shrinkage* estimator. We illustrate some of these problems with specific examples.

***Example* 4.2.1.** Let $\boldsymbol{X}$ has be a random vector following a $\mathcal{N}_n(\theta\mathbf{1}_n, \sigma^2\boldsymbol{I}_n)$ distribution, where $\theta \in \mathbb{R}$ and $\sigma > 0$. Consider the squared error loss function for an estimator of θ. For the sample mean $\overline{X}_n$, the risk is a constant for all $\theta \in \mathbb{R}$ (Exercise 4.2.1). Thus, we have a translation-invariant risk function. As a matter of fact, for any translation-equivariant estimator of θ, this risk is independent of θ. Exercise 4.2.2 is set to show that the MLE is the MRE. □

***Example* 4.2.2.** Let $\boldsymbol{X} = (X_1, \dots, X_p)'$, for some $p \geq 3$, have a multivariate normal distribution with mean vector $\theta\mathbf{1}$ and covariance matrix $\boldsymbol{I}_p$. Consider the estimator of θ under a squared error loss function. We can show that the MLE of θ is the mean $\overline{X}_p = p^{-1}\boldsymbol{X}'\mathbf{1}$ and its risk is equal to p^{-1}, a constant for all θ. Following James and Stein (1961), consider the estimator

$$T_{JS} = [1 - c(\boldsymbol{X}'\boldsymbol{X})^{-1})]\overline{X}_p, \tag{4.2.8}$$

where c is a nonnegative constant $[0 < c < 2(p-2)]$. Note that whereas the MLE is translation-equivariant, T_{JS} is not. Nevertheless, it can be shown that (Exercise 4.2.6)

$$\mathbb{E}(T_{JS} - \theta)^2 \leq \mathbb{E}(\overline{X}_p - \theta)^2, \quad \forall\theta \in \mathbb{R}, \tag{4.2.9}$$

so that T_{JS} dominates $\overline{X}_p$ in quadratic risk, and as a result, the MLE is inadmissible. Consider a plot of the two risk functions, the first one is a constant $(= p^{-1})$ and the second one depends on θ only through $\Delta = \theta^2$; at $\Delta = 0$, it is a minimum when $c = p - 2$, and as Δ increases, the risk, still being smaller than p^{-1}, approaches the upper asymptote. Furthermore, as p increases, the risk dominance picture becomes more prominent.

This is depicted in Figure 4.2.2, where the relative risk is the ratio of the two, that is p times the risk of the shrinkage estimator. □

There are many variants of such shrinkage estimators, adapted to more complex schemes, and having many other interpretations too. A direct generalization relates to i.i.d. random vectors $\boldsymbol{X}_i = (X_{i1}, \dots, X_{ip})', i = 1, \dots, n$ each one having a p-variate normal distribution with mean vector $\theta\mathbf{1}$, $\theta \in \mathbb{R}$ and covariance matrix $\sigma^2\boldsymbol{I}_p$ where both θ and $\sigma^2 > 0$ are unknown parameters. Also, instead of choosing the null pivot, we may take an arbitrary

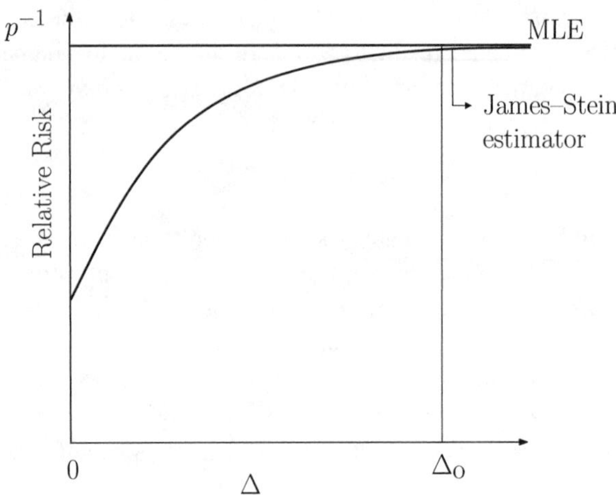

Figure 4.2.2. Relative risk of the shrinkage estimator in (4.2.8).

pivot θ_0 and consider a shrinkage estimator $\theta_{S,n}$ around the pivot θ_0. In the same vein as in Example 4.2.2, one may then consider a shrinkage estimator

$$\theta_{S,n} = \theta_0 + \left[1 - c_n \left(\sum_{i=1}^{n} ||\boldsymbol{X}_i - \mathbf{1}\theta_0||^{-2} \right) \right] (\overline{X}_n - \theta_0), \tag{4.2.10}$$

where $\overline{X}_n = (np)^{-1} \sum_{i=1}^{n} \sum_{j=1}^{p} X_{ij}$ is the grand mean and the shrinkage factor depends on n as well as p. The term within brackets, [ ], in (4.2.10) may be negative with a positive probability, in which case, there could be overshrinking. To avoid this undesirable phenomenon, this term is taken to be 0 whenever it is negative, and the resulting estimator is called a *positive-rule shrinkage* estimator. These estimators will be briefly discussed in the next section.

4.3 Bayes Estimation Methods

In SDT, Bayes procedures play a basic role. Some of the concepts outlined in the preceding section have been further generalized in the light of the well-known Bayes theorem. In a Bayesian decision rule setup, in addition to the basic random element $\boldsymbol{X}$ following a density $p(\boldsymbol{X}|\theta)$ it is also assumed that the "parameter" θ is itself stochastic with a *prior distribution function* $\Pi(\theta)$ or, simply, *prior*, defined on Θ. Note that θ is unobservable but, its distribution function $\Pi(\theta)$ is assumed to be given, although taken to be independent of $\boldsymbol{X}$. Thus, the joint density function (when it exists) of $(\boldsymbol{X}, \theta)$ is given by

$$p(\boldsymbol{X}|\theta)\pi(\theta), \quad \boldsymbol{X} \in \mathcal{X}, \theta \in \Theta, \tag{4.3.1}$$

where $p(\boldsymbol{X}|\theta)$[1] is the conditional density of $\boldsymbol{X}$, given θ. Therefore, the marginal density of $\boldsymbol{X}$ is

$$p^*(\boldsymbol{X}) = \int_{\Theta} p(\boldsymbol{X}|\theta)\pi(\theta)\mathrm{d}\theta, \quad \boldsymbol{X} \in \mathcal{X}. \tag{4.3.2}$$

[1] In this section, we write $p(\boldsymbol{X}|\theta)$ instead of $p(\boldsymbol{X};\theta)$ to highlight that it represents the "conditional" density of $\boldsymbol{X}$ given θ.

As a result, the conditional density of θ, given $\boldsymbol{X}$, is

$$p^{\pi}_{\theta|X} = \frac{\pi(\theta)p(\boldsymbol{X}|\theta)}{\int_{\Theta} p(\boldsymbol{X}|\theta)\pi(\theta)\mathrm{d}\theta}, \quad \theta \in \Theta, \quad \boldsymbol{X} \in \mathcal{X}. \tag{4.3.3}$$

This is called the *posterior density* of (the unobservable) θ, given the observable $\boldsymbol{X}$. This posterior density is in conformity with the usual probability law that given two events A and $B_i \in \{B_j, j \in I\}$,

$$\begin{aligned} P(B_i|A) &= P(A \cap B_i)/P(A) \\ &= P(A|B_i)P(B_i)/P(A) \\ &= P(A|B_i)P(B)/\sum_{j\in I} P(B_j)P(A|B_j), \end{aligned} \tag{4.3.4}$$

known as the *Bayes Theorem*. Therefore, by analogy, statistical decision theory based on such posterior laws is called *Bayes decision theory*. In a frequentist setup, as discussed in the preceding chapters, θ, though unknown, is treated as nonstochastic, while in a Bayesian context, given the observable random elements, the posterior density of θ provides the natural clue to draw conclusions on it in the light of the observations. Often, the two approaches may not yield identical decision rules.

Under a Bayesian formulation, we must extend the notion of risk and soforth, to suit such more general setups. The definition of a loss function $L(a, b)$ is the same as in the preceding section. However, since here the sample-based decision rule $d(\boldsymbol{X})$ and θ are both treated as stochastic, the *Bayes risk*, incorporating both of these stochastic elements, is defined as

$$\begin{aligned} R(\pi, d) = \mathbb{E}^{\pi}[R(\theta, \delta)] &= \int_{\Theta}\int_{\mathcal{X}} L[\theta, \mathrm{d}(\boldsymbol{X})]\mathrm{d}P(\boldsymbol{X}, \theta)\mathrm{d}\Pi(\theta) \\ &= \int_{\mathcal{X}}\left\{\int_{\Theta} L[\theta, d(\boldsymbol{X})]\mathrm{d}P(\theta|\boldsymbol{X})\right\}\mathrm{d}P^*(\boldsymbol{X}) = \mathbb{E}\{\mathbb{E}[L(\theta, d(\boldsymbol{X}))|\boldsymbol{X}]\}. \end{aligned} \tag{4.3.5}$$

An estimator d^{π}_{B} which minimizes this *posterior risk* $R(\pi, d)$ over the class of estimators $\{d : d \in \mathcal{D}\}$ is called the *minimum Bayes risk estimator* [with respect to the prior $\pi(\theta)$], that is,

$$R[\pi, d^{\pi}_{B}] = \inf\{R[\pi, d] : d \in \mathcal{D}\}. \tag{4.3.6}$$

Unfortunately, for a general loss function and an arbitrary prior π, a *minimum posterior risk* (or Bayes) estimator may not exist in a closed form and may even be difficult to formulate. However, if there exists an estimator [say, $d^*(\boldsymbol{X})$], such that $R[\pi, d^*(\boldsymbol{X})] = \inf\{R[\pi, d(\boldsymbol{X})] : \mathrm{d} \in \mathcal{D}\}$, then clearly $d^*(\boldsymbol{X})$ is a Bayes estimator. For the squared error loss function $L(a, b) = (a - b)^2$, we note that

$$\begin{aligned} \mathbb{E}\{\mathbb{E}[(\theta - d(\boldsymbol{X}))^2|\boldsymbol{X}]\} &= \mathbb{E}\{[(\theta - E(\theta|\boldsymbol{X})]^2\} + \mathbb{E}\{[\mathbb{E}(\theta|\boldsymbol{X}) - d(\boldsymbol{X})]^2\} \\ &\geq \mathbb{E}[\theta - \mathbb{E}(\theta|\boldsymbol{X})]^2, \quad \forall d \in \mathcal{D}, \end{aligned} \tag{4.3.7}$$

where the equality sign holds only when $d(\boldsymbol{X}) = E(\theta|\boldsymbol{X})$, a.e. $\pi(\theta)$. Whenever there is no confusion, this is called the *Bayes estimator*. In summary, $d^{\pi}_{B}(\boldsymbol{X}) = \mathbb{E}(\theta|\boldsymbol{X})$, the *posterior mean* is the minimum Bayes risk estimator, under the squared error loss function and the prior $\pi(\theta)$. For the absolute error loss we have, on parallel lines,

$$\begin{aligned}\mathbb{E}\{\mathbb{E}|\theta - d(\boldsymbol{X})||\boldsymbol{X}\} &= \int_{\mathcal{X}} \left[\int_{\boldsymbol{\theta}} |\theta - d(\boldsymbol{X})| \mathrm{d}\Pi(\theta)\right] \mathrm{d}P^*(\boldsymbol{X}) \\ &\geq \int_{\mathcal{X}} \left[\int_{\Theta} |\theta - m(\boldsymbol{X})| \mathrm{d}\Pi(\theta)\right] \mathrm{d}P^*(\boldsymbol{X}), \end{aligned} \tag{4.3.8}$$

where $m(\boldsymbol{X})$ is the posterior median of θ, given $\boldsymbol{X}$, and the equality sign in (4.3.8) holds only when $d(\boldsymbol{X}) = m(\boldsymbol{X})$ a.e. Thus, the *posterior median* is the minimum Bayes risk estimator under the absolute error loss function. If the posterior distribution of θ, given $\boldsymbol{X}$, is symmetric then its mean and median are the same. However, in general, this symmetry can not be taken for granted, and hence, depending on the choice of the loss function, the optimal Bayes estimator has to be derived. We consider some simple examples.

***Example* 4.3.1.** Let $X_1, \ldots, X_n$ be i.i.d. random variables having, for a given θ, a normal density with mean θ and variance σ^2, and let θ have a normal density with mean μ and variance τ^2. The posterior density of θ is normal with mean $\mu\sigma^2(\sigma^2 + n\tau^2)^{-1} + \overline{X}_n n\tau^2(\sigma^2 + n\tau^2)^{-1}$ and variance $\sigma^2\tau^2(\sigma^2 + n\tau^2)^{-1}$, where $\overline{X}_n$ is the sample mean. Hence, or otherwise, it can be shown (Exercise 4.3.1) that under both the squared error and absolute error loss functions, when μ, σ and τ are specified, we have the same minimum risk Bayes estimator given by

$$d_B(\boldsymbol{X}) = \mu\left(\frac{\sigma^2}{n\tau^2 + \sigma^2}\right) + \overline{X}_n\left(\frac{n\tau^2}{n\tau^2 + \sigma^2}\right). \tag{4.3.9}$$

We let $B = \sigma^2/(n\tau^2 + \sigma^2)$ such that $0 \leq B \leq 1$. Then,

$$d_B(\boldsymbol{X}) = B\mu + (1 - B)\overline{X}_n.$$

Thus, B represents the impact of the prior in the estimation of θ. B is called the *Bayes factor*.

Furthermore, under the squared error loss function, the Bayes risk is

$$(\tau^2\sigma^2)/(n\tau^2 + \sigma^2) = \tau^2 B, \tag{4.3.10}$$

the mean squared error under the posterior density of θ. Also, as the posterior density is normal, it is easily seen (Exercise 4.3.1) that the Bayes risk under an absolute error loss function [i.e., $\mathbb{E}|d_B(\boldsymbol{X}) - \theta|$] is

$$\sqrt{(2/\pi)}[(\tau^2\sigma^2)/(n\tau^2 + \sigma^2)]^{1/2}. \tag{4.3.11}$$

□

***Example* 4.3.2.** Let $X \sim$Bin(n, p) and let $p \in (0, 1)$ have the beta distribution with parameters (a, b), that is, with density

$$\pi(p) = \frac{\Gamma(a + b)}{\Gamma(a)\Gamma(b)} p^{a-1}(1 - p)^{b-1}, \quad a, b > 0. \tag{4.3.12}$$

The marginal distribution of X, and hence, the posterior density of p are, respectively,

$$f(x) = \binom{n}{x} \frac{\Gamma(a + b)\Gamma(a + x)\Gamma(b + n + x)}{\Gamma(a)\Gamma(b)\Gamma(a + b + n)}$$

and

$$f(p|x) = p^{x+a-1}(1-p)^{n-x+b-1}\frac{\Gamma(a+b+n)}{\Gamma(a+x)\Gamma(b+n-x)}.$$

Exercise 4.3.2 is set to verify this to obtain the Bayes estimator under a squared error loss function. □

Consider next the case of vector-valued parameter $\boldsymbol{\theta}$. Under a quadratic error loss function $L(\boldsymbol{a}, \boldsymbol{b}) = (\boldsymbol{a} - \boldsymbol{b})'\boldsymbol{Q}(\boldsymbol{a} - \boldsymbol{b})$ with $\boldsymbol{Q}$ denoting a p.d. matrix, we may use the Rao–Blackwell Theorem 2.3.2 to verify that the Bayes estimator is the mean vector of the posterior distribution of $\boldsymbol{\theta}$. There could be, in general, technical problems in defining the L_1-norm Bayes estimator, stemming mainly from the difficulties in formulating multivariate medians of the posterior density. This is outlined in Exercise 4.3.4.

An important and sometimes controversial theme in the formulation of a Bayes estimator or, more generally, a Bayes decision rule is the choice of an appropriate prior density $\pi(\boldsymbol{\theta})$. Usually, this is done from extraneous prior information, and often, from past studies of coherent nature. However, by choosing $\pi(\boldsymbol{\theta})$ as a *bonafide* distribution it is tacitly assumed that the measure associated with $\boldsymbol{\Theta}$ is finite (namely, equal to one). In principle, a prior distribution can be chosen so as to capture the maximum information available. However, in practice, often the prior distribution is so chosen that the posterior distribution of $\boldsymbol{\theta}$ belongs to the same family housing the specific conditional density $p(\boldsymbol{X}|\boldsymbol{\theta})$. Such a prior is called a *conjugate prior*. For example, if $p(\boldsymbol{X}|\boldsymbol{\theta})$ belongs to an exponential family then there exists a conjugate prior for which the posterior distribution belongs to the same family. As an illustration, we consider the multinomial distribution

$$p(\boldsymbol{X}|\boldsymbol{\theta}) = \frac{(\boldsymbol{X}'\boldsymbol{1})!}{\prod_{j=1}^{k}(X_j)!}\prod_{j=1}^{k}\theta_j{}^{X_j}, \tag{4.3.13}$$

where $\boldsymbol{\theta} = (\theta_1, \ldots, \theta_k)'$ lies in the simplex $S_{k-1} = \{\boldsymbol{a} \in \mathbb{R}^k : \boldsymbol{a} \geq \boldsymbol{0},\ \boldsymbol{a}'\boldsymbol{1} = 1\}$ and the random vector $\boldsymbol{X}$ has nonnegative integer valued coordinates satisfying the constraint $\boldsymbol{X}'\boldsymbol{1} = n$. Here we may either interpret $\boldsymbol{\theta}$ as parameters or even we may take the particular probability law governing $\boldsymbol{X}$, generated by a specific $\boldsymbol{\theta}$, as a member of the entire class of multinomial probability laws on the simplex S_{k-1}, the latter formulation having a nonparametric flavor. As a multivariate extension of the beta density, used in Example 4.3.2, consider the *Dirichlet density*, given by

$$\pi(\boldsymbol{\theta}) = \frac{\Gamma\left(\sum_{j=1}^{k}\alpha_j\right)}{\prod_{j=1}^{k}\Gamma(\alpha_j)}\prod_{j=1}^{k}\theta_j^{\alpha_j-1}, \tag{4.3.14}$$

where $\boldsymbol{\alpha} = (\alpha_1, \ldots, \alpha_k)' > \boldsymbol{0},\ j = 1, \ldots, k$, and $\boldsymbol{\theta} \in S_{k-1}$. For this conjugate prior, the posterior density of $\boldsymbol{\theta}$ is

$$\frac{\Gamma\left[1 + \sum_{j=1}^{k}(X_j + \alpha_j)\right]}{\prod_{j=1}^{k}\Gamma(1 + X_j + \alpha_j)}\prod_{j=1}^{k}\theta_j^{X_j+\alpha_j-1}, \tag{4.3.15}$$

that is, a Dirichlet density with parameters $(X_j + \alpha_j) > 0,\ j = 1, \ldots, k$ and $\boldsymbol{\theta} \in S_{k-1}$. This posterior density leads to routine computation of Bayes estimators. Often, such conjugate priors run into interpretational difficulties, and for nonexponential families of densities, may not even exist in a closed form. In some studies, lacking such precise prior information, a (Bayesian) statistician may still like to use some *vague* or imprecisely defined prior, so that it is given a somewhat less dominating role in the choice of a decision rule. For instance, in Exercise 4.3.4, we may assume a measure ξ for $\boldsymbol{\theta}$, for which the finiteness of $\xi(\mathbb{R}^p)$ is not tenable. Still, under such an *improper prior*, as long as the marginal distribution of $\boldsymbol{X}$ is properly defined, that is, for almost all $\boldsymbol{X}$,

$$\int_{\boldsymbol{\Theta}} p(\boldsymbol{X}|\boldsymbol{\theta})\mathrm{d}\xi(\boldsymbol{\theta}) < \infty, \tag{4.3.16}$$

Bayes decision theory works out well.

***Example* 4.3.3.** Let $\boldsymbol{X} = (X_1, \ldots, X_n)'$ have i.i.d. coordinates each following a density $f(x-\theta)$ for which the form does not involve θ. Furthermore, assume that θ has the uniform (improper) prior distribution over $\mathbb{R}$. It can be shown (Exercise 4.3.5) that the MRE within this class is of the form

$$T(\boldsymbol{X}) = X_{n:1} - \mathbb{E}(X_{n:1}|X_{n:j} - X_{n:1},\ 2 \le j \le n), \tag{4.3.17}$$

where $X_{n:1} < \cdots < X_{n:n}$ are the order statistics corresponding to $(X_1, \ldots, X_n)$. This characterization is based on the fact that for the group of translation transformations (from the X_i to $X_i + a,\ a \in \mathbb{R}$), the maximal invariants are $X_{n:j} - X_{n:1},\ j = 2, \ldots, n$. As such, it can be shown that the translation-equivariant MRE is of form

$$T(\boldsymbol{X}) = X_{n:1} - \frac{\int_{\mathbb{R}} tf(t)\prod_{j=2}^{n} f(t + X_{n:j} - X_{n:1})\mathrm{d}t}{\int_{\mathbb{R}} f(t)\prod_{j=2}^{n} f(t + X_{n:j} - X_{n:1})\mathrm{d}t}. \tag{4.3.18}$$

This estimator is known as the *Pitman estimator of location.* □

***Example* 4.3.4.** In Exercise 4.3.4, assume that $\boldsymbol{\theta}$, instead of a normal prior, has an improper prior with uniform measure (say, ξ) on $\mathbb{R}^p$. Then, the marginal distribution of $\boldsymbol{X}$ is properly defined, and hence, the posterior density of $\boldsymbol{\theta}$ can be obtained and its mean vector considered to formulate the corresponding Bayes estimator under the quadratic error loss function (see Exercise 4.3.6). □

If for the sample point $\boldsymbol{X}$, the joint density $p(\boldsymbol{X}, \theta)$ admits a sufficient statistic (say, T) for θ in the conventional way as in Chapter 2, then $p(\boldsymbol{X}, \theta) = p_1(T|\theta)p_2(\boldsymbol{X}|T)$, where the second factor does not depend on θ. If we use this factorization in the definition of the marginal distribution $p^*(\boldsymbol{X})$, we may show that the posterior density of θ, given $\boldsymbol{X}$ depends on the sample only through the density of the sufficient statistic T [i.e., the factor $p_2(\boldsymbol{X}, T)$ cancels from both the numerator and denominator]. Then it is possible to consider the posterior density based solely on the density of the sufficient statistic (whenever it exists) and the same prior on θ, leading to the original posterior density based on the joint density $p(\boldsymbol{X}, \theta)$ (Exercise 4.3.7). Therefore, whenever a sufficient statistic T exists, Bayes decision rules will depend on the sample through T alone. This prescription covers general loss functions and general SDT problems. Even when there is no sufficient statistic, but we use a constant (possibly, improper) prior, the posterior density of θ is proportional to the likelihood $p(\theta|\boldsymbol{X})$, so that the conventional maximum likelihood estimator is the mode of the posterior density, whenever such a mode is uniquely defined (i.e., the posterior is unimodal). In this sense, the mode of the posterior density in (4.3.3), whenever it exists uniquely, is called the *generalized MLE*.

There is another variant of a Bayes decision rule, known as *empirical Bayes* (EB) *rule*, where the prior may not be totally prespecified. It is adapted, at least, partly, from the observable random element in conjunction with some priors for which only the form is specified. We illustrate this EB rule with Example 4.3.1, where σ^2 and τ^2 are both treated as unknown. Then, the Bayes estimator is given by (4.3.9), but it involves the nuisance σ^2 and τ^2. Note that $(n-1)s_n^2/\sigma^2$ has the χ_{n-1}^2 distribution so that

$$\mathbb{E}\left(\frac{n-3}{n-1}\frac{\sigma^2}{s_n^2}\right) = \mathbb{E}(\chi_{n-1}^{-2})(n-3) = 1, \tag{4.3.19}$$

providing an estimator of σ^2 as $\widehat{\sigma}^2 = \sum_{i=1}^n (X_i - \overline{X})^2/(n-3)$. Also, for the marginal distribution of $\boldsymbol{X}$, we have

$$(\overline{X}_n - \mu)^2/(\sigma^2/n + \tau^2) \sim \chi_1^2, \tag{4.3.20}$$

providing the estimator, $n(\overline{X}_n - \mu)^2$ for $(\sigma^2 + n\tau^2)$. From these two estimators we arrive at a plug-in estimator of the Bayes factor $B = \sigma^2/(\sigma^2 + n\tau^2)$.

4.4 Bayes Hypothesis Testing

In a Bayesian setup, hypothesis testing problems are generally formulated in terms of two orthogonal prior distributions on $\boldsymbol{\theta}$ pertaining to the two hypotheses. This prescription goes over smoothly to the *multiple decision theory*, which will be also outlined here.

As in Chapter 3, consider a partitioning of the parameter space $\boldsymbol{\Theta}$ into two disjoint subspaces $\boldsymbol{\Theta}_0$ and $\boldsymbol{\Theta}_1$, and formulate the null and alternative hypotheses, respectively, as

$$H_0 : \boldsymbol{\theta} \in \boldsymbol{\Theta}_0 \quad \text{and} \quad H_1 : \boldsymbol{\theta} \in \boldsymbol{\Theta}_1. \tag{4.4.1}$$

In a Bayesian setup, as $\boldsymbol{\theta}$ is treated as random, we take two priors $\Lambda_0(\boldsymbol{\theta})$ and $\Lambda_1(\boldsymbol{\theta})$ for $\boldsymbol{\theta}$ under H_0 and H_1, respectively. Since $\Lambda_0(\boldsymbol{\theta})$ and $\Lambda_1(\boldsymbol{\theta})$ are defined on two disjoint subspaces (i.e., they are null on the complementary subsets), they are *orthogonal*. Then, we may write

$$\Lambda(\boldsymbol{\theta}) = \Lambda_0(\boldsymbol{\theta})I(\boldsymbol{\theta} \in \boldsymbol{\Theta}_0) + \Lambda_1(\boldsymbol{\theta})I(\boldsymbol{\theta} \in \boldsymbol{\Theta}_1), \tag{4.4.2}$$

and the marginal law for the observable $\boldsymbol{X}$ is

$$\begin{aligned} p^*(\boldsymbol{X}) &= \int_{\boldsymbol{\Theta}} p(\boldsymbol{X}|\boldsymbol{\theta})\mathrm{d}\Lambda(\boldsymbol{\theta}) \\ &= \int_{\boldsymbol{\Theta}_0} p(\boldsymbol{X}|\boldsymbol{\theta})\mathrm{d}\Lambda_0(\boldsymbol{\theta}) + \int_{\boldsymbol{\Theta}_1} p(\boldsymbol{X}|\boldsymbol{\theta})\mathrm{d}\Lambda_1(\boldsymbol{\theta}) \\ &= p_0^*(\boldsymbol{X}) + p_1^*(\boldsymbol{X}), \quad \text{say,} \end{aligned} \tag{4.4.3}$$

where

$$p_j^*(\boldsymbol{X}) = \int_{\boldsymbol{\Theta}_j} p(\boldsymbol{X}|\boldsymbol{\theta})\mathrm{d}\Lambda_j(\boldsymbol{\theta}), \quad j = 0, 1. \tag{4.4.4}$$

The posterior distribution of $\boldsymbol{\theta}$ is

$$g(\boldsymbol{\theta}|\boldsymbol{X}) = p(\boldsymbol{X}|\boldsymbol{\theta})d\Lambda(\boldsymbol{\theta})/p^*(\boldsymbol{X}), \quad \boldsymbol{\theta} \in \boldsymbol{\Theta}. \tag{4.4.5}$$

Let us define

$$\alpha_0 = \int_{\mathbf{\Theta}_0} g(\boldsymbol{\theta}|\boldsymbol{X})\mathrm{d}\boldsymbol{\theta} = \frac{p_0^*(\boldsymbol{X})}{p_0^*(\boldsymbol{X}) + p_1^*(\boldsymbol{X})} \tag{4.4.6}$$

and

$$\alpha_1 = \int_{\mathbf{\Theta}_1} g(\boldsymbol{\theta}|\boldsymbol{X})\mathrm{d}\boldsymbol{\theta} = \frac{p_1^*(\boldsymbol{X})}{p_0^*(\boldsymbol{X}) + p_1^*(\boldsymbol{X})}. \tag{4.4.7}$$

We may visualize α_0 and α_1 as the *posterior confidence* or evidence in favor of H_0 and H_1, respectively, and as such, it is quite intuitive to consider the following *Bayes test* procedure for testing H_0 versus H_1: *Reject H_0 in favor of H_1 when $\alpha_0 < \alpha_1$, or equivalently, when* $p_0^*(\boldsymbol{X}) < p_1^*(\boldsymbol{X})$; otherwise, *reject H_1 in favor of H_0.*

Let $\mathbf{\Theta}_0$ and $\mathbf{\Theta}_1$ consist of single mass points $\boldsymbol{\theta}_0$ and $\boldsymbol{\theta}_1$, respectively. In the Neyman–Pearson setup we consider a test function $\tau(\boldsymbol{X})$ that takes on the values 1 and 0 when $p(\boldsymbol{X}|\boldsymbol{\theta}_1)/p(\boldsymbol{X}|\boldsymbol{\theta}_0)$ is $>$ or $\leq K_\alpha$, where the constant K_α is chosen so that

$$P[\tau(\boldsymbol{X}) = 1|H_0] = \alpha, \tag{4.4.8}$$

for some prespecified significance level $0 < \alpha < 1$. Now, denote the corresponding prior probabilities for θ_0 and θ_1 as π_0 and π_1, respectively; then the Bayes test function reduces to the following: *reject H_0 in favor of H_1 when the likelihood ratio*

$$p(\boldsymbol{X}|\boldsymbol{\theta}_1)/p(\boldsymbol{X}|\boldsymbol{\theta}_0) > \pi_0/\pi_1; \tag{4.4.9}$$

otherwise, *reject H_1 in favor of H_0*. The left-hand side of (4.4.9) is also known as the Bayes factor [Kass and Raftery (1995)]. Thus, it is possible to characterize the Neyman–Pearson test as a Bayes test whenever the priors π_0 and π_1 are so chosen that the size of the test is equal to some preassigned α. We pose some specific cases in Exercises 4.4.1 and 4.4.2.

Now, we formulate some multiple hypotheses testing problems in a Bayesian setup along the same line as above. Consider a set of $K \geq 2$ hypotheses $H_j : \boldsymbol{\theta} \in \mathbf{\Theta}_j,\ j = 1, \ldots, K$ where the $\mathbf{\Theta}_j$ are all disjoint, and as above, we introduce prior distributions given by the distribution functions $\Lambda_j(\boldsymbol{\theta}),\ j = 1, \ldots, K$ which are mutually orthogonal. Let then

$$\Lambda(\boldsymbol{\theta}) = \sum_{j=1}^{K} \Lambda_j(\boldsymbol{\theta}). \tag{4.4.10}$$

This leads us to the following marginal law for $\boldsymbol{X}$:

$$p^*(\boldsymbol{X}) = \sum_{j=1}^{K} p_j^*(\boldsymbol{X}), \tag{4.4.11}$$

where

$$p_j^*(\boldsymbol{X}) = \int_{\mathbf{\Theta}} p(\boldsymbol{X}|\boldsymbol{\theta})\mathrm{d}\Lambda_j(\boldsymbol{\theta}), \quad j = 1, \ldots, K. \tag{4.4.12}$$

Define then

$$g_j(\boldsymbol{\theta}) = \frac{p(\boldsymbol{X}|\boldsymbol{\theta})\lambda_j(\boldsymbol{\theta})}{p^*(\boldsymbol{X})}, \quad j = 1, \ldots, K, \tag{4.4.13}$$

where $\lambda_j(\boldsymbol{\theta}) = \mathrm{d}\Lambda_j(\boldsymbol{\theta})/\mathrm{d}\boldsymbol{\theta}$. Analogously to the two-decision problem, let

$$\alpha_j = \int_{\boldsymbol{\Theta}_j} g_j(\boldsymbol{\theta})\mathrm{d}\boldsymbol{\theta} = p_j^*(\boldsymbol{X})/p^*(\boldsymbol{X}), \quad j = 1, \ldots, K \tag{4.4.14}$$

Then, regarding the α_j as the posterior confidence or evidence in favor of H_j, for $j = 1, \ldots, K$, at least intuitively, we may consider the following test: *among the K hypotheses $H_1, \ldots, H_K$, accept H_j (and reject the others), if*

$$\alpha_j > \alpha_k, \quad \forall k (\neq j). \tag{4.4.15}$$

In (4.4.9) and (4.4.15) it is tacitly assumed that the strict inequality sign holds a.e. $(\boldsymbol{\theta})$ – a case that requires some modifications when the posterior density is not continuous a.e. We clarify some of these problems through Exercise 4.4.3.

4.5 Confidence Sets

In the frequentist approach, the setting of a confidence interval, particularly in the presence of nuisance parameter(s), tacitly assumes that there exists a *pivotal statistic* involving only the parameter of interest [along with estimates of the nuisance parameter(s)] whose probability law is free from all the parameters. This enables us to use the known distribution of the pivotal statistic to set a confidence interval for the parameter(s) of interest. As an example, consider the case of n i.i.d. random variables following a normal distribution with mean μ and variance σ^2, both unknown. We are interested in setting a confidence interval for μ with coverage probability $1 - \alpha$, for some prespecified $0 < \alpha < 1$, usually taken as 0.05 or 0.01. If $\overline{X}_n$ and s_n^2, respectively, denote the sample mean and (unbiased) variance, then,

$$t_n = \sqrt{n}(\overline{X}_n - \mu)/s_n \tag{4.5.1}$$

has the Student t-distribution with $n - 1$ degrees of freedom, and we construct the following confidence interval for μ

$$P[\overline{X}_n - n^{-1/2}s_n t_{n-1,1-\alpha/2} \leq \mu \leq \overline{X}_n + n^{-1/2}s_n t_{n-1,1-\alpha/2}] = 1 - \alpha. \tag{4.5.2}$$

In the same vein, if we want to set a confidence interval for σ^2, we use the pivotal statistic $Q_n = (n-1)s_n^2/\sigma^2$, that follows the chi-squared distribution with $n - 1$ degrees of freedom, and use its percentile points to setup a confidence interval for σ^2. We leave the details to Exercise 4.5.1.

We consider now a second multiparameter model: let $\boldsymbol{X}_1, \ldots, \boldsymbol{X}_n$ be i.i.d. random vectors having a multivariate normal distribution with mean vector $\boldsymbol{\mu}$ and covariance matrix $\boldsymbol{\Sigma}$, both unspecified, the latter is assumed to be positive definite. We are interested in constructing a confidence set for $\boldsymbol{\mu}$. Let $\overline{\boldsymbol{X}}_n$ and $\boldsymbol{S}_n$ be the sample mean vector and (unbiased) covariance matrix, respectively, and consider the pivotal statistic

$$T_n^2 = n(\overline{\boldsymbol{X}}_n - \boldsymbol{\mu})'\boldsymbol{S}_n^{-1}(\overline{\boldsymbol{X}}_n - \boldsymbol{\mu}), \tag{4.5.3}$$

which follows the Hotelling T^2-distribution with $n - 1$ degrees of freedom. Hence, letting $T_{n,\alpha}^2$ denote its upper α-quantile, we obtain $P(T_n^2 \leq T_{n,\alpha}^2) = \alpha$, yielding the confidence set

$$\mathcal{I}_n(\boldsymbol{\mu}) = \{\boldsymbol{\mu} : T_n^2 \leq T_{n,1-\alpha}^2\}, \tag{4.5.4}$$

an ellipsoid with center $\overline{\boldsymbol{X}}_n$. Exercise 4.5.2 is set to verify that this confidence set can be equivalently expressed as

$$P[\sqrt{n}|\boldsymbol{a}'(\boldsymbol{\mu} - \overline{\boldsymbol{X}}_n)| \leq (\boldsymbol{a}'\boldsymbol{S}_n\boldsymbol{a})^{1/2}T_{n,1-\alpha}, \forall \boldsymbol{a} \in \mathbb{R}^p] = 1 - \alpha. \tag{4.5.5}$$

Let us consider a less favorable situation, where $X_n \sim \text{Bin}(n, \theta)$. We wish to set a confidence interval for θ. Unlike the previous examples, here $(X_n - n\theta)$ has a distribution with mean 0, but its form as well as its variance depend on the unknown θ. Thus, setting an exact confidence interval for θ is troublesome. Exercise 4.5.3 is set to work out various possibilities to handle this problem.

A point estimator usually is judged by its performance characteristics (e.g., risk in the SDT setup); the same is true for confidence intervals (or sets), and we outline specific perspectives here. If for a confidence set $\mathfrak{I}(\boldsymbol{\theta})$ at confidence level $1 - \alpha$, we have

$$P_{\boldsymbol{\theta}}[\boldsymbol{\theta} \in \mathfrak{I}(\boldsymbol{\theta})] = 1 - \alpha, \quad \forall \boldsymbol{\theta}, \tag{4.5.6}$$

and further, for every other $\boldsymbol{\theta}'$,

$$P_{\boldsymbol{\theta}'}[\boldsymbol{\theta} \in \mathfrak{I}(\boldsymbol{\theta})] \leq 1 - \alpha, \tag{4.5.7}$$

then $\mathfrak{I}(\boldsymbol{\theta})$ is called an *unbiased confidence set*; if part of $\boldsymbol{\theta}$ relates to nuisance parameters, then the above two equations should be true for all nuisance parameter(s). In a Bayesian setup, the above two equations should hold also for the chosen prior parameters. Furthermore, if for a class of confidence sets $\{\mathfrak{I}_r(\boldsymbol{\theta}), r \geq 0\}$, at confidence level $1 - \alpha$, there exists a set $\mathfrak{I}_0(\boldsymbol{\theta})$ such that

$$P_{\boldsymbol{\theta}'}[\boldsymbol{\theta} \in \mathfrak{I}_0(\boldsymbol{\theta})] \leq P_{\boldsymbol{\theta}'}[\boldsymbol{\theta} \in \mathfrak{I}_r(\boldsymbol{\theta})], \tag{4.5.8}$$

for all $r \geq 1, \boldsymbol{\theta}'$ and $\boldsymbol{\theta}$, then $\mathfrak{I}_0(\boldsymbol{\theta})$ is the *uniformly most accurate* (UMA) confidence set for $\boldsymbol{\theta}$ in that class. It is not difficult to see a connection between UMPU tests and UMA unbiased confidence sets. The concept of invariance (and equivariance) arising in point-estimation theory extends directly to confidence sets. We pose this as Exercise 4.5.4.

In a Bayesian setup, incorporating the information contained in the prior distribution and given the data, the posterior distribution of $\boldsymbol{\theta}$ captures all the information on the parameter(s) contained in the prior distribution and in the data. Hence, if the assumed prior distribution is conjugate, the Bayes setup is quite interpretable and often convenient to work with. On the other hand, if the chosen prior is different from the true (unknown) one, we could end up with unreasonable statistical resolutions. However, as the chosen prior also includes nuisance parameter(s), by proper integration over this set, the posterior distribution for the parameters of interest can be conveniently used to obtain a confidence interval (set) for the parameter(s) of interest. To illustrate this point, let us consider the binomial model discussed above. We refer to (4.3.12) and (4.3.15), for the associate prior and posterior densities. In particular, the posterior density of θ based on the Beta(a, b) prior is

$$p(\theta|X_n, a, b) = \frac{\Gamma(a + b + n + 1)}{\Gamma(a + X_n + 1)\Gamma(b + n + 1 - X_n)}\theta^{a+X_n-1}(1 - \theta)^{b+n-X_n-1}, \tag{4.5.9}$$

for $\theta \in (0, 1)$. Thus, for given a, b and observed X_n, the above completely specified posterior density provides a confidence interval for θ; we leave the details to Exercise 4.5.5.

In some cases, the posterior distribution of $\boldsymbol{\theta}$ has some standard functional form, involving parameters, which are thus determined from the observed $\boldsymbol{X}$ and the incorporated prior. This was the case with the betabinomial model discussed above. If, in particular, the posterior density is (multi)normal allowing possible transformation on $\boldsymbol{\theta}$ (such as the logit), or even a member of the exponential family, then such Bayesian confidence intervals (sets) can be conveniently constructed in terms of the associated parameters, using their posterior laws. In a general multiparameter setup, one may use the *maximum posterior likelihood* principle based on the posterior density $g(\boldsymbol{\theta})$ and seek to have a lower cutoff point x_α, such that

$$\int_{g(\boldsymbol{\theta})\geq x_\alpha} g(\boldsymbol{\theta})\mathrm{d}\boldsymbol{\theta} = 1-\alpha. \tag{4.5.10}$$

As such, we may consider a (Bayesian) confidence set

$$\mathfrak{I}(\boldsymbol{\theta}) = \{\boldsymbol{\theta} \in \boldsymbol{\Theta} : g(\boldsymbol{\theta}) \geq x_\alpha\}. \tag{4.5.11}$$

If the posterior distribution is unimodal, the above confidence set would be a closed subspace of $\boldsymbol{\Theta}$; for multimodal posterior densities, such confidence sets may be the union of some disjoint closed subsets. We pose Exercise 4.5.6 for illustration. In the above development, it is tacitly assumed that the prior density on $\boldsymbol{\Theta}$ is of known form. In the event that this is not available, improper priors are used and analogous solutions are advocated.

4.6 Concluding Notes

For SDT and Bayesian inference, in particular, prior distributions are incorporated to accommodate existing information on the parameters. For small sample sizes, the impact of the prior is quite visible, while as the sample size increases, the prior plays a decreasing role. In fact, for large-sample sizes, Bayes estimators and MLE are almost the same; if we look in (4.3.9), we may observe that for any (fixed) σ^2 and τ^2, as $n \to \infty$, the Bayes factor $\sigma^2/(n\tau^2+\sigma^2) \to 0$ at a rate $O(n^{-1})$, so that $d_B(\boldsymbol{X}) - \overline{X}_n = O_p(n^{-1})$. This feature holds for general Bayes estimators in a asymptotic setup.

For confidence intervals or sets, finding an exact confidence level, specially, in the presence of nuisance parameters, is generally difficult. In this sense, the usual likelihood ratio statistics-based procedures work out well for large sample sizes. In the same vein, the posterior likelihood plays a vital role in the construction of Bayesian confidence sets, and excepting some simple cases, asymptotics provide great relief in deriving good approximations to such intervals and sets. Again, this will be briefly discussed in Chapter 8.

Many computational techniques have been recently developed for the computation of Bayes estimators. Among them we mention *Gibbs sampling*, *Metropolis Hastings*, and *MCMC*. We do not address these issues here; see Robert and Casella (2004).

4.7 Exercises

4.2.1 Show that for $\overline{X}_n$ in Example 4.2.1 the risk is a constant for all $\theta \in \mathbb{R}$.

4.2.2 Show that the MLE in Example 4.2.1 is the MRE.

4.2.3 For the same model as in Example 4.2.1, consider the absolute error loss function for an estimator of θ. Examine the risk for translation invariant estimators of θ, and find the MRE. Is it the same as in the preceding problem?

4.2.4 Let $X_1, \ldots, X_n$ be i.i.d. random variables having (for a fixed θ) the Laplace density, that is, $f(x;\theta) = (1/2)\exp[-|x-\theta|]$, $x \in \mathbb{R}$. Considering the absolute error loss function and within the class of translation-equivariant estimators of θ, find the MRE.

4.2.5 Let $\boldsymbol{X} = (X_1, \ldots, X_n)'$ with each X_i having the exponential density $f(x;\theta) = \theta^{-1}\exp(-x/\theta)I(x \geq 0)$, where $\theta \in \mathbb{R}^+$. Consider the class of scale-equivariant estimators and a loss function of the form $L(a, b) = |a/b - 1|^p$ where p is either 1 or 2. Obtain the scale-equivariant MRE under both loss functions.

4.2.6 Refer to Example 4.2.2 and show that

$$E(T_{JS} - \theta)^2 \leq E(\overline{X} - \theta)^2, \quad \forall \theta \in \mathbb{R},$$

so that T_{JS} dominates $\overline{X}$ in quadratic risk, and as a result, conclude that the MLE is inadmissible.

4.3.1 Work out the details for obtaining expressions (4.3.9)–(4.3.11) in Example 4.3.1.

4.3.2 Using (4.3.12) show that the marginal distribution of X and the posterior density of p are given by $f(x)$ and $f(p|x)$, respectively, defined after (4.3.12). Hence, obtain the mean of the posterior density $f(p|x)$ and show that the Bayes estimator of p is given by $(a + x)/(a + b + n)$.

4.3.3 Let $X_1, \ldots, X_n$ be i.i.d. random variables having the normal density with mean 0 and variance $\theta > 0$. Also, let $\pi(\theta)$ have the gamma density with scale parameter σ^2 and shape parameter ν. Obtain the posterior density of θ, and examine its mean and median. Are these the same? Examine any stochastic ordering of the two Bayes estimators under the squared error and absolute error loss.

4.3.4 Let $\boldsymbol{X}_1, \ldots, \boldsymbol{X}_n$ be i.i.d. random variables having a multivariate normal distribution with mean vector $\boldsymbol{\theta}$ and covariance matrix $\boldsymbol{I}_p$. Also, let $\boldsymbol{\theta}$ have a multivariate normal distribution with mean vector $\boldsymbol{\mu}$ and covariance matrix $\tau^2\boldsymbol{I}_p$, for some $\tau > 0$. Obtain the Bayes estimator under a quadratic error loss. Examine the difficulties in deriving the Bayes estimator under absolute error loss function.

4.3.5 For the situation considered in Example 4.3.3, obtain the posterior density of θ. Considering the class of translation-equivariant estimators, show that the MRE class is of the form depicted in (4.3.17) and then obtain (4.3.18).

4.3.6 In Example 4.3.4, show that the marginal distribution of $\boldsymbol{X}$ is properly defined and, hence, obtain the posterior density of $\boldsymbol{\theta}$. Considering quadratic error loss function, obtain the Bayes estimator.

4.3.7 Let $X_1, \ldots, X_n$, given θ, be i.i.d. $\mathcal{N}(\theta, \sigma^2)$ variables and let θ have the prior $\mathcal{N}(\mu, \tau^2)$ (see Example 4.3.1). Given (θ, σ^2), $\overline{X}_n$ and s_n^2 are jointly sufficient for (θ, σ^2). Then show that the posterior density of θ depends on $(X_1, \ldots, X_n)$ only through $(\overline{X}_n, s_n^2)$. Use this result to justify an empirical Bayes estimator as formulated in (4.3.19) and (4.3.20), treating both σ^2 and τ^2 as nuisance.

4.4.1 Consider the loss functions $L_0(\theta, a)$ and $L_1(\theta, a)$, where $L_0(\theta, a) = 0$ or K_0 according as $\theta \in \Theta_0$ or Θ_1 and $L_1(\theta, a) = 0$ or K_2 according as $\theta \in \Theta_1$ or Θ_2. Then work out the Bayes test for testing $H_0 : \theta \in \Theta_0$ versus $H_1 : \theta \in \Theta_1$ and show its relation with the Neyman–Pearson test.

4.4.2 For testing $H_0 : \theta \leq 0$ versus $H_1 : \theta > 0$ (in the $\mathcal{N}(\theta, 1)$ setup), let $\Lambda_0(\theta)$ be the folded $\mathcal{N}(0, \tau_0)$ distribution over $(-\infty, 0]$, while $\Lambda_1(\theta)$ be the folded $\mathcal{N}(0, \tau_1)$ distribution over $(0, \infty)$. Obtain the Bayes test for H_0 versus H_1 and appraise its relation with the Neyman–Pearson test.

4.4.3 In the $\mathcal{N}(\theta, 1)$ setup, consider the hypotheses $H_0 : \theta = 0$, $H_1 : \theta > 0$ and $H_2 : \theta < 0$. Let $\Lambda_0(\theta)$ be a prior that has masspoint π_0 at 0 and a pdf $(1 - \pi_0)\varphi(\tau\theta)$ over $\theta \in \mathbb{R}$. Similarly, let $\Lambda_1(\theta)$ be a folded $\mathcal{N}(0, \gamma_1)$ distribution over $(0, \infty)$ and $\Lambda_2(\theta)$ be a folded $\mathcal{N}(0, \gamma_2)$ distribution over $(-\infty, 0)$. Work out the Bayes test in (4.4.15).

4.5.1 Consider a random sample of size n from a $\mathcal{N}(\mu, \sigma^2)$ distribution and the pivotal statistic $Q_n = (n-1)s_n^2/\sigma^2$. Use the quantiles of its distribution to construct a confidence interval for σ^2.

4.5.2 Consider a random sample of size n from a multinormal distribution $\mathcal{N}(\boldsymbol{\mu}, \boldsymbol{\Sigma})$. Let T_n^2 be as in (4.5.3) and provide the details to show that (4.5.4) is equivalent to (4.5.5).

4.5.3 Let $X_n \sim \text{Bin}(n, \theta)$. Work out various possibilities to construct a confidence set for θ.

4.5.4 Let $\boldsymbol{X}_i = (X_{i1}, \ldots, X_{ip}) \sim \mathcal{N}(\boldsymbol{\theta}, \boldsymbol{\Sigma})$, $i = 1, \ldots, n$. Obtain the confidence set $\mathcal{I}(\boldsymbol{\theta})$ in (4.5.4) and (4.5.5). Show that (4.5.7) holds, that is, $\mathcal{I}(\boldsymbol{\theta})$ is an unbiased confidence set. Use the affine invariance property and show that $\mathcal{I}(\boldsymbol{\theta})$ is UMA confidence set within the class of all affine-invariant confidence sets for $\boldsymbol{\theta}$.

4.5.5 In the binomial setup, for given a, b and observed X_n, use (4.5.9) to obtain a confidence interval for θ.

4.5.6 Let $\boldsymbol{X} = (X_1, \ldots, X_n)$ have the Cauchy distribution, so that $p(\boldsymbol{x}|\theta) = \pi^{-n}\prod_{i=1}^n[1 + (x_i - \theta)^2]^{-1}$. Also, let $\pi(\theta)$ be a noninformative prior on $\mathbb{R}$, so that

$$p^*(\boldsymbol{x}) = \int_{\mathbb{R}} p(\boldsymbol{x}|\theta)\mathrm{d}\theta = \pi^{-n}\int_{\mathbb{R}}\prod_{i=1}^{n}[1 + (x_i - \theta)^2]^{-1}\mathrm{d}\theta.$$

The posterior density $g(\theta|\boldsymbol{x})$ is given by

$$\prod_{i=1}^{n}[1 + (x_i - \theta)^2]^{-1}\bigg/\int_{\mathbb{R}}\prod_{i=1}^{n}[1 + (x_i - \theta)^2]^{-1}\mathrm{d}\theta.$$

Consider the case of $n = 3$ and let $X_{(1)} < X_{(2)} < X_{(3)}$ be the ordered values of X_1, X_2, X_3. Show that depending on the spacings $X_{(2)} - X_{(1)}$ and $X_{(3)} - X_{(2)}$, $g(\theta|\boldsymbol{x})$ may not be unimodal (in θ). Hence, or otherwise, comment on the (Bayesian) confidence set for θ.

5

Stochastic Processes: An Overview

5.1 Introduction

In preceding chapters, collections of independent random variables (or vectors) constituting random samples were considered for drawing statistical conclusions on the underlying populations. In many practical problems independence may be questionable, most notably in data arising as realizations of collections of random elements that are not generally independent. To deemphasize independence we may consider more general sets of random variables, called *stochastic processes*. Formally, a stochastic process $\boldsymbol{X} = \{X_t,\ t \in \mathcal{T}\}$ is a collection of random variables, indexed by the set $\mathcal{T}$, called the *space-* or *time-parameter set*. The time-parameter t may be discrete or continuous, univariate or multivariate. Also, associated with $\boldsymbol{X}$, there is a (joint) probability law $P(\boldsymbol{x})$. When t is discrete, it is the joint distribution; when t is continuous, there are some technical complications that we will discuss in Section 5.4. We motivate the need for such more complex sets of random variables with some illustrative examples.

***Example* 5.1.1.** The daily maximum temperature (Y) in a given region is recorded over a period of time with the objective of verifying the existence of a linear trend and seasonality. A suitable model for the data is

$$Y_t = \alpha + \beta t + e_t,\ t = 1, \ldots, n,$$

where α and β are unknown constants, $e_t = \rho e_{t-1} + u_t$, $e_0 = 0$, $|\rho| < 1$, and $u_t \sim \mathcal{N}(0, \sigma^2)$. The variables in the set $Y = \{Y_t,\ t = 1, \ldots, n\}$ are clearly dependent and constitute a stochastic process called a *time series* process with a discrete time parameter t. □

***Example* 5.1.2.** Mortality, morbidity, and other vital events are recorded in a given period of time, a year, say, for a sample of counties in a given state with the objective of evaluating the spatial distribution of the corresponding prevalences. Here, a convenient model is

$$\boldsymbol{Y}_s = \boldsymbol{\mu}_s + \boldsymbol{e}_s,\ \boldsymbol{s} = (s_1, s_2)' \in \boldsymbol{S} \subset \mathbb{R}^2,$$

where $\boldsymbol{Y}_s$ denotes the observed counts for a county with coordinates $\boldsymbol{s}$, $\boldsymbol{\mu}_s$ denotes the expected vector of prevalences, and $\boldsymbol{e}_s$ represents a random term for which $\mathbb{C}\mathrm{orr}(\boldsymbol{e}_{s_1}, \boldsymbol{e}_{s_2})$ decreases as the distance between $\boldsymbol{s}_1$ and $\boldsymbol{s}_2$ increases. Again, the elements in the set $\boldsymbol{Y} = \{\boldsymbol{Y}_s, \boldsymbol{s} \in \boldsymbol{S} \subset \mathbb{R}^2\}$ are clearly dependent and relates to a stochastic process commonly known as a *spatial* stochastic process, with a discrete spatial-parameter set $\boldsymbol{S}$. □

***Example* 5.1.3.** For a metropolitan area consisting of a number of municipalities, records of the daily number of automobile accidents, $N(s,t)$ over a period of time, say, one year, may be modeled as in Examples 5.1.1 and 5.1.2 by letting $N = \{N(s,t);\ s \in \mathcal{S},\ t \in \mathcal{T}\}$, where $\mathcal{S}$ is the set if municipalities ad $\mathcal{T}$ is the time-index set. Then, the set N constitutes a *spatiotemporal* process. □

***Example* 5.1.4.** Suppose that a pesticide is sprayed on a colony with n insects with the objective of evaluating its effectiveness. Let n_t be the number of insects alive at time $t > 0$. Clearly, n_t is nonincreasing, and n_t and n_r, $r \neq t$ are dependent random variables. The set $N = \{n_t,\ t \geq 0\}$ is a stochastic process with a continuous time-parameter, commonly known as a *death process*. □

***Example* 5.1.5.** At a given generation, a bacterium either splits into two or dies. Let X_n be the number of bacteria at the nth generation and assume that $X_0 = 1$. If the splitting and death occur with positive probabilities, then the process $\{X_n\}$ is called a *branching process*. One may be interested in the eventual extinction probability. More general branching processes relating to more than two offspring may also be considered. □

***Example* 5.1.6.** Let X_n, $n \geq 1$ be independent random variables taking on the values -1 and $+1$ with probability $1/2$. Assume that $X_0 = 0$ and $S_0 = 0$ and let $S_n = X_0 + \cdots + X_n$, for $n \geq 1$. Thus, S_n may be interpreted as the position of a particle at step $n \geq 1$. Note that $-n \leq S_n \leq n$ and that S_n can only take on the values $-n, -n+2, \ldots, n-2, n$, for $n \geq 1$. Furthermore, observe that $S_{n+1} = S_n + X_{n+1}$, $n \geq 0$, so that given $S_n = s$, it follows that S_{n+1} can only take on the values $s+1$ and $s-1$ with probabilities $1/2$. Then,

$$\frac{1}{2}(n + S_n) = T_n \sim \text{Bin}(n, 1/2), \quad \forall n \geq 1,$$

although the S_n are not independent. However, for every $k \geq 1$, $n \geq 0$, it follows that $S_{n+k} - S_n = X_{n+1} + \cdots + X_{n+k}$ is independent of S_n and has the same distribution as S_k. Therefore, the stochastic process $\{S_n,\ n \geq 0\}$ is said to have *independent* and *homogeneous* increments, and $\mathbb{E}(S_n) = 0$, $\forall n \geq 0$. It is the *simple random walk*. If, however, we have $P(X_n = +1) = 1 - P(X_n = -1) = p$, then $\mathbb{E}(S_n) = n(2p-1)$, but the independent and homogeneous increment property remains intact and the process is known as a *drifted random walk* with a drift parameter given by $n(2p-1)$, $n \geq 1$. Moreover, even for the general drifted case, for any $n \geq 1$,

$$P(S_{n+1} = k | S_n, \ldots, S_0) = P(S_{n+1} = k | S_n), \quad \forall k, n \geq 1$$

is a property known as *Markov dependence*. Furthermore, if we let $S_n^* = S_n - n(2p-1)$, $n \geq 0$, we obtain

$$\mathbb{E}(S_{n+1}^* | S_n^* = q_1, S_{n-1}^* = q_2, \ldots, S_0^* = q_n) = \mathbb{E}(S_{n+1}^* | S_n^* = q_1) = q_1,$$

$\forall n \geq 1$, $q \in \mathbb{R}$. Then, $\{S_n^*;\ n \geq 0\}$ is called a (zero-mean) *martingale*, which will be formally defined in (5.2.7). □

***Example* 5.1.7.** Let X_n, $n \geq 1$, be i.i.d. random variables with mean μ and finite variance σ^2 and let $S_n = X_1 + \cdots + X_n$, $\forall n \geq 1$. Then, the stochastic process $\{S_n;\ n \geq 0\}$, usually known as a *partial sum process*, has the Markov dependence property. □

***Example* 5.1.8.** Let $X_1, \ldots, X_n$ be i.i.d. random variables with a distribution function F, defined on $\mathbb{R}$ and let F_n be the corresponding empirical distribution function defined in (2.4.18). Then, $\{F_n(x) - F(x);\ x \in \mathbb{R}\}$ is a (zero-mean) continuous time parameter stochastic process with time parameter set $\mathbb{R}$. This definition extends to the multivariate case in a natural way. □

***Example* 5.1.9.** Let $X_1, \ldots, X_n$ be i.i.d. random variables having the exponential density $f(x;\theta) = \theta^{-1}\exp(-x/\theta)I(x \geq 0)$, where $\theta > 0$; let the associated order statistics $X_{n:0}(= 0) < X_{n:1} < \cdots < X_{n:n+1}(= \infty)$. If these n items are put in a *life testing* context, then the *total time on test* (TTT) up to a point $x > 0$, for $X_{n:k} \leq x < X_{n:k+1}$ is

$$\begin{aligned}
T_n(x) &= \sum_{i=1}^{n} X_{n:i} I(X_{n:i} < x) + x\left[n - \sum_{i=1}^{n} I(X_{n:i} > x)\right] \\
&= (n-k)(x - X_{n:k}) + (n-k+1)(X_{n:k} - X_{n:k-1}) + \cdots + n(X_{n:1} - 0) \\
&= (n-k)(x - X_{n:k}) + \sum_{j=1}^{k}(n-j+1)(X_{n:j} - X_{n:j-1}) \\
&= (n-k)(x - X_{n:k}) + \sum_{j=1}^{k} d_{nj}, \quad 0 \leq k \leq n,
\end{aligned}$$

where $d_{nj} = (n-j+1)(X_{n:j} - X_{n:j-1})$ are the *normalized spacings*. The set $T_n = \{T_n(x), x \geq 0\}$ is a stochastic process with continuous time parameter; note that $T_n(x) = T_n(X_{n:n})$, $\forall x \geq X_{n:n}$. □

***Example* 5.1.10.** For workers in a tobacco factory, consider the following *health states*: A_1: respiratory disease free, A_2: possibly infected but not diagnosed, A_3: diagnosed respiratory problem, A_4: ambulatory care needed, A_5: death due to respiratory problems, and A_6: discharged from the factory (or death due to other causes). Let X_n denote the health status of a worker on month n and assume that $X_0 = A_1$. Thus, given $X_n = A_{i_n}$ (for some $i_n = 1, \ldots, 6$), X_{n+1} may remain at the same state or it may transit from X_n to one of the other states. The $36(= 6 \times 6)$ possible transitions may be denoted by

$$A_{i_n} \rightarrow A_{i_{n+1}} \quad i_n, i_{n+1} = 1, \ldots, 6.$$

Once a transition occurs to state A_5 or A_6, no further transition occurs – so these are called *absorbing states*. Here $\{X_n,\ n \geq 0\}$ is a *multistate* stochastic process with two absorbing states. It may or may not have a Markov property. □

5.2 Processes with Markov Dependencies

Let $X = \{X_1, \ldots, X_n\}$ be a stochastic process with discrete time parameter, indexed by $\mathcal{J} = \{1, 2, \ldots, n, \ldots\}$. For any $n \geq 1$, let $F(x_1, \ldots, x_n)$ be the joint distribution function (or probability law) of $(X_1, \ldots, X_n)$. If the X_i were independent, $F(x_1, \ldots, x_n)$ would be the product of the marginal probabilities, but, in general, we may write

$$F(x_1, \ldots, x_n) = F_1(x_1) \prod_{j=2}^{n} F_j(x_j | x_1, \ldots, x_{j-1}). \tag{5.2.1}$$

If the conditional distribution $F_j(x_j|x_1, \ldots, x_{j-1})$ depends only on x_{j-1}, that is, $F_j(x_j|x_1, \ldots, x_{j-1}) = F_j(x_j|x_{j-1})$, $\forall j \geq 2$, then (5.2.1) simplifies to

$$F(x_1, \ldots, x_n) = F_1(x_1) \prod_{j=2}^{n} F_j(x_j | x_{j-1}), \quad \forall n \geq 2. \tag{5.2.2}$$

In this case, $\boldsymbol{X}$ is said to have *Markov dependence*. Examples 5.1.6 and 5.1.7 relate to this Markov dependence pattern. If X_i can assume values in a discrete state space $\mathcal{S}$, then $\boldsymbol{X}$, with the Markov dependence pattern, is called a *Markov chain*. In this case, if the conditional probability function $P(x_j|x_{j-1})$ does not depend on j (≥ 1) but only on the states x_j, x_{j-1}, $\boldsymbol{X}$ is called a *stationary Markov chain*. Example 5.1.6, thus, relates to a stationary Markov chain. If the X_i have density functions, we may write (5.2.2) as

$$f(x_1, \ldots, x_n) = f_1(x_1) \prod_{j=2}^{n} f_j(x_j | x_{j-1}), \tag{5.2.3}$$

and $\boldsymbol{X}$ is called a *Markov process*. If additionally, $f_j(y|x) = f(y|x)$ $\forall j \geq 1$, it would be called a *stationary Markov process*. Example 5.1.7 pertains to this scheme.

For Markov chains, $P_j(X_j = b|X_{j-1} = a) = p_{ab}^{(j)}$, $j \geq 1$, are the *transition probabilities*. For a stationary Markov chain, $p_{ab}^{(j)} = p_{ab}$, for all $j \geq 1$ and all $a, b \in \mathcal{S}$. The matrix $\boldsymbol{P} = ((p_{ab}))$ is the *transition matrix*. Using the Kolmogorov–Chapman equation [see, for example, Chiang (1980)]

$$p_{ab} = P(X_{j+2} = a|X_j = b) = \sum_{c \in \mathcal{S}} P(X_{j+2} = a|X_{j+1} = c)P(X_{j+1} = c|X_j = a), \tag{5.2.4}$$

we may show that the two-step transition matrix is given by

$$\boldsymbol{P}^{(2)} = \boldsymbol{P} \times \boldsymbol{P} = \boldsymbol{P}^2. \tag{5.2.5}$$

In general, by induction, we have $\boldsymbol{P}^{(m)} = \boldsymbol{P}^m$ $\forall m \geq 1$.

Note that Markov dependence holds for qualitative as well as quantitative state spaces (see Examples 5.1.6 and 5.1.10). In the quantitative case, if the underlying distributions admit first-order moments and if $\{X_n,\ n \geq 1\}$ is a Markov process such that the conditional distribution of X_n given X_j, $j \leq n-1$, depends only on X_{n-1}, we may write

$$\mathbb{E}(X_n - X_{n-1}|X_j,\ j \leq n-1) = h_{n-1}(X_{n-1}),\ n \geq 1. \tag{5.2.6}$$

As such, if we write $H_n = \sum_{j \leq n} h_{j-1}(X_{j-1})$, $n \geq 1$, and let $X_n^* = X_n - H_n$, $\forall n \geq 1$, then from (5.2.6), we obtain that

$$\begin{aligned} \mathbb{E}(X_n^*|X_j^*,\ j \leq n-1) &= X_{n-1} + h_{n-1}(X_{n-1}) - H_n = X_{n-1} - H_{n-1} \\ &= X_{n-1}^*, \quad \forall n \geq 1. \end{aligned} \tag{5.2.7}$$

This property [i.e., $\mathbb{E}(X_n^*|X_j^*,\ 1 \leq j \leq n-1) = X_{n-1}^*$ a.e., $\forall n \geq 1$] is known as the *martingale* property.

A natural step to eliminate the assumption of independence is to bring in some weaker structure on conditional expectations. In this context, we introduce first some basic definitions and illustrate them by suitable examples.

Let $\{T_n,\ n \geq 1\}$ be a stochastic process for which $\mathbb{E}(T_n)$ exists for every $n \geq 1$. If

$$\mathbb{E}(T_n|T_j,\ j \leq n-1) = T_{n-1} \quad \text{a.e.}, \quad n > 1, \tag{5.2.8}$$

then $\{T_n,\ n \geq 1\}$ is called a *martingale*. For each $n \geq 1$, let $\{T_{n,k},\ k \geq 1\}$ be a stochastic process, assume that $\mathbb{E}(T_{n,k})$ exists for every n, k, and that

$$\mathbb{E}(T_{n,k}|T_{n,j},\ j \leq k-1) = T_{n,k-1} \quad \text{a.e.}, \quad k \geq 1,\ n \geq 1. \tag{5.2.9}$$

Then $\{T_{n,k},\ k \geq 1, n \geq 1\}$ is called a *martingale array*.

We consider the following examples.

***Example* 5.2.1.** Let X_k, $k \geq 1$, be independent random variables with means μ_k. Also, let $T_k = \sum_{j \leq k}(X_j - \mu_j)$, for $k \geq 1$. Then $T_n = T_{n-1} + (X_n - \mu_n)$. Since X_n is independent of $X_1, \ldots, X_{n-1}$ it is also independent of T_{n-1}. Therefore, (5.2.9) holds, so that $\{T_n\}$ is a martingale. Note that the X_k need not be identically distributed for (5.2.9) to hold. Similarly, if $\{X_{n,k},\ k \geq 1,\ \geq 1\}$ is a *triangular scheme* of (row-wise) independent random variables, then (5.2.9) holds for $T_{n,k} = \sum_{j \leq k}[X_{n,j} - \mathbb{E}(X_{n,j})]$, $k \geq 1$. Thus, $\{T_{n,k}\}$ forms a martingale array. □

Example* 5.2.2 *(Likelihood ratio statistics). Let $X_1, \ldots, X_n$ be a set of i.i.d. random variables with the density function $f(x;\theta)$, $x \in \mathbb{R}$ and $\theta \in \Theta$. For testing a null hypothesis $H_0 : \theta = \theta_0$ against an alternative $H_1 : \theta = \theta_1$, the likelihood ratio test statistic is

$$\lambda_n = \frac{L_n(\theta_1)}{L_n(\theta_0)} = \lambda_{n-1}\frac{f(X_n;\theta_1)}{f(X_n;\theta_0)} = \lambda_{n-1}h_n, \quad n \geq 1, \tag{5.2.10}$$

where $h_n = f(X_n;\theta_1)/f(X_n;\theta_0)$ and, conventionally, we take $\lambda_0 = 1$. Note that h_n is independent of $X_1, \ldots, X_{n-1}$ (and, hence, of λ_j, $j \leq n-1$), so that by (5.2.10), we have for all $n \geq 1$,

$$\mathbb{E}(\lambda_n|\lambda_j \leq n-1) = \lambda_{n-1}\mathbb{E}(h_n|\lambda_j,\ j \leq n-1) = \lambda_{n-1}\mathbb{E}(h_n). \tag{5.2.11}$$

Also, by the definition of h_n,

$$\mathbb{E}_{\theta_0}(h_n) = \int h_n f(x_n;\theta_0)\mathrm{d}x_n = \int f(x;\theta_1)\mathrm{d}x = 1, \tag{5.2.12}$$

so that from (5.2.11) and (5.2.12), we obtain

$$\mathbb{E}_{\theta_0}[\lambda_n|\lambda_j,\ j \leq n-1] = \lambda_{n-1} \quad \text{a.e.}, \quad n \geq 1, \tag{5.2.13}$$

and, hence, under H_0, $\{\lambda_n\}$ is a martingale. In this context, we may note that $\lambda_n - \lambda_{n-1} = \lambda_{n-1}(h_n - 1)$ is not necessarily independent of λ_{n-1}, so that the likelihood ratio statistic may not have independent summands, but it has the martingale structure under H_0. It may be noted further that if we consider the log-likelihood ratio statistic

$$W_n = \log \lambda_n = \sum_{i=1}^{n} \log h_i = \sum_{i=1}^{n} \log\left[\frac{f(X_i;\theta_1)}{f(X_i;\theta_0)}\right], \tag{5.2.14}$$

then the summands are i.i.d. random variables (regardless of whether H_0 holds or not), and, hence, $\{W_n - \mathbb{E}_\theta(W_n),\ n \geq 1\}$ is a martingale, whenever $\mathbb{E}_\theta\{\log[f(X_1;\theta_1)/f(X_1;\theta_0)]\}$ exists. Compare this with (5.2.10), where the martingale property holds for $\theta = \theta_0$. Next, to illustrate the martingale array structure, consider the order statistics $X_{n:1}, \ldots, X_{n:n}$. Note that for any $n \geq 1$, the joint density of $X_{n:1}, \ldots, X_{n:k}$, for $k \leq n$, is

$$L_{n,k}(\theta) = n(n-1)\cdots(n-k+1)f(X_{n:1};\theta)\cdots f(X_{n:k};\theta)[1 - F(X_{n:k};\theta)]^{n-k}, \quad (5.2.15)$$

where $F(x, \theta)$ is the distribution function corresponding to the density $f(x, \theta)$. Note that by (5.2.15), we have

$$L_{n,k+1}(\theta) = L_{n,k}(\theta)\left\{(n-k)f(X_{n:k+1};\theta)\frac{[1 - F(X_{n:k+1};\theta)]^{n-k-1}}{[1 - F(X_{n:k};\theta)]^{n-k}}\right\}, \quad (5.2.16)$$

where defining the likelihood ratio statistics (based on $X_{n:1}, \ldots, X_{n:k}$) as in (5.2.10) and denoting it by $\lambda_{n,k}$, $k = 1, \ldots, n$ (and the corresponding h_n by $h_{n,k}$), we obtain on parallel lines, that for $k = 1, \ldots, n-1$,

$$\begin{aligned}
\mathbb{E}_\theta[h_{n,k+1}|L_{n,j},\ j \leq k] &= \mathbb{E}_\theta[h_{n,k+1}|X_{n:k}] \\
&= \mathbb{E}_\theta\left[\frac{\lambda_{n,k+1}(\theta_1)}{\lambda_{n,k+1}(\theta_0)}|X_{n:k}\right] \\
&= \int \lambda_{n,k+1}(\theta_1)\lambda_{n,k+1}(\theta)[\lambda_{n,k+1}(\theta_0)]^{-1}\mathrm{d}x_{n:k+1} \\
&= \int \lambda_{n,k+1}(\theta_1)\mathrm{d}x_{n:k+1} = 1, \quad \text{whenever} \quad \theta = \theta_0, \quad (5.2.17)
\end{aligned}$$

so that under $\theta = \theta_0$, for every integer $n(\geq 1)$, $\{\lambda_{n,k},\ k \leq n\}$ is a martingale array. □

A similar martingale array characterization holds for the partial sequence of log-likelihood ratio statistics based on the order statistics [defined as in (5.2.14)]. The likelihood function can be formulated in a more general manner and used for a wider class of random variables having a Markov dependence structure, not necessarily stationary. In such a case, the decomposition in (5.2.16) (in terms of the conditional density functions) or its natural extensions provide the martingale structure. In passing, we may remark that for the definition of the likelihood function we have taken $f(x, \theta)$ as the probability density function of the random variable X_1 (under the assumption that the distribution function F admits a density). In case F is itself a discrete distribution (e.g., Poisson, binomial, hypergeometric), we may as well work with the corresponding probability function and define the likelihood function accordingly. This will result in a summation (instead of an integral) in (5.2.13) or (5.2.17). We will elaborate on this point in a later chapter.

***Example* 5.2.3.** Consider a sequence $\{X_i;\ i \geq 1\}$ of i.i.d. random variables, such that $P(X_1 = 1) = P(X_1 = -1) = 1/2$. Let $A_0 > 0$ be a positive number, and consider the model

$$Y_n = Y_{n-1} + X_n c_{n-1}, \quad n \geq 1; \quad Y_0 = A_0, \quad (5.2.18)$$

where the bet $c_{n-1} = c_{n-1}(A_0, X_j;\ j \leq n-1)$ depends on the initial fortune A_0 as well as the outcome X_j, $j \leq n-1$, up to the $(n-1)$th stage, for $n \geq 1$. If $c_{n-1} = c$,

$n \geq 1$, then $Y_n - Y_{n-1}$ $(= cX_n)$ is independent of Y_{n-1}, and, hence, the characterization in Example 5.2.1 applies. In the general case, where the c_n are possibly random variables, we may note that as the X_i are independent,

$$\begin{aligned}\mathbb{E}(X_n c_{n-1}|X_j;\ j \leq n-1) &= \mathbb{E}[\mathbb{E}(X_n c_{n-1}|X_j;\ j \leq n-1)]\\ &= \mathbb{E}[\mathbb{E}(X_n)c_{n-1}] = 0, \quad n \geq 1. \end{aligned} \tag{5.2.19}$$

Hence, we have $\mathbb{E}[Y_n|Y_j,\ j \leq n-1] = Y_{n-1}$ a.e., for every $n \geq 1$. Then, $\{Y_n\}$ is a martingale, although the increments are not generally independent. Typically, in a gambling context where the model in (5.2.18) arises, c_{n-1} may be a proportion of the fortune (Y_{n-1}) at the $(n-1)$th stage, whenever Y_{n-1} is positive, and 0 otherwise. □

A sequence $\{T_n\}$ of random variables is called a *submartingale* (or *supermartingale*) if

$$\mathbb{E}[T_n|T_j,\ j \leq n-1] \geq (\leq)\ T_{n-1} \quad \text{a.e.,} \quad n \geq 1. \tag{5.2.20}$$

Before we present some examples of submartingales or supermartingales, we may note that if $\{Y_n\}$ is a martingale and g is a convex function, then, for $T_n = g(Y_n)$, we may use a version of the Jensen Inequality (1.5.43) adapted to the conditional expectations and obtain that

$$\begin{aligned}\mathbb{E}[T_n|T_j,\ j \leq n-1] &= \mathbb{E}[g(Y_n)|g(Y_j),\ j \leq n-1]\\ &= \mathbb{E}[g(Y_n)|Y_1, \ldots, Y_{n-1}]\\ &\geq g(\mathbb{E}[Y_n|Y_j, j \leq n-1])\\ &= g(Y_{n-1}) \quad \text{a.e., } n \geq 1. \end{aligned} \tag{5.2.21}$$

Thus, for a martingale $\{Y_n\}$, and for any convex function g, $\{T_n = g(Y_n)\}$ is a submartingale. Similarly, if g is a concave function, we will have a supermartingale. In particular, if X_i, $i \geq 1$, are independent random variables with zero means and finite pth order (absolute) moments, for some $p \geq 1$, then, on letting $Y_n = X_1 + \cdots + X_n$, $n \geq 1$, it follows that $\{Y_n\}$ is a martingale. Let $T_n = |Y_n|^p$, $n \geq 1$. Then, $\{T_n\}$ is a submartingale. Specifically, in the case $p = 2$, for zero-mean independent random variables X_i and setting $Y_0 = 0$, we get

$$\mathbb{E}(Y_n^2|Y_j,\ j \leq n-1) \geq Y_{n-1}^2 \quad \text{a.e.,} \quad n \geq 1. \tag{5.2.22}$$

***Example* 5.2.4.** Let $X_i, i \geq 1$ be i.i.d. random variables, and, for every $n \geq 1$, let $X_{n:n}$ $(X_{n:1})$ denote the maximum (minimum) of the X_j, $j \leq n$. Then, note that $X_{n:n}$ cannot be smaller than $X_{n-1:n-1}$, so that letting $T_n = X_{n:n}$, $n \geq 1$, we have $\mathbb{E}[T_n|T_1, \ldots, T_{n-1}] \geq T_{n-1}$ a.e., for every $n > 1$. Similarly, $\mathbb{E}[X_{n:1}|X_{j:1},\ j \leq n-1] \leq X_{n-1:1}$, for every $n > 1$. It may be noted that in the above discussion we have tacitly assumed that $\mathbb{E}(X_{n:n})$ exists (which may not always be true); if $\mathbb{E}(X_{n:n}) = \infty$, the inequality may still be interpreted in a meaningful way. A similar modification holds for $X_{n:1}$, where we take $\mathbb{E}(X_{n:1}) = -\infty$ whenever it does not exist. □

***Example* 5.2.5.** In Example 5.2.2, for the sequence $\{\lambda_n\}$ of likelihood ratio statistics in (5.2.10), define $T_{n,p} = \lambda_n^p$, for some $p > 0$. Note that by the Jensen Inequality (1.5.43),

$$\mathbb{E}_{\theta_0}(h_n^p) \geq [\mathbb{E}_{\theta_0}(h_n)]^p, \quad p \geq 1, \tag{5.2.23}$$

and that the opposite inequality holds for $p \in (0, 1)$. Therefore, proceeding as in (5.2.10)–(5.2.13), but using (5.2.23) instead of (5.2.12), we obtain that

(i) for every (fixed) $p \geq 1$, $\{T_{n,p}\}$ is a submartingale,
(ii) for $p \in (0, 1)$, $\{T_{n,p}\}$ is a supermartingale. □

The definition of submartingales or supermartingales may also be extended to triangular schemes. However, we do not repeat the details. A useful decomposition is given in the following theorem.

***Theorem* 5.2.1.** *A submartingale $\{T_n\}$ can be decomposed into a martingale $\{T_{n0}\}$ and a residual $\{T_{n1}\}$, where T_{n1} is a.s. nonnegative and nondecreasing in n. (Note that T_{n1} depends only on the T_j, $j \leq n-1$.)*

Proof. Note that

$$T_n = \sum_{j=0}^{n-1}\{[T_{j+1} - \mathbb{E}(T_{j+1}|T_r, r \leq j)] + [\mathbb{E}(T_{j+1}|T_r, r \leq j) - T_j]\}$$

$$= \sum_{j \leq n-1} Y_{j+1} + \sum_{j<n} U_j = T_{n0} + T_{n1}, \quad n \geq 1, \tag{5.2.24}$$

where $T_{n0} = \sum_{j \leq n-1} Y_{j+1}$ and $T_{n1} = \sum_{j<n} U_j$. By definition, $\{T_{n0}\}$ is a martingale, whereas, by (5.2.20), the T_{n1} are nonnegative and nondecreasing in n. ■

A sequence $\{T_n\}$ of random variables (or vectors or more general elements) is called a *reverse martingale* (or *reverse submartingale*) if for every n

$$\mathbb{E}(T_n|T_{n+1}, T_{n+2}, \ldots) = (\geq)\, T_{n+1} \quad \text{a.e.} \tag{5.2.25}$$

Here also, by using the Jensen Inequality (1.5.43), we may conclude that if $\{T_n\}$ is a reverse martingale, and if g is a convex function, then $\{g(T_n)\}$ is a reverse submartingale.

A more precise definition of the stochastic processes discussed above may be given in terms of appropriate subsigma fields. However, for simplicity of presentation, we will sacrifice some of these refinements. Next, we consider the following example.

***Example* 5.2.6.** Let $\{X_i,\ i \geq 1\}$ be a sequence of i.i.d. random variables with finite mean μ and let $\boldsymbol{X}_{(n+1)} = (X_1, \ldots, X_{n+1})$ denote the *collection* of the first (unordered) $n+1$ observations (not necessarily distinct). Also, let $\mathcal{C}_{n+1} = \{\boldsymbol{X}_{(n+1)}, \boldsymbol{X}_{(n+j)},\ j \geq 2\}$ be the *tail sigmafield*, containing all information generated by the collections $\boldsymbol{X}_{(n+1)}, \boldsymbol{X}_{(n+j)},\ j \geq 2$. Because of the nested nature of $\mathcal{C}_{n+1}$, $\mathcal{C}_{n+2}$, and soforth, the tail sigma field $\mathcal{C}_{n+1}$ may also be expressed as $\mathcal{C}_{n+1} = \{\boldsymbol{X}_{(n+1)}, X_{n+j},\ j \geq 2\}$. If we assume that $\mathcal{C}_{n+1}$ is given, the order of $X_{n+j},\ j \geq 2$ is fixed, but the order of the elements of $(X_1, \ldots, X_{n+1})$ is not, so that all $(n+1)!$ possible permutations $(X_{n+1:i_1}, \ldots, X_{n+1:i_{n+1}})$ of the order statistics $(X_{n+1:1}, \ldots, X_{n+1:n+1})$ are equally likely. Therefore, for all $j, k = 1, \ldots, n+1$, we have

$$P(X_k = X_{n+1:j}|\mathcal{C}_{n+1}) = P(X_1 = X_{n+1:j}|\mathcal{C}_{n+1}) = (n+1)^{-1}, \tag{5.2.26}$$

$k = 1, \ldots, n+1$, and since given $\mathcal{C}_{n+1}$, the random element $(X_1, \ldots, X_n)$ may assume values corresponding to all $n+1$ subsets of cardinality n of the set of order statistics

$(X_{n+1:1}, \dots, X_{n+1:n+1})$, it follows that for every $n \geq 1$,

$$\begin{aligned}\mathbb{E}(\overline{X}_n|\mathcal{C}_{n+1}) &= n^{-1}\sum_{k=1}^{n}\mathbb{E}(X_k|\mathcal{C}_{n+1}) = \mathbb{E}(X_1|\mathcal{C}_{n+1}) \\ &= (n+1)^{-1}\sum_{j=1}^{n+1} X_{n+1:j} = (n+1)^{-1}\sum_{i=1}^{n+1} X_i \\ &= \overline{X}_{n+1} \text{ a.e.} \end{aligned} \tag{5.2.27}$$

Thus, $\{\overline{X}_n, n \geq 1\}$ is a reverse martingale. Note that the same treatment holds for $\overline{X}_n - a$, $n \geq 1$, for an arbitrary a, and, hence,

$$\{\overline{X} - \mu,\ n \geq 1\} \quad \text{is a reverse martingale.}$$

This result, in conjunction with the Jensen Inequality (1.5.43), implies that for every (fixed) $p \geq 1$,

$$\{|\overline{X}_n - \mu|^p,\ n \geq 1\} \quad \text{is a reverse submartingale.}$$

If the original random variables are vector-valued, then replacing the order statistics by the collection matrix, a similar argument holds. Note that here $\boldsymbol{X}_{(n+j)}$ is a $[p \times (n+1)]$ matrix. □

Example* 5.2.7 *(U-statistic). In Chapter 2 we considered the U-statistic (2.4.28) given by

$$U_n = \binom{n}{m}^{-1} \sum_{\{1 \leq i_1 < \cdots < i_m \leq n\}} g(X_{i_1}, \dots, X_{i_m}).$$

When g defines the functional $\theta(F) = \mu$ [see (2.4.27)], it is easy to verify that $U_n = \overline{X}_n$, $n \geq 1$. Also, for the case of $\theta(F) = \sigma^2$, on using $g(a, b) = (a - b)^2/2$, we obtain that $U_n = s_n^2$.

As can be easily verified for s_n^2 when $m \geq 2$, the summands in U_n are not all independent random variables. However, we may note that $(X_{i_1}, \dots, X_{i_m})$ may assume values corresponding to all $(n+1)!/[(n+1-m)!m!]$ subsets of cardinality m of the set of order statistics $X_{n+1:1}, \dots, X_{n+1:n+1}$. Therefore, as in Example 5.2.6, we may define a similar tail sigma field $\mathcal{C}_{n+1}$, and observe that conditionally on such sigma field and using (2.4.30) we have, for every $n \geq m$,

$$\begin{aligned}\mathbb{E}(U_n|\mathcal{C}_{n+1}) &= \mathbb{E}[g(X_1, \dots, X_m)|\mathcal{C}_{n+1}] \\ &= \binom{n+1}{m}^{-1} \sum_{\{1 \leq i_1 < \cdots < i_m \leq n+1\}} g(X_{n+1:i_1}, \dots, X_{n+1:i_m}) \\ &= \binom{n+1}{m}^{-1} \sum_{\{1 \leq i_1 < \cdots < i_m \leq n+1\}} g(X_{i_1}, \dots, X_{i_m}) \\ &= U_{n+1} \quad \text{a.e.}\end{aligned} \tag{5.2.28}$$

Hence, $\{U_n,\ n \geq m\}$ is a reverse martingale. Along the same lines of the results for the sample mean, we have, for general U-statistics and a constant C, that

$$\{U_n - C,\ n \geq m\} \quad \text{is a reverse martingale,}$$

and

$$\left\{|U_n - C|^p,\ n \geq m\right\} \quad \text{is a reverse submartingale for every } p \geq 1.$$

In particular, we prefer to take $C = \theta(F)$ so that U_n is centered. We may remark that by virtue of the reverse martingale property, for any $N > n$, $\mathbb{E}(U_n|\mathcal{C}_N) = U_N$ a.e., and, hence,

$$\begin{aligned}\mathbb{E}(U_n - U_N)^2 &= \mathbb{E}[(U_n - \theta(F)]^2 + \mathbb{E}[U_N - \theta(F)]^2 - 2E\{[U_n - \theta(F)][U_N - \theta(F)]\}\\ &= E[(U_n - \theta(F)]^2 - E[U_N - \theta(F)]^2.\end{aligned}$$

This result is useful in finite population sampling, as is treated in Example 5.2.9.

In passing, we may remark that the arguments used in (5.2.26) in favor of the order statistics easily extends to the vector case, where the order statistics are to be replaced by the collection of vectors, leading to the same permutation distribution of the indices $(s_{n+1,1}, \ldots, s_{n+1,n+1})$ over the set $\{1, \ldots, n+1\}$. Hence, the reverse martingale (and submartingale) results for U-statistics remain valid even in the case where the $\boldsymbol{X}_i$ are random vectors and/or the kernel g is a vector-valued function. □

Example 5.2.8 (Empirical distribution function). Let $X_1, \ldots, X_n$ be i.i.d. random variables with distribution function F, defined on $\mathbb{R}$ and F_n be the corresponding empirical distribution function defined in (1.5.82). Note that for all $x, y \in \mathbb{R}$,

$$\mathbb{E}[F_n(x)] = F(x) \quad \text{and} \quad \mathbb{E}[I(X_1 \leq x)I(X_1 \leq y)] = F[\min(x, y)]. \tag{5.2.29}$$

Letting $g_x(X_i) = I(X_i \leq x)$, for $x \in \mathbb{R}$, we obtain, as in (5.2.28), that for every (fixed) $x \in \mathbb{R}$,

$$\mathbb{E}[F_n(x)|\mathcal{C}_{n+1}] = F_{n+1}(x) \quad \text{a.e.},\ n \geq 1, \tag{5.2.30}$$

because of the one-to-one correspondence between the order statistics and the empirical distribution function discussed in Chapter 2; thus, the tail sequence defined in terms of $F_{n+1}, F_{n+2}, \ldots$ is equivalent to the one defined in terms of the order statistics $X_{n+1:j}$, $j = 1, \ldots, n+1$, and by the X_{n+j}, $j \geq 2$, as considered in Example 5.2.6. Consequently, the pointwise reverse martingale structure in (5.2.30) extends readily to the entire line, that is, for all $n \geq 1$, we have

$$F\{[F_n(x),\ x \in \mathbb{R}]|\mathcal{C}_{n+1}\} = \{F_{n+1}(x),\ x \in \mathbb{R}\} \quad \text{a.e.} \tag{5.2.31}$$

Thus, the empirical distribution function is a reverse martingale process. In fact, $\{F_n(x) - F(x),\ x \in \mathbb{R}\}$, $n \geq 1$, is a reverse martingale process, and $\{|F_n(x) - F(x)|,\ x \in \mathbb{R}\}$, $n \geq 1$, is a reverse submartingale process. Moreover, we have that $\sup\{f(x) : x \in \mathbb{R}\}$ and $\sup\{|f(x)| : x \in \mathbb{R}\}$ are both convex functions, and, hence, using (5.2.31) and the Jensen Inequality (1.5.43), we obtain that $\sup_{x\in\mathbb{R}}[F_n(x) - F(x)]$ and $\sup_{x\in\mathbb{R}} |F_n(x) - F(x)|$ are reverse submartingales.

Similar characterizations also hold for multivariate random variables. □

Example 5.2.9 (Finite population sampling). Let $A_N = \{a_1, \ldots, a_N\}$ stand for the vector of N observations in a finite population. A simple random sample of size n is drawn without replacement, and the sample observations are denoted by $X_1, \ldots, X_n$. Note that we may write $X_i = a_{R_i}$, $i = 1, \ldots, n$, where $R_1, \ldots, R_n$ can assume values in any subset of n integers out of the N natural numbers $1, \ldots, N$ with the common probability $(N^{[n]})^{-1}$, where $N^{[n]} = N \cdots (N - n + 1)$. Thus, we may formally write, for $1 \leq R_1 \neq \cdots \neq R_n \leq N$,

$$P(X_1 = a_{R_1}, \ldots, X_n = a_{R_n}) = [N \cdots (N - n + 1)]^{-1}. \tag{5.2.32}$$

Clearly, the X_i are not independent – they are, however, exchangeable random variables. From the sample, we may compute the sample mean $\overline{X}_n$ or the variance s_n^2, and we may want to study their stochastic convergence properties. If we examine (5.2.32) carefully, we will see that it leads to the same conditional law given in (5.2.26); therefore, in simple random sampling without replacement, U-statistics have the reverse martingale property (parallel to the case of i.i.d. random variables). Thus, the results of the previous examples remain valid in this case. If the sample units are drawn with replacement, then, of course, $X_1, \ldots, X_n$ are i.i.d. random variables with $P(X_i = a_k) = N^{-1}$, for every k $(= 1, \ldots, N)$ and $i \geq 1$, and, hence, that characterization remains valid. □

5.3 Discrete Time-Parameter Processes

First, we consider the random walk model illustrated in Example 5.1.6. For independent $X_n, n \geq 1$, with $P(X_n = 1) = p = 1 - P(X_n = 0), n \geq 1$, and $X_0 = 0$, the position at step n is defined by

$$S_n = \sum_{j \leq n} X_j, \; n = 0, 1, \ldots.$$

Thus, $\mathbb{E}(S_n) = n(2p - 1)$ and $\mathbb{V}\text{ar}(S_n) = np(1 - p)$. Moreover, $Y_n = (n + S_n)/n$ is $\text{Bin}(n, p)$, so that

$$P(S_n = r) = P\left(Y_n = \frac{n+r}{2}\right) = \begin{cases} 0, & \text{if } n + r \text{ is odd;} \\ \binom{n}{(n+r)/2} p^{\frac{n+r}{2}} (1 - p)^{\frac{n-r}{2}}, & \text{if } n + r \text{ is even.} \end{cases}$$

As a result, if n is even, S_n can only take on even positions $\{0, \pm 2, \ldots, \pm n\}$, while for odd n, the (odd) realizable values of S_n are $-n, -n + 2, \ldots, n - 2, n$. If we look into the transition at time point n, we note that $S_n = S_{n-1} + X_n$, so that

$$p_{n,ij} = P(S_n = j | S_{n-1} = i) = P(X_n = j - i) = \begin{cases} p & \text{if } j = i + 1 \\ 1 - p & \text{if } j = i - 1 \\ 0 & \text{otherwise.} \end{cases}$$

Thus,

$$P(S_n = j) = P(S_{n-1} = j - 1)p + P(S_{n-1} = j + 1)(1 - p), \quad \forall j.$$

Similarly,

$$\begin{aligned} P(S_{n+1} = j | S_{n-1} = j - 1) &= \sum_{-n \leq k \leq n} P(S_{n+1} = j, S_n = k | S_{n-1} = i) \\ &= \sum_{-n \leq k \leq n} P(S_{n+1} = j | S_n = k) P(S_n = k | S_{n-1} = i) \\ &= \sum_k p_{n+1,kj}\, p_{n,ik}, \; \forall i, j. \end{aligned}$$

Note that in this setup, $P(S_{n+1} = j|S_{n-1} = i) = 0$, unless $j = i + 2, i$, or $i - 2$, and this can be easily verified by analysing the following relationship between the transition matrices

$$\boldsymbol{P}_n^{(2)} = \boldsymbol{P}_n^{(1)} \times \boldsymbol{P}_{n+1}^{(1)}, \ \forall n \geq 0, \tag{5.3.1}$$

where $\boldsymbol{P}_n^{(1)} = ((p_{n,ij}))$. We leave the details to Exercise 5.3.1.

Recall that the probability of returning to the origin at time n is

$$P(S_n = 0) = \begin{cases} 0, & \text{if } n = 2m+1, m = 1, 2, \ldots, \\ \binom{n}{n/2}[p(1-p)]^{n/2}, & \text{if } n = 2m, m = 1, 2, \ldots. \end{cases}$$

Exercise 5.3.2 is set to verify the behavior of the above probability when $n \to \infty$ for both the cases $p = 1/2$ and $p \neq 1/2$. Exercise 5.3.3 is set to verify whether or not $\{S_n, \ n \geq 0\}$ has independent and homogeneous increments.

Next, we consider a branching process, also known as the *Galton–Watson process*, in a little more generality than in Example 5.1.5. Let $Z_0 = 1$ be a organism, which gives rise to Z_1 offspring, $Z_1 = 0, 1, 2, \ldots$ with respective probabilities $p_0, p_1, \ldots$. If $Z_1 = 0$, the process ends, otherwise, each of the Z_1 units gives rise to a number of offsprings, say, according to the same probability law $P = (p_0, p_1, \ldots)$, and the process continues along these lines. Let Z_n be the number of units at the nth generation. For $n \geq 1$, we may write

$$Z_n = \begin{cases} 0 & \text{if } Z_{n-1} = 0, \\ X_{n1} + \cdots + X_{nZ_{n-1}}, & \text{if } Z_{n-1} > 0, \end{cases}$$

where the X_{nj} are independent random variables following the probability law $P = (p_0, p_1, p_2, \ldots)$ and are also independent of the random variables Z_n. Here we note that the independence of the X_{ni}, $i \geq 1$, $n \geq 1$, and their homogeneity are the basic assumptions that may not hold for human populations. Also, it is tacitly assumed that $0 < p_0 < 1$; if $p_0 = 0$, Z_n is increasing in n with probability 1, while if $p_0 = 1$, $Z_n = 0, \forall n \geq 1$.

For $Z_1 = X_{11}$, the probability generating function (1.5.70) is

$$g_1(s) = g(s) = \sum_{k \geq 0} s^k p_k.$$

Let $g_n(s) = \mathbb{E}(s^{Z_n})$ be the generating function for Z_n, $n \geq 1$. Observe that $Z_2 = 0$ when $Z_1 = 0$; otherwise, $Z_2 = X_{21} + \cdots + X_{2Z_1}$. Hence,

$$\begin{aligned} g_2(s) = \mathbb{E}(s^{Z_2}) &= \mathbb{E}[s^{Z_2} I(Z_1 = 0)] + \mathbb{E}[s^{Z_2} I(Z_1 > 0)] \\ &= P(Z_1 = 0) + \sum_{k \geq 1} P(Z_1 = k) \mathbb{E}(s^{X_{21} + \cdots + X_{2Z_1}} | Z_1 = k) \\ &= g_1(0) + \sum_{k \geq 1} P(Z_1 = k) [\mathbb{E}(s^{X_{21}})]^k \\ &= p_o + \sum_{k \geq 1} p_k [g_1(s)]^k = g[g_1(s)] = g_1[g(s)]. \end{aligned}$$

Exercise 1.5.4 is set to show that for every $n \geq 1$,

$$g_n(s) = \mathbb{E}(s_n^{Z_n}) = g_{n-1}[g(s)] = g[g_{n-1}(s)], \quad 0 \leq s \leq 1,$$

so that the probability of $Z_n = k$ as well as the moments of Z_n can be conveniently expressed in terms of the original probability law.

5.4 Continuous Time-Parameter Processes

Let $\{X_t,\ t \in \mathcal{T}\}$ be a stochastic process where the time parameter t varies continuously over an interval $\mathcal{T}$ although X_t can itself be either a discrete or a continuous random variable. First, we consider some simple cases where a Markovian structure prevails, and where for a continuous $t \in \mathcal{T}$, the classical Kolmogorov–Chapman stochastic differential equations work out well. For example, we take Poisson processes where X_t assumes only nonnegative integer values and is increasing in $t \in \mathcal{T}$. To illustrate this, we consider the number of traffic accidents occurring in a given location during the interval $(0, t]$, $t > 0$, and we assume that the probability of an accident in a small period of time is very small. We also assume that in a small interval $\Delta > 0$ the probability of 2 or more occurrences is $o(\Delta)$[1], while the probability of 1 occurrences is $\lambda\Delta + o(\Delta)$ for some $\lambda > 0$; thus the probability of no occurrence is $1 - \lambda\Delta + o(\Delta)$. Let $p_k(t) = P(X_t = k)$, for $k = 0, 1, \ldots, t \geq 0$. Then,

$$\begin{aligned}
p_k(t + \Delta_t) = P(X_{t+\Delta t} = k) &= \sum_{q \geq 0} P(X_t = q, X_{t+\Delta_t} = k) \\
&= \sum_{q \geq 0} p_q(t) P(X_{t+\Delta_t} = k | X_t = q) \\
&= \sum_{q \geq 0} p_q(t) P(X_{t+\Delta_t} - X_t = k - q | X_t = q) \\
&= \sum_{0 \leq q \leq k} p_q(t) P(X_{t+\Delta_t} - X_t = k - q | X_t = q). \qquad (5.4.1)
\end{aligned}$$

Next we assume that $X_{t+\Delta_t} - X_t$ takes on nonnegative integer values with probabilities identical to those of X_{Δ_t}, that is, that the independence and homogeneity assumptions, fundamental to this model hold. Thus,

$$P(X_{t+\Delta_t} - X_t = k - q | X_t = q) = \begin{cases} 1 - \lambda\Delta_t + o(\Delta_t), & \text{if } k = q, \\ \lambda\Delta_t + o(\Delta_t), & \text{if } k = q + 1, \\ o(\Delta_t), & \text{if } k \geq q + 2. \end{cases} \qquad (5.4.2)$$

As a result,

$$p_k(t + \Delta_t) = p_k(t)[1 - \lambda\Delta_t + o(\Delta_t) + p_{k-1}(t)\lambda\Delta_t + o(\Delta_t)]$$

for all $k \geq 0$, where $P_{-m}(t) = 0$, $\forall m \geq 1$, $t \geq 0$, and $p_0(t) = 1$ or 0 according as $t = 0$ or $t > 0$. Thus,

$$\frac{1}{\Delta_t}[p_0(t + \Delta_t) - p_o(t)] = -\lambda p_0(t) + o(1), \qquad (5.4.3)$$

and

$$\frac{1}{\Delta_t}[p_k(t + \Delta_t) - p_k(t)] = -\lambda p_k(t) + \lambda p_{k-1}(t), \quad k \geq 1.$$

Therefore, $(\mathrm{d}/\mathrm{d}t)p_0(t) = -\lambda p_0(t)$, $\forall t \geq 0$, so that $p_0(t) = \exp(-\lambda t)$, for all $t \geq 0$. Furthermore,

$$\frac{\mathrm{d}}{\mathrm{d}t} p_1(t) = -\lambda p_1(t) + \lambda p_0(t) = -\lambda p_1(t) + \lambda e^{-\lambda t},$$

[1] For details on the $O(\cdot)$ and $o(\cdot)$ notations, see (6.2.7) and (6.2.8).

which can be rewritten as

$$e^{\lambda t}\frac{\mathrm{d}}{\mathrm{d}t}p_1(t)+\lambda e^{\lambda t}p_1(t)=\lambda$$

or

$$\frac{\mathrm{d}}{\mathrm{d}t}[e^{\lambda t}p_1(t)]=\lambda \Rightarrow p_1(t)=(\lambda t)e^{-\lambda t}.$$

It then follows that (Exercise 5.4.1)

$$p_k(t)=(\lambda t)^k e^{-\lambda t}/k!, \quad \text{for } k=0,1,2,\ldots, \tag{5.4.4}$$

which is the usual Poisson distribution with parameter λt. Exercise 5.4.2 is set to show that for every $\tau \geq 0$, $X_{\tau+t}-X_\tau$ has the same distribution as X_t, and furthermore, that $X_{\tau+1}-X_\tau$ is independent of X_τ. Thus, the process $\{X_t,\ t\geq 0\}$ has independent and homogeneous increments.

Consider next the case where in (5.4.2), λ is possibly time dependent, that is, $\lambda=\lambda(t)$. Let $\Lambda(t)=\int_0^t \lambda(u)\mathrm{d}u$ be the *cumulative intensity function*. Then, analogously to (5.4.3), we have

$$\frac{\mathrm{d}}{\mathrm{d}t}p_k(t)=-\lambda(t)p_k(t)+\lambda(t)p_{k-1}(t), \quad k\geq 0,\ t>0, \tag{5.4.5}$$

which in turn leads to (see Exercise 5.4.3)

$$p_k(t)=\frac{[\Lambda(t)]^k}{k!}e^{-\Lambda(t)}, \quad k\geq 0,\ t\geq 0. \tag{5.4.6}$$

Note that in this general case, $\Lambda(\tau+t)-\Lambda(\tau)$ may not be equal to λt for all $\tau\geq 0, t\geq 0$, so that we will have a *nonhomogeneous Poisson process*, still preserving the independence of the increment.

Let us next consider some variations of Poisson processes. Let X_t denote the number of units subject to a failure process at time t, that is, one in which the probability that an individual alive at time t will fail during $(t, t+\Delta_t)$ is $\mu(t)\Delta_t+o(\Delta_t)$, $t\geq 0$. Also, let $X_0=n$ and $H(t)=\int_0^t \mu(u)\mathrm{d}u$, $t\geq 0$. Then, similarly to the case of the Poisson process (Exercise 5.4.4), we have

$$P(X_t=k|X_o=n)=\binom{n}{k}e^{-kH(t)}[1-e^{-H(t)}]^{n-k}, \tag{5.4.7}$$

for $k=0,1,\ldots,n$. The set $\{X_t,\ t\geq 0\}$, with the nonnegative integers between 0 and n as sample paths, is called a *pure death process*. Here, X_t is decreasing in t (≥ 0).

Consider next a *pure birth process*. Here, if $X_t=k$ at time t, the probability that a new event (i.e., a birth) will occur in the interval $(t, t+\Delta_t)$ is $\lambda_k\Delta_t+o(\Delta_t)$; also, the probability that two or more new events will occur in this interval is $o(\Delta_t)$ and that none will occur is $1 - \lambda_k \Delta_t + o(\Delta_t)$. This will lead to

$$\begin{aligned} p_k(t + \Delta_t) &= P(X_{t+\Delta_t} = k) \\ &= \sum_{q \geq 0} P(X_t = q) P(X_{t+\Delta_t} = k | X_t = q) \\ &= \sum_{0 \leq q \leq k} p_q(t) P(X_{t+\Delta_t} - X_t = k - q | X_t = q) \\ &= p_k(t)\{1 - \lambda_k \Delta_t\} + p_{k-1}(t)\lambda_{k-1}\Delta_t + o(\Delta_t), \quad \forall k,\ t \geq 0, \end{aligned}$$

so that

$$\frac{\mathrm{d}}{\mathrm{d}t} p_k(t) = -\lambda_k p_k(t) + \lambda_{k-1} p_{k-1}(t), \quad k \geq 0,\ t \geq 0, \tag{5.4.8}$$

where starting with $X_o = m$, we have $p_k(0) = 1$ or 0 according as $k = m$ or $k \neq m$. A general solution of $p_k(t)$ in (5.4.8) depends on the specific λ_k, $k \geq 0$. In the particular case of $\lambda_k = k\lambda$, $\lambda > 0$, (5.4.8) simplifies to

$$\frac{\mathrm{d}}{\mathrm{d}t} p_k(t) = -k\lambda p_k(t) + (k-1)\lambda p_{k-1}(t), \quad k \geq 0,\ t \geq 0, \tag{5.4.9}$$

where $p_k(0) = 1$ or 0 according as $k = m$ or not. Thus,

$$p_k(t) = \binom{k-1}{k-m} e^{-m\lambda t}(1 - e^{-\lambda t})^{k-m}, \quad \forall k \geq m,\ t \geq 0. \tag{5.4.10}$$

This is known as the *Yule process*; it relates to the classical negative binomial probability law with $p = \exp(-\lambda t)$.

There are many generalizations of these pure birth and death processes, and for the sake of brevity, we will not enter into their formulation. An excellent introduction may be found in Chiang (1980).

Let us now consider some *renewal processes* with continuous time parameter. As in the previous section, we let $T_r = t_1 + \cdots + t_r$, $r \geq 1$, $T_0 = 0$ where the t_j are i.i.d. nonnegative random variables with a continuous distribution function $F(t)$. Then noting that t_j are nonnegative, and denoting the distribution function of T_r by $G_r(t)$, $t \geq 0$, we obtain $G_1(t) = F(t)$, $t \geq 0$, $G_2(t) = \int_0^t G_1(t-u)\mathrm{d}G_1(u)$, $t \geq 0$, and, in general, that (see Exercise 5.4.5)

$$G_r(t) = \int_0^t G_{r-1}(t-u)\mathrm{d}G_1(u), \quad t \geq 0,\ r \geq 1. \tag{5.4.11}$$

As a notable example, consider the case of $F(t) = 1 - \exp(-\lambda t)$, $t \geq 0$ (the exponential distribution function). Then,

$$\begin{aligned} G_2(t) &= \int_0^t [1 - e^{-\lambda(t-u)}]\lambda e^{-\lambda u} \mathrm{d}u = \lambda \int_0^t (e^{-\lambda u} - e^{-\lambda t})\mathrm{d}u \\ &= 1 - e^{-\lambda t} - e^{-\lambda t}\lambda t, \quad t \geq 0. \end{aligned}$$

In general, (see Exercise 5.4.6), we would have

$$G_r(t) = 1 - e^{-\lambda t}\left[1 + \lambda t + \cdots + \frac{(\lambda t)^{r-1}}{(r-1)!}\right]$$
$$= \sum_{k\geq r} e^{-\lambda t}\frac{(\lambda t)^k}{k!}, \quad \text{for } r = 1, 2, \ldots. \tag{5.4.12}$$

Associated with the T_r, $r \geq 0$, we have a *counting process* $\{N_t,\ t \geq 0\}$ defined by

$$N_t = r \quad \text{if} \quad T_r \leq t < T_{r+1}, \quad r = 0, 1, 2, \ldots. \tag{5.4.13}$$

Thus, $\{N_t \geq r\} \Leftrightarrow \{T_r \leq t\}$, and

$$P(N_t = r) = P(N_t \geq r) - P(N_t \geq r+1)$$
$$= P(T_r \leq t) - P(T_{r+1} \leq t) \tag{5.4.14}$$
$$= G_r(t) - G_{r+1}(t), \quad t \geq 0,\ r \geq 0. \tag{5.4.15}$$

As a result,

$$\sum_{r\geq 0} P(N_t = r) = \sum_{r\geq 0}[G_r(t) - G_{r+1}(t)] = G_o(t) = 1, \quad \forall t \geq 0,$$

so that N_t is a proper random variable for all $t \geq 0$. Furthermore,

$$U(t) = E(N_t) = \sum_{r\geq 0} r[G_r(t) - G_{r+1}(t)] = \sum_{r\geq 1} G_r(t). \tag{5.4.16}$$

The function $U(t)$, $t \geq 0$ is known as the *renewal function*. If $U(t)$ is absolutely continuous, $u(t) = (\mathrm{d}/\mathrm{d}t)U(t)$ is called the *renewal density* although it is not a proper density function. For the exponential case in (5.4.2) we may obtain explicit expressions for $U(t)$ and $u(t)$. This is the objective of Exercise 5.4.7.

Counting processes may emerge in many other contexts, even without a strict renewal setup. The Poisson processes, discussed earlier, may also be looked upon as counting processes, by noting that $X_t = N_t$, $\forall t \geq 0$. Suppose that the interevent times are i.i.d. random variables with distribution function $F(x) = 1 - \exp[-\Lambda(x)]$, $x \geq 0$, where $\Lambda(x)$ is defined in (5.4.6). Then, using (5.4.12) and (5.4.14) and extending the results to the nonhomogeneous case (see Exercise 5.4.8), we obtain that

$$G_r(t) = e^{-\Lambda(t)}\sum_{k\geq r}\frac{1}{k!}[\Lambda(t)]^k,$$

so that

$$P(X_t = r) = G_r(t) - G_{r+1}(t) = e^{-\Lambda(t)}\frac{[\Lambda(t)]^r}{r!}, \quad r \geq 0. \tag{5.4.17}$$

Note that for the interevent time t^o, with distribution function $F(t) = 1 - \exp[-\Lambda(t)]$, $t \geq 0$, the first two moments are

$$\mathbb{E}(t^o) = \int_0^\infty e^{-\Lambda(t)}\mathrm{d}t \quad \text{and} \quad \mathbb{E}(t^{o2}) = 2\int_0^\infty te^{-\Lambda(t)}\mathrm{d}t. \tag{5.4.18}$$

For the simple case, $\Lambda(t) = t\lambda$, we have from (5.4.18), $\mathbb{E}(t^o) = 1/\lambda$, $\mathbb{E}(t^{o2}) = 2/\lambda^2$, so that $\mathbb{V}\text{ar}(t^o) = 1/\lambda^2$. Thus, in this special case, $\mathbb{E}(X_t) = \lambda t = t/\mathbb{E}(t^o)$ and $\mathbb{V}\text{ar}(X_t) = \lambda t = t\mathbb{V}\text{ar}(t^o)/[\mathbb{E}(t^o)]^3$, $\forall t > 0$. Although this simple relation may not hold for general renewal processes, it can be shown that

$$\lim_{t\to\infty} t^{-1}\mathbb{E}(N_t) = 1/\mathbb{E}(t^o) \quad \text{and} \quad \lim_{t\to\infty} t^{-1}\mathbb{V}\text{ar}(N_t) = \frac{\mathbb{V}\text{ar}(t^o)}{[\mathbb{E}(t^o)]^3}.$$

Counting processes allowing for formal derivations of small- and large-sample properties of estimators and test statistics occur naturally in Survival Analysis. Initially envisioned by Aalen (1978), where nonparametric inference based on counting process is described in a more structured way, this methodology is well-suited for situations such as the one described in Example 1.3.5. To clarify this, we start by considering that n items are followed over time, up to the occurrence of one (or more) events of interest. Let $T_1, \ldots, T_n$ be failure times associated to each of the items, as outlined in Example 1.3.5. In most real-life applications, some of the T_i may not be completely observed due to the occurrence of another event (not of primary interest), taking place at time C_i, designated as *censoring time*. The C_i are possibly random and are usually assumed to be independent of T_i. Note that the main interest remains on the characterization of the failure times T_i, usually done via the hazard function presented in Example 1.3.5. This may be formally defined as

$$h_i(t) = \lim_{\Delta\to 0} \Delta^{-1} P\{T_i \in [t, t+\Delta) | T_i \geq t\} \quad i = 1, \ldots, n. \tag{5.4.19}$$

Alternatively, one may consider the *cumulative hazard function*

$$H_i(t) = \int_0^t h_i(u)\mathrm{d}u, \quad i = 1, \ldots, n. \tag{5.4.20}$$

The data actually observed may be represented by the random variables

$$Z_i = \min(T_i, C_i) \quad \text{and} \quad \delta_i = I(T_i \leq C_i),$$

$i = 1, \ldots, n$. Aalen (1978) considers the empirical estimation of $H(t)$ using the theory of time-continuous martingales and stochastic integrals and such approach leads to a unified way to deal with several methods in Survival Analysis. Based on $\{(Z_i, \delta_i),\ 1 \leq i \leq n\}$, right-continuous multivariate counting processes $\boldsymbol{N}(t) = \{N_1(t), \ldots, N_n(t)\}$, $t \geq 0$ are defined, such that

$$N_i(t) = I(Z_i \leq t, \delta_i = 1), \quad i = 1, \ldots, n.$$

The path of each one of such processes is zero before the event is observed for each item, and jumps to one as soon as the event takes place. Note that if the ith observation is censored, no jump is observed for N_i. We also need the information regarding whether an item is still *at risk* (i.e., whether no censoring occurred and the event has not been observed for that item up to time t), so that a left-continuous process may be defined as

$$Y_i(t) = I(Z_i \geq t), \quad t \geq 0, \tag{5.4.21}$$

for $i = 1, \ldots, n$. All the information contained in $\{(T_i, \delta_i),\ i = 1, \ldots, n\}$ is also contained in $\{[N_i(t), Y_i(t)],\ i = 1, \ldots, n,\ t \geq 0\}$. In order to obtain the intensity process for the counting process N_i, we follow the heuristic approach presented by Andersen, Borgan, Gill, and Keiding (1996) or Fleming and Harrington (2005). Consider a small time interval $[t, t + \mathrm{d}t)$; given the information on $\boldsymbol{N}$ available up to time t^- (i.e., just prior to t), the probability that N_i will jump at $[t, t+\mathrm{d}t)$ [i.e., $\mathrm{d}N_i(t) = N_i[(t+\mathrm{d}t)^-] - N_i(t^-) = 1$] will be zero if either the event has already occurred or the item has been censored prior to t (in which case $Y_i(t) = 0$); otherwise, such probability will be approximately (for $\mathrm{d}t \to 0$) equal to $h(t)\mathrm{d}t$ if the unit is still at risk [in which case $Y_i(t) = 1$.] Therefore, we may combine both situations to write

$$P[\mathrm{d}N_i(t) = 1|\mathcal{H}_t] = h_i(t)Y_i(t)\mathrm{d}t,$$

where $\mathcal{H}_t$ represents the history of $\{N(s),\ s < t\}$. Since for fixed t the "jump" $\mathrm{d}N_i(t)$ is a 0–1 random variable, we may write

$$\mathbb{E}[\mathrm{d}N_i(t)|\mathcal{H}_t] = h_i(t)Y_i(t)\mathrm{d}t = \lambda_i(t)\mathrm{d}t.$$

Therefore, λ_i is the intensity process associated to N_i. We may then consider the *integrated intensity process*

$$\Lambda_i(t) = \int_0^t \lambda_i(u)\mathrm{d}u$$

and define the process

$$M_i(t) = N_i(t) - \Lambda_i(t). \tag{5.4.22}$$

If we know at instant t^- whether item i is or not at risk at instant t [and, hence, that the value of $Y_i(t)$ is known at instant t^-], then it is possible to show that

$$\mathbb{E}[\mathrm{d}M_i(t)|\mathcal{H}_t] = 0, \tag{5.4.23}$$

and, hence, that M_i is a time-continuous (local) martingale (see Exercise 5.4.9). The decomposition in (5.4.22) is often used in Survival Analysis to represent estimators and test statistics as martingales.

Since $M_i(0) = 0$, the result in (5.4.23) implies that $\{M_i(t),\ t \geq 0\}$ is a zero-mean martingale. From (5.4.22) we may also write $N_i(t) = \Lambda_i(t) + M_i(t)$, which may be intuitively interpreted as a decomposition of the counting process N_i (an observable quantity) into a systematic part [the process Λ_i] and a pure noise process [the martingale M].

As an illustration, we consider an estimator of $H(t)$ defined in (5.4.20). Since that expression may be interpreted as the ratio between the accumulation of the number of events in perspective to the number of items still in risk at time t, we define $\overline{N}(t) = \sum_{1\leq i\leq n} N_i(t)$ as the number of observed events up to time t and $\overline{Y}(t) = \sum_{1\leq i\leq n} Y_i(t)$ as the number of items at risk at instant t^-, so that a natural estimator for $H(t)$ is

$$\int_0^t \frac{\mathrm{d}\overline{N}(u)}{\overline{Y}(u)}.$$

Since when $\overline{Y}(t) = 0$ such a quantity is undefined, an (asymptotically) equivalent estimator, proposed by Aalen (1978) is

$$\widehat{H}(t) = \int_0^t I[\overline{Y}(u) > 0]\frac{\mathrm{d}\overline{N}(u)}{\overline{Y}(u)}.$$

This estimator is known as the *Nelson–Aalen estimator* and is based on the estimator proposed by Nelson (1969).

Due to the lack of space, formal derivations of the results discussed above will not be presented. We refer to Fleming and Harrington (2005) and Andersen, Borgan, Gill, and Keiding (1996) for further details.

5.5 Exercises

5.2.1 Consider the first-order (stationary) Markov process $X_k = \rho X_{k-1} + \varepsilon_k$, $k \geq 1$, where the ε_k are i.i.d. random variables with zero mean and (finite) variance σ^2 and ε_k is independent of X_{k-1}, for every $k \geq 1$. Set $U_n = \sum_{k=1}^{n-1} X_k^2$ and $\widehat{\rho}_n = U_n/V_n$. Show that $\{V_n(\widehat{\rho}_n - \rho),\ n \geq 2\}$ is a zero-mean martingale. Hence, or otherwise, verify that $\widehat{\rho}_n \overset{a.s.}{\longrightarrow} \rho$.

5.3.1 Consider the two-step transition matrix (5.3.1) for a random walk based on n i.i.d. random variables X_j, such that $P(X_j = 1) = 1 - P(X_j = 0) = p$, $j = 1, \ldots, n$, and show that

$$\boldsymbol{P}_n^{(2)} = \boldsymbol{P}_n^{(1)} \times \boldsymbol{P}_{n+1}^{(1)}, \ \forall n \geq 0,$$

where $\boldsymbol{P}^{(1)}$ is the (one-step) transition matrix associated to the process.

5.3.2 In the context of the random walk described in Example 5.1.6, show that the probability of returning to the origin at time n is

$$P(S_n = 0) = \begin{cases} 0, & \text{if } n = 2m+1, m = 1, 2, \ldots, \\ \binom{n}{n/2}[p(1-p)]^{n/2}, & \text{if } n = 2m, m = 1, 2, \ldots; \end{cases}$$

also, using the Stirling approximation (1.5.33), show that as $n \to \infty$, $P(S_n = 0)$ behaves like $(2/\sqrt{\pi n})\rho^{n/2}$ with $\rho = 1$ if $\pi = 1/2$ and $\rho < 1$ if $p \neq 1/2$.

5.3.3 Under the setup of Exercise 5.3.2, verify whether or not $\{S_n,\ n \geq 0\}$ has independent and homogeneous increments.

5.4.1 Using (5.4.3), show by induction that for homogeneous Poisson processes,

$$p_k(t) = (\lambda t)^k e^{-\lambda t}/k! \quad \text{for } k = 0, 1, 2, \ldots.$$

5.4.2 Consider the Poisson process and show that for every $\tau \geq 0$, $X_{\tau+1} - X_\tau$ has the same distribution as X_t. Furthermore, show that $X_{\tau+1} - X_\tau$ is independent of X_τ.

5.4.3 Using (5.4.5) show by induction that (5.4.6) holds.

5.4.4 Identifying that the probability of failure of a unit before time t is $1 - \exp[-H(t)]$, show that expression (5.4.7) holds. State all of the assumptions you need in this context.

5.4.5 Derive the expression (5.4.11)

5.4.6 Derive the expression (5.4.12)

5.4.7 For a renewal process with exponential distribution function F, obtain the expressions for $U(t)$ and $u(t)$.

5.4.8 Start with $G_1(x) = F(x) = 1 - \exp[-\Lambda(x)]$, $x \geq 0$ and, using (5.4.11) with $r = 2$, show that

$$G_2(x) = e^{-\Lambda(x)} \sum_{k \geq 2} \frac{1}{k!}[\Lambda(x)]^k.$$

Hence, by induction, show that for every $r \geq 1$,

$$G_r(x) = e^{-\Lambda(x)} \sum_{k \geq r} \frac{1}{k!}[\Lambda(x)]^k.$$

5.4.9 Discuss (5.4.23).

6

Stochastic Convergence and Probability Inequalities

6.1 Introduction

Unbiasedness, efficiency, sufficiency, and ancillarity, as outlined in Chapters 2 and 3, are essentially finite-sample concepts, but consistency refers to indefinitely increasing samples sizes, and thus has an asymptotic nature. In general, finite-sample optimality properties of estimators and tests hold basically for a small class of probability laws, mostly related to the exponential family of distributions; consistency, however, holds under much less restricted setups as we will see. Moreover, even when finite-sample optimal statistical procedures exist, they may not lead to closed-form expressions and/or be subject to computational burden. These problems are not as bothersome when we adopt an asymptotic point of view and use the corresponding results to obtain good approximations of such procedures for large (although finite) samples. This is accomplished with the incorporation of *probability inequalities*, *limit theorems*, and other tools that will be developed in this and subsequent chapters.

In this context, a minimal requirement for a good statistical decision rule is its increasing reliability with increasing sample sizes (consistency). For an estimator, consistency relates to an increasing closeness to its population counterpart as the sample sizes become larger. In view of its stochastic nature, this closeness needs to incorporate its fluctuation around the parameter it estimates and thus requires an appropriate adaptation of the definitions usually considered in nonstochastic setups. Generally, a distance function or norm of this stochastic fluctuation is incorporated in the formulation of this closeness, and consistency refers to the convergence of this norm to 0 in some well-defined manner. Similarly, when dealing with tests of statistical hypotheses, consistency refers to the increasing degree of confidence in their ability to reject the null hypothesis when it is not true as the sample size increases.

Basically, the consistency of a sequence of statistics as discussed in Section 2.2 rests on some *convergence properties*, which are intimately related to *stochastic convergence* in the classical theory of probability.

We say that a sequence $\{a_n\}$ of real numbers converges to a limit a as $n \to \infty$, if for every positive ε, there exists a positive integer n_0 $[= n_0(\varepsilon)]$, such that

$$|a_n - a| < \varepsilon, \quad n \geq n_0. \tag{6.1.1}$$

Equivalently, we may state this in terms of the *Cauchy property*

$$|a_{n+m} - a_n| < \varepsilon, \quad n \geq n_0, \ m \geq 1. \tag{6.1.2}$$

In a more general case, the a_n may be q-vectors, for some $q > 1$, denoted here by $\boldsymbol{a}_n$. In this case, if $\boldsymbol{a} = (a_1, \ldots, a_q)' \in \mathbb{R}^q$ we may consider the *Euclidean norm*

$$\|\boldsymbol{a}\| = (\boldsymbol{a}'\boldsymbol{a})^{1/2} = \left(\sum_{i=1}^{q} a_i^2\right)^{1/2}, \tag{6.1.3}$$

and both (6.1.1) and (6.1.2) remain intact if we replace $|a_n - a|$ and $|a_{n+m} - a_n|$ by $\|\boldsymbol{a}_n - \boldsymbol{a}\|$ and $\|\boldsymbol{a}_{n+m} - \boldsymbol{a}_n\|$, respectively. Instead of the simple Euclidean norm, we may even consider other norms, such as

$$\textit{max-norm}: \|\boldsymbol{a}\|_\infty = \max\{|a_j| : 1 \le j \le q\}, \tag{6.1.4}$$

or

$$\textit{quadratic norm}: \|\boldsymbol{a}\|_{\boldsymbol{W}} = (\boldsymbol{a}'\boldsymbol{W}\boldsymbol{a})^{1/2}, \tag{6.1.5}$$

where $\boldsymbol{W}$ is a given p.d. matrix; the particular choice $\boldsymbol{W} = \boldsymbol{I}$ leads to the Euclidean norm. We may even proceed one step further and consider a more general situation where $a_n = \{a_n(t), t \in \mathcal{T}\}$ is a function defined on an interval $\mathcal{T}$ (*e.g.*, $\mathcal{T} = [0, 1]$ or $[0, \infty)$). In such a case, for a function $x = \{x(t), t \in \mathcal{T}\}$, an appropriate norm may be

$$\|x\|_{\sup} = \sup\{|x(t)| : t \in \mathcal{T}\} \tag{6.1.6}$$

or some ramifications of it. When $x(t)$ is a q-vector, we may as well replace $|x(t)|$ in (6.1.6) by one of the norms in (6.1.4) or (6.1.5). Again, the definition of the convergence of $\{a_n\}$ to a limit a $[= \{a(t), t \in \mathcal{T}\}]$, given in (6.1.1) or (6.1.2), stands valid when we replace the simple norm in (6.1.1) by (6.1.6). The notion of convergence, as has been outlined above, can also be formulated in terms of a general *metric space*. However, we will refrain ourselves from such a relatively more abstract treatment.

In dealing with a sequence $\{T_n\}$ of random elements (statistics), we may, however, note that no matter how large n is, the inequality $|T_n - T| < \varepsilon$ or $\|\boldsymbol{T}_n - \boldsymbol{T}\| < \varepsilon$ may not hold with probability equal to one, but may do so with probability arbitrarily close to one when n is indefinitely large. Moreover, in the case of nonstochastic elements, (6.1.1) may also be equivalently written as

$$\sup_{N \ge n} |T_N - T| < \varepsilon, \quad n \ge n_0, \tag{6.1.7}$$

where again $|T_N - T|$ may be replaced by $\|\boldsymbol{T}_N - \boldsymbol{T}\|$, or other norms too. On the other hand, for the case of stochastic elements, convergence can be defined in terms of the marginal event $\|\boldsymbol{T}_n - \boldsymbol{T}\|$ or in terms of the simultaneous events in (6.1.7), and this subtle difference in the mode of convergence leads us to formulate two allied concepts: *weak convergence*, also known as *convergence in probability*, and *strong* (*almost sure* or *almost certain*) *convergence*; these will be elaborated in the subsequent sections of this chapter. Dealing with nonstochastic elements, both these modes of convergence, (6.1.1) and (6.1.7), are equivalent. However, in the case of stochastic elements, weak convergence may not necessarily ensure strong convergence and, generally, more stringent regularity conditions may be necessary for the latter to hold. There are some other modes of convergence (notably, *complete convergence* and *convergence in the r.t.h. mean*, for $r > 0$), which will also be considered. The interrelationship of these modes of convergence will be studied too.

Typically, in a statistical estimation problem, T_n refers to an estimator (possibly vector-valued), so that it is a random element, whereas T, the corresponding parameter, is nonstochastic. But, in a more general setup, we may encounter situations where $\{T_n\}$ is a stochastic sequence converging to a (possibly) stochastic T (in some suitable sense). Hence, to preserve generality, we will consider the various modes of stochastic convergence treating both $\{T_n\}$ and T as possibly random elements. These concepts are systematically presented in Section 6.2. Mutual implications of these modes of convergence are also discussed in the same section. Probability inequalities and the classical *Laws of Large Numbers* (LLN) play a vital role in this context, and these are treated in full generality in Section 6.3. These LLN and probability inequalities are usually posed for sums or averages of independent (and, often, identically distributed) random variables (or vectors). In many statistical problems, however, we may have nonstandard situations where the statistics may not be expressible as linear functions of independent random variables (i.e., they may be nonlinear and may have nonindependent components). To cope with such situations, some extensions of the LLN and probability inequalities to possibly dependent random elements, including martingales, reverse martingales and related sequences are considered in Section 6.4. A handful of examples has also been set to illustrate their potential use in problems of statistical inference. The last section deals with some rather isolated topics having some use in statistical inference.

6.2 Modes of Stochastic Convergence

The concepts of stochastic convergence as outlined in the previous section may be formalized in the following definitions. In this context, we consider a set $\mathcal{A}$ of elements $\{a_i\}$ along with a distance function $d(a_i, a_j)$ with the properties that

(i) $d(a_i, a_i) = 0$,
(ii) $d(a_i, a_j) > 0$ if $a_i \neq a_j$,
(iii) $d(a_i, a_j) = d(a_j, a_i)$,
(iv) $d(a_i, a_j) + d(a_j, a_k) \geq d(a_i, a_k)$,

for all i, j, k. Note that the norms introduced in (6.1.1)–(6.1.5) satisfy these conditions.

Convergence in probability. A sequence $\{T_n\}$ of random elements is said to converge in probability to a (possibly degenerate) random element T, if for every (given) positive numbers ε and η, there exists a positive integer $n_0 = n_0(\varepsilon, \eta)$, such that

$$P[d(T_n, T) > \varepsilon] < \eta, \quad n \geq n_0. \tag{6.2.1}$$

This mode of convergence is usually denoted by

$$T_n - T \xrightarrow{P} 0. \tag{6.2.2}$$

In the case where T is nonstochastic, we may also write

$$T_n \xrightarrow{P} T.$$

Almost sure (or strong) convergence. A sequence $\{T_n\}$ of random elements is said to converge almost surely (a.s.) or strongly to a (possibly degenerate) random element T, if for every (given) positive numbers ε and η, there exists a positive integer $n_0 = n_0(\varepsilon, \eta)$, such that

$$P[d(T_N, T) > \varepsilon \text{ for some } N \geq n] < \eta, \quad n \geq n_0. \tag{6.2.3}$$

In symbols, we write this as

$$T_n - T \xrightarrow{a.s.} 0. \tag{6.2.4}$$

If T is nonstochastic, (6.2.4) may also be written as

$$T_n \xrightarrow{a.s.} T.$$

To clarify the difference between (6.2.1) and (6.2.4), let $A_n = [d(T_n, T) > \varepsilon]$, for $n \geq n_0$; then the stochastic convergence in (6.2.1) entails that

$$P(A_n) \to 0 \quad \text{as } n \to \infty. \tag{6.2.5}$$

On the other hand, the a.s. convergence in (6.2.3) leads us to

$$P\left(\bigcup_{N \geq n} A_N\right) \to 0 \quad \text{as } n \to \infty \tag{6.2.6}$$

and, hence, (6.2.3) is a stronger mode of convergence than (6.2.1). Since $P(\bigcup_{N \geq n} A_N) \geq P(A_n)$, it is quite clear that (6.2.4) implies (6.2.1); however, the converse may not be true. We will make additional comments later in this section.

To illustrate these modes of convergence, let us consider some simple examples. Let $X_1, X_2, \ldots$ be a sequence of i.i.d. random variables having a distribution function F, defined on $\mathbb{R}$, such that the mean μ and the variance σ^2 are finite. Based on a sample $(X_1, \ldots, X_n)$ of size n, let $\overline{X}_n$, s_n^2 and $F_n(x)$, $x \in \mathbb{R}$ denote the sample mean, the sample variance and the empirical distribution function, respectively. Now consider the following:

(i) $T_n = \overline{X}_n$, $T = \mu$ and $d(x, y) = |x - y|$;
(ii) $\boldsymbol{T}_n = (\overline{X}_n, s_n^2)'$, $\boldsymbol{T} = (\mu, \sigma^2)'$, and $d(\boldsymbol{x}, \boldsymbol{y}) = \{(\boldsymbol{x} - \boldsymbol{y})'(\boldsymbol{x} - \boldsymbol{y})\}^{1/2}$;
(iii) $T_n = \{F_n(x),\ x \in \mathbb{R}\}$, $T = \{F(x),\ x \in \mathbb{R}\}$ and $d(x, y) = \|x - y\|_{\text{sup}}$.

In the first case, $\{T_n\}$ corresponds to a sequence of real-valued random variables, in the second case to a sequence of vector-valued random variables and, in the last, to a sequence of random functions or stochastic processes. In the first two cases, we have the usual Euclidean norm for d, whereas in the last case, in view of the infinite dimensionality, the Euclidean distance is not applicable and we have adapted a more meaningful metric introduced in (6.1.6). As we will see in subsequent sections, in all three cases, $T_n - T \to 0$ in probability or almost surely, as $n \to \infty$. In passing, we may remark that unless the functional form of F is assumed to be known (and it involves T_n as a sufficient statistic), the empirical distribution function F_n contains more information than the statistic T_n. Thus, one may be interested in the empirical distribution function itself for a better feeling of the underlying distribution function, and hence, may like to deal with such a stochastic process rather than finite dimensional random vectors. This subtle point will be made clearer in Chapter 11 where the significance of nonparametric methods relatively to their parametric counterparts will be discussed.

We take this opportunity to introduce some notations that will be consistently used in this as well as in later chapters. Consider two sequences $\{a_n\}$ and $\{b_n\}$ of real numbers.

Then, we say that $a_n = O(b_n)$ if there exists a finite positive number K and a positive integer $n(K)$, such that

$$|a_n/b_n| \le K, \quad n \ge n(K). \tag{6.2.7}$$

In particular, $a_n = O(1)$ means that $|a_n| \le K$ for n sufficiently large, where K is some finite, positive constant, that is, that $\{a_n\}$ is essentially bounded.

Also, we say that $a_n = o(b_n)$ if for any $\varepsilon > 0$, there exists a positive integer $n(\varepsilon)$, such that

$$|a_n/b_n| < \varepsilon, \quad n \ge n(\varepsilon). \tag{6.2.8}$$

In particular $a_n = o(1)$ means that $a_n \to 0$ as $n \to \infty$.

Similarly, for a sequence $\{X_n\}$ of random variables, if for every $\eta > 0$ there exists a positive constant $K(\eta)$ ($< \infty$) and a positive integer $n(\eta)$, such that

$$P\left[|X_n| \le K(\eta)\right] \ge 1 - \eta, \quad n \ge n(\eta), \tag{6.2.9}$$

then we say that $X_n = O_p(1)$. We also say that if (6.2.9) holds, then $\{X_n\}$ is *bounded in probability*.

The definition in (6.2.9) readily extends to the case of random vectors $\{\boldsymbol{X}_n\}$; all we need is to replace $|X_n|$ with the Euclidean norm $\|\boldsymbol{X}_n\|$. We will write $\boldsymbol{x}_n = O_p(\mathbf{1})$ to designate the vector case.

If for a sequence $\{X_n\}$ of random variables and another sequence $\{b_n\}$ (of real numbers or possibly of random variables), for every $\eta > 0$, $\varepsilon > 0$, there exists a positive integer $n(\varepsilon, \eta)$, such that

$$P\left[|X_n/b_n| \le K(\eta)\right] \ge 1 - \eta, \quad n \ge n(\eta), \tag{6.2.10}$$

then we say that $X_n = O_p(b_n)$. The definition immediately extends to the vector case by replacing the norm in (6.2.10) by the usual Euclidean norm. Note that (6.2.9) is a special case of (6.2.10) when we allow b_n to be bounded in probability. Furthermore, with the metric in (6.1.6), all these notations can also be extended to random functions.

For a sequence $\{X_n\}$ of random variables, if for every $\eta > 0$, $\varepsilon > 0$, there exists a positive integer $n(\varepsilon, \eta)$, such that

$$P\left(|X_n| > \eta\right) < \varepsilon, \quad n \ge n(\varepsilon, \eta), \tag{6.2.11}$$

then we say that $X_n = o_p(1)$. In other words, $X_n = o_p(1)$ is equivalent to saying that $X_n \xrightarrow{P} 0$. The definition extends directly to the vector case by adapting the Euclidean norm. Here we write $\boldsymbol{X}_n = o_p(\mathbf{1})$. Also, the metric in (6.1.6) may be used to extend it to the case of random functions.

If for a sequence $\{X_n\}$ of random variables and another sequence $\{b_n\}$ (of real numbers or possibly random variables), for every $\eta > 0$, $\varepsilon > 0$, there exists a positive integer $n(\varepsilon, \eta)$, such that

$$P\left(|X_n/b_n| > \eta\right) < \varepsilon, \quad n \ge n(\varepsilon, \eta), \tag{6.2.12}$$

then we say that $X_n = o_p(b_n)$; the definition also extends to the vector case under the Euclidean norm, and to the functional case, under (6.1.6).

Now, note that

$$X_n = o_p(b_n) \text{ and } b_n = O_p(c_n) \Rightarrow X_n = o_p(c_n). \tag{6.2.13}$$

Thus, (6.2.11) can be obtained from (6.2.12) if (6.2.9) holds for the $\{b_n\}$.

If, for any sequence of real numbers $\{x_n\}$, satisfying $x_n \to L$ as $n \to \infty$, we have $a(x_n) = O[b(x_n)]$, in the sense of (6.2.10), then we say that $a(x) = O[b(x)]$ as $x \to L$. This definition can be fitted to (6.2.10) by letting $a_n = a(x_n)$ and $b_n = b(x_n)$, and allowing $x_n \to L$.

If, for any sequence of real numbers $\{x_n\}$, satisfying $x_n \to L$ as $n \to \infty$, we have $a(x_n) = o[b(x_n)]$, in the sense of (6.2.12), then we say that $a(x) = o[b(x)]$, as $x \to L$.

Furthermore, if in the setup of (6.2.9), we have

$$P\left[|X_N| > K(\eta) \text{ for some } N \geq n\right] < \eta, \quad n \geq n(\eta), \tag{6.2.14}$$

then we say that $X_n = O(1)$ a.s. Similarly, if we replace (6.2.10) by

$$P\left[|X_N/b_N| > K(\eta) \text{ for some } N \geq n\right] < \eta, \quad n \geq n(\eta), \tag{6.2.15}$$

then we say that $X_n = O(b_n)$ a.s.

If (6.2.11) is replaced by

$$P\left[|X_N| > \eta \text{ for some } N \geq n\right] < \varepsilon, \quad n \geq n(\varepsilon, \eta), \tag{6.2.16}$$

then we say that $X_n = o(1)$ a.s. Similarly, replacing (6.2.12) by

$$P\left[|X_N/b_N| > \eta \text{ for some } N \geq n\right] < \varepsilon, \quad n \geq n(\varepsilon, \eta), \tag{6.2.17}$$

we say that $X_n = o(b_n)$ a.s.

Finally, we may note that (6.2.14)–(6.2.20) all extend to the vector case under the adaptation of the Euclidean norm, and also to the case of random functions under the metric in (6.1.6).

By virtue of the notations introduced above, we have the following results:

$$T_n - T \xrightarrow{P} 0 \Leftrightarrow d(T_n, T) = o_p(1), \tag{6.2.18}$$

$$T_n - T \xrightarrow{a.s.} 0 \Leftrightarrow d(T_n, T) = o(1) \text{ a.s.} \tag{6.2.19}$$

Thus, both convergence in probability and almost sure convergence results can be stated in the framework of *stochastic orders* of the distance $d(T_n, T)$, and this also holds in the vector case as well as for random functions. Consequently, one may reduce the problem to that involving the sequence $\{d(T_n, T)\}$ of real-valued random variables and to verify (6.2.11) or (6.2.16) for this sequence.

There is a third mode of convergence, which has interest of its own and that also provides a simpler way of establishing (6.2.2). It will be advantageous for us to introduce this mode of stochastic convergence for the real-valued or vector-valued statistics first, and then to extend the definition to the more general case.

Convergence in the rth mean ($r > 0$). A sequence $\{\boldsymbol{T}_n\}$ of random variables (or vectors) is said to converge in the rth mean (for some $r > 0$) to a (possibly degenerate) random variable (or vector) $\boldsymbol{T}$, if for any given positive ε, there exists a positive integer $n_0 = n_0(\varepsilon)$, such that $\mathbb{E}(\|\boldsymbol{T}_n - \boldsymbol{T}\|^r) < \varepsilon$, for every $n \geq n_0$, or in other words, if

$$\mathbb{E}(\|\boldsymbol{T}_n - \boldsymbol{T}\|^r) \to 0 \text{ as } n \to \infty, \text{ that is, } \mathbb{E}(\|\boldsymbol{T}_n - \boldsymbol{T}\|^r) = o(1). \tag{6.2.20}$$

Here, we adopt the Euclidean distance in (6.1.3). In symbols, we put

$$\boldsymbol{T}_n - \boldsymbol{T} \xrightarrow{r\text{th}} 0. \tag{6.2.21}$$

To cover the general case of random functions, we conceive of a metric d, such as the one in (6.1.6), and we say that $T_n - T \xrightarrow{r\text{th}} 0$ (in the topology d) if for every (given) $\varepsilon > 0$, there exists a positive integer $n_0 = n_0(\varepsilon)$, such that

$$\mathbb{E}\{[d(T_n, T)]^r\} < \varepsilon, \quad n \geq n_0 \quad (r > 0). \tag{6.2.22}$$

Here also, if T is nonstochastic, we may write $T_n \xrightarrow{r\text{th}} T$.

Before we proceed to study the mutual implications of these modes of convergence, we present another concept that plays a fundamental role in large sample theory.

Convergence in distribution or *weak convergence.* A sequence $\{T_n\}$ of random variables with distribution functions $\{F_n\}$ is said to converge in distribution (or in law) to a (possibly degenerate) random variable T with a distribution function F, if for every $\varepsilon > 0$, there exists an integer $n_0 = n_0(\varepsilon)$, such that at every point of continuity (x) of F,

$$|F_n(x) - F(x)| < \varepsilon, \quad n \geq n_0. \tag{6.2.23}$$

This mode of convergence is denoted by either

$$F_n \xrightarrow{w} F, \quad T_n \xrightarrow{D} T \quad \text{or} \quad \mathcal{L}(T_n) \to \mathcal{L}(T). \tag{6.2.24}$$

If the $\boldsymbol{T}_n$ are vectors, we may, similarly, work with the multivariate distribution functions [i.e., $F_n(\boldsymbol{x}) = P(\boldsymbol{T}_n \leq \boldsymbol{x})$ and $F(\boldsymbol{x}) = P(\boldsymbol{T} \leq \boldsymbol{x})$ and (6.2.23) still works out neatly. However, when the T_n are random functions, this definition may run into some technical difficulties, and additional conditions may be needed to ensure (6.2.24). These will be outlined in Chapter 11.

Recall that, in (6.2.23), T_n and T need not be defined on the same probability space. Furthermore, $d(T_n, T)$ may not make any sense, because, in fact, T_n may not converge to T in any mode; rather, the distribution of T_n converges to that of T, and in (6.2.24), $\mathcal{L}(T_n) \to \mathcal{L}(T)$ actually signifies that. It may be noted further that if T is a degenerate random variable (at the point c), then

$$F(t) = \begin{cases} 0, & \text{if } t < c \\ 1, & \text{if } t \geq c \end{cases}. \tag{6.2.25}$$

Therefore, if (6.2.23) holds and if F is a degenerate distribution function at a point c, then it follows that F_n converges to the degenerate distribution function F, and hence, as $n \to \infty$,

$$F_n(t) \to \ 0 \text{ or } 1 \text{ accordingly as } t \text{ is } < \text{ or } > c, \tag{6.2.26}$$

so that $T_n \xrightarrow{P} c$. This leads us to the following result in Theorem 6.2.1.

***Theorem* 6.2.1.** *For a sequence $\{T_n\}$ of random variables or vectors, convergence in law (or distribution) to a random variable or vector T, degenerate at a point c, ensures that $T_n \to c$, in probability, as $n \to \infty$.*

We will find this result very useful in what follows. We proceed now to study the mutual implications of the various modes of convergence; in practice, this gives us a variety of tools to establish a particular mode of convergence. We first consider the following (probability) inequality, which has also numerous other applications.

Chebyshev inequality. Let U be a nonnegative random variable with a finite mean $\mu = \mathbb{E}(U)$. Then, for every $t > 0$,

$$P(U > t\mu) \le t^{-1}. \tag{6.2.27}$$

Note that

$$\begin{aligned}\mu = \mathbb{E}(U) &= \int_0^{+\infty} u\mathrm{d}P(U \le u)\\ &= \int_0^{t\mu} u\mathrm{d}P(U \le u) + \int_{t\mu}^{+\infty} u\mathrm{d}P(U \le u)\\ &\ge \int_{t\mu}^{+\infty} u\mathrm{d}P(U \le u)\\ &\ge t\mu \int_{t\mu}^{+\infty} \mathrm{d}P(U \le u) = t\mu P\{U > t\mu\},\end{aligned} \tag{6.2.28}$$

and hence, dividing both sides by $t\mu$, we obtain (6.2.27).

We now consider a series of results on the interrelationship between the different modes of stochastic convergence.

***Theorem* 6.2.2.** *Almost sure convergence implies convergence in probability.*

Proof. The result follows directly from (6.2.5) and (6.2.6). ■

***Theorem* 6.2.3.** *Convergence in the rth mean for some $r > 0$ implies convergence in probability.*

Proof. Let $U_n = |T_n - T|^r$, where $r > 0$. Then by (6.2.27), it follows that for every $t > 0$, $P[U_n > t\mathbb{E}(U_n)] \le t^{-1}$, that is,

$$P\{|T_n - T| > t^{1/r}[\mathbb{E}(|T_n - T|^r)]^{1/r}\} \le t^{-1}. \tag{6.2.29}$$

Thus, for given $\varepsilon > 0$ and $\eta > 0$, choose n_0 and t, such that

$$t^{-1} < \eta, \quad \text{and} \quad t^{1/r}[\mathbb{E}(|T_n - T|^r)]^{1/r} < \varepsilon, \quad n \ge n_0, \tag{6.2.30}$$

which is always possible, since $\mathbb{E}(|T_n - T|^r) \to 0$ as $n \to \infty$. Thus, from (6.2.29), we obtain that $(|T_n - T| > \varepsilon) \le \eta$, for every $n \ge n_0$, and hence, $T_n - T \xrightarrow{P} 0$. ■

Note that for Theorem 6.2.3, r is arbitrary (but positive). To show that the converse implication may not be always true, let $X_n \sim \text{Bin}(n, \pi)$ and $T_n = n^{-1}X_n$. Note that for all $\pi \in (0, 1)$,

$$\mathbb{E}(T_n) = \pi \quad \text{and} \quad \mathbb{E}(T_n - \pi)^2 = n^{-1}\pi(1 - \pi) \le (4n)^{-1}. \tag{6.2.31}$$

Thus, using Theorem 6.2.3, we immediately obtain that T_n converges to π in the second mean, and hence, that $T_n \xrightarrow{P} \pi$. Consider next the parameter $\theta = \pi^{-1}$ and its natural estimator $Z_n = T_n^{-1}$. Since $T_n \xrightarrow{P} \pi$ and θ is a continuous function of π (except at $\pi = 0$), it immediately follows that $Z_n \xrightarrow{P} \theta$. However, $P(X_n = 0) = (1-\pi)^n > 0$ for every $n \geq 1$ (although it converges to 0 as $n \to \infty$) and, hence, for any $r > 0$, $\mathbb{E}(Z_n^r)$ is not finite. Therefore, Z_n does not converge to θ, in the rth mean, for any $r > 0$. This simple example also shows that the tails of a distribution may be negligible, but their contributions to its moments may not be so. We will discuss more about this in connection with *uniform integrability* in Section 6.5. In some simple situations, a sufficient condition under which convergence in probability implies convergence in rth mean may be identified, and we present this below.

***Theorem* 6.2.4.** *Convergence in probability for almost surely bounded random variables implies convergence in the* rth *mean, for every* $r > 0$.

Proof. A random variable X is a.s. bounded, if there exists a positive constant $c < \infty$, such that $P(|X| \leq c) = 1$. Suppose now that $T_n - T \xrightarrow{P} 0$ and that $P(|T_n - T| \leq c) = 1$, for some $0 < c < \infty$. Then observe that for any $\varepsilon > 0$, we may write

$$
\begin{aligned}
\mathbb{E}(|T_n - T|^r) &= \int_{|t|\leq c} |t|^r \mathrm{d}P(|T_n - T| \leq t) \\
&= \int_{|t|\leq \varepsilon} |t|^r \mathrm{d}P(|T_n - T| \leq t) + \int_{c \geq |t| > \varepsilon} |t|^r \mathrm{d}P(|T_n - T| \leq t) \\
&\leq \varepsilon^r P(|T_n - T| \leq \varepsilon) + c^r P(|T_n - T| > \varepsilon). \qquad (6.2.32)
\end{aligned}
$$

Since $T_n - T \xrightarrow{P} 0$ and ε is arbitrary, the right-hand side of (6.2.32) can be made arbitrarily small by choosing ε small and n adequately large. This ensures that $|T_n - T| \xrightarrow{r\text{th}} 0$. Since $r > 0$ is arbitrary, the proof is complete. ■

As an illustration, we go back to the binomial example cited previously. Note that $P(0 \leq T_n \leq 1) = 1$, for every $n \geq 1$, whereas $\pi \in [0, 1]$. Hence, it is easy to verify that $P(|T_n - \pi| \leq 1) = 1$ for every $n \geq 1$. Also, as we have seen earlier, $T_n \xrightarrow{P} \pi$. Thus, by Theorem 6.2.4, we conclude that $T_n - \pi \xrightarrow{r\text{th}} 0$ for every $r > 0$. Incidentally, in proving convergence in probability, here, we started with convergence in the second mean, but because of the boundedness of $T_n - \pi$, we ended up with convergence in rth mean for every $r > 0$.

Next, we explore the situation where the basic random variables may not be a.s. bounded. Toward this, we may note that a transformation $|T_n - T|$ to $W_n = g(|T_n - T|)$, for some monotone and bounded function g, may lead to an affirmative answer. In this context, we present the following theorem.

***Theorem* 6.2.5.** *Let* w *be a monotone nondecreasing, continuous and bounded function defined on* $0 \leq t < \infty$ *and such that* $w(0) = 0$. *Let* $W_n = w(|T_n - T|)$, $n \geq 1$. *Then,* $T_n - T \xrightarrow{P} 0$ *if and only if* $W_n \xrightarrow{r\text{th}} 0$, *for some* $r > 0$.

Proof. Note that under the assumed regularity conditions on w, for every $\varepsilon > 0$ there exists an $\eta > 0$ such that $\eta \to 0$ as $\varepsilon \to 0$ and $0 \leq t \leq \varepsilon$ is equivalent to $w(t) < \eta$. Now, suppose that $T_n - T \xrightarrow{P} 0$. Then, $W_n = w(|T_n - T|) \xrightarrow{P} 0$. Further, by construction, W_n is a.s. bounded and, hence, by Theorem 6.2.4, $W_n \xrightarrow{P} 0$, which implies that $W_n \xrightarrow{r\text{th}} 0$ for any $r > 0$. Alternatively, if $W_n \xrightarrow{r\text{th}} 0$ for some $r > 0$, then by Theorem 6.2.3, $W_n \xrightarrow{P} 0$ and this, in turn, implies that $|T_n - T| \xrightarrow{P} 0$, completing the proof. ■

In practice, we may choose

$$W_n = W_n^r = |T_n - T|^r/[1 + |T_n - T|^r]$$

for an appropriate $r > 0$; a typical choice for r is 2. Note that W_n is then a.s. bounded by 1, and, for $r = 2$, the evaluation of the moments of W_n may also be quite convenient. We illustrate this point with the estimation of $\theta = \pi^{-1}$ related to the binomial example cited previously in this section. In the context of (6.2.31), we define $Z_n = T_n^{-1} = n/X_n$. Then,

$$\begin{aligned} W_n^2 &= \frac{(Z_n - \theta)^2}{1 + (Z_n - \theta)^2} = \frac{(n - \pi^{-1}X_n)^2}{X_n^2 + (n - \pi^{-1}X_n)^2} \\ &= \frac{(n\pi - X_n)^2}{\pi^2 X_n^2 + (n\pi - X_n)^2} = \frac{(T_n - \pi)^2}{(T_n - \pi)^2 + \pi^2 T_n^2} \leq 1. \end{aligned} \tag{6.2.33}$$

Thus, $(T_n - \pi) \xrightarrow{2\text{nd}} 0$ implies $W_n \xrightarrow{2\text{nd}} 0$, and a direct application of Theorem 6.2.5 yields that $Z_n - \theta \xrightarrow{P} 0$.

We should not, however, overemphasize the role of Theorem 6.2.5. In general, it may not be very easy to show that for an arbitrary $r > 0$, $W_n \xrightarrow{r\text{th}} 0$, and simpler proofs of the stochastic convergence may be available. We will continue to explore these tools in what follows.

Next, we study the mutual implications of convergence in rth mean for various ordered r.

***Theorem* 6.2.6.** *Convergence in the* rth *mean, for some* $r > 0$ *implies convergence in the* sth *mean, for every* s *such that* $0 < s \leq r$.

Proof. Suppose that for some $r > 0$, $T_n - T \xrightarrow{r\text{th}} 0$. Letting $U_n = |T_n - T|^r$ and $V_n = |T_n - T|^s$ for some s, $0 < s < r$, it follows that $U_n = (V_n)^t$, $t = r/s > 1$. Since x^t is a convex function of x ($\in \mathbb{R}^+$), for every $t \geq 1$ (and is strictly convex, for $t > 1$) we may employ the Jensen Inequality (1.5.43), to show that

$$\mathbb{E}(|T_n - T|^r) = \mathbb{E}(U_n) \geq [\mathbb{E}(V_n)]^{r/s} = [\mathbb{E}(|T_n - T|^s)]^{r/s}, \quad r \geq s, \tag{6.2.34}$$

and this implies that

$$[\mathbb{E}(|T_n - T|^s)]^{1/s} \text{ is nondecreasing in } s > 0. \tag{6.2.35}$$

Therefore, we conclude that $\lim_{n\to\infty} \mathbb{E}(|T_n - T|^r) = 0$, for some $r > 0$, implies $\lim_{n\to\infty} \mathbb{E}(|T_n - T|^s) = 0$, for every $s \leq r$, and this completes the proof. ■

To show a.s. convergence, we may need to verify (6.2.6), and this may not be a simple task in all cases. In fact, convergence in rth mean by itself may not convey much information toward this. There is another mode of convergence (stronger than a.s. convergence) that is often used to simplify (6.2.6), and we consider this next.

Complete convergence. A sequence $\{T_n\}$ of random elements is said to converge completely to a (possibly degenerate) random element T, if for every $\varepsilon > 0$,

$$\sum_{n \geq 1} P[d(T_n, T) > \varepsilon] < \infty. \tag{6.2.36}$$

This mode of convergence is denoted by $T_n - T \xrightarrow{c} 0$. The metric d defined in the beginning of this section.

Note that the convergence of the series in expression (6.2.36) implies that the tail sum $\sum_{n \geq n_0} P[d(T_n, T) > \varepsilon]$ converges to 0 as $n_0 \to \infty$; as such, noting that for arbitrary events $\{A_k\}$, $P(\bigcup A_k) \leq \sum_k P(A_k)$, we conclude from (6.2.6) and (6.2.36) that

$$T_n - T \xrightarrow{c} 0 \quad \Rightarrow \quad T_n - T \xrightarrow{a.s.} 0. \tag{6.2.37}$$

The converse, however, may not be true. In practice, to verify (6.2.36), all we need is to verify that for every $\varepsilon > 0$, $P[d(T_n, T) > \varepsilon]$ converges to 0 as $n \to \infty$ at a rate such that the series $\sum_{n \geq n_0} P[d(T_n, T) > \varepsilon]$ converges, and, in this context, suitable probability inequalities may be used. We will discuss these inequalities in detail in the next section.

Let us next consider the implications of convergence in probability, almost sure convergence, convergence in the rth mean and complete convergence on weak convergence.

***Theorem* 6.2.7.** *Convergence in probability implies convergence in distribution.*

Proof. Suppose that $T_n - T \xrightarrow{P} 0$ and that T has the distribution function F; for simplicity, we restrict ourselves to real-valued random variables. Let x be a point of continuity of F, and let $\varepsilon > 0$ be an arbitrary small number. Then,

$$\begin{aligned} F(x - \varepsilon) &= P(T \leq x - \varepsilon) \\ &= P(T \leq x - \varepsilon, |T_n - T| \leq \varepsilon) + P(T \leq x - \varepsilon, |T_n - T| > \varepsilon) \\ &\leq P(T_n \leq x) + P(|T_n - T| > \varepsilon) \end{aligned} \tag{6.2.38}$$

and

$$F(x + \varepsilon) \geq P(T_n \leq x) - P(|T_n - T| > \varepsilon). \tag{6.2.39}$$

Given that the last term on the right-hand side of (6.2.38) and (6.2.39) converges to 0 as $n \to \infty$, we have

$$F(x - \varepsilon) \leq \liminf_{n \to \infty} F_n(x) \leq \limsup_{n \to \infty} F_n(x) \leq F(x + \varepsilon). \tag{6.2.40}$$

Since x is a point of continuity of F, $F(x + \varepsilon) - F(x - \varepsilon)$ can be made arbitrarily small by letting $\varepsilon \downarrow 0$, and, hence, we conclude that (6.2.23) holds, completing the proof. ■

The diagram in Figure 6.2.1 summarizes the relationship between the different modes of convergence.

The picture is, however, incomplete. Because of their complexity, the relationships between complete (or a.s.) convergence and the convergence in the rth mean ($r > 0$) are not indicated.

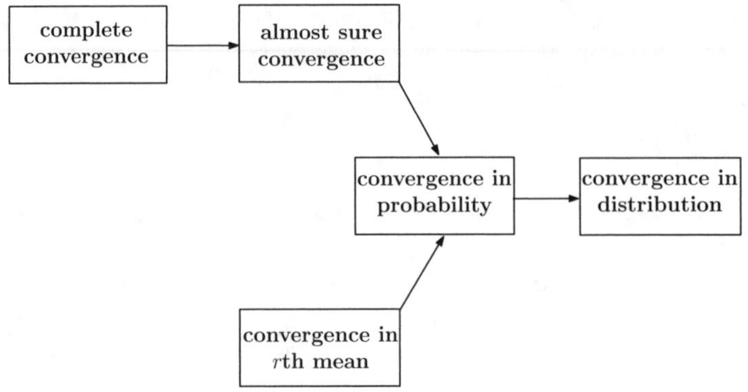

Figure 6.2.1. Convergence implication diagram.

We conclude this section with an important observation on the role of characteristic functions defined in (1.5.73) in weak convergence; this will also be useful in the study of stochastic convergence in the subsequent sections. For a distribution function F, defined on $\mathbb{R}$, the characteristic function $\phi_F(t)$, for real t, is

$$\phi_F(t) = \int_{\mathbb{R}} e^{itx} \mathrm{d}F(x), \quad t \in \mathbb{R}.$$

In general, the characteristic function $\phi_F(t)$ is a complex-valued function. If the distribution function F is degenerate at the point $c \in \mathbb{R}$, then, $\phi_F(t) = \exp(itc)$, for every real t. The classical Lévy Theorem [Cramér (1946)] asserts that if two distribution functions F and G both have the same characteristic function $\phi_F(t)$, $t \in \mathbb{R}$, then F and G are the same, and the converse is also true. A more useful result in this context, stated without proof, is the following theorem.

Theorem 6.2.8 (Lévy–Cramér). *Let $\{F_n, n \geq 1\}$ be a sequence of distribution functions (defined on $\mathbb{R}$) with corresponding characteristic functions $\{\phi_{F_n}(t), n \geq 1\}$. Then, a necessary and sufficient condition for $F_n \xrightarrow{w} F$ is that, for every real t, $\phi_{F_n}(t)$ converges to a limit $\phi_F(t)$ which is continuous at $t = 0$ and is the characteristic function of the distribution function F.*

Theorem 6.2.1 is a direct corollary of the Lévy–Cramér Theorem. This result is quite useful in practice, and it provides the useful consistency results under minimal regularity conditions. To illustrate this point, we consider a typical sequence $\{T_n\}$ of (real-valued) estimators of a real parameter θ. We may consider the Implication Diagram in Figure 6.2.1 and directly verify either the convergence in probability criterion or any of the other three criteria that imply convergence in probability. For any one of these, we may need to verify certain regularity conditions. On the other hand, we may as well use Theorem 6.2.1 and verify the same under (usually) less restrictive conditions. As a specific case, consider $T_n = \overline{X}_n$, where the X_i are i.i.d. random variables with mean θ. By using the Lévy–Cramér Theorem (6.2.8), it can easily be shown that $T_n \xrightarrow{D} T$, where T is degenerate at the point θ. Thus, by Theorem 6.2.1, we may claim that $T_n \xrightarrow{P} \theta$, and this does not need any moment condition on the distribution function of X_i other than the finiteness of θ. On the other hand, using convergence in the rth mean (to ensure convergence in probability), we may arrive at the same conclusion, but that generally needs the existence of the moment of order r, for some $r > 1$; the simplest way of doing this is to appeal to (6.2.27) on letting $U = (\overline{X}_n - \theta)^2$, so that $\mathbb{E}(U) = n^{-1}\sigma^2 \downarrow 0$ as $n \to \infty$, provided the second central moment (σ^2) of X_i is finite. Evaluation of $\mathbb{E}(|\overline{X}_n - \theta|^r)$ for a non integer $r > 0$ or for r other than 2 may require more tedious manipulations and, hence, may not be generally advocated. If, however, by any other means, it is possible to obtain a probability inequality of the form

$$P(|T_n - T| > \varepsilon) \leq \psi_\varepsilon(n), \quad n \geq n_0, \quad \varepsilon > 0, \tag{6.2.41}$$

where $\psi_\varepsilon(n) \downarrow 0$ as $n \to \infty$, then, of course, the desired result follows directly. If $\sum_{m \geq n} \psi_\varepsilon(m) \to 0$ as n$\to \infty$, then the a.s. (or complete) convergence also follows from (6.2.41). Thus, refined probability inequalities may provide information beyond convergence in probability (or a.s. convergence). Alternatively, we do not need (6.2.41) with a specific rate for $\psi_\varepsilon(n)$, for the stochastic (or a.s.) convergence of a sequence of random elements; often, these may be achieved by the use of the *laws of large numbers*. Hence, before we consider some deeper results on the modes of convergence of random elements, we present some basic probability inequalities and laws of large numbers in the next section.

6.3 Probability Inequalities and Laws of Large Numbers

The Chebyshev Inequality (6.2.27) is the precursor of other forms of more general probability inequalities. Actually, if we have a sequence $\{T_n\}$ of random variables, and we put $\sigma_n^2 = \mathbb{E}[(T_n - \theta)^2]$, then on letting $U = (T_n - \theta)^2$, we may use (6.2.27) to obtain the following alternative form of such inequality:

$$P(|T_n - \theta| > t\sigma_n) \leq t^{-2}, \quad n \geq 1, \quad t \in \mathbb{R}^+. \tag{6.3.1}$$

Note that (6.3.1) depends on the distribution function of T_n only through σ_n and, hence, may be termed a *distribution-free inequality*.

The simplicity of (6.3.1) is its main appeal. Whenever σ_n^2 converges to 0 as $n \to \infty$, it follows that T_n converges in probability to θ. However, there are two points we should keep in mind here. First, for the stochastic convergence of T_n to θ, it is not necessary to assume that the mean squared error (given here by σ_n^2) exists nor that it converges to 0 as $n \to \infty$. Second, if we assume that the mean squared error exists, we may even obtain stronger convergence results in a more general setup. These will be explored throughout this section. We may also mention that (6.3.1), although quite simple, may not be generally very sharp. To illustrate this point, we consider the classical case of i.i.d. random variables $\{X_i\}$ having a normal distribution with mean θ and variance σ^2 and take $T_n = \overline{X}_n$. Then, by (6.3.1), we get

$$P(|T_n - \theta| \geq t\sigma/\sqrt{n}) \leq t^{-2}, \quad t \in \mathbb{R}^+. \tag{6.3.2}$$

On the other hand, $n^{1/2}(\overline{X}_n - \theta)/\sigma$ has the standard normal distribution, and, hence,

$$P(|T_n - \theta| > t\sigma/\sqrt{n}) = 2[1 - \Phi(t)], \quad t \in \mathbb{R}^+, \tag{6.3.3}$$

where $\Phi(t)$, $t \in \mathbb{R}$, is the standard normal distribution function. Letting $t = 2$, the probability in (6.3.3) is equal to 0.0455, whereas the upper bound in (6.3.2) is 0.25 (quite higher). Similarly, for $t = 3$, (6.3.3) yields 0.0029, whereas the corresponding value obtained via

(6.3.2) is 1/9 (= 0.1111). Also, for $t^2 \leq 1$, (6.3.2) is of no use, although (6.3.3) can be computed for every real t.

Better rates of convergence (or sharper bounds) can be obtained (at the price of assuming the existence of higher order moments) via the application of the following inequality.

Markov Inequality. Let U be a nonnegative random variable with finite rth moment $\mu^{(r)} = \mathbb{E}(U^r)$, for some $r > 0$. Then, for every $\varepsilon > 0$, we have

$$P(U > \varepsilon) \leq \varepsilon^{-r}\mu^{(r)}. \tag{6.3.4}$$

To show this, first let $V = U^r$, so that $\mathbb{E}(V) = \mu^{(r)}$. Then, note that $U > \varepsilon$ is equivalent to $V > \varepsilon^r$, so that (6.3.4) follows directly from (6.2.27).

As an illustration of the utility of this inequality, we consider the binomial example treated in Section 6.2 and let $nT_n \sim \text{Bin}(n, \pi)$. From (6.2.31) and (6.3.2), it follows that for every $\varepsilon > 0$,

$$P(|T_n - \pi| > \varepsilon) \leq \frac{\pi(1-\pi)}{n\varepsilon^2} = O(n^{-1}), \tag{6.3.5}$$

whereas, from Exercise 6.3.1 and (6.3.4), for $r = 4$, we have

$$P(|T_n - \pi| > \varepsilon) \leq n^{-2}\frac{3}{16\varepsilon^4} = O(n^{-2}). \tag{6.3.6}$$

Thus, for large n, at least, (6.3.6) yields a much better rate of convergence than (6.3.5). In fact, using values of r even greater than 4, one may still obtain better rates of convergence. This is possible because T_n is a bounded random variable, and hence all moments of finite order exist. Taking the lead from this observation, we may go even further and obtain a sharper form of a probability inequality when, in fact, the moment generating function $M(t) = \mathbb{E}[\exp(tX)]$ exists for t belonging to a suitable domain, insuring the existence of moments of all finite order. In this direction, we consider the following.

Bernstein Inequality. Let U be a random variable, such that $M_U(t) = \mathbb{E}[\exp(tU)]$ exists for all $t \in [0, K]$, for some $K > 0$. Then, for every real u, we have

$$P(U \geq u) \leq \inf\{e^{-tu}M_U(t) : t \in [0, K]\}. \tag{6.3.7}$$

To show this, first let $g(x)$, $x \in \mathbb{R}$, be a nonnegative and nondecreasing function. Using the Chebyshev Inequality (6.2.27), we have

$$P(U \geq u) = P[g(U) \geq g(u)] \leq \mathbb{E}[g(U)]/g(u). \tag{6.3.8}$$

Letting $g(x) = \exp(tx)$, $t > 0$, $x \in \mathbb{R}$, we obtain from (6.3.8) that

$$P(U > u) \leq \frac{\mathbb{E}\left(e^{tU}\right)}{e^{tu}} = \frac{M_U(t)}{e^{tu}} = e^{-tu}M_U(t), \quad t > 0. \tag{6.3.9}$$

Since (6.3.9) holds for every $t \in [0, K]$, minimizing the right-hand side over t, (6.3.7) follows.

Note that, in (6.3.7), U need not be a nonnegative random variable. Also, note that by the same arguments,

$$\begin{aligned} P(U \leq u) &= P(-U \geq -u) \\ &\leq \inf\{e^{tu} M_{(-U)}(t) : t \in [0, K']\} \\ &= \inf\{e^{tu} M_U(-t) : t \in [0, K^*]\}, \end{aligned}$$

where K^* is a suitable positive number, such that $M_U(-t)$ exists for every $t \leq K^*$. Hence, we may write

$$\begin{aligned} P(|U| \geq u) &= P(U \geq u) + P(U \leq -u) \\ &\leq \inf\{e^{-tu} M_U(t) : t \in [0, K]\} + \inf\{e^{-tu} M_U(-t) : t \in [0, K^*]\} \\ &\leq \inf\{e^{-tu}(M_U(t) + M_U(-t)) : 0 < t \leq \min(K, K^*)\}. \end{aligned} \tag{6.3.10}$$

We illustrate the effectiveness of the Bernstein Inequality (6.3.7) with the same binomial example treated earlier. Note that $nT_n \sim \text{Bin}(n, \pi)$, so that the moment generating function of nT_n is

$$M_{nT_n}(t) = \sum_{r=0}^{n} \binom{n}{r} e^{tr} \pi^r (1-\pi)^{n-r} = [e^t \pi + (1-\pi)]^n.$$

Thus, taking $U = nT_n - n\pi$, we obtain from (6.3.10) that for all $\varepsilon > 0$ and $\pi \in (0, 1)$,

$$\begin{aligned} P(|T_n - \pi| > \varepsilon) &= P(|nT_n - n\pi| > n\varepsilon) \\ &= P(U < -n\varepsilon) + P(U > n\varepsilon) \\ &\leq \inf_{t \geq 0}\{e^{-n(1-\pi)t - n\varepsilon t}[1 + (1-\pi)(e^t - 1)]^n\} \\ &\quad + \inf_{t > 0}\{e^{-n\pi t - n\varepsilon t}[1 + \pi(e^t - 1)]^n\} \\ &= [\rho(\varepsilon, \pi)]^n + [\rho(\varepsilon, 1-\pi)]^n, \end{aligned} \tag{6.3.11}$$

where, for $\pi \in (0, 1)$, $(\pi + \varepsilon) \in (0, 1)$,

$$\rho(\varepsilon, \pi) = \inf_{t > 0}\{e^{-(\pi+\varepsilon)t}[1 + \pi(e^t - 1)]\}. \tag{6.3.12}$$

Note that T_n lies between 0 and 1 with probability 1, so that $\rho(\varepsilon, \pi) = 0$ for $\pi + \varepsilon > 1$. Next, to obtain an explicit expression for (6.3.12), we set $w = \exp(t)$ and $g(w) = w^{-(\pi+\varepsilon)}[1 + \pi(w-1)]$, so that $(\partial/\partial w) \log g(w) = -(\pi+\varepsilon)/w + \pi/[1 + \pi(w-1)]$, and equating this to 0, we obtain that the point w^* at which $g(w)$ is an extremum is

$$w^* = [(\pi+\varepsilon)(1-\pi)]/[\pi(1-\pi-\varepsilon)].$$

Since

$$\frac{\partial^2}{\partial w^2} \log g(w)|_{w=w^*} = [(\pi+\varepsilon)^{-1} - 1]\pi^2/(1-\pi)^2 > 0,$$

it follows that w^* corresponds to a minimum. Therefore, we may write

$$\rho(\varepsilon,\pi) = g(w^*) = \left[\frac{(\pi+\varepsilon)(1-\pi)}{\pi(1-\pi-\varepsilon)}\right]^{-(\pi+\varepsilon)}\left[1+\frac{\pi\varepsilon}{\pi(1-\pi-\varepsilon)}\right]$$
$$= \left(\frac{\pi}{\pi+\varepsilon}\right)^{\pi+\varepsilon}\left(\frac{1-\pi}{1-\pi-\varepsilon}\right)^{1-\pi-\varepsilon} > 0. \quad (6.3.13)$$

Next, we note that by (6.3.13), $\rho(0,\pi) = 1$, for every $\pi \in (0, 1)$. Furthermore,

$$\frac{\partial}{\partial\varepsilon}\log\rho(\varepsilon,\pi) = \log\left[\frac{\pi(1-\pi-\varepsilon)}{(1-\pi)(\pi+\varepsilon)}\right] < 0, \quad \varepsilon > 0. \quad (6.3.14)$$

Hence $\rho(\varepsilon,\pi) < 1$, for every (ε,π) such that $0 < \pi+\varepsilon < 1$. Since both $\rho(\varepsilon,\pi)$ and $\rho(\varepsilon, 1-\pi)$ are positive fractions that are less than 1, by (6.3.11), we may conclude that for every $\varepsilon > 0$, $P(|T_n-\pi| > \varepsilon)$ converges to 0 at an exponential rate as n increases. In fact, we may set

$$\rho^*(\varepsilon) = \sup\{\rho(\varepsilon,\pi) : 0 < \pi < 1-\varepsilon\}, \quad \varepsilon > 0, \quad (6.3.15)$$

and obtain from (6.3.11) and (6.3.15) that for every $\varepsilon > 0$ and $n \geq 1$,

$$P(|T_n-\pi| > \varepsilon) \leq 2[\rho^*(\varepsilon)]^n, \quad \pi \in (0, 1). \quad (6.3.16)$$

Interestingly, it may be noted that this powerful inequality is not confined to the binomial case alone. The same exponential rate of convergence appears in a wider class of problems involving a.s. bounded random variables (for which the moment generating functions exist), and then it may not be necessary to assume that these random variables are identically distributed. In this context, we present the following elegant probability inequality due to Hoeffding (1963).

Hoeffding Inequality. Let X_k, $k \geq 1$, be independent (but not necessarily identically distributed) random variables, such that $P(0 \leq X_k \leq 1) = 1$ for every $k \geq 1$. Let $\mu_k = \mathbb{E}(X_k)$, $k \geq 1$, and $\overline{\mu}_n = n^{-1}\sum_{k=1}^n \mu_k$, $n \geq 1$. Also, let $\overline{X}_n = n^{-1}\sum_{k=1}^n X_k$, $n \geq 1$. Then, for every $\varepsilon > 0$,

$$P(|\overline{X}_n - \overline{\mu}_n| > \varepsilon) \leq [\rho(\varepsilon,\overline{\mu}_n)]^n + [\rho(\varepsilon, 1-\overline{\mu}_n)]^n, \quad (6.3.17)$$

where $\rho(\varepsilon, a)$ is defined as in (6.3.12). Thus, $\overline{X}_n - \overline{\mu}_n$ converges in probability to 0 at an exponential rate.

To prove the result, first observe that for every $x \in [0, 1]$,

$$g_t(x) = e^{tx} \leq 1 + x(e^t - 1), \quad t > 0$$

(see Figure 6.3.1). As the $X_k \in [0, 1]$ with probability 1, this ensures that for every k,

$$M_{X_k}(t) = \mathbb{E}(e^{tX_k}) \leq 1 + (e^t-1)\mathbb{E}(X_k) = 1 + (e^t-1)\mu_k. \quad (6.3.18)$$

Moreover, $X_1, \ldots, X_n$ are independent, so that for every real t,

$$\mathbb{E}\{\exp[t(X_1+\cdots+X_n)]\} = \prod_{k=1}^n \mathbb{E}[\exp(tX_k)] = \prod_{k=1}^n M_{X_k}(t). \quad (6.3.19)$$

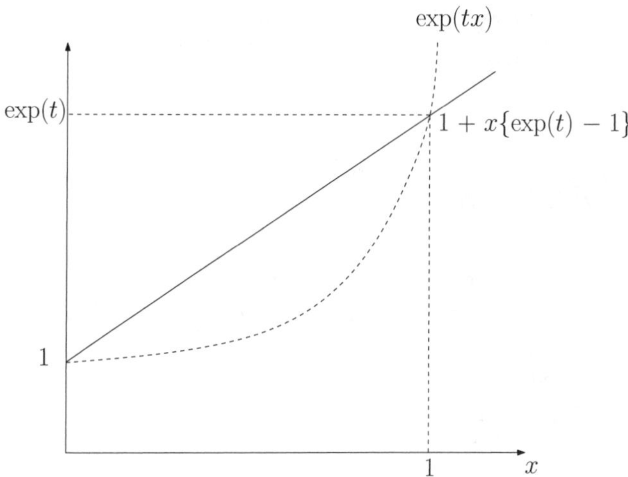

Figure 6.3.1. Upper bound for e^{tx}, $x \in [0, 1]$.

Therefore, using (6.3.10) for $U = (X_1 + \cdots + X_n) - (\mu_1 + \cdots + \mu_n) = n(\overline{X}_n - \overline{\mu}_n)$, we obtain that for every $\varepsilon > 0$,

$$\begin{aligned} P(|\overline{X}_n - \overline{\mu}_n| > \varepsilon) &= P(|U| > n\varepsilon) \\ &\leq \inf_{t>0} \left\{ \exp[-tn(\varepsilon + \overline{\mu}_n)] \prod_{k=1}^{n} M_{X_k}(t) \right\} \\ &\quad + \inf_{t>0} \left\{ \exp[-tn(\varepsilon - \overline{\mu}_n)] \prod_{k=1}^{n} M_{X_k}(-t) \right\}. \end{aligned} \tag{6.3.20}$$

At this stage, we make use of (6.3.18) along with the Arithmetic/Geometric Mean/Harmonic Mean Inequality (1.5.40) and obtain that

$$\prod_{k=1}^{n} M_{X_k}(t) \leq \left[n^{-1} \sum_{k=1}^{n} M_{X_k}(t) \right]^n \leq \left[1 + (e^t - 1)\overline{\mu}_n \right]^n. \tag{6.3.21}$$

As such, the first term on the right-hand side of (6.3.20) is bounded from above by the first term on the right-hand side of (6.3.11), where π has to be replaced by $\overline{\mu}_n$. A very similar inequality holds for the second term on the right-hand side of (6.3.20). As such, following (6.3.11)–(6.3.13) we obtain (6.3.17), concluding the proof.

The above result may easily be extended to the case of random variables belonging to an arbitrary compact interval $[a, b]$, for some $-\infty < a < b < +\infty$. In this context, if $P(a \leq X_k \leq b) = 1$ for every $k \geq 1$, we may set $Z_k = (X_k - a)/(b - a)$, so that $P(0 \leq Z_k \leq 1) = 1$, for every $k \geq 1$. Note that

$$\begin{aligned} &(X_1 + \cdots + X_n) - (\mu_1 + \cdots + \mu_n) \\ &\quad = (b - a)[Z_1 + \cdots + Z_n - \mathbb{E}(Z_1) - \cdots - \mathbb{E}(Z_n)], \end{aligned} \tag{6.3.22}$$

so that the only change in (6.3.17) would be to replace ε by $\varepsilon^* = \varepsilon/(b - a)$ and $\overline{\mu}_n$ by $(\overline{\mu}_n - a)/(b - a) = \overline{\mu}_n^*$, say. For some other forms of probability inequalities for a.s. bounded random variables, we refer to Hoeffding (1963); for brevity, we do not consider them here. In passing, we may remark that by virtue of (6.3.14),

$$\frac{\partial}{\partial \varepsilon} \log \rho(\varepsilon, \pi)|_{\varepsilon=0} = 0$$

and

$$\begin{aligned}\frac{\partial^2}{\partial \varepsilon^2} \log \rho(\varepsilon, \pi) &= -(1 - \pi - \varepsilon)^{-1} - (\pi + \varepsilon)^{-1} \\ &= -[(\pi + \varepsilon)(1 - \pi - \varepsilon)]^{-1} \leq -4, \quad 0 \leq \pi + \varepsilon \leq 1.\end{aligned}$$

As such, by a second-order Taylor expansion, we have $\log \rho(\varepsilon, \pi) \leq -2\varepsilon^2$, so that $\rho(\varepsilon, \pi) \leq \exp(-2\varepsilon^2)$, for every π such that $0 \leq \pi + \varepsilon \leq 1$. Then, in (6.3.11) and (6.3.17), the right-hand side can be bounded from above by $2\exp(-2n\varepsilon^2) = 2[\exp(-2\varepsilon^2)]^n$, and this specifies an upper bound for the exponential rate of convergence (to 0) of the tail probabilities in (6.3.11) or (6.3.17). We may also remark that for such an exponential rate of convergence, it may not be necessary to assume that the random variables are a.s. bounded. Although for unbounded random variables, (6.3.18) may not hold [see Figure 6.3.1], an exponential rate may often be achieved under the finiteness of the moment generating functions. We then present the following inequality.

Corollary to the Bernstein Inequality. Suppose that X_k, $k \geq 1$ are i.i.d. random variables and, without any loss of generality, assume that $\mathbb{E}(X_k) = 0$ and $M(t) = \mathbb{E}[\exp(tX_1)] < \infty$, for t in some neighborhood of 0. For $\varepsilon > 0$, let

$$\rho_+(\varepsilon) = \inf\{e^{-t\varepsilon} M(t) : t > 0\} \text{ and } \rho_-(\varepsilon) = \inf\{e^{-t\varepsilon} M(-t) : t > 0\}.$$

Then, for every $\varepsilon > 0$ and $n \geq 1$,

$$P(|\overline{X}_n| > \varepsilon) \leq [\rho_+(\varepsilon)]^n + [\rho_-(\varepsilon)]^n. \tag{6.3.23}$$

This result is a direct consequence of (6.3.20) and the definitions of $\rho_+(\varepsilon)$ and $\rho_-(\varepsilon)$. In fact, we do not need to assume that the X_k are identically distributed; all that is needed is that $\prod_{k=1}^n M_{X_k}(t) \leq [\overline{M}_n(t)]^n$, where $\overline{M}_n(t)=n^{-1}\sum_{i=1}^n M_{X_i}(t)$ satisfies the condition ensuring (6.3.23) for every $n \geq n_0$.

To illustrate this, we consider the following examples.

***Example* 6.3.1.** Let $X_k \sim \mathcal{N}(\mu, \sigma^2), k \geq 1$, and let $T_n = \overline{X}_n - \mu, n \geq 1$. Then, for $X_k - \mu$, the moment generating function is $M(t) = \exp(t^2\sigma^2/2)$, $t \in \mathbb{R}$. Thus, it is easy to verify that $\rho_+(\varepsilon) = \rho_-(\varepsilon) = \exp(-\varepsilon^2/2\sigma^2)$, so that along lines parallel to (6.3.23), we have $P(|\overline{X} - \mu| > \varepsilon) \leq 2[\exp(-\varepsilon^2/2\sigma^2)]^n$, for every $\varepsilon > 0$. It is also possible to avoid the assumption of homogeneity of the means and the variances of these random variables. For possibly different means, we need to take the average mean for μ in the definition of T_n, whereas as long as the variances are bounded from above by a constant γ^2, we have the same bound with σ^2 replaced by γ^2. □

***Example* 6.3.2.** Let X_k be i.i.d. random variables with the double exponential distribution [i.e., with a density function given by $f(x) = (1/2)\exp(-|x|)$, $x \in \mathbb{R}$], where, for simplicity, we take the mean equal to 0. For every t, such that $|t| < 1$, $M(t)$ is finite, follows (6.3.23). The proof is the object of Exercise 6.3.4. □

As has been stated earlier, these probability inequalities can be fruitfully incorporated in the proofs of various modes of convergence of suitable random elements. In this context, we first present a preliminary result that will lead to the *Borel–Cantelli Lemma*.

***Theorem* 6.3.1.** *Let $\{a_n\}$ be a sequence of real numbers, such that $0 < a_n < 1$, for all $n \geq 1$. Then $\sum_{n\geq 1} a_n = +\infty$ implies $\prod_{k\geq 1}(1 - a_k) = 0$.*

Proof. Let

$$K_n = \prod_{k=1}^{n}(1 - a_k) \text{ and } L_n = \sum_{k=1}^{n} a_k, n \geq 1. \tag{6.3.24}$$

Since the a_k lie in the unit interval, we have for every $n \geq 1$,

$$\begin{aligned} \log K_n &= \sum_{k=1}^{n} \log(1 - a_k) \\ &= -\sum_{k=1}^{n}\left(a_k + \frac{1}{2}a_k^2 + \frac{1}{3}a_k^3 + \cdots\right) \\ &\leq -\sum_{k=1}^{n} a_k = -L_n. \end{aligned} \tag{6.3.25}$$

Allowing $n \to \infty$, it follows that $-L_n \to -\infty$ and, hence, that $K_n \to 0$. ■

Theorem* 6.3.2 *(Borel–Cantelli Lemma). *Let $\{A_n\}$ be a sequence of events and let A denote the event that the A_n occur infinitely often. Then,*

$$\sum_{n\geq 1} P(A_n) < \infty \Rightarrow P(A) = 0, \tag{6.3.26}$$

no matter whether the A_n are independent or not. If the A_n are independent, then,

$$\sum_{n\geq 1} P(A_n) = \infty \Rightarrow P(A) = 1. \tag{6.3.27}$$

Proof. Let $B_n = \bigcup_{N\geq n} A_N,\ n \geq 1$. Then, note that

$$A = \bigcap_{n\geq 1}\bigcup_{N\geq n} A_N = \bigcap_{n\geq 1} B_n \subseteq B_n, \quad n \geq 1. \tag{6.3.28}$$

Thus, we have

$$\begin{aligned} P(A) = P\left(\bigcap_{n\geq 1} B_n\right) &\leq P(B_m) = P\left(\bigcup_{N\geq m} A_N\right) \\ &\leq \sum_{N\geq m} P(A_N), \quad m \geq 1. \end{aligned} \tag{6.3.29}$$

In the last step, we use the basic inequality that $P(\bigcup A_r) \leq \sum_r P(A_r)$ without necessarily requiring that the A_r be independent. Thus, whenever $\sum_{n\geq 1} P(A_n) < \infty$, by definition, $\sum_{n\geq m} P(A_n) \to 0$, and this proves (6.3.26). Next, we note by (6.3.28) that

$A^c = \bigcup_{n\geq 1}\bigcap_{N\geq n} A_N^c$, so that incorporating the independence of the A_n (or A_n^c, equivalently), we have

$$\begin{aligned} 1 - P(A) = P(A^c) &\leq \sum_{n\geq 1} P\left(\bigcap_{N\geq n} A_N^c\right) \\ &= \sum_{n\geq 1}\prod_{N\geq n} P(A_N^c) = \sum_{n\geq 1}\prod_{N\geq n}\{1 - P(A_N)\}. \end{aligned} \tag{6.3.30}$$

By Theorem 6.3.1, the right-hand side of (6.3.30) is equal to 0 when $\sum_{n\geq 1} P(A_n) = \infty$, and this completes the proof. ■

The Borel–Cantelli Lemma (Theorem 6.3.2) occupies a central position in Probability Theory and has a variety of uses in Asymptotic Theory. One of these uses relates to the verification of complete (and, hence, a.s.) convergence by using the Markov (or some other) Inequality (6.3.4) along with (6.3.26). Before we illustrate this point with the help of some examples, we consider the following theorem.

***Theorem* 6.3.3.** *For every $p > 1$, the series $\sum_{n\geq 1} n^{-p}$ converges.*

Proof. Note that for $p > 1$, $2^{-(p-1)} < 1$, and

$$\begin{aligned} \sum_{n\geq 1} n^{-p} &= 1 + \sum_{k\geq 1}\left\{\sum_{j=2^{k-1}+1}^{2^k} j^{-p}\right\} \\ &\leq 1 + \sum_{k\geq 1}(2^{-(k-1)})^p\{2^k - 2^{k-1}\} \\ &= 1 + \sum_{k\geq 1} 2^{-(k-1)(p-1)} \\ &= 1 + (1 - 2^{-(p-1)})^{-1} < \infty. \end{aligned} \tag{6.3.31}$$

■

***Example* 6.3.3.** Let $X_k, k \geq 1$ be i.i.d. random variables with $\mathbb{E}(X) = \mu, \mathbb{E}[(X-\mu)^2] = \sigma^2$ and $\mathbb{E}[(X-\mu)^4] = \mu_4 < \infty$. Then, let $T_n = \overline{X}_n$ and $A_n = [|\overline{X}_n - \mu| > \varepsilon]$, for $n \geq 1$, where $\varepsilon > 0$ is arbitrary. Considering the Markov Inequality (6.3.4) with $r = 4$, we have

$$\begin{aligned} P(A_n) = P(|\overline{X}_n - \mu| > \varepsilon) &\leq \varepsilon^{-4}\mathbb{E}[(\overline{X}_n - \mu)^4] \\ &= \frac{n^{-2}3\sigma^4 + n^{-3}(\mu_4 - 3\sigma^4)}{\varepsilon^4}, \quad n \geq 1. \end{aligned} \tag{6.3.32}$$

Consequently, using (6.3.31) for $p = 2$ and the first part of the Borel–Cantelli Lemma (6.3.26), we immediately obtain that

$$P(A) = P\left(\bigcap_{n\geq 1}\bigcup_{N\geq n} A_N\right) = 0,$$

so that the A_n do not occur infinitely often, or $\overline{X}_n - \mu \xrightarrow{a.s.} 0$. In fact, in this case, $P(\bigcup_{N\geq n} A_N) \leq \sum_{N\geq n} P(A_N) \to 0$, as $n \to \infty$, and, hence, $\overline{X}_n \xrightarrow{c} \mu$. Note that the conclusions derived here remain true for the entire class of distributions for which the fourth moment is finite. Later on, we will see that this fourth moment condition may also be relaxed. □

***Example* 6.3.4.** Consider now the binomial example treated in (6.3.11)–(6.3.13). Here, letting $A_n = [|T_n - \pi| > \varepsilon]$, $n \geq 1$, $\varepsilon > 0$ arbitrary, we have

$$\sum_{n\geq 1} P(A_n) \leq \sum_{n\geq 1} \{[\rho(\varepsilon, \pi)]^n + [\rho(\varepsilon, 1-\pi)]^n\}$$
$$= \frac{\rho(\varepsilon, \pi)}{1-\rho(\varepsilon, \pi)} + \frac{\rho(\varepsilon, 1-\pi)}{1-\rho(\varepsilon, 1-\pi)} < \infty, \tag{6.3.33}$$

and, hence, the Borel–Cantelli Lemma (Theorem 6.3.2) leads us to the conclusion that $T_n - \pi \xrightarrow{a.s.} 0$. We could have obtained the same result by using the previous example wherein we take $X_k = 1$ with probability π and $X_k = 0$, with probability $1 - \pi$. This result is known in the literature as the *Borel Strong Law of Large Numbers* □

Essentially, the use of the Borel–Cantelli Lemma (Theorem 6.3.2) to establish the a.s. or complete convergence is based on the specification of a sequence of nonnegative functions $h(n)$, defined in such a way that $P(A_n) = P\{|T_n - T| > \varepsilon\} \leq c_\varepsilon h(n)$, for all $n \geq n_0$ and $\sum_{n\geq n_0} h(n) < \infty$, with c_ε denoting a constant, which may depend on ε. Theorem 6.3.3 provides a general class of such $h(n)$. In most of the practical problems, a choice of a suitable $h(n)$ is made by using the Markov (or some other) probability inequality (6.3.4), which generally deems the existence of positive order (absolute) moments of $T_n - T$. However, this is not necessary, and there are numerous examples where such a moment condition may not actually hold, but the desired order of $h(n)$ can be achieved by alternative methods. To illustrate this point, we consider the following example.

Example* 6.3.5 *(CMRR procedure). Consider the problem of estimating the number N of fish in a lake. The *Capture-Mark-Release and Recapture (CMRR)* procedure consists of (i) drawing an initial sample of n_1 fish at random from the lake, marking them suitably (e.g., on the fin) and releasing them in the lake; (ii) drawing a second random sample of size n_2 fishes from the lake and counting the number (r_2) of marked fish in this sample, and (iii) estimating N by maximum likelihood method as

$$\widehat{N} = [n_1 n_2 / r_2]. \tag{6.3.34}$$

Note that given n_1 and n_2, the probability law for r_2 is

$$P(r_2 = r | n_1, n_2, N) = \binom{n_1}{r}\binom{N-n_1}{n_2-r}\bigg/\binom{N}{n_2}, \quad r = 0, 1, \ldots, \min(n_1, n_2), \tag{6.3.35}$$

so that

$$\frac{P(r|N, n_1, n_2)}{P(r|N-1, n_1, n_2)} = \frac{(N-n_1)(N-n_2)}{N(N-n_1-n_2+r)} \geq (\leq)\ 1 \tag{6.3.36}$$

according as $n_1 n_2 / r$ is $\geq (\leq) N$, and this characterizes the estimator in (6.3.34). Since

$$P(0|N, n_1, n_2) = \frac{(N-n_1)!(N-n_2)!}{N!(N-n_1-n_2)!} > 0,$$

the estimator $\widehat{N}$ in (6.3.34) may assume the value $+\infty$ with a positive probability, and, hence, any positive-order moment of its distribution fails to be finite. Therefore, we are not in a position to use the probability inequalities considered earlier, which rest on the finiteness of the sth moment for some $s > 0$. On the other hand, we may note that by (6.3.34)

$$N/\widehat{N} \simeq (r_2/n_2)/(n_1/N), \tag{6.3.37}$$

where (r_2/n_2) converges in probability (or in the sth mean, for any $s > 0$) to n_1/N, so that $N/\widehat{N}$ converges in the same mode to 1. Thus, $\widehat{N}/N$ converges in probability to 1 although it does not converge in the rth mean, for any $r > 0$. □

Motivated by this example, we may consider the following useful tool to bypass such a moment existence problem through suitable transformations of statistics.

***Theorem* 6.3.4.** *Let $\{T_n\}$ be a sequence of random variables, such that for some real θ, T_n converges to θ, in probability, or a.s. or completely. Also, let $g : \mathbb{R} \to \mathbb{R}$, be a continuous function at $x = \theta$. Then, $g(T_n)$ converges to $g(\theta)$, in the same mode as T_n converges to θ.*

Proof. By virtue of the assumed continuity of g at θ, for every $\varepsilon > 0$, there exists a $\delta = \delta(\varepsilon) > 0$, such that

$$|t - \theta| < \varepsilon \Rightarrow |g(t) - g(\theta)| < \delta. \tag{6.3.38}$$

Therefore, for every (fixed) n and $\varepsilon > 0$,

$$P[|g(T_n) - g(\theta)| \leq \delta] \geq P(|T_n - \theta| \leq \varepsilon);$$
$$P\left[\bigcap_{N \geq n} |g(T_N) - g(\theta)| \leq \delta\right] \geq P\left[\bigcap_{N \geq n} |T_N - \theta| \leq \varepsilon\right]. \tag{6.3.39}$$

Thus, if $T_n \xrightarrow{P} \theta$, then $P(|T_n - \theta| \leq \varepsilon) \to 1$ as $n \to \infty$, and, hence, the result follows from the first inequality in (6.3.39); if $T_n \to \theta$ a.s. (or completely), then $P(\bigcap_{N \geq n} |T_N - \theta| \leq \varepsilon) \to 1$ [or $\sum_{N \geq n} P(|T_N - \theta| > \varepsilon) \to 0$], as $n \to \infty$, so that the result follows from the second inequality. ■

The theorem remains valid when T_n and/or g are vector-valued; the proof is essentially the same, provided the appropriate distance function is used. If, we want to maintain the same generality as in (6.2.1) or (6.2.4), that is, replacing θ by a (possibly degenerate) random variable T, then we may need to strengthen the continuity of g to *uniform continuity* in order to derive a parallel result. This is presented below.

***Theorem* 6.3.5.** *If $T_n - T \to 0$, in probability, a.s. or completely, and if g is uniformly continuous (a.e. T), then $g(T_n) - g(T) \to 0$ in the same mode, as $n \to \infty$.*

Proof. If g is uniformly continuous (a.e. T), then (6.3.38) holds with θ and $g(\theta)$ replaced by T and $g(T)$, respectively, almost everywhere in T, and, hence, the rest of the proof follows as in Theorem 6.3.4. If $T = \theta$ with probability 1, then uniform continuity of g (a.e. T) is equivalent to the continuity of g at $t = \theta$, so that Theorem 6.3.4 is a special case of Theorem 6.3.5. ■

We illustrate the use of the last two theorems with the following examples.

***Example* 6.3.6.** With reference to the binomial model in Example 6.3.4, consider now the estimation of $\theta = \pi^{-1}$, so that $\theta \in (0, \infty)$. Let $T_n = n^{-1}X_n$ so that $nT_n \sim \text{Bin}(n, \pi)$, and $T_n \to \pi$, in probability or a.s. or completely as well as in the rth mean for any $r > 0$. A natural estimator of θ is $g(T_n) = T_n^{-1}$. As we have seen, $P(T_n = 0) = (1 - \pi)^n > 0$, so that T_n^{-1} assumes the value $+\infty$ with a positive probability (which converges to 0 exponentially in n), and, hence, $g(T_n) \to \pi^{-1} = \theta$, in probability or a.s. or completely, as $n \to \infty$ (although the rth mean convergence does not hold, for any $r > 0$). □

***Example* 6.3.7.** In the setup of Example 6.3.5, assume that N, n_1, n_2 are all large and $n_1/N \to \alpha, n_2/N \to \beta$, where $0 < \alpha, \beta < 1$. Let $T_{n_2} = r_2/n_2$, so that by (6.3.34), $\widehat{N}/N = (n_1/N)T_{n_2}^{-1} = g(T_{n_2})$, where $g(t) = (n_1/N)/t, t \geq 0$. Again, $g(t)$ is continuous, except at $t = 0$, whereas, by the Chebyshev Inequality (6.2.27), we have for all $\varepsilon > 0$,

$$P\left(\left|T_{n_2} - \frac{n_1}{N}\right| > \varepsilon\right) \leq (n_2\varepsilon^2)^{-1} n_1 \frac{(N - n_1)(N - n_2)}{N^2(N - 1)}, \tag{6.3.40}$$

so that $T_{n_2} \to n_1/N$, in probability, as $N \to \infty$. Thus, by Theorem 6.3.4, we conclude that $\widehat{N}/N \xrightarrow{P} 1$, as $N \to \infty$. □

Example* 6.3.8 *(Ratio estimator). Let $(X_1, Y_1), \ldots, (X_n, Y_n)$ be i.i.d. random variables with mean vector (μ, ν) and dispersion matrix $\boldsymbol{\Sigma} = ((\sigma_{ij}))$, $i, j = 1, 2$; assume that all these parameters are finite and that $\nu \neq 0$. Our concern is to estimate $\theta = \mu/\nu$. A natural estimator of θ is $\widehat{\theta}_n = \overline{X}_n/\overline{Y}_n$. By the use of the Chebyshev Inequality (6.2.27) or other inequalities, we may easily verify that $(\overline{X}_n, \overline{Y}_n)$ stochastically converges to (μ, ν), as $n \to \infty$. However, $\widehat{\theta}_n$ may not have any finite positive-order moment, so that such a probability inequality may not be directly applicable to $\widehat{\theta}_n$. We note that $g(t_1, t_2) = t_1/t_2$ is a continuous function of (t_1, t_2), except on the line $t_2 = 0$, and, hence, whenever $\nu \neq 0$, we conclude that $(\overline{X}_n, \overline{Y}_n) \to (\mu, \nu)$, in probability or almost surely or completely, implies that $\widehat{\theta}_n \to \theta$, in the same mode. □

Example* 6.3.9 *(ANOVA model). Let X_{ij}, $j = 1, \ldots, n$, be *i.i.d.* random variables with mean μ_i and finite variance σ^2, for $i = 1, \ldots, k, k \geq 2$; all these k samples are assumed to be independent. Let $\overline{X}_i = n^{-1}\sum_{j=1}^n X_{ij}$, $i = 1, \ldots, k$, $\overline{X} = k^{-1}\sum_{i=1}^k \overline{X}_i$, $s_B^2 = \sum_{i=1}^k(\overline{X}_i - \overline{X})^2$, $s_n^2 = [k(n-1)]^{-1}\sum_{i=1}^k(X_{ij} - \overline{X}_i)^2$. For testing the null hypothesis of homogeneity of $\mu_1, \ldots, \mu_k$, the usual analysis of variance test statistic is

$$F_{k-1,k(n-1)} = s_B^2/[(k-1)s_n^2]. \tag{6.3.41}$$

Side by side, we introduce the statistic $F_{k-1}^* = s_B^2/[(k-1)\sigma^2]$. Computation of the actual central moments of $F_{k-1,k(n-1)}$ may be quite cumbersome (when the actual distribution of the X_{ij} is not known), and for possibly nonnormal distributions, the mean, variance or other central moments of $F_{k,k(n-1)}$ may not exist. Thus, a direct use of the Chebyshev Inequality (6.2.27) or other moment-based probability inequality may not be feasible. We will see in Example 6.3.11 that whenever σ^2 is finite, $s_n^2 \xrightarrow{a.s.} \sigma^2$. Consequently, whenever $0 < \sigma^2 < \infty$, by using Theorem 6.3.5 we conclude that

$$F_{k-1,k(n-1)} - F_{k-1}^* \xrightarrow{P} 0. \tag{6.3.42}$$

For F^*_{k-1} , as we will see in Chapter 7, a suitable limiting distribution exists under quite general conditions, and, hence, we may use the same limiting distribution for $F_{k-1,k(n-1)}$ as well; in this context, we use the implications depicted in Figure 6.2.1. □

In most of the cases dealing with simple statistical models, we encounter suitable statistics that can be expressed as averages over independent random variables (or vectors) or as functions of such averages. In many other situations, by suitable expansions, we may also be able to express a statistic as an average over independent random variables plus a remainder term, which converges to 0 in a certain mode. As such, the study of the stochastic convergence of averages of independent random variables occupies a central position in asymptotic theory. Although the probability inequalities studied earlier (or some others to follow) may be used in this context, a more direct approach (requiring generally less-restrictive regularity conditions) has been worked out; conventionally, such results are categorized under the denomination of Laws of Large Numbers. We may start with the following theorem.

Theorem 6.3.6 (Khintchine Weak Law of Large Numbers). *Let $\{X_k, k \geq 1\}$ be a sequence of i.i.d. random variables with $\mathbb{E}(X_1) = \theta$, and let $\overline{X}_n = n^{-1}\sum_{k=1}^n X_k$, $n \geq 1$. Then $\overline{X}_n \xrightarrow{P} \theta$.*

Proof. Let $F(x)$, $x \in \mathbb{R}$, be the distribution function of X_1, and let $\phi_F(t)$, $t \in \mathbb{R}$, be the characteristic function of X_1. Also, note that as $n \to \infty$,

$$\begin{aligned}\phi_{\overline{X}_n}(t) &= \mathbb{E}\left(\prod_{k=1}^n e^{itX_k/n}\right) = [\phi_F(t/n)]^n \\ &= \{1 + itn^{-1}\mathbb{E}(X_1) + o(n^{-1})t\}^n \to e^{it\theta}. \end{aligned} \tag{6.3.43}$$

Since $\exp(it\theta)$ is the characteristic function of a random variable Z, which is degenerate at the point θ [i.e., $P(Z = \theta) = 1$], the proof can be completed by appealing to Theorem 6.2.1. ■

The simple treatment in (6.3.43) may not work out that well when the X_k are independent but not identically distributed; if we denote by $\phi_{F_k}(t)$ the characteristic function of X_k, then, of course, we have

$$\phi_{\overline{X}_n}(t) = \prod_{k=1}^n \phi_{F_k}(t/n) \Rightarrow \log \phi_{\overline{X}_n}(t) = \sum_{k=1}^n \log \phi_{F_k}(t/n), \tag{6.3.44}$$

but in order for the right-hand side to converge to a limiting characteristic function of a degenerate random variable, we may need additional (uniformity) conditions on the $\phi_{F_k}(t)$, and, these may, in turn, require higher-order moment conditions on the X_k. In addition, if the X_k are not independent, then the factorization in (6.3.44) may not hold, and the treatment of the convergence of $\phi_{\overline{X}_n}(t)$ to a degenerate characteristic function may become still more complicated. Finally, if instead of the convergence in probability, we seek an a.s. convergence, then, even in the i.i.d. case, Theorem 6.2.1 fails to provide the desired tool, and a more involved proof is, therefore, in order. Keeping this spectrum in mind, we may appreciate the need for the different LLNs, which address different degrees of dependence structure and/or more conditions on the underlying random variables. Before we present some of these LLNs, we may note that for independent random variables, if moments up to the order r exist for some $r \geq 2$, then by the use of the Chebyshev Inequality (6.2.27) or the Markov Inequality (6.3.4), the stochastic convergence of $\overline{X}_n - \mathbb{E}(\overline{X}_n)$ to 0 can be easily established. On the other hand, for the i.i.d. case, the first moment suffices. Thus, there is a natural question: for random variables that are not necessarily identically distributed, do we still need the existence of the second moment for some LLN to hold, or we may relax this condition to a certain extent? Toward this, we consider the following example.

***Example* 6.3.10.** Consider a sequence $\{X_k\}$ of independent random variables, such that, for $k \geq 1$,

$$X_k = \begin{cases} -2^{k(1-\varepsilon)} & \text{with probability } 2^{-k-1} \\ 0 & \text{with probability } 1 - 2^{-k} \\ 2^{k(1-\varepsilon)} & \text{with probability } 2^{-k-1}, \end{cases} \tag{6.3.45}$$

where $0 < \varepsilon < 1$. Clearly, $\mathbb{E}(X_k) = 0$ and $\mathbb{V}\mathrm{ar}(X_k) = \mathbb{E}(X_k^2) = 2^{k(1-2\varepsilon)} < \infty$, for every $k \geq 1$. Thus, we have $\mathbb{E}(\overline{X}_n) = 0$, for every $n \geq 1$, and

$$\begin{aligned} \mathbb{V}\mathrm{ar}(\overline{X}_n) = \gamma_n^2 &= n^{-2} \sum_{k=1}^{n} (2^{1-2\varepsilon})^k \\ &= n^{-2} 2^{1-2\varepsilon} \left[\frac{2^{n(1-2\varepsilon)} - 1}{2^{(1-2\varepsilon)} - 1} \right]. \end{aligned} \tag{6.3.46}$$

Thus, for $\varepsilon > 1/2$, of course, $\gamma_n^2 \to 0$ as $n \to \infty$, and, hence, the Chebyshev Inequality (6.2.27) provides the access to the desired goal. A similar picture holds for the Markov Inequality (6.3.4) involving moments of order $r \geq 2$. However, none of these inequalities relates to necessary conditions, and the LLN may hold without either of them. This will be made clear from the following LLN. We will come back to this example later. □

Theorem* 6.3.7 *(Markov Weak Law of Large Numbers). *Let X_k, $k \geq 1$, be independent random variables, such that $\mu_k = \mathbb{E}(X_k)$, $k \geq 1$ exist, and for some δ such that $0 < \delta \leq 1$, along with the existence of $\mathbb{E}(|X_k - \mu_k|^{1+\delta})$, suppose that*

$$n^{-1-\delta} \sum_{k=1}^{n} \mathbb{E}(|X_k - \mu_k|^{1+\delta}) = \rho_n(\delta) \to 0, \quad \text{as } n \to \infty. \tag{6.3.47}$$

Then $\overline{X}_n - \mathbb{E}(\overline{X}_n) \xrightarrow{P} 0$.

Proof. Expression (6.3.47) is known as the *Markov condition*. Note that without any loss of generality we may take $0 < \delta < 1$, since for $\delta \geq 1$, the Chebyshev Inequality (6.2.27) applies. We let $U_k = X_k - \mu_k$, so that $\mathbb{E}(U_k) = 0$, $k \geq 1$. Then $\overline{X}_n - \mathbb{E}(\overline{X}_n) = \overline{U}_n$, $n \geq 1$. We denote the characteristic functions of U_k and $\overline{U}_n$ by $\phi_{U_k}(t)$ and $\phi_{\overline{U}_n}(t)$, respectively. Since the U_k are independent, as in (6.3.44), we have for all $t \in \mathbb{R}$,

$$\phi_{\overline{U}_n}(t) = \prod_{k=1}^{n} \phi_{U_k}(t/n) \Rightarrow \log \phi_{\overline{U}_n}(t) = \sum_{k=1}^{n} \log \phi_{U_k}(t/n). \tag{6.3.48}$$

At this stage, we recall Theorem 1.5.2 and note that under the assumed moment condition, for every $k \geq 1$ and real θ,

$$\phi_{U_k}(\theta) = 1 + i\theta\mathbb{E}(U_k) + R_k(\theta), \tag{6.3.49}$$

where

$$|R_k(\theta)| \leq c|\theta|^{1+\delta}\mathbb{E}(|U_k|^{1+\delta}), \quad c < \infty.$$

Also, note that for every $t \in \mathbb{R}$,

$$\begin{aligned} \max_{1\leq k\leq n} |R_k(t/n)| &\leq \sum_{k=1}^{n} |R_k(t/n)| \\ &\leq cn^{-1-\delta}|t|^{1+\delta}\sum_{k=1}^{n}\mathbb{E}(|U_k|^{1+\delta}), \end{aligned} \tag{6.3.50}$$

where, by (6.3.47), the right-hand side of (6.3.50) converges to 0 as $n \to \infty$. Finally, note that $\mathbb{E}(U_k) = 0$ for every $k \geq 1$, and if $a_{n1}, \ldots, a_{nn}$ are numbers (real or imaginary), such that $\sum_{k=1}^{n} |a_{nk}| \to 0$ as $n \to \infty$, then,

$$\left|\sum_{k=1}^{n} \log(1 + a_{nk}) - \sum_{k=1}^{n} a_{nk}\right| \to 0, \quad \text{as } n \to \infty.$$

Combining these with (6.3.48), we immediately obtain that

$$\left|\log \phi_{\overline{U}_n}(t) - \sum_{k=1}^{n} R_k(t/n)\right| \to 0 \quad \text{as } n \to \infty, \tag{6.3.51}$$

so that, by (6.3.50) and (6.3.51), under (6.3.47) we obtain that, for every real t, $\log \phi_{\overline{U}_n}(t) \to 0$ as $n \to \infty$, and, hence, that $\phi_{\overline{U}_n}(t) \to 1$, as $n \to \infty$, for every $t \in \mathbb{R}$. This ensures that $\overline{U}_n$ converges in law to a degenerate random variable having the entire probability mass at the point 0, and the proof of the theorem follows by using Theorem 6.2.1. ■

We illustrate the use of the Markov LLN with Example 6.3.10. First note that

$$\mathbb{E}(|X_k|^{1+\delta}) = 2^{-k+k(1-\varepsilon)(1+\delta)} \leq 1 \quad \text{if} \quad (1+\delta)(1-\varepsilon) \leq 1. \tag{6.3.52}$$

In particular, if we choose $\delta = \varepsilon$, then $\mathbb{E}(|X_k|^{1+\delta}) < 1$, for every $k \geq 1$, and, hence, (6.3.47) is bounded by $n^{-\varepsilon}$ ($\downarrow 0$, as $n \to \infty$). Thus, the weak LLN applies to the $\overline{X}_n$ under the Markov condition (6.3.47), for every $\varepsilon > 0$, although we have seen earlier that for $\varepsilon < 1/2$, the Chebyshev Inequality (6.2.27) or the Markov Inequality (6.3.4) fail to yield the desired result.

For the stochastic convergence of averages of independent (but not necessarily identically distributed) random variables, the Markov weak LLN is the most general result. However, this is not enough to ensure a.s. convergence in the same setup and an a.s. convergence result can be established under a second moment condition on the U_k. In this context, we consider first a very useful probability inequality which has a variety of uses in Asymptotic Theory.

Theorem 6.3.8 (Kolmogorov Maximal Inequality). *Let $X_1, \ldots, X_n$ be independent random variables, such that $\mathbb{E}(X_i) = \mu_i$ and $\mathbb{E}[(X_i - \mu_i)^2] = \sigma_i^2$ exist for every $i = 1, \ldots, n$.*

Then let $T_k = (X_1 - \mu_1) + \cdots + (X_k - \mu_k)$, for $k = 1, \ldots, n$, and, conventionally, we let $T_0 = 0$. Then, for every $t > 0$,

$$P\left(\max_{0 \le k \le n} |T_k| > t\right) \le t^{-2}\mathbb{V}ar(T_n) = t^{-2}\sum_{k=1}^{n} \sigma_k^2. \tag{6.3.53}$$

[Note the analogy with the Chebyshev Inequality in (6.3.1) where we have to take $T_n = \overline{X}_n$. For both, the right-hand sides are the same, but compared to the marginal event $\{|T_n| > t\}$, here we have the union of the events $\{|T_k| > t\}$, for $k = 1, \ldots, n$, and, hence, the statement on the left-hand side of (6.3.53) is much stronger than the one in (6.3.1).]

Proof. Without loss of generality, we set $\mu_i = 0$, for $i = 1, \ldots, n$ and define the following mutually disjoint events

$$\begin{aligned} A_0 &= \{|T_k| \le t,\ k = 0, 1, \ldots, n\}, \\ A_k &= \{|T_j| \le t,\ j \le k-1,\ |T_k| > t\},\ k = 1, \ldots, n. \end{aligned} \tag{6.3.54}$$

Then, we may observe that

$$P\left(\max_{1 \le k \le n} |T_k| > t\right) = P\left(\bigcup_{k=1}^{n} A_k\right) = \sum_{k=1}^{n} P(A_k) = 1 - P(A_0). \tag{6.3.55}$$

Also, we have

$$\sum_{k=1}^{n} \sigma_k^2 = \mathbb{E}(T_n^2) = \sum_{k=0}^{n} \mathbb{E}[T_n^2 I(A_k)] \ge \sum_{k=1}^{n} \mathbb{E}[T_n^2 I(A_k)]. \tag{6.3.56}$$

Therefore, to establish (6.3.53), it suffices to show that for every $k = 1, \ldots, n$,

$$\mathbb{E}[T_n^2 I(A_k)] \ge t^2 P(A_k). \tag{6.3.57}$$

For this purpose, we write $T_n = T_k + (T_n - T_k)$ and note that $T_n - T_k$ is independent of T_j, $j \le k$, for every $k = 1, \ldots, n-1$. Moreover, the event A_k in (6.3.54) depends only on the T_j, $j \le k$, so that on the set A_k, $T_n - T_k$ has expectation 0. Thus, we have

$$\begin{aligned} \mathbb{E}[T_n^2 I(A_k)] &= \mathbb{E}[T_k^2 I(A_k)] + \mathbb{E}[(T_n - T_k)^2 I(A_k)] + 2\mathbb{E}[T_k(T_n - T_k)I(A_k)] \\ &= \mathbb{E}[T_k^2 I(A_k)] + \mathbb{E}[(T_n - T_k)^2 I(A_k)] + 0 \\ &\ge \mathbb{E}[T_k^2 I(A_k)] \\ &\ge t^2 P(A_k), \end{aligned} \tag{6.3.58}$$

where the last step follows because of A_k, $|T_k| > t$, for $k = 1, \ldots, n$. ■

We will consider various extensions of the Kolmogorov Maximal Inequality (6.3.53) for suitable sequences of dependent random variables in the next section. We consider another extension of this inequality that plays a vital role in the subsequent developments.

Theorem 6.3.9 (Hájek–Rènyi Inequality). *Let $X_1, \ldots, X_n$ be independent random variables with means $\mu_1, \ldots, \mu_n$ and finite variances $\sigma_1^2, \ldots, \sigma_n^2$, respectively. Let $T_k = \sum_{i=1}^{k}(X_i - \mu_i)$, $k = 1, \ldots, n$. Then, if $c_1 \ge c_2 \ge \cdots \ge c_n$ are positive constants, we have* *for every* $t > 0$,

$$P\left(\max_{1\le k\le n} c_k|T_k| > t\right) \le t^{-2}\sum_{k=1}^{n} c_k^2\sigma_k^2. \tag{6.3.59}$$

Proof. As in the proof of the Kolmogorov Inequality (6.3.53), we let $T_0 = 0$, and define

$$A_0 = \{c_k|T_k| \le t,\ k = 0, \ldots, n\} \quad (c_0 = c_1),$$
$$A_k = \left\{c_j|T_j| \le t,\ j \le k-1;\ c_k|T_k| > t\right\}, \quad k = 1, \ldots, n. \tag{6.3.60}$$

Then, by the definition of the A_k, we have

$$\sum_{k=1}^{n} \mathbb{E}\left[c_k^2 T_k^2 I(A_k)\right] \ge \sum_{k=1}^{n} \mathbb{E}[t^2 I(A_k)]$$
$$= t^2 \sum_{k=1}^{n} P(A_k) = t^2 P\left(\max_{1\le k\le n} c_k|T_k| > t\right). \tag{6.3.61}$$

Thus, to prove (6.3.59), it suffices to show that

$$\sum_{k=1}^{n} c_k^2 \mathbb{E}\left[T_k^2 I(A_k)\right] \le \sum_{k=1}^{n} c_k^2\sigma_k^2. \tag{6.3.62}$$

For this purpose, we let $B_0 = \Omega$ and $B_k = \left(\bigcup_{i=1}^k A_i\right)^c = \bigcap_{i=1}^k A_i^c$, for $k = 1, \ldots, n$. Then, noting that $I(A_k) = I(B_{k-1}) - I(B_k)$ and $c_k \ge c_{k+1}$, $k = 1, \ldots,$ we may write

$$\sum_{k=1}^{n} c_k^2 \mathbb{E}\left[T_n^2 I(A_k)\right] = \sum_{k=1}^{n} c_k^2\left\{\mathbb{E}\left[T_k^2 I(B_{k-1})\right] - \mathbb{E}[T_k^2 I(B_k)]\right\}$$
$$\le c_1^2\mathbb{E}\left(T_1^2\right) + \sum_{k=2}^{n} c_k^2\mathbb{E}\left\{\left[T_k^2 - T_{k-1}^2\right] I(B_{k-1})\right\} - c_n^2\mathbb{E}\left[T_n^2 I(B_n)\right]$$
$$\le c_1^2\mathbb{E}\left(T_1^2\right) + \sum_{k=2}^{n} c_k^2\left\{\mathbb{E}\left[(X_k - \mu_k)^2 I(B_{k-1})\right]\right.$$
$$\left. + \mathbb{E}\left[(X_k - \mu_k)T_{k-1} I(B_{k-1})\right]\right\}$$
$$\le c_1^2\sigma_1^2 + \sum_{k=2}^{n} c_k^2\sigma_k^2 + 0 = \sum_{k=1}^{n} c_k^2\sigma_k^2, \tag{6.3.63}$$

where, the penultimate step follows because X_k is independent of B_{k-1}, for every $k = 1, \ldots, n$. ■

We will consider some extensions of this inequality for dependent random variables in the next section. In passing, we may remark that in (6.3.53) and (6.3.59), the X_i are not necessarily identically distributed. As such, if we define $Y_1 = \cdots = Y_{M-1} = 0$ and $Y_k = T_k = \sum_{i=1}^k (X_i - \mu_i)$, for $k \ge M$, where the X_i are independent random variables with means μ_i and finite variances σ_i^2, $i \ge 1$, we obtain, on letting $c_k = k^{-1}$, for $k \ge M$

and $n = N$, that for all $t > 0$,

$$P\left(\max_{M\le k\le N} k^{-1}|T_k| > t\right) \le t^{-2}\left(M^{-2}\mathbb{V}\text{ar}(Y_M) + \sum_{k=M+1}^{N} k^{-2}\sigma_k^2\right)$$
$$\le t^{-2}\left[M^{-2}\left(\sum_{i=1}^{M}\sigma_i^2\right) + \sum_{k=M+1}^{N} k^{-2}\sigma_k^2\right]. \qquad (6.3.64)$$

The last inequality provides the key to the proof of the following result.

Theorem 6.3.10 (Kolmogorov Strong Law of Large Numbers). *Let X_i, $i \ge 1$, be independent random variables, such that $\mathbb{E}(X_i) = \mu_i$ and $\mathbb{V}ar(X_i) = \sigma_i^2$ exist for every $i \ge 1$. Also, let $\overline{\mu}_n = n^{-1}\sum_{i=1}^{n}\mu_i$, for $n \ge 1$. Then,*

$$\sum_{k\ge 1} k^{-2}\sigma_k^2 < \infty \Rightarrow \overline{X}_n - \overline{\mu}_n \xrightarrow{a.s.} 0. \qquad (6.3.65)$$

Proof. Let $D_k = \sum_{n\ge k} n^{-2}\sigma_n^2$, for $k \ge 1$ and note that $D_1 < \infty$ ensures that D_k is nonincreasing in k and $\lim_{k\to\infty} D_k = 0$. Also, note that for every $M \ge 1$,

$$\frac{1}{M^2}\sum_{k\le M}\sigma_k^2 = \frac{1}{M^2}\sum_{k\le M} k^2[D_k - D_{k+1}]$$
$$\le \frac{1}{M^2}\sum_{k=1}^{M}(2k-1)D_k, \qquad (6.3.66)$$

where the right-hand side converges to 0 as $M \to \infty$. Now, to prove that $\overline{X}_n - \overline{\mu}_n \xrightarrow{a.s.} 0$, we make use of (6.3.64) and obtain that for every $N > M$, $\varepsilon > 0$,

$$P\left(\max_{M\le k\le N}|\overline{X}_n - \overline{\mu}_n| > \varepsilon\right) = P\left(\max_{M\le k\le N} k^{-1}|T_k| > \varepsilon\right)$$
$$\le \varepsilon^{-2}\left[M^{-2}\left(\sum_{k=1}^{M}\sigma_k^2\right) + \sum_{k=M+1}^{N} k^{-2}\sigma_k^2\right]. \qquad (6.3.67)$$

Thus, for any given M, first allowing $N \to \infty$, we obtain from (6.3.67) that

$$P\left(\sup_{k\ge M}|\overline{X}_k - \overline{\mu}_k| > \varepsilon\right) \le \varepsilon^{-2}\left(M^{-2}\sum_{k=1}^{M}\sigma_k^2 + D_{M+1}\right), \qquad (6.3.68)$$

and, hence, allowing M to go to $+\infty$, it follows that the right-hand side of (6.3.68) converges to 0. ■

In passing, we may remark that if the X_i in Theorem 6.3.10 are i.i.d. random variables, then $\sigma_i^2 = \sigma^2$, $i \ge 1$, whereas $\sum_{k\ge 1} k^{-2} = \pi^2/6 < \infty$. Hence, $\overline{X}_n - \mathbb{E}(\overline{X}_n) \xrightarrow{a.s.} 0$, whenever $\sigma^2 < \infty$. However, as we will see later in this section, for the i.i.d. case we do not require the second moment condition, and the finiteness of the first moment suffices. The interesting point is that the Kolmogorov condition in (6.3.65) allows the σ_i^2 to be varying with i, but at a rate slower than i, and, in particular, if the σ_i^2 are all bounded (but not necessarily equal), then this condition holds automatically. In order to consider better results for the i.i.d. case, we first consider the following theorem.

***Theorem* 6.3.11.** *Let X be a real-valued random variable Then*

$$\mathbb{E}(|X|) < \infty \Leftrightarrow \sum_{k\geq 1} kP(k \leq |X| < k+1) < \infty. \tag{6.3.69}$$

Proof. First, note that for every $k \geq 0$,

$$\begin{aligned} kP(k \leq |X| < k+1) &\leq \int_k^{k+1} x\mathrm{d}P(|X| \leq x) \\ &\leq (k+1)P(k \leq |X| < k+1). \end{aligned} \tag{6.3.70}$$

Hence, noting that $\mathbb{E}(|X|) = \sum_{k\geq 0} \int_k^{k+1} x\mathrm{d}P(|X| \leq x)$, summing over k in (6.3.70), the first inequality yields that

$$\mathbb{E}(|X|) < \infty \Rightarrow \sum_{k\geq 1} kP(k \leq |X| < k+1) < \infty,$$

whereas from the last inequality we obtain

$$\sum_{k\geq 1} kP(k \leq |X| < k+1) < \infty \Rightarrow \mathbb{E}(|X|) < \infty. \qquad \blacksquare$$

Theorem* 6.3.12 *(Khintchine Equivalence Lemma). *Let $\{X_n\}$ and $\{Y_n\}$ be two arbitrary sequences of random variables. Then, if $\sum_{i\geq 1} P(X_i \neq Y_i) < \infty$, the Strong Law of Large Numbers (SLLN) holds for both sequences or for none.*

Proof. Note that

$$n^{-1}\sum_{i\leq n} X_i = n^{-1}\sum_{i\leq n} Y_i + n^{-1}\sum_{i\leq n}(X_i - Y_i), \quad n \geq 1.$$

If $\sum_{i\geq 1} P\{X_i \neq Y_i\} < \infty$, if follows from the Borel–Cantelli Lemma (Theorem 6.3.2) that only finitely many of the events $\{X_i \neq Y_i\}$ may occur and, hence, $n^{-1}\sum_{i\leq n}(X_i - Y_i) \xrightarrow{a.s.} 0$. Therefore, $n^{-1}\sum_{i\leq n} X_i$ and $n^{-1}\sum_{i\leq n} Y_i$ are *convergence equivalent*, that is, as $n \to \infty$, either both series converge or none. ■

We are now in a position to prove the following SLLN.

Theorem* 6.3.13 *(Khintchine Strong Law of Large Numbers). *Let X_i, $i \geq 1$ be i.i.d. random variables . Then $\overline{X}_n \xrightarrow{a.s.} c$, if and only if $\mathbb{E}(X_1)$ exists and $c = \mathbb{E}(X_1)$.*

Proof. First assume that $\mu = \mathbb{E}(X_1)$ exists, and let $U_i = X_i - \mu$, $i \geq 1$. Then, let us show that $\overline{U}_n \xrightarrow{a.s.} 0$. For this purpose, define

$$Y_i = U_i I(|U_i| \leq i), \quad i \geq 1. \tag{6.3.71}$$

Since the U_i are i.i.d. random variables, we have

$$P(Y_i \neq U_i) = P(|U_i| > i) = P(|U_1| > i) = \sum_{k \geq i} P(k < |U_i| \leq k+1), \tag{6.3.72}$$

for every $i \geq 1$, so that, by Theorems 6.3.11 and (6.3.72),

$$\sum_{i \geq 1} P(Y_i \neq U_i) = \sum_{i \geq 1} \sum_{k \geq i} P(k < |U_1| \leq k+1) = \sum_{k \geq 1} k P(k < |U_1| \leq k+1) < \infty. \tag{6.3.73}$$

Using the Khintchine Equivalence Lemma (Theorem 6.3.12) along with (6.3.73), it suffices to show that $\overline{Y}_n \xrightarrow{a.s.} 0$. Toward this, first, we observe that, for $n \geq 1$,

$$\begin{aligned} \mathbb{E}(\overline{Y}_n) &= n^{-1} \sum_{i=1}^{n} \mathbb{E}[U_i I(|U_i| \leq i)] = -n^{-1} \sum_{i=1}^{n} \mathbb{E}[U_1 I(|U_1| > i)] \\ &= -n^{-1} \sum_{i=1}^{n-1} i \mathbb{E}[U_1 I(i < |U_1| \leq i+1)] - \mathbb{E}[U_1 I(|U_1| > n)]. \end{aligned} \tag{6.3.74}$$

Now, $\mathbb{E}(U_1)$ exists and, hence, $|\mathbb{E}[U_1 I(|U_1| > n)]| \to 0$ as $n \to \infty$. Similarly,

$$\begin{aligned} &\left| n^{-1} \sum_{i=1}^{n-1} i \mathbb{E}[U_1 I(i < |U_1| \leq i+1)] \right| \\ &\leq k^2 n^{-1} P(|U_1| \leq k) + \sum_{i=k+1}^{n-1} i P(i < |U_1| \leq i+1) \\ &\leq k^2 n^{-1} + \mathbb{E}[|U_1| I(|U_1| \geq k+1)], \quad k \in (1, n-1). \end{aligned} \tag{6.3.75}$$

Therefore, by (6.3.74) and (6.3.75), we conclude [on letting $k = k_n = o(n^{1/2})$] that

$$\mathbb{E}(\overline{Y}_n) \to 0 \text{ as } n \to \infty. \tag{6.3.76}$$

Thus, to establish that $\overline{Y}_n \xrightarrow{a.s.} 0$, it suffices to show that $\overline{Y}_n - \mathbb{E}(\overline{Y}_n) \xrightarrow{a.s.} 0$, and for this purpose, we may as well use the Kolmogorov SLLN (Theorem 6.3.10). In this setup, note that for all $i \geq 1$,

$$\mathbb{V}\mathrm{ar}(Y_i) \leq \mathbb{E}(Y_i^2) = \mathbb{E}[U_1^2 I(|U_1| \leq i)] \leq \sum_{k=0}^{i-1} (k+1)^2 P(k < |U_1| \leq k+1). \tag{6.3.77}$$

Consequently,

$$\begin{aligned}\sum_{k\geq 1}\frac{\mathbb{V}\mathrm{ar}(Y_k)}{k^2} &\leq \sum_{k\geq 1}k^{-2}\sum_{i=0}^{k-1}(i+1)^2P(i<|U_1|\leq i+1)\\ &\leq \sum_{i=0}^{\infty}(i+1)^2P(i<|U_1|\leq i+1)\sum_{k=i+1}^{\infty}k^{-2}\\ &\leq P(|U_1|\leq 1)\sum_{i=1}^{\infty}i^{-2}\\ &\quad+\sum_{k=1}^{\infty}P(k<|U_1|\leq k+1)(k+1)^2\sum_{i=k+1}^{\infty}i^{-2}\\ &\leq \left(\frac{\pi^2}{6}\right)P(|U_1|\leq 1)+4\sum_{k=1}^{\infty}kP(k<|U_1|\leq k+1). \end{aligned}\tag{6.3.78}$$

Since $\sum_{i\geq 1}i^{-2}=\pi^2/6$ and, for every $k\geq 1$,

$$(k+1)^2\sum_{i\geq k+1}i^{-2}\leq 2(k+1)^2\sum_{i\geq k+1}[i(i+1)]^{-1}=2(k+1),$$

the right-hand side of (6.3.78) converges, and hence, by the Kolmogorov SLLN (Theorem 6.3.10), $\overline{Y}_n-\mathbb{E}(\overline{Y}_n)\xrightarrow{a.s.}0$. Thus, if $\mathbb{E}(X_1)$ exists, then $\overline{X}_n-\mathbb{E}(X_1)\xrightarrow{a.s.}0$.

Alternatively, suppose now that $\overline{X}_n\xrightarrow{a.s.}c$, for some finite c. Then we show that $c=\mathbb{E}(X)$. Let $U_i=X_i-c$, $i\geq 1$, so that by the definition of a.s. convergence, we claim that for every $\varepsilon>0$,

$$P\left[\bigcup_{N\geq n}(|\overline{U}_N|>\varepsilon)\right]\to 0,\quad \text{as } n\to\infty. \tag{6.3.79}$$

Note that $(N+1)\overline{U}_{n+1}=N\overline{U}_N+U_{N+1}$, for every N, so that $|\overline{U}_N|<\varepsilon$, $N\geq n$, implies that $|N^{-1}U_N|\leq 2\varepsilon$, for every $N\geq n$. Consequently, we have

$$P\left(\bigcap_{N\geq n}(N^{-1}|U_N|\leq 2\varepsilon)\right)\geq P\left(\bigcap_{N\geq n}(|\overline{U}_N|\leq\varepsilon)\right), \tag{6.3.80}$$

which in conjunction with (6.3.79) implies that, for every positive ε, as $n\to\infty$,

$$P\left[\bigcup_{N\geq n}(N^{-1}|U_N|>2\varepsilon)\right]\leq P\left[\bigcup_{N\geq n}(|\overline{U}_N|>\varepsilon)\right]\to 0. \tag{6.3.81}$$

Since the events $A_N=\left(N^{-1}|U_N|>\varepsilon\right)$, $N\geq 1$, are independent and $\varepsilon>0$ is arbitrary, on letting $\varepsilon=1/2$ and using the second part of the Borel–Cantelli Lemma (Theorem 6.3.2) it follows that (6.3.81) implies

$$\sum_{N\geq 1}P(N^{-1}|U_N|>1)<\infty. \tag{6.3.82}$$

On the other hand, the left-hand side of (6.3.82) is equal to

$$\begin{aligned}\sum_{N\geq 1} P(N^{-1}|U_N| > 1) &= \sum_{N\geq 1}\sum_{k\geq N} P(k < |U_N| \leq k+1)\\ &= \sum_{N\geq 1}\sum_{k\geq N} P\,(k < |U_1| \leq k+1)\\ &= \sum_{k\geq 1} kP\,(k < |U_1| \leq k+1)\\ &\geq \mathbb{E}(|U_1|) - 1,\end{aligned}$$

by Theorem 6.3.11. Therefore, $\mathbb{E}(U_1)$ exists. As $\mathbb{E}(U_1) = \mathbb{E}(X_1) - c$, it follows that $\mathbb{E}(X_1)$ also exists. If $c \neq \mathbb{E}(X_1)$, then by the first part of the theorem, $\overline{X}_n \xrightarrow{a.s.} \mathbb{E}(X_1) = \mu$, which is a contradiction. Therefore, $c = \mathbb{E}(X_1)$. ■

The Khintchine SLLN (Theorem 6.3.13) provides the a.s. convergence for the average of i.i.d. random variables under minimal conditions. The theorem extends directly to the vector case by treating the average of each coordinate separately. In fact, we may treat a somewhat more general result as follows. Suppose that $g(\boldsymbol{y})$ is a real-valued function of a vector $\boldsymbol{y} \in \mathbb{R}^p$, for some $p \geq 1$. Suppose further that there exists a sequence $\{\boldsymbol{Z}_i, i \geq 1\}$ of i.i.d. p-vectors, such that $\mathbb{E}(\boldsymbol{Z}_1) = \boldsymbol{\theta}$ exists. Finally, suppose that g is continuous at $\boldsymbol{\theta}$. Then, $g(\overline{\boldsymbol{Z}}_n) \xrightarrow{a.s.} g(\boldsymbol{\theta})$. A parallel result holds when g is itself a q-vector, for some $q \geq 1$. We invite the reader to complete the proof of this result by using the definition of continuity of g in the multivariate case and the Khintchine SLLN (Theorem 6.3.13) for the $\boldsymbol{Z}_i$. To illustrate the utility of the above results, we consider the following examples.

***Example* 6.3.11.** Let $\{X_i, i \geq 1\}$ be i.i.d. random variables, such that $\mathbb{E}(X_1) = \theta$ and $\mathbb{V}\text{ar}(X_1) = \sigma^2$ where both θ and σ^2 are assumed to be finite. A natural estimator of σ^2 is the sample variance s_n^2. Note that

$$n^{-1}(n-1)s_n^2 = n^{-1}\sum_{i=1}^{n}(X_i - \theta)^2 - (\overline{X}_n - \theta)^2. \tag{6.3.83}$$

Thus, if we let $Z_{1i} = (X_i - \theta)^2$, $Z_{2i} = (X_i - \theta)$ and $\boldsymbol{Z}_i = (Z_{1i}, Z_{2i})'$, $i \geq 1$, then by the Khintchine SLLN (Theorem 6.3.13) (in the vector case), $\overline{\boldsymbol{Z}}_n \xrightarrow{a.s.} (\sigma^2, 0)'$. On the other hand, $(n-1)/n \to 1$ as $n \to \infty$, whereas $g(t_1, t_2) = t_1 - t_2^2$ is a continuous function of (t_1, t_2). Hence, we conclude that

$$s_n^2 \xrightarrow{a.s.} \sigma^2 \quad \text{whenever } \mathbb{E}(X_1^2) < \infty. \tag{6.3.84}$$

Note that for (6.3.84), the finiteness of the second moment suffices. Had we used the Kolmogorov SLLN (Theorem 6.3.10) for each component, the finiteness of $\mathbb{E}(X_1^4)$ would be required and that would have been more restrictive than the second moment condition in (6.3.84). If, however, we use a slightly more general model, where the X_i have the same variance σ^2 (but possibly different distributions), then to establish the a.s. convergence result for s_n^2, we may not be able to use the Khintchine SLLN (Theorem 6.3.13), and some other conditions may be necessary. □

We now use some probability inequalities to show the stochastic convergence of sample quantiles and extreme order statistics.

***Theorem* 6.3.14.** *Let $\{X_1, \ldots, X_n\}$ be a random sample corresponding to a random variable with distribution function F and assume that its pth quantile, ξ_p, $0 < p < 1$, is uniquely defined, i.e., for all $\eta > 0$, $F(\xi_p - \eta) < F(\xi_p) = p < F(\xi_p + \eta)$. Then, for the pth sample quantile $X_{n:k}$, with $k = k_p$ such that (1.5.99) or (1.5.100) holds, we have $X_{n:k} \xrightarrow{P} \xi_p$ $(X_{n:k} \xrightarrow{a.s.} \xi_p)$.*

Proof. For every $\varepsilon > 0$, let $p_\varepsilon = F(\xi_p + \varepsilon)$ and note that $p_\varepsilon > p$. Then observe that

$$\begin{aligned} Q_{n,k}(\xi_p + \varepsilon) &= P(X_{n:k} \le \xi_p + \varepsilon) \\ &= P(k \text{ or more among } (X_1, \ldots, X_n) \text{ are } \le \xi_p + \varepsilon) \\ &= P\left(\sum_{i=1}^n I(X_i \le \xi_p + \varepsilon) \ge k\right) \\ &= P\left(\frac{1}{n}\sum_{i=1}^n I(X_i \le \xi_p + \varepsilon) - p_\varepsilon \ge \frac{k}{n} - p_\varepsilon\right) \\ &= P(U_n \ge a_n) \end{aligned} \tag{6.3.85}$$

where $U_n = n^{-1}\sum_{i=1}^n I(X_i \le \xi_p + \varepsilon) - p_\varepsilon$ is such that $U_n \xrightarrow{P} 0$ by the Khintchine (Borel) Weak Law of Large Numbers (Theorem 6.3.6), and $a_n = k/n - p_\varepsilon$ is such that $a_n \to a = p - p_\varepsilon < 0$ as $n \to \infty$. Thus, given $\eta > 0$, there exists $n_0 = n_0(\eta)$, such that as $n \to \infty$

$$\begin{aligned} P(U_n \ge a_n) &= P(U_n - a \ge a_n - a) \\ &\ge P(U_n - a \ge -\eta) \\ &= P(U_n \ge a - \eta) \\ &\ge P(a - \eta \le U_n \le -a + \eta) \to 1. \end{aligned} \tag{6.3.86}$$

From (6.3.85) and (6.3.86), we obtain

$$Q_{n,k}(\xi_p + \varepsilon) = P(X_{n:k} - \xi_p \le \varepsilon) \to 1 \quad \text{as } n \to \infty. \tag{6.3.87}$$

Now, letting $p_\varepsilon^* = F(\xi_p - \varepsilon)$ we have $p_\varepsilon^* < p$ and proceeding analogously we may show that

$$Q_{n,k}(\xi_p - \varepsilon) = P(X_{n:k} - \xi_p \le -\varepsilon) \to 0 \quad \text{as } n \to \infty. \tag{6.3.88}$$

From (6.3.87) and (6.3.88) it follows that $X_{n:k} \xrightarrow{P} \xi_p$.

To show strong consistency, first let

$$\begin{aligned} A_n = (X_{n:k} \ge \xi_p + \varepsilon) &= \left[\sum_{i=1}^n I(X_i \le \xi_p + \varepsilon) \le k\right] \\ &= [nF_n(\xi_p + \varepsilon) \le k], \end{aligned} \tag{6.3.89}$$

and

$$B_n = (X_{n:k} \leq \xi_p - \varepsilon) = \left[\sum_{i=1}^{n} I(X_i \leq \xi_p - \varepsilon) \geq k\right]$$
$$= (nF_n(\xi_p - \varepsilon) \geq k) \tag{6.3.90}$$

and note that since $nF_n(\xi_p \pm \varepsilon) \sim \text{Bin}[n, F(\xi_p \pm \varepsilon)]$, we may use the Bernstein Inequality (6.3.16) for binomial events to see that

$$P(A_n) \leq \{\rho[\varepsilon, F(\xi_p)]\}^n$$

and

$$P(B_n) \leq \{\rho^*[\varepsilon, F(\xi_p)]\}^n,$$

where $0 \leq \rho[\varepsilon, F(\xi_p)]$, $\rho^*[\varepsilon, F(\xi_p)] < 1$. Therefore, we have $\sum_{i=1}^{\infty} P(A_n) < \infty$ and $\sum_{n=1}^{\infty} P(B_n) < \infty$ and using the Borel–Cantelli Lemma (Theorem 6.3.2), it follows that both $P(\cup_{n\geq m} A_n) \to 0$ and $P(\cup_{n\geq m} B_n) \to 0$ as $m \to \infty$. The desired result is then a consequence of

$$P[\cup_{n\geq m}(|X_{n:k_p} - \xi_p| \geq \varepsilon)] = P(\cup_{n\geq m} A_n) + P(\cup_{n\geq m} B_n). \tag{6.3.91}$$

■

It is noteworthy to remark that by the very use of the Bernstein Inequality we are able to achieve an exponential rate of convergence. Also, note that the above theorem holds for any $0 < p < 1$.

***Theorem* 6.3.15.** *Let* $\{X_1, \ldots, X_n\}$ *be a random sample corresponding to a random variable with distribution function* F. *If* F *has a finite upper end point or lower end point* ξ_1 (ξ_0), *then* $X_{n:n} \xrightarrow{\text{P,a.s.}} \xi_1$ $\quad (X_{n:1} \xrightarrow{\text{P,a.s.}} \xi_0)$.

Proof. First note that if ξ_1 is a finite upper end point of F, then given $\varepsilon > 0$, there exists $0 < \eta = \eta(\varepsilon) < 1$ such that $F(\xi_1 - \varepsilon) = 1 - \eta$. Therefore, given $\varepsilon > 0$,

$$P(X_{n:n} < \xi_1 - \varepsilon) = P[\text{all of } (X_1, \ldots, X_n) < \xi_1 - \varepsilon]$$
$$\leq [F(\xi_1 - \varepsilon)]^n = (1 - \eta)^n.$$

Thus,

$$\sum_{n=1}^{\infty} P(X_{n:n} < \xi_1 - \varepsilon) \leq \sum_{n=1}^{\infty} (1 - \eta)^n = (1 - \eta)/\eta,$$

and by the Borel–Cantelli Lemma (Theorem 6.3.2), it follows that

$$P\left[\lim_{n\to\infty} (X_{n:n} < \xi_1 - \varepsilon)\right] = 0. \tag{6.3.92}$$

Now observe that

$$P(X_{n:n} > \xi_1 + \varepsilon) = 1 - P(X_{n:n} \leq \xi_1 + \varepsilon) = 1 - F(\xi_1 + \varepsilon) = 0.$$

Thus $\sum_{n=1}^{\infty} P(X_{n:n} > \xi_1 + \varepsilon) = 0$, and by the Borel–Cantelli Lemma (Theorem 6.3.2), we obtain

$$P\Big[\lim_{n\to\infty}(X_{n:n} > \xi_1 + \varepsilon)\Big] = 0. \quad (6.3.93)$$

Now, since $(|X_{n:n} - \xi_1| > \varepsilon) = (X_{n:n} < \xi_1 - \varepsilon) \cup (X_{n:n} > \xi_1 + \varepsilon)$, it follows from (6.3.92) and (6.3.93) that $X_{n:n} \xrightarrow{a.s.} \xi_1$; the result $X_{n:n} \xrightarrow{P} \xi_1$ follows then as a direct consequence of Theorem 6.2.2.

In order to show that $X_{n:1} \to \xi_0$, in probability or almost surely, observe that given $\varepsilon > 0$, there exists $0 < \eta = \eta(\varepsilon) < 1$ such that $F(\xi_0 + \varepsilon) = \eta$; therefore, given $\varepsilon > 0$, we have

$$\begin{aligned} P(X_{n:1} > \xi_0 + \varepsilon) &= P[\text{all of } (X_1, \ldots, X_n) > \xi_0 + \varepsilon] \\ &= [1 - F(\xi_0 + \varepsilon)]^n = (1 - \eta)^n. \end{aligned}$$

Proceeding as in the first part of the proof, the result follows. ■

Now let us consider the case of infinite end points. In this direction let us introduce the following definition conventionally used in extreme value theory.

Characteristic largest (smallest) observation. Let X be a random variable with distribution function F. The characteristic largest (smallest) observation of F is defined as a solution ξ_{1n}^* (ξ_{0n}^*) to $F(\xi) = 1 - 1/n$ $(= 1/n)$.

Given a random sample $\{X_1, \ldots, X_n\}$ corresponding to F, we know that because F_n is left-continuous, $F_n(X_{n:n}^-) = 1 - 1/n$; therefore, it seems reasonable to expect that $X_{n:n} - \xi_{1n}^* \xrightarrow{P} 0$; similarly, since $F_n(X_{n:1}^+) = 1/n$, we expect that $X_{n:1} - \xi_{0n}^* \xrightarrow{P} 0$. As we will see through some examples, this is not always true.

Note that if x_n is such that $F(x_n) = 1 - cn^{-1}\log n$ for some $c > 1$, we have $x_n \to \infty$ as $n \to \infty$; furthermore, observe that

$$P(X_{n:n} < x_n) = (1 - cn^{-1}\log n)^n \approx n^{-c},$$

and, hence, by the Borel–Cantelli Lemma (Theorem 6.3.2), $P(X_{n:n} < x_n$ infinitely often) $\to 0$ as $n \to \infty$. Therefore $X_{n:n} \xrightarrow{a.s.} \infty$. In order to show that $X_{n:n} - \xi_{1n}^* \xrightarrow{P} 0$, we must prove that for all $\varepsilon > 0$

$$P(X_{n:n} \le \xi_{1n}^* + \varepsilon) \to 1 \quad \text{and} \quad P(X_{n:n} \le \xi_{1n}^* - \varepsilon) \to 0 \quad \text{as } n \to \infty,$$

which is equivalent to

$$[F(\xi_{1n}^* + \varepsilon)]^n \to 1 \quad \text{and} \quad [F(\xi_{1n}^* - \varepsilon)]^n \to 0 \quad \text{as } n \to \infty,$$

which, in turn, can be equivalently stated as

$$n \log F(\xi_{1n}^* + \varepsilon) \to 0 \quad \text{and} \quad n \log F(\xi_{1n}^* - \varepsilon) \to -\infty \quad \text{as} \quad n \to \infty. \quad (6.3.94)$$

Analogous conditions are required to show that $X_{n:1} - \xi_{0n}^* \xrightarrow{P} 0$.

***Example* 6.3.12.** Let X be a random variable with the logistic distribution, that is, with distribution function $F(x) = \{1 + \exp(-x)\}^{-1}$, $x \in \mathbb{R}$. Then,

$$F(\xi_{1n}^*) = (1 + e^{-\xi_{1n}^*})^{-1} = \frac{n-1}{n}$$

so that $n \exp(\xi_{1n}^*) = n \exp(\xi_{1n}^*) - \exp(\xi_{1n}^*) + n - 1$, which implies $\exp(\xi_{1n}^*) = n - 1$; thus $\xi_{1n}^* = \log(n-1)$. Now note that

$$F(\xi_{1n}^* + \varepsilon) = (1 + e^{-\xi_{1n}^* - \varepsilon})^{-1} = [1 + e^{-\varepsilon}(n-1)^{-1}]^{-1}.$$

Thus, as $n \to \infty$

$$\begin{aligned} n \log F(\xi_n^* + \varepsilon) &= n \log \left(1 + \frac{e^{-\varepsilon}}{n-1}\right)^{-1} \\ &= -n \log \left(1 + \frac{e^{-\varepsilon}}{n-1}\right) \\ &= -n \left[\frac{e^{-\varepsilon}}{n-1} - \frac{e^{-2\varepsilon}}{2(n-1)^2} + o(n^{-2})\right] \to -e^{-\varepsilon}, \end{aligned}$$

and in view of (6.3.94) we may not conclude that $X_{n:n} - \xi_{1n}^* \xrightarrow{P} 0$. Along similar lines we may show that $\xi_{0n}^* = -\log(n-1)$ and that $n \log[1 - F(\xi_{0n}^* + \varepsilon)] \to -\exp(\varepsilon)$ as $n \to \infty$, so we cannot conclude that $X_{n:1} - \xi_{0n}^* \xrightarrow{P} 0$. □

***Example* 6.3.13.** Let $X \sim \mathcal{N}(0, 1)$. Writing

$$1 - \Phi(x) = \int_x^\infty \frac{1}{\sqrt{2\pi}} \exp(-t^2/2)\mathrm{d}t,$$

integrating by parts and letting $\varphi = \Phi'$, we can show that as $x \to \infty$

$$1 - \Phi(x) = \frac{\varphi(x)}{x}[1 + O(x^{-2})].$$

Therefore, the corresponding characteristic largest observation ξ_{1n}^* is defined as the solution to

$$1 - \Phi(\xi_{1n}^*) = \frac{1}{\xi_{1n}^* \sqrt{2\pi}} \exp\left(-\frac{1}{2}\xi_{1n}^{*2}\right) [1 + o(1)] = \frac{1}{n}. \tag{6.3.95}$$

Re-expressing (6.3.95) as

$$e^{-\frac{1}{2}\xi_{1n}^{*2}} = \sqrt{2\pi}\xi_{1n}^* n^{-1}[1 + o(1)]$$

and taking logarithms of both sides, we obtain

$$-\frac{1}{2}\xi_{1n}^{*2} = \log\sqrt{2\pi} + \log \xi_{1n}^* - \log n + \log[1 + o(1)],$$

which implies

$$\begin{aligned} {\xi_{1n}^*}^2 &= 2\log n - 2\log\sqrt{2\pi} - 2\log \xi_{1n}^* + o(1) \\ &= 2\log n\left[1 - \frac{\log\sqrt{2\pi}}{\log n} - \frac{\log \xi_{1n}^*}{\log n} + o\left(\frac{1}{\log n}\right)\right]. \end{aligned} \tag{6.3.96}$$

Now, from (6.3.95), for sufficiently large n we have

$$1 - \Phi(\xi_{1n}^*) = \frac{1}{n} < \frac{1}{\sqrt{2\pi}\xi_{1n}^*} e^{-\frac{1}{2}\xi_{1n}^{*2}} < e^{-\frac{1}{2}\xi_{1n}^{*2}}$$

and then we may write

$$-\log n < -\frac{1}{2}\xi_{1n}^{*\,2} \Rightarrow 2\log n > \xi_{1n}^{*\,2} \Rightarrow \log(2\log n) > 2\log \xi_{1n}^*,$$

which, in turn, implies that

$$\log \xi_{1n}^* < \frac{\log(2\log n)}{2} = \frac{\log(\log n^2)}{2}.$$

Thus,

$$0 < \sqrt{\log n}\frac{\log \xi_{1n}^*}{\log n} = \frac{\log \xi_{1n}^*}{\sqrt{\log n}} < \frac{\log(\log n^2)}{2\sqrt{\log n}}. \tag{6.3.97}$$

Applying the l'Hôpital rule, we may show that

$$\lim_{n\to\infty} \log(\log n^2)/2\sqrt{\log n} = 0,$$

and it follows from (6.3.97) that

$$\log \xi_{1n}^* / \log n = o[(\log n)^{-1/2}].$$

Using this in (6.3.96) we get

$$\xi_{1n}^{*2} = 2\log n\left[1 + o\left(\frac{1}{\sqrt{\log n}}\right)\right],$$

which implies that

$$\xi_{1n}^* = \sqrt{2\log n}\left[1 + o\left(\frac{1}{\sqrt{\log n}}\right)\right]^{1/2}.$$

Observing that for $x \geq -1$,

$$(1+x)^{1/2} = 1 + x/2 + o(x),$$

we have

$$\left[1 + o\left(\frac{1}{\sqrt{\log n}}\right)\right]^{1/2} = 1 + o\left(\frac{1}{\sqrt{\log n}}\right)$$

and then

$$\xi_{1n}^* = \sqrt{2\log n}\left[1 + o\left(\frac{1}{\sqrt{\log n}}\right)\right] = \sqrt{2\log n} + o(1).$$

Now, note that

$$
\begin{aligned}
n \log \Phi(\xi_{1n}^* + \varepsilon) n \log \left\{1 - [1 - \Phi(\xi_{1n}^* + \varepsilon)]\right\} \\
&= n \log \left\{1 - \frac{1}{\sqrt{2\pi}(\xi_{1n}^* + \varepsilon)} e^{-\frac{1}{2}(\xi_{1n}^* + \varepsilon)^2}[1 + o(1)]\right\} \\
&= n \log \left\{1 - \frac{\xi_{1n}^*}{\xi_{1n}^* + \varepsilon} \frac{1}{\sqrt{2\pi}\xi_{1n}^*} e^{-\frac{1}{2}\xi_{1n}^{*\,2} - \xi_{1n}^*\varepsilon - \frac{1}{2}\varepsilon^2}[1 + o(1)]\right\} \\
&= n \log \left\{1 - \frac{\xi_{1n}^*}{\xi_{1n}^* + \varepsilon} \frac{1}{n} e^{-\xi_{1n}^*\varepsilon - \frac{1}{2}\varepsilon^2}[1 + o(1)]\right\} \\
&= -n \left\{\frac{\xi_{1n}^*}{\xi_{1n}^* + \varepsilon} \frac{1}{n} e^{-\xi_{1n}^*\varepsilon - \frac{1}{2}\varepsilon^2} + o(n^{-2})\right\} \\
&= -\frac{\xi_{1n}^*}{\xi_{1n}^* + \varepsilon} e^{-\xi_{1n}^*\varepsilon - \frac{1}{2}\varepsilon^2} + o(n^{-1}) \to 0 \quad \text{as } n \to \infty.
\end{aligned}
$$

Along the same lines, we may show that

$$
n \log \Phi(\xi_{1n}^* - \varepsilon) = -\frac{\xi_{1n}^*}{\xi_{1n}^* - \varepsilon} e^{\xi_{1n}^*\varepsilon - \frac{1}{2}\varepsilon^2} + o(n^{-1}) \to 0 \quad \text{as } n \to \infty.
$$

Thus, in view of (6.3.96) we may conclude that $X_{n:n} - \xi_{1n}^* \xrightarrow{P} 0$. Similarly, one can prove that $\xi_{0n}^* = -\sqrt{2 \log n} + o(1)$ and that $X_{n:n} - \xi_{0n}^* \xrightarrow{P} 0$. □

An important result that extends the concepts of almost sure convergence of real-valued or vector-valued random variables to random processes is given in the next theorem.

Theorem* 6.3.16 *(Glivenko–Cantelli). *Let $\{X_1, \ldots, X_n\}$ be a random sample corresponding to a random variable X with distribution function F. Then, if F_n denotes the associated empirical distribution function, we have*

$$
\sup_{x \in \mathbb{R}} |F_n(x) - F(x)| \xrightarrow{a.s.} 0.
$$

Proof. Assume that F is continuous. Then $Y_i = F(X_i) \sim \text{Uniform}(0, 1)$, which implies $G(t) = P\{Y_i \le t\} = t, 0 \le t \le 1$. Let $x = F^{-1}(t)$; then $F(x) = t$ and

$$
\begin{aligned}
F_n(x) &= \frac{1}{n} \sum_{i=1}^n I(X_i \le F^{-1}(t)) = \frac{1}{n} \sum_{i=1}^n I(F(X_i) \le t) \\
&= \frac{1}{n} \sum_{i=1}^n I(Y_i \le t) = G_n(t). \qquad (6.3.98)
\end{aligned}
$$

The function G_n is known as the reduced empirical distribution function. Consequently,

$$
\sup_{x \in \mathbb{R}} |F_n(x) - F(x)| = \sup_{0 \le t \le 1} |G_n(t) - t|. \qquad (6.3.99)
$$

Now note that G_n is nondecreasing; therefore, given $m > 0$, it follows that for all $t \in [(k-1)/m,\ k/m]$, $k = 1, \ldots, m$, we have

$$\begin{aligned}
&G_n\left(\frac{k-1}{m}\right) \le G_n(t) \le G_n\left(\frac{k}{m}\right)\\
&\Rightarrow G_n\left(\frac{k-1}{m}\right) - \frac{k}{m} \le G_n(t) - t \le G_n\left(\frac{k}{m}\right) - \frac{k-1}{m}\\
&\Rightarrow G_n\left(\frac{k-1}{m}\right) - \frac{k-1}{m} - \frac{1}{m} \le G_n(t) - t \le G_n\left(\frac{k}{m}\right) - \frac{k}{m} + \frac{1}{m}\\
&\Rightarrow |G_n(t) - t| \le \max_{s=k-1,k} \left| G_n\left(\frac{s}{m}\right) - \frac{s}{m}\right| + \frac{1}{m}\\
&\Rightarrow \sup_{0\le t\le 1} |G_n(t) - t| \le \max_{1\le k\le m} \sup_{\frac{k-1}{m}\le t\le \frac{k}{m}} |G_n(t) - t|\\
&\le \max_{1\le k\le m} \left| G_n\left(\frac{k}{m}\right) - \frac{k}{m}\right| + \frac{1}{m}.
\end{aligned}$$

Given $\varepsilon > 0$, there exists m such that $m^{-1} < \varepsilon/2$, and then

$$\sup_{0\le t\le 1} |G_n(t) - t| \le \max_{1\le k\le m} \left| G_n\left(\frac{k}{m}\right) - \frac{k}{m}\right| + \frac{\varepsilon}{2}. \tag{6.3.100}$$

Now recall that $nG_n(k/m) = \sum_{i=1}^n I(Y_i \le k/m) \sim \text{Bin}(n, k/m)$, and apply the Bernstein Inequality for binomial events (6.3.16) to see that, given $\eta > 0$,

$$P\left[\left| G_n\left(\frac{k}{m}\right) - \frac{k}{m}\right| > \eta\right] \le 2[\rho(\eta)]^n, \quad \text{where } 0 < \rho(\eta) < 1.$$

Choose $\eta = \varepsilon/2$ and note that

$$\begin{aligned}
P\left[\max_{0\le k\le m} \left| G_n\left(\frac{k}{m}\right) - \frac{k}{m}\right| > \frac{\varepsilon}{2}\right] &\le \sum_{k=1}^{m-1} P\left[\left| G_n\left(\frac{k}{m}\right) - \frac{k}{m}\right| > \frac{\varepsilon}{2}\right]\\
&\le 2(m-1)[\rho(\varepsilon/2)]^n
\end{aligned}$$

so that

$$\begin{aligned}
&P\left[\bigcup_{N\ge n} \max_{0\le k\le m} \left| G_N\left(\frac{k}{m}\right) - \frac{k}{m}\right| > \frac{\varepsilon}{2}\right]\\
&\le \sum_{N\ge n} P\left[\max_{0\le k\le m} \left| G_N\left(\frac{k}{m}\right) - \frac{k}{m}\right| > \frac{\varepsilon}{2}\right]\\
&\le 2(m-1) \sum_{N\ge n} \left[\rho\left(\frac{\varepsilon}{2}\right)\right]^N\\
&= \frac{2(m-1)[\rho(\varepsilon/2)]^n}{1-\rho(\varepsilon/2)} \to 0 \quad \text{as } n \to \infty,
\end{aligned}$$

which implier that

$$\max_{1\le k\le m} \left| G_n\left(\frac{k}{m}\right) - \frac{k}{m}\right| \xrightarrow{a.s.} 0.$$

In view of (6.3.99) and (6.3.100) the result is proven for F continuous.

Now suppose that F has a finite number of jump points at $a_1 < a_2 < \cdots < a_M$. Then, given $\varepsilon > 0$, let M_ε denote the number of jump points with jumps $> \varepsilon$. Excluding the neighborhoods of the jump points, the first part of the proof holds. For the jump point at a_j, observe that $nF_n(a_j) \sim \text{Bin}(n, F_n(a_j))$ and then apply the Bernstein Inequality for binomial events (6.3.16) to see that

$$P[|F_n(a_j) - F(a_j)| > \varepsilon] \leq 2[\rho(\varepsilon)]^n, \quad \text{where } 0 < \rho(\varepsilon) < 1.$$

Finally note that because there are only finitely many jumps of magnitude $> \varepsilon > 0$, we obtain

$$P\left(\bigcup_{N \geq n} \max_{1 \leq j \leq M} |F_N(a_j) - F(a_j)| > \varepsilon\right) \to 0 \quad \text{as } n \to \infty$$

and the result follows. ■

To extend the above result to the bivariate case, consider a random sample $\{(X_1, Y_1)', \ldots, (X_n, Y_n)'\}$ from random vector $(X, Y)'$ with distribution function F and let

$$F_n(x, y) = \frac{1}{n} \sum_{i=1}^{n} I(X_i \leq x) I(Y_i \leq y), \quad x, y \in \mathbb{R}.$$

Assuming F continuous, let $W_i = F(X_i, \infty)$ and $Z_i = F(\infty, Y_i)$ then write $X_i = F^{-1}(W_i, \infty)$ and $Y_i = F^{-1}(\infty, Z_i)$, so that

$$\begin{aligned}
F_n(x, y) &= \frac{1}{n} \sum_{i=1}^{n} I[X_i \leq F^{-1}(w, \infty)] I[Y_i \leq F^{-1}(\infty, z)] \\
&= \frac{1}{n} \sum_{i=1}^{n} I[F(X_i, \infty) \leq w] I[F(\infty, Y_i) \leq z] \\
&= \frac{1}{n} \sum_{i=1}^{n} I(W_i \leq w) I(Z_i \leq z) = G_n(w, z).
\end{aligned}$$

So, letting $G(w, z) = P(W_i \leq w, Z_i \leq z)$, we obtain

$$\sup_{(x,y)' \in \mathbb{R}^2} |F_n(x, y) - F(x, y)| = \sup_{(w,z)' \in [0,1] \times [0,1]} |G_n(w, z) - G(w, z)|.$$

Now, since G_n and G are nondecreasing functions, it follows that

$$\begin{aligned}
&G_n\left(\frac{k-1}{m}, \frac{l-1}{m}\right) - G\left(\frac{k}{m}, \frac{l}{m}\right) \\
&\leq G_n(w, z) - G(w, z) \\
&\leq G_n\left(\frac{k}{m}, \frac{l}{m}\right) - G\left(\frac{k-1}{m}, \frac{l-1}{m}\right),
\end{aligned} \tag{6.3.101}$$

for all $(w, z) \in [\frac{k-1}{m}, \frac{k}{m}) \times [\frac{l-1}{m}, \frac{l}{m})$, $k, l = 1, \ldots, m$, where m is a positive integer. Then, given $\varepsilon > 0$, we may choose $m^{-1} < \varepsilon/4$ so that

$$
\begin{aligned}
G_n&\left(\frac{k}{m}, \frac{l}{m}\right) - G\left(\frac{k-1}{m}, \frac{l-1}{m}\right) \\
&= \left[G_n\left(\frac{k}{m}, \frac{l}{m}\right) - G\left(\frac{k}{m}, \frac{l}{m}\right)\right] + \left[G\left(\frac{k}{m}, \frac{l}{m}\right) - G\left(\frac{k-1}{m}, \frac{l-1}{m}\right)\right] \\
&= G_n\left(\frac{k}{m}, \frac{l}{m}\right) - G\left(\frac{k}{m}, \frac{l}{m}\right) + \left[G\left(\frac{k}{m}, \frac{l}{m}\right) - G\left(\frac{k}{m}, \frac{l-1}{m}\right)\right] \\
&\quad + \left[G\left(\frac{k}{m}, \frac{l-1}{m}\right) - G\left(\frac{k-1}{m}, \frac{l-1}{m}\right)\right] \\
&\leq G_n\left(\frac{k}{m}, \frac{l}{m}\right) - G\left(\frac{k}{m}, \frac{l}{m}\right) + \left[G\left(1, \frac{l}{m}\right) - G\left(1, \frac{l-1}{m}\right)\right] \\
&\quad + \left[G\left(\frac{k}{m}, 1\right) - G\left(\frac{k-1}{m}, 1\right)\right] \\
&\leq \left[G_n\left(\frac{k}{m}, \frac{l}{m}\right) - G\left(\frac{k}{m}, \frac{l}{m}\right)\right] + \frac{\varepsilon}{4} + \frac{\varepsilon}{4}, \qquad k, l = 1, \ldots, m.
\end{aligned}
\tag{6.3.102}
$$

Using a similar argument we may show that

$$
\begin{aligned}
G_n&\left(\frac{k-1}{m}, \frac{l-1}{m}\right) - G\left(\frac{k}{m}, \frac{l}{m}\right) \\
&\leq \left[G_n\left(\frac{k-1}{m}, \frac{l-1}{m}\right) - G\left(\frac{k-1}{m}, \frac{l-1}{m}\right)\right] + \frac{\varepsilon}{2}
\end{aligned}
\tag{6.3.103}
$$

so that from (6.3.101)–(6.3.103), we may write

$$
\begin{aligned}
&\sup_{(w,z)' \in [0,1]\times[0,1]} |G_n(w, z) - G(w, z)| \\
&\quad \leq \max_{k,l=1,\ldots,m} \left|G_n\left(\frac{k}{m}, \frac{l}{m}\right) - G\left(\frac{k}{m}, \frac{l}{m}\right)\right| + \frac{\varepsilon}{2}.
\end{aligned}
$$

Recalling that $nG_n(k/m, l/m) \sim \text{Bin}[n, G(k/m, l/m)]$ and following the same steps as in Theorem 6.3.16 we may readily complete the proof.

6.4 Extensions to Dependent Variables

In the last section, we considered some basic probability inequalities and laws of large numbers for sums (or averages) of independent random variables. Because, in many inferential problems, the relevant statistics may not be expressed in that form, it is of natural interest to explore the adaptability of the results of Section 6.3 to cases where the underlying random variables are dependent. Fortunately this is possible if we consider the martingale, submartingale and related processes discussed in Section 5.2. We start with the following theorem.

Theorem 6.4.1 (Extension of the Kolmogorov Maximal Inequality). *Let $\{T_n, n \geq 1\}$ be a (zero-mean) martingale such that $\mathbb{E}(T_n^2)$ exists, for every $n \geq 1$. Then, for every $t > 0$,* *we have*

$$P\left(\max_{1\le k\le n}|T_k|>t\right)\le t^{-2}\mathbb{E}(T_n^2). \tag{6.4.1}$$

If $\{T_n,\ n\ge m\}$ *is a (zero-mean) reverse martingale, then, for every* $N\ge n\ge m$,

$$P\left(\max_{n\le k\le N}|T_k|>t\right)\le t^{-2}\mathbb{E}(T_n^2),\quad t>0. \tag{6.4.2}$$

Proof. First, consider the case of martingales. We follow the proof of the Kolmogorov Maximal Inequality (6.3.8) and we need to verify only that (6.3.57) holds for the martingales. For this, it suffices to show that $\mathbb{E}(T_n^2|T_j,\ j\le k)\ge T_k^2$ a.e., for every $k\le n$, and this has already been proved in (5.2.22). Next, consider the case of reverse martingales. We write $S_1=T_N, S_2=T_{N-1},\dots,S_{N-n+1}=T_n$ and observe that the reverse martingale property of $\{T_n\}$ yields that the $\{S_k\}$ form a forward martingale. As such, (6.4.2) follows from (6.4.1) as adapted to the partial sequence $\{S_1,\dots,S_{N-n+1}\}$. ■

We do not need the $\{T_n\}$ to be a zero-mean sequence for the inequalities in (6.4.1) and (6.4.2). However, in the case of a zero-mean sequence, we may as well replace $\mathbb{E}(T_n^2)$ by $\mathbb{V}\text{ar}(T_n)$. In actual statistical applications, with suitable choice of t, the right-hand side of (6.4.1) or (6.4.2) can be made small when T_n has zero mean, but for nonzero mean, the contribution of $[\mathbb{E}(T_n)/t]^2$ may make the bound of little utility. Furthermore, it is not necessary to assume that $\{T_n\}$ is a martingale (or reverse martingale) sequence; we may as well assume that $\{T_n\}$ is a submartingale (or a reverse submartingale); this follows simply by noting that $[\max_{1\le k\le n}|T_k|>t]$ is equivalent to $[\max_{1\le k\le n}T_k^2>t^2]$, and by noting that the $\{T_k^2\}$ form a submartingale. In fact, if $\{T_n\}$ is a submartingale and g is any nonnegative (nondecreasing) convex function, such that $\mathbb{E}[g(T_n)]$ exists, then, by (5.2.21), $\{g(T_n)\}$ is a submartingale, and, hence, using Theorem 6.4.1 on these $g(T_k)$, we immediately obtain that for any positive t,

$$P\left(\max_{1\le k\le n}T_k>t\right)\le\frac{\mathbb{E}[g(T_n)]}{g(t)}. \tag{6.4.3}$$

In particular, if T_n has a finite-moment generating function $M_n(\theta)=\mathbb{E}[\exp(\theta T_n)]$, for all $\theta\in(0,\theta_0)$, $\theta_0>0$, then, letting $g(x)=\exp(\theta x)$, we obtain from (6.4.3) the following inequality [compare with the Bernstein Inequality (6.3.7)].

Bernstein Inequality for Submartingales. Let $\{T_n\}$ be a submartingale and assume that $M_n(\theta)=\mathbb{E}[\exp(\theta T_n)]$ exists for all $\theta\in(0,\theta_0)$, $\theta_0>0$. Then, for every $t>0$,

$$P\left(\max_{1\le k\le n}T_k>t\right)\le\inf_{\theta>0}[e^{-\theta t}M_n(\theta)]. \tag{6.4.4}$$

If $\{T_n\}$ is a reverse submartingale such that $M_n(\theta)$ exists for all $\theta\in(0,\theta_0)$, then, for every $t>0$,

$$P\left(\sup_{N\ge n}T_N>t\right)\le\inf_{\theta>0}[e^{-\theta t}M_n(\theta)]. \tag{6.4.5}$$

***Example* 6.4.1.** Let X_i, $i\ge1$ be i.i.d. random variables with $\mathbb{E}(X_1)=0$ without loss of generality, and assume that $\mathbb{E}(|X_1|^p)<\infty$, for some $p\ge1$. Then let $T_n=\overline{X}_n, n\ge1$. Note that $\{\overline{X}_n, n \geq 1\}$ is a reverse martingale, so that $\{|\overline{X}_n|^p, n \geq 1\}$ is a reverse submartingale. Consequently, by a parallel version of (6.4.3) for reverse submartingales, we immediately obtain that for every $\varepsilon > 0$

$$P\left(\sup_{N \geq n} |\overline{X}_n| > \varepsilon\right) \leq \varepsilon^{-p}\mathbb{E}(|\overline{X}_n|^p). \tag{6.4.6}$$

On the other hand, $\mathbb{E}(|X_1|^p) < \infty$ implies $\mathbb{E}(|\overline{X}_n|^p) \to 0$ as $n \to \infty$, and, hence, the right-hand side of (6.4.6) converges to 0 as $n \to \infty$. Thus, $\overline{X}_n \xrightarrow{a.s.} 0$. In this context, we may note that for $p = 2$, $\mathbb{E}(\overline{X}_n^2) = \mathbb{V}\mathrm{ar}(\overline{X}_n) = n^{-1}\mathbb{V}\mathrm{ar}(X_1) = n^{-1}\mathbb{E}(X_1^2)$, whereas, for $p \in (1, 2)$, $\mathbb{E}(|\overline{X}_n|^p) = o(n^{1-p}) \to 0$, as $n \to \infty$; we may refer to Pyke and Root (1968) for a detailed treatment of these related L_p-convergence results. □

***Example* 6.4.2.** Let X_i be i.i.d. random variables with $P(X_1 = 1) = 1 - P(X_1 = 0) = \pi$ $(0 < \pi < 1)$. Then, the moment-generating function of $T_n = \overline{X}_n$ exists, as we have noticed in Section 6.3. The reverse martingale structure of $\{\overline{X}_n\}$ follows from the preceding example. Hence, we readily conclude that the Bernstein Inequality (6.4.5) actually applies to the entire tail, and, hence, $\overline{X}_n \xrightarrow{a.s.} \pi$ at an exponential rate. □

Hájek–Rènyi–Chow Inequality. Let $\{T_n, n \geq 1\}$ be a submartingale and let $\{c_n^*, n \geq 1\}$ be a nonincreasing sequence of positive numbers. Let $T_n^+ = \max\{T_n, 0\}$, $n \geq 1$, and assume that $\mathbb{E}(T_n^+)$ exists for every $n \geq 1$. Then, for every $\varepsilon > 0$,

$$P\left(\max_{1 \leq k \leq n} c_k^* T_k > \varepsilon\right) \leq \varepsilon^{-1}\left[c_1^*\mathbb{E}(T_1^+) + \sum_{k=2}^{n} c_k^*\mathbb{E}(T_k^+ - T_{k-1}^+)\right]. \tag{6.4.7}$$

If $\{T_n, n \geq 1\}$ is a reverse submartingale and the c_k^* are nondecreasing, we have for every $N \geq n$,

$$P\left(\max_{n \leq k \leq N} c_k^* T_k > \varepsilon\right) \leq \varepsilon^{-1}\left[c_n^*\mathbb{E}(T_n^+) + \sum_{k=n+1}^{N} (c_k^* - c_{k-1}^*)\mathbb{E}(T_k^+)\right]. \tag{6.4.8}$$

Note that if we take X_i, $i \geq 1$, as independent random variables with zero mean, then $T_n = (X_1 + \cdots + X_n)^2$ is a nonnegative submartingale (so that $T_n^+ = T_n$), and then (6.4.7) reduces to (6.3.59), with $c_k^* = c_k^2$, $k \geq 1$. The proof of (6.4.7) consists in verifying (6.3.63) with the s_k^2 being replaced by T_k^+ and c_k^2 by c_k^*. We omit the details in view of this similarity and the basic definition in (5.2.20). The case of reverse submartingales can be treated as in the forward submartingale case after reversing the order of the index set (as we have done in the proof of Theorem 6.4.1).

For independent random variables, the Kolmogorov SLLN (Theorem 6.3.10) provides the a.s. convergence result under a simple condition that $\sum_{k \geq 1} k^{-2}\sigma_k^2$ converges [see (6.3.65)]. Actually, the second moment condition in (6.3.65) can be relaxed and the independence assumption may as well be replaced by a martingale structure. Toward this, we present the following theorem.

Theorem* 6.4.2 *(Kolmogorov SLLN for martingales). *First consider a martingale* $\{T_n = \sum_{k \leq n} Y_k, n \geq 1\}$, *such that for some* $1 \leq p \leq 2$, $\mathbb{E}(|Y_k|^p)$ *exists for every* $k \geq 1$, *that is,* $\{T_n, n \geq 1\}$ *is a zero mean* L_p *martingale. Also assume that there is a sequence* $\{b_n, n \geq 1\}$

of increasing positive numbers such that $b_n \to \infty$ *as* $n \to \infty$, *and*

$$\sum_{n \geq 2} b_n^{-p} \mathbb{E}[|Y_n|^p | Y_j, j < n] < \infty \quad a.s. \tag{6.4.9}$$

Then, $b_n^{-1} T_n \xrightarrow{a.s.} 0$.

We may note that if the Y_k are independent, then $\mathbb{E}[|Y_k|^p | Y_j,\ j < k] = \mathbb{E}[|Y_k|^p]$, for every $k \geq 2$, so that (6.4.9) reduces to (6.3.65) when we take $p = 2$ and $b_n = n$. For martingales, the conditional expectations are generally random variables, and, hence, we need the series in (6.4.9) to converge a.s. It may also be noted that if the Y_k are independent or, more generally, if for some $p > 1$, $\mathbb{E}[|Y_k|^p | Y_j,\ j < k] \leq c < \infty$, for every $k \geq 2$, we obtain that $n^{-1} T_n \xrightarrow{a.s.} 0$, by noting that $\sum_{n \geq 2} n^{-p} < \infty$, for every $p > 1$. Thus, the second moment condition in (6.3.65) is not that crucial for the SLLN to hold. For the i.i.d. case, a slightly less stringent condition, such as the one in (6.4.9), suffices.

Looking back at Example 6.4.1 and the reverse submartingale inequality, we may conclude that the following general result holds.

***Theorem* 6.4.3.** *For a reverse submartingale* $\{T_n,\ n \geq m\}$, *there exists an (possibly random variable)* T, *such that* $T_n - T \xrightarrow{a.s.} 0$. *If* $\{T_n\}$ *is a reverse martingale, then* $T_n - T \to 0$ *a.s., as well as in the first mean.*

In simple random sampling (with or without replacement) from a finite population, or for i.i.d. random variables, we have observed earlier that U-statistics form a reverse martingale. As such, we are in a position to use (6.4.8), that is, the Hájek–Rènyi–Chow Inequality for reverse martingales. In order to make use of the submartingale structure in (6.4.8), we may take $T_n = [U_n - \mathbb{E}(U_n)]^2$, $n \geq m$, and $c_k^* = c_k^2$. We may, of course, use $T_n = |T_n - \mathbb{E}(U_n)|^p$, for some $p \geq 1$. Thus, allowing N to be indefinitely large in (6.4.8), we arrive at the a.s. convergence of U_n to $\theta = \mathbb{E}(U_n)$. For $p > 1$, we may use the L_p-convergence of U-statistics and claim that $\mathbb{E}(|U_n - \theta|^p) = O(n^{-r(p)})$, where $r(p) = p - 1$ if $p \in (1, 2]$, and $p/2$ for $p \geq 2$, so that the right hand side of (6.4.8) converges to 0 as $n \to \infty$. On the other hand, the treatment is a bit more delicate for $p = 1$. In this context, we first note that a tail event is an event the probability of which remains the same when a finite number of the X_n are changed, that is, the event belongs to the tail sigma field of the X_n. Similarly, a tail function does not depend on any finite segment of the X_n. The celebrated *Zero-One Law* can be presented as follows.

Theorem* 6.4.4 *(Kolmogorov Zero-One Law). *Let* $\{X_n\}$ *be a sequence of independent random variables. Then, the probability of any tail event is either 0 or 1, and any tail function is a constant with probability 1.*

To conceive of such tail functions, consider for example for an arbitrary function g,

(i) $\limsup_{n\to\infty} g(X_n)$,
(ii) $\liminf_{n\to\infty} g(X_n)$,
(iii) $\limsup_{n\to\infty} n^{-1} \sum_{i=1}^{n} g(X_i)$,

We recall that the U-statistics are symmetric functions of the sample observations (i.e., they remain invariant under any permutation of the indices $1, \ldots, n$ of the random variables $X_1, \ldots, X_n$). Such a function is called on *exchangeable* function.

Theorem 6.4.5 (Hewitt–Savage Zero-One Law). *Let $\{X_i, i \geq 1\}$ be i.i.d. random variables. Then every exchangeable event has probability either equal to 0 or 1.*

Coming back to the case of U-statistics, we observe that by virtue of Theorem 6.4.3, $U_n - U \xrightarrow{a.s.} 0$ where by the Hewitt–Savage Zero-One Law, we claim that $U = \theta$ with probability 1. Hence, $U_n \xrightarrow{a.s.} \theta$.

6.5 Miscellaneous Convergence Results

Let us first consider the convergence of a series of random variables. For a sequence $\{a_n\}$ of real numbers, we say that the series $\sum_{n\geq 1} a_n$ converges if for every $\varepsilon > 0$, there exists a positive integer $n_0 = n_0(\varepsilon)$, such that

$$\left| \sum_{n=m+1}^{m+N} a_n \right| \leq \varepsilon, \quad m \geq n_0, \ N \geq 1. \tag{6.5.1}$$

If instead of the sequence $\{a_n\}$ of real numbers, we have a sequence of random variables, say $\{X_n\}$, and we define the partial sums as

$$T_n = X_1 + \cdots + X_n, \quad n \geq 1, \tag{6.5.2}$$

then a natural question may arise: Does T_n converge (in a meaningful sense) as $n \to \infty$? In this context, we may note that the T_n are themselves random variables, and, hence, T_n may not converge to a nonstochastic limit as $n \to \infty$; rather, it may have some nondegenerate distribution. Thus, we may need to interpret the convergence of a series of random variables in a different manner. It seems quite logical to incorporate the basic requirement in (6.5.1) in a stochastic sense and to define the desired convergence result as follows.

Almost sure convergence of series. A series T_n, $n \geq 1$, defined as in (6.5.2), converges almost surely if for every given positive numbers ε and η, there exists a positive integer $n_0 = n_0(\varepsilon, \eta)$, such that

$$P\left(\max_{m+1 \leq n \leq m+N} \left| \sum_{i=m+1}^{n} X_i \right| > \varepsilon \right) < \eta, \quad m \geq n_0, \ N \geq 1. \tag{6.5.3}$$

Note that this definition allows the T_n to retain their stochastic nature but implies that the tail contribution becomes a.s. negligible as n is increased. Note that if the X_i are i.i.d. random variables with a nondegenerate distribution F, then the series T_n does not converge a.s. in the sense of (6.5.3) although $n^{-1}T_n$ converges a.s. to $\mathbb{E}(X)$ whenever the latter exists. However, as we see in the following theorems, there are situations involving independent but possibly nonidentically distributed random variables where this mode of convergence holds and has suitable applications in other problems too.

***Theorem* 6.5.1.** *Let $\{X_i,\ i \ge 1\}$ be a sequence of independent random variables, such that $\mathbb{E}(X_k) = 0$ and $\mathbb{V}ar(X_k) = \sigma_k^2 < \infty$, for all $k \ge 1$. If $\sum_{k\ge1} \sigma_k^2 < \infty$, then $\sum_{n\ge1} X_n$ converges a.s.*

Proof. It follows from the Kolmogorov Maximal Inequality in (6.3.53) that for every $\eta > 0$ and all $m \ge 1$, $N \ge 1$,

$$P\left(\max_{m+1\le n\le m+N}\left|\sum_{k=m+1}^{n} X_k\right| > \eta\right) \le \eta^{-2}\sum_{k=m+1}^{m+N}\sigma_k^2 \le \eta^{-2}\sum_{k>m}\sigma_k^2. \tag{6.5.4}$$

Since the series $\sum_{k\ge1}\sigma_k^2$ converges, for every $\varepsilon > 0$, there exists an $n_0 = n_0(\varepsilon, \eta)$, such that $\sum_{k>n_0}\sigma_k^2 \le \eta^2\varepsilon$, and, hence, the result follows from (6.5.4). ■

***Example* 6.5.1.** Let $Z_i, i \ge 1$ be i.i.d. random variables having the standard normal distribution and let $X_k = k^{-1}Z_k$, $k \ge 1$. Then, $\mathbb{E}(X_k) = 0$ and $\mathbb{E}(X_k^2) = \mathbb{V}ar(X_k) = \sigma_k^2 = k^{-2}$, $k \ge 1$. Thus, $\sum_{k\ge1}\sigma_k^2 = (\pi^2/6) < \infty$, and, hence, by Theorem 6.5.1 we may conclude that the series $\sum_{k\ge1} X_k$ converges a.s. In fact, here $\sum_{k\ge1} X_k$ has a normal distribution with zero mean and variance $\pi^2/6$. □

Before we consider the converse of Theorem 6.5.1, we present the following probability inequality; a proof of this may be found in most books in Probability Theory [it runs along the same lines as in the proof of (6.3.53)] and is omitted.

Kolmogorov Maximal Inequality (Lower Bound). Let $X_1, \ldots, X_n$ be independent random variables, such that for some finite positive c, $P(|X_i| \le c) = 1$, for every $i \ge 1$. Let $T_k = \sum_{i\le k}[X_i - \mathbb{E}(X_i)]$, $k \ge 1$. Then, for every $\varepsilon > 0$,

$$P\left(\max_{1\le k\le n}|T_k| \ge \varepsilon\right) \ge \frac{1-(\varepsilon+2c)^2}{\sum_{i=1}^{n}\sigma_i^2 + (\varepsilon+2c)^2 - \varepsilon^2}. \tag{6.5.5}$$

It is of interest to compare (6.3.53) and (6.5.5). The former provides an upper bound, whereas the latter provides a lower bound under an additional condition that the random variables are all bounded. We use (6.5.5) to prove the following result.

***Theorem* 6.5.2.** *Let $\{X_i,\ i \ge 1\}$ be a sequence of independent random variables, such that $P(|X_k| \le c) = 1$, for all k, for some finite $c > 0$. Then, the series $\sum_{k\ge1}[X_k - \mathbb{E}(X_k)]$ converges a.s., if and only if $\sum_{k\ge1}\mathbb{V}ar(X_k) < \infty$.*

Proof. Writing $Y_k = X_k - \mathbb{E}(X_k)$, $k \ge 1$, the "if" part of the theorem follows directly from Theorem 6.5.1. For the "only if" part, we make use of (6.5.5) so that

$$P\left(\max_{m+1\le n\le m+N}\left|\sum_{i=m+1}^{n} Y_i\right| > \varepsilon\right) \ge \frac{1-(\varepsilon+2c)^2}{\sum_{i=m+1}^{m+N}\sigma_i^2 + (\varepsilon+2c)^2 - \varepsilon^2}. \tag{6.5.6}$$

If $\sum_{k\ge1}\sigma_k^2 = +\infty$, (6.5.6) would imply (on letting $N \to \infty$) that

$$P\left(\max_{m+1\le n\le m+N}\left|\sum_{i=m+1}^{n} Y_i\right| > \varepsilon\right) = 1,$$

contradicting the hypothesis of a.s. convergence of $\sum_{k>1} Y_k$. Therefore, we must have $\sum_{k\geq 1}\sigma_k^2 < \infty$. ■

***Theorem* 6.5.3.** *Let $\{X_k; k \geq 1\}$ be independent random variables, such that for some finite $c > 0$, $P(|X_k| \leq c) = 1$, for all k. Then $\sum_{k\geq 1} X_k$ converges a.s., if and only if, $\sum_{k\geq 1}\mathbb{E}(X_k) < \infty$ and $\sum_{k\geq 1}\mathbb{V}ar(X_k) < \infty$.*

Proof. If $\sum_{k\geq 1}\mathbb{V}\text{ar}(X_k) < \infty$, by Theorem 6.5.2, $\sum_{k\geq 1}[X_k - \mathbb{E}(X_k)]$ converges a.s., so that the convergence of the series $\sum_{k\geq 1}\mathbb{E}(X_k)$ ensures the a.s. convergence of $\sum_{k\geq 1} X_k$. Suppose next that $\sum_{k\geq 1} X_k$ converges a.s. Consider another sequence $\{X_k^*; k \geq 1\}$ of independent random variables, such that for every $k \geq 1$, X_k^* has the same distribution as $(-1)X_k$. Let $Z_k = X_k + X_k^*$, $k \geq 1$; then $\mathbb{E}(Z_k) = 0$ and $\mathbb{V}\text{ar}(Z_k) = 2\mathbb{V}\text{ar}(X_k)$, for $k \geq 1$. Since $|\sum_{k\geq 1} Z_k| \leq |\sum_{k\geq 1} X_k| + |\sum_{k\geq 1} X_k^*|$, it follows that the series $\sum_{k\geq 1} Z_k$ converges a.s. whenever $\sum_{k\geq 1} X_k$ does so. Therefore by Theorem 6.5.2, we conclude that $\sum_{k\geq 1}\mathbb{V}\text{ar}(Z_k) = 2\sum_{k\geq 1}\mathbb{V}\text{ar}(X_k) < \infty$. Also, this, in turn [by Theorem 6.5.1], implies that $\sum_{k\geq 1}[X_k - \mathbb{E}(X_k)]$ converges a.s., and the a.s. convergence of $\sum_{k\geq 1} X_k$ and $\sum_{k\geq 1}[X_k - \mathbb{E}(X_k)]$ imply that the series $\sum_{k\geq 1}\mathbb{E}(X_k)$ converges. ■

We proceed now to formulate a basic theorem on the a.s. convergence of a series of random variables. In this context, for every $c > 0$, we define

$$X_i^c = X_i I(|X_i| \leq c), \quad i \geq 1. \tag{6.5.7}$$

Theorem* 6.5.4 *(Kolmogorov Three-Series Criterion). *Let X_i, $i \geq 1$ be independent random variables. Then $\sum_{n=1}^{\infty} X_n$ converges a.s. if and only if the following three conditions hold:*

$$\sum_{n\geq 1} P\{|X_n| > c\} < \infty, \tag{6.5.8}$$

$$\sum_{n\geq 1}\mathbb{E}(X_n^c) < \infty, \tag{6.5.9}$$

$$\sum_{n\geq 1}\mathbb{V}ar(X_n^c) < \infty. \tag{6.5.10}$$

Proof. First assume that (6.5.8) and (6.5.9) hold. Since

$$P\left(X_n \neq X_n^c\right) = P\left(|X_n| > c\right),$$

by (6.5.8) and the Khintchine Equivalence Lemma (Theorem 6.3.12), it follows that $\sum_{n\geq 1} X_n$ and $\sum_{n\geq 1} X_n^c$ are convergence-equivalent. Since the X_n^c are bounded random variables, by Theorem 6.5.3, we obtain that under (6.5.9) and (6.5.10), $\sum_{n\geq 1} X_n$ converges a.s. Next, suppose that $\sum_{n\geq 1} X_n$ converges a.s.; then $|\sum_{n=n_1+1}^{n_1+n_2} X_n| \to 0$ a.s., and hence $X_n \xrightarrow{a.s.} 0$. Since the X_n are independent (and so are the X_n^c), by the second part of the Borel–Cantelli Lemma (Theorem 6.3.2), we conclude that $X_n \xrightarrow{a.s.} 0$ implies $\sum_{n\geq 1} P(|X_n| > c) < \infty$, for all $c > 0$, that is, (6.5.8) holds. Again (6.5.8) ensures the convergence equivalence of $\sum_{n\geq 1} X_n$ and $\sum_{n\geq 1} X_n^c$, so that Theorem 6.5.3 leads to the satisfaction of (6.5.9) and (6.5.10). ■

There are some other convergence theorems having fruitful applications in Asymptotic Theory, among which the celebrated *Law of Iterated Logarithm* is the most notable one. See Chow and Teicher (1978) for details. We conclude this section with a brief introduction to the situation in *Sequential Sampling Schemes*, where the sample size is itself a nonnegative integer-valued random variable. To this point, we consider first the following example.

Example 6.5.2 (Inverse sampling). Often, to estimate the probability π of a rare event, instead of the binomial sampling scheme, we adopt the following alternative. Units are drawn one by one with replacement until a specified number, say, m, of the event occurs. Thus, N, the total number of units drawn to yield exactly m occurrences, is itself a positive random variable with $P\{n \geq m\} = 1$. For $\pi \in (0, 1)$, the probability law for n is given by the negative binomial distribution with a probability function (1.5.50), which can also be written as

$$P(n|m, \pi) = \binom{n-1}{m-1} \pi^m (1-\pi)^{n-m}, \; n = m, m+1, \ldots. \tag{6.5.11}$$

An estimator of π is $T_n = m/n$, where m is a given positive integer and n is stochastic. For the stochastic convergence of T_n to π, the results developed in this chapter may not be directly applicable. Furthermore, in such a case, in view of the stochastic nature of n, we may like to formulate the stochastic convergence in terms of the nonstochastic variable m, that is, we may pose the problem as follows: as m is made to increase, does T_n converges a.s. or in probability to π? The answer is, of course, affirmative, and this can be posed in a more general framework as follows.

Consider a sequence $\{N_m \cdot m \geq 1\}$ of nonnegative integer-valued random variables, and let $T_{(m)} = T_{N_m}, m \geq 1$, be a sequence of (sequential) estimators of some parameter θ. Then note that if (i) $T_n \xrightarrow{a.s.} \theta$ and (ii) $N_m \to \infty$ a.s., as $m \to \infty$, we would have $T_{N_m} \to \theta$ a.s., as $m \to \infty$. Thus, under the additional condition (ii) the a.s. convergence of a sequential estimator follows from that of the classical version. On the other hand, suppose that there exists a sequence $\{a_m\}$ of positive numbers, such that

$$N_m/a_m \xrightarrow{P} 1, \quad \text{as } m \to \infty. \tag{6.5.12}$$

Then, for every positive ε and η, we have

$$\begin{aligned} P\left(|T_{N_m} - \theta| > \varepsilon\right) &= P\left(|T_{N_m} - \theta| > \varepsilon, \left|\frac{N_m}{a_m - 1}\right| \leq \eta\right) \\ &\quad + P\left(|T_{N_m} - \theta| > \varepsilon, \left|\frac{N_m}{a_m - 1}\right| > \eta\right) \\ &\leq P\left(\max_{k:|k-a_m|\leq \eta a_m} |T_k - \theta| > \varepsilon\right) + P\left(\left|\frac{N_m}{a_m - 1}\right| > \eta\right). \end{aligned} \tag{6.5.13}$$

By (6.5.12), the second term on the right-hand side of (6.5.13) converges to 0 as $m \to \infty$. On the other hand, the first term on the right-hand side of (6.5.13) is more stringent than the usual $P(|T_{[a_m]} - \theta| > \varepsilon) \to 0$ (but is less stringent than $T_n \xrightarrow{a.s.} \theta$). Fortunately, the Kolmogorov Maximal Inequality (Theorem 6.3.8) or its various extensions may often be used to verify that the first term on the right hand side of (6.5.13) goes to 0 as $m \to \infty$. Thus, toward the proof of weak consistency of estimators based on stochastic sample sizes, the Kolmogorov or related Maximal inequalities may be conveniently used. In general, for the

SLLN, the stochastic nature of the sample size makes no difference (provided the sample size goes to ∞ a.s.), but for the weak LLN, the stochastic convergence of the classical T_n may not be enough, and a slightly more stringent condition, such as the one in (6.5.13), is usually needed. □

***Example* 6.5.3.** Let X_i, $i \geq 1$, be i.i.d. random variables with finite mean θ and variance $\sigma^2 < \infty$. Let $T_n = \overline{X}_n$, for $n \geq 1$ and N_m, $m \geq 1$ be a sequence of positive integer-valued random variables, such that (6.5.12) holds. For example, letting $s_n^2 = (n-1)^{-1}\sum_{i=1}^n (X_i - \overline{X}_n)^2$, $n \geq 2$, we may define

$$N_m = \min\left(n \geq 2 : \frac{S_n}{\sqrt{n}} \leq m^{-1/2}\right), \quad m > 0, \tag{6.5.14}$$

then (6.5.12) holds with $a_m \sim m\sigma^2$, since $s_n^2 \xrightarrow{a.s.} \sigma^2$. Actually, here $N_m \to \infty$ a.s., as $m \to \infty$, whereas by the Khintchine SLLN (Theorem 6.3.13), $\overline{X}_n \xrightarrow{a.s.} \theta$. Consequently, $\overline{X}_{N_m} \to \theta$ a.s., as $m \to \infty$. Alternatively, we could have used the Kolmogorov Maximal Inequality (Theorem 6.3.8) for the $\overline{X}_k$ and verified that the first term on the right-hand side of (6.5.13) converges to 0 as $m \to \infty$. This yields the stochastic convergence of the sequential estimator $\overline{X}_{N_m}$ under the usual second moment condition. □

Example* 6.5.4 *(Renewal Theorem). Consider a sequence $\{X_i,\ i \geq 1\}$ of i.i.d. (usually nonnegative) random variables with $\mu = \mathbb{E}(X)$, $0 < \mu < \infty$, and let $T_n = \sum_{i \leq n} X_i$, $n \geq 1$, $T_0 = 0$. For $d > 0$, define $\tau_d = \inf\{n \geq 1 : T_n > d\}$. Then, τ_d takes the role of N_m while $a_m = m/\mu$. Exercise 6.5.5 is set to verify the details. □

We conclude this chapter with a brief introduction to other useful convergence concepts.

Uniform integrability. Let $\{G_n,\ n \geq n_0\}$ be a sequence of distribution functions, defined on $\mathbb{R}$, and let $h(y)$, $y \in \mathbb{R}$, be a real-valued continuous function. If

$$\sup_{n \geq n_0} \int_{\{|y| \geq a\}} |h(y)| \mathrm{d}G_n(y) \to 0 \quad \text{as} \quad a \to \infty, \tag{6.5.15}$$

then h is called *uniformly* (in n) *integrable* relatively to $\{G_n\}$. This definition extends to $\mathbb{R}^p$ (for G_n) and/or $\mathbb{R}^q$ (for $h(\cdot)$), for $p, q \geq 1$, by replacing the norm $|\cdot|$ by the Euclidean norm $\|\cdot\|$.

We recall that if $G_n(y) = P\{T_n \leq y\}$ stands for the distribution function of a statistic T_n, such that $T_n \to \theta$, in probability or a.s., as $n \to \infty$, we are not automatically permitted to conclude that $T_n \to \theta$ in the first (or rth) mean (see Example 6.3.6). The uniform integrability condition (6.5.15) provides this access (and this is not true in the situation considered in Example 6.3.6). This concept is strengthened with the following result.

Theorem 6.5.5 (Lebesgue dominated convergence theorem). *For a sequence $\{X_n\}$ of measurable functions, suppose that there exists a random variable Y, such that*

$$|X_n| \leq Y \quad \textit{a.e., where} \quad \mathbb{E}(Y) < \infty, \tag{6.5.16}$$

and either $X_n \to X$ a.e. or $X_n \xrightarrow{D} X$, for a suitable X. Then,

$$\mathbb{E}(|X_n - X|) \to 0, \quad \textit{as } n \to \infty. \tag{6.5.17}$$

A related version of (6.5.17) wherein the role of the dominating random variable Y is de-emphasized is the following:

Let $\{X, X_n, n \geq n_0\}$ be a sequence of random variables, such that $X_n - X \xrightarrow{P} 0$ and

$$\mathbb{E}\left(\sup_{n \geq n_0} |X_n|\right) < \infty. \tag{6.5.18}$$

Then (6.5.17) holds, and, hence, $\mathbb{E}(X_n) \to \mathbb{E}(X)$.

It may be remarked that the uniform integrability condition in (6.5.15) is implicit in (6.5.16) or (6.5.18), and, hence, the proof of the theorem follows by some standard steps; we may refer to Chow and Teicher (1978), among others, for details. For clarification of ideas, we refer to Exercises 6.5.2 and 6.5.3. Exercise 6.5.4 is set to verify that in Exercise 5.2.1, $\widehat{\rho}_n \to \rho$ also in the first mean.

6.6 Concluding Notes

With due emphasis on the scope of applications, a variety of examples has been worked out throughout this chapter. However, a few of the basic results (e.g., the Lévy–Cramér Theorem, law of iterated logarithm, and lower bound to the Kolmogorov Maximal Inequality, among others) have been presented without derivation. Some supplementary reading of these "left-out" proofs is strongly recommended for a methodology-oriented reader. Nevertheless, for the rest of the book, this little omission will not create any problems in the understanding of the outlined methodology as well as the contemplated applications. The probability inequalities and the laws of large numbers have primarily been developed for sums of independent random variables and/or martingale/reverse martingale type of dependent sequences. Although these sequences cover a broad range of statistical models, there may be some other situations [e.g., unequal probability sampling (without replacement) from a finite population] where, at best, one can approximate (in a meaningful way) the actual sequence of statistics by a (sub)martingale or reverse (sub)martingale. Furthermore, the role of the sample size n, in a large sample context, may be replaced by some other characteristic. Example 6.5.2 and the sequential scheme, following it, pertain to this extended mode. There are numerous other examples of this type, some of which will be considered in the subsequent chapters. For these reasons, at this stage, we present (below) only a few selected exercises (some of which are rather artificial) for further clarification of ideas. We will, of course, encourage applications-oriented readers to look into various applied problems and to see to what extent the current chapter provides insight.

6.7 Exercises

6.1.1 Construct a sequence $\{a_n\}$ for which (6.1.1) holds, but a_n is not equal to a, for any finite n.

6.1.2 Construct a sequence $\{a_n\}$ for which (6.1.1) does not hold, but there exists a subsequence $\{a_{n_j}\}$ for which (6.1.1) holds. Can different subsequences have different limits? Construct suitable $\{a_n\}$ to support your answer.

6.2.1 Let denote the empirical distribution function F_n as in (2.4.18). Suppose that the true distribution function F is concentrated on three mass points a, b and c with respective probability masses p_a, p_b and p_c (where $p_a + p_b + p_c = 1$). Verify that $\|F_n - F\| \to 0$ a.s., as $n \to \infty$.

6.2.2 Extend the result in the previous exercise to the case where F is discrete and (i) has finitely many mass points and (ii) has countably infinite number of mass points.

6.2.3 Let $X_1, \ldots, X_n$ be n i.i.d. random variable from the Uniform$(0, \theta)$ distribution function, $\theta > 0$. Let $T_n = \max(X_1, \ldots, X_n)$. Show that (6.2.41) holds with $\psi_\varepsilon(n) = O(n^{-r})$ for $r \geq 2$. Hence, or otherwise, show that $T_n \to \theta$ a.s., as well as completely, as $n \to \infty$.

6.2.4 Let $X_1, \ldots, X_n$ be i.i.d. random variable from the negative exponential distribution for which the density is (1.5.49), and let $T_n = \min(X_1, \ldots, X_n)$. Show that (6.2.41) holds with $T = 0$ and $\psi_\varepsilon(n) = \exp(-n\varepsilon/\theta)$.

6.2.5 A *system* has two components connected in series, so that it *fails* when at least one of the components fail. Suppose that each component has a negative exponential life distribution [defined by (1.5.59)] with mean θ. What is the life distribution of the system? Suppose that there are n copies of this system, and let their lifetimes be denoted by $Y_1, \ldots, Y_n$, respectively. Show that $\overline{Y}_n \xrightarrow{a.s.} \theta/2$.

6.2.6 In the previous exercise, let the system have two components in parallel; verify the a.s. convergence of $\overline{Y}_n$ and work out the limit.

6.2.7 Consider the Bin(n, p) distribution. Let $T = \pi$ and consider the estimator $T_n = n^{-1}X_n$. Note that for all $\pi \in [0, 1]$,

$$\mathbb{E}(T_n - \pi)^4 \leq n^{-2}\frac{3}{16}.$$

Hence, using Chebyshev Inequality (6.3.1) with $U_n = (T_n - \pi)^4$, show that for every $\varepsilon > 0$,

$$P(\|T_n - \pi\| > \varepsilon) \leq n^{-2}\frac{3}{16\varepsilon^4},$$

so that the series in (6.2.36) converges. Therefore, conclude that $T_n \xrightarrow{c} \pi$.

6.3.1 Let $nT_n \sim$ Bin(n, θ), $0 < \theta < 1$. Compare (6.3.2) and (6.3.3) [assuming that the normality in (6.3.3) holds closely], when $n = 50$, $t = 1$, and $\theta = 1/2, 1/4$, and $1/10$.

6.3.2 In the previous exercise, let $\varepsilon = 0.05$, $n = 50$, and $\theta = 1/2(= \pi)$ and compare (6.3.11) with (6.3.2).

6.3.3 Verify (6.3.14).

6.3.4 Verify (6.3.23) in the context of Example 6.3.2.

6.3.5 Consider the simple logistic density function $f(x) = \exp(-x)[1 + \exp(-x)]^{-2}, x \in \mathbb{R}$. Verify that here $M(t)$ is finite for every t such that $|t| < 1$ and, hence, obtain (6.3.23) with explicit forms for $\rho_+(\varepsilon)$ and $\rho_-(\varepsilon)$.

6.3.6 Consider the t-distribution with m degrees of freedom. Show that for this distribution, the moment generating function $M(t)$ is not finite for any real t, and, hence, comment on the rate of convergence for the t-statistic. Another counterexample is the classical Cauchy distribution for which the density function is given by $f(x) = \pi^{-1}\lambda[\lambda^2 + (x - \theta)^2]^{-1}, x \in \mathbb{R}$. Show that for this distribution, no positive moment of order 1 or more is finite, and, hence, none of the probability inequalities discussed so far apply.

6.3.7 Obtain explicit forms for $\rho_+(\varepsilon)$ and $\rho_-(\varepsilon)$ in the context of Exercise 6.3.5.

6.3.8 Consider the Pareto density f, given by $f(x; a, \nu) = \frac{1}{a}\nu(x/a)^{-\nu-1}I(x \geq a)$, $a > 0$, $\nu > 0$. Show that for this density, the moment generating function $M(t)$ does not exist for any $t > 0$. For what range of values of ν, does $\mathbb{E}(X^k)$ exist, for a given $k > 0$? Hence, or otherwise, show that if $\nu > 1$, the Khintchine SLLN (Theorem 6.3.13) applies to samples from this distribution.

6.3.9 In the context of Examples 6.3.5 (CMRR procedure) and 6.3.7, consider the asymptotic situation where n_1/N and n_2/N both converge to 0 as $N \to \infty$, but $n_1 n_2/N \to \lambda$, for some $\lambda > 0$. Use (6.3.35) to verify that r_2 has closely a Poisson distribution with parameter λ. Hence, or otherwise, find out the asymptotic moment generating function of r_2 and show that $r_2/\lambda \xrightarrow{P} 1$, as $\lambda \to \infty$.

6.3.10 Verify that the Khintchine SLLN [Theorem 6.3.13] holds when the random variables X_i have a common distribution F, defined on $\mathbb{R}^p$, for some $p \geq 1$. Hence, or otherwise, show that the case of vector-valued or matrix-valued i.i.d. random variables is covered by this extension.

6.3.11 Let (X_i, Y_i), $i \geq 1$, be i.i.d. random variables with a bivariate distribution function having finite moments up to the second order. Let r_n be the sample correlation coefficient (for a sample of size n) and ρ be the population counterpart. Show that $r_n \xrightarrow{a.s.} \rho$.

6.4.1 Consider a *bundle* of n parallel filaments whose individual strengths are denoted by $X_1, \ldots, X_n$ (nonnegative random variables). Also, let $X_{n:1} \leq \cdots \leq X_{n:n}$ be the ordered values of $X_1, \ldots, X_n$. The *bundle strength* is then defined by Daniels (1945) as

$$B_n = \max\,[(n - i + 1)X_{n:i} : 1 \leq i \leq n]\,.$$

Consider the empirical distribution function F_n as in (2.4.18) and show that

$$Z_n = n^{-1} B_n = \sup\,\{x[1 - F_n(x)] : x \geq 0\}\,.$$

Hence, or otherwise, use (5.2.31), to verify that $\{Z_n\}$ is a nonnegative reverse submartingale. Hence, use Theorem 6.4.3 to show that as $n \to \infty$

$$Z_n \xrightarrow{a.s.} \theta = \sup\{x[1 - F(x)] : x \geq 0\}.$$

6.4.2 In the context of the previous exercise, show that as $n \to \infty$

$$n^{1/2}|\mathbb{E}(Z_n) - \theta| \to 0,$$

and $\mathbb{E}(Z_n) \geq \theta$, for all $n \geq 1$. [Sen, Bhattacharyya, and Suh (1973)].

6.4.3 Show that in Exercise 5.2.1, $\widehat{\rho}_n \to \rho$ in the first mean.

6.5.1 For the negative binomial law in (1.5.50), index the random variable n by n_m, and rewrite m/n_m as T_m^*. Then show that T_m^* is a bounded random variable for every $m \geq 1$. Moreover, using (6.5.11), compute the mean and variance of T_m^* and, hence, use the Chebyshev Inequality (6.3.1) to show that $T_m^* \xrightarrow{P} \pi$ as $m \to \infty$. Can you claim that $T_m^* \xrightarrow{r\text{th}} \pi$, as $m \to \infty$?

6.5.2 Show that for every $m \geq 1$, n_m as defined in Exercise 6.5.1 can be expressed as $Z_1 + \cdots + Z_m$, where the Z_i are i.i.d. random variables, each having a negative binomial law (with the index $m = 1$). Hence, use Theorem 6.3.13 to show that $n_m/m \xrightarrow{a.s.} \mathbb{E}(Z_1)$ as $m \to \infty$, and conclude that $T_m^* \xrightarrow{a.s.} \pi$.

6.5.3 Define N_m as in (6.5.14) and show that

$$s^2_{N_m - 1} > \frac{1}{m}(N_m - 1) \geq s^2_{N_m} - \frac{1}{m}, \quad m \geq 1.$$

Also verify that N_m is monotonically nondecreasing in m, and, for every fixed m, N_m is finite a.s. Furthermore, by (6.3.84), $s_n^2 \xrightarrow{a.s.} \sigma^2$. Thus, on letting $a_m = [m\sigma^2]$, verify that $N_m/a_m \xrightarrow{a.s.} 1$, as $m \to \infty$.

6.5.4 In Exercise 5.2.1, show that $\widehat{\rho}_n \to \rho$ in the first mean.

6.5.5 (Renewal Theorem). Let $\{X_i, i \geq 1\}$ be a sequence of i.i.d. random variable with $0 < \mathbb{E}(X) = \mu < \infty$, and set

$$T_n = \sum_{i=1}^{n} X_i, \quad n \geq 1, \text{ and } \tau_d = \min(n : T_n > d), d > 0.$$

Note that T_d is nondecreasing in $d > 0$ and, further,

$$T_{\tau_d} > d \geq T_{\tau_d - 1}, \quad d > 0.$$

Moreover by Theorem 6.3.13, $n^{-1}T_n \xrightarrow{a.s.} \mu$. Then, show that $d^{-1}\tau_d \xrightarrow{a.s.} 1/\mu$, as $d \to \infty$. Also, $\mathbb{E}(\tau_d)/d \to 1/\mu$, as $d \to \infty$.

7

Asymptotic Distributions

7.1 Introduction

Although desirable, consistency (whether of an estimator or a test statistic) may not convey the full statistical information contained in the data set at hand, and, hence, by itself, may not provide efficient statistical conclusions. To illustrate this point, suppose that we have n independent random variables $X_1, \ldots, X_n$ drawn from an unknown distribution with mean μ and finite variance σ^2 (both unknown). The sample mean $\overline{X}_n$ is a natural estimator of μ, and from the results of Chapter 6, we may conclude that as n increases, $\overline{X}_n$ converges to μ in some well-defined manner (e.g., in probability/almost surely/second mean). This certainly endows us with an increasing confidence on the estimator as the sample size increases. Suppose now that we desire to set a confidence interval (L_n, U_n) for μ with a prescribed coverage probability or confidence coefficient $1 - \alpha$ $(0 < \alpha < 1)$, that is, we intend to determine suitable statistics L_n and U_n $(L_n \leq U_n)$, such that

$$P(L_n \leq \mu \leq U_n) \geq 1 - \alpha. \tag{7.1.1}$$

If σ is known, we can use the Chebyshev Inequality (6.3.1) to obtain that, for every $n \geq 1$,

$$P(|\overline{X}_n - \mu| > t) \leq \sigma^2/(nt^2), \tag{7.1.2}$$

so that, on setting $\alpha = \sigma^2/(nt^2)$, that is, $t = \sigma/(n\alpha)^{1/2}$, we have

$$P[L_n = \overline{X}_n - \sigma/(n\alpha)^{1/2} \leq \mu \leq \overline{X}_n + \sigma/(n\alpha)^{1/2} = U_n] \geq 1 - \alpha. \tag{7.1.3}$$

We may even get sharper bounds (for L_n and U_n) using the Markov Inequality (6.3.4), if we assume that higher-order moments of the underlying random variables exist, or the Bernstein Inequality (6.3.7), if we assume that the moment generating function is finite. On the other hand, if we assume further that $X_1, \ldots, X_n$ follow normal distributions with mean μ and variance σ^2, we have, for every real x and every $n \geq 1$,

$$P[n^{1/2}(\overline{X}_n - \mu)/\sigma \leq x] = \Phi(x). \tag{7.1.4}$$

It follows from (7.1.4) that, for every $n \geq 1$ and $0 < \alpha < 1$,

$$P(L_n^* = \overline{X}_n - n^{-1/2}\sigma z_{\alpha/2} \leq \mu \leq \overline{X}_n + n^{-1/2}\sigma z_{\alpha/2} = U_n^*) = 1 - \alpha. \tag{7.1.5}$$

Comparing the width of the two bounds in (7.1.3) and (7.1.5), we get

$$(U_n^* - L_n^*)/(U_n - L_n) = \alpha^{1/2} z_{\alpha/2}. \tag{7.1.6}$$

The right-hand side of (7.1.6) is strictly less than 1 for every $0 < \alpha < 1$. For $\alpha = 0.01$, 0.025, 0.05, and 0.10, it is equal to 0.258, 0.354, 0.438, and 0.519, respectively. A similar picture holds for the case of the Markov or Bernstein Inequality–based bounds. This simple example illustrates that the knowledge of the actual sampling distribution of a statistic generally leads to more precise confidence bounds than those obtained by using some of the probability inequalities considered in Chapter 6. For an estimator T_n of a parameter θ, more precise studies of its properties may be achieved when its sampling distribution is known. A similar situation arises when one wants to test for the null hypothesis (H_0) that μ is equal to some specified value μ_0 against one-sided or two-sided alternatives. In order that the test has a specified margin $0 < \alpha < 1$ for the Type I error, one may use the probability inequalities from Chapter 6 for the demarcation of the critical region. Knowledge of the distribution of the test statistic (under H_0) provides a test that has generally better power properties.

In small samples, a complete determination of the sampling distribution of a statistic T_n may not only demand a knowledge of the functional form of the underlying distribution function F but also cumbrous algebraic manipulations (especially when T_n is not linear). The situation changes drastically as n increases. Under fairly general regularity conditions, it is possible to approximate the actual sampling distribution of the normalized version of a statistic T_n by some simple ones (such as the normal law), and in this way, one may have more flexibility to study suitable estimators and/or test statistics, which possess some asymptotic optimality properties that may serve as good approximations when the sample size is large. To illustrate this point, let us go back to the same example we considered before but now assume that both μ and σ are unknown. The sample variance s_n^2 is a consistent estimator of σ^2 (as has already been established in Chapter 6), and, hence, it may be quite tempting to use the normalized Student t-statistic

$$t_n = n^{1/2}(\overline{X}_n - \mu)/s_n, \quad n \geq 2. \tag{7.1.7}$$

It is well-known that for $n > 30$, the actual sampling distribution of t_n may be very well-approximated by the normal law. Incidentally, to use some of the probability inequalities on t_n we need extra manipulations, and the simplicities of the Chebyshev/Markov/Bernstein inequalities are no longer attainable. Actually, even if the underlying distribution function F is not normal but possesses a finite second-order moment, (7.1.4) extends in a natural way to the following:

$$\lim_{n\to\infty} P[n^{1/2}(\overline{X}_n - \mu)/\sigma \leq x] = \Phi(x), \quad x \in \mathbb{R} \tag{7.1.8}$$

and, in the literature, this is referred to as the classical *Central Limit Theorem* (CLT). Moreover, (7.1.8) holds even when σ is replaced by s_n, so that we need not assume that σ^2 is known in providing a confidence interval for μ or to test for a null hypothesis on μ. Presumably, one may have to pay a little penalty: the rate of convergence may be generally slower for (7.1.7). Numerous other examples of this type will be considered in this chapter and in subsequent ones too.

Keeping the case of $n^{1/2}(\overline{X}_n - \mu)/\sigma$ or $n^{1/2}(\overline{X}_n - \mu)/s_n$ in mind, in a typical statistical inference problem, we encounter a sequence $\{T_n; n \geq n_0\}$ of statistics, such that for suitable normalizing constants $\{a_n, b_n;\ n \geq n_0\}$, $(T_n - a_n)/b_n$ has a limiting distribution, which is typically nondegenerate, and our goal is to formulate this in a simple manner.

Let

$$F_{(n)}(t) = P[(T_n - a_n)/b_n \leq t], \quad t \in \mathbb{R}, \quad n \geq n_0. \tag{7.1.9}$$

Our contention is to approximate $F_{(n)}$ by some simple distribution F (e.g., normal, Poisson, chi-squared) for large sample sizes. Associated with such an F, we have a random variable, say Z, such that $P(Z \leq t) = F(t)$, for all t. Recalling (6.2.24) we say that $(T_n - a_n)/b_n$ converges in distribution (or law) to Z, symbolically,

$$(T_n - a_n)/b_n \xrightarrow{D} Z. \tag{7.1.10}$$

Some points of clarification are in order here. First, in (7.1.10), we really mean that $F_{(n)}(t) \to F(t)$ as $n \to \infty$, for almost all t. Now, the distribution function $F_{(n)}$ may not be continuous everywhere (e.g., the binomial case), nor necessarily is the distribution function F continuous everywhere (e.g., the Poisson distribution). As such, if t is a jump point of F, $F_{(n)}(t)$ may not converge to $F(t)$. We eliminate this problem by saying that $F_{(n)} \to F$ weakly if $F_{(n)}(t) \to F(t)$ as $n \to \infty$ at all points of continuity of F. In symbols, we write

$$F_{(n)} \to F \quad \text{if} \quad F_{(n)}(t) \to F(t), \quad \forall t \in J, \quad \text{as} \quad n \to \infty, \tag{7.1.11}$$

where J is the set of continuity points t of F. Second, if we let $Z_n = (T_n - a_n)/b_n, n \geq n_0$, then Z_n and Z need not be defined on the same probability space, and $Z_n - Z$ (even if properly defined) may not converge to 0 in probability or in some other mode (as mentioned in Section 6.2) when n becomes large. For example, let nT_n be $\text{Bin}(n, \pi)$, $0 < \pi < 1$, and $Z_n = n^{1/2}(T_n - \pi)$, $n \geq 1$. Then, for every finite n, Z_n has a discrete distribution with positive probability mass attached to the points $n^{-1/2}(k - n\pi)$, $k = 0, 1, \ldots, n$, while we will see in Section 7.3 that here (7.1.8) holds with $\sigma^2 = \pi(1 - \pi)$, so that $F(= \Phi)$ is an absolutely continuous distribution function In the same example, if $\pi = \pi_n$ is allowed to depend on n in such a way that as $n \to \infty$, $n\pi_n \to \lambda$, $0 < \lambda < \infty$, then the binomial law converges to a Poisson law with parameter λ. In this case, the Poisson distribution has jump points 0,1,2,. . . . , at which the right-hand and left-hand side limits may not agree. Thus, we need to eliminate these jump points from the convergence criterion, and this reflects the necessity of the set J in (7.1.11).

Whenever the statistics $\boldsymbol{T}_n, n \geq 1$, are random p-vectors, they have distribution functions defined on $\mathbb{R}^p$, for some $p \geq 1$. Then we may consider a real-valued random element $Z_n = h(\boldsymbol{T}_n; n)$ and apply the result in (7.1.10). A very common problem arising in this context involves a linear combination of the elements of $\boldsymbol{T}_n$, in which case, we may have typically a normal asymptotic distribution. In a variety of other cases, we would also consider a quadratic form in the elements of $\boldsymbol{T}_n$, and these may relate to asymptotic (central/noncentral) chi-squared and allied distributions. In a general multivariate setup, we consider a sequence of (normalized) stochastic vectors $\{\boldsymbol{T}_n, n \geq n_0\}$ and let $F_{(n)}(\boldsymbol{t}) = P(\boldsymbol{T}_n \leq \boldsymbol{t}), \boldsymbol{t} \in \mathbb{R}^p$. Also, we conceive of a distribution function F defined on $\mathbb{R}^p$ and define the continuity-point set J $(\subset \mathbb{R}^p)$ in a similar manner. Let F stand for the distribution function of a random vector $\boldsymbol{T}$. Then, we say that $\boldsymbol{T}_n$ converges in distribution (or law) to $\boldsymbol{T}$ ($\boldsymbol{T}_n \xrightarrow{D} \boldsymbol{T}$) or $F_{(n)}$ weakly converges to F, if

$$F_{(n)}(\boldsymbol{t}) \to F(\boldsymbol{t}), \quad \forall \boldsymbol{t} \in J \ (\subset \mathbb{R}^p), \quad \text{as} \quad n \to \infty. \tag{7.1.12}$$

The concept of weak convergence is not confined to the finite-dimensional vector case. Its extension to the case of general probability spaces enables one to include infinite-dimensional vectors or stochastic processes. However, to be consistent with our intended (intermediate) level of presentation, we present this in Chapter 11.

The normal distribution (in univariate as well as multivariate setups) occupies a central position in asymptotic theory or large-sample methods. The *De Moivre–Laplace Theorem* (on the convergence of the binomial to the normal law) is one of the classical examples of the weak convergence of $\{F_{(n)}\}$. For binomial, multinomial, hypergeometric, and other discrete distributions, earlier work on weak convergence have exploited direct expansions of the associated probability functions (by the use of the *Stirling approximations* to factorials of natural integers), resulting in a mode of convergence stronger than in (7.1.12). However, in most of the cases, the algebraic manipulations are heavy and specifically dependent on the particular underlying model. Moreover, they are not that necessary for actual applications. For sums (or averages) of independent random variables, the weak convergence in (7.1.11) or (7.1.12) has been proven under increasing generality by a host of workers (Liapounov, Lindeberg, Feller, and others). As we will see in Section 7.3, for most of the statistical applications, this weak convergence suffices, and the other local limit theorems available for some discrete distributions are not that necessary.

The normal (or Gaussian) distribution has many interesting properties. Among these, the structure of its central moments has been widely used in weak convergence studies. Note that the $(2k)$th central moment of a standard normal distribution is $(2^k k!)^{-1}(2k)!$ and the $(2k+1)$th moment is 0, for every $k \geq 1$. This inspired a lot of workers to evaluate the first four moments of a statistic T_n and to show that these converge to the corresponding ones of a suitable normal distribution. From this moment convergence, they attempted to conclude that the distribution of T_n (for the normalized version) also converges to a normal one. Although this conclusion remains acceptable in some cases, it cannot be justified theoretically. If the convergence of the asymptotic distribution of T_n to a normal law has to be based on the convergence of moments, then one needs to consider moments of all finite orders (not just the first four), and, moreover, the limiting distribution must be uniquely defined by its moments. The evaluation of moments of all finite orders may indeed be a monumental task, and, therefore, this approach will not be stressed here too.

The characteristic functions introduced in Chapter 1 play a fundamental role in the weak convergence studies to be pursued in here. For example, for sums or averages or linear statistics, the approach based on characteristic functions yields general theorems, which are much less model-dependent, and these remain vastly adaptable in most of the statistical applications; they depend on some basic results that will be discussed in Section 7.2. We will consider a general review of these broad results, because they facilitate the extensions of the main theorems to triangular schemes of (row-wise independent) random variables and vectors and to multivariate central limit theorems.

The central limit theorems will be considered in Section 7.3. It is important to note that they are no longer confined to sums of independent random variables (vectors). During the past decades, these central limit theorems have been successfully extended for various dependent random variables. In particular, *martingales*, reverse martingales, and related sequences have been annexed in these developments, and they have some important applications too. We will also review some of these developments.

Section 7.4 is essentially included for the sake of completeness and deals with the rates of convergence to normality. This is a very specialized topic and the reader may skip it and still understand of the rest of the book.

The Slutsky Theorem and some related *projection results* [mostly, due to Hoeffding (1948) and Hájek (1968)] are considered in Section 7.5; their applications are stressed too. Transformations of statistics and variables play an important role in large sample theory. In particular, the *variance-stabilizing transformations* are interesting on their own. Interestingly enough, such transformations often accelerate the rate of convergence to the limiting distributions. Related asymptotic theory is considered in the same section.

Section 7.6 deals with the asymptotic properties of quadratic forms. There the chi-squared distribution replaces the Gaussian distribution as the limiting law. The *Cochran Theorem* (on quadratic forms) in a multinormal setup has many important uses in statistical inference. In this setup, too, it may not be necessary to assume multinormality of the underlying distribution function, and incorporating the usual multidimensional central limit theorems, extensions of the Cochran Theorem will be considered too. The asymptotic behavior of sample quantiles and empirical distribution functions play an important role in the study of the properties of many estimators based on U-, L-, M-, and R-statistics. This is the topic covered in Section 7.7. More special attention is given to the limiting behavior of extreme order statistics. In Section 7.8, we consider the following natural question: is the weak convergence result in (7.1.10) accompanied by the convergence of the moments of Z_n (up to a certain order) to the corresponding ones of Z? Some additional regularity conditions pertain to this problem, and these will be briefly presented.

7.2 Some Important Tools

In this section, we present some important results on weak convergence of distribution functions, that serve as the basis for many useful applications

Theorem 7.2.1 (Helly–Bray Lemma). *Let $\{F_{(n)}\}$ be a sequence of distribution functions; also, let g be a continuous function on $[a, b]$ where $-\infty < a < b < \infty$ are continuity points of a distribution function F. Then, if $F_{(n)} \xrightarrow{w} F$, we have*

$$\int_a^b g(x)\,dF_{(n)}(x) \to \int_a^b g(x)\,dF(x) \quad as \quad n \to \infty.$$

Proof. Let

$$I_n = \int_a^b g(x)\,\mathrm{d}F_{(n)}(x) \quad \text{and} \quad I = \int_a^b g(x)\,\mathrm{d}F(x);$$

now, given $\varepsilon > 0$, define a partition of $[a, b]$ by taking

$$a = x_0 < x_1 < \cdots < x_m < x_{m+1} = b,$$

where

(i) $x_j, 0 \le j \le m+1$, are points of continuity of F;
(ii) $x_{j+1} - x_j < \varepsilon, 0 \le j \le m$.

Then, for $x_l \le x \le x_{l+1}$, $0 \le l < m$, define

$$g_m(x) = g[(x_l + x_{l+1})/2] = \sum_{l=0}^{m} g[(x_l + x_{l+1})/2] I(x_l < x < x_{l+1}),$$

$$I_n(m) = \int_a^b g_m(x)\, \mathrm{d}F_{(n)}(x),$$

$$I(m) = \int_a^b g_m(x)\, \mathrm{d}F(x),$$

and observe that

$$|I_n - I| \le |I_n - I_n(m)| + |I_n(m) - I(m)| + |I(m) - I|.$$

Analyzing each term of this inequality, we have:

(a) since g is continuous, for some $\delta = \delta(\varepsilon)$,

$$\begin{aligned}|I(m) - I| &\le \sup_{x\in[a,b]} |g_m(x) - g(x)| \int_a^b \mathrm{d}F(x) \\ &\le \sup_{x\in[a,b]} |g_m(x) - g(x)| < \delta;\end{aligned}$$

(b) since $F_{(n)} \overset{w}{\to} F$ and x_l, $0 \le l \le m+1$, are points of continuity of F,

$$I_n(m) - I(m) = \sum_{l=0}^{m} g\left(\frac{x_l + x_{l+1}}{2}\right)[\Delta F_{(n)}(x_l) - \Delta F(x_l)],$$

where $\Delta F_{(n)}(x_l) = F_{(n)}(x_{l+1}) - F_{(n)}(x_l)$ and $\Delta F(x_l) = F(x_{l+1}) - F(x_l)$ converges to 0 as $n \to \infty$; and

(c) since g is continuous, for some $\delta = \delta(\varepsilon)$,

$$\begin{aligned}|I_n - I_n(m)| &\le \sup_{x\in[a,b]} |g(x) - g_m(x)| \int_a^b \mathrm{d}F_{(n)}(x) \\ &\le \sup_{x\in[a,b]} |g(x) - g_m(x)| < \delta.\end{aligned}$$

Therefore, given $\delta > 0$ it is possible to choose a convenient $\varepsilon > 0$ and a sufficiently large n such that $|I_n - I| < 3\delta$, which completes the proof. ■

Theorem 7.2.2 (Extension of the Helly–Bray Lemma). *Let $g(x)$ be a continuous and bounded function and let $\{F_{(n)}\}$ be a sequence of distribution functions, such that $F_{(n)} \overset{w}{\longrightarrow} F$. Then,*

$$\int_{-\infty}^{+\infty} g(x)\, dF_{(n)}(x) \to \int_{-\infty}^{+\infty} g(x)\, dF(x) \quad as \quad n \to \infty.$$

Proof. Given $\varepsilon' > 0$, let $-\infty < a < b < \infty$ be points of continuity of F such that $\int_{-\infty}^{a} \mathrm{d}F(x) < \varepsilon'$ and $\int_b^{\infty} \mathrm{d}F(x) < \varepsilon'$. Then let $\varepsilon = g^*\varepsilon'$ where $g^* = \sup_{x\in\mathbb{R}} |g(x)| < \infty$ and observe that as $n \to \infty$, we have

(i) $|\int_{-\infty}^{a} g(x)\,\mathrm{d}F_{(n)}(x)| \le g^* F_{(n)}(a) \to g^* F(a) < g^*\varepsilon' = \varepsilon$ and
(ii) $|\int_{b}^{\infty} g(x)\,\mathrm{d}F_{(n)}(x)| \le g^*[1 - F_{(n)}(b)] \to g^*[1 - F(b)] < g^*\varepsilon' = \varepsilon$.

Now write $I_n = \int_{-\infty}^{\infty} g(x)\,\mathrm{d}F_{(n)}(x)$ and $I = \int_{-\infty}^{\infty} g(x)\,\mathrm{d}F(x)$ and note that

$$\begin{aligned}|I_n - I| \le &\left|\int_{-\infty}^{a} g(x)\,\mathrm{d}F_{(n)}(x) - \int_{-\infty}^{a} g(x)\,\mathrm{d}F(x)\right| \\ &+ \left|\int_{a}^{b} g(x)\,\mathrm{d}F_{(n)}(x) - \int_{a}^{b} g(x)\,\mathrm{d}F(x)\right| \\ &+ \left|\int_{b}^{\infty} g(x)\,\mathrm{d}F_{(n)}(x) - \int_{b}^{\infty} g(x)\,\mathrm{d}F(x)\right|. \qquad (7.2.1)\end{aligned}$$

Applying the triangular inequality (1.5.39) to each relation (i) and (ii), it follows that for sufficiently large n

$$\left|\int_{-\infty}^{a} g(x)\,\mathrm{d}F_{(n)}(x) - \int_{-\infty}^{a} g(x)\,\mathrm{d}F(x)\right| < 2\varepsilon, \qquad (7.2.2)$$

and

$$\left|\int_{b}^{\infty} g(x)\,\mathrm{d}F_{(n)}(x) - \int_{b}^{\infty} g(x)\,\mathrm{d}F(x)\right| < 2\varepsilon. \qquad (7.2.3)$$

Furthermore, an application of the Helly–Bray Lemma 7.2.1 yields

$$\left|\int_{a}^{b} g(x)\,\mathrm{d}F_{(n)}(x) - \int_{a}^{b} g(x)\,\mathrm{d}F(x)\right| < \varepsilon. \qquad (7.2.4)$$

Then, from (7.2.1)–(7.2.4) we may conclude that for sufficiently large n, $|I_n - I| < 5\varepsilon$, which completes the proof. ■

Note that the above theorem holds even in cases where g is not real valued. For example, let $g(x) = \exp(itx) = \cos(tx) + i\sin(tx)$, $t \in \mathbb{R}$, and consider a sequence of distributions functions $\{F_{(n)}\}$, such that $F_{(n)} \xrightarrow{w} F$. Applying Theorem 7.2.2 to $g_1(x) = \cos(tx)$ and $g_2(x) = \sin(tx)$ and using the triangular inequality, we have

$$\phi_{F_{(n)}}(t) = \int_{-\infty}^{\infty} e^{itx}\,\mathrm{d}F_{(n)}(x) \to \int_{-\infty}^{\infty} e^{itx}\,\mathrm{d}F(x) = \phi_F(t) \quad \text{as} \quad n \to \infty.$$

Actually, the converse proposition is also true, and together they constitute one of the most powerful tools in the field of weak convergence, the well-known Lévy–Cramér Theorem 6.2.8.

***Example* 7.2.1.** Consider a sequence $\{R_n\}$ of random variable such that $R_n \sim \text{Bin}(n, \pi)$ and let $n \to \infty$, $\pi \to 0$ in such a way that $n\pi \to \lambda < \infty$. We know that

$$\phi_{R_n}(t) = \mathbb{E}(e^{itR_n}) = [1 - \pi(1 - e^{it})]^n = \left[1 - \frac{\lambda}{n}(1 - e^{it})\right]^n.$$

Then, $\lim_{n\to\infty} \phi_{R_n}(t) = \exp\{-\lambda[1 - \exp(it)]\}$, which we recognize as the characteristic function of a random variable R following a Poisson(λ) distribution. Therefore, we may conclude that $R_n \xrightarrow{D} R$. □

The following result, parallel to that of Theorem 6.3.4, constitutes an important tool in the search for the asymptotic distribution of transformations of statistics with known asymptotic distributions.

Theorem 7.2.3 (Sverdrup). *Let $\{T_n\}$ be a sequence of random variables such that $T_n \xrightarrow{D} T$ and $g : \mathbb{R} \to \mathbb{R}$ be a continuous function. Then $g(T_n) \xrightarrow{D} g(T)$.*

Proof. For all $u \in \mathbb{R}$, let

$$\begin{aligned}\phi_{g(T_n)}(u) &= \mathbb{E}\{\exp[iug(T_n)]\}\\ &= \int_{-\infty}^{+\infty} \cos[ug(t)]\,\mathrm{d}F_{T_n}(u) + i\int_{-\infty}^{\infty} \sin[ug(t)]\,\mathrm{d}F_{T_n}(u).\end{aligned}$$

Now, since $\cos(x)$, $\sin(x)$, and $g(x)$ are continuous and bounded functions, it follows from Theorem 7.2.2 that as $n \to \infty$

$$\begin{aligned}\phi_{g(T_n)}(u) &\to \int_{-\infty}^{+\infty} \cos[ug(t)]\,\mathrm{d}F_T(u) + i\int_{-\infty}^{+\infty} \sin[ug(t)]\,\mathrm{d}F_T(u)\\ &= \mathbb{E}\{\exp[iug(T)]\} = \phi_{g(T)}(u).\end{aligned}$$

Thus, from the Lévy–Cramér Theorem 6.2.8 the result follows. ■

Extensions to random vectors may be easily obtained via the following theorem, which is the most important tool for the generalization of many univariate results to the multivariate case.

Theorem 7.2.4 (Cramér–Wold). *Let $\boldsymbol{X}, \boldsymbol{X}_1, \boldsymbol{X}_2, \ldots$ be random vectors in $\mathbb{R}^p$; then $\boldsymbol{X}_n \xrightarrow{D} \boldsymbol{X}$ if and only if, for every fixed $\boldsymbol{\lambda} \in \mathbb{R}^p$, we have $\boldsymbol{\lambda}'\boldsymbol{X}_n \xrightarrow{D} \boldsymbol{\lambda}'\boldsymbol{X}$.*

Proof. Suppose that $\boldsymbol{X}_n \xrightarrow{D} \boldsymbol{X}$; then, for every $\alpha \in \mathbb{R}$ and $\boldsymbol{\lambda} \in \mathbb{R}^p$, it follows from Theorem 7.2.2 that as $n \to \infty$

$$\phi_{\boldsymbol{\lambda}'\boldsymbol{X}_n}(\alpha) = \phi_{\boldsymbol{X}_n}(\alpha\boldsymbol{\lambda}) \to \phi_{\boldsymbol{X}}(\alpha\boldsymbol{\lambda}) = \phi_{\boldsymbol{\lambda}'\boldsymbol{X}}(\alpha).$$

Using the Lévy–Cramér Theorem 6.2.8 we conclude that $\boldsymbol{\lambda}'\boldsymbol{X}_n \xrightarrow{D} \boldsymbol{\lambda}'\boldsymbol{X}$. The converse follows by similar arguments. ■

Finally, another result that is continuously used in the remainder of the text is the object of the following theorem.

Theorem 7.2.5. *Let $\{X_n\}$ be a sequence of random variables, such that $X_n \xrightarrow{D} X$, where X has a nondegenerate distribution function F. Then $X_n = O_p(1)$.*

Proof. From the definition of weak convergence it follows that given $\gamma > 0$, there exists $n_0 = n_0(\gamma)$, such that for all points of continuity of $F(x) = P(X \leq x)$, we have

$$|P(X_n \leq x) - P(X \leq x)| < \gamma, \quad \forall n \geq n_0. \tag{7.2.5}$$

Now, from the definition of $F(x)$, we know that given $\eta > 0$, there exists $M_1 = M_1(\eta)$ and $M_2 = M_2(\eta)$, such that $P(X \leq x) > 1 - \eta$, for every $x \geq M_1$, and $P(X \leq x) < \eta$, for every $x \leq M_2$; thus, taking $M = \max(|M_1|, |M_2|)$, we may conclude that $P(X \leq x) > 1 - \eta$, for every $x \geq M$ and $P(X \leq x) < \eta$, for every $x \leq -M$. Then, from (7.2.5) we have that for all $n \geq n_0$

$$
\begin{aligned}
P(X_n \leq x) > P(X \leq x) - \gamma &\Rightarrow P(X_n \leq M) > 1 - \eta - \gamma \\
&\Rightarrow P(X_n > M) < \eta + \gamma, \qquad (7.2.6) \\
P(X_n \leq x) < P(X \leq x) + \gamma &\Rightarrow P(X_n \leq -M) < \eta + \gamma. \qquad (7.2.7)
\end{aligned}
$$

Now, given $\varepsilon > 0$, it suffices to take $\eta = \gamma = \varepsilon/4$ and use (7.2.5)–(7.2.7) to show that for all $n \geq n_0$ we have

$$
P(|X_n| > M) \leq P(X_n > M) + P(X_n < -M) < \varepsilon,
$$

concluding the proof. ■

7.3 Central Limit Theorems

Here we essentially investigate the convergence of sequences of distribution functions $\{F_{(n)}\}$ of sums of random variables to the normal distribution function Φ. The theorems considered in this section may be classified under the denomination of *Central Limit Theorems* and although they constitute special cases of more general results, they are sufficient for most situations of practical interest. We start with the simplest form of such a theorem and proceed to extend the results to cases with less-restrictive assumptions.

For the sake of simplicity, we will use the notation $X_n \xrightarrow{D} \mathcal{N}(\mu, \sigma^2)$ or $X_n \xrightarrow{D} \chi^2_q$ to indicate that $X_n \xrightarrow{D} X$, where $X \sim \mathcal{N}(\mu, \sigma^2)$ or $X_n \xrightarrow{D} X$, where $X \sim \chi^2_q$, respectively.

Theorem 7.3.1 (Classical Central Limit Theorem). *Let X_k, $k \geq 1$ be i.i.d. random variables with mean μ and finite variance σ^2. Also let*

$$
Z_n = (T_n - n\mu)/\sigma\sqrt{n},
$$

where $T_n = X_1 + \cdots + X_n$. Then, $Z_n \xrightarrow{D} \mathcal{N}(0, 1)$.

Proof. Let $\phi_{Z_n}(t) = \mathbb{E}[\exp(itZ_n)]$ and note that

$$
\begin{aligned}
\phi_{Z_n}(t) &= \mathbb{E}\left\{\exp\left[\frac{it}{\sigma\sqrt{n}}\left(\sum_{k=1}^{n} X_k - n\mu\right)\right]\right\} \\
&= \prod_{k=1}^{n} \mathbb{E}\left\{\exp\left[\frac{it}{\sigma\sqrt{n}}(X_k - \mu)\right]\right\} \\
&= \left\{\mathbb{E}\left[\exp\left(\frac{it}{\sigma\sqrt{n}}U\right)\right]\right\}^n = \left[\phi_U\left(\frac{t}{\sqrt{n}}\right)\right]^n,
\end{aligned}
$$

where $U = (X_1 - \mu)/\sigma$ and $\phi_U(t) = \mathbb{E}[\exp(itU)]$. Considering a Taylor expansion of $\phi_U(t/\sqrt{n})$ and observing that $\mathbb{E}(U) = 0$ and $\mathbb{E}(U^2) = 1$, we may write

$$\begin{aligned}\phi_U\left(\frac{t}{\sqrt{n}}\right) &= 1 + i\mathbb{E}(U)\frac{t}{\sqrt{n}} + i^2\frac{t^2}{2n}\mathbb{E}(U^2) + o\left(\frac{t^2}{n}\right)\\ &= 1 - \frac{t^2}{2n} + o\left(\frac{t^2}{n}\right).\end{aligned}$$

Therefore, as $n \to \infty$, we have

$$\phi_{Z_n}(t) = \left[1 - \frac{t^2}{2n} + o\left(\frac{t^2}{n}\right)\right]^n \to \exp\left(\frac{-t^2}{2}\right),$$

which is the characteristic function of the standard normal distribution, completing the proof. ■

***Example* 7.3.1.** Let $T_n \sim \text{Bin}(n, \pi)$ and write $T_n = \sum_{k=1}^n X_k$, where

$$X_k = \begin{cases} 1, & \text{with probability } \pi \\ 0, & \text{with probability } 1 - \pi. \end{cases}$$

Since $\mathbb{E}(X_k) = \pi$ and $\mathbb{V}\text{ar}(X_k) = \pi(1 - \pi)$, a direct application of Theorem 7.3.1 implies that $(T_n - n\pi)/[n\pi(1 - \pi)]^{1/2}$ is asymptotically distributed as $\mathcal{N}(0, 1)$. This result is known as the *De Moivre–Laplace Theorem*. □

An extension of Theorem 7.3.1 to cover the case of sums of independent but not identically distributed random variables is given by Theorem 7.3.2.

Theorem* 7.3.2 *(Liapounov). *Let X_k, $k \geq 1$ be independent random variables, such that $\mathbb{E}(X_k) = \mu_k$ and $\mathbb{V}ar(X_k) = \sigma_k^2$, and for some $0 < \delta \leq 1$,*

$$\nu_{2+\delta}^{(k)} = \mathbb{E}(|X_k - \mu_k|^{2+\delta}) < \infty, \quad k \geq 1.$$

Also let $T_n = \sum_{k=1}^n X_k$, $\xi_n = \mathbb{E}(T_n) = \sum_{k=1}^n \mu_k$, $\tau_n^2 = \mathbb{V}ar(T_n) = \sum_{k=1}^n \sigma_k^2$, $Z_n = (T_n - \xi_n)/\tau_n$ and $\rho_n = \tau_n^{-(2+\delta)}\sum_{k=1}^n \nu_{2+\delta}^{(k)}$. Then, if $\lim_{n\to\infty} \rho_n = 0$, we have $Z_n \xrightarrow{D} \mathcal{N}(0, 1)$.

Proof. First, for every $k \geq 1$ apply the Jensen Inequality (1.5.40) to conclude that

$$\sigma_k^2 = \mathbb{E}[(X_k - \mu_k)^2] \leq \{\mathbb{E}[|X_k - \mu_k|^{2+\delta}]\}^{\frac{2}{2+\delta}} = [\nu_{2+\delta}^{(k)}]^{\frac{2}{2+\delta}},$$

which implies

$$\sigma_k^2/\tau_n^2 \leq \left[\nu_{2+\delta}^{(k)}/\tau_n^{2+\delta}\right]^{\frac{2}{2+\delta}}.$$

Consequently, we have

$$\begin{aligned}\max_{1\leq k\leq n}\left(\frac{\sigma_k^2}{\tau_n^2}\right) &\leq \max_{1\leq k\leq n}\left[\frac{\nu_{2+\delta}^{(k)}}{\tau_n^{2+\delta}}\right]^{\frac{2}{2+\delta}} = \left\{\max_{1\leq k\leq n}\left[\frac{\nu_{2+\delta}^{(k)}}{\tau_n^{2+\delta}}\right]\right\}^{\frac{2}{2+\delta}}\\ &\leq \left[\sum_{k=1}^n \frac{\nu_{2+\delta}^{(k)}}{\tau_n^{2+\delta}}\right]^{\frac{2}{2+\delta}} = \rho_n^{\frac{2}{2+\delta}}.\end{aligned}$$

Since $\lim_{n\to\infty} \rho_n = 0$, it follows that

$$\max_{1\leq k\leq n} \left(\frac{\sigma_k^2}{\tau_n^2}\right) \to 0 \quad \text{as} \quad n \to \infty. \tag{7.3.1}$$

Next, let $U_k = (X_k - \mu_k)/\sigma_k$. Then, $\mathbb{E}(U_k) = 0$, $\mathbb{E}(U_k^2) = 1$ and $\mathbb{E}(|U_k|^{2+\delta}) = \nu_{2+\delta}^{(k)}/\sigma_k^{2+\delta}$; furthermore, $Z_n = \sum_{k=1}^n \sigma_k U_k/\tau_n$. Now, writing $\phi_{Z_n}(t) = \mathbb{E}[\exp(itZ_n)]$ and $\phi_{U_k}(t) = \mathbb{E}[\exp(itU_k)]$, we have

$$\begin{aligned}
\phi_{Z_n}(t) &= \mathbb{E}\left[\exp\left(it\sum_{k=1}^n \frac{\sigma_k}{\tau_n} U_k\right)\right] \\
&= \prod_{k=1}^n \mathbb{E}\left[\exp\left(it\frac{\sigma_k}{\tau_n} U_k\right)\right] \\
&= \prod_{k=1}^n \phi_{U_k}\left(\frac{\sigma_k}{\tau_n} t\right),
\end{aligned}$$

which implies

$$\log \phi_{Z_n}(t) = \sum_{k=1}^n \log \phi_{U_k}\left(\frac{\sigma_k}{\tau_n} t\right). \tag{7.3.2}$$

Now, for $\left|1 - \phi_{U_k}(\sigma_k t/\tau_n)\right| < 1$, consider the expansion

$$\begin{aligned}
\log \phi_{U_k}\left(\frac{\sigma_k}{\tau_n} t\right) &= \log\left\{1 - \left[1 - \phi_{U_k}\left(\frac{\sigma_k}{\tau_n} t\right)\right]\right\} \\
&= -\left[1 - \phi_{U_k}\left(\frac{\sigma_k}{\tau_n} t\right)\right] - \sum_{r=2}^{\infty} \frac{1}{r}\left[1 - \phi_{U_k}\left(\frac{\sigma_k}{\tau_n} t\right)\right]^r
\end{aligned} \tag{7.3.3}$$

and let us examine the terms on the right-hand side; using the Taylor expansion

$$e^x = 1 + x + \frac{1}{2}x^2 e^{hx}, \quad 0 < h < 1,$$

we get

$$\begin{aligned}
&\left|1 - \phi_{U_k}\left(\frac{\sigma_k}{\tau_n} t\right)\right| = \left|\int_{-\infty}^{+\infty}\left[1 - \exp\left(it\frac{\sigma_k}{\tau_n} u\right)\right] \mathrm{d}P(U_k \leq u)\right| \\
&= \left|-it\frac{\sigma_k}{\tau_n}\int_{-\infty}^{+\infty} u\, \mathrm{d}P(U_k \leq u) + \frac{1}{2}t^2\frac{\sigma_k^2}{\tau_n^2}\int_{-\infty}^{+\infty} u^2 \exp\left(it\frac{\sigma_k}{\tau_n} uh\right) \mathrm{d}P(U_k \leq u)\right| \\
&\leq \frac{1}{2}t^2\frac{\sigma_k^2}{\tau_n^2}\int_{-\infty}^{+\infty} u^2 \left|\exp\left(it\frac{\sigma_k}{\tau_n} uh\right)\right| \mathrm{d}P(U_k \leq u) \\
&= \frac{1}{2}t^2\frac{\sigma_k^2}{\tau_n^2}\int_{-\infty}^{+\infty} u^2\, \mathrm{d}P(U_k \leq u) = \frac{1}{2}t^2\frac{\sigma_k^2}{\tau_n^2}.
\end{aligned} \tag{7.3.4}$$

Therefore, using (7.3.1), it follows that

$$\max_{1\leq k\leq n}\left|1 - \phi_{U_k}\left(\frac{\sigma_k}{\tau_n} t\right)\right| \leq \frac{1}{2}t^2 \max_{1\leq k\leq n}\left(\frac{\sigma_k^2}{\tau_n^2}\right) = o(1) \quad \text{as} \quad n \to \infty. \tag{7.3.5}$$

Furthermore, observe that

$$\begin{aligned}
\left|\sum_{r=2}^{\infty}\left\{-\frac{1}{r}\left[1-\phi_{U_k}\left(\frac{\sigma_k}{\tau_n}t\right)\right]^r\right\}\right| &\le \sum_{r=2}^{\infty}\frac{1}{r}\left|1-\phi_{U_k}\left(\frac{\sigma_k}{\tau_n}t\right)\right|^r \\
&\le \frac{1}{2}\sum_{r=2}^{\infty}\left|1-\phi_{U_k}\left(\frac{\sigma_k}{\tau_n}t\right)\right|^r = \frac{\frac{1}{2}|1-\phi_{U_k}(\sigma_k t/\tau_n)|^2}{1-|1-\phi_{U_k}(t\sigma_k/\tau_n)|} \\
&\le \left|1-\phi_{U_k}\left(\frac{\sigma_k}{\tau_n}t\right)\right|^2, \quad n \ge n_0. \qquad (7.3.6)
\end{aligned}$$

From (7.3.3)–(7.3.6) we obtain that, for $n \ge n_0$,

$$\begin{aligned}
\sum_{k=1}^{n}\left|\log\phi_{U_k}\left(\frac{\sigma_k}{\tau_n}t\right)+\left[1-\phi_{U_k}\left(\frac{\sigma_k}{\tau_n}t\right)\right]\right| &\le \sum_{k=1}^{n}\sum_{r=2}^{\infty}\frac{1}{r}\left|1-\phi_{U_k}\left(\frac{\sigma_k}{\tau_n}t\right)\right|^r \\
&\le \sum_{k=1}^{n}\left|1-\phi_{U_k}\left(\frac{\sigma_k}{\tau_n}t\right)\right|^2 \\
&\le \max_{1\le k\le n}\left|1-\phi_{U_k}\left(\frac{\sigma_k}{\tau_n}t\right)\right|\sum_{k=1}^{n}\left|1-\phi_{U_k}\left(\frac{\sigma_k}{\tau_n}t\right)\right| \\
&\le o(1)\sum_{k=1}^{n}\frac{1}{2}t^2\frac{\sigma_k^2}{\tau_n^2} = o(1)\frac{t^2}{2} = o(1) \quad \text{as} \quad n\to\infty,
\end{aligned}$$

which implies that as $n\to\infty$,

$$\begin{aligned}
\log\phi_{Z_n}(t) &= \sum_{k=1}^{n}\log\phi_{U_k}\left(\frac{\sigma_k}{\tau_n}t\right) \\
&= -\sum_{k=1}^{n}\left[1-\phi_{U_k}\left(\frac{\sigma_k}{\tau_n}t\right)\right]+o(1). \qquad (7.3.7)
\end{aligned}$$

Then, using Theorem 1.5.2, expression (7.3.7) enables us to write

$$\log\phi_{Z_n}(t) = \sum_{k=1}^{n}\left[-\frac{t^2}{2}\frac{\sigma_k^2}{\tau_n^2}+R_{2k}\left(\frac{\sigma_k}{\tau_n}t\right)\right]+o(1) \quad \text{as} \quad n\to\infty, \qquad (7.3.8)$$

where

$$|R_{2k}(t\sigma_k/\tau_n)| \le c|t|^{2+\delta}\nu_{2+\delta}^{(k)}/\tau_n^{2+\delta}, \qquad \text{for some} \quad c>0.$$

Now, note that

$$\begin{aligned}
\left|\sum_{k=1}^{n}R_{2k}\left(\frac{\sigma_k}{\tau_n}t\right)\right| &\le \sum_{k=1}^{n}\left|R_{2k}\left(\frac{\sigma_k}{\tau_n}t\right)\right| \\
&\le c|t|^{2+\delta}\sum_{k=1}^{n}\frac{\nu_{2+\delta}^{(k)}}{\tau_n^{2+\delta}} = c|t|^{2+\delta}\rho_n \to 0 \quad \text{as} \quad n\to\infty.
\end{aligned}$$

Thus, from (7.3.8) we get $\log\phi_{Z_n}(t) = -t^2/2+o(1)$, and it follows that as $n\to\infty$, $\phi_{Z_n}(t)\to\exp(-t^2/2)$, which is the characteristic function of the standard normal distribution. ■

In the original formulation, Liapounov used $\delta = 1$, but even the existence of $\nu_{2+\delta}^{(k)}$, $0 < \delta \leq 1$ is not a necessary condition, as we may see from the following theorem.

Theorem 7.3.3 (Lindeberg–Feller). *Let X_k, $k \geq 1$, be independent random variables, such that $\mathbb{E}(X_k) = \mu_k$ and $\mathbb{V}ar(X_k) = \sigma_k^2$, $k \geq 1$; also let $T_n = \sum_{k=1}^n X_k$, $\xi_n = \mathbb{E}(T_n) = \sum_{k=1}^n \mu_k$, $\tau_n^2 = \mathbb{V}ar(T_n) = \sum_{k=1}^n \sigma_k^2$ and $Z_n = (T_n - \xi_n)/\tau_n = \sum_{k=1}^n Y_{nk}$, where $Y_{nk} = (X_k - \mu_k)/\tau_n$. Consider the following conditions:*

(a) Uniform asymptotic negligibility (UAN) condition:

$$\max_{1\leq k\leq n}\left(\frac{\sigma_k^2}{\tau_n^2}\right) \to 0 \quad as \quad n \to \infty.$$

Note that this condition implies that the random variables Y_{nk} are infinitesimal, that is, $\max_{1\leq k\leq n} P(|Y_{nk}| > \varepsilon) \to 0$ as $n \to \infty$ for every $\varepsilon > 0$. To see this, observe that for every $\varepsilon > 0$

$$\begin{aligned}
0 \leq \max_{1\leq k\leq n} P(|Y_{nk}| > \varepsilon) &= \max_{1\leq k\leq n} \int_{\{|y|\geq\varepsilon\}} dP(Y_{nk} \leq y) \\
&\leq \frac{1}{\varepsilon^2} \max_{1\leq k\leq n} \int_{\{|y|\geq\varepsilon\}} y^2\, dP(Y_{nk} \leq y) \\
&\leq \frac{1}{\varepsilon^2} \max_{1\leq k\leq n} \frac{\sigma_k^2}{\tau_n^2} \to 0 \quad as \quad n \to \infty.
\end{aligned}$$

In other words, the UAN condition states that the random variables Y_{nk}, $1 \leq k \leq n$, are uniformly in k, asymptotically in n, negligible.

(b) Asymptotic normality condition:

$$P(Z_n \leq z) \to \frac{1}{\sqrt{2\pi}} \int_{-\infty}^{z} e^{-t^2/2}\, dt = \Phi(z).$$

(c) Lindeberg–Feller condition (uniform integrability):

$$\forall \varepsilon > 0, \quad \frac{1}{\tau_n^2} \sum_{k=1}^{n} \mathbb{E}[(X_k - \mu_k)^2 I(|X_k - \mu_k| > \varepsilon\tau_n)] \to 0 \quad as \quad n \to \infty.$$

Then, (a) and (b) hold simultaneously if and only if (c) holds.

Proof. First suppose that (c) holds and let us show that this implies (a) and (b). Note that, for all $k \geq 1$,

$$\begin{aligned}
\sigma_k^2 &= \mathbb{E}[(X_k - \mu_k)^2 I(|X_k - \mu_k| \leq \varepsilon\tau_n)] + \mathbb{E}[(X_k - \mu_k)^2 I(|X_k - \mu_k| > \varepsilon\tau_n)] \\
&\leq \varepsilon^2\tau_n^2 + \mathbb{E}[(X_k - \mu_k)^2 I(|X_k - \mu_k| > \varepsilon\tau_n)].
\end{aligned}$$

Thus, we have

$$\begin{aligned}
\max_{1\leq k\leq n}\left(\frac{\sigma_k^2}{\tau_n^2}\right) &\leq \varepsilon^2 + \max_{1\leq k\leq n}\left\{\frac{1}{\tau_n^2}\mathbb{E}[(X_k - \mu_k)^2 I(|X_k - \mu_k| > \varepsilon\tau_n)]\right\} \\
&\leq \varepsilon^2 + \frac{1}{\tau_n^2} \sum_{k=1}^{n} \mathbb{E}[(X_k - \mu_k)^2 I(|X_k - \mu_k| > \varepsilon\tau_n)].
\end{aligned}$$

From condition (c) and the fact that $\varepsilon > 0$ can be made arbitrarily small, it follows that (a) holds. Now let us show that (c) and (a) imply (b). In this direction we may follow the lines of the proof of Theorem 7.3.2 up to expression (7.3.7). Now, let

$$R_{2k}(t) = \int_{-\infty}^{+\infty} (e^{itu} - 1 - itu + t^2u^2/2)\,\mathrm{d}P(U_k \le u)$$

and consider the expansion

$$\begin{aligned}\phi_{U_k}(t) &= 1 + it\int_{-\infty}^{+\infty} u\,\mathrm{d}P(U_k \le u) - \frac{t^2}{2}\int_{-\infty}^{+\infty} u^2\,\mathrm{d}P(U_k \le u) + R_{2k}(t)\\ &= 1 - \frac{t^2}{2} + R_{2k}(t).\end{aligned}$$

Therefore, using (7.3.7) we may write

$$\log\phi_{Z_n}(t) = -\frac{1}{2}t^2 + \sum_{k=1}^{n} R_{2k}\left(\frac{\sigma_k}{\tau_n}t\right) + o(1) \quad \text{as} \quad n \to \infty, \tag{7.3.9}$$

and all we have to show is that $\sum_{k=1}^{n} |R_{2k}(t\sigma_k/\tau_n)| < \varepsilon$, for all $\varepsilon > 0$. Write

$$g(u) = \exp\left(it\frac{\sigma_k}{\tau_n}u\right) - 1 - it\frac{\sigma_k}{\tau_n}u + \frac{t^2}{2}\frac{\sigma_k^2}{\tau_n^2}u^2$$

and note that

$$\begin{aligned}\left|R_{2k}\left(\frac{\sigma_k}{\tau_n}t\right)\right| \le &\left|\int_{\{|u|\le\varepsilon\tau_n/\sigma_k\}} g(u)\,\mathrm{d}P(U_k \le u)\right|\\ &+ \left|\int_{\{|u|>\varepsilon\tau_n/\sigma_k\}} g(u)\,\mathrm{d}P(U_k \le u)\right|.\end{aligned} \tag{7.3.10}$$

Using a third-order Taylor expansion for $\exp(itu\sigma_k/\tau_n)$, we have

$$g(u) = \frac{1}{3!}\left(\frac{itu\sigma_k}{\tau_n}\right)^3 \exp\left(\frac{it\sigma_k uh}{\tau_n}\right),$$

for some $0 < h < 1$, and it follows that

$$\begin{aligned}&\left|\int_{\{|u|\le\varepsilon\tau_n/\sigma_k\}} g(u)\,\mathrm{d}P(U_k \le u)\right|\\ &\quad\le \int_{\{|u|\le\varepsilon\tau_n/\sigma_k\}} \frac{1}{3!}\left|\left(\frac{itu\sigma_k}{\tau_n}\right)^3\right|\left|\exp\left(\frac{it\sigma_k uh}{\tau_n}\right)\right|\mathrm{d}P\{U_k \le u\}\\ &\quad\le |t|^3\frac{\sigma_k^3}{\tau_n^3}\int_{\{|u|\le\varepsilon\tau_n/\sigma_k\}} |u|^3\,\mathrm{d}P(U_k \le u)\\ &\quad\le \varepsilon|t|^3\frac{\sigma_k^2}{\tau_n^2}\int_{\{|u|\le\varepsilon\tau_n/\sigma_k\}} u^2\,\mathrm{d}P(U_k \le u)\\ &\quad\le \varepsilon|t|^3\frac{\sigma_k^2}{\tau_n^2}.\end{aligned} \tag{7.3.11}$$

Using a second-order Taylor expansion for $\exp(itu\sigma_k/\tau_n)$, we have

$$g(u) = \frac{1}{2}(tu\sigma_k/\tau_n)^2(1 - \exp(it\sigma_k uh/\tau_n)),$$

for some $0 < h < 1$, and it follows that

$$\begin{aligned}
&\left| \int_{\{|u|>\varepsilon\tau_n/\sigma_k\}} g(u)\,\mathrm{d}P(U_k \leq u) \right| \\
&\quad \leq \frac{1}{2}\left(\frac{t\sigma_k}{\tau_n}\right)^2 \int_{\{|u|>\varepsilon\tau_n/\sigma_k\}} u^2 |1 - e^{(it\sigma_k uh/\tau_n)}|\,\mathrm{d}P(U_k \leq u) \\
&\quad \leq \frac{t^2}{\tau_n^2} \int_{\{|u|>\varepsilon\tau_n/\sigma_k\}} u^2 \sigma_k^2\,\mathrm{d}P(U_k \leq u) \\
&\quad = \frac{t^2}{\tau_n^2} \mathbb{E}[(X_k - \mu_k)^2 I(|X_k - \mu_k| > \varepsilon\tau_n)]. \qquad (7.3.12)
\end{aligned}$$

From (7.3.10)–(7.3.12) we obtain

$$\left| R_{2k}\left(\frac{\sigma_k}{\tau_n} t\right) \right| \leq \varepsilon |t|^3 \frac{\sigma_k^2}{\tau_n^2} + \frac{t^2}{\tau_n^2} \mathbb{E}[(X_k - \mu_k)^2 I(|X_k - \mu_k| > \varepsilon\tau_n)],$$

and summing over k we may write

$$\sum_{k=1}^{n} \left| R_{2k}\left(\frac{\sigma_k}{\tau_n} t\right) \right| \leq \varepsilon |t|^3 + \frac{t^2}{\tau_n^2} \sum_{k=1}^{n} \mathbb{E}[(X_k - \mu_k)^2 I(|X_k - \mu_k| > \varepsilon\tau_n)],$$

Since the first term on the right-hand side can be made arbitrarily small by choosing $\varepsilon > 0$ sufficiently small and the second term converges to zero by condition (c), it follows from (7.3.9) that as $n \to \infty$, $\phi_{Z_n}(t) \to \exp(-t^2/2)$, which is characteristic function of a standard normal distribution and, therefore, (b) holds.

Now let us prove that (a) and (b) imply (c). First recall that condition (a) implies (7.3.7). Then, we may write

$$\begin{aligned}
\log \phi_{Z_n}(t) = &-\sum_{k=1}^{n} \mathbb{E}\left[1 - \cos\left(\frac{t\sigma_k}{\tau_n} U_k\right)\right] \\
&+ i \sum_{k=1}^{n} \mathbb{E}\left[\sin\left(\frac{t\sigma_k}{\tau_n} U_k\right)\right] + o(1) \quad \text{as} \quad n \to \infty.
\end{aligned}$$

Since (b) holds, we may conclude that the coefficient of the imaginary term must converge to zero; furthermore, for the real-valued term we must have

$$\sum_{k=1}^{n} \int_{-\infty}^{+\infty} \left[1 - \cos\left(\frac{t\sigma_k}{\tau_n} u\right)\right] \mathrm{d}P(U_k \leq u) = \frac{t^2}{2} + o(1) \quad \text{as} \quad n \to \infty.$$

Thus, as $n \to \infty$,

$$\begin{aligned}
&\frac{t^2}{2} - \sum_{k=1}^{n} \int_{\{|u|\leq\varepsilon\tau_n/\sigma_k\}} \left[1 - \cos\left(\frac{t\sigma_k}{\tau_n} u\right)\right] \mathrm{d}P(U_k \leq u) \\
&\quad = \sum_{k=1}^{n} \int_{\{|u|>\varepsilon\tau_n/\sigma_k\}} \left[1 - \cos\left(\frac{t\sigma_k}{\tau_n} u\right)\right] \mathrm{d}P(U_k \leq u) + o(1). \qquad (7.3.13)
\end{aligned}$$

Now observe that for $|y| < 1$ we have

$$\cos(y) = 1 - \frac{y^2}{2} + \frac{y^4}{4!} - \frac{y^6}{6!} + \cdots,$$

which implies $1 - \cos(y) \le y^2/2$. Thus, for the left-hand side of (7.3.13) we get

$$\begin{aligned}
&\frac{t^2}{2} - \sum_{k=1}^{n} \int_{\{|u| \le \varepsilon\tau_n/\sigma_k\}} \left[1 - \cos\left(\frac{t\sigma_k}{\tau_n} u\right)\right] \mathrm{d}P(U_k \le u) \\
&\quad \ge \frac{t^2}{2} - \frac{t^2}{2} \sum_{k=1}^{n} \frac{\sigma_k^2}{\tau_n^2} \int_{\{|u| \le \varepsilon\tau_n/\sigma_k\}} u^2 \, \mathrm{d}P(U_k \le u) \\
&\quad = \frac{t^2}{2} \left\{1 - \frac{1}{\tau_n^2} \sum_{k=1}^{n} \mathbb{E}[(X_k - \mu_k)^2 I(|X_k - \mu_k| \le \varepsilon\tau_n)]\right\}.
\end{aligned} \tag{7.3.14}$$

Observing that $1 - \cos(y) \le 2$ and that $u^2\sigma_k^2/\varepsilon^2\tau_n^2 > 1$ in the region $\{|u| > \varepsilon\tau_n/\sigma_k\}$, for the term on the right-hand side of (7.3.13) we may write:

$$\begin{aligned}
&\sum_{k=1}^{n} \int_{\{|u| > \varepsilon\tau_n/\sigma_k\}} \left[1 - \cos\left(\frac{t\sigma_k}{\tau_n} u\right)\right] \mathrm{d}P(U_k \le u) \\
&\quad \le 2 \sum_{k=1}^{n} \int_{\{|u| > \varepsilon\tau_n/\sigma_k\}} \mathrm{d}P(U_k \le u) \\
&\quad \le \frac{2}{\varepsilon^2\tau_n^2} \sum_{k=1}^{n} \sigma_k^2 \int_{\{|u| > \varepsilon\tau_n/\sigma_k\}} u^2 \, \mathrm{d}P(U_k \le u) \\
&\quad = \frac{2}{\varepsilon^2\tau_n^2} \sum_{k=1}^{n} \mathbb{E}[(X_k - \mu_k)^2 I(|X_k - \mu_k| > \varepsilon\tau_n)].
\end{aligned} \tag{7.3.15}$$

Then, from (7.3.13)–(7.3.15) we have

$$\begin{aligned}
&\frac{t^2}{2} \left\{1 - \frac{1}{\tau_n^2} \sum_{k=1}^{n} \mathbb{E}[(X_k - \mu_k)^2 I(|X_k - \mu_k| \le \varepsilon\tau_n)]\right\} \\
&\quad \le \frac{2}{\varepsilon^2\tau_n^2} \sum_{k=1}^{n} \mathbb{E}[(X_k - \mu_k)^2 I(|X_k - \mu_k| > \varepsilon\tau_n)] + o(1) \quad \text{as} \quad n \to \infty.
\end{aligned}$$

Dividing both members by $t^2/2$, it follows that

$$\begin{aligned}
&1 - \frac{1}{\tau_n^2} \sum_{k=1}^{n} \mathbb{E}[(X_k - \mu_k)^2 I(|X_k - \mu_k| \le \varepsilon\tau_n)] \\
&\quad \le \frac{4}{\varepsilon^2 t^2 \tau_n^2} \sum_{k=1}^{n} \mathbb{E}[(X_k - \mu_k)^2 I(|X_k - \mu_k| > \varepsilon\tau_n)] + o(t^{-2}) \\
&\quad \le \frac{4}{\varepsilon^2 t^2} \sum_{k=1}^{n} \frac{\sigma_k^2}{\tau_n^2} + o(t^{-2}) \\
&\quad = \frac{4}{\varepsilon^2 t^2} + o(t^{-2}) \quad \text{as} \quad n \to \infty.
\end{aligned}$$

Since this must hold for all t, we may let $t \to \infty$ as well, to conclude that the right-hand side converges to zero. Because the left-hand side is positive, it must also converge to zero.

Hence, we have

$$1 - \frac{1}{\tau_n^2}\sum_{k=1}^{n}\mathbb{E}[(X_k-\mu_k)^2 I(|X_k-\mu_k|\le\varepsilon\tau_n)]$$
$$= \frac{1}{\tau_n^2}\left\{\tau_n^2 - \sum_{k=1}^{n}\mathbb{E}[(X_k-\mu_k)^2 I(|X_k-\mu_k|\le\varepsilon\tau_n)]\right\}$$
$$= \frac{1}{\tau_n^2}\sum_{k=1}^{n}\mathbb{E}[(X_k-\mu_k)^2 I(|X_k-\mu_k|>\varepsilon\tau_n)] = o(1) \quad \text{as} \quad n\to\infty$$

and the proof is complete. ■

The first part of the proof, (c) implies (a, b) is according to Lindeberg and the second part (a, b) implies (c) is due to Feller (1971). It is interesting to note that one of (a) or (b) may hold if (c) does not hold; in this direction, consider the following example:

***Example* 7.3.2.** Let $Y_k, k\ge 1$, be independent $\mathcal{N}(0,1)$ random variables and let $X_k = a^k Y_k$, $k\ge 1$, with $a\ne 1$. Then, it follows that X_k, $k\ge 1$, are independent $\mathcal{N}(0, a^{2k})$ random variables; also, denoting $\mathbb{V}\text{ar}(X_k)=\sigma_k^2$, we have $\tau_n^2=\sum_{k=1}^n\sigma_k^2 = a^2(a^{2n}-1)/(a^2-1)$. Hence, for $a>1$,

$$\max_{1\le k\le n}\left(\frac{\sigma_k^2}{\tau_n^2}\right) = \frac{a^{2n}(a^2-1)}{a^2(a^{2n}-1)} \to 1-\frac{1}{a^2}>0 \quad \text{as} \quad n\to\infty,$$

and for $a<1$,

$$\max_{1\le k\le n}\left(\frac{\sigma_k^2}{\tau_n^2}\right) = \frac{a^2(a^2-1)}{a^2(a^{2n}-1)} \to 1-a^2>0 \quad \text{as} \quad n\to\infty.$$

Thus, (a) does not hold and from the first part of the Lindeberg–Feller Theorem 7.3.3 it follows that (c) does not hold. However, $\sum_{k=1}^n X_k \sim \mathcal{N}(0,\tau_n^2)$, which implies that $\tau_n^{-1}\sum_{k=1}^n X_k\sim\mathcal{N}(0,1)$ and, consequently, that (b) holds. □

It is also noteworthy to point that the Liapounov Theorem 7.3.2 is a corollary of the Lindeberg–Feller Theorem 7.3.3; this follows directly from

$$\frac{1}{\tau_n^2}\sum_{k=1}^{n}\mathbb{E}[(X_k-\mu_k)^2 I(|X_k-\mu_k|>\varepsilon\tau_n)]$$
$$\le \frac{1}{\varepsilon^\delta\tau_n^{2+\delta}}\sum_{k=1}^{n}\mathbb{E}[|X_k-\mu_k|^{2+\delta} I(|X_k-\mu_k|>\varepsilon\tau_n)]$$
$$\le \frac{1}{\varepsilon^\delta\tau_n^{2+\delta}}\sum_{k=1}^{n}\mathbb{E}(|X_k-\mu_k|^{2+\delta}) = \frac{\rho_n}{\varepsilon^\delta}.$$

If the random variables under consideration are bounded, the following theorem holds.

***Theorem* 7.3.4.** *Let X_k, $k\ge 1$ be independent random variables such that $P(a\le X_k\le b)=1$, for some finite scalars $a<b$. Also let $\mathbb{E}(X_k)=\mu_k$, $\mathbb{V}ar(X_k)=\sigma_k^2$, $T_n=\sum_{k=1}^n X_k$,*

$\xi_n = \sum_{k=1}^n \mu_k$ *and* $\tau_n^2 = \sum_{k=1}^n \sigma_k^2$. *Then,* $Z_n = (T_n - \xi_n)/\tau_n \xrightarrow{D} \mathcal{N}(0,1)$, *if and only if,* $\tau_n \to \infty$ *as* $n \to \infty$.

Proof. First, suppose that $s_n \to \infty$ as $n \to \infty$. Then, note that

$$|X_k - \mu_k|^3 = |X_k - \mu_k|(X_k - \mu_k)^2 \leq (b-a)(X_k - \mu_k)^2,$$

which implies

$$\mathbb{E}(|X_k - \mu_k|^3) \leq (b-a)\sigma_k^2.$$

Therefore,

$$\rho_n = \sum_{k=1}^n \mathbb{E}(|X_k - \mu_k|^3)/\tau_n^3 \leq (b-a)/\tau_n \to 0 \quad \text{as} \quad n \to \infty$$

and the result follows from Liapounov Theorem 7.3.2. Now suppose that $\tau_n \to \tau < \infty$ as $n \to \infty$ and write

$$Z_n = (T_n - \xi_n)/\tau_n = (X_1 - \mu_1)/\tau_n + \sum_{k=2}^n (X_k - \mu_k)/\tau_n.$$

Then, if $Z_n \xrightarrow{D} \mathcal{N}(0,1)$ each of the terms on the right-hand side must also converge to normal random variables; this is absurd since $(X_1 - \mu_1)/\tau$ is a bounded random variable. Therefore, $\tau_n \to \infty$ as $n \to \infty$. ■

Up to this point we have devoted our attention to the weak convergence of sequences of statistics $\{T_n, n \geq 1\}$ constructed from *independent* underlying random variables $X_1, X_2, \ldots$. Now we consider some extensions of the central limit theorems where such restriction may be relaxed. The first of such extensions holds for sequences of (possibly dependent) random variables, which may be structured as a *double array* of the form

$$\begin{pmatrix} X_{11}, & X_{12}, & \cdots, & X_{1k_1} \\ X_{21}, & X_{22}, & \cdots, & X_{2k_2} \\ \vdots & \vdots & \ddots & \vdots \\ X_{n1}, & X_{n2}, & \cdots, & X_{nk_n} \end{pmatrix},$$

where the X_{nk} are row-wise independent. The case where $k_n = n$, $n \geq 1$, is usually called a *triangular array* of random variables. As we will see in Section 7.6, such a result is very useful in the field of order statistics.

Theorem 7.3.5. *Consider a double array of random variables* $\{X_{nk}, 1 \leq k \leq k_n, n \geq 1\}$, *where* $k_n \to \infty$ *as* $n \to \infty$ *and such that for each* n, $\{X_{nk}, 1 \leq k \leq k_n\}$ *are independent. Then,*

(*i*) $\{X_{nk}, 1 \leq k \leq k_n, n \geq 1\}$ *is an infinitesimal system of random variables, that is, for every* $\varepsilon > 0$, $\max_{1 \leq k \leq k_n} P(|X_{nk}| > \varepsilon) \to 0$ *as* $n \to \infty$ *(the UAN condition), and*

(*ii*) $Z_n = \sum_{k=1}^{k_n} X_{nk} \xrightarrow{D} \mathcal{N}(0,1)$

hold simultaneously, if and only if, for every $\varepsilon > 0$, *as* $n \to \infty$

(a) $\sum_{k=1}^{k_n} P(|X_{nk}| > \varepsilon) \to 0$,

(b) $\sum_{k=1}^{k_n} \left\{ \int_{\{|x|\le\varepsilon\}} x^2 \, dP(X_{nk} \le x) - \left[\int_{\{|x|\le\varepsilon\}} x \, dP(X_{nk} \le x) \right]^2 \right\} \to 1$.

Proof. Here we prove that (a) and (b) are sufficient conditions; see Tucker (1967, p. 197) for a proof of the necessity part, that follows the same lines of Theorem 7.3.3. First, note that $\max_{1\le k\le k_n} P(|X_{nk}| > \varepsilon) \le \sum_{k=1}^{k_n} P(|X_{nk}| > \varepsilon)$; thus (i) follows from (a). Then, let $Y_{nk} = X_{nk} I(|X_{nk}| \le \varepsilon)$, $1 \le k \le k_n$, $n \ge 1$, and $Z_n^* = \sum_{k=1}^{k_n} Y_{nk}$. Now observe that

$$\begin{aligned} P(Z_n \neq Z_n^*) &\le P(X_{nk} \neq Y_{nk} \text{ for some } k = 1, \ldots, k_n) \\ &\le \sum_{k=1}^{k} P(|X_{nk}| > \varepsilon). \end{aligned}$$

Therefore from, (a), it follows that $P(Z_n \neq Z_n^*) \to 0$ as $n \to \infty$, and we may restrict ourselves to the limiting distribution of Z_n^*. Next note that since $P(|Y_{nk}| \le \varepsilon) = 1$, we may write

$$\begin{aligned} \rho_n = \sum_{k=1}^{k_n} \frac{\mathbb{E}(|Y_{nk} - \mathbb{E}(Y_{nk})|^3)}{\left[\sum_{k=1}^{k_n} \mathbb{V}\mathrm{ar}(Y_{nk})\right]^{3/2}} &\le \frac{2\varepsilon \sum_{k=1}^{k_n} \mathbb{E}[Y_{nk} - \mathbb{E}(Y_{nk})]^2}{\left[\sum_{k=1}^{k_n} \mathbb{V}\mathrm{ar}(Y_{nk})\right]^{3/2}} \\ &= 2\varepsilon \left[\sum_{k=1}^{k_n} \mathbb{V}\mathrm{ar}(Y_{nk}) \right]^{-1/2}. \end{aligned}$$

Then, since ε can be made arbitrarily small and (b) implies $\sum_{k=1}^{k_n} \mathbb{V}\mathrm{ar}(Y_{nk}) \to 1$ as $n \to \infty$, we may conclude that $\rho_n \to 0$ as $n \to \infty$. The result follows from the Liapounov Theorem 7.3.2. ∎

An application of the above theorem, which is especially useful in Regression Analysis, is given by the following result.

Theorem 7.3.6 (Hájek–Šidak). *Let* $\{Y_n\}$ *be a sequence of i.i.d. random variables with mean* μ *and finite variance* σ^2*; let* $\{\mathbf{c}_n\}$ *be a sequence of real vectors* $\mathbf{c}_n = (c_{n1}, \ldots, c_{nn})'$. *Then, if*

$$\max_{1\le i\le n} \left[c_{ni}^2 \Big/ \sum_{i=1}^{n} c_{ni}^2 \right] \to 0 \quad \text{as} \quad n \to \infty,$$

it follows that

$$Z_n = \left[\sum_{i=1}^{n} c_{ni}(Y_i - \mu) \right] \Big/ \left[\sigma^2 \sum_{i=1}^{n} c_{ni}^2 \right]^{1/2} \xrightarrow{D} \mathcal{N}(0, 1).$$

Proof. Let $X_{ni} = c_{ni} Y_i$ and write $T_n = \sum_{i=1}^{n} X_{ni}$, $\mathbb{E}(T_n) = \mu \sum_{i=1}^{n} c_{ni}$. Without loss of generality, we take $\sum_{i=1}^{n} c_{ni}^2 = 1$ so that $\tau_n^2 = \mathbb{V}\mathrm{ar}(T_n) = \sigma^2$. Then, note that for all $\varepsilon > 0$, as $n \to \infty$,

$$\frac{1}{\tau_n^2}\sum_{i=1}^{n}\mathbb{E}[(X_{ni}-c_{ni}\mu)^2 I(|X_{ni}-c_{ni}\mu|>\varepsilon\tau_n)]$$
$$=\frac{1}{\sigma^2}\sum_{i=1}^{n}c_{ni}^2\mathbb{E}[(Y_i-\mu)^2 I(|Y_i-\mu|>\varepsilon\sigma/c_{ni})]$$
$$\leq\frac{1}{\sigma^2}\mathbb{E}\big[(Y_1-\mu)^2 I\big[(Y_1-\mu)^2>\varepsilon^2\sigma^2/\max_{1\leq i\leq n}c_{ni}^2\big]\big]\to 0, \qquad (7.3.16)$$

since $\sum_{i=1}^{n}c_{ni}^2/\max_{1\leq i\leq n}c_{ni}^2\to\infty$ as $n\to\infty$. All we have to do is to verify (a) and (b) of Theorem 7.3.5, which follows trivially from (7.3.16). ■

It is still possible to relax further the independence assumption on the underlying random variables. The following theorems, stated without proofs, constitute examples of central limit theorems for dependent random variables having a martingale (or reverse martingale) structure. For further details, the reader is referred to Brown (1971a), Dvoretzky (1971), Loynes (1970), or McLeish (1974).

Theorem 7.3.7 (Martingale Central Limit Theorem). *Consider a sequence $\{X_k, k\geq 1\}$ of random variables such that $\mathbb{E}(X_k)=0$, $\mathbb{E}(X_k^2)=\sigma_k^2<\infty$ and $\mathbb{E}\{X_k|X_1,\ldots,X_{k-1}\}=0$, $(X_0=0)$. Also let $T_n=\sum_{k=1}^{n}X_k$, $\tau_n^2=\sum_{k=1}^{n}\sigma_k^2$, $v_k^2=\mathbb{E}(X_k^2|X_1,\ldots,X_{k-1})$ and $w_n^2=\sum_{k=1}^{n}v_k^2$. If*

(a) $w_n^2/\tau_n^2\xrightarrow{P}1$ as $n\to\infty$ and

(b) for every $\varepsilon>0$, $\tau_n^{-2}\sum_{k=1}^{n}\mathbb{E}[X_k^2 I(|X_k|>\varepsilon\tau_n)]\to 0$ as $n\to\infty$ (the Lindeberg–Feller condition),

then the sequence $\{X_k, k\geq 1\}$ is infinitesimal and

$$Z_n=T_n/\tau_n\xrightarrow{D}\mathcal{N}(0,1).$$

It is of interest to remark that:

(i) The v_k^2s are random variables (they depend on $X_1,\ldots,X_{k-1}$); condition (a) essentially states that all the information about the variability in the X_k is contained in $X_1,\ldots,X_{k-1}$.

(ii) $\{T_n, n\geq 1\}$ is a *zero mean martingale*, since

$$\mathbb{E}(T_n|T_1,\ldots,T_{n-1})=\mathbb{E}(T_n|X_1,\ldots,X_{n-1})$$
$$=\mathbb{E}(T_{n-1}+X_n|X_1,\ldots,X_{n-1})$$
$$=T_{n-1}$$

and $\mathbb{E}(T_n)=0$; furthermore, since, for all $j>i$,

$$\mathbb{E}(X_iX_j)=\mathbb{E}[X_i\mathbb{E}(X_j|X_1,\ldots,X_i)]=0,$$

it follows that

$$\mathbb{E}(T_n^2)=\mathbb{E}\left(\sum_{k=1}^{n}X_k\right)^2=\sum_{k=1}^{n}\mathbb{E}(X_k^2)+2\sum_{1\leq i\leq j\leq n}\mathbb{E}(X_iX_j)=\sum_{k=1}^{n}\sigma_k^2=\tau_n^2.$$

Theorem 7.3.8 (Reverse Martingale Central Limit Theorem). *Consider a sequence $\{T_k, k \geq 1\}$ of random variables, such that*

$$\mathbb{E}(T_n|T_{n+1}, T_{n+2}, \ldots) = T_{n+1} \quad \textit{and} \quad \mathbb{E}(T_n) = 0,$$

that is, $\{T_k, k \geq 1\}$ is a zero-mean reverse martingale. Assume that $\mathbb{E}(T_n^2) < \infty$ and let $Y_k = T_k - T_{k+1}, k \geq 1$, $v_k^2 = \mathbb{E}(Y_k^2|T_{k+1}, T_{k+2}, \ldots)$ and $w_n^2 = \sum_{k=n}^{\infty} v_k^2$. If

(a) $w_n^2/\mathbb{E}(w_n^2) \xrightarrow{a.s.} 1$

(b) $w_n^{-2} \sum_{k=n}^{\infty} \mathbb{E}\big[Y_k^2 I(|Y_k| > \varepsilon w_n)\big|T_{k+1}, T_{k+2}, \ldots\big] \xrightarrow{P} 0, \quad \varepsilon > 0$

or $w_n^{-2} \sum_{k=n}^{\infty} Y_k^2 \xrightarrow{a.s.} 1$,

then, $T_n/\sqrt{\mathbb{E}(w_n^2)} \xrightarrow{D} \mathcal{N}(0, 1)$.

***Example* 7.3.3.** Let $\{X_k,\ k \geq 1\}$ be a sequence of i.i.d. random variables such that $\mathbb{E}(X_k) = 0$ and $\mathbb{V}\text{ar}(X_k) = \sigma^2 < \infty$. We know from previous results that $n^{1/2}\overline{X}_n/\sigma \xrightarrow{D} \mathcal{N}(0, 1)$. We now indicate how the same conclusion may be reached by an application of the above theorem. First, recall from Example 5.2.6 that $\{\overline{X}_n, n \geq 1\}$ is a *zero-mean reverse martingale*, that is, $\mathbb{E}(\overline{X}_k|\overline{X}_{k+1}, \overline{X}_{k+2}, \ldots) = \overline{X}_{k+1}$, $\mathbb{E}(\overline{X}_k) = 0$. Then, note that for every $k \leq n$ we have:

$$\begin{aligned}\mathbb{E}(\overline{X}_k\overline{X}_n) &= \mathbb{E}[\mathbb{E}(\overline{X}_k\overline{X}_n|\overline{X}_n, \overline{X}_{n+1}, \ldots)] \\ &= \mathbb{E}[\overline{X}_n\mathbb{E}(\overline{X}_k|\overline{X}_n, \overline{X}_{n+1}, \ldots)] = \mathbb{E}\big(\overline{X}_n^2\big).\end{aligned}$$

Now, let $Y_k = \overline{X}_k - \overline{X}_{k+1}$; this implies $\mathbb{E}(Y_k|\overline{X}_n, \overline{X}_{n+1}, \ldots) = 0$, for every $k \leq n$. Furthermore, it follows that for every $j \geq 1$

$$\begin{aligned}\mathbb{E}(Y_kY_{k+j}) &= \mathbb{E}[(\overline{X}_k - \overline{X}_{k+1})(\overline{X}_{k+j} - \overline{X}_{k+j+1})] \\ &= \mathbb{E}(\overline{X}_k\overline{X}_{k+j}) - \mathbb{E}(\overline{X}_k\overline{X}_{k+j+1}) - \mathbb{E}(\overline{X}_{k+1}\overline{X}_{k+j}) + \mathbb{E}(\overline{X}_{k+1}\overline{X}_{k+j+1}) \\ &= \mathbb{E}\big(\overline{X}_{k+j}^2\big) - \mathbb{E}\big(\overline{X}_{k+j+1}^2\big) - \mathbb{E}\big(\overline{X}_{k+j}^2\big) + \mathbb{E}\big(\overline{X}_{k+j+1}^2\big) = 0.\end{aligned}$$

Write

$$v_k^2 = \mathbb{E}\big(Y_k^2|\overline{X}_{k+1}, \overline{X}_{k+2}, \ldots\big) = \mathbb{E}\big[(\overline{X}_k - \overline{X}_{k+1})^2\big|\overline{X}_{k+1}, \overline{X}_{k+2}, \ldots\big]$$

and note that since $\overline{X}_k - \overline{X}_{k+1} = \frac{1}{k}(\overline{X}_{k+1} - X_{k+1})$, we have

$$\begin{aligned}v_k^2 &= \frac{1}{k^2}\mathbb{E}\big[\big(\overline{X}_{k+1}^2 - 2\overline{X}_{k+1}X_{k+1} + X_{k+1}^2\big)\big|\overline{X}_{k+1}, \overline{X}_{k+2}, \ldots\big] \\ &= \frac{1}{k^2}\big[\overline{X}_{k+1}^2 - 2\overline{X}_{k+1}\mathbb{E}(X_{k+1}\big|\overline{X}_{k+1}, \overline{X}_{k+2}, \ldots) \\ &\quad + \mathbb{E}\big(X_{k+1}^2|\overline{X}_{k+1}, \overline{X}_{k+2}, \ldots\big)\big] \\ &= \frac{1}{k^2}\big[\overline{X}_{k+1}^2 - 2\overline{X}_{k+1}^2 + \mathbb{E}\big(X_{k+1}^2|\overline{X}_{k+1}, \overline{X}_{k+2}, \ldots\big)\big]. \qquad (7.3.17)\end{aligned}$$

Now observe that

$$\begin{aligned}
&\mathbb{E}(X_{k+1}|\overline{X}_{k+1}, \overline{X}_{k+2}, \ldots) \\
&= \binom{k+1}{k}^{-1} \sum_{\{1\le\alpha_1<\cdots<\alpha_k\le k+1\}} \left(\sum_{j=1}^{k+1} X_j - \sum_{j=1}^{k} X_{\alpha_j}\right) \\
&= \frac{1}{k+1}\left[(k+1)\sum_{j=1}^{k+1} X_j - \sum_{\{1\le\alpha_1<\cdots<\alpha_k\le k+1\}} \sum_{j=1}^{k} X_{\alpha_j}\right] \\
&= \frac{1}{k+1}\left[(k+1)\sum_{j=1}^{k+1} X_j - k\sum_{j=1}^{k+1} X_j\right] \\
&= \overline{X}_{k+1}. \qquad (7.3.18)
\end{aligned}$$

Following a similar argument we have

$$\begin{aligned}
&\mathbb{E}\big(X_{k+1}^2|\overline{X}_{k+1}, \overline{X}_{k+2}, \ldots\big) \\
&= \binom{k+1}{k}^{-1} \sum_{\{1\le\alpha_1<\cdots<\alpha_k\le k+1\}} \left[\sum_{j=1}^{k+1} X_j^2 - \sum_{j=1}^{k} X_{\alpha_j}^2\right] \\
&= \frac{1}{k+1}\left[(k+1)\sum_{j=1}^{k+1} X_j^2 - k\sum_{j=1}^{k+1} X_j^2\right] \\
&= \frac{1}{k+1}\sum_{j=1}^{k+1} X_j^2. \qquad (7.3.19)
\end{aligned}$$

From (7.3.17)–(7.3.19) it follows that

$$\begin{aligned}
v_k^2 &= \frac{1}{k^2}\left[\frac{1}{k+1}\sum_{j=1}^{k+1} X_j^2 - \overline{X}_{k+1}^2\right] = \frac{1}{k^2}\left[\frac{1}{k+1}\sum_{j=1}^{k+1}(X_j - \overline{X}_{k+1})^2\right] \\
&= \frac{1}{k(k+1)}\left[\frac{1}{k}\sum_{j=1}^{k+1}(X_j - \overline{X}_{k+1})^2\right] = \left(\frac{1}{k} - \frac{1}{k+1}\right) s_{k+1}^2,
\end{aligned}$$

where $s_{k+1}^2 = \frac{1}{k}\sum_{j=1}^{k+1}(X_j - \overline{X}_{k+1})^2$. Now, using (6.3.84), it follows that $s_{k+1}^2 \xrightarrow{a.s.} \sigma^2$ as $k \to \infty$. Then, given $\varepsilon > 0$, there exists $n_0 = n_0(\varepsilon)$ such that for all $k \ge n_0$, $\sigma^2 - \varepsilon \le s_{k+1}^2 \le \sigma^2 + \varepsilon$. Thus, using the definition of w_n^2, for all $n \ge n_0$ we have

$$\begin{aligned}
nw_n^2 = n\sum_{k=n}^{\infty} v_k^2 &= n\sum_{k=n}^{\infty}\left(\frac{1}{k} - \frac{1}{k+1}\right) s_{k+1}^2 \\
&\le (\sigma^2 + \varepsilon)n\sum_{k=n}^{\infty}\left(\frac{1}{k} - \frac{1}{k+1}\right) = \sigma^2 + \varepsilon.
\end{aligned}$$

Similarly, we may show that for all $n \ge n_0$, $nw_n^2 \ge \sigma^2 - \varepsilon$, and since ε is arbitrary, we may conclude that $nw_n^2 \xrightarrow{a.s.} \sigma^2$. Then, observing that $\mathbb{E}(w_n^2) = \sum_{k=n}^{\infty}(\frac{1}{k} - \frac{1}{k+1})\mathbb{E}(s_{k+1}^2) = \sigma^2/n$, it follows that condition (a) of Theorem 7.3.8 holds.

The proof of condition (b) is somewhat more elaborate; the reader is referred to Loynes (1970) for details. □

The central limit theorems discussed above may be generalized to cover the multivariate case. Given a sequence $\{\boldsymbol{X}_n\}$ of random vectors in $\mathbb{R}^p$, with mean vectors $\boldsymbol{\mu}_n$ and covariance matrices $\boldsymbol{\Sigma}_n$, $n \geq 1$, to show that $n^{-1/2} \sum_{i=1}^n (\boldsymbol{X}_i - \boldsymbol{\mu}_i) \xrightarrow{D} \mathcal{N}_p(\boldsymbol{0}, \boldsymbol{\Sigma})$ with $\boldsymbol{\Sigma} = \lim_{n\to\infty} n^{-1} \sum_{i=1}^n \boldsymbol{\Sigma}_i$ (here we abuse notation as in the univariate case), one generally proceeds according to the following strategy:

(i) Use one of the univariate central limit theorems to show that for every fixed $\boldsymbol{\lambda} \in \mathbb{R}^p$, $n^{-1/2} \sum_{i=1}^n \boldsymbol{\lambda}'(\boldsymbol{X}_i - \boldsymbol{\mu}_i) \xrightarrow{D} \mathcal{N}(0, \gamma^2)$ with $\gamma^2 = \lim_{n\to\infty} n^{-1} \boldsymbol{\lambda}'(\sum_{i=1}^n \boldsymbol{\Sigma}_i)\boldsymbol{\lambda}$.
(ii) Use the Cramér–Wold Theorem 7.2.4 to complete the proof.

As an example consider the following result in Theorem 7.3.9.

Theorem 7.3.9. *Let $\{\boldsymbol{X}_n\}$ be a sequence of random vectors in $\mathbb{R}^p$ with mean vectors $\boldsymbol{\mu}_n$ and finite covariance matrices $\boldsymbol{\Sigma}_n$, $n \geq 1$, such that*

$$\max_{1\leq i\leq n} \max_{1\leq j\leq p} \mathbb{E}(|X_{ij} - \mu_{ij}|^{2+\delta}) < \infty, \quad \textit{for some} \quad 0 < \delta < 1,$$

and

$$\boldsymbol{\Sigma} = \lim_{n\to\infty} n^{-1} \sum_{i=1}^n \boldsymbol{\Sigma}_i$$

exist. Then, $n^{-1/2} \sum_{i=1}^n (\boldsymbol{X}_i - \boldsymbol{\mu}_i) \xrightarrow{D} \mathcal{N}_p(\boldsymbol{0}, \boldsymbol{\Sigma})$.

Proof. For every $\boldsymbol{\lambda} \in \mathbb{R}^p$ define $\gamma_n^2 = n^{-1} \sum_{i=1}^n \boldsymbol{\lambda}'\boldsymbol{\Sigma}_i\boldsymbol{\lambda}$. Now, using the C_r Inequality 1.5.38 we have

$$\mathbb{E}[|\boldsymbol{\lambda}'(\boldsymbol{X}_i - \boldsymbol{\mu}_i)|^{2+\delta}] \leq p^{1+\delta}\big(\max_{1\leq j\leq p} |\boldsymbol{\lambda}_j|\big) \sum_{i=1}^p \mathbb{E}(|X_{ij} - \mu_{ij}|^{2+\delta}).$$

Then, for all $\boldsymbol{\lambda} \in \mathbb{R}^p$ such that $\liminf \gamma_n > 0$, we have

$$\begin{aligned}\rho_n &= \sum_{i=1}^n \frac{\mathbb{E}(|\boldsymbol{\lambda}'(\boldsymbol{X}_i - \boldsymbol{\mu}_i)|^{2+\delta})}{(\sqrt{n}\gamma_n)^{2+\delta}} \\ &\leq \frac{p^{1+\delta}}{\gamma_n^{2+\delta}} \big(\max_{1\leq j\leq p} |\boldsymbol{\lambda}_j|\big) \left(\sum_{i=1}^n \sum_{j=1}^p \frac{\mathbb{E}(|X_{ij} - \mu_{ij}|^{2+\delta})}{n^{1+\delta/2}} \right).\end{aligned}$$

Since

$$\begin{aligned}&\sum_{i=1}^n \sum_{j=1}^p n^{-(1+\delta/2)} \mathbb{E}(|X_{ij} - \mu_{ij}|^{2+\delta}) \\ &\quad \leq np/n^{1+\delta/2} \max_{1\leq i\leq n} \max_{1\leq j\leq p} \mathbb{E}(|X_{ij} - \mu_{ij}|^{2+\delta}) = O(n^{-\delta/2}),\end{aligned}$$

it follows that $\rho_n \to 0$ as $n \to \infty$; thus, from Liapounov Theorem 7.3.2 we may conclude that

$$n^{-1/2}\gamma_n^{-1}\sum_{i=1}^{n}(\boldsymbol{\lambda}'\boldsymbol{X}_i - \boldsymbol{\lambda}'\boldsymbol{\mu}) \xrightarrow{D} \mathcal{N}(0, 1).$$

Finally, a direct application of Cramér–Wold Theorem 7.2.4 completes the proof. ■

To finalize this section we want to present the central limit theorem related to the elementary Renewal Theorem introduced via Example 6.5.4. Let $\{X_i,\ i \geq 1\}$ be a sequence of i.i.d. nonnegative random variables with mean $\mu \in (0, \infty)$ and variance $\sigma^2 < \infty$. For every $n \geq 1$ let

$$T_n = X_1 + \cdots + X_n, \quad T_0 = 0. \tag{7.3.20}$$

Also, for every $t > 0$, let

$$N_t = \max(k : T_k \leq t). \tag{7.3.21}$$

Thus, $N_0 = 0$ and N_t is increasing in t, $t \geq 0$. Note that, by definition,

$$T_{N_t} \leq t \leq T_{N_t+1}, \quad t \geq 0. \tag{7.3.22}$$

From every $a \in \mathbb{R}$, consider the event

$$\{t^{-1/2}(N_t - t/\mu) < a\} = \{N_t < t/\mu + t^{1/2}a\}. \tag{7.3.23}$$

Then, let $\{r_t; t \geq 0\}$ be a sequence of nonnegative integers, defined by

$$r_t \leq t/\mu + t^{1/2}a < r_t + 1, \quad t > 0, \tag{7.3.24}$$

where a is held fixed. Recall that $0 < \mu < \infty$, so that for any fixed a,

$$\lim_{t\to\infty} r_t(t/\mu) = 1, \quad \text{that is,} \quad r_t \to \infty \text{ as } t \to \infty. \tag{7.3.25}$$

Consequently, by (7.3.22) and (7.3.23), we have

$$\begin{aligned}
\lim_{t\to\infty} P[t^{-1/2}(N_t - t/\mu) < \alpha] &= \lim_{r_t\to\infty} P(T_{r_t+1} > t) \\
&= \lim_{r_t\to\infty} P[T_{r_t+1} - (r_t+1)\mu > t - (r_t+1)\mu] \\
&= \lim_{r_t\to\infty} P\left\{(r_t+1)^{-1/2}[T_{r_t+1} - \mu(r_t+1)] > \frac{t-(r_t+1)\mu}{\sqrt{r_t+1}}\right\}.
\end{aligned} \tag{7.3.26}$$

Now, from (7.3.24), we obtain

$$\lim_{t\to\infty}\frac{t-(r_t+1)\mu}{\sqrt{r_t+1}} = \lim_{t\to\infty}\frac{-a\mu\sqrt{t}}{\sqrt{t/\mu + \sqrt{t}/a}} = -a\mu^{3/2}. \tag{7.3.27}$$

On the other hand, by Theorem 7.3.1, as $n \to \infty$,

$$n^{-1/2}(T_n - n\mu)/\sigma \xrightarrow{D} \mathcal{N}(0, 1) \tag{7.3.28}$$

so that by (7.3.25), (7.3.27), and (7.3.28), for $a \in \mathbb{R}$ and $Z \sim \mathcal{N}(0, 1)$, the right-hand side of (7.3.26) reduces to

$$P(Z > -a\mu^{3/2}/\sigma) = 1 - \Phi(-a\mu^{3/2}/\sigma) = \Phi(a\mu^{3/2}/\sigma), \tag{7.3.29}$$

so that

$$t^{-1/2}(N_t - t/\mu) \xrightarrow{D} \mathcal{N}(0, \sigma^2/\mu^3). \tag{7.3.30}$$

In this context, we may note that if $\{T_n - n\mu;\ n \geq 0\}$ is a zero-mean martingale, (7.3.28) follows from Theorem 7.3.7, so that for the Renewal Central Limit Theorem in (7.3.30), the independence of the X_is may also be replaced by the aforesaid martingale structure.

7.4 Rates of Convergence to Normality

Let $\{X_n\}$ be a sequence of random variables such that $\mathbb{E}(X_i) = \mu_i$ and $\mathbb{V}\text{ar}(X_i) = \sigma_i^2, i \geq 1$. Let $T_n = \sum_{i=1}^n X_i$, $\xi_n = \sum_{i=1}^n \mu_i$, $s_n^2 = \sum_{i=1}^n \sigma_i^2$ and suppose that the conditions required by some version of the Central Limit Theorem are satisfied. Then, for every $x \in \mathbb{R}$ we may write

$$F_{(n)}(x) = P\left(\frac{T_n - \xi_n}{s_n} \leq x\right) \to \int_{-\infty}^{x} \frac{1}{\sqrt{2\pi}} e^{-t^2/2}\, dt = \Phi(x) \quad \text{as} \quad n \to \infty.$$

A question of both theoretical and practical interest concerns the speed with which the above convergence takes place. In other words, how large should n be so that $\Phi(x)$ may be considered as a good approximation for $F_{(n)}(x)$. Although there are no simple answers to this question, the following results may serve as general guidelines in that direction.

Theorem* 7.4.1 *(Polya). *Let $\{F_{(n)}\}$ be a sequence of distribution functions weakly converging to a continuous distribution function F. Then,*

$$\lim_{n\to\infty} \sup_{x\in\mathbb{R}} |F_{(n)}(x) - F(x)| = 0.$$

Proof. Since F is continuous, given $\varepsilon > 0$, there exists $n_0 = n_0(\varepsilon)$ such that for all $x \in \mathbb{R}$, $|F_{(n)}(x) - F(x)| < \varepsilon$. Therefore, $\sup_{x\in\mathbb{R}} |F_{(n)}(x) - F(x)| \leq \varepsilon$ and the result follows from the definition of limit. ■

Theorem* 7.4.2 *(Berry–Esséen). *Let $\{X_n\}$ be a sequence of i.i.d. random variables with $\mathbb{E}(X_1) = \mu$, $\mathbb{V}ar(X_1) = \sigma^2$ and suppose that $\mathbb{E}(|X_1 - \mu|^{2+\delta}) = \nu_{2+\delta} < \infty$ for some $0 < \delta \leq 1$. Also let $T_n = \sum_{i=1}^n X_i$ and $F_{(n)}(x) = P[(T_n - n\mu)/(\sigma\sqrt{n}) \leq x]$, $x \in \mathbb{R}$. Then, there exists a constant C such that*

$$\Delta_n = \sup_{x\in\mathbb{R}} |F_{(n)}(x) - \Phi(x)| \leq C \frac{\nu_{2+\delta} n^{-\delta/2}}{\sigma^{2+\delta}}.$$

The proof is out of the scope of this text and the reader is referred to Feller (1971) for details. Berry (1941) proved the result for $\delta = 1$ and a sharper limit with $C = 0.41$ has been obtained by Zolotarev (1967) and van Beeck (1972). The usefulness of the theorem, however, is limited, since the rates of convergence attained are not very sharp. Suppose, for example, that $\delta = 1, n = 100$ and $\gamma_3 = \nu_3/\sigma^3 = 1.5$; then $\Delta_n \leq C\gamma_3 n^{-1/2} = 0.41 \times 1.5 \times 1/10 = 0.0615$.

Alternatively, for the rates of convergence of the sequence of distribution functions $F_{(n)}$ to Φ or of the density functions $f_{(n)}$ (when they exist) to φ, the density function of the standard normal distribution may be assessed by *Gram–Charlier* or *Edgeworth expansions*.

Essentially, the idea behind the Gram–Charlier expansions is based on the fact that any probability density function g may be formally expressed as

$$g(x) = \sum_{j=0}^{\infty} \frac{c_j}{j!} \varphi^{(j)}(x) = \frac{1}{\sqrt{2\pi}} \sum_{j=0}^{\infty} \frac{c_j}{j!} \frac{\mathrm{d}^j}{\mathrm{d}x^j} e^{-x^2/2}$$
$$= \varphi(x) \sum_{j=0}^{\infty} (-1)^j \frac{c_j}{j!} H_j(x), \qquad (7.4.1)$$

where

$$H_j(x) = (-1)^j e^{-x^2/2}\, \mathrm{d}^j/\mathrm{d}x^j e^{-x^2/2}, \quad j = 0, 1, \ldots$$

are orthogonal polynomials associated with the normal distribution, known in the literature as *Hermite polynomials*, and φ is the density function for the normal distribution. In particular, they satisfy the relations

$$\frac{1}{\sqrt{2\pi}} \int_{-\infty}^{+\infty} H_m(x) H_n(x) e^{-x^2/2}\, \mathrm{d}x = \begin{cases} m! & \text{if } m = n \\ 0 & \text{if } m \neq n. \end{cases}$$

Now, from (7.4.1), it follows that for $m = 0, 1, \ldots,$ we have

$$\int_{-\infty}^{+\infty} H_m(x) g(x) dx = \sum_{j=0}^{\infty} (-1)^j \frac{c_j}{j!} \frac{1}{\sqrt{2\pi}} \int_{-\infty}^{+\infty} H_m(x) H_j(x) e^{-x^2/2}\, \mathrm{d}x,$$
$$= (-1)^m c_m$$

which implies that

$$c_m = (-1)^m \int_{-\infty}^{+\infty} g(x) H_m(x)\, \mathrm{d}x. \qquad (7.4.2)$$

For example, if X is a random variable with $\mathbb{E}(X) = 0$, $\mathbb{V}\mathrm{ar}(X) = 1$ and probability density function g we may use (7.4.2) to obtain

$$c_0 = 1,$$
$$c_1 = -\int_{-\infty}^{+\infty} x g(x)\, \mathrm{d}x = 0,$$
$$c_2 = (-1)^2 \int_{-\infty}^{+\infty} (x^2 - 1) g(x)\, \mathrm{d}x = 1 - 1 = 0,$$
$$c_3 = (-1)^3 \int_{-\infty}^{+\infty} (x^3 - 3x) g(x)\, \mathrm{d}x = -\mu_3,$$
$$\vdots$$

so that from (7.4.1) we may write

$$g(x) = \varphi(x) \left[1 + \sum_{j=3}^{\infty} (-1)^j \frac{c_j}{j!} H_j(x) \right]. \qquad (7.4.3)$$

Note that we are not really interested in the convergence of the series in (7.4.3); our main objective is to know whether a small number of terms (say, one or two) are sufficient to guarantee a good approximation to $g(x)$. In particular, we may consider

$$g(x) = \varphi(x)[1 + \mu_3(x^3 - 3x)/6] + r(x),$$

and the problem is to find out whether $r(x)$ is small enough.

Recalling that from the definition of the Hermite polynomials,

$$\Phi^{(j)}(x) = \frac{1}{\sqrt{2\pi}}(-1)^{j-1}H_{j-1}(x)e^{-x^2/2}, \quad j \geq 1,$$

a formal integration of (7.4.3) yields

$$\int_{-\infty}^{x} g(t)\,\mathrm{d}t = \int_{\infty}^{x} \varphi(t)\,\mathrm{d}t + \sum_{j=3}^{\infty}(-1)^j \frac{c_j}{j!} \int_{-\infty}^{x} \frac{1}{\sqrt{2\pi}} e^{-t^2/2} H_j(t)\,\mathrm{d}t. \tag{7.4.4}$$

But, since

$$\begin{aligned}\int_{-\infty}^{x} \frac{1}{\sqrt{2\pi}} e^{-t^2/2} H_j(t)\,\mathrm{d}t &= \int_{-\infty}^{x} (-1)^j \Phi^{(j+1)}(t)\,\mathrm{d}t \\ &= (-1)^j \Phi^{(j)}(t)\Big|_{-\infty}^{x} = (-1)^j \Phi^{(j)}(x),\end{aligned}$$

it follows that

$$\begin{aligned}G(x) &= \Phi(x)\left[1 + \sum_{j=3}^{\infty} \frac{c_j}{j!}\Phi^{(j)}(x)\right] \\ &= \Phi(x) - \varphi(x)\sum_{j=3}^{\infty}(-1)^j \frac{c_j}{j!} H_{j-1}(x),\end{aligned} \tag{7.4.5}$$

where G is the distribution function of X. In particular we may write

$$G(x) = \Phi(x) + \varphi(x)\mu_3(x^2 - 1)/6 + R(x).$$

In order to verify how the above results may be applied in the context of large samples, let $\{X_n\}$ denote a sequence of i.i.d. random variables with $\mathbb{E}(X_1) = 0$, $\mathbb{V}\mathrm{ar}(X_1) = \sigma^2 < \infty$ and characteristic function ϕ. Let $Z_n = \sum_{i=1}^{n} X_i/\sigma\sqrt{n}$ and suppose that the corresponding probability density function $f_{(n)}$ exists; also let $F_{(n)}$ and ϕ_n respectively denote the distribution function and the characteristic function associated with $f_{(n)}$. In order to specify an expansion of the form (7.4.3) for $f_{(n)}$, we must obtain the coefficients c_j, $j \geq 1$. In this direction, we first consider the following formal Taylor series expansions:

$$\phi(t) = \sum_{j=0}^{\infty} \phi^{(j)}(0)\frac{t^j}{j!} = 1 + \sum_{j=1}^{\infty} \frac{\mu_j}{j!}(it)^j, \tag{7.4.6}$$

where $\mu_j = i^{-j}\phi^{(j)}(0)$, $j \geq 1$, are the moments of X_1, and

$$\log\phi(t) = \sum_{j=0}^{\infty} [\log\phi(t)]_{t=0}^{(j)} \frac{t^j}{j!} = \sum_{j=0}^{\infty} \frac{\lambda_j}{j!}(it)^j, \tag{7.4.7}$$

where $\lambda_j = i^{-j}[\log\phi(t)]^{(j)}_{t=0}$, $j \geq 1$, are the cumulants of X_1. From (7.4.6) and (7.4.7) it follows that

$$\log\phi(t) = \log\left[1 + \sum_{j=1}^{\infty}\frac{\mu_j}{j!}(it)^j\right] = \sum_{j=0}^{\infty}\frac{\lambda_j}{j!}(it)^j,$$

which implies

$$\sum_{h=1}^{\infty}\frac{(-1)^{h+1}}{h}\left[\sum_{j=1}^{\infty}\frac{\mu_j}{j!}(it)^j\right]^h = \sum_{j=0}^{\infty}\frac{\lambda_j}{j!}(it)^j. \tag{7.4.8}$$

Expanding both members of (7.4.8) in powers of (it) and equating the coefficients of terms with the same power, we have $\lambda_1 = \mu_1$, $\lambda_2 = \mu_2 - \mu_1^2$, $\lambda_3 = \mu_3 - 3\mu_1\mu_2 + 2\mu_1^3$, $\lambda_4 = \mu_4 - 3\mu_2^2 - 4\mu_1\mu_3 + 12\mu_1^2\mu_2 - 6\mu_4, \ldots$ and, consequently, $\mu_1 = \lambda_1$, $\mu_2 = \lambda_2 - \lambda_1^2$, $\mu_3 = \lambda_3 + 3\lambda_1\lambda_2 + \lambda_1^3$, $\mu_4 = \lambda_4 + 3\lambda_2^2 + 4\lambda_1\lambda_3 + 6\lambda_1^2\lambda_2 + \lambda_1^4 \cdots$.

Now observe that

$$\phi_n(t) = [\phi(t/\sigma\sqrt{n})]^n \rightarrow \log\phi_n(t) = n\log\phi(t/\sigma\sqrt{n});$$

expanding both members via (7.4.7) we obtain

$$\sum_{j=1}^{\infty}\lambda_j^{(n)}\frac{(it)^j}{j!} = \sum_{j=1}^{\infty}n\lambda_j\frac{(it/\sigma\sqrt{n})^j}{j!} = \sum_{j=1}^{\infty}\frac{\lambda_j}{j!\sigma^j n^{3/2-1}}(it)^j,$$

where $\lambda_j^{(n)}$, $j \geq 1$, are the cumulants of Z_n; equating the coefficients of the terms with the same power of (it) we arrive at

$$\begin{aligned}
\lambda_1^{(n)} &= \lambda_1 = \mu_1 = 0,\\
\lambda_2^{(n)} &= \lambda_2/\sigma^2 = \sigma^2/\sigma^2 = 1,\\
\lambda_3^{(n)} &= \lambda_3/\sigma^3\sqrt{n} = \mu_3/\sigma^3\sqrt{n},\\
\lambda_4^{(n)} &= \lambda_4/\sigma^4 n = (\mu_4 - 3\sigma^4)/\sigma^4 n,\\
&\vdots
\end{aligned}$$

From the relation between cumulants and moments we get

$$\begin{aligned}
\mu_1^{(n)} &= 0,\\
\mu_2^{(n)} &= 1,\\
\mu_3^{(n)} &= \mu_3/\sigma^3\sqrt{n},\\
\mu_4^{(n)} &= \frac{\mu_4 - 3\sigma^4}{\sigma^4 n} + 3,\\
&\vdots
\end{aligned}$$

Finally, from (7.4.2), we may conclude that

$$
\begin{aligned}
c_1 &= c_2 = 0, \\
c_3 &= -\int_{-\infty}^{+\infty} (x^3 - 3x) f_{(n)}(x)\,\mathrm{d}x = -\mu_3^{(n)} + 3\mu_1^{(n)} = -\mu_3/\sigma^3\sqrt{n}, \\
c_4 &= \int_{-\infty}^{+\infty} (x^4 - 6x^2 + 3) f_{(n)}(x)\,\mathrm{d}x = \mu_4^{(n)} - 6\mu_2^{(n)} + 3 = \frac{\mu_4 - 3\sigma^4}{\sigma^4 n}, \\
&\vdots
\end{aligned}
$$

so that the Gram–Charlier expansion (7.4.3) is

$$
f_{(n)}(x) = \varphi(x)\left[1 + \frac{\mu_3}{6\sigma^3\sqrt{n}}(x^3 - 3x) + \frac{\mu_4 - 3\sigma^4}{24\sigma^4 n}(x^4 - 6x^2 + 3) + \cdots\right]. \quad (7.4.9)
$$

Analogously, we may obtain an expansion for the distribution function:

$$
F_{(n)}(x) = \Phi(x) + \varphi(x)\left[\frac{\mu_3}{6\sigma^3\sqrt{n}}(1 - x^2) + \frac{\mu_4 - 3\sigma^4}{24\sigma^4 n}(3x - x^3) + \cdots\right]. \quad (7.4.10)
$$

Although the above expansions offer a better insight to the problem of evaluation of the rates of convergence to normality than that provided by Berry–Esséen Theorem 7.4.2, they have the disadvantage of requiring the knowledge of the moments of the parent distribution. Others alternatives, such as the Edgeworth expansions are available, but they are out of the scope of this text. See Cramér (1946) for a detailed description and other references.

7.5 Projections and Variance-Stabilizing Transformations

In this section, we are mainly concerned with the asymptotic normality of statistics that may not be expressed as sums of independent random variables. For example, let $\{X_n\}$ be a sequence of i.i.d. random variables with mean μ and variance σ^2, and suppose that we are interested in verifying whether the sample variance s_n^2 follows an asymptotic normal distribution.

A convenient strategy in that direction consists of decomposing the statistic of interest in two terms, one of which is a sum of i.i.d. random variables and the other converges in probability to zero and then uses one of the results presented in this section.

***Theorem* 7.5.1.** *Consider the statistic $T_n = T(X_1, \ldots, X_n)$ where the X_i are i.i.d. random variables. Let $T_n = G_n + R_n$, where $G_n = \sum_{i=1}^n g(X_i)$ and $n^{-1/2}R_n \xrightarrow{P} 0$. Furthermore, let $\mathbb{E}[g(X_i)] = \xi$ and $\mathbb{V}\mathrm{ar}[g(X_i)] = \nu^2 < \infty$ and suppose that $(G_n - n\xi)/\sqrt{n}\nu \xrightarrow{D} \mathcal{N}(0, 1)$. Then, $(T_n - n\xi)/\sqrt{n}\nu \xrightarrow{D} \mathcal{N}(0, 1)$.*

Proof. First observe that given $\varepsilon > 0$:

$$\begin{aligned}
&P[(T_n - n\xi)/\nu\sqrt{n} \le x] \\
&\quad = P[(G_n - n\xi)/\sqrt{n}\nu + R_n/\sqrt{n}\nu \le x] \\
&\quad = P[(G_n - n\xi)/\nu\sqrt{n} + R_n/\sqrt{n}\nu \le x; |R_n|/\nu\sqrt{n} \le \varepsilon] \\
&\qquad + P[(G_n - n\xi)/\nu\sqrt{n} + R_n/\sqrt{n}\nu \le x; |R_n|/\nu\sqrt{n} > \varepsilon] \\
&\quad \le P[(G_n - n\xi)/\nu\sqrt{n} \le x + \varepsilon; |R_n|/\nu\sqrt{n} \le \varepsilon] \\
&\qquad + P(|R_n|/\nu\sqrt{n} > \varepsilon) \\
&\quad \le P[(G_n - n\xi)/\nu\sqrt{n} \le x + \varepsilon] \\
&\qquad + P(|R_n|/\nu\sqrt{n} > \varepsilon) \to \Phi(x+\varepsilon) \quad \text{as} \quad n \to \infty.
\end{aligned}$$

Similarly,

$$\begin{aligned}
&P[(T_n - n\xi)/\nu\sqrt{n} \le x] \\
&\quad = P[(G_n - n\xi)/\nu\sqrt{n} + R_n/\sqrt{n}\nu \le x] \\
&\quad \ge P[(G_n - n\xi)/\nu\sqrt{n} \le x - \varepsilon; |R_n|/\nu\sqrt{n} \le \varepsilon] \\
&\quad \ge P[(G_n - n\xi)/\nu\sqrt{n} \le x - \varepsilon] \\
&\qquad - P[(G_n - n\xi)/\nu\sqrt{n} \le x - \varepsilon; |R_n|/\nu\sqrt{n} > \varepsilon] \\
&\quad \ge P[(G_n - n\xi)/\nu\sqrt{n} \le x - \varepsilon] \\
&\qquad - P(|R_n|/\nu\sqrt{n} > \varepsilon) \to \Phi(x-\varepsilon) \quad \text{as} \quad n \to \infty.
\end{aligned}$$

From the two expressions above we have,

$$\Phi(x-\varepsilon) \le \lim_{n\to\infty} P\left(\frac{T_n - n\xi}{\nu\sqrt{n}} \le x\right) \le \Phi(x+\varepsilon).$$

Now, since for all $\varepsilon > 0$,

$$\Phi(x+\varepsilon) - \Phi(x-\varepsilon) = (2\pi)^{-1/2}\int_{x-\varepsilon}^{x+\varepsilon} \exp(-t^2/2)\,\mathrm{d}t \le 2\varepsilon/\sqrt{2\pi},$$

it follows that $\lim_{n\to\infty} P[(T_n - n\xi)/\nu\sqrt{n} \le x] = \Phi(x)$. ■

***Example* 7.5.1.** Let $\{X_n\}$ be a sequence of i.i.d. random variables with mean μ and variance σ^2. Assume that $\mathbb{E}(X_i^4) < \infty$ and write $\mathbb{E}(X_i - \mu)^4 = \mu_4 < \infty$ and $\mathbb{V}\mathrm{ar}[(X_i - \mu)^2] = \mu_4 - \sigma^4 = \gamma^2 > 0$. Then, observe that

$$\frac{\sqrt{n}}{\gamma}(s_n^2 - \sigma^2) = \frac{n}{n-1}\frac{1}{\sqrt{n}\gamma}[T_n - (n-1)\sigma^2], \tag{7.5.1}$$

where

$$\begin{aligned}
T_n &= \sum_{i=1}^{n}(X_i - \overline{X}_n)^2 = \sum_{i=1}^{n}\left[(X_i - \mu)^2 - \frac{\sigma^2}{n}\right] - n(\overline{X}_n - \mu)^2 + \sigma^2 \\
&= G_n + R_n,
\end{aligned}$$

$G_n = \sum_{i=1}^n g(X_i)$ with $g(X_i) = (X_i - \mu)^2 - \sigma^2/n$ and $R_n = -n(\overline{X}_n - \mu)^2 + \sigma^2$. Note that

$$\mathbb{E}(|R_n/\sqrt{n}|) = \sqrt{n}\mathbb{E}[(\overline{X}_n - \mu)^2] + \sigma^2/\sqrt{n} = 2\sigma^2/\sqrt{n} \to 0 \quad \text{as} \quad n \to \infty;$$

thus, from Theorem 6.2.3, it follows that $n^{-1/2}R_n \xrightarrow{P} 0$. Observing that $\mathbb{E}[g(X_i)] = (1 - 1/n)\sigma^2$ and $\mathbb{V}\text{ar}[g(X_i)] = \gamma^2$, a direct application of Theorem 7.5.1 in connection with (7.5.1) permits us to conclude that $\sqrt{n}(s_n^2 - \sigma^2)/\gamma \xrightarrow{D} \mathcal{N}(0, 1)$. □

This type of decomposition technique may be generalized further along the lines of the following well-known result.

Theorem* 7.5.2 *(Slutsky). *Let $\{X_n\}$ and $\{Y_n\}$ be sequences of random variables such that $X_n \xrightarrow{D} X$ and $Y_n \xrightarrow{P} c$, where c is a constant. Then, it follows that*

(i) $X_n + Y_n \xrightarrow{D} X + c$,

(ii) $Y_n X_n \xrightarrow{D} cX$,

(iii) $X_n / Y_n \xrightarrow{D} X/c$ *if* $c \neq 0$.

Proof. Let F denote the distribution function of X; then, for every $\varepsilon > 0$ and every continuity point of F, x, we may write

$$\begin{aligned}
&P(X_n + Y_n \le x)\\
&\quad = P(X_n + Y_n \le x; |Y_n - c| \le \varepsilon) + P(X_n + Y_n \le x; |Y_n - c| > \varepsilon)\\
&\quad \le P(X_n \le x - c + \varepsilon; |Y_n - c| \le \varepsilon) + P(|Y_n - c| > \varepsilon)\\
&\quad \le P(X_n \le x - c + \varepsilon) + P(|Y_n - c| > \varepsilon.
\end{aligned}$$

Therefore,

$$\begin{aligned}
&\limsup_{n\to\infty} P(X_n + Y_n \le x)\\
&\quad \le \lim_{n\to\infty} P(X_n \le x - c + \varepsilon) + \lim_{n\to\infty} P(|Y_n - c| > \varepsilon)\\
&\quad = P(X \le x - c + \varepsilon) = F(x - c + \varepsilon)
\end{aligned} \tag{7.5.2}$$

Also,

$$\begin{aligned}
P(X_n + Y_n \le x) &\ge P(X_n + Y_n \le x; |Y_n - c| \le \varepsilon)\\
&\ge P(X_n \le x - c - \varepsilon; |Y_n - c| \le \varepsilon)\\
&= P(X_n \le x - c - \varepsilon) - P(X_n \le x - c - \varepsilon; |Y_n - c| > \varepsilon)\\
&\ge P(X_n \le x - c - \varepsilon) - P(|Y_n - c| > \varepsilon).
\end{aligned}$$

Therefore,

$$\begin{aligned}
&\liminf_{n\to\infty} P(X_n + Y_n \le x)\\
&\quad \ge \lim_{n\to\infty} P(X_n \le x - c - \varepsilon) - \lim_{n\to\infty} P(|Y_n - c| > \varepsilon)\\
&\quad = P(X \le x - c - \varepsilon)\\
&\quad = F(x - c - \varepsilon).
\end{aligned} \tag{7.5.3}$$

From (7.5.2) and (7.5.3), we have

$$\begin{aligned} F(x-c-\varepsilon) &\le \liminf_{n\to\infty} P(X_n + Y_n \le x) \\ &\le \limsup_{n\to\infty} P(X_n + Y_n \le x) \\ &\le F(x-c+\varepsilon). \end{aligned}$$

Letting $\varepsilon \to 0$, it follows that

$$\lim_{n\to\infty} P(X_n + Y_n \le x) = F(x-c)$$

and the proof of (i) is completed. To prove (ii), note that, for $x \ge 0$, without loss of generality,

$$\begin{aligned} P(X_nY_n \le x) &= P\left(X_nY_n \le x;\ \left|\frac{Y_n}{c} - 1\right| \le \varepsilon\right) + P\left(X_nY_n \le x;\ \left|\frac{Y_n}{c} - 1\right| > \varepsilon\right) \\ &\le P\left(X_n \le \frac{x}{c(1-\varepsilon)};\ \left|\frac{Y_n}{c} - 1\right| \le \varepsilon\right) + P\left(\left|\frac{Y_n}{c} - 1\right| > \varepsilon\right) \\ &\le P\left(X_n \le \frac{x}{c(1-\varepsilon)}\right) + P\left(\left|\frac{Y_n}{c} - 1\right| > \varepsilon\right). \end{aligned}$$

Therefore, for $x \ge 0$,

$$\begin{aligned} &\limsup_{n\to\infty} P(X_nY_n \le x) \\ &\quad \le \lim_{n\to\infty} P\left[X_n \le \frac{x}{c(1-\varepsilon)}\right] + \lim_{n\to\infty} P\left(\left|\frac{Y_n}{c} - 1\right| > \varepsilon\right) \\ &\quad = F[x/c(1-\varepsilon)]. \end{aligned} \tag{7.5.4}$$

Similarly, for $x \ge 0$,

$$\begin{aligned} P(X_nY_n \le x) &\ge P\left[X_n \le \frac{x}{c(1+\varepsilon)};\ \left|\frac{Y_n}{c} - 1\right| \le \varepsilon\right] \\ &= P\left[X_n \le \frac{x}{c(1+\varepsilon)}\right] - P\left[X_n \le \frac{x}{c(1+\varepsilon)};\ \left|\frac{Y_n}{c} - 1\right| > \varepsilon\right] \\ &\ge P\left[X_n \le \frac{x}{c(1+\varepsilon)}\right] - P\left(\left|\frac{Y_n}{c} - 1\right| > \varepsilon\right). \end{aligned}$$

Therefore, for $x \ge 0$,

$$\begin{aligned} &\liminf_{n\to\infty} P(X_nY_n \le x) \\ &\quad \ge \lim_{n\to\infty} P\left[X_n \le \frac{x}{c(1+\varepsilon)}\right] - \lim_{n\to\infty} P\left(\left|\frac{Y_n}{c} - 1\right| > \varepsilon\right) \\ &\quad = F\{x/c(1+\varepsilon)\}. \end{aligned} \tag{7.5.5}$$

From (7.5.4) and (7.5.5), we have, for $x \ge 0$,

$$\begin{aligned} F[x/c(1+\varepsilon)] &\le \liminf_{n\to\infty} P(X_nY_n \le x) \le \limsup_{n\to\infty} P(X_nY_n \le x) \\ &\le F[x/c(1-\varepsilon)]. \end{aligned}$$

Letting $\varepsilon \to 0$, it follows that $\lim_{n\to\infty} P(X_n Y_n \le x) = F(x/c)$ and the proof of (ii) is complete; (iii) is proved along similar arguments. ■

***Example* 7.5.2.** Let $\{X_n\}$ be a sequence of i.i.d. random variables with mean μ and variance $\sigma^2 < \infty$. We are interested in the asymptotic distribution of the statistic $t_n = \sqrt{n}(\overline{X}_n - \mu)/s_n$. First, write

$$t_n = \frac{\sqrt{n}(\overline{X} - \mu)/\sigma}{\sqrt{s_n^2/\sigma^2}}$$

and note that

(i) the Khintchine Weak Law of Large Numbers (Theorem 6.3.6) implies that

$$n^{-1}\sum_{i=1}^{n}(X_i - \mu)^2 \xrightarrow{P} \mathbb{E}[(X_i - \mu)^2] = \sigma^2 \quad \text{and} \quad \overline{X}_n \xrightarrow{P} \mu;$$

consequently, from the definition of convergence in probability, it follows that

$$s_n^2 = \frac{n}{n-1}\left\{\frac{1}{n}\sum_{i=1}^{n}(X_i - \mu)^2 - (\overline{X}_n - \mu)^2\right\} \xrightarrow{P} \sigma^2,$$

which, in turn, implies that $s_n \xrightarrow{P} \sigma$;

(ii) $Z_n = \sqrt{n}(\overline{X}_n - \mu)/\sigma \xrightarrow{D} \mathcal{N}(0, 1)$ by the Classical Central Limit Theorem 7.3.1.

From (i) and (ii), a direct application of the Slutsky Theorem 7.5.2 implies that $t_n \xrightarrow{D} \mathcal{N}(0, 1)$. □

A multivariate version of the Slutsky Theorem may be stated as follows.

Theorem 7.5.3. *Let $\{\boldsymbol{X}_n\}$ and $\{\boldsymbol{Y}_n\}$ be sequences of random p-vectors such that $\boldsymbol{X}_n \xrightarrow{D} \boldsymbol{X}$ and $\boldsymbol{Y}_n \xrightarrow{P} \boldsymbol{0}$; also let $\{\boldsymbol{W}_n\}$ be a sequence of random $(w \times p)$ matrices, such that $\mathrm{tr}\{(\boldsymbol{W}_n - \boldsymbol{W})'(\boldsymbol{W}_n - \boldsymbol{W})\} \xrightarrow{P} 0$ where $\boldsymbol{W}$ is a nonstochastic matrix. Then,*

(i) $\boldsymbol{X}_n \pm \boldsymbol{Y}_n \xrightarrow{D} \boldsymbol{X}$,

(ii) $\boldsymbol{W}_n\boldsymbol{X}_n \xrightarrow{D} \boldsymbol{W}\boldsymbol{X}$.

Proof. Given $\boldsymbol{\lambda} \in \mathbb{R}^p$, arbitrary but fixed, and applying the Lévy–Cramér Theorem 6.2.8, it follows that for all $r \in \mathbb{R}$

$$\begin{aligned}\phi_{\boldsymbol{\lambda}'\boldsymbol{X}_n}(r) &= \mathbb{E}[\exp(ir\boldsymbol{\lambda}'\boldsymbol{X}_n)] \\ &= \phi_{\boldsymbol{X}_n}(r\boldsymbol{\lambda}) \to \phi_{\boldsymbol{X}}(r\boldsymbol{\lambda}) \\ &= \mathbb{E}[\exp(ir\boldsymbol{\lambda}'\boldsymbol{X})] = \phi_{\boldsymbol{\lambda}'\boldsymbol{X}}(r).\end{aligned}$$

Thus, $\boldsymbol{\lambda}'\boldsymbol{X}_n \xrightarrow{D} \boldsymbol{\lambda}'\boldsymbol{X}$. Observing that $\boldsymbol{\lambda}'\boldsymbol{Y}_n \xrightarrow{P} 0$, it follows from the Slutsky Theorem 7.5.2 that $\boldsymbol{\lambda}'(\boldsymbol{X}_n \pm \boldsymbol{Y}_n) \xrightarrow{D} \boldsymbol{\lambda}'\boldsymbol{X}$. Since this is true for all $\boldsymbol{\lambda} \in \mathbb{R}^p$, (i) follows from the Cramér–Wold Theorem 7.2.4.

To prove (ii) first write

$$\boldsymbol{W}_n\boldsymbol{X}_n = \boldsymbol{W}\boldsymbol{X}_n + (\boldsymbol{W}_n - \boldsymbol{W})\boldsymbol{X}_n. \tag{7.5.6}$$

Then, use an argument similar to the one above to show that for all $\boldsymbol{\lambda} \in \mathbb{R}^p$ (arbitrary, but fixed)

$$\boldsymbol{\lambda}'\boldsymbol{W}\boldsymbol{X}_n \xrightarrow{D} \boldsymbol{\lambda}'\boldsymbol{W}\boldsymbol{X} \tag{7.5.7}$$

and note that

$$\begin{aligned}|\boldsymbol{\lambda}'(\boldsymbol{W}_n - \boldsymbol{W})\boldsymbol{X}_n| &\leq \|\boldsymbol{\lambda}\|\|(\boldsymbol{W}_n - \boldsymbol{W})\boldsymbol{X}_n\| \\ &= \|\boldsymbol{\lambda}\||\boldsymbol{X}_n'(\boldsymbol{W}_n - \boldsymbol{W})'(\boldsymbol{W}_n - \boldsymbol{W})\boldsymbol{X}_n|^{1/2} \\ &\leq \|\boldsymbol{\lambda}\||\text{ch}_1[(\boldsymbol{W}_n - \boldsymbol{W})'(\boldsymbol{W}_n - \boldsymbol{W})]\boldsymbol{X}_n'\boldsymbol{X}_n|^{1/2} \\ &\leq \|\boldsymbol{\lambda}\||\text{tr}[(\boldsymbol{W}_n - \boldsymbol{W})'(\boldsymbol{W}_n - \boldsymbol{W})]\boldsymbol{X}_n'\boldsymbol{X}_n|^{1/2}.\end{aligned}$$

From Theorem 7.2.3 it follows that $\boldsymbol{X}_n'\boldsymbol{X}_n \xrightarrow{D} \boldsymbol{X}'\boldsymbol{X}$ and an application of the Slutsky Theorem 7.5.2 yields

$$\boldsymbol{\lambda}'(\boldsymbol{W}_n - \boldsymbol{W})\boldsymbol{X}_n \xrightarrow{P} \boldsymbol{0}, \tag{7.5.8}$$

which in conjunction with (7.5.6) and (7.5.7) implies (ii) via a subsequent application of the Slutsky Theorem 7.5.2. ■

In Example 7.5.1 we have shown that the sample variance s_n^2 is such that $\sqrt{n}(s_n^2 - \sigma^2)/\gamma \xrightarrow{D} \mathcal{N}(0, 1)$ using a decomposition result. Alternatively, we could have used the fact that $s_n^2 = \binom{n}{2}^{-1} \sum_{1 \leq i < j \leq n} \psi(X_i, X_j)$ with $\psi(a, b) = (a - b)^2/2$ is a *reverse martingale* to prove the same result via Theorem 7.3.8, although some rather tedious algebraic manipulation would have been necessary to show condition (a).

Let $\{X_k, k \geq 1\}$ be a sequence of i.i.d. random variables with distribution function F and consider a statistic $T_n = T_n(X_1, \ldots, X_n)$; suppose we are interested in verifying whether $[T_n - \mathbb{E}(T_n)]/\sqrt{\mathbb{V}\text{ar}(T_n)}$ is asymptotically normal. First, note that $T_{ni} = \mathbb{E}(T_n|X_i) = f_n(X_i)$, $i = 1, \ldots, n$ are i.i.d. random variables, such that $\mathbb{E}(T_{ni}) = \mathbb{E}(T_n)$. Then, letting $v_n = \sum_{i=1}^n [T_{ni} - \mathbb{E}(T_n)]$ we may use the Central Limit Theorem for Triangular Schemes 7.3.5 to show the asymptotic normality of $\sqrt{n}v_n/\sqrt{\mathbb{V}\text{ar}(v_n)}$; if we can show that $[T_n - \mathbb{E}(T_n)]/\sqrt{\mathbb{V}\text{ar}(T_n)} - v_n/\sqrt{\mathbb{V}\text{ar}(v_n)} \xrightarrow{P} 0$, then by the Slutsky Theorem 7.5.2 the asymptotic distribution of $[T_n - \mathbb{E}(T_n)]/\sqrt{\mathbb{V}\text{ar}(T_n)}$ will be the same as that of $v_n/\sqrt{\mathbb{V}\text{ar}(v_n)}$. Another way of obtaining similar results, especially important for U-statistics, relies on the *H-Projection Technique* proposed by Hoeffding (1948), as we detail in the sequence.

Let us denote by

$$g_c(x_1, \ldots, x_c) = \mathbb{E}_F[g(x_1, \ldots, x_c, X_{c+1}, \ldots, X_m)], \tag{7.5.9}$$

for $c = 0, 1, \ldots, m$. Note that $g_0 = \theta(F)$ and $g_m = g$. Then, let

$$Y_n = \frac{m}{n} \sum_{i=1}^n [g_1(X_i) - \theta(F)] \tag{7.5.10}$$

and

$$\xi_c = \mathbb{E}_F\left[g_c^2(X_1, \ldots, X_c)\right] - \theta^2(F), \quad 0 \leq c \leq m;\ \xi_0 = 0. \tag{7.5.11}$$

Then, by the Lindeberg–Feller Central Limit Theorem 7.3.3,

$$\sqrt{n}Y_n \xrightarrow{D} \mathcal{N}(0, m^2\xi_1) \quad \text{whenever} \quad 0 < \xi_1 < \infty. \tag{7.5.12}$$

Let us also note that for each $i = 1, \dots, n$,

$$\begin{aligned}
\mathbb{E}[U_n - \theta(F)|X_i] &= \binom{n}{m}^{-1} \sum_{\{1 \le i_1 < \cdots < i_m \le n\}} \mathbb{E}\left[g(X_{i_1}, \dots, X_{i_m}) - \theta(F)|X_i\right] \\
&= \binom{n}{m}^{-1} \left\{ \binom{n-1}{m-1} [g_1(X_i) - \theta(F)] + \binom{n-1}{m} [\theta(F) - \theta(F)] \right\} \\
&= \frac{m}{n} [g_1(X_i) - \theta(F)]. \qquad (7.5.13)
\end{aligned}$$

Therefore,

$$\sum_{i=1}^{n} \mathbb{E}[U_n - \theta(F)|X_i] = Y_n, \tag{7.5.14}$$

which shows that Y_n is the projection of $U_n - \theta(F)$; also, (7.5.10) shows that Y_n has independent summands. Furthermore, by (7.5.14),

$$\begin{aligned}
\mathbb{E}\{[U_n - \theta(F)]Y_n\} &= \frac{m}{n} \sum_{i=1}^{n} \mathbb{E}\{[U_n - \theta(F)][g_1(X_i) - \theta(F)]\} \\
&= \frac{m}{n} \sum_{i=1}^{n} \mathbb{E}\left\{[g_1(X_i) - \theta(F)]\mathbb{E}([U_n - \theta(F)]|X_i)\right\} \\
&= m^2 n^{-2} \sum_{i=1}^{n} \mathbb{E}\{[g_1(X_i) - \theta(F)]^2\} \\
&= m^2 n^{-1} \mathbb{E}[g_1(X_1) - \theta(F)]^2 \\
&= m^2 n^{-1} \xi_1 = \mathbb{E}(Y_n^2). \qquad (7.5.15)
\end{aligned}$$

Thus,

$$n\mathbb{E}[U_n - \theta(F) - Y_n]^2 = n\mathbb{E}[U_n - \theta(F)]^2 - n\mathbb{E}(Y_n^2), \tag{7.5.16}$$

where, by (2.4.28) and (7.5.9)–(7.5.11), we have

$$\begin{aligned}
\mathbb{E}[U_n - \theta(F)]^2 &= \binom{n}{m}^{-1} \sum_{c=1}^{m} \binom{m}{c}\binom{n-m}{m-c} \xi_c \\
&= m^2 n^{-1} \xi_1 + O(n^{-2}), \qquad (7.5.17)
\end{aligned}$$

so that, by (7.5.15)–(7.5.17), we obtain

$$n\mathbb{E}[U_n - \theta(F) - Y_n]^2 = O(n^{-1}), \tag{7.5.18}$$

and, hence, by the Chebyshev Inequality (6.2.27), it follows that

$$\sqrt{n}|U_n - \theta(F) - Y_n| \xrightarrow{P} 0. \tag{7.5.19}$$

By (7.5.12), (7.5.19), and the Slutsky Theorem 7.5.2, we have

$$\sqrt{n}\,[U_n - \theta(F)] \xrightarrow{D} \mathcal{N}(0, m^2\xi_1). \tag{7.5.20}$$

If, in addition to $\mathbb{E}_F[g^2(X_1, \ldots, X_m)] < \infty$, we assume that

$$\mathbb{E}_F[|g_F(X_{i_1}, \ldots, X_{i_m})|] < \infty, \quad 1 \le i_1 \le i_2 \le \cdots \le i_m \le m, \tag{7.5.21}$$

by using the decomposition

$$V_n = U_n + n^{-1} \left\{ \frac{1}{n} \sum_{i=1}^{n} [g(X_i, X_i) - U_n] \right\},$$

for $m = 2$ (or a similar one for $m \ge 2$), we obtain

$$\mathbb{E}[|U_n - V_n|] = O(n^{-1}), \tag{7.5.22}$$

so that $\sqrt{n}|U_n - V_n| \xrightarrow{P} 0$. Hence, under (7.5.21) and recalling that $\mathbb{E}_F(g^2) < \infty$, we have

$$\sqrt{n}\,[V_n - \theta(F)] \xrightarrow{D} \mathcal{N}(0, m^2\xi_1). \tag{7.5.23}$$

It may be remarked that (7.5.20), (7.5.22), and (7.5.23) reflect the asymptotic equivalence of U-statistics and V-statistics. Therefore, for large sample sizes, they share the common properties (i.e., consistency, asymptotic normality, because). In many practical applications, it may be necessary to estimate ξ_1 from the sample in a comparable way because the estimation of $\sigma^2 = \mathbb{V}\text{ar}(X)$ is necessary for the estimation of $\theta = \mathbb{E}(X)$]. Note that by (7.5.11),

$$\begin{aligned} \xi_1 = \mathbb{E}_F\{g(X_1, \ldots, X_m)g(X_m, \ldots, X_{2m-1}) \\ - \mathbb{E}_F[g(X_1, \ldots, X_m)g(X_{m+1}, \ldots, X_{2m})]\} \end{aligned} \tag{7.5.24}$$

and is, therefore, an estimable parameter of degree $2m$. However, the computation of the unbiased estimator of ξ_1 (i.e., the U-statistic corresponding to ξ_1), for $m \ge 2$, may be generally very cumbersome. This can be largely avoided by using a *jackknifing* method, which in this special case, reduces to the following one suggested by earlier by Sen (1960). For each $i = 1, \ldots, n$, let

$$U_{n(i)} = \binom{n-1}{m-1}^{-1} \sum_{S_{n,i}} g(X_i, X_{i_2}, \ldots, X_{i_m}), \tag{7.5.25}$$

where the summation $S_{n,i}$ extends over all $1 \le i_2 < \cdots < i_m \le n$ with $i_j \ne i$ for $j = 2, \ldots, m$. Then, note that by (2.4.28) and (7.5.25), $U_n = n^{-1} \sum_{i=1}^{n} U_{n(i)}$, so that

$$s_n^2 = \frac{1}{(n-1)} \sum_{i=1}^{n} \left[U_{n(i)} - U_n\right]^2. \tag{7.5.26}$$

It is easy to verify that s_n^2 is a (strongly) consistent estimator of ξ_1. Thus, for large sample sizes, the standardized form

$$m^{-1}s_n^{-1}n^{1/2}[U_n - \theta(F)] \tag{7.5.27}$$

can be used to test suitable hypotheses for $\theta(F)$ or to provide a confidence interval for $\theta(F)$. By virtue of (7.5.22) and (7.5.23), U_n may as well be replaced by V_n in (7.5.27). Intuitively speaking, in (2.4.29), writing $F_n = F + (F_n - F)$, we end up with 2^m terms that can be gathered into $(m+1)$ subsets, where the kth subset contains $\binom{m}{k}$ terms, for

$k = 0, 1, \ldots, m$. This leads us to

$$V_n - \theta(F) = \sum_{k=1}^{m} \binom{m}{k} V_n^{(k)}, \tag{7.5.28}$$

where for $k = 1, \ldots, m$,

$$V_n^{(k)} = \int \cdots \int g_k(x_1, \ldots, x_k)\,\mathrm{d}[F_n(x_1) - F(x_1)] \cdots \mathrm{d}[F_n(x_k) - F(x_k)].$$

A very similar decomposition holds for U_n, that is,

$$U_n - \theta(F) = \sum_{k=1}^{m} \binom{m}{k} U_n^{(k)}, \qquad \binom{m}{1} U_n^{(1)} = \binom{m}{1} V_n^{(1)} = Y_n. \tag{7.5.29}$$

In the literature this is known as the *Hoeffding decomposition* of U-statistics. It may be remarked that $\mathbb{E}(U_n^{(k)}) = 0$ and $\mathbb{E}[(U_n^{(k)})^2] = O(n^{-k})$, $k \geq 1$, so that the successive terms in the decomposition in (7.5.29) are of stochastically smaller order of magnitude, and they are pairwise orthogonal too. Furthermore, as we have seen in Chapter 5, $\{U_n, U_n^{(1)}, \ldots, U_n^{(m)}; n \geq m\}$ form a reverse martingale, so that the corresponding asymptotic properties (such as a.s. convergence, asymptotic distribution, etc.) can be incorporated to study similar results for $\{U_n\}$ or $\{V_n\}$. Detailed study of these asymptotic properties is somewhat beyond the scope of our text; we may, however, refer to Sen (1981, chapter 3) for some general accounts. The U-statistics have also been extended to multisample models where the parameter θ is viewed as a functional of several distribution functions. For example, in a two-sample model, with two distribution functions F and G (say), we may write $\theta = \theta(F, G)$ as

$$\theta = \int \cdots \int g(x_1, \ldots, x_r; y_1, \ldots, y_s)\,\mathrm{d}F(x_1) \cdots \mathrm{d}F(x_r)\,\mathrm{d}G(y_1) \cdots \mathrm{d}G(y_s), \tag{7.5.30}$$

where $g(\cdot)$ is a kernel of degree (r, s) and $r \geq 1, s \geq 1$. For two samples $X_1, \ldots, X_{n_1}$, from F and $Y_1, \ldots, Y_{n_2}$ from G, we define the (generalized) U-statistic as

$$U_{n_1,n_2} = \binom{n_1}{r}^{-1} \binom{n_2}{s}^{-1} \sum g(X_{i_1}, \ldots, X_{i_r}; Y_{i_1}, \ldots, Y_{i_s}) \tag{7.5.31}$$

where the summation extends over all $1 \leq i_1 < \cdots < i_r \leq n_1$ and $1 \leq j_1 < \cdots < j_s \leq n_2$. The Hoeffding projection and Hoeffding decomposition for U_{n_1,n_2} work out neatly, and the asymptotic theory of such generalized U-statistics has been developed on parallel lines; we again refer to Sen (1981, Chapter 3) for some of these details.

Example 7.5.3 (Rank weighted mean). [Sen (1964)] Let $X_{n:1} \leq \cdots \leq X_{n:n}$ be the order statistics in a sample of size n from a distribution F with mean θ. For every $k \geq 1$, define

$$T_{n,k} = \binom{n}{2k+1}^{-1} \sum_{i=1}^{n} \binom{i-1}{k} \binom{n-i}{k} X_{n:i}, \quad n \geq 2k+1. \tag{7.5.32}$$

Note that $T_{n,0}$ $(= \overline{X}_n)$ is the sample mean and $T_{n,[(n+1)/2]}$ is the sample median. On the ground of robustness, often $T_{n,1}$ or $T_{n,2}$ are preferred to $T_{n,0}$. For every k (≥ 0), $T_{n,k}$ is a linear combination of order statistics, and its population counterpart is

$$\begin{aligned}\theta_k(F) &= \frac{\Gamma(2k+2)}{[\Gamma(k+1)]^2}\int_{-\infty}^{\infty}\{F(x)[1-F(x)]\}^k x\,\mathrm{d}F(x)\\ &= \frac{\Gamma(2k+2)}{[\Gamma(k+1)]^2}\int_0^1 [u(1-u)]^k F^{-1}(u)\,\mathrm{d}u,\end{aligned} \tag{7.5.33}$$

where $F^{-1}(u) = \inf\{x : F(x) \geq u\}$, $0 \leq u \leq 1$. Thus, whenever F is symmetric about θ, we have $\theta_k = \theta$, for all $k \geq 0$. For $k = 0$, the results of Chapter 6 and the preceeding sections of this chapter apply directly, and, hence, we will consider only the case of $k \geq 1$. Note that for $T_{n,0}$ $(= \overline{X}_n)$, one needs to assume that $\sigma^2 = \mathbb{V}\text{ar}(X) < \infty$, so that the asymptotic normality of $n^{1/2}(T_{n,0} - \theta)$ holds. However, for $k \geq 1$, less stringent regularity conditions suffice. Note that whenever $\mathbb{E}_F(|X|^r) < \infty$ for some $r > 0$, $|x|^r F(x)[1 - F(x)]$ is bounded for all $x \in \mathbb{R}$ and it converges to zero as $x \to \pm\infty$. Thus, for $k \geq 1$, $|x|F(x)[1 - F(x)]^k = \{|x|^{1/k}F(x)[1 - F(x)]\}^k$ is bounded (and smooth) whenever $\mathbb{E}_F(|X|^r) < \infty$ for some $r \geq 1/k$. This explains why, for $k \geq 1$, $T_{n,k}$ is more robust than $\overline{X}_n$. Thus, whenever $F(u)$ is absolutely continuous and strictly monotone, we have $[u(1-n)]^k F^{-1}(u) = J(u)$, say, a bounded and continuous function on [0, 1], provided $\mathbb{E}_F(|X|^{1/k}) < \infty$.Interestingly enough, $T_{n,k}$ may also be expressed as a U-statistic. Let

$$g(X_1, \ldots, X_{2k+1}) = \text{med}(X_1, \ldots, X_{2k+1}). \tag{7.5.34}$$

Then, it is easy to verify that

$$T_{n,k} = \binom{n}{2k+1}^{-1} \sum_{\{1 \leq i_1 < \cdots < i_{2k+1} \leq n\}} g(X_{i_1}, \ldots, X_{i_{2k+1}}). \tag{7.5.35}$$

As such, the asymptotic properties of U-statistics are all transmittable to $T_{n,k}$, for any fixed $k \geq 0$. On the other hand, if $k = k_n$ is made to depend on n, then the kernel in (7.5.34) also depends on n, and, hence, we may need some additional regularity conditions. These may not be necessary if we use the functional approach to be discussed in Chapter 11. □

In many practical situations one might be interested in obtaining the asymptotic distribution of statistics, which may not be decomposed in a suitable form to apply Theorem 7.5.1 or the Slutsky Theorem 7.5.2 as in the case of the sample standard deviation. If, however, the statistic of interest is a well-behaved function of some other statistic known to be asymptotically normal, its asymptotic distribution may be determined via the following result known in the literature as the *Delta method*.

Theorem 7.5.4. *Suppose that $\sqrt{n}(T_n - \theta)/\sigma \xrightarrow{D} \mathcal{N}(0, 1)$ and let g be a continuous function such that $g'(\theta)$ exists and $g'(\theta) \neq 0$. Then, it follows that*

$$\sqrt{n}[g(T_n) - g(\theta)]/\sigma g'(\theta) \xrightarrow{D} \mathcal{N}(0, 1).$$

Proof. Let $Z_n = \sqrt{n}(T_n - \theta)/\sigma$ and $G_n = [g(T_n) - g(\theta)]/(T_n - \theta)$ and observe that

$$\frac{\sqrt{n}}{\sigma}[g(T_n) - g(\theta)] = \frac{1}{\sigma}G_n Z_n. \tag{7.5.36}$$

Now, using the fact that $Z_n \xrightarrow{D} Z$ and Theorem 7.2.5 we may conclude that $\sqrt{n}\sigma^{-1}(T_n - \theta) = O_p(1)$, which implies $(T_n - \theta) = O_p(1)O_p(n^{-1/2}) = O_p(n^{-1/2}) = o_p(1)$, that is, $T_n - \theta \xrightarrow{P} 0$; consequently it follows that $G_n \xrightarrow{P} g'(\theta)$. Then, applying the Slutsky Theorem 7.5.2 to (7.5.9), we have $\sqrt{n}\sigma^{-1}[g(T_n) - g(\theta)] \xrightarrow{D} g'(\theta)Z$ and the result follows. ■

***Example* 7.5.4.** Let $\{X_n\}$ be a sequence of i.i.d. random variables with mean μ, variance σ^2 and such that $\mu_4 = \mathbb{E}[(X-\mu)^4] < \infty$. We have seen in Example 7.5.1 that $\sqrt{n}(s_n^2 - \sigma^2) \xrightarrow{D} \mathcal{N}(0, \gamma^2)$, but now we are interested in the asymptotic distribution of $\sqrt{n}(s_n - \sigma)$. Note that $s_n = \{(n-1)^{-1}\sum_{i=1}^n (X_i - \overline{X})^2\}^{1/2}$ is not expressible as a sum of independent random variables as in the case of s_n^2. However, since $s_n = \sqrt{s_n^2}$, we may take $g(x) = \sqrt{x}$ [which implies $g'(x) = (2\sqrt{x})^{-1}$] and apply Theorem 7.5.4 to conclude that

$$\frac{\sqrt{n}(s_n - \sigma)}{(2\sigma)^{-1}\gamma} \xrightarrow{D} \mathcal{N}(0, 1). \qquad \square$$

***Example* 7.5.5.** Let $R_n \sim \text{Bin}(n, \pi)$ and write $T_n = R_n/n$. We have seen in Example 7.3.1 that $\sqrt{n}(T_n - \pi)/[\pi(1-\pi)]^{1/2} \xrightarrow{D} \mathcal{N}(0, 1)$. Suppose we are interested in the asymptotic distribution of $\sqrt{n}(G_n - \pi^{-1})$, where $G_n = T_n^{-1}$ is an estimate of π^{-1}. Taking $g(x) = 1/x$, which implies $g'(x) = -1/x^2$, it follows from Theorem 7.5.4 that

$$\frac{\sqrt{n}(G_n - \pi^{-1})}{\pi^{-2}\sqrt{\pi(1-\pi)}} = \frac{\sqrt{n}(G_n - \pi^{-1})}{\sqrt{(1-\pi)/\pi^3}} \xrightarrow{D} \mathcal{N}(0, 1). \qquad \square$$

***Example* 7.5.6.** Let $R_n \sim \text{Bin}(n, \pi)$ and suppose we are interested in the asymptotic distribution of $\sqrt{n}[G_n - \pi(1-\pi)]$ where $G_n = R_n(n - R_n)/n^2$ is an estimator of $\pi(1-\pi)$. Observe that $G_n = g(R_n/n)$ where $g(x) = x(1-x)$ [which implies $g'(x) = 1 - 2x$] and apply Theorem 7.5.4 to see that, for $\pi \neq 1/2$,

$$\frac{\sqrt{n}[G_n - \pi(1-\pi)]}{(1-2\pi)\sqrt{\pi(1-\pi)}} \xrightarrow{D} \mathcal{N}(0, 1). \qquad \square$$

The generalization of the above results to the multivariate case is of great practical importance and will be considered next. Let $\{\boldsymbol{T}_n\}$ be a sequence of p-dimensional random vectors and suppose that $\sqrt{n}(\boldsymbol{T}_n - \boldsymbol{\theta}) \xrightarrow{D} \mathcal{N}_p(\boldsymbol{0}, \boldsymbol{\Sigma})$; essentially, we are interested in the asymptotic distribution of $\sqrt{n}[g(\boldsymbol{T}_n) - g(\boldsymbol{\theta})]$, where $g(\cdot)$ is a real-valued function of $\boldsymbol{T}_n$. This is the object of the following theorem.

***Theorem* 7.5.5.** *Let $\{\boldsymbol{T}_n\}$ be a sequence of random p-vectors such that $\sqrt{n}(\boldsymbol{T}_n - \boldsymbol{\theta}) \xrightarrow{D} \mathcal{N}(\boldsymbol{0}, \boldsymbol{\Sigma})$ and consider a real-valued function $g(\boldsymbol{T}_n)$ such that $\dot{\boldsymbol{g}}(\boldsymbol{\theta}) = \partial g(\boldsymbol{x})/\partial \boldsymbol{x}|_{\boldsymbol{\theta}}$ is nonnull and continuous in a neighborhood of $\boldsymbol{\theta}$. Then*

$$\sqrt{n}[g(\boldsymbol{T}_n) - g(\boldsymbol{\theta})] \xrightarrow{D} N(0, \gamma^2) \quad \textit{with} \quad \gamma^2 = [\dot{\boldsymbol{g}}(\boldsymbol{\theta})]'\boldsymbol{\Sigma}[\dot{\boldsymbol{g}}(\boldsymbol{\theta})].$$

Proof. First, note that for $\boldsymbol{x}, \boldsymbol{y} \in \mathbb{R}^p$ we may write

$$\begin{aligned} g(\boldsymbol{x}) - g(\boldsymbol{y}) &= g(x_1, \ldots, x_p) - g(y_1, \ldots y_p) \\ &= g(x_1, \ldots, x_p) - g(x_1, \ldots, x_{p-1}, y_p) \\ &\quad + g(x_1, \ldots, x_{p-1}, y_p) - g(x_1, \ldots, x_{p-2}, y_{p-1}, y_p) \\ &\quad + g(x_1, \ldots, x_{p-2}, y_{p-1}, y_p) \\ &\quad - g(x_1, \ldots, x_{p-3}, y_{p-2}, y_{p-1}, y_p) \\ &\quad + \cdots \\ &\quad + g(x_1, y_2, \ldots, y_p) - g(y_1, y_2, \ldots, y_p) \\ &= \sum_{i=1}^{p} (x_i - y_i)\widetilde{g}_i(\boldsymbol{x}, \boldsymbol{y}), \end{aligned} \tag{7.5.37}$$

where

$$\widetilde{g}_i(\boldsymbol{x}, \boldsymbol{y}) = \{g(x_1, \ldots, x_i, y_{i+1}, \ldots, y_p) - g(x_1, \ldots, x_{i-1}, y_i, \ldots, y_p)/(x_i - y_i)\}$$

is such that

$$\lim_{x_i \to y_i} \widetilde{g}_i(\boldsymbol{x}, \boldsymbol{y}) = \partial g(\boldsymbol{x})/\partial x_i|_{\boldsymbol{y}} = \dot{g}_i(\boldsymbol{y}).$$

Now let

$$\widetilde{\boldsymbol{g}}(T_n, \boldsymbol{\theta}) = [\widetilde{g}_1(\boldsymbol{T}_n, \boldsymbol{\theta}), \ldots, \widetilde{g}_p(\boldsymbol{T}_n, \boldsymbol{\theta})]'$$

and observe that, from (7.5.37), we may write

$$\sqrt{n}[g(\boldsymbol{T}_n) - g(\boldsymbol{\theta})] = \sqrt{n}(\boldsymbol{T}_n - \boldsymbol{\theta})'\widetilde{\boldsymbol{g}}(\boldsymbol{T}_n, \boldsymbol{\theta}). \tag{7.5.38}$$

From a direct extension of Theorem 7.2.5 it follows that $\sqrt{n}(\boldsymbol{T}_n - \boldsymbol{\theta}) = O_p(\boldsymbol{1})$, which implies $(\boldsymbol{T}_n - \boldsymbol{\theta}) \xrightarrow{P} \boldsymbol{0}$ and, therefore, $\widetilde{\boldsymbol{g}}(\boldsymbol{T}_n, \boldsymbol{\theta}) \xrightarrow{P} \dot{\boldsymbol{g}}(\boldsymbol{\theta})$. Then, applying the Slutsky Theorem 7.5.2 to (7.5.38) we have

$$\sqrt{n}(\boldsymbol{T}_n - \boldsymbol{\theta})'\widetilde{\boldsymbol{g}}_n \xrightarrow{D} \boldsymbol{T}'\dot{\boldsymbol{g}}(\boldsymbol{\theta}) \sim \mathcal{N}(0, \gamma^2).$$

■

Example 7.5.7. Let $\{X_n\}$ be a sequence of i.i.d. random variables with mean μ, variance σ^2 and $\mathbb{E}(X_1^4) < \infty$. Let $\nu = \sigma/\mu$ denote the coefficient of variation and $v_n = s_n/\overline{X}_n$ the corresponding sample counterpart. We are interested in the asymptotic distribution of $\sqrt{n}(v_n - \nu)$. First, let us show that the vector $\sqrt{n}\boldsymbol{T}_n$ where $\boldsymbol{T}_n = [\overline{X}_n - \mu, s_n^2 - \sigma^2]'$ is asymptotically normally distributed. In this direction, observe that we may write

$$\begin{aligned} \left(s_n^2 - \sigma^2\right) &= \frac{n-1}{n}\left(s_n^2 - \sigma^2\right) + \frac{1}{n}\left(s_n^2 - \sigma^2\right) \\ &= \frac{1}{n}\sum_{i=1}^{n}[(X_i - \mu)^2 - \sigma^2] - (\overline{X}_n - \mu)^2 + \frac{\sigma^2}{n} + \frac{1}{n}\left(s_n^2 - \sigma^2\right) \\ &= \frac{1}{n}\sum_{i=1}^{n} Y_i + R_n, \end{aligned}$$

where $Y_i = (X_i - \mu)^2 - \sigma^2$ is such that $\mathbb{E}(Y_i) = 0$ and $\mathbb{V}\mathrm{ar}(Y_i) = \mu_4 - \sigma^4 = \gamma^2$ and $R_n = (s_n^2 - \sigma^2)/n + \sigma^2/n - (\overline{X}_n - \mu)^2$ is such that $\sqrt{n}R_n = o_p(1)$. Then, let $\lambda_1, \lambda_2 \in \mathbb{R}$ be arbitrary but fixed and let

$$\begin{aligned}
Z_n &= \sqrt{n}\big[\lambda_1(\overline{X}_n - \mu) + \lambda_2(s_n^2 - \sigma^2)\big] \\
&= \sqrt{n}\left\{\frac{1}{n}\sum_{i=1}^{n}[\lambda_1(X_i - \mu) + \lambda_2 Y_i]\right\} + \sqrt{n}\lambda_2 R_n \\
&= \sqrt{n}\left(\frac{1}{n}\sum_{i=1}^{n} W_i\right) + o_p(1), \qquad (7.5.39)
\end{aligned}$$

where $W_i = \lambda_1(X_i - \mu) + \lambda_2 Y_i$, $i = 1, 2, \ldots, n$ are independent random variables such that $\mathbb{E}(W_i) = 0$ and $\mathbb{V}\mathrm{ar}(W_i) = \lambda_1^2\sigma^2 + \lambda_2^2(\mu_4 - \sigma^4) + 2\mu_3\lambda_1\lambda_2$ with $\mu_4 = \mathbb{E}[(X_i - \mu)^4]$ and $\mu_3 = \mathbb{E}[(X_i - \mu)^3]$. Applying the Classical Central Limit Theorem 7.3.1 in conjunction with the Slutsky Theorem 7.5.2 to (7.5.39), we may conclude that Z_n is asymptotically normally distributed. Since λ_1 and λ_2 are arbitrary, it follows from the Cramér–Wold Theorem 7.2.4 that

$$\sqrt{n}\boldsymbol{T}_n \xrightarrow{D} \mathcal{N}(\mathbf{0}, \boldsymbol{\Sigma}), \quad \text{where} \quad \boldsymbol{\Sigma} = \begin{pmatrix} \sigma^2 & \mu_3 \\ \mu_3 & \mu_4 - \sigma^4 \end{pmatrix}.$$

Now, let $\boldsymbol{\theta} = (\mu,\ \sigma^2)'$ and $g(x, y) = \sqrt{y}/x$; thus, $g(\overline{X}_n, s_n^2) = s_n/\overline{X}_n = v_n$ and $g(\boldsymbol{\theta}) = \sigma/\mu = \nu$. Furthermore, $\overset{\bullet}{g}_1(\boldsymbol{\theta}) = \partial g(x, y)/\partial x|_{\boldsymbol{\theta}} = -\sigma/\mu^2$ and $\overset{\bullet}{g}_2(\theta) = \partial g(x, y)/\partial y|_{\boldsymbol{\theta}} = 1/2\mu\sigma$ exist; therefore, applying Theorem 7.5.5 it follows that

$$\sqrt{n}(v_n - \nu) \xrightarrow{D} \mathcal{N}\left(0, \frac{\sigma^4}{\mu^4} - \frac{\mu_3}{\mu^3} + \frac{\mu_4}{4\mu^2\sigma^2} - \frac{\sigma^2}{4\mu^2}\right). \qquad \square$$

***Example* 7.5.8.** Let $\{X_n, Y_n\}$ be a sequence of independent random variables following a bivariate distribution with mean vector $\boldsymbol{\mu} = (\mu_X, \mu_Y)'$ and p.d. covariance matrix:

$$\boldsymbol{\Sigma} = \begin{pmatrix} \sigma_X^2 & \rho\sigma_X\sigma_Y \\ \rho\sigma_X\sigma_Y & \sigma_Y^2 \end{pmatrix}.$$

We are interested in the asymptotic distribution of the sample coefficient of correlation

$$R_n = \sum_{i=1}^{n}(X_i - \overline{X}_n)(Y_i - \overline{Y}_n) \bigg/ \left[\sum_{i=1}^{n}(X_i - \overline{X}_n)^2 \sum_{i=1}^{n}(Y_i - \overline{Y}_n)^2\right]^{1/2}.$$

First, observe that

$$\begin{aligned}
s_{nXY} &= n^{-1}\sum_{i=1}^{n}(X_i - \overline{X}_n)(Y_i - \overline{Y}_n) \\
&= n^{-1}\sum_{i=1}^{n}(X_i - \mu_X)(Y_i - \mu_Y) - (\overline{X}_n - \mu_X)(\overline{Y}_n - \mu_Y) \\
&= n^{-1}\sum_{i=1}^{n}(X_i - \mu_X)(Y_i - \mu_Y) + O_p(n^{-1}),
\end{aligned}$$

since from Theorem 7.2.5 it follows that $\overline{X}_n - \mu_X = O_p(n^{-1/2})$ and $(\overline{Y}_n - \mu_Y) = O_p(n^{-1/2})$. Similarly we may conclude that

$$s_{nX}^2 = n^{-1}\sum_{i=1}^{n}(X_i - \overline{X}_n)^2 = n^{-1}\sum_{i=1}^{n}(X_i - \mu_X)^2 + O_p(n^{-1}),$$

$$s_{nY}^2 = n^{-1}\sum_{i=1}^{n}(Y_i - \overline{Y}_n)^2 = n^{-1}\sum_{i=1}^{n}(Y_i - \mu_Y)^2 + O_p(n^{-1}).$$

Then, consider $\boldsymbol{t} = (t_1, t_2, t_3)' \in \mathbb{R}^3$, fixed but arbitrary and write

$$\begin{aligned}\sqrt{n}&\big[t_1\big(s_{nX}^2 - \sigma_X^2\big) + t_2\big(s_{nY}^2 - \sigma_Y^2\big) + t_3(s_{nXY} - \rho\sigma_X\sigma_Y)\big]\\ &= \frac{1}{\sqrt{n}}\sum_{i=1}^{n}\big\{t_1\big[(X_i - \mu_X)^2 - \sigma_X^2\big] + t_2\big[(Y_i - \mu_Y)^2 - \sigma_Y^2\big]\\ &\quad + t_3[(X_i - \mu_X)(Y_i - \mu_Y) - \rho\sigma_X\sigma_Y]\big\} + O_p(n^{-1})\\ &= \frac{1}{\sqrt{n}}\sum_{i=1}^{n}U_i + O_p(n^{-1}),\end{aligned}$$

where

$$\begin{aligned}U_i &= t_1\big[(X_i - \mu_X)^2 - \sigma_X^2\big] + t_2[(Y_i - \mu_Y)^2 - \sigma_Y^2]\\ &\quad + t_3[(X_i - \mu_X)(Y_i - \mu_Y) - \rho\sigma_X\sigma_Y],\end{aligned}$$

$i = 1, \ldots, n$ are i.i.d. random variables with $\mathbb{E}(U_1) = 0$ and $\mathbb{V}\text{ar}(U_1) = \mathbb{E}(U_1^2) = \boldsymbol{t}'\boldsymbol{\Gamma}\boldsymbol{t}$ with $\boldsymbol{\Gamma} = \mathbb{V}\text{ar}(W_1, W_2, W_3)$, $W_1 = (X_1 - \mu_X)^2 - \sigma_X^2$, $W_2 = (Y_1 - \mu_Y)^2 - \sigma_Y^2$ and $W_3 = (X_1 - \mu_X)(Y_1 - \mu_Y) - \rho\sigma_X\sigma_Y$. Thus, from the Classical Central Limit Theorem 7.3.1 and the Cramér–Wold Theorem 7.2.4, it follows that

$$\sqrt{n}\big[\big(s_{nX}^2, s_{nY}^2, s_{nXY}\big)' - \big(\sigma_X^2, \sigma_Y^2, \rho\sigma_X\sigma_Y\big)'\big] \xrightarrow{D} \mathcal{N}(\boldsymbol{0}, \boldsymbol{\Gamma}).$$

Now, since $R_n = s_{nXY}/s_{nX}s_{nY}$, if we define $g : \mathbb{R}_+ \times \mathbb{R}_+ \times \mathbb{R} \to \mathbb{R}$ by $g(\boldsymbol{x}) = x_3/\sqrt{x_1x_2}$, we may apply Theorem 7.5.5 to obtain the desired result. More specifically, letting $\boldsymbol{\theta} = (\sigma_X^2, \sigma_Y^2, \rho\sigma_X\sigma_Y)'$ we have

$$\begin{aligned}\overset{\bullet}{g}_1(\boldsymbol{x}) &= -x_3/(2x_1\sqrt{x_1x_2}),\\ \overset{\bullet}{g}_2(\boldsymbol{x}) &= -x_3/(2x_2\sqrt{x_1x_2}),\\ \overset{\bullet}{g}_3(\boldsymbol{x}) &= (x_1x_2)^{-1}\end{aligned}$$

and

$$\overset{\bullet}{\boldsymbol{g}}(\boldsymbol{\theta}) = \rho\left(-\frac{1}{2\sigma_X^2}, -\frac{1}{2\sigma_Y^2}, \frac{1}{\rho\sigma_X\sigma_Y}\right)'$$

so that $\sqrt{n}(R_n - \rho) \xrightarrow{D} \mathcal{N}(0, \gamma^2)$ with $\gamma^2 = \overset{\bullet}{\boldsymbol{g}}(\boldsymbol{\theta})'\boldsymbol{\Gamma}\overset{\bullet}{\boldsymbol{g}}(\theta)$. □

***Example* 7.5.9.** Consider the Pearsonian system of distributions, that is, the family characterized by (at most) the first four moments, $\boldsymbol{\tau} = \boldsymbol{\tau}(\boldsymbol{\theta}) = [\mu_1(\boldsymbol{\theta}), \mu_2(\boldsymbol{\theta}), \mu_3(\boldsymbol{\theta}), \mu_4(\boldsymbol{\theta})]$, where

$\boldsymbol{\theta} = [\theta_1, \theta_2, \theta_3, \theta_4]'$ are unknown parameters. The Method of Moments estimator of $\boldsymbol{\theta}$ based on a random sample $X_1, \ldots, X_n$ is defined as a solution $\widehat{\boldsymbol{\theta}}_n$ to the equation $\boldsymbol{T}_n = \boldsymbol{\tau}(\boldsymbol{\theta})$, where $\boldsymbol{T}_n = [m_{n1}, m_{n2}, m_{n3}, m_{n4}]'$ with $m_{nk} = n^{-1} \sum_{i=1}^{n} X_i^k$. In the case of the normal distribution, we have $\boldsymbol{\theta} = [\mu, \sigma^2]'$, $\boldsymbol{\tau} = [\mu, \mu^2 + \sigma^2]'$, $\boldsymbol{T}_n = [\overline{X}_n, n^{-1} \sum_{i=1}^{n} (X_i - \overline{X}_n)^2]'$, which implies $\widehat{\boldsymbol{\theta}}_n = [\widehat{\mu}_n, \widehat{\sigma}_n^2]'$ with $\widehat{\mu}_n = \overline{X}_n$ and $\widehat{\sigma}_n^2 = n^{-1} \sum_{i=1}^{n} (X_i - \overline{X}_n)^2$. Using some version of the Central Limit Theorem we may show that $\sqrt{n}[\boldsymbol{T}_n - \boldsymbol{\tau}(\boldsymbol{\theta})] \xrightarrow{D} \mathcal{N}(\boldsymbol{0}, \boldsymbol{\Sigma})$, where $\boldsymbol{\Sigma}$ is a p.d. matrix, provided the moment of order eight exists. The question is whether we can obtain the asymptotic distribution of $\sqrt{n}(\widehat{\boldsymbol{\theta}}_n - \boldsymbol{\theta})$. Now, since the correspondence between $\boldsymbol{\theta}$ and $\boldsymbol{\tau}$ is one-to-one, the inverse function $\boldsymbol{\tau}^{-1}$ exists, and we may write $\sqrt{n}(\widehat{\boldsymbol{\theta}}_n - \boldsymbol{\theta}) = \boldsymbol{\tau}^{-1} \sqrt{n}[\boldsymbol{T}_n - \boldsymbol{\tau}(\boldsymbol{\theta})]$. The desired result is a consequence of the following generalization of the previous theorem to the case of vector-valued functions. □

Theorem 7.5.6. *Let $\{\boldsymbol{T}_n\}$ be a sequence of random p-vectors such that $\sqrt{n}(\boldsymbol{T}_n - \boldsymbol{\theta}) \xrightarrow{D} \mathcal{N}(\boldsymbol{0}, \boldsymbol{\Sigma})$ and consider a vector-valued function $\boldsymbol{g} : \mathbb{R}^p \to \mathbb{R}^q$ such that $\dot{\boldsymbol{g}}(\boldsymbol{\theta}) = (\partial/\partial \boldsymbol{x}')\boldsymbol{g}(\boldsymbol{x})\big|_{\boldsymbol{\theta}}$ exists. Then,*

$$\sqrt{n}[\boldsymbol{g}(\boldsymbol{T}_n) - \boldsymbol{g}(\boldsymbol{\theta})] \xrightarrow{D} \mathcal{N}[\boldsymbol{0}, \dot{\boldsymbol{G}}(\boldsymbol{\theta})\boldsymbol{\Sigma}\dot{\boldsymbol{G}}(\boldsymbol{\theta})'].$$

Proof. Let $\dot{\boldsymbol{G}}(\boldsymbol{\theta}) = [\dot{\boldsymbol{g}}_1(\boldsymbol{\theta}), \ldots, \dot{\boldsymbol{g}}_q(\boldsymbol{\theta})]$ and write for each coordinate j, $1 \leq j \leq q$,

$$\sqrt{n}[g_j(\boldsymbol{T}_n) - g_j(\boldsymbol{\theta})] = \sqrt{n}(\boldsymbol{T}_n - \boldsymbol{\theta})'\widetilde{\boldsymbol{g}}_j(\boldsymbol{T}_n, \boldsymbol{\theta}),$$

where $\widetilde{\boldsymbol{g}}_j(\boldsymbol{T}_n, \boldsymbol{\theta})$ as in the previous theorem is such that $\widetilde{\boldsymbol{g}}_j(\boldsymbol{T}_n, \boldsymbol{\theta}) \xrightarrow{P} \dot{\boldsymbol{g}}_j(\boldsymbol{\theta})$. Then, let $\boldsymbol{\lambda} \in \mathbb{R}^q$ be arbitrary and write

$$G_n(\boldsymbol{\lambda}) = \sqrt{n}\boldsymbol{\lambda}'[\boldsymbol{g}(\boldsymbol{T}_n) - \boldsymbol{g}(\boldsymbol{\theta})] = \sqrt{n}(\boldsymbol{T}_n - \boldsymbol{\theta})'\boldsymbol{\lambda}'\widetilde{\boldsymbol{G}}(\boldsymbol{T}_n, \boldsymbol{\theta}),$$

where $\widetilde{\boldsymbol{G}}(\boldsymbol{T}_n, \boldsymbol{\theta}) = [\widetilde{\boldsymbol{g}}_1(\boldsymbol{T}_n, \boldsymbol{\theta}), \ldots, \widetilde{\boldsymbol{g}}_q(\boldsymbol{T}_n, \boldsymbol{\theta})]$. Now apply Theorem 7.5.5 to conclude that $G_n(\boldsymbol{\lambda}) \xrightarrow{D} \mathcal{N}[0, \boldsymbol{\lambda}'\dot{\boldsymbol{g}}(\boldsymbol{\theta})\boldsymbol{\Sigma}\dot{\boldsymbol{G}}'(\boldsymbol{\theta})\boldsymbol{\lambda}]$. Finally, since $\boldsymbol{\lambda}$ is arbitrary, the result follows from the Cramér–Wold Theorem 7.2.4. ■

The results on the asymptotic distribution of functions of asymptotically normal statistics are of special interest for the *variance-stabilizing transformations*. Consider, for example, the problem of obtaining an asymptotic confidence interval for σ^2 in the setup of Example 7.5.1. Since

$$\sqrt{n}\left(s_n^2 - \sigma^2\right) \xrightarrow{D} \mathcal{N}(0, (\mu_4 - \sigma^4),$$

we have that as $n \to \infty$,

$$P\left[s_n^2 - z_{\alpha/2}\sqrt{(\mu_4 - \sigma^4)/n} \leq \sigma^2 \leq s_n^2 + z_{\alpha/2}\sqrt{(\mu_4 - \sigma^4)/n}\right] \to 1 - \alpha.$$

Now, since the asymptotic variance depends on the unknown parameter σ^2 this statement is of no practical use. The question is whether we may consider a suitable transformation of s_n^2 for which the asymptotic variance does not depend on the unknown parameter. More generally, let

$$\sqrt{n}(T_n - \theta) \xrightarrow{D} \mathcal{N}[0, h(\theta)].$$

We want to consider a transformation $g(T_n)$ such that

$$\sqrt{n}[g(T_n) - g(\theta)] \xrightarrow{D} \mathcal{N}(0, c^2),$$

c independent of θ. Note that when such *variance-stabilizing transformations* exists we may write

$$P[g(T_n) - z_{\alpha/2}c/\sqrt{n} \le g(\theta) \le g(T_n) + z_{\alpha/2}c/\sqrt{n}] \to 1 - \alpha \quad \text{as} \quad n \to \infty.$$

Furthermore, if g is monotone, the above statement may be "inverted" to produce an asymptotic confidence interval for θ, that is, as $n \to \infty$,

$$P\{g^{-1}[g(T_n) - z_{\alpha/2}c/\sqrt{n}] \le \theta \le g^{-1}[g(T_n) + z_{\alpha/2}c/\sqrt{n}]\} \to 1 - \alpha.$$

In the setup of Example 7.5.1, letting $g(x) = \log x$, we have $g'(x) = x^{-1}$ and $g'(\sigma^2) = 1/\sigma^2 > 0$; thus, by Theorem 7.5.4 it follows that

$$\sqrt{n}\big[\log s_n^2 - \log \sigma^2\big] \xrightarrow{D} \mathcal{N}(0, c^2)$$

with $c^2 = \mu_4 - \sigma^{-4} = \beta_2 - 1$, $\beta_2 = \mu_4/\sigma^4$. Thus, as $n \to \infty$

$$P\big[\log s_n^2 - z_{\alpha/2}\sqrt{(\beta_2 - 1)/n} \le \log \sigma^2 \le \log s_n^2 + z_{\alpha/2}\sqrt{(\beta_2 - 1)/n}\big]$$

converges to $1 - \alpha$ which implies that as $n \to \infty$

$$P\big\{\exp[\log s_n^2 - z_{\alpha/2}\sqrt{(\beta_2 - 1)/n}] \le \sigma^2 \le \exp[\log s_n^2 + z_{\alpha/2}\sqrt{(\beta_2 - 1)/n}]\big\}$$

converges to $1 - \alpha$. Now observe that if g is continuous and $0 < |g'(\theta)| < \infty$, it follows from Theorem 7.5.4 that

$$\sqrt{n}[g(T_n) - g(\theta)] \xrightarrow{D} \mathcal{N}(0, [g'(\theta)]^2 h(\theta)\}.$$

Thus, to obtain a variance-stabilizing transformation we need to define g such that $[g'(\theta)]^2 h(\theta) = c^2$, which implies

$$g'(\theta) = \frac{c}{\sqrt{h(\theta)}}$$

so that

$$g(\theta) = \int g'(\theta)\, d\theta = \int \frac{c}{\sqrt{h(\theta)}}\, d\theta.$$

Returning to the setup of Example 7.5.1, letting $\theta = \sigma^2$, we have $h(\theta) = (\beta_2 - 1)\theta^2$; then, $g(\theta) = c(\beta_2 - 1)^{-1/2} \int \theta^{-1}\, d\theta = c^* \log \theta$, where $c^* = c/\sqrt{\beta_2 - 1}$. Choosing $c^* = 1$, we get $g(T_n) = \log(T_n)$ as the appropriate transformation.

***Example* 7.5.10.** Let $R_n \sim \text{Bin}(n, \pi)$ and consider the statistic $T_n = R_n/n$; we have seen that $\sqrt{n}(T_n - \pi) \to Z$ where $Z \sim \mathcal{N}[0, \pi(1 - \pi)]$. Here $h(\pi) = \pi(1 - \pi)$ and the variance-stabilizing transformation may be obtained from $g(\pi) = c \int [\pi(1 - \pi)]^{-1/2}\, d\pi$. Letting $\pi = [\sin(x)]^2$, we have $1 - \pi = [\cos(x)]^2$ and $d\pi = 2\sin(x)\cos(x)\, dx$, which yields

$$g(\pi) = c \int \frac{2\sin(x)\cos(x)}{\sin(x)\cos(x)}\, dx = 2cx = 2c \sin^{-1}(\sqrt{\pi}).$$

Choosing $c = 1/2$, we get $g(T_n) = \sin^{-1}(\sqrt{T_n})$, $0 < T_n < 1$. More explicitly, we may write

$$\sqrt{n}[\sin^{-1}(\sqrt{T_n}) - \sin^{-1}(\sqrt{\pi})] \xrightarrow{D} \mathcal{N}(0, 1/4) \quad \text{for} \quad 0 < \pi < 1.$$

An approximate $100(1-\alpha)\%$ confidence interval for π has lower and upper limits respectively given by $L_n = \sin^2[\sin^{-1}(\sqrt{T_n}) - z_{\alpha/2}/2\sqrt{n}]$ and $U_n = \sin^2[\sin^{-1}(\sqrt{T_n}) + z_{\alpha/2}/2\sqrt{n}]$. It is interesting to note the approximation is fairly good, even for $n \cong 10$. Anscombe (1948) suggested the alternative transformation $g(T_n) = \sin^{-1}[\sqrt{(R_n + 3/8)/(n + 3/4)}]$, which achieves a better approximation for moderate values of n. □

***Example* 7.5.11.** Let $\{X_\theta\}$ be a sequence of independent random variables with a Poisson(θ) distribution. Then, $\theta^{-1/2}(\overline{X}_\theta - \theta) \xrightarrow{D} \mathcal{N}(0, \theta)$. Letting $h(\theta) = \theta$ we get $g(\theta) = c\int d\theta/\sqrt{\theta} = 2c\sqrt{\theta}$ and choosing $c = 1/2$ we conclude that the appropriate variance-stabilizing transformation is $g(X_\theta) = \sqrt{X_\theta}$. More explicitly, we may write

$$\theta^{-1/2}(\sqrt{X_\theta} - \sqrt{\theta}) \xrightarrow{D} \mathcal{N}(0, 1/4).$$

Anscombe (1948) suggested an alternative transformation, where $g(X_\theta) = \sqrt{X_\theta + 3/8}$, which works out better for moderate θ. □

***Example* 7.5.12.** Consider the setup of Example 7.5.8 and suppose that (X_n, Y_n), $n \geq 1$, follow a bivariate normal distribution. In this case, we have

$$\begin{aligned}\gamma^2 = {} & \frac{\rho^2}{4}\left\{\mathbb{E}\left(\frac{X_1 - \mu_X}{\sigma_X}\right)^4 + \mathbb{E}\left(\frac{Y_1 - \mu_Y}{\sigma_Y}\right)^4\right\} \\ & + \rho^2\left(\frac{1}{\rho^2} + \frac{1}{2}\right)\mathbb{E}\left(\frac{X_1 - \mu_X}{\sigma_X}\right)^2\left(\frac{Y_1 - \mu_Y}{\sigma_Y}\right)^2 \\ & - \rho\left\{\mathbb{E}\left(\frac{X_1 - \mu_X}{\sigma_X}\right)^3\left(\frac{Y_1 - \mu_Y}{\sigma_Y}\right) + \mathbb{E}\left(\frac{X_1 - \mu_X}{\sigma_X}\right)\left(\frac{Y_1 - \mu_Y}{\sigma_Y}\right)^3\right\}.\end{aligned} \tag{7.5.40}$$

Since $\mathbb{E}[(X_1 - \mu_X)/\sigma_X]^2 \sim \chi_1^2$, we have

$$\mathbb{E}\left(\frac{X_1 - \mu_X}{\sigma_X}\right)^4 = \mathrm{Var}\left(\frac{X_1 - \mu_X}{\sigma_X}\right)^2 + \left[\mathbb{E}\left(\frac{X_1 - \mu_X}{\sigma_X}\right)^2\right]^2 = 3.$$

Similarly, it follows that $\mathbb{E}[(Y_1 - \mu_Y)/\sigma_Y]^4 = 3$. On the other hand, we know that the conditional distribution of Y_1 given $X_1 = x$ is

$$N\left[\mu_Y + \rho\frac{\sigma_Y}{\sigma_X}(x - \mu_X), \sigma_Y^2(1 - \rho^2)\right];$$

therefore,

$$\begin{aligned}
\mathbb{E}\left(\frac{X_1-\mu_X}{\sigma_X}\right)^2\left(\frac{Y_1-\mu_Y}{\sigma_Y}\right)^2 &= \mathbb{E}\left\{\mathbb{E}\left[\left(\frac{X_1-\mu_X}{\sigma_X}\right)^2\left(\frac{Y_1-\mu_Y}{\sigma_Y}\right)^2\Bigg|X_1\right]\right\}\\
&= \mathbb{E}\left\{\left(\frac{X_1-\mu_X}{\sigma_X}\right)^2\left[(1-\rho^2)+\rho^2\left(\frac{X_1-\mu_X}{\sigma_X}\right)^2\right]\right\}\\
&= (1-\rho^2)+\rho^2\mathbb{E}\left(\frac{X_1-\mu_X}{\sigma_X}\right)^4\\
&= 1-\rho^2+3\rho^2\\
&= 1+2\rho^2
\end{aligned}$$

and also,

$$\begin{aligned}
\mathbb{E}\left(\frac{Y_1-\mu_Y}{\sigma_Y}\right)\left(\frac{X_1-\mu_X}{\sigma_X}\right)^3 &= \mathbb{E}\left\{\mathbb{E}\left[\left(\frac{Y_1-\mu_Y}{\sigma_Y}\right)\left(\frac{X_1-\mu_X}{\sigma_X}\right)^3\Bigg|X_1\right]\right\}\\
&= \mathbb{E}\left\{\left(\frac{X_1-\mu_X}{\sigma_X}\right)^3\rho\left(\frac{X_1-\mu_X}{\sigma_X}\right)\right\}\\
&= 3\rho.
\end{aligned}$$

Similarly, we have $\mathbb{E}[(Y_1-\mu_Y)/\sigma_Y]^3[(X_1-\mu_X)/\sigma_X]=3\rho$; substituting these results in (7.5.40) we obtain $\gamma^2=(1-\rho^2)^2$, so that

$$\sqrt{n}(R_n-\rho)\xrightarrow{D}\mathcal{N}[0,(1-\rho^2)^2].$$

To obtain a variance stabilizing transformation, we let $h(p)=(1-\rho^2)^2$ and

$$\begin{aligned}
g(\rho)=c\int\frac{\mathrm{d}\rho}{1-\rho^2} &= \frac{c}{2}\int\left[\frac{1}{1-\rho}+\frac{1}{1+\rho}\right]\mathrm{d}\rho\\
&= \frac{c}{2}\int\frac{\mathrm{d}\rho}{1-\rho}+\frac{c}{2}\int\frac{\mathrm{d}\rho}{1+\rho}\\
&= \frac{c}{2}\log\frac{1+\rho}{1-\rho}.
\end{aligned}$$

Choosing $c=1$ and recalling that $\tanh(x)=[\exp(x)-\exp(-x)]/[\exp(x)+\exp(-x)]$, it follows that $g(R_n)=\tanh^{-1}(R_n)$. Therefore, we may write $\sqrt{n}[\tanh^{-1}(R_n)-\tanh^{-1}(\rho)]\xrightarrow{D}\mathcal{N}(0,1)$. Gayen (1951) showed that a better approximation is obtained if we replace $\sqrt{n}$ by $\sqrt{n-3}$ in the above expression; this produces reasonable approximations for $n\geq 10$. □

7.6 Quadratic Forms

Consider now a sequence of random p-vectors $\{\boldsymbol{T}_n\}$ such that $\sqrt{n}(\boldsymbol{T}_n-\boldsymbol{\theta})\xrightarrow{D}\mathcal{N}(\boldsymbol{0},\boldsymbol{\Sigma})$. A problem of great practical interest concerns the determination of the asymptotic distribution of quadratic forms of the type $Q_n=n(\boldsymbol{T}_n-\boldsymbol{\theta})'\boldsymbol{A}_n(\boldsymbol{T}_n-\boldsymbol{\theta})$ where $\{\boldsymbol{A}_n\}$ is a sequence of p.s.d. matrices. Theorem 7.5.5 may not be applied here, since $\partial Q_n(\boldsymbol{x})/\partial\boldsymbol{x}\big|_{\boldsymbol{\theta}}=\boldsymbol{0}$; however, the following results may be considered instead.

Theorem* 7.6.1 *(Cochran). *Let* $\{\boldsymbol{T}_n\}$ *be a sequence of random p-vectors such that* $\sqrt{n}(\boldsymbol{T}_n - \boldsymbol{\theta}) \xrightarrow{D} \mathcal{N}(\boldsymbol{0}, \boldsymbol{\Sigma})$, *with* $\text{rank}(\boldsymbol{\Sigma}) = q \le p$, *and* $\{\boldsymbol{A}_n\}$ *be a sequence of p.s.d. nonstochastic matrices such that* $\boldsymbol{A}_n \to \boldsymbol{A}$, *where rank*$(\boldsymbol{A}) \ge q$. *Then,*

$$Q_n = n(\boldsymbol{T}_n - \boldsymbol{\theta})'\boldsymbol{A}_n(\boldsymbol{T}_n - \boldsymbol{\theta}) \xrightarrow{D} \chi^2_q,$$

if and only if, $\boldsymbol{A}$ *is a generalized inverse* $\boldsymbol{\Sigma}$.

Proof. Since $\boldsymbol{\Sigma}$ is symmetric p.s.d., there exists a $(p \times q)$ matrix $\boldsymbol{B}$ of rank q such that $\boldsymbol{\Sigma} = \boldsymbol{B}\boldsymbol{B}'$. Now, let $\boldsymbol{Z}_n = (\boldsymbol{B}'\boldsymbol{B})^{-1}\boldsymbol{B}'\sqrt{n}(\boldsymbol{T}_n - \boldsymbol{\theta})$ and note that $\boldsymbol{Z}_n \xrightarrow{D} \boldsymbol{Z}$, where $\boldsymbol{Z} \sim \mathcal{N}(\boldsymbol{0}, \boldsymbol{I}_q)$. Then take $\sqrt{n}(\boldsymbol{T}_n - \boldsymbol{\theta}) = \boldsymbol{B}\boldsymbol{Z}_n$ and observe that, using a multivariate version of the Sverdrup Theorem 7.2.3 it follows that

$$Q_n = \boldsymbol{Z}_n'\boldsymbol{B}'\boldsymbol{A}_n\boldsymbol{B}\boldsymbol{Z}_n \xrightarrow{D} \boldsymbol{Z}'\boldsymbol{B}'\boldsymbol{A}\boldsymbol{B}\boldsymbol{Z}.$$

A necessary and sufficient condition for $\boldsymbol{Z}'\boldsymbol{B}'\boldsymbol{A}\boldsymbol{B}\boldsymbol{Z}$ to be distributed as χ^2_q is that $\boldsymbol{B}'\boldsymbol{A}\boldsymbol{B}$ be idempotent, that is,

$$\boldsymbol{B}'\boldsymbol{A}\boldsymbol{B}\boldsymbol{B}'\boldsymbol{A}\boldsymbol{B} = \boldsymbol{B}'\boldsymbol{A}\boldsymbol{\Sigma}\boldsymbol{A}\boldsymbol{B} = \boldsymbol{B}'\boldsymbol{A}\boldsymbol{B},$$

which occurs if and only if $\boldsymbol{A}$ is a generalized inverse of $\boldsymbol{\Sigma}$, that is, $\boldsymbol{\Sigma}\boldsymbol{A}\boldsymbol{\Sigma} = \boldsymbol{\Sigma}$ or equivalently $\boldsymbol{A}\boldsymbol{\Sigma}\boldsymbol{A} = \boldsymbol{A}$. Note that $\boldsymbol{B}'\boldsymbol{A}\boldsymbol{\Sigma}\boldsymbol{A}\boldsymbol{B} = \boldsymbol{B}'\boldsymbol{A}\boldsymbol{B}$ if and only if $\boldsymbol{B}\boldsymbol{B}'\boldsymbol{A}\boldsymbol{\Sigma}\boldsymbol{A}\boldsymbol{B}\boldsymbol{B}' = \boldsymbol{B}\boldsymbol{B}'\boldsymbol{A}\boldsymbol{B}\boldsymbol{B}'$, that is, $\boldsymbol{\Sigma}\boldsymbol{A}\boldsymbol{\Sigma}\boldsymbol{A}\boldsymbol{\Sigma} = \boldsymbol{\Sigma}\boldsymbol{A}\boldsymbol{\Sigma}$, or, equivalently, $\boldsymbol{A}\boldsymbol{\Sigma}\boldsymbol{A} = \boldsymbol{A}$. ■

***Theorem* 7.6.2.** *Let* $\{\boldsymbol{T}_n\}$ *be a sequence of random p-vectors such that* $\sqrt{n}(\boldsymbol{T}_n - \boldsymbol{\theta}) \xrightarrow{D} \mathcal{N}(\boldsymbol{0}, \boldsymbol{\Sigma})$, *with* $\text{rank}(\boldsymbol{\Sigma}) = q \le p$ *and* $\{\boldsymbol{A}_n\}$ *be a sequence of random matrices such that* $\boldsymbol{A}_n \xrightarrow{P} \boldsymbol{A}$ *where* $\boldsymbol{A}\boldsymbol{\Sigma}\boldsymbol{A} = \boldsymbol{A}$, *and rank*$(\boldsymbol{A}) = r$. *Then,*

$$Q_n = n(\boldsymbol{T}_n - \boldsymbol{\theta})'\boldsymbol{A}_n(\boldsymbol{T}_n - \boldsymbol{\theta}) \xrightarrow{D} \chi^2_r.$$

Proof. We will prove the result for $\text{rank}(\boldsymbol{\Sigma}) = p$. First, note that since $\boldsymbol{A}_n \xrightarrow{P} \boldsymbol{A}$ it follows that $\boldsymbol{x}'\boldsymbol{A}_n\boldsymbol{x}/\boldsymbol{x}'\boldsymbol{A}\boldsymbol{x} \xrightarrow{P} 1$, for all $\boldsymbol{x} \in \mathbb{R}^p$. Therefore, given $\varepsilon > 0$ we may write

$$P\left(1 - \varepsilon < \frac{\boldsymbol{x}'\boldsymbol{A}_n\boldsymbol{x}}{\boldsymbol{x}'\boldsymbol{A}\boldsymbol{x}} < 1 + \varepsilon\right) \to 1 \quad \text{as} \quad n \to \infty. \tag{7.6.1}$$

Then, let $\lambda_{np} = \sup_{\boldsymbol{x} \neq \boldsymbol{0}} \boldsymbol{x}'\boldsymbol{A}_n\boldsymbol{x}/\boldsymbol{x}'\boldsymbol{A}\boldsymbol{x}$ and observe that given $\varepsilon > 0$, there exists $\boldsymbol{x}_0 \in \mathbb{R}^p$ such that

(i) $\boldsymbol{x}_0'\boldsymbol{A}_n\boldsymbol{x}_0/\boldsymbol{x}_0'\boldsymbol{A}\boldsymbol{x}_0 > \lambda_{np} - \varepsilon \Rightarrow \lambda_{np} < \boldsymbol{x}_0'\boldsymbol{A}_n\boldsymbol{x}_0/\boldsymbol{x}_0'\boldsymbol{A}\boldsymbol{x}_0 + \varepsilon$,
(ii) $\boldsymbol{x}_0'\boldsymbol{A}_n\boldsymbol{x}_0/\boldsymbol{x}_0'\boldsymbol{A}\boldsymbol{x}_0 \le \lambda_{np} \Rightarrow \lambda_{np} \ge \boldsymbol{x}_0'\boldsymbol{A}_n\boldsymbol{x}_0/\boldsymbol{x}_0'\boldsymbol{A}\boldsymbol{x}_0 - \varepsilon$.

Thus, given $\varepsilon > 0$, it follows that

$$\left\{1 - \varepsilon < \frac{\boldsymbol{x}_0'\boldsymbol{A}_n\boldsymbol{x}_0}{\boldsymbol{x}_0'\boldsymbol{A}\boldsymbol{x}_0} < 1 + \varepsilon\right\} \subset \{1 - 2\varepsilon < \lambda_{np} < 1 + 2\varepsilon\},$$

which implies that $\lambda_{np} \xrightarrow{P} 1$ by (7.6.1). Using a similar argument we may show that $\lambda_{n1} = \inf_{\boldsymbol{x} \neq \boldsymbol{0}} \boldsymbol{x}'\boldsymbol{A}_n\boldsymbol{x}/\boldsymbol{x}'\boldsymbol{A}\boldsymbol{x} \xrightarrow{P} 1$. Now use the previous theorem to conclude that

$$Q_n^* = n(\boldsymbol{T}_n - \boldsymbol{\theta})'\boldsymbol{A}(\boldsymbol{T}_n - \boldsymbol{\theta}) \xrightarrow{D} \chi^2_r,$$

and let $Q_n = Q_n^* W_n$, where

$$W_n = (\boldsymbol{T}_n - \boldsymbol{\theta})'\boldsymbol{A}_n(\boldsymbol{T}_n - \boldsymbol{\theta})/(\boldsymbol{T}_n - \boldsymbol{\theta})'\boldsymbol{A}(\boldsymbol{T}_n - \boldsymbol{\theta}).$$

Since $\lambda_{n1} \le W_n \le \lambda_{np}$ we have $W_n \xrightarrow{P} 1$ and the result follows from a direct application of the Slutsky Theorem 7.5.2. ■

***Example* 7.6.1.** Let $\{\boldsymbol{X}_n\}$ be a sequence of i.i.d. random p-vectors with means $\boldsymbol{\mu}$ and p.d. covariance matrices $\boldsymbol{\Sigma}$. We know from the multivariate version of the Classical Central Limit Theorem 7.3.1 that $\sqrt{n}(\overline{\boldsymbol{X}}_n - \boldsymbol{\mu}) \xrightarrow{D} \mathcal{N}(\boldsymbol{0}, \boldsymbol{\Sigma})$; we also know that

$$\boldsymbol{S}_n = (n-1)^{-1}\sum_{i=1}^{n}(\boldsymbol{X}_i - \overline{\boldsymbol{X}}_n)(\boldsymbol{X}_i - \overline{\boldsymbol{X}}_n)' \xrightarrow{P} \boldsymbol{\Sigma}.$$

Thus, from the above result it follows that Hotelling's T^2-statistic, namely,

$$T_n^2 = n(\overline{\boldsymbol{X}}_n - \boldsymbol{\mu}_0)'\boldsymbol{S}_n^{-1}(\overline{\boldsymbol{X}}_n - \boldsymbol{\mu}_0),$$

is such that $T_n^2 \xrightarrow{D} \chi_p^2$ under the hypothesis that $\boldsymbol{\mu} = \boldsymbol{\mu}_0$. □

We now consider a more general version of Theorem 7.6.2.

***Theorem* 7.6.3.** *Let $\{\boldsymbol{T}_n\}$ be a sequence of random p-vectors such that $\sqrt{n}(\boldsymbol{T}_n - \boldsymbol{\theta}) \xrightarrow{D} \mathcal{N}(\boldsymbol{0}, \boldsymbol{\Sigma})$, with rank$(\boldsymbol{\Sigma}) = q \le p$, and let $\{\boldsymbol{A}_n\}$ be a sequence of p.s.d. random matrices such that $\boldsymbol{A}_n \xrightarrow{P} \boldsymbol{A}$. Then,*

$$Q_n = n(\boldsymbol{T}_n - \boldsymbol{\theta})'\boldsymbol{A}_n(\boldsymbol{T}_n - \boldsymbol{\theta}) \xrightarrow{D} \sum_{j=1}^{q} w_j Z_j^2,$$

where the w_j are constants and the Z_j^2 follow independent χ_1^2 distributions.

Proof. Since $\boldsymbol{\Sigma}$ is p.s.d., we can obtain an orthogonal matrix $\boldsymbol{B}$, such that $\boldsymbol{\Sigma} = \boldsymbol{B}\boldsymbol{B}'$ and $\boldsymbol{D}_n = \boldsymbol{B}'\boldsymbol{A}_n\boldsymbol{B} = \text{diag}(w_{n1}, \ldots, w_{np})$, $w_{nj} \ge 0$. Now let $\sqrt{n}(\boldsymbol{T}_n - \boldsymbol{\theta}) = \boldsymbol{B}\boldsymbol{Z}_n$, where $\boldsymbol{Z}_n \xrightarrow{D} \mathcal{N}(\boldsymbol{0}, \boldsymbol{I}_q)$. Then

$$Q_n = \boldsymbol{Z}_n'\boldsymbol{B}'\boldsymbol{A}_n\boldsymbol{B}\boldsymbol{Z}_n = \boldsymbol{Z}'\boldsymbol{D}_n\boldsymbol{Z}_n = \sum_{j=1}^{q} w_{nj} Z_j^2.$$

Since $\boldsymbol{A}_n \xrightarrow{P} \boldsymbol{A}$ then $\boldsymbol{D}_n \xrightarrow{P} \boldsymbol{D}$, where $\boldsymbol{D} = \text{diag}(w_1, \ldots, w_q)$, so that

$$Q_n \xrightarrow{D} \sum_{j=1}^{q} w_j Z_j^2,$$

and the result follows.

If the w_j are 0 or 1 then $\boldsymbol{B}\boldsymbol{A}\boldsymbol{B}$ is idempotent and Q_n follows a χ_r^2 distribution, where $r = \text{rank}(\boldsymbol{B}\boldsymbol{A}\boldsymbol{B})$. ■

***Example* 7.6.2.** Consider a one-way layout under which the random variables X_{ij}, $i = 1, \ldots, p$, $j = 1, \ldots, m$ follow independent distributions with means μ_i and variances σ_i^2.

To test for $H_0 : \mu_1 = \cdots = \mu_p$, we may consider the statistic

$$Q_m = \sum_{i=1}^{p} v_i(\overline{X}_i - \overline{X}_v)^2,$$

where $\overline{X}_i = \sum_{j=1}^{m} X_{ij}/m$, $v_i = \sigma_i^{-2}/\sum_{k=1}^{p}\sigma_k^{-2}$ and $\overline{X}_v = \sum_{i=1}^{p} v_i\overline{X}_i$. When H_0 holds, we expect Q_m to be close to zero.

Letting $\overline{\boldsymbol{X}} = (\overline{X}_1, \ldots, \overline{X}_p)'$, it follows from the CLT that, under H_0, $\sqrt{m}(\overline{\boldsymbol{X}} - \mathbf{1}_p\overline{X}_v) \xrightarrow{D} \mathcal{N}_{p-1}(\mathbf{0}, \boldsymbol{\Sigma})$, where rank$(\boldsymbol{\Sigma}) = p - 1$. Then, we may write

$$Q_m = m(\overline{\boldsymbol{X}} - \mathbf{1}_p\overline{X}_v)'\boldsymbol{A}_m(\overline{\boldsymbol{X}} - \mathbf{1}\overline{X}_v),$$

where $\boldsymbol{A}_m = m^{-1}\text{diag}\{v_1, \ldots, v_p\}$ and using Theorem 7.6.3 we may conclude that the asymptotic distribution of Q_m is a weighted sum of $(p - 1)$ variables following a χ_1^2 distribution. Under homoskedasticity, that it, $\sigma_i^2 = \sigma^2$, $i = 1, \ldots, p$, all the weights are equal, and the distribution of Q_n may be approximated by a chi-squared distribution with $p - 1$ degrees of freedom. □

7.7 Order Statistics and Empirical Distributions

In this section we apply the techniques considered so far to study the asymptotic distributions of empirical distributions, sample quantiles, and order statistics. Most of the results we discuss are intrinsically important because of their usefulness in nonparametric inference, reliability, and life-testing; many are also important, because they serve as paradigms to similar results developed for other types of statistics, like U-, M-, L-, and R-statistics.

Consider a sample of i.i.d. real-valued random variables $X_1, \ldots, X_n$ drawn from the distribution function F. Let F_n be the corresponding empirical distribution function (1.5.82). Note that from (1.5.84) if follows that $\mathbb{V}\text{ar}[F_n(x)] \to 0$ as $n \to \infty$. Then, using the Classical Central Limit Theorem 7.3.1 we may conclude that that for each fixed x

$$\sqrt{n}[F_n(x) - F(x)] \xrightarrow{D} \mathcal{N}\{0,\ F(x)[1 - F(x)]\}. \tag{7.7.1}$$

It is also easy to verify that for every $x \le y$

$$\mathbb{E}[I(X_i \le x)I(X_i \le y)] = F(x) \tag{7.7.2}$$

so that for all $x, y \in \mathbb{R}$

$$\begin{aligned} &\mathbb{E}\{[F_n(x) - F(x)]\}\{F_n(y) - F(y)\} \\ &= \mathbb{C}\text{ov}\{F_n(x), F_n(y)\} \\ &= n^{-1}F[\min(x, y)]\{1 - F[\max(x, y)]\}. \end{aligned} \tag{7.7.3}$$

On the other hand, $F_n - F = \{F_n(x) - F(x)\colon x \in \mathbb{R}\}$ is a *random function* defined on $\mathbb{R}$, and, hence, to study its various properties we may need more than (7.7.1); this topic will be considered partly in the present chapter and partly in Chapter 11.

We now focus our attention on the asymptotic normality of the sample quantiles. The first authors to deal with this topic, like Cramér (1946) or Mosteller (1946), showed the convergence of the density function (1.5.111) to the normal density function. Their proofs were essentially based on taking

$$x = \xi_p + n^{-1/2}t, \quad |t| \le K, \ K > 0,$$

and using the Stirling approximation (1.5.33) in (1.5.111). We will consider a weaker result, which is sufficient for most statistical applications, however.

***Theorem* 7.7.1.** *Let $\{X_1, \ldots, X_n\}$ be a random sample corresponding to a random variable with distribution function F and density function f, continuous at ξ_p, $0 < p < 1$, and such that $f(\xi_p) > 0$; define k in such a way that $k = np + o(n^{1/2})$. Then,*

$$\sqrt{n}(X_{n:k} - \xi_p) \xrightarrow{D} \mathcal{N}[0, p(1-p)/f^2(\xi_p)]. \tag{7.7.4}$$

Proof. Let $Z_n = \sqrt{n}(X_{n:k} - \xi_p)$ and note that

$$\begin{aligned} P(Z_n \le x) &= P\left(X_{n:k} \le \xi_p + n^{-1/2}x\right) \\ &= P\left[\sum_{i=1}^n I(X_i \le \xi_p + n^{-1/2}x) \ge k\right] \\ &= P\left(\frac{1}{\sqrt{n}}\sum_{i=1}^n Y_{ni} \ge t_n\right), \end{aligned} \tag{7.7.5}$$

where

$$Y_{ni} = I(X_i \le \xi_p + n^{-1/2}x) - F(\xi_p + n^{-1/2}x), \quad i = 1, \ldots, n,$$

and

$$t_n = [k - nF(\xi_p + n^{-1/2}x)]/\sqrt{n}.$$

Now, since $Y_{ni} = 1 - F(\xi_p + n^{-1/2}x)$ with probability $F(\xi_p + n^{-1/2}x)$ and $Y_{ni} = -F(\xi_p + n^{-1/2}x)$ with probability $1 - F(\xi_p + n^{-1/2}x)$, and $F(\xi_p + n^{-1/2}x) = F(\xi_p) + n^{-1/2}xf(\xi_p) + o(n^{-1/2})$, it follows that $\mathbb{E}(Y_{ni}) = 0$, $\mathbb{E}(Y_{ni}^2) = p(1-p) + O(n^{-1/2})$ and that

$$t_n = \frac{1}{\sqrt{n}}[k - nF(\xi_p) - n^{1/2}xf(\xi_p) + o(n^{-1/2})] = -xf(\xi_p) + o(1).$$

Therefore, using the Central Limit Theorem 7.3.4 and the Slutsky Theorem 7.5.2 in (7.7.5), we may conclude that as $n \to \infty$

$$\begin{aligned} P(Z_n \le x) &= P\left[\frac{1}{\sqrt{np(1-p)}}\sum_{i=1}^n Y_{ni} \ge \frac{t_n}{\sqrt{p(1-p)}}\right] \\ &\to 1 - \Phi\left[\frac{-xf(\xi_p)}{\sqrt{p(1-p)}}\right] = \Phi\left[\frac{xf(\xi_p)}{\sqrt{p(1-p)}}\right], \end{aligned} \tag{7.7.6}$$

which implies that as $n \to \infty$

$$P\left[\frac{f(\xi_p)}{\sqrt{p(1-p)}}\sqrt{n}(X_{n:k} - \xi_p) \le x\right] \to \Phi(x), \tag{7.7.7}$$

proving the result. ■

The next theorem generalizes the preceding one by examining the asymptotic joint distribution of two sample quantiles.

Theorem 7.7.2. *Let $\{X_1, \ldots, X_n\}$ be a random sample corresponding to a random variable with distribution function F and density function f continuous at ξ_{p_j}, $j = 1, 2$, $0 < p_1 < p_2 < 1$, and such that $f(\xi_{p_j}) > 0$, $j = 1, 2$; define k_j in such a way that $k_j = np_j + o(n^{1/2})$, $j = 1, 2$. Then,*

$$\sqrt{n}(X_{n:k_1} - \xi_{p_1}, X_{n:k_2} - \xi_{p_2})' \xrightarrow{D} \mathcal{N}_2(\mathbf{0}, \boldsymbol{\Sigma})$$

where $\boldsymbol{\Sigma} = ((\sigma_{ij}))$ with $\sigma_{11} = p_1(1-p_1)/f^2(\xi_{p_1})$, $\sigma_{12} = \sigma_{21} = p_1(1-p_2)/[f(\xi_{p_1})f(\xi_{p_2})]$ and $\sigma_{22} = p_2(1-p_2)/f^2(\xi_{p_2})$.

Proof. Let $\boldsymbol{Z}_n = (Z_{n1}, Z_{n2})'$ where $Z_{nj} = \sqrt{n}(X_{n:k_j} - \xi_{p_j})$, $j = 1, 2$, and $\boldsymbol{x} = (x_1, x_2)'$, and use arguments similar to the ones considered in the previous theorem to see that

$$P(\boldsymbol{Z}_n \le \boldsymbol{x}) = P\left(\frac{1}{\sqrt{n}}\sum_{i=1}^{n} \boldsymbol{Y}_{ni} \ge \boldsymbol{t}_n\right), \tag{7.7.8}$$

where $\boldsymbol{Y}_{ni} = (Y_{ni1}, Y_{ni2})'$ with

$$Y_{nij} = I(X_i \le \xi_{pj} + n^{-1/2}x_j) - F(\xi_{p_j} + n^{-1/2}x_j), \quad j = 1, 2,$$

and $\boldsymbol{t}_n = (t_{n1}, t_{n2})'$ with

$$t_{nj} = [k_j - nF(\xi_{p_j} + n^{-1/2}x_j)]/\sqrt{n}, \quad j = 1, 2.$$

Also, note that $\mathbb{E}(Y_{nij}) = 0$ and $\mathbb{E}(Y_{nij}^2) = p_j(1-p_j) + O(n^{-1/2})$, $j = 1, 2$. Furthermore, since we are assuming $p_1 < p_2$, it follows that

$$\begin{aligned}
\mathbb{E}(Y_{ni1}Y_{ni2}) &= \mathbb{E}[I(X_i \le \xi_{p_1} + n^{-1/2}x_1)I(X_i \le \xi_{p_2} + n^{-1/2}x_2)] \\
&\quad - F(\xi_{p_1} + n^{-1/2}x_1)F(\xi_{p_2} + n^{-1/2}x_2) \\
&= F(\xi_{p_1} + n^{-1/2}x_1) - F(\xi_{p_1} + n^{-1/2}x_1)F(\xi_{p_2} + n^{-1/2}x_2) \\
&= p_1(1-p_2) + O(n^{-1/2}).
\end{aligned}$$

Thus, putting

$$\boldsymbol{\Gamma} = \begin{bmatrix} p_1(1-p_1) & p_1(1-p_2) \\ p_1(1-p_2) & p_2(1-p_2) \end{bmatrix}$$

we may write $\mathbb{V}\text{ar}(\boldsymbol{Y}_{ni}) = \boldsymbol{\Gamma} + \boldsymbol{\Delta}_n$ where $\boldsymbol{\Delta}_n = O(\boldsymbol{n}^{-1/2})$. Finally, observing that

$$\boldsymbol{t}_n = -\boldsymbol{D}\boldsymbol{x} + o(\boldsymbol{1}),$$

where $\boldsymbol{D} = \text{diag}\{f(\xi_{p_1}), f(\xi_{p_2})\}$, we may apply the multivariate versions of the Central Limit Theorem 7.3.1 and the Slutsky Theorem 7.5.2 to (7.7.8), to conclude that as $n \to \infty$

$$P(\boldsymbol{Z}_n \leq \boldsymbol{x}) = P\left(n^{-1/2}\boldsymbol{\Gamma}^{-1/2}\sum_{i=1}^{n}\boldsymbol{Y}_{ni} \geq \boldsymbol{\Gamma}^{-1/2}\boldsymbol{t}_n\right)$$
$$\to 1 - \Phi_2(-\boldsymbol{\Gamma}^{-1/2}\boldsymbol{D}\boldsymbol{x}) = \Phi_2(\boldsymbol{\Gamma}^{-1/2}\boldsymbol{D}\boldsymbol{x}),$$

where Φ_2 denotes the bivariate standard normal distribution function. Therefore, we may conclude that

$$\boldsymbol{Z}_n \xrightarrow{D} \mathcal{N}_2(\boldsymbol{0}, \boldsymbol{D}^{-1}\boldsymbol{\Gamma}\boldsymbol{D}^{-1}), \qquad (7.7.9)$$

and since $\boldsymbol{\Sigma} = \boldsymbol{D}^{-1}\boldsymbol{\Gamma}\boldsymbol{D}^{-1}$, the proof is completed. ■

Note that this result may be easily extended to cover the joint asymptotic distribution of $q \geq 2$ sample quantiles. In this direction, let $0 < p_1 < \cdots < p_q < 1$, $\boldsymbol{Z}_n = (Z_{n1}, \ldots, Z_{nq})'$, where $Z_{nj} = \sqrt{n}(X_{n:k_1} - \xi_{p_j})$, $j = 1, \ldots, q$ and $\boldsymbol{x} = (x_1, \ldots, x_q)'$. Then, (7.7.9) holds with $\boldsymbol{D} = \text{diag}\{f(\xi_{p_1}), \ldots, f(\xi_{p_q})\}$ and $\boldsymbol{\Gamma} = ((\gamma_{ij}))$ with $\gamma_{ij} = \min(p_i, p_j) - p_i p_j, i, j = 1, \ldots, q$. An extension to the case of sample quantiles of multivariate distributions was considered by Mood (1941) using elaborate analysis; the method of proof outlined in Theorem 7.7.1 is much simpler and goes through without much change. Without loss of generality we state the result for bivariate distributions.

Theorem 7.7.3. *Let $\{X_1, \ldots, X_n\}$ be a random sample corresponding to a bivariate random variable $\boldsymbol{X} = (X^{(1)}, X^{(2)})'$; let F, F_1 and F_2 respectively denote the joint and marginal distribution functions of $\boldsymbol{X}$ and assume that the corresponding joint and marginal density functions, f, f_1, and f_2 are such that f_j is continuous at $\xi_{p_j}^{(j)}$ and $f_j(\xi_{p_j}^{(j)}) > 0$, $0 < p_j < 1$, $j = 1, 2$. Define k_j in such a way that $k_{p_j} = np_j + o(n^{1/2})$, $j = 1, 2$. Then*

$$\sqrt{n}\big(X_{n:k_1}^{(1)} - \xi_{p_1}^{(1)}, X_{n:k_2}^{(2)} - \xi_{p_2}^{(2)}\big)' \xrightarrow{D} \mathcal{N}(\boldsymbol{0}, \boldsymbol{\Sigma}),$$

where $\boldsymbol{\Sigma} = ((\sigma_{ij}))$ with $\sigma_{11} = p_1(1 - p_1)/f_1^2(\xi_{p_1}^{(1)})$, $\sigma_{22} = p_2(1 - p_2)/f_2^2(\xi_{p_2}^{(2)})$ and $\sigma_{12} = [F(\xi_{p_1}^{(1)}, \xi_{p_2}^{(2)}) - p_1 p_2]/[f_1(\xi_{p_1}^{(1)})f_2(\xi_{p_2}^{(2)})]$.

Proof. Let $\boldsymbol{Z}_n = (Z_{n1}, Z_{n2})'$ with $Z_{nj} = \sqrt{n}(X_{n:k_j}^{(j)} - \xi_{p_j}^{(j)})$, $j = 1, 2$, and note that

$$P(\boldsymbol{Z}_n \leq \boldsymbol{x}) = P\left(\frac{1}{\sqrt{n}}\sum_{i=1}^{n}\boldsymbol{Y}_{ni} \geq \boldsymbol{t}_n\right),$$

where $\boldsymbol{Y}_{ni} = (Y_{ni1}, Y_{ni2})'$, with

$$Y_{nij} = I\big(X_i^{(j)} \leq \xi_{p_j}^{(j)} + n^{-1/2}x_j\big) - F_j\big(\xi_{p_j}^{(j)} + n^{-1/2}x_j\big), \quad j = 1, 2,$$

and $\boldsymbol{t}_n = (t_{n1}, t_{n2})'$ with

$$t_{nj} = n^{-1/2}\big[k_j - nF_j\big(\xi_{p_j}^{(j)} + n^{-1/2}x_j\big)\big], \quad j = 1, 2.$$

Now, recalling that for $a, b \in \mathbb{R}$

$$\mathbb{E}\big[I(X_i^{(1)} \leq a)I(X_i^{(2)} \leq b)\big] = P\big(X_i^{(1)} \leq a, X_i^{(2)} \leq b\big) = F(a, b),$$

it follows that

$$\begin{aligned}\mathbb{E}(Y_{ni1}Y_{ni2}) &= \mathbb{E}\big[I(X_i^{(1)} \le \xi_{p_1}^{(1)} + n^{-1/2}x_1)I\big(X_i^{(2)} \le \xi_{p_2} + n^{-1/2}x_2\big)\big] \\ &\quad F_1\big(\xi_{p_1}^{(1)} + n^{-1/2}x_1\big)F_2\big(\xi_{p_2}^{(2)} + n^{-1/2}x_2\big) \\ &= F\big(\xi_{p_1}^{(1)}, \xi_{p_2}^{(2)}\big) - p_1p_2 + O(n^{-1/2}).\end{aligned}$$

The rest of the proof follows the same arguments as those in Theorem 7.7.2. ■

A multivariate generalization of this theorem is straightforward.

An interesting application of the above results is in the estimation of the parameters of the *location-scale* families of distributions. Consider a random sample $X_1, \ldots, X_n$ corresponding to a random variable X with distribution function F and density function f such that $f(y) = \lambda^{-1} f_0[(y - \mu)/\lambda]$ where $\mu \in \mathbb{R}$ and $\lambda > 0$ are unknown location and scale parameters, respectively, f_0 is a known density function, free of μ and λ and F_0 is the corresponding distribution function In many cases the solution to the likelihood equations is difficult to obtain and alternative methods to estimate μ and λ must be employed. In this direction let $Y_i = (X_i - \mu)/\lambda$, $i = 1, \ldots, n$, and note that the density function of Y_i is f_0. Now consider the order statistics $X_{n:1} \le X_{n:2} \le \cdots \le X_{n:n}$, which imply $X_{n:i} = \mu + \lambda Y_{n:i}$, $i = 1, \ldots, n$, where $Y_{n:1}$ come from the known density f_o, free from μ and λ. Let $\boldsymbol{\xi}_n^0 = \mathbb{E}(\boldsymbol{Y}_n)$, where $\boldsymbol{Y}_{(n)} = (Y_{n:1}, \ldots, Y_{n:n})'$ and $\Gamma_{n0} = \mathbb{C}\mathrm{ov}(\boldsymbol{Y}_{(n)}) = ((\gamma_{n0ij}))$. The quantities in $\boldsymbol{\xi}_n^0$ and (more pressingly) in $\boldsymbol{\Gamma}_{n0}$ though known for a given f_0, can be computationally intensive. As such, we may use $\boldsymbol{\xi}_n$, with $\xi_{ni} = F_0^{-1}[i/(n+1)]$ and $\boldsymbol{\Gamma}_n$ with

$$n\gamma_{nij} = \frac{i(n+1-j)}{(n+1)^2} \frac{1}{f_0(\xi_{ni})f_0(\xi_{nj})}$$

for $i \le j$, providing good asymptotic approximations. Let then $\boldsymbol{W}_n = (\mathbf{1}_n, \boldsymbol{\xi}_n)$ and $\boldsymbol{W}_n^0 = (\mathbf{1}_n, \boldsymbol{\xi}_n^0)$. Then, if $\boldsymbol{\xi}_n^0$ and $\boldsymbol{\Gamma}_{n0}$ are given, we construct

$$Q_n^0(\boldsymbol{\theta}) = \big(\boldsymbol{X}_n - \boldsymbol{W}_n^0\boldsymbol{\theta}\big)'\boldsymbol{\Gamma}_{n0}^{-1}\big(\boldsymbol{X}_n - \boldsymbol{W}_n^0\boldsymbol{\theta}\big).$$

In a more likely case, we use the asymptotic norm

$$Q_n(\boldsymbol{\theta}) = (\boldsymbol{X}_n - \boldsymbol{W}_n\boldsymbol{\theta})'\boldsymbol{\Gamma}_n^{-1}(\boldsymbol{X}_n - \boldsymbol{W}_n\boldsymbol{\theta}), \tag{7.7.10}$$

yielding

$$\widehat{\boldsymbol{\theta}}_n = (\boldsymbol{W}_n'\boldsymbol{\Gamma}_n^{-1}\boldsymbol{W}_n)^{-1}\boldsymbol{W}_n'\boldsymbol{\Gamma}_n^{-1}\boldsymbol{X}_n \tag{7.7.11}$$

with covariance matrix

$$\mathbb{V}\mathrm{ar}(\widehat{\boldsymbol{\theta}}_n) = \lambda^2(\boldsymbol{W}_n'\boldsymbol{\Gamma}_n^{-1}\boldsymbol{W}_n)^{-1}. \tag{7.7.12}$$

Now, since

$$\boldsymbol{W}_n'\boldsymbol{\Gamma}_n^{-1}\boldsymbol{W}_n = \begin{pmatrix}\mathbf{1}_n'\boldsymbol{\Gamma}_n^{-1}\mathbf{1}_n & \mathbf{1}_n'\boldsymbol{\Gamma}_n^{-1}\boldsymbol{\xi}_n \\ \boldsymbol{\xi}_n'\boldsymbol{\Gamma}_n^{-1}\mathbf{1}_n & \boldsymbol{\xi}_n'\boldsymbol{\Gamma}_n^{-1}\boldsymbol{\xi}_n\end{pmatrix} \tag{7.7.13}$$

it follows from (7.7.11) that $\widehat{\mu}_n = -\boldsymbol{\xi}_n'\boldsymbol{a}_n\boldsymbol{X}_n$ and $\widehat{\lambda}_n = \mathbf{1}_n'\boldsymbol{a}_n\boldsymbol{X}_n$ where $\boldsymbol{a}_n = \Delta_n^{-1}[\boldsymbol{\Gamma}_n^{-1}(\mathbf{1}_n\boldsymbol{\xi}_n' - \boldsymbol{\xi}_n\mathbf{1}_n')\boldsymbol{\Gamma}_n^{-1}]$ and $\Delta_n = |\boldsymbol{W}_n'\boldsymbol{\Gamma}_n^{-1}\boldsymbol{W}_n|$, or, in other words, that $\widehat{\mu}_n$ and $\widehat{\lambda}_n$ are linear combinations of the order statistics $\boldsymbol{X}_n$, with coefficients depending only on the known function f_0. Also, from the generalized version of the Gauss–Markov Theorem confined to the class of linear estimators based on order statistics, $\widehat{\mu}_n$ and $\widehat{\lambda}_n$ are asymptotically *best linear unbiased estimators* (ABLUE) of μ and λ, respectively. This ABLUE property is retained by the convergence of the mean vector $\boldsymbol{\xi}_n$ and the covariance matrix $\boldsymbol{\Gamma}_n$ to their respective exact forms. From (7.7.12) and (7.7.13) it follows that

$$\mathbb{V}\text{ar}(\widehat{\mu}_n) = \lambda^2 \Delta_n^{-1} \boldsymbol{\xi}_n' \boldsymbol{\Gamma}_n^{-1} \boldsymbol{\xi}_n,$$

$$\mathbb{V}\text{ar}(\widehat{\lambda}_n) = \lambda^2 \Delta_n^{-1} \mathbf{1}_n' \boldsymbol{\Gamma}_n^{-1} \mathbf{1}_n,$$

and

$$\mathbb{C}\text{ov}(\widehat{\mu}_n, \widehat{\lambda}_n) = -\lambda^2 \Delta_n^{-1} \mathbf{1}_n' \boldsymbol{\Gamma}_n^{-1} \boldsymbol{\xi}_n.$$

For further details, see Chapter 10.

The same approach may be employed with censored data, that is, when only a subset of the order statistics is available. Suppose that the only order statistics available correspond to $\boldsymbol{X}_n(J) = (X_{n:j_1}, \ldots, X_{n:j_m})'$, where $J = \{j_1, \ldots, j_m\} \subseteq \{1, \ldots, n\}$; then, defining $\boldsymbol{\xi}_n(J) = (\xi_{p_{j_1}}, \ldots, \xi_{p_{j_m}})'$ and $\boldsymbol{\Gamma}_n(J) = ((\gamma_{j_k j_l}))$ with $\gamma_{j_k j_l} = n\mathbb{C}\text{ov}(X_{n:j_k}, X_{n:j_l})$, $j_k, j_l \in J$, the procedure consists of minimizing

$$Q_n(\boldsymbol{\theta}, J) = [\boldsymbol{X}_n(J) - \boldsymbol{W}_n(J)\boldsymbol{\theta}]' \boldsymbol{\Gamma}_n^{-1}(J)[\boldsymbol{X}_n(J) - \boldsymbol{W}_n(J)\boldsymbol{\theta}],$$

where $\boldsymbol{W}_n(J) = (\mathbf{1}_{\#J}, \ \boldsymbol{\xi}_n(J))$ and $\#J$ denotes the cardinality of J. A nice account of these topics is given in Sarhan and Greenberg (1962) or David (1970), for example. We will consider some related results in Chapter 8.

We now turn our attention to the asymptotic distribution of extreme order statistics. The basis for the related asymptotic distribution theory is laid down by the following theorem.

Theorem 7.7.4. *Let $\{X_1, \ldots, X_n\}$ be a random sample corresponding to a random variable with a continuous distribution function F and let $X_{n:1} < \cdots < X_{n:n}$ denote the set of associated order statistics. Also let $V_n = n[1 - F(X_{n:n})]$. Then, $V_n \xrightarrow{D} V$, where $V \sim$* Exp(1).

Proof. Since $Y_i = F(X_i) \sim \text{Unif}(0, 1)$, it follows that for $v \geq 0$

$$\begin{aligned} G_n(v) = P(V_n \leq v) &= P(n[1 - F(X_{n:n})] \leq v) = P(1 - Y_{n:n} \leq v/n) \\ &= P(Y_{n:n} \geq 1 - v/n) = 1 - P(Y_{n:n} \leq 1 - v/n) \\ &= 1 - (1 - v/n)^n \to 1 - e^{-v} \quad \text{as} \quad n \to \infty. \end{aligned}$$

■

Our next task is to determine conditions on the underlying distribution function F under which $X_{n:n}$ can be normalized by sequences of constants $\{a_n\}$ and $\{b_n\}$ so that $(X_{n:n} - a_n)/b_n$ converges to some nondegenerate distribution. In this direction we first consider the following definitions:

Terminal contact of order m. Let X be a random variable with distribution function F and a finite upper end point ξ_1. We say that F has a terminal contact of order m at ξ_1 if $1 - F(x)$ and the left-hand derivatives $F^{(j)}(x)$, $j = 1, \ldots, m$, vanish at ξ_1, whereas $F^{(m+1)}(\xi_1) \neq 0$.

Exponential type distribution. Let X be a random variable with distribution function F with an infinite upper end point and such that $F^{(j)}(x)$, $j = 1, 2, \ldots,$ exists. We say that F is of the exponential type if for large x

$$-\frac{F^{(1)}(x)}{1 - F(x)} \cong \frac{F^{(2)}(x)}{F^{(1)}(x)} \cong \frac{F^{(3)}(x)}{F^{(2)}(x)} \cong \cdots .$$

Cauchy type distribution. Let X be a random variable with distribution function F. We say that F is of the Cauchy type if for some $k > 0$ and $c > 0$, we have

$$x^k[1 - F(x)] \to c \quad \text{as} \quad x \to \infty.$$

Distributions of the exponential type have finite moments of all orders and include those commonly employed in statistical methods, such as the normal, exponential, or gamma distributions. Cauchy-type distributions, on the other hand, have no finite moments of order $\geq k$ and are named after the typical member, the Cauchy distribution, for which $k = 1$. As we will see in the sequel, the asymptotic behavior of sample minima or maxima depends on the type of the underlying distribution.

Theorem 7.7.5. *Let $\{X_1, \ldots, X_n\}$ be a random sample corresponding to a random variable X with distribution function F having an mth order terminal contact at the upper end point ξ_1. Then there exist sequences of constants $\{a_n\}$ and $\{b_n\}$, such that as $n \to \infty$,*

$$P[(X_{n:n} - \xi_1)/b_n \leq t] \to \begin{cases} \exp[-(-t)^{m+1}], & t \leq 0 \\ 1, & t > 0. \end{cases}$$

Proof. Consider the following Taylor expansion:

$$\begin{aligned} F(\xi_1 - s) &= F(\xi_1) - sF^{(1)}(\xi_1) + \cdots + \frac{(-1)^m}{m!} s^m F^{(m)}(\xi_1) \\ &\quad + \frac{(-1)^{m+1}}{(m+1)!} s^{m+1} F^{(m+1)}(\xi_1 - \theta s) \\ &= F(\xi_1) + \frac{(-1)^{m+1}}{(m+1)!} s^{m+1} F^{(m+1)}(\xi_1 - \theta s) \end{aligned} \tag{7.7.14}$$

for some $0 < \theta < 1$. Then, note that

$$\begin{aligned} V_n = n[1 - F(X_{n:n})] &= n[1 - F(\xi_1)] + n[F(\xi_1) - F(X_{n:n})] \\ &= n[F(\xi_1) - F(X_{n:n})] \end{aligned}$$

and let $s = \xi_1 - X_{n:n}$ in (7.7.14) to see that

$$\begin{aligned} V_n &= \frac{(-1)^m}{(m+1)!} n F^{(m+1)}(\xi_1)(\xi_1 - X_{n:n})^{m+1} \frac{F^{(m+1)}[(1-\theta)\xi_1 + \theta X_{n:n}]}{F^{(m+1)}(\xi_1)} \\ &= \left(\frac{\xi_1 - X_{n:n}}{b_n}\right)^{m+1} W_n = \left[-\left(\frac{X_{n:n} - \xi_1}{b_n}\right)\right]^{m+1} W_n, \end{aligned} \tag{7.7.15}$$

where

$$b_n = \left[(-1)^m (m+1)!/nF^{(m+1)}(\xi_1)\right]^{\frac{1}{m+1}},$$

and

$$W_n = F^{(m+1)}[(1-\theta)\xi_1 + \theta X_{n:n}]/F^{(m+1)}(\xi_1) \xrightarrow{P} 1.$$

Therefore, from (7.7.15) and the Slutsky Theorem 7.5.2, it follows that the asymptotic distribution of $\{-[(X_{n:n}-\xi_1)/b_n]\}^{m+1}$ is the same as that of V_n. Finally, using Theorem 7.7.4 we may conclude that, as $n \to \infty$

$$P\left(\frac{X_{n:n}-\xi_1}{b_n} \le t\right) = P\left\{\left[-\frac{X_{n:n}-\xi_1}{b_n}\right]^{m+1} \ge (-t)^{m+1}\right\}$$

so that

$$P\left(\frac{X_{n:n}-\xi_1}{b_n}\right) \to \begin{cases} \exp[-(-t)^{m+1}], & \text{if } t \le 0 \\ 1, & \text{if } t > 0. \end{cases} \tag{7.7.16}$$

■

Note that (7.7.16) is known as the *extreme value distribution of the first type.*

***Example* 7.7.1.** Let $\{X_1, \ldots, X_n\}$ be a random sample corresponding to a random variable with the Uniform$(0, \theta)$ distribution. Note that since $F(x) = x/\theta, 0 \le x \le \theta$, and $F^{(1)}(x) = \theta^{-1}, 0 \le x \le \theta$, it follows that F has terminal contact of order $m = 0$ and in view of Theorem 7.7.5, we may write

$$P\left[\frac{n}{\theta}(X_{n:n}-\theta) \le t\right] \to e^t, \quad t \le 0 \quad \text{as} \quad n \to \infty.$$

It can be shown that:

(i) $\mathbb{E}(X_{n:n}) = n\theta/(n+1)$,

(ii) $\mathbb{V}\text{ar}(X_{n:n}) = \theta^2 n/[(n+2)(n+1)^2]$.

Although the sample maximum $X_{n:n}$ is a biased estimate of θ, its variance is smaller than that of the sample median, $X_{n:k}$ with $k = [n/2]+1$, an alternative estimate. In this direction, recall that from Theorem 7.7.1 we have

$$\sqrt{n}\left(X_{n:k} - \frac{\theta}{2}\right) \xrightarrow{D} \mathcal{N}(0, \theta^2/4),$$

which implies that for large n, $n\mathbb{V}\text{ar}(X_{n:k}) \sim \theta^2/4n$. □

***Example* 7.7.2.** Let $\{X_1, \ldots, X_n\}$ be a random sample corresponding to a random variable with the Triangular$(0, \theta)$ distribution. Then,

$$F(x) = \begin{cases} 2x^2/\theta^2, & 0 \le x \le \theta/2 \\ 4x/\theta - 2x^2/\theta^2 - 1, & \theta/2 \le x \le \theta, \end{cases}$$

$$F^{(1)}(x) = \begin{cases} 4x/\theta^2, & 0 \le x \le \theta/2 \\ 4/\theta - 4x/\theta^2, & \theta/2 \le x \le \theta \end{cases},$$

which imply $F^{(1)}(\theta) = 0$ and $F^{(2)}(\theta) = -4/\theta^2$ so that F has terminal contact of order $m = 1$. In this case $b_n = \theta/\sqrt{2n}$ and $\mathbb{V}\text{ar}(X_{n:n}) = \theta^2/2n$, which is of the same order as that of the variance of the corresponding sample median. We note that for $m \ge 2$ the rate of convergence of the distribution of the (standardized) sample maximum becomes slow and in such cases this statistic is of little practical interest. □

Theorem 7.7.6. *Let $\{X_1, \ldots, X_n\}$ be a random sample corresponding to a random variable with distribution function F of the Cauchy type; that is, such that as $x \to \infty$, $x^k[1 - F(x)] \to c$, for some $k > 0$ and $c > 0$; also let ξ_{1n}^* denote the characteristic largest observation of F. Then, as $n \to \infty$*

$$P\left(\frac{X_{n:n}}{\xi_{1n}^*} \le t\right) \to \exp(-t^{-k}), \quad t \ge 0.$$

Proof. Since $1 - F(\xi_{1n}^*) = n^{-1}$ we may write

$$\begin{aligned} V_n = n\{1 - F(X_{n:n})\} &= \frac{1 - F(X_{n:n})}{1 - F(\xi_{1n}^*)} \\ &= \left(\frac{\xi_{1n}^*}{X_{n:n}}\right)^k \left[\frac{X_{n:n}^k[1 - F(X_{n:n})]}{\xi_{1n}^{*\,k}[1 - F(\xi_{1n}^*)]}\right]. \end{aligned}$$

Observing that the term within the braces, { }, converges in probability to 1 and using the Slutsky Theorem 7.5.2, it follows that the asymptotic distribution of $(\xi_{1n}^*/X_{n:n})^k$ is the same as that of V_n. Thus, from Theorem 7.7.4, as $n \to \infty$,

$$P\left(\frac{X_{n:n}}{\xi_{1n}^*} \le t\right) = P\left[\left(\frac{\xi_{1n}^*}{X_{n:n}}\right)^k \ge t^{-k}\right] \to \exp(-t^{-k}), \quad t \ge 0. \tag{7.7.17}$$

■

The distribution in (7.7.17) is known as the *extreme value distribution of the second type*.

Theorem 7.7.7. *Let $\{X_1, \ldots, X_n\}$ be a random sample corresponding to random variable with distribution function F of the exponential type; also let ξ_{1n}^* denote the characteristic largest observation of F. Then there exists a sequence of constants $\{b_n\}$, such that, as $n \to \infty$,*

$$P\left(\frac{X_{n:n} - \xi_{1n}^*}{b_n} \le t\right) \to \exp[-\exp(-t)], \quad t \in \mathbb{R}. \tag{7.7.18}$$

Proof. Let $f(x) = F^{(1)}(x)$, $\gamma_n = nf(\xi_{1n}^*)$ and $X_{n:n} = \xi_{1n}^* + h/\gamma_n$, for some $h \in \mathbb{R}$. Then observe that

$$\begin{aligned} V_n &= n[1 - F(X_{n:n})] \\ &= n[1 - F(\xi_{1n}^*)] - n[F(X_{n:n}) - F(\xi_{1n}^*)] \\ &= 1 - n[F(X_{n:n}) - F(\xi_{1n}^*)] \end{aligned} \tag{7.7.19}$$

and consider the following Taylor expansion:

$$\begin{aligned} F(X_{n:n}) &= F(\xi_{1n}^*) + (X_{n:n} - \xi_{1n}^*)f(\xi_{1n}^*) \\ &\quad + \frac{1}{2!}(X_{n:n} - \xi_{1n}^*)^2 F^{(2)}(\xi_{1n}^*) + \cdots . \end{aligned} \tag{7.7.20}$$

Substituting (7.7.19) into (7.7.20) we obtain

$$V_n = 1 - n(X_{n:n} - \xi_{1n}^*)f(\xi_{1n}^*) - \frac{n}{2!}(X_{n:n} - \xi_{1n}^*)^2 F^{(2)}(\xi_{1n}^*) - \cdots . \tag{7.7.21}$$

Recalling that $1 - F(\xi_{1n}^*) = n^{-1}$, the typical term of (7.7.21) is given by

$$\begin{aligned}
\frac{n}{k!}(X_{n:n} - \xi_{1n}^*)^k F^{(k-1)}(\xi_{1n}^+) &= \frac{n}{k!}\frac{h^k}{\gamma_n^k} F^{(k-1)}(\xi_{1n}^*) \\
&= \frac{h^k}{k!}\frac{n}{[nf(\xi_{1n}^*)]^k}\frac{F^{(k-1)}(\xi_{1n}^*)}{F^{(k-2)}(\xi_{1n}^*)}\frac{F^{(k-2)}(\xi_{1n}^*)}{F^{(k-3)}(\xi_{1n}^*)}\cdots\frac{F^2(\xi_{1n}^*)}{f(\xi_{1n}^*)}f(\xi_{1n}^*) \\
&= \frac{h^k}{k!}\Bigg\{\left[\frac{1 - F(\xi_{1n}^*)}{f(\xi_{1n}^*)}\right]^k \\
&\quad \frac{F^{(k-1)}(\xi_{1n}^*)}{F^{(k-2)}(\xi_{1n}^*)}\frac{F^{(k-2)}(\xi_{1n}^*)}{F^{(k-3)}(\xi_{1n}^*)}\cdots\frac{F^2(\xi_{1n}^*)}{f(\xi_{1n}^*)}\frac{f(\xi_{1n}^*)}{[1 - F(\xi_{1n}^*)]}\Bigg\} \\
&\cong (-1)^k\frac{h^k}{k!},
\end{aligned}$$

since the assumption that F is of the exponential type implies that we may approximate the term within the braces, { }, by $(-1)^k$ for large n. Then, from (7.7.21) we get

$$V_n = 1 - h + \frac{h^2}{2!} - \frac{h^3}{3!} + \cdots = \exp(-h) = \exp\{-\gamma_n(X_{n:n} - \xi_{1n}^*)\}$$

and taking $b_n = \gamma_n^{-1}$, we may recal, Theorem 7.7.4 to write, for all $t \in \mathbb{R}$,

$$\begin{aligned}
P\left(\frac{X_{n:n} - \xi_{1n}^*}{b_n} \le t\right) &= P\{\exp[-\gamma_n(X_{n:n} - \xi_{1n}^*)] \ge \exp(-t)\} \\
&\cong P[V_n \ge \exp(-t)] \to \exp(-e^{(-t)})
\end{aligned} \tag{7.7.22}$$

as $n \to \infty$. ■

The distribution in (7.7.22) is known as the *extreme value distribution of the third type*.

The above results extend directly to the case of the sample minimum $X_{n:1}$. More specifically, consider a random sample $\{X_1, \ldots, X_n\}$ corresponding to a random variable X having distribution function F; then

(i) if F has an mth order terminal contact at the lower end point ξ_0, it follows that, as $n \to \infty$,

$$P[(X_{n:1} - \xi_0)/b_n \le t] \to \begin{cases} 1 - \exp(-t)^{m+1}, & t \ge 0 \\ 0, & t < 0 \end{cases}$$

where $b_n = [(m+1)!/nF^{(m+1)}(\xi_0)]^{\frac{1}{m+1}}$;

(ii) if F is of the Cauchy type, it follows that , as $n \to \infty$,

$$P(X_{n:1}/\xi_{0n}^* \le t) \to 1 - \exp(-t^{-k}), \quad t \in \mathbb{R}; \quad \text{and}$$

(iii) if F is of the exponential type, it follows that, as $n \to \infty$,

$$P[(X_{n:1} - \xi_{0n}^*)/b_n \le t] \to 1 - \exp[-\exp(t)], \quad t \in \mathbb{R},$$

where $b_n = [nf(\xi_{0n}^*)]^{-1}$.

Gnedenko (1943) showed that the limiting distributions presented above are the only possible limiting distributions (domains of attraction) for the sample extreme values.

The class of exponential type distributions is particularly important since its members are commonly employed for practical purposes. If the interest lies in approximating the largest (smallest) characteristic observation based on a (large) sample of size n, Theorem 7.7.7 allows us to obtain limits I_α and U_α such that

$$P\left(\frac{I_\alpha}{\gamma_n} \leq X_{n:n} - \xi_{1n}^* \leq \frac{U_\alpha}{\gamma_n}\right) \cong 1 - \alpha, \tag{7.7.23}$$

where $\gamma_n = nf(\xi_{1n}^*)$ is known as the *extremal intensity function* (extremal failure rate). Note that for practical applications, (7.7.23) is useful only if $\gamma_n \to \infty$ as $n \to \infty$, since, in such a case, $X_{n:n} - \xi_{1n}^* \xrightarrow{P} 0$. In view of this fact, the members of the exponential class of distributions may be classified into three categories according to the behavior of the extremal intensity function:

(i) *convex exponential type*, when $\gamma_n \to \infty$ as $n \to \infty$ (e.g., normal distribution);
(ii) *simple exponential type*, when $\gamma_n \to c > 0$ as $n \to \infty$ (e.g., exponential or double exponential distributions); and
(iii) *concave exponential type*, when $\gamma_n \to 0$ as $n \to \infty$ (e.g., Laplace or logistic distributions).

For simple or concave exponential type distributions, the extreme observations do not provide much information for estimation of the extreme characteristic observations and may be discarded with little loss.

***Example* 7.7.3.** Consider the standard normal distribution and note that

$$\gamma_n = n\varphi(\xi_{1n}^*) = \varphi(\xi_{1n}^*)/[1 - \Phi(\xi_{1n}^*)] \approx \xi_{1n}^* \approx \sqrt{2\log n}$$

for large n, in view of Example 6.3.13 Now, using Theorems 7.7.5 and 7.5.5, it follows that $\gamma_n(X_{n:n} - \xi_{1n}^*) = O_p(1)$, which, in turn, implies that $X_{n:n} - \xi_{1n}^* = O_p[(\log n)^{-1/2}]$. Although the convergence is slow when compared to the pth-quantile case $(0 < p < 1)$, where $X_{n:k_p} - \xi_p = O_p(n^{-1/2})$, we may still use the information contained in the extreme observations. For the Exp(θ) distribution, we have $\gamma_n = f(\xi_{1n}^*)/[1 - F(\xi_{1n}^*)] = \theta$; following the same argument as above, we may only conclude that $X_{n:n} - \xi_{1n}^* = O_p(1)$, indicating that here the largest observation is of no utility for approximating the largest characteristic observation. □

Extensions to the case of the lth largest (or sth smallest) order statistics have been considered by many authors; see Gumbel (1958), among others, for details. In particular, to obtain the asymptotic joint distribution of the sth smallest and the lth largest order statistics of samples of size n corresponding to an absolutely continuous distribution function F, first consider $V_n = nF(X_{n:s})$ and $W_n = n[1 - F(X_{n:l})]$ and note that from (1.5.114), the exact joint density function for V_n and W_n is given by

$$g_{n,l,s}(v, w) = \frac{n!}{n^2(s-1)!(n-l-s)!(l-1)!} \times \left(\frac{v}{n}\right)^{s-1}\left(1 - \frac{w}{n} - \frac{v}{n}\right)^{n-l-s}\left(\frac{w}{n}\right)^{l-1} \tag{7.7.24}$$

since the Jacobian of the transformation

$$(X_{n:s}, X_{n:n-l}) \to (V_n, W_n) \tag{7.7.25}$$

is $[n^2 f(t) f(u)]^{-1}$. Then, if we let $n \to \infty$ while maintaining s and l fixed, we obtain the asymptotic joint density function

$$g_{l,s}(v, w) = \frac{v^{s-1}e^{-v}}{\Gamma(s)} \frac{w^{l-1}e^{-w}}{\Gamma(l)}, \tag{7.7.26}$$

which implies that V_n and W_n are asymptotically independent. If we impose the restriction that F is monotonically increasing, then (7.7.25) defines a one-to-one transformation and, therefore, we may conclude also that $X_{n:s}$ and $X_{n:n-l}$ are asymptotically independent. In particular, for $s = 1$ and $l = n$, it follows that the sample minimum and maximum are asymptotically independent.

7.8 Concluding Notes

As we have seen, the normal distribution occupies a central position in large sample theory. In the first place, it may be considered as an approximation to the distributions of many statistics (conveniently standardized) commonly used in practice. In many other cases, the corresponding asymptotic distributions (such as the central or noncentral chi-squared or f distributions) are basically related to suitable functions of normal variables. There are some problems, however, where normality, albeit being embedded, may not provide the appropriate results. A notable example of this kind is the U-statistic U_n defined by (2.4.28) when the parameter $\theta(F)$, given by (2.4.27), is stationary of order 1, in the sense that

$$\zeta_1(F) = \mathbb{E}_F[g^2(X_1)] - \theta^2(F) = 0 < \zeta_2(F) = \mathbb{E}_F[g^2(X_1, X_2)] - \theta^2(F),$$

where

$$g_c(x_1, \dots, x_c) = \mathbb{E}[g(X_1, \dots, X_m)|X_1 = x_1, \dots, X_c = x_c].$$

In such a case $n[U_n - \theta(F)] \xrightarrow{D} \sum_{j\geq 1} \lambda_j (Z_j^2 - 1)$, where λ_j are unknown coefficients and Z_j are i.i.d. random variables following the $\mathcal{N}(0, 1)$ law. A similar problem will be encountered when dealing with Kolmogorov–Smirnov-type statistics. The techniques developed in this chapter do not suffice to handle these situations and a survey of some modern probability tools more appropriate for such nonregular cases will be presented in Chapter 11.

Another related problem deals with the convergence of the binomial law (1.5.48) to the Poisson distribution (1.5.52) when $n \to \infty$, $\pi \to 0$ but $n\pi \to \lambda > 0$; this convergence to a Poisson law may also arise in a variety of other problems (see Exercise 7.1.1). For various reasons, we will not enter into details of this Poisson convergence results. We like to refer to Tucker (1967), Chow and Teicher (1978) and other standard texts in Probability Theory for some nice accounts of these developments. For applications of martingale central limit theorems in nonparametrics, see to Sen (1981). Further results are posed in the form of exercises.

A final but important topic in this section is that of the *convergence of moments* of functions of statistics that are asymptotically normal. In the setup of Example 7.5.7 we have seen that the sample coefficient of variation $v_n = s_n/\overline{X}_n$ is such that $\sqrt{n}(v_n - \nu) \xrightarrow{D} \mathcal{N}(0, \gamma^2)$; the question is whether we may approximate the moments of $\sqrt{n}(v_n - \nu)$ by appropriate functions of the moments of $\overline{X}_n$ and s_n^2. More specifically, let $\{T_n\}$ be a sequence of statistics such that $\sqrt{n}(T_n - \theta) \xrightarrow{D} \mathcal{N}(0, \sigma^2)$ and let g be a function such that $g'(\theta) \neq 0$ exists. We have seen above that

$$\sqrt{n}[g(T_n) - g(\theta)] \xrightarrow{D} \mathcal{N}\{0, [g'(\theta)]^2\sigma^2\}.$$

Furthermore, we also know that $(T_n - \theta) = O_p(n^{-1/2})$ and that the Taylor expansion formula $\sqrt{n}[g(T_n) - g(\theta)] = g'(\theta)\sqrt{n}(T_n - \theta) + o_p(1)$. Then, if r is a positive integer, we have

$$\begin{aligned}\{\sqrt{n}[g(T_n) - g(\theta)]\}^r &= \sum_{i=0}^{r} \binom{r}{i} [g'(\theta)\sqrt{n}(T_n - \theta)]^i [o_p(1)]^{r-i} \\ &= [g'(\theta)]^r[\sqrt{n}(T_n - \theta)]^r + \sum_{i=0}^{r-1} \binom{r}{i} [O_p(1)]^i [o_p(1)]^{r-i} \\ &= [g'(\theta)]^r[\sqrt{n}(T_n - \theta)]^r + o_p(1).\end{aligned}$$

Thus, we may write

$$\mathbb{E}\{\sqrt{n}[g(T_n) - g(\theta)]\}^r = [g'(\theta)]^r\mathbb{E}[\sqrt{n}(T_n - \theta)]^r + \mathbb{E}[o_p(1)].$$

The question is whether we may claim that $\mathbb{E}[o_p(1)] = o(1)$. In general, the answer is negative, but under some further conditions on g, the result holds. In this direction we consider the following theorem.

Theorem 7.8.1 (Cramér). *Let $\{T_n\}$ be a sequence of random variables with mean θ, r a positive integer and g a function satisfying the following conditions:*

(i) $|g(T_n)| \leq cn^p$ where c and p are finite constants;

(ii) there exists $k > r(p + 1)$ and $n_0 = n_0(k)$ such that for every $n \geq n_0$

$$\mu_{2k} = \mathbb{E}\{[\sqrt{n}(T_n - \theta)]^{2k}\} < \infty; \quad \textit{and}$$

(iii) $g''(x)$ is continuous and bounded in some neighborhood of θ.

Then,

$$\mathbb{E}\{[\sqrt{n}[g(T_n) - g(\theta)]\}^r = [g'(\theta)]^r\mathbb{E}[\sqrt{n}(T_n - \theta)]^r + O(n^{-1/2})$$

or equivalently

$$\mathbb{E}[g(T_n) - g(\theta)]^r = [g'(\theta)]^r\mathbb{E}(T_n - \theta)^r + O(n^{-(r+1)/2}).$$

Proof. Let $Z_n = \sqrt{n}[g(T_n) - g(\theta)]$ and $U_n = \sqrt{n}(T_n - \theta)$. Then, for all $\varepsilon > 0$

$$\mathbb{E}\big(Z_n^r\big) = \mathbb{E}\big[Z_n^r I(|U_n| < \varepsilon\sqrt{n})\big] + \mathbb{E}\big[Z_n^r I(|U_n| \geq \varepsilon\sqrt{n})\big]. \tag{7.8.1}$$

Now, note that

$$\begin{aligned}|Z_n| \leq \sqrt{n}[|g(T_n)| + |g(\theta)|] &\leq cn^{p+1/2} + n^{1/2}|g(\theta)| \\ &< n^{p+1/2}[c + |g(\theta)|] \\ &= C_1 n^{p+1/2},\end{aligned}$$

where $C_1 = c + |g(\theta)|$. Then, considering the second term on the right-hand side of (7.8.1), we have

$$\begin{aligned}|\mathbb{E}\big[Z_n^r I(|U_n| \geq \varepsilon\sqrt{n})\big]| &\leq \mathbb{E}\big[|Z_n^r| I(|U_n| \geq \varepsilon\sqrt{n})\big] \\ &< C_1^r n^{r(p+1/2)} P(|U_n| \geq \varepsilon\sqrt{n}) \\ &\leq C_1^r n^{r(p+1/2)} \frac{\mathbb{E}(|T_n - \theta|^{2k})}{\varepsilon^{2k}} \\ &= C_1^r n^{r(p+1/2)} \frac{\mu_{2k}}{\varepsilon^{2k} n^k} \\ &= C_2 n^{r(p+1/2)-k},\end{aligned}$$

where $C_2 = C_1 \mu_{2k}/\varepsilon^{2k}$. Now, for all $n \geq n_0$ we have from (ii)

$$r(p+1/2) - k < r(p+1/2) - r(p+1) = -r/2 \leq -1/2$$

so it follows that, for all $n \geq n_0$

$$\big|\mathbb{E}[Z_n^r I(|U_n| \geq \varepsilon\sqrt{n})]\big| \leq C_2 n^{-1/2} = O(n^{-1/2}). \tag{7.8.2}$$

In order to deal with the first term on the right-hand side of (7.8.1), note that

$$\begin{aligned}Z_n^r I(|U_n| < \varepsilon\sqrt{n}) &= n^{r/2}[g(T_n) - g(\theta)]^r I(|U_n| < \varepsilon\sqrt{n}) \\ &= n^{r/2}\left[g'(\theta)(T_n - \theta) + \frac{g''(\xi_n)}{2}(T_n - \theta)^2\right]^r I(|U_n| < \varepsilon\sqrt{n}),\end{aligned}$$

where $\theta - \varepsilon < \xi_n < \theta + \varepsilon$. Then,

$$\begin{aligned}Z_n^r I(|U_n| < \varepsilon\sqrt{n}) &= \left[g'(\theta)U_n + \frac{g''(\xi_n)}{2\sqrt{n}}U_n^2\right]^r I(|U_n| < \varepsilon\sqrt{n}) \\ &= \sum_{i=0}^{r}\binom{r}{i}\left[\frac{1}{2\sqrt{n}}g''(\xi_n)U_n^2\right]^i [g'(\theta)U_n]^{r-i} I(|U_n| < \varepsilon\sqrt{n}) \\ &= [g'(\theta)]^r U_n^r I[|T_n - \theta| < \varepsilon\sqrt{n}] \\ &\quad + \sum_{i=1}^{r}\binom{r}{i}\left(\frac{1}{2\sqrt{n}}\right)^i [g'(\theta)]^{r-i}[g''(\xi_n)]^i U_n^{r+i} I(|T_n - \theta| < \varepsilon\sqrt{n}).\end{aligned}$$

Therefore,

$$\begin{aligned}&\mathbb{E}\big[Z_n^r I(|U_n| < \varepsilon\sqrt{n})\big] \\ &\quad = [g'(\theta)]^r\{\mathbb{E}[(\sqrt{n}(T_n - \theta))^r] - \mathbb{E}[(\sqrt{n}(T_n - \theta))^r I(|U_n| \geq \varepsilon\sqrt{n})]\} \\ &\quad + \sum_{i=1}^{r}\binom{r}{i}\left(\frac{1}{2\sqrt{n}}\right)^i [g'(\theta)]^{r-i}\mathbb{E}\{[g''(\xi_n)]^i U_n^{r+i} I(|U_n| \geq \varepsilon\sqrt{n})\}.\end{aligned} \tag{7.8.3}$$

Now observe that

$$
\begin{aligned}
&|\mathbb{E}\{(\sqrt{n}(T_n-\theta))^r I(|U_n| \geq \varepsilon\sqrt{n})\}| \\
&\quad \leq \mathbb{E}\{[\sqrt{n}(T_n-\theta)]^{2r}\}^{1/2}[P(|U_n| \geq \varepsilon\sqrt{n})]^{1/2} \\
&\quad \leq \mathbb{E}[n^r(T_n-\theta)^{2r}]^{1/2}\left[\frac{\mathbb{E}(|T_n-\theta|^{2k})}{\varepsilon^k}\right]^{1/2} \\
&\quad = \mu_{2r}^{1/2}\frac{\mu_{2k}^{1/2}}{\varepsilon^k n^{k/2}} \\
&\quad = C_3 n^{-k/2} < C_3 n^{-1/2} = O(n^{-1/2}),
\end{aligned} \tag{7.8.4}
$$

where $C_3 = (\mu_{2r}\mu_{2k}/\varepsilon^{2k})^{1/2}$.

Let A_n denote the second term on the right-hand side of (7.8.3) and note that

$$
|A_n| \leq \sum_{i=1}^{r}\binom{r}{i}\left(\frac{1}{2\sqrt{n}}\right)^i |g'(\theta)|^{r-i}\mathbb{E}[|g''(\xi_n)|^i|U_n|^{r+i}I(|U_n| < \varepsilon\sqrt{n})].
$$

Using condition (iii), it follows that there exists $\varepsilon > 0$ such that $A = \sup\{|g''(X)| : \theta-\varepsilon \leq X \leq \theta+\varepsilon\} < \infty$; then we have

$$
\begin{aligned}
|A_n| &\leq \sum_{i=1}^{r}\binom{r}{i}\left(\frac{1}{2\sqrt{n}}\right)^i A^i|g'(\theta)|^{r-i}\mathbb{E}[|U_n^{r+i}|I(|U_n| < \varepsilon\sqrt{n})] \\
&\leq \sum_{i=1}^{r}\binom{r}{i}\left(\frac{1}{2\sqrt{n}}\right)^i A^i|g'(\theta)|^{r-i}\mathbb{E}(|U_n|^{r+i}) \\
&= \sum_{i=1}^{r}\binom{r}{i}\left(\frac{1}{2\sqrt{n}}\right)^i A^i|g'(\theta)|^{r-i}\mu_{r+i} \\
&\leq \frac{1}{\sqrt{n}}\left[\sum_{i=1}^{r}\binom{r}{i}A^i|g'(\theta)|^{r-i}\mu_{r+i}\right] = O(n^{-1/2}).
\end{aligned} \tag{7.8.5}
$$

Thus, from (7.8.3)–(7.8.5) it follows that

$$
\mathbb{E}\big[Z_n^r I(|U_n| < \varepsilon\sqrt{n})\big] = [g'(\theta)]^r\mathbb{E}\{[\sqrt{n}(T_n-\theta)]^r\} + O(n^{-1/2}),
$$

which in connection with (7.8.1) completes the proof. ■

A useful generalization of the above result is given by:

Theorem 7.8.2 (Cramér). *Let $\{\boldsymbol{T}_n\}$ be a sequence of random q-vectors with mean vector $\boldsymbol{\theta}$, r a positive integer and $g : \mathbb{R}^q \to \mathbb{R}$ satisfying the following conditions:*

(i) $|g(\boldsymbol{T}_n)| \leq cn^p$ where c and p are finite positive constants;

(ii) there exists $k > r(p+1)$ and $n_0 = n_0(k)$ such that for all $n \geq n_0$,

$$
\mu_{2k}^{(j)} = \mathbb{E}\{[\sqrt{n}(T_{nj}-\theta_j)]^{2k}\} < \infty, \quad j = 1,\ldots,q; \quad \textit{and}
$$

(iii) $g''_{ij}(\boldsymbol{x}) = (\partial^2/\partial x_i\partial x_j)g(\boldsymbol{x})$ is a continuous and bounded in some neighborhood of $\boldsymbol{\theta}$, $i, j = 1,\ldots,q$.

Then,

$$\mathbb{E}\{[\sqrt{n}(g(\boldsymbol{T}_n) - g(\boldsymbol{\theta}))]^r\} = \mathbb{E}\left\{\left[\sqrt{n}\sum_{j=1}^{q} g_j'(\boldsymbol{\theta})(T_{nj} - \theta_j)\right]^r\right\} + O(n^{-1/2}).$$

***Example* 7.8.1.** Consider the setup of Example 7.5.7 and let $v_n = s_n/\overline{X}_n = g(\overline{X}_n, s_n^2)$. Now note that

$$v_n^2 = \frac{\frac{1}{n-1}\left[\sum_{i=1}^n X_i^2 - n\overline{X}_n^2\right]}{\overline{X}_n^2} \le \frac{\frac{1}{n-1}\left[\sum_{i=1}^n X_i^2\right]}{\frac{1}{n}\left[\sum_{i=1}^n X_i\right]^2}$$
$$\le \frac{n}{n-1}\sum_{i=1}^{n}\left[X_i \Big/ \sum_{i=1}^{n} X_i\right]^2 \le n$$

and then $|v_n| = |g(\overline{X}_n, s_n^2)| \le n^{1/2}$, which implies that (i) in Theorem 7.8.2 holds with $c = 1$ and $p = 1/2$. If $\mu > 0$ and $\sigma^2 < \infty$, then (iii) is also satisfied. Therefore, if we want to obtain an approximation for the second moment of $\sqrt{n}(v_n - \nu)$, we must have $\mu_{2k}^{(j)} < \infty$ for some $k > r(p+1) = 3$. This implies that $\mathbb{E}[(s_n^2 - \sigma^2)^8] < \infty$, for all $n \ge n_0$ and consequently that $\mathbb{E}[|X_1|^{16}] < \infty$.

Suppose, for example, that the distribution function of X_1 is given by $F(x) = 1 - x^{-5}$ if $x \ge 1$ and $F(x) = 0$ for $x < 1$. Then,

$$\mathbb{E}(X^4) = \int_1^\infty x^4 5x^{-6}\,\mathrm{d}x = 5\int_1^\infty x^{-2}\,\mathrm{d}x = 5[-x^{-1}]_1^\infty = 5 < \infty.$$

Therefore, from Example 7.8.1 it follows that

$$\sqrt{n}(v_n - \nu) \xrightarrow{D} \mathcal{N}(0, \gamma^2).$$

However, $\mathbb{E}(X^{16}) = \int_1^\infty x^{16}5x^{-6}\,\mathrm{d}x = 5\int_1^\infty x^{10}\,\mathrm{d}x = \infty$, so we may not use the previous theorem to approximate the moments of $\sqrt{n}(v_n - \nu)$. □

Finally, we point that although the assumptions required by the Cramér Theorem 7.8.1 are sufficient, they are not necessary. If, for example, $g(T_n) = T_n = \overline{X}_n$, (i) may not hold (i.e., when the distribution function F has infinite support). Nevertheless, we may still be able to obtain convenient moment convergence, provided some extra regularity conditions are met. In this direction we state, without proof, the following result due to von Bahr (1965).

***Theorem* 7.8.3.** *Let $X_1, \ldots, X_n$ be i.i.d. random variables with zero means and unit variances. If $\mathbb{E}(|X_i|^k) < \infty$ for an integer $k \ge 3$, we have*

$$\mathbb{E}[(\sqrt{n}\overline{X}_n)^k] = \int_{-\infty}^{\infty} x^k\, d\Phi(x) + O(n^{-1/2}).$$

7.9 Exercises

7.1.1 Let X_{ni}, $i \ge 1$, be independent random variables such that $P(X_{ni} = 1) = 1 - P(X_{ni} = 0) = \pi_{ni}$, $i \ge 1$. If the π_{ni} are strictly positive and as $n \to \infty$, (i) $\sum_{i=1}^n \pi_{ni} = \pi_n \to \lambda$ $(0 < \lambda < \infty)$

and (ii) $\max_{i\le i\le n}(\pi_n^{-1}\pi_{ni}) \to 0$, show that $T_n = \sum_{i=1}^n X_{ni}$ has, asymptotically, a Poisson (λ) distribution.

7.1.2 In the above exercise, show that as λ increases, $(T_n - \lambda)/\sqrt{\lambda}$ converges in law to a normal variate.
Hint: For both Exercises 7.1.1 and 7.1.2 you may use characteristic functions.

7.1.3 Consider the setup of Exercise 7.1.1. Show that $\mathbb{V}\text{ar}(T_n) = \pi_n(1-\overline{\pi}_n) - \sum_{i=1}^n(\pi_{ni} - \overline{\pi}_n)^2$, where $\overline{\pi}_n = n^{-1}\pi_n$. Hence, or otherwise, show that $\mathbb{V}\text{ar}(T_n) \to \lambda$. Also, show that $\mathbb{V}\text{ar}(T_n)/[n\overline{\pi}_n(1-\overline{\pi}_n)] \le 1$, where the upper bound is attained when $\sum_{i=1}^n(\pi_{ni} - \overline{\pi}_n)^2/n\overline{\pi}_n(1-\overline{\pi}_n) \to 0$.

7.3.1 In the same setup of Exercise 7.1.1 show that the divergence of the series $\sum_{i=1}^n \pi_{ni}(1-\pi_{ni})$ is a necessary and sufficient condition for the asymptotic normality of $(T_n - \pi_n)/\sqrt{n}$.

7.3.2 Let T_n be a random variable having a noncentral chi-squared distribution with n degrees of freedom and noncentrality parameter Δ_n. Show T_n as the same asymptotic distribution as $X_1 + \cdots + X_n$, for all $n \ge 1$, where the X_j are i.i.d. random variables following a noncentral chi-squared distribution with 1 degree of freedom and noncentrality parameter $\delta_n = n^{-1}\Delta_n$. Hence, use Theorem 7.3.5 and show that $n^{1/2}(T_n - n - \Delta_n)$ is asymptotically normal. In particular, for $\Delta_n = 0$, verify that $(2n)^{-1/2}(T_n - n) \xrightarrow{D} \mathcal{N}(0,1)$.f

7.3.3 Let $X_1, \ldots, X_n$ be i.i.d. random variables with a continuous distribution function F and let $c_1, \ldots, c_n$ be n given constants, not all equal to zero. Let $a_n(i) = \mathbb{E}[\psi(U_{ni})]$, $1 \le i \le n$, where $\psi(u)$, $u \in (0,1)$ is square integrable and $U_{n1} < \cdots < U_{nn}$ are the ordered random variables of a sample of size n from the Uniform(0, 1) distribution. Consider the linear rank statistic $L_n = \sum_{i=1}^n(c_i - \overline{c}_n)a_n(R_{ni})$ where $\overline{c}_n = n^{-1}\sum_{i=1}^n c_i$ and R_{ni} is the rank of X_i among $X_1, \ldots, X_n$, for $i = 1, \ldots n$.

(i) Show that $\{L_n; n \ge 1\}$ is a zero mean martingale sequence.

(ii) Apply Theorem 7.3.7 to show that $C^{-1}L_n \xrightarrow{D} \mathcal{N}(0, \gamma^2)$ where $C_n^2 = \sum_{i=1}^n(c_i - \overline{c}_n)^2$ and $\gamma^2 = \int_0^1 \psi^2(u)\,du - \overline{\psi}^2$ with $\overline{\psi} = \int_0^1 \psi(u)\,du$.

(iii) Let $L_n^0 = \sum_{i=1}^n(c_i - \overline{c}_n)\psi\{F(X_i)\}$, so that $\mathbb{E}(L_n^0|R_{n1}, \ldots, R_{nn}) = L_n$ and $C_n^{-2}\mathbb{E}[(L_n^0 - L_n)^2] \to \infty$. Use Theorem 7.3.6 to verify that $C_n^{-1}L_n^0 \xrightarrow{D} \mathcal{N}(0, \gamma^2)$, and, hence, use the Projection Technique discussed in Section 7.5 to obtain the asymptotic normality of $C_n^{-1}L_n$ from that of $C_n^{-1}L_n^0$.

7.3.4 Let X_i, $i \ge 1$, be i.i.d. random variables with a distribution function F symmetric about 0 [i.e., $F(x) + F(-x) = 1, \forall x$]. Also let $\psi(u)$ be skew symmetric [i.e., $\psi(1/2+u) + \psi(1/2-u) = 0$, $\forall u \in (0, 1/2)$] and $\psi^+(u) = \psi[(1+u)/2]$, $0 < u < 1$. Furthermore, define $T_n = \sum_{i=1}^n \text{sign}(X_i)a_n^+(R_{ni}^+)$, $n \ge 1$, where $a_n^+(i) = \mathbb{E}[\psi^+(U_{ni})]$, $i = 1, \ldots, n$, $U_{n1} < \cdots < U_{nn}$ are ordered random variables of a sample of size n from the Uniform(0, 1) distribution and R_{ni}^+ denotes the rank of $|X_i|$ among $|X_1|, \ldots, |X_n|$, for $i = 1, \ldots, n$.

(i) Show that $\{T_n; n \ge 1\}$ is a zero mean martingale sequence. Hence, use Theorem 7.3.7 to verify the asymptotic normality of T_n (conveniently standardized).

(ii) Define $T_n^0 = \sum_{i=1}^n \psi[F(X_i)]$, $n \ge 1$, and show that

$$\mathbb{E}\big(T_n^0|\text{sign}(X_i),\quad R_{ni}^+,\ 1 \le i \le n\big) = T_n$$

and, hence, use Theorem 7.3.6 (on T_n^0) and the Projection Technique discussed in Section 7.5 to verify the asymptotic normality of T_n in an alternative manner.

7.3.5 For the CMRR procedure in Example 6.3.5, let $n_1/N \to \alpha_1$ and $n_2/N \to \alpha_2$ $(0 < \alpha_1, \alpha_2 < 1)$ as $n_i \to \infty$, $i = 1, 2$ and $N \to \infty$. Then show that $N^{-1/2}(\widehat{N} - N) \xrightarrow{D} \mathcal{N}[0, \sigma^2(\alpha_1, \alpha_2)]$ where $\sigma^2(\alpha_1, \alpha_2) = (1-\alpha_1)(1-\alpha_2)/\alpha_1\alpha_2$. Verify that for a given α $(= \alpha_1 + \alpha_2)$, $\sigma^2(\alpha_1, \alpha_2)$ is minimized when $\alpha_1 = \alpha_2 = \alpha/2$.

7.5.1 Let X be a random variable with finite moments; let μ_i denote the corresponding central moments and $\beta_1 = \mu_3^2/\mu_2^3$ and $\beta_2 = \mu_4/\mu_2^2$ be the conventional measures of *skewness* and *kurtosis*, respectively. Consider their sample counterparts $\widehat{\beta}_{1n} = m_{n3}^2/m_{n2}^3$ and $\widehat{\beta}_{2n} = m_{n4}/m_{n2}^2$ where $m_{nj} = n^{-1}\sum_{i=1}^{n}(X_i - \overline{X}_n)^j$, $j \geq 1$. Obtain the asymptotic normality of $\sqrt{n}(\widehat{\beta}_{jn} - \beta_j)$, $j = 1, 2$, as well as the joint one. (Use Theorems 7.5.1 and 7.5.4.)

7.6.1 Let $\boldsymbol{X}_n \sim N_p(\boldsymbol{0}, \boldsymbol{\Sigma})$ and consider the quadratic form $A_n = \boldsymbol{X}_n'\boldsymbol{A}_n\boldsymbol{X}_n$, where $\boldsymbol{A}_n$ $(p \times p)$ is p.s.d. Let $\alpha_0 = \lim_{n\to\infty} \text{ch}_p(\boldsymbol{A}_n\boldsymbol{\Sigma})$ and $\alpha^0 = \lim_{n\to\infty} \text{ch}_1(\boldsymbol{A}_n\boldsymbol{\Sigma})$. Show that $\alpha_0\boldsymbol{X}_n'\boldsymbol{\Sigma}^{-1}\boldsymbol{X}_n \leq Q_n \leq \alpha^0\boldsymbol{X}_n'\boldsymbol{\Sigma}^{-1}\boldsymbol{X}_n$. Hence, or otherwise, verify that, for every real t (≥ 0), we have

$$P\big(\chi_p^2 \leq t/\alpha^0\big) \leq \lim_{n\to\infty} P(Q_n \leq t) \leq P\big(\chi_p^2 \leq t/\alpha_0\big).$$

Use these inequalities to show that $\alpha_0 = \alpha^0$ implies $\lim_{n\to\infty} P(Q_n \leq t) = P(\chi_p^2 \leq t/\alpha_0)$.

7.6.2 Extend the results in the previous exercise to the case where $\boldsymbol{A}_n$ is a stochastic matrix, by making $\alpha_{0n} = \text{ch}_p(\boldsymbol{A}_n\boldsymbol{\Sigma}_n) \xrightarrow{P} \alpha_0$ and $\alpha_{1n} = \text{ch}_1(\boldsymbol{A}_n\boldsymbol{\Sigma}_n) \xrightarrow{P} \alpha^0$ $(\geq \alpha_0)$.

7.7.1 Let X_i, $i \geq 1$ be i.i.d. random variables with density

$$f(x;\theta) = c(x-\theta)^2 I(\theta - 1 \leq x \leq \theta + 1), \quad c > 0.$$

Define the sample median $\widetilde{X}_n$ by $X_{n:m}$ where $m = [n/2]$. Show that $\widetilde{X}_n$ is a strongly consistent estimator of θ. Also, let $Z_n = n^{1/6}(\widetilde{X}_n - \theta)$. Show that Z_n^3 has asymptotically a normal distribution with zero mean and a finite variance.

7.7.2 Let $X_1, \ldots, X_n$ be i.i.d. random variables with a density function $f(x;\theta) = f(x-\theta)$, where f is symmetric about 0. Let then $\widehat{\xi}_{p,n} = [X_{n:[np]} + X_{n:n-[np]+1}]/2$, for $0 < p < 1$.

(i) What is the asymptotic variance of $n^{1/2}(\widehat{\xi}_{p,n} - \theta)$?

(ii) Under what condition on f, does it have a minimum at $p = 1/2$?

(iii) Verify (ii) when f is (a) normal and (b) double exponential.

7.7.3 Suppose that $F(x)$ has the following (scale-perturbed) form:

$$F(x) = \begin{cases} \Phi((x-\theta)/a), & x > \theta, \quad a > 0 \\ \Phi((x-\theta)/b), & x < \theta, \quad b > 0, \quad b \neq a, \end{cases}$$

where Φ is the standard normal distribution function Show that F does not have a density function at θ, although the right-hand side and left-hand side derivatives exist at 0. Verify that the median of F is still uniquely defined and the sample median converges a.s. to the population median, as $n \to \infty$. What can be said about the asymptotic law for $n^{1/2}(\widetilde{X}_n - \theta)$?

7.7.4 Let θ be the median of F [i.e., $F(\theta) = 1/2$], and use (2.4.24) to verify that for every $n > 1$ and $0 \leq r < s \leq n+1$ (where $X_{n:0} = -\infty$ and $X_{n:n+1} = +\infty$),

$$P(X_{n:r} \leq \theta \leq X_{n:s}) = 2^{-n}\sum_{i=r}^{s-1}\binom{n}{i}.$$

Choose $s = n - r + 1, r \leq n/2$, and show that the above formula provides a distribution-free confidence interval for θ. What can be said about the confidence coefficient (i.e., coverage probability) when n is not large.

7.7.5 Let $X_1, \ldots, X_n$ be i.i.d. random variables with density function

$$f(x) = (1/2)\exp(-|x|), \quad x \in \mathbb{R}.$$

Define $M_n = (X_{n:1} + X_{n:n})/2$ and $Y_n = (X_{n:[np]} + X_{n:n-[np]+1})/2$, where $p \in (0, 1)$. Obtain the asymptotic distribution of the normalized form of M_n and, hence, show that as $n \to \infty$, M_n does not converge to 0, in probability. Obtain the asymptotic distribution of the normalized form of Y_n, and show that $Y_n \xrightarrow{P} 0$, as $n \to \infty$, for every $p \in (0, 1)$. Compare the asymptotic variance of Y_n for $p \in (0, 1.2)$ with that of $p = 1/2$ (i.e., the median), and comment on their asymptotic relative efficiency.

7.7.6 In the previous exercise, instead of the Laplace density, consider a logistic density function and conclude on the parallel results on M_n and Y_n.

7.7.7 Consider a distribution F, symmetric about θ and having a finite support $[\theta - \delta, \theta + \delta]$, where $0 < \delta < \infty$. Suppose that F has a terminal contact of order m. Define the kth midrange $M_{nk} = \frac{1}{2}(X_{n:k} + X_{n:n-k+1})$, for $k = 1, 2, \ldots$. Use (7.5.18)–(7.5.19) and Theorem 7.7.5 to derive the asymptotic distribution of the normalized form of M_{nk}. Compute the variance from this asymptotic distribution and denote this by $V_m(k)$, for $k > 1$. Show that, for $m = 1$, $V_1(k)$ is nondecreasing in k, whereas for $m \geq 2$, an opposite inequality may hold.

7.7.8 Consider the Cramér–von Mises statistic

$$C_n = n \int_{-\infty}^{\infty} [F_n(x) - F(x)]^2 \, \mathrm{d}F(x).$$

Derive the expression for the following: (i) $\mathbb{E}(C_n)$, (ii) $\mathbb{V}\mathrm{ar}(C_n)$, (iii) third and fourth central moments of C_n, and (iv) the usual Pearsonian measures of skewness and kurtosis of C_n. Hence, comment on the asymptotic distribution of C_n.

7.7.9 Comment on the difficulties you may encounter if you want to study the Cramér–von Mises statistic in the bivariate (or multivariate) case where the distribution function $F(x)$ is defined on $\mathbb{R}^2$ (or $\mathbb{R}^p$, $\rho \geq 2$) and is not necessarily expressible as the product of its marginal distributions.

7.7.10 Consider the Kolmogorov–Smirnov statistic for the one-sample and two-sample problems, and comment why you do not expect the asymptotic normality to be tenable in such cases. (See Chapter 11 for more details on their asymptotic distributions.)

8

Asymptotic Behavior of Estimators and Tests

8.1 Introduction

In Chapters 6 and 7 we developed tools to study the stochastic convergence and the asymptotic distributions of a general class of statistics. These tools can be directly incorporated in the study of asymptotic properties of a variety of (point as well as interval) estimators and tests. However, many of the methods discussed in Section 2.4 rest on suitable *estimating equations* with solutions that, in general, do not lead to estimators that can be expressed as explicit functions of the sample observations. Even when such closed-form expressions are available, they are not exactly in the form of the statistics treated in Chapters 6 and 7. Similarly, for composite hypotheses, specially when the underlying distributions do not belong to the exponential family, test statistics may suffer from the same drawbacks. The classical maximum likelihood estimators and likelihood ratio tests generally have this undesirable feature, albeit they may have well defined asymptotic optimality properties. In this chapter, we consider a general method to obtain asymptotic properties for a very broad class of statistics that fall in this category. Basically, we provide a viable link to the topics considered in the preceding two chapters borrowing strength from the methodology outlined there.

In Section 8.2, using a *uniform asymptotic linearity* property we obtain the asymptotic properties of estimators generated by estimating equations like (2.4.21). Although the MLE is a special member of this family of estimators, we provide related specialized details in Section 8.3, because of its important role in statistical inference. The case of M-estimators and other alternatives is treated in Section 8.4. The concept of *asymptotic efficiency* is presented in Section 8.5 along with its relation to the finite sample counterpart discussed in Chapter 2. Section 8.6 deals with asymptotic considerations for tests of hypotheses with due emphasis on likelihood ratio tests and their usual contenders, namely, Rao score tests and Wald tests. Asymptotic theory for resampling methods like the bootstrap and jackknife are dealt with in Section 8.7

8.2 Estimating Equations and Local Asymptotic Linearity

We motivate this section with a couple of examples and then proceed to formulate a general methodology useful in a broad context that includes parametric inference as well as beyond parametrics.

***Example* 8.2.1.** Let $X_1, \ldots, X_n$ be i.i.d. random variables following the Cauchy distribution, that is, having density function $f(x;\theta) = \{\pi[1 + (x - \theta)^2]\}^{-1}$. Assume that we are interested in estimating the location parameter θ. The likelihood function is $L_n(\theta) = \prod_{i=1}^n \{\pi[1+(X_i-\theta)^2]\}^{-1}$ and the corresponding estimating function for the MLE is

$$U_n(\theta) = 2\sum_{i=1}^n \{(X_i-\theta)/[1+(X_i-\theta)^2]\}.$$

This behaves as a polynomial of degree $2n-1$ in θ, and consequently, the estimating equation $U_n(\widehat{\theta}_n) = 0$ has $2n-1$ roots, some leading to (local) maxima and others to (local) minima; the MLE of θ, even if exists uniquely, is one of these multiple roots. Thus, the methodology developed in Chapters 6 and 7 cannot be directly applied here. □

***Example* 8.2.2.** Let $X_1, \ldots, X_n$ be i.i.d. random variables with density function $f(x;\theta) = \theta\exp(-\theta x)$, $x \geq 0$. The method of moments estimator of θ is based on $E(X_i) = 1/\theta$ and the corresponding estimating equation is

$$\overline{X}_n = 1/\widehat{\theta}_n.$$

Here, $\widehat{\theta}_n = 1/\overline{X}_n$ and the methodology developed in Chapters 6 and 7 can be easily amended. □

We now exploit the asymptotic linearity of estimating equations to derive the asymptotic distribution of the estimator obtained as their solution. Still confining ourselves to the single parameter case, we consider estimating equations of the form[1]:

$$M_n(\widehat{\theta}_n) = n^{-1/2}\sum_{i=1}^n \psi(X_i, \widehat{\theta}_n) = 0, \tag{8.2.1}$$

where $\psi(t,\theta)$, $t \in \mathbb{R}$ is a suitable score function satisfying

$$\int_{\mathbb{R}} \psi(x,\theta)\mathrm{d}F(x;\theta) = 0, \tag{8.2.2}$$

and

$$\int_{\mathbb{R}} \psi^2(x,\theta)\mathrm{d}F(x;\theta) = \sigma^2(\psi,\theta) < \infty, \tag{8.2.3}$$

with F denoting the distribution function of the underlying random variables. We are interested in the asymptotic distribution of the solution $\widehat{\theta}_n$ to (8.2.1). First, we let

$$\lambda(\psi,\theta,\delta) = \mathbb{E}_\theta[\psi(X,\theta+\delta)], \tag{8.2.4}$$

and assume it to exist for all δ in a small neighborhood of 0. By (8.2.2) we have $\lambda(\psi,\theta,0) = 0$, and, hence,

$$\begin{aligned}\lambda(\psi,\theta,\delta) &= \int_{\mathbb{R}} [\psi(x,\theta+\delta) - \psi(x,\theta)]\mathrm{d}F(x;\theta)\\ &= \delta\int_{\mathbb{R}} \delta^{-1}[\psi(x,\theta+\delta) - \psi(x,\theta)]\mathrm{d}F(x;\theta).\end{aligned} \tag{8.2.5}$$

[1] Note that the least squares estimating equations (2.4.5), the likelihood estimating equations (2.4.15) or the M-estimating equations (2.4.21) constitute special cases of (8.2.1).

As such, if ψ is absolutely continuous in θ in a neighborhood of the true parameter, then

$$\nu(\psi, \theta) = \lim_{\delta \to 0} \delta^{-1} \lambda(\psi, \theta, \delta) = \int_{\mathbb{R}} \psi'_\theta(x, \theta) \mathrm{d}F(x; \theta), \tag{8.2.6}$$

where

$$\psi'_\theta(x, \theta) = \lim_{\delta \to 0} [\psi(x, \theta + \delta) - \psi(x, \theta)]/\delta.$$

If the distribution F has an absolutely continuous (in θ) density function $f(x;\theta)$, we may write

$$\lambda(\psi, \theta, \delta) = \int_{\mathbb{R}} \psi(x, \theta)[f(x;\theta) - f(x;\theta - \delta)]\mathrm{d}x$$

so that $\nu(\psi, \theta)$ can also be expressed as

$$\nu(\psi, \theta) = \int_{\mathbb{R}} \psi(x, \theta) f'_\theta(x;\theta)\mathrm{d}x, \tag{8.2.7}$$

where $f'_\theta(x, \theta) = \lim_{\delta \to 0} \delta^{-1}[f(x;\theta + \delta) - f(x;\theta)]$. We may use (8.2.6) and (8.2.7) interchangeably, depending on the nature of the distribution F and the score function ψ. Also note that the Fisher score function (as it appears in the likelihood score statistic) is

$$\psi_f(x, \theta) = f'_\theta(x;\theta)/f(x;\theta), \tag{8.2.8}$$

while the Fisher information on θ (per observation) is

$$I_f(\theta) = \int_{\mathbb{R}} [f'_\theta(x;\theta)/f(x;\theta)]^2 f(x;\theta)\mathrm{d}x, \tag{8.2.9}$$

so that by (8.2.2) and (8.2.8), we may write (8.2.7) as

$$\begin{aligned} \nu(\psi, \theta) &= \int_{\mathbb{R}} \psi(x, \theta)\psi_f(x, \theta) f(x;\theta)\mathrm{d}x \\ &= \langle \psi(\cdot, \theta), \psi_f(\cdot, \theta) \rangle \\ &= \sigma(\psi, \theta)[I_f(\theta)]^{1/2} \rho[\psi(\cdot, \theta), \psi_f(\cdot, \theta)] \end{aligned} \tag{8.2.10}$$

where ρ is the correlation coefficient of the two score functions [and is thereby confined to the interval $(-1, 1)$].

The following theorem provides a basic result that may be employed to obtain the required asymptotic distribution.

***Theorem* 8.2.1.** *Let $X_1, \ldots, X_n$ be i.i.d. random variables with absolutely continuous distribution function F. Let ψ denote a score function satisfying (8.2.2) and (8.2.3) and assume that it also satisfies the following uniform continuity condition*

$$\xi_\delta = \mathbb{E}\left\{ \sup_{|u| \le \delta} [|\psi'_\theta(X, \theta + u) - \psi'_\theta(X, \theta)|] \right\} \to 0 \textit{ as } \delta \to 0. \tag{8.2.11}$$

Then, letting $M_n(\theta) = n^{-1/2} \sum_{i=1}^m \psi(X_i, \theta)$, for any fixed $K < \infty$,

$$\sup_{|u| \le \delta} [|M_n(\theta + n^{-1/2}u) - M_n(\theta) - u\nu(\psi, \theta)|] \xrightarrow{P} 0. \tag{8.2.12}$$

Proof. Using a first-order Taylor expansion, we obtain that for any $|u| \le K < \infty$, there exists a $\gamma \in (0, 1)$ such that

$$\begin{aligned} M_n(\theta + n^{-1/2}u) &= M_n(\theta) + n^{-1/2}uM_n'(\theta) + n^{-1/2}u[M_n'(\theta + \gamma n^{-1/2}u) - M_n'(\theta)] \\ &= M_n(\theta) + uV_n(\theta) + uR_n(u), \end{aligned} \tag{8.2.13}$$

where

$$V_n(\theta) = n^{-1}\sum_{i=1}^{n} \psi_\theta'(X_i, \theta), \tag{8.2.14}$$

and

$$R_n(u) = V_n(\theta + \gamma n^{-1/2}u) - V_n(\theta).$$

By the Khintchine SLLN (Theorem 6.3.13) and the definition in (8.2.7),

$$V_n(\theta) \xrightarrow{a.s.} \nu(\psi, \theta). \tag{8.2.15}$$

Also,

$$\sup_{|u|\le K} [|R_n(u)|] \le n^{-1}\sum_{i=1}^{n} Z_{ni}, \tag{8.2.16}$$

where

$$Z_{ni} = \sup_{\{|u|\le K\}} |\psi_\theta'(X_i, \theta + n^{-1/2}u) - \psi_\theta'(X_i, \theta)|$$

are i.i.d. nonnegative random variables with mean $\xi_{K/\sqrt{n}}$, which, by (8.2.11), converges to 0, as $n \to \infty$. Then, again by the Khintchine SLLN (Theorem 6.3.13), we have

$$\sup_{|u|\le K} [|R_n(u)|] \xrightarrow{a.s.} 0. \tag{8.2.17}$$

From (8.2.13), (8.2.15), and (8.2.17), it follows that, uniformly in $u \in [-K, K]$,

$$M_n(\theta + n^{-1/2}u) = M_n(\theta) + u\nu(\psi, \theta) + o_p(1), \tag{8.2.18}$$

completing the proof. ■

We are now ready to consider the main theorem of this section.

Theorem 8.2.2. *Under the regularity conditions of Theorem 8.2.1 it follows that the solution $\widehat{\theta}_n$ to the estimating equation (8.2.1) is such that*

$$\sqrt{n}(\widehat{\theta}_n - \theta) \xrightarrow{D} \mathcal{N}\{0, [I_f(\theta)\rho^2(\psi, \psi_f)]^{-1}\}. \tag{8.2.19}$$

Proof. Letting

$$\widehat{u} = \sqrt{n}(\widehat{\theta}_n - \theta) \tag{8.2.20}$$

it follows that $\widehat{\theta}_n = \theta + n^{-1/2}\widehat{u}$, so that $M_n(\widehat{\theta}_n) = M_n(\theta + n^{-1/2}\widehat{u}) = 0$. Then, recalling that $M_n = n^{-1/2}\sum_{i=1}^{n} \psi(X_i, \theta)$ and in view of (8.2.3)–(8.2.4), we may use the Central

Limit Theorem 7.3.1, to conclude that

$$M_n(\theta) \xrightarrow{D} \mathcal{N}[0, \sigma^2(\psi, \theta)]. \tag{8.2.21}$$

As a result, $|M_n(\theta)|/\sigma_\psi = O_p(1)$. From (8.2.18), we obtain

$$\widehat{u} = -\frac{M_n(\theta)}{\nu(\psi, \theta)} + o_p(1)$$

and using this in conjunction with (8.2.20) the proof is completed by noting that

$$\sigma^2(\psi, \theta)/\nu^2(\psi, \theta) = [I_f(\theta)\rho^2(\psi, \psi_f)]^{-1}. \tag{8.2.22}$$

■

Although (8.2.11) can be replaced by a less stringent assumption, it is easier to verify the former in specific applications. Also, for discrete F, the absolute continuity of the probability function $f(x;\theta)$ in θ suffices, and all we need is to replace the integral signs by summation signs in the appropriate expressions.

Let us now consider the general case of $\boldsymbol{\theta}$ being a p-vector for some $p \geq 1$. In such a case, the generalized score statistic (multiplied by $n^{-1/2}$) is a p-vector

$$\boldsymbol{M}_n(\boldsymbol{\theta}) = n^{-1/2} \sum_{i=1}^{n} \boldsymbol{\psi}(X_i, \boldsymbol{\theta}), \tag{8.2.23}$$

and the system of estimating equations is $\boldsymbol{M}_n(\widehat{\boldsymbol{\theta}}_n) = \mathbf{0}$. Here we assume that the conditions (8.2.2) and (8.2.3) hold for each coordinate of $\boldsymbol{\psi}$ and, instead of $\sigma^2(\psi, \theta)$, we have

$$\boldsymbol{\Sigma}(\boldsymbol{\psi}, \theta) = \int_{\mathbb{R}} [\boldsymbol{\psi}(x, \boldsymbol{\theta})][\boldsymbol{\psi}(x, \boldsymbol{\theta})]' \mathrm{d}F(x; \boldsymbol{\theta}). \tag{8.2.24}$$

Similarly, we replace (8.2.5) by the $p \times p$ matrix

$$\boldsymbol{\nu}(\boldsymbol{\psi}, \boldsymbol{\theta}) = \int_{\mathbb{R}} \frac{\partial}{\partial \boldsymbol{\theta}} \boldsymbol{\psi}(x, \boldsymbol{\theta}) \mathrm{d}F(x; \boldsymbol{\theta}) \tag{8.2.25}$$

and the Fisher score (vector) is defined as

$$\boldsymbol{\psi}_f(x, \boldsymbol{\theta}) = \frac{\partial}{\partial \boldsymbol{\theta}} \log f(x; \boldsymbol{\theta}). \tag{8.2.26}$$

Thus,

$$\boldsymbol{\nu}(\boldsymbol{\psi}, \boldsymbol{\theta}) = \int_{\mathbb{R}} \boldsymbol{\psi}(x, \boldsymbol{\theta})[\boldsymbol{\psi}_f(x, \boldsymbol{\theta})]' \mathrm{d}F(x; \boldsymbol{\theta}), \tag{8.2.27}$$

and

$$\boldsymbol{I}_f(\boldsymbol{\theta}) = \int_{\mathbb{R}} \boldsymbol{\psi}_f(x, \boldsymbol{\theta})[\boldsymbol{\psi}_f(x, \boldsymbol{\theta})]' \mathrm{d}F(x; \boldsymbol{\theta}) \tag{8.2.28}$$

is the *Fisher information matrix*. A generalization of Theorem 8.2.1 follows.

***Theorem* 8.2.3.** *Assume that the conditions (8.2.2) and (8.2.3) hold for each coordinate of the score function* $\boldsymbol{\psi}$ *in (8.2.23) and that*

$$\xi_\delta = \mathbb{E}\left\{ \sup_{||\boldsymbol{u}|| \leq K} [||\dot{\boldsymbol{\psi}}(X, \boldsymbol{\theta} + \boldsymbol{u}) - \dot{\boldsymbol{\psi}}(X, \boldsymbol{\theta})||] \right\} \to 0 \text{ as } \delta \to 0, \tag{8.2.29}$$

where $||\cdot||$ *stands for the Euclidean norm, and* $\overset{\bullet}{\boldsymbol{\psi}}(X, \boldsymbol{\theta})$ *is coordinatewise defined as in (8.2.6). Then for any fixed* $K < \infty$,

$$\sup_{||\boldsymbol{u}|| \leq K} ||\boldsymbol{M}_n(\boldsymbol{\theta} + n^{-1/2}\boldsymbol{u}) - \boldsymbol{M}_n(\boldsymbol{\theta}) - \boldsymbol{\nu}(\boldsymbol{\psi}, \boldsymbol{\theta})'\boldsymbol{u}|| \xrightarrow{P} 0. \tag{8.2.30}$$

The proof follows the lines of that of Theorem 8.2.1 and is left as Exercise 8.2.1.

Note further that, as $n \to \infty$,

$$\boldsymbol{M}_n(\boldsymbol{\theta}) \xrightarrow{D} \mathcal{N}_p[\boldsymbol{0}, \boldsymbol{I}_f(\boldsymbol{\theta})]. \tag{8.2.31}$$

Therefore, setting $\widehat{\boldsymbol{u}} = \sqrt{n}(\widehat{\boldsymbol{\theta}}_n - \boldsymbol{\theta})$ and noting that $||\boldsymbol{M}_n(\boldsymbol{\theta})|| = O_p(1)$, with $\mathbb{E}[\boldsymbol{M}_n(\boldsymbol{\theta})] = \boldsymbol{0}$, we obtain the main result of this section for vector valued parameters from the estimating equations generated by equating (8.2.23) to zero, namely,

$$\sqrt{n}(\widehat{\boldsymbol{\theta}}_n - \boldsymbol{\theta}) \xrightarrow{D} \mathcal{N}_p[\boldsymbol{0}, \boldsymbol{\Gamma}(\boldsymbol{\psi}, \boldsymbol{\theta})], \tag{8.2.32}$$

where $\boldsymbol{\Gamma}(\boldsymbol{\psi}, \boldsymbol{\theta}) = [\boldsymbol{\nu}(\boldsymbol{\psi}, \boldsymbol{\theta})]^{-1}\boldsymbol{I}_f(\boldsymbol{\theta})[\boldsymbol{\nu}(\boldsymbol{\psi}, \boldsymbol{\theta})']^{-1}$.

8.3 Asymptotics for MLE

As we have noted earlier, if $f(x; \boldsymbol{\theta})$ denotes the density of the underlying random variables, the MLE score function is

$$\boldsymbol{U}_n(x, \boldsymbol{\theta}) = \frac{\partial}{\partial \boldsymbol{\theta}} \log f(x; \boldsymbol{\theta}). \tag{8.3.1}$$

In Section 2.3, we showed that under the regularity conditions of the Cramér–Rao–Fréchet Theorem (2.3.1),

$$\mathbb{E}_{\boldsymbol{\theta}}[\boldsymbol{U}_n(X, \boldsymbol{\theta})] = \boldsymbol{0} \quad \text{and} \quad \mathbb{E}_{\boldsymbol{\theta}}[\boldsymbol{U}_n(X, \boldsymbol{\theta})][\boldsymbol{U}_n(X, \boldsymbol{\theta})]' = \boldsymbol{I}_f(\boldsymbol{\theta}),$$

where $\boldsymbol{I}_f(\boldsymbol{\theta})$ is the per unit observation Fisher information matrix on $\boldsymbol{\theta}$. If in addition to the Cramér–Rao–Fréchet regularity conditions, we have that, for $\delta \downarrow 0$,

$$\mathbb{E}_{\boldsymbol{\theta}}\left[\sup_{||\boldsymbol{u}|| < \delta} \left\| \frac{\partial^2}{\partial \boldsymbol{\theta} \partial \boldsymbol{\theta}'} \log f(X; \boldsymbol{\theta} + \boldsymbol{u}) - \frac{\partial^2}{\partial \boldsymbol{\theta} \partial \boldsymbol{\theta}'} \log f(X; \boldsymbol{\theta}) \right\| \right] = \xi(\delta) \to 0, \tag{8.3.2}$$

a specific p-variate version of Theorem 8.2.2 may be formalized for the MLE as follows.

***Theorem* 8.3.1.** *Under the Cramér–Rao–Fréchet regularity conditions and the uniform continuity condition (8.3.2), the solution to the estimating equations with the score function (8.3.1) is a* $\sqrt{n}$*-consistent estimator* $\widehat{\boldsymbol{\theta}}_n$ *of* $\boldsymbol{\theta}$ *such that*

$$\sqrt{n}(\widehat{\boldsymbol{\theta}}_n - \boldsymbol{\theta}) \xrightarrow{D} \mathcal{N}\{\boldsymbol{0}, [\boldsymbol{I}_f(\boldsymbol{\theta})]^{-1}\}. \tag{8.3.3}$$

The proof is left for the exercises. More specifically, Exercise 8.3.1 is set to verify (8.3.2) and the objective of Exercise 8.3.2 is to verify that for the score function in (8.3.1), we have

$$\boldsymbol{\nu}(\boldsymbol{U}_n, \boldsymbol{\theta}) = \boldsymbol{I}_f(\boldsymbol{\theta}). \tag{8.3.4}$$

In the sequel, to simplify expressions, we may use the notation $\boldsymbol{U}_n(\boldsymbol{\theta})$ and $L_n(\boldsymbol{\theta})$ instead of $\boldsymbol{U}_n(\boldsymbol{X}, \boldsymbol{\theta})$ and $L_n(\boldsymbol{\theta}; \boldsymbol{X})$, respectively.

Theorem 8.3.1 does not necessarily imply that the MLE is unique nor that all solutions to the estimating equations have the same asymptotic distribution. Rather, it shows that in a $\sqrt{n}$-neighborhood of the true parameter $\boldsymbol{\theta}$, a $\sqrt{n}$-consistent solution exists and that it is a MLE of $\boldsymbol{\theta}$. The uniqueness of the MLE may generally demand extra regularity conditions. In the one-parameter case, if the Fisher score function is monotone (as is the case with densities having the monotone likelihood property), then the MLE is unique. In the multiparametric case, Theorem 8.2.3 asserts a quadratic surface of the likelihood function in an $\sqrt{n}$-neighborhood of $\boldsymbol{\theta}$. Thus, whenever for large K and $\varepsilon > 0$, as $n \to \infty$,

$$\sup_{\boldsymbol{\theta}^*}(L_n(\boldsymbol{\theta}) : ||\boldsymbol{\theta}^* - \boldsymbol{\theta}|| > Kn^{-1/2}) \leq L_n(\boldsymbol{\theta}) - \varepsilon, \quad \text{in probability,} \tag{8.3.5}$$

then all nonlocal solutions of the estimating equations can be neglected in probability, while all local solutions will converge to a common one for which Theorem 8.3.1 holds. In practice, it is much easier to verify (8.3.2) instead of the uniform convergence of the observed curvature of $\log L_n(\boldsymbol{\theta})$ [e.g., Foutz (1977)]. We consider some examples to illustrate these features.

***Example* 8.3.1.** Let $X_1, \ldots, X_n$ be i.i.d. random variables following the Cauchy distribution, that is, with density function $f(x;\theta) = \{\pi[1 + (x - \theta)^2]\}^{-1}$, $x \in \mathbb{R}$. Then, it can be shown (Exercise 8.3.3) that since the estimating equation for θ has $(2n - 1)$ roots, there may be $n - 1$ (or n) MLE of θ. □

***Example* 8.3.2.** Let $X_1, \ldots, X_n$ be i.i.d. random variables with density function $f(x;\theta) = \exp[-(x - \theta)]I(x \geq \theta)$, $\theta \in \mathbb{R}$. In this case, the smallest observation in the sample, $X_{n:1}$, is the MLE and is also sufficient for θ (Exercise 8.3.4). Furthermore, from Section 7.7, we know that $n(X_{n:1} - \theta)$ has the simple exponential density function. Thus, here the rate of convergence is n (not $\sqrt{n}$) and the (asymptotic as well as exact) distribution is exponential (not normal). Note that the Cramér–Rao–Fréchet regularity conditions do not hold in this case. □

***Example* 8.3.3.** Let $X_1, \ldots, X_n$ be i.i.d. random variables with density function $f(x;\theta) = \theta^{-1}I(0 \leq x \leq \theta)$, $\theta > 0$. Here also the Cramér–Rao–Fréchet regularity conditions do not hold (Exercise 8.3.5). Also, the largest sample observation, $X_{n:n}$, is sufficient and is the MLE of θ. Furthermore, $n(\theta - X_{n:n})$ has asymptotically the simple exponential distribution. □

***Example* 8.3.4.** Let $X_1, \ldots, X_n$ be i.i.d. random variables with density function $f(x;\boldsymbol{\theta}) = (1/\sqrt{2\pi\theta_2})\exp[-(x - \theta_1)^2/(2\theta_2)]$, $x \in \mathbb{R}$ and let $\boldsymbol{\theta} = (\theta_1, \theta_2)'$. Then,

$$\log f(x;\boldsymbol{\theta}) = \frac{1}{2}\log(2\pi) - \frac{1}{2}\log\theta_2 - \frac{1}{2\theta_2}(x - \theta_1)^2.$$

Exercise 8.3.6 is set to verify that (8.3.2) holds in this case. □

***Example* 8.3.5.** Let $X_1, \ldots, X_n$ be i.i.d. random variables with density function $f(x;\theta) = c(x)\exp[a(\theta)t(x) - b(\theta)]$, where $a(\theta)$ and $b(\theta)$ are twice continuously differentiable and

$t(X)$ has finite moments up to the second order. Then, (8.3.2) holds. This is the object of Exercise 8.3.7. □

Although the analysis of qualitative data based on the multinomial model will be discussed in detail in Chapter 9, we consider a simple setup here to show how the asymptotic results discussed so far also hold in this case.

Recall from Example 2.4.1, the MLE for the parameter vector $\boldsymbol{\pi}$ is

$$\widehat{\boldsymbol{\pi}} = \frac{1}{n}\sum_{i=1}^{n} \boldsymbol{X}_i, \tag{8.3.6}$$

where $\boldsymbol{X}_i = (X_{i1}, \ldots, X_{ic})'$, $i = 1, \ldots, n$ are i.i.d. random vectors with $\mathbb{E}(\boldsymbol{X}_i) = \boldsymbol{\pi}$, $\mathbb{V}\mathrm{ar}(\boldsymbol{X}_i) = \boldsymbol{D}_\pi - \boldsymbol{\pi}\boldsymbol{\pi}'$ and $\boldsymbol{D}_\pi = \mathrm{diag}(\pi_1, \ldots, \pi_c)$. Thus, as we will detail in Chapter 9, the uniform continuity condition (8.3.2) holds and Theorem 8.3.1 leads to

$$\sqrt{n}(\widehat{\boldsymbol{\pi}}_n - \boldsymbol{\pi}) \xrightarrow{D} \mathcal{N}_k(\boldsymbol{0}, \boldsymbol{D}_\pi - \boldsymbol{\pi}\boldsymbol{\pi}'). \tag{8.3.7}$$

Two features of (8.3.7) are noteworthy. First, because of the constraint $\boldsymbol{\pi}'\boldsymbol{1} = 1$, we need to use Lagrange multipliers to formulate the estimating equations and also, to verify (8.3.2). Second, $\boldsymbol{D}_\pi - \boldsymbol{\pi}\boldsymbol{\pi}'$ is not of full rank and

$$\sqrt{n}\boldsymbol{D}_\pi^{-1/2}(\widehat{\boldsymbol{\pi}}_n - \boldsymbol{\pi}) \xrightarrow{D} \mathcal{N}\big(\boldsymbol{0}, \boldsymbol{I}_k - \boldsymbol{D}_\pi^{-1}\boldsymbol{\pi}\boldsymbol{\pi}'\big) \tag{8.3.8}$$

relates to a singular multinormal law.

In Example 8.3.1, we observed that the MLE of $\boldsymbol{\theta}$ cannot be expressed in a closed algebraic form, and therefore needs some iterative approach for its numerical evaluation. This characteristic is shared by many other estimating equations of the type (8.2.1), where $\widehat{\theta}_n$ can not be obtained in a closed form, as a simple function of the sample observations, that is, we have implicit solutions. Even for MLE, this implicit equation prevails for distributions not belonging to the *exponential family of densities* – logistic, Laplace, and Cauchy are important cases. Looking at (8.2.1) from this perspective, an iterative approach can be advocated, albeit we are in an stochastic environment, where the classical Newton–Raphson method needs some crucial appraisal. In a Newton–Raphson algorithm, the success hinges on a initial solution being located in a neighborhood of the true solution. Here, the true solution, the M-estimator, is itself a random variable. Hence, a natural requirement is that an initial solution be in a neighborhood of this stochastic solution. This delicate aspect along with the convergence prospects have been formulated in the following manner–it is generally referred to in the literature as the (Fisher–Rao) *Method of Scoring*.

Consider some alternative estimator $\widetilde{\theta}_n$ which is $\sqrt{n}$-consistent, that is, $\sqrt{n}|\widetilde{\theta}_n - \theta| = O_p(1)$. For example, in the Cauchy location model, $\widetilde{\theta}_n$ could be the sample median. Next, note that $M_n(\theta + a/\sqrt{n})$ can be written as $M_n(\theta) + aV_n(\theta) + aR_n(a)$ [see (8.2.13)], where $\sup[|R_n(a)| : |a| \leq k] \xrightarrow{P} 0$, as $n \to \infty$. Note further that $M_n(\theta)$, being asymptotically normal, is $O_p(1)$, while $V_n(\theta) \xrightarrow{P} \nu(\psi, \theta)$ [see (8.2.15)]. Thus, letting $\widehat{u}_n = n^{1/2}(\widehat{\theta} - \theta)$ and noting that $M_n(\widehat{\theta}) = o_p(1)$, we get a first-step estimator

$$\widehat{\theta}_n^{(1)} = \widetilde{\theta}_n - n^{-1/2}M_n(\widetilde{\theta}_n)[V_n(\widetilde{\theta}_n)]^{-1}.$$

This allow us to conceive an iterative scheme:

$$\widehat{\theta}_n^{(k)} = \widehat{\theta}_n^{(k-1)} - n^{-1/2}M_n(\widehat{\theta}_n^{(k-1)})[V_n(\widehat{\theta}_n^{k-1})]^{-1}, \quad k \geq 1. \tag{8.3.9}$$

Then, (8.2.12) insures the convergence of $\widehat{\theta}_n^{(k)}$ to $\widehat{\theta}_n$ and, in fact, one or two steps may generally suffice for this convergence. In dealing particularly with MLE, $\nu(\psi, \theta)$ reduces to the Fisher information $I(\theta)$, so that in (8.3.9) we may use $I(\widehat{\theta}_n^{k-1})$ whenever $I(\theta)$ has a closed expression. The Fisher scoring method pertains to that scheme, while the Rao scoring method applies to more general cases as reported here.

We illustrate this first with Example 8.3.1 pertaining to the Cauchy distribution. For the estimation of the corresponding location parameter, the sample median, $\widetilde{\theta}_n$, is a $\sqrt{n}$-consistent estimator for which Theorem 8.2.1 applies. Furthermore, here

$$U_n(\theta) = \frac{\partial}{\partial\theta} \log L_n(\theta) = \sum_{i=1}^{n} \frac{2(\theta - X_i)}{1 + (X_i - \theta)^2},$$

and

$$\begin{aligned} V_n(\theta) &= \frac{1}{n}\frac{\partial^2}{\partial\theta^2} \log L_n(\theta) = \frac{1}{n}\sum_{i=1}^{n}\left\{\frac{2}{1 + (X_i - \theta)^2} - \frac{4(X_i - \theta)^2}{[1 + (X_i - \theta)]^2}\right\} \\ &= \frac{2}{n}\sum_{i=1}^{n} \frac{1 - (X_i - \theta)^2}{[1 + (X_i - \theta)^2]^2}. \end{aligned}$$

Then, at the MLE, $\widehat{\boldsymbol{\theta}}_n$, we have

$$0 = n^{-1}U_n(\theta)|_{\widehat{\theta}_n} = n^{-1}U_n(\theta)|_{\widetilde{\theta}_n} + (\widehat{\theta}_n - \widetilde{\theta}_n)V_n(\theta_n^*),$$

where $\theta^* \in (\widehat{\theta}_n, \widetilde{\theta}_n)$. Appealing to (8.3.2), we may set

$$\widehat{\theta}_n \doteq \widetilde{\theta}_n - [V_n(\widetilde{\theta}_n)]^{-1} n^{-1} U_n(\theta)|_{\widetilde{\theta}_n}.$$

Thus, evaluating $U_n(\theta)$ and $V_n(\theta)$ at $\widetilde{\theta}_n$, we obtain the *one-step MLE*, $\widehat{\theta}_n^{(1)}$, of θ. We may repeat this process with $\widetilde{\theta}_n$ replaced by $\widehat{\theta}_n^{(1)}$, and so on, until convergence is achieved. Note that by (8.2.15) and (8.3.4), $V_n(\widetilde{\theta}_n)$ converges in probability to $I_f(\theta)$, ensuring that the iterative process terminates in probability. Also, in this context, $V_n(\theta)$ may be replaced by $I_f(\theta)$ and one may use $I_f(\widetilde{\theta}_n)$ instead of $V_n(\widetilde{\theta}_n)$ in the iterative process.

In the multivariate case a similar scheme may be employed. To see this, consider the following first-order multivariate Taylor expansion around some initial guess $\boldsymbol{\theta}_n^{(0)}$;

$$\begin{aligned} \mathbf{0} &= \frac{\partial}{\partial\boldsymbol{\theta}} \log L_n(\boldsymbol{\theta})\bigg|_{\widehat{\boldsymbol{\theta}}_n} \\ &= \frac{\partial}{\partial\boldsymbol{\theta}} \log L_n(\boldsymbol{\theta})\bigg|_{\boldsymbol{\theta}_n^{(0)}} + \frac{\partial^2}{\partial\boldsymbol{\theta}\partial\boldsymbol{\theta}'} \log L_n(\boldsymbol{\theta})\bigg|_{\boldsymbol{\theta}_n^*} (\widehat{\boldsymbol{\theta}}_n - \boldsymbol{\theta}_n^{(0)}), \end{aligned}$$

where $\boldsymbol{\theta}_n^*$ lies between $\widehat{\boldsymbol{\theta}}_n$ and $\boldsymbol{\theta}_n^{(0)}$. Therefore, we have

$$\widehat{\boldsymbol{\theta}}_n = \boldsymbol{\theta}_n^{(0)} - \left[\frac{\partial^2}{\partial\boldsymbol{\theta}\partial\boldsymbol{\theta}'} \log L_n(\theta)\bigg|_{\boldsymbol{\theta}_n^*}\right]^{-1} \left[\frac{\partial}{\partial\boldsymbol{\theta}} \log L_n(\boldsymbol{\theta})\bigg|_{\boldsymbol{\theta}_n^{(0)}}\right]. \tag{8.3.10}$$

If we choose $\boldsymbol{\theta}_n^{(0)}$ in a neighborhood of $\widehat{\boldsymbol{\theta}}_n$ (i.e., if $\boldsymbol{\theta}_n^{(0)}$ is based on some consistent estimator of $\boldsymbol{\theta}$), we may use (8.3.10) to consider a first-step estimator

$$\widehat{\boldsymbol{\theta}}_n^{(1)} = \boldsymbol{\theta}_n^{(0)} - \left[\frac{\partial^2}{\partial\boldsymbol{\theta}\partial\boldsymbol{\theta}'} \log L_n(\boldsymbol{\theta})\bigg|_{\boldsymbol{\theta}_n^{(0)}}\right]^{-1} \left[\frac{\partial}{\partial\boldsymbol{\theta}} \log L_n(\boldsymbol{\theta})\bigg|_{\boldsymbol{\theta}_n^{(0)}}\right], \tag{8.3.11}$$

and then proceed with the Newton–Raphson algorithm, with $\boldsymbol{\theta}_n^{(0)}$ replaced by the value of $\widehat{\boldsymbol{\theta}}_n^{(i)}$ obtained at the previous step.

The method of scoring is easy to interpret and operate under the regularity conditions of Theorem 8.2.1. However, if those conditions are not tenable, the picture might be completely different, and we must require additional conditions to guarantee its convergence.

The method of scoring describer above works out well for the regular case where the Cramér–Rao–Fréchet regularity conditions are satisfied. As a counterexample, consider a nonregular case where the conditions may not hold. Let X_i have the Uniform$[0, \theta]$ density, for some $\theta > 0$. The MLE of θ is $X_{(n)} = \max\{X_1, \ldots, X_n\}$, and it is a sufficient statistic, so that, by the Rao–Blackwell Theorem 2.3.2, $[(n+1)/n]X_{(n)}$ is an unbiased sufficient statistic for θ. There is no need for the method of scoring. However, the rate of convergence is n^{-1} (not $n^{-1/2}$), and the MLE is not asymptotically normal. Consider a second example where $f(x;\theta)$ is exponential with parameter $\theta > 0$. However, there is a (left-)censoring and the observed elements are the order statistics $X_{n:n-k+1}, \ldots, X_{n:n}$ for some fixed $k \geq 1$. Thus, we need to estimate θ based on $X_{n:j}$, $j \geq n-k+1$. The score statistic in this case may not be asymptotically normal nor its rate of convergence is $O(n^{-1/2})$ – it is typically much slower [e.g., $(\log n)^{-1}$]. Here, one can use the method of scoring with a preliminary estimator based on $X_{n:n-k+1}$ alone and then iterate the scoring process. The scheme works out but has slower rate of convergence.

There are other variants of the method of scoring. Among these, the *EM-algorithm* [Dempster, Laird, and Rubin (1977)] has often been advocated, specially for incomplete or partially observable data models. In the presence of sufficient statistics, the EM-algorithm provides a computational scheme that is easy to interpret and convenient to adapt to other nonstandard models. However, the EM algorithm may not provide the most efficient computational scheme. It is often claimed (but not necessarily true) that the EM algorithm is less sensitive to poor initial points. The relative advantages of the EM algorithm in nonstandard cases may have to be contrasted with its slow convergence. Also, it may not provide an estimator of the mean square error (or its multiparameter counterparts). We will briefly outline the EM-algorithm in Chapter 10.

8.4 Asymptotics for Other Classes of Estimators

Maximum Likelihood is, perhaps, the most frequently used method for parametric estimation when the functional form of the underlying distribution is known. In this section we deal with the asymptotic properties of some alternative estimation procedures which are usually employed in situations where only limited knowledge about the parent distribution is available. First we consider the *Method of Moments*.

Let $X_1, \ldots, X_n$ be i.i.d. random variables with density function $f(x;\boldsymbol{\theta})$, $\boldsymbol{\theta} \in \boldsymbol{\Theta} \subset \mathbb{R}^q$, $q \geq 1$, $x \in \mathbb{R}$, and assume that $\mathbb{E}(X_1^k) = \mu^{(k)} = h_k(\boldsymbol{\theta}) < \infty$, $k = 1, \ldots, q$; also let $m_n^{(k)} = n^{-1}\sum_{i=1}^n X_i^k$, $k = 1, \ldots, q$. The method of moments estimator (MME) $\widetilde{\boldsymbol{\theta}}_n$ of $\boldsymbol{\theta}$ is a solution (in $\boldsymbol{\theta}$) to the equations $h_k(\widetilde{\boldsymbol{\theta}}) = m_n^{(k)}$, $k = 1, \ldots, q$. The following theorem establishes conditions under which the asymptotic distribution of $\widetilde{\boldsymbol{\theta}}_n$ may be obtained.

Theorem 8.4.1. *Under the setup described above, assume that $\mu^{(k)} < \infty$, $k = 1, \ldots, 2q$. Also let $\boldsymbol{h}(\boldsymbol{\theta}) = [h_1(\boldsymbol{\theta}), \ldots, h_q(\boldsymbol{\theta})]'$ and $\boldsymbol{H}(\boldsymbol{\theta}) = (\partial/\partial\boldsymbol{\theta})\boldsymbol{h}(\boldsymbol{\theta})$ be such that $r[\boldsymbol{H}(\boldsymbol{\theta})] = q$ and* *that its elements, $H_{ij}(\boldsymbol{\theta}) = (\partial/\partial\theta_j)h_i(\boldsymbol{\theta})$, $i, j = 1, \ldots, q$, are continuous in $\boldsymbol{\theta}$. Then,*

$$\sqrt{n}(\widetilde{\boldsymbol{\theta}}_n - \boldsymbol{\theta}) \xrightarrow{D} \mathcal{N}_q\{\mathbf{0}, [\boldsymbol{H}(\boldsymbol{\theta})]^{-1}\boldsymbol{\Sigma}\{[\boldsymbol{H}(\boldsymbol{\theta})]^{-1}\}'\}$$

with $\boldsymbol{\Sigma}$ denoting a $q \times q$ matrix with the element (j, k) given by $\mu^{(j+k)} - \mu^{(j)}\mu^{(k)}$.

Proof. Consider the Taylor expansion:

$$\boldsymbol{h}(\boldsymbol{\theta} + n^{-1/2}\boldsymbol{u}) = \boldsymbol{h}(\boldsymbol{\theta}) + n^{-1/2}\boldsymbol{H}(\boldsymbol{\theta})\boldsymbol{u} + n^{-1/2}[\boldsymbol{H}(\boldsymbol{\theta}^*) - \boldsymbol{H}(\boldsymbol{\theta})]\boldsymbol{u}, \tag{8.4.1}$$

where $\boldsymbol{\theta}^* = \boldsymbol{\theta} + n^{-1/2}h\boldsymbol{u}, 0 \le h \le 1$. In view of the assumption on the continuity of $\boldsymbol{H}(\boldsymbol{\theta})$, (8.4.1) may be re-expressed as

$$\sqrt{n}[\boldsymbol{h}(\boldsymbol{\theta} + n^{-1/2}\boldsymbol{u}) - \boldsymbol{h}(\boldsymbol{\theta})] = \boldsymbol{H}(\boldsymbol{\theta})\boldsymbol{u} + o(\mathbf{1}). \tag{8.4.2}$$

Letting $\boldsymbol{u} = \sqrt{n}(\widetilde{\boldsymbol{\theta}}_n - \boldsymbol{\theta})$, we get $\boldsymbol{h}(\boldsymbol{\theta} + n^{-1/2}\boldsymbol{u}) = \boldsymbol{h}(\widetilde{\boldsymbol{\theta}}_n) = \boldsymbol{m}_n$. Thus, (8.4.2) may be written as

$$\sqrt{n}\,[\boldsymbol{m}_n - \boldsymbol{h}(\boldsymbol{\theta})] = \boldsymbol{H}(\boldsymbol{\theta})\sqrt{n}(\widetilde{\boldsymbol{\theta}}_n - \boldsymbol{\theta}) + o_p(\mathbf{1}). \tag{8.4.3}$$

Now let $\boldsymbol{\lambda} \in \mathbb{R}^q$, $\boldsymbol{\lambda} \neq \mathbf{0}$, be an arbitrary, but fixed vector, and set

$$\begin{aligned}
\sqrt{n}\boldsymbol{\lambda}'[\boldsymbol{m}_n - \boldsymbol{h}(\boldsymbol{\theta})] &= \sqrt{n}[(\lambda_1 m_n^{(1)} + \cdots + \lambda_q m_n^{(q)}) - (\lambda_1\mu^{(1)} + \cdots + \lambda_q\mu^{(q)})] \\
&= \frac{1}{\sqrt{n}}\sum_{i=1}^{n}\left[\sum_{j=1}^{q}\lambda_j[X_i^j - \mathbb{E}(X_i^j)]\right] = \frac{1}{\sqrt{n}}\sum_{i=1}^{n}U_i,
\end{aligned}$$

where $U_i = \sum_{i=1}^{q}\lambda_j[X_i^j - \mathbb{E}(X_i^j)]$ is such that $\mathbb{E}(U_i) = 0$ and

$$\begin{aligned}
\mathbb{E}(U_i^2) &= \sum_{j=1}^{q}\sum_{k=1}^{q}\lambda_j\lambda_k\mathbb{E}(X_i^j - \mu^{(j)})(X_i^k - \mu^{(k)}) \\
&= \sum_{j=1}^{q}\sum_{k=1}^{q}\lambda_j\lambda_k[\mu^{(j+k)} - \mu^{(j)}\mu^{(k)}] = \boldsymbol{\lambda}'\boldsymbol{\Sigma}\boldsymbol{\lambda} < \infty.
\end{aligned}$$

Since the U_i are i.i.d. random variables, it follows from the Central Limit Theorem 7.3.1 that

$$\sqrt{n}\sum_{i=1}^{n}U_i = \sqrt{n}\boldsymbol{\lambda}'[\boldsymbol{m}_n - \boldsymbol{h}(\boldsymbol{\theta})] \xrightarrow{D} \mathcal{N}(0, \boldsymbol{\lambda}'\boldsymbol{\Sigma}\boldsymbol{\lambda});$$

using the Cramér-Wold Theorem 7.2.4 we may conclude that

$$\sqrt{n}[\boldsymbol{m}_n - \boldsymbol{h}(\boldsymbol{\theta})] \xrightarrow{D} \mathcal{N}_q(\mathbf{0}, \boldsymbol{\Sigma}).$$

Finally, from (8.4.3), we get

$$\sqrt{n}(\widetilde{\boldsymbol{\theta}}_n - \boldsymbol{\theta}) \xrightarrow{D} \mathcal{N}_q\{\mathbf{0}, [\boldsymbol{H}^{-1}(\boldsymbol{\theta})]\boldsymbol{\Sigma}[\boldsymbol{H}^{-1}(\boldsymbol{\theta})]'\}.$$

■

***Example* 8.4.1.** If $X_1, \ldots, X_n$ are i.i.d. random variables following a $\mathcal{N}(\mu, \sigma^2)$ distribution, we have (see Exercise 2.4.1)

$$\boldsymbol{m}_n = \frac{1}{n}\left(\sum_{i=1}^n X_i, \sum_{i=1}^n X_i^2\right)', \qquad \boldsymbol{h}(\boldsymbol{\theta}) = (\mu, \mu^2 + \sigma^2)'$$

and

$$\boldsymbol{H}(\boldsymbol{\theta}) = \begin{pmatrix} 1 & 0 \\ 2\mu & 1 \end{pmatrix};$$

since $\mu^{(3)} = \mathbb{E}(X_1^3) = \mu^3 + 3\mu\sigma^2$ and $\mu^{(4)} = \mathbb{E}(X_1^4) = \mu^4 + 3\sigma^4 + 6\mu^2\sigma^2$ a direct application of Theorem 8.4.1 leads us to the conclusion that

$$n^{-1/2}\left\{\left[\sum_{i=1}^n X_i, \sum_{i=1}^n X_i^2 - n^{-1}\left(\sum_{i=1}^n X_i\right)^2\right] - (\mu, \sigma^2)\right\}' \xrightarrow{D} \mathcal{N}_2(\boldsymbol{0}, \boldsymbol{\Gamma})$$

with

$$\boldsymbol{\Gamma} = \begin{pmatrix} \sigma^2 & 0 \\ 0 & 2\sigma^4 \end{pmatrix}.$$

In addition, if we consider $\boldsymbol{T}_n = n^{-1}\left\{\sum_{i=1}^n X_i, [n/(n-1)]\sum_{i=1}^n (X_i - \overline{X}_n)^2\right\}'$, then $\boldsymbol{h}(\boldsymbol{\theta}) = \mathbb{E}(\boldsymbol{T}_n) = (\mu, \sigma^2)'$ and, hence, $\boldsymbol{H}(\boldsymbol{\theta}) = \boldsymbol{I}_2$. Then, it follows that

$$\sqrt{n}[(\overline{X}_n, s_n^2) - (\mu, \sigma)]' \xrightarrow{D} \mathcal{N}_2(\boldsymbol{0}, \boldsymbol{\Gamma}). \qquad \square$$

Observe that the method described above may be used to obtain the asymptotic distribution of the statistic $\boldsymbol{T}_n = n^{-1}\sum_{i=1}^n \boldsymbol{g}(X_i)$, where $\boldsymbol{g}(x) = [g_1(x), \ldots, g_q(x)]'$ is a vector-valued function, by making $\boldsymbol{h}(\boldsymbol{\theta}) = \mathbb{E}(\boldsymbol{T}_n)$.

For location problems, M-estimators were introduced in Section 2.4 and correspond to solutions of special cases of (8.2.1) when $\psi(x, \theta) = \psi(x - \theta)$. The following theorem illustrates a useful technique for proving the consistency and asymptotic normality of M-estimators.

***Theorem* 8.4.2.** *Let $X_1, \ldots, X_n$ be i.i.d. random variables with distribution function F and $\psi(x - \theta)$ be nonincreasing in θ. Also, let θ_0 be an isolated root of*

$$M(\theta) = \int \psi(x - \theta)dF(x) = 0.$$

Assume that (i) $M'(\theta) \neq 0$, for θ in a neighborhood of θ_0; (ii) $\int \psi^2(x - \theta)dF(x) < \infty$ and is continuous at θ_0. Then,

(a) if (i) holds, $\widehat{\theta}_n$ is consistent for θ_0;

(b) if (i) and (ii) hold,

$$\sqrt{n}(\widehat{\theta}_n - \theta_0) \xrightarrow{D} \mathcal{N}\{0, \sigma^2(\theta_0)/[M'(\theta_0)]^2\}$$

with $\sigma^2(\theta_0) = \mathbb{V}ar[\psi(X_1 - \theta_0)]$.

Proof. First note that since $\psi(x-\theta)$ is nonincreasing in θ, so are $M(\theta)$ and $M_n(\theta)$; thus, given $\varepsilon > 0$, we have $\widehat{\theta}_n > \theta_0 + \varepsilon \Rightarrow M_n(\theta_0+\varepsilon) \geq M_n(\widehat{\theta}_n) = 0$, so that

$$\begin{aligned}
P(\widehat{\theta}_n - \theta_0 > \varepsilon) &\leq P[M_n(\theta_0+\varepsilon) \geq 0] \\
&= P\left[\frac{1}{n}\sum_{i=1}^{n}\psi(X_i-\theta_0-\varepsilon) \geq 0\right] \\
&= P\left(\frac{1}{n}\sum_{i=1}^{n}\{\psi(X_i-\theta_0-\varepsilon) - \mathbb{E}[\psi(X_1-\theta_0-\varepsilon)]\} \geq -\mathbb{E}[\psi(X_1-\theta_0-\varepsilon)]\right) \\
&= P\left\{\frac{1}{n}\sum_{i=1}^{n}Z_i \geq -\mathbb{E}[\psi(X_1-\theta_0-\varepsilon)]\right\}, \qquad (8.4.4)
\end{aligned}$$

where $Z_i = \psi(X_i-\theta_0-\varepsilon) - \mathbb{E}[\psi(X_1-\theta_0-\varepsilon)], i = 1,\ldots,n$ are i.i.d. random variables with zero means. Now consider the following first-order Taylor expansion of $M(\theta_0+\varepsilon) = \mathbb{E}[\psi(X_1-\theta_0-\varepsilon)]$ around θ_0:

$$\mathbb{E}[\psi(X_1-\theta_0-\varepsilon)] = M(\theta_0+\varepsilon) = M(\theta_0) + \varepsilon M'(\theta_0+\varepsilon\gamma),$$

where $0 \leq \gamma \leq 1$. Since M is nonincreasing and $M(\theta_0) = 0$, we get $M(\theta_0+\varepsilon) = \varepsilon M'(\theta_0 + \varepsilon\gamma) < 0$. Therefore, given $\varepsilon > 0$, there exists $\delta = \delta(\varepsilon) = -\varepsilon M'(\theta_0+\varepsilon\gamma) > 0$ such that (8.4.4) may be expressed as

$$P(\widehat{\theta}_n - \theta_0 > \varepsilon) \leq P\left(\frac{1}{n}\sum_{i=1}^{n}Z_i \geq \delta\right),$$

which converges to zero as $n \to \infty$ by the Khintchine WLLN (Theorem 6.3.6). Repeating a similar argument for the term $P(\widehat{\theta}_n - \theta_0 < -\varepsilon)$ it is possible to show that $P\{|\widehat{\theta}_n - \theta_0| > \varepsilon\} \to 0$ as $n \to \infty$, proving (a).

To prove part (b), first note that for all $u \in \mathbb{R}$, we may write

$$\begin{aligned}
P[\sqrt{n}(\widehat{\theta}_n - \theta_0) \leq u] &= P\left(\widehat{\theta}_n \leq \theta_0 + n^{-1/2}u\right) \\
&= P\left[M_n\left(\theta_0 + n^{-1/2}u\right) < 0\right] \\
&= P\left\{\frac{1}{\sqrt{n}}\sum_{i=1}^{n}\left[\frac{\psi(X_i-\theta_0-u/\sqrt{n}) - \mathbb{E}[\psi(X_1-\theta_0-u/\sqrt{n})]}{\sqrt{\mathbb{V}\text{ar}[\psi(X_1-\theta_0-u/\sqrt{n})]}}\right]\right. \\
&\qquad \left. \leq -\frac{\sqrt{n}\mathbb{E}[\psi(X_1-\theta_0-u/\sqrt{n})]}{\sqrt{\mathbb{V}\text{ar}[\psi(X_1-\theta_0-u/\sqrt{n})]}}\right\} \\
&= P\left\{\frac{1}{\sqrt{n}}\sum_{i=1}^{n}Z_{ni} \leq -\frac{\sqrt{n}\mathbb{E}[\psi(X_1-\theta_0-u/\sqrt{n})]}{\sqrt{\mathbb{V}\text{ar}[\psi(X_1-\theta_0-u/\sqrt{n})]}}\right\}, \qquad (8.4.5)
\end{aligned}$$

where

$$Z_{ni} = \frac{1}{\sqrt{n}}\left\{\frac{\psi(X_i-\theta_0-u/\sqrt{n}) - \mathbb{E}[\psi(X_1-\theta_0-u/\sqrt{n})]}{\sqrt{\mathbb{V}\text{ar}[\psi(X_1-\theta_0-u/\sqrt{n})]}}\right\}, \qquad i = 1,\ldots,n,$$

are i.i.d. random variables with $\mathbb{E}(Z_{ni}) = 0$ and $\mathbb{V}\text{ar}(Z_{ni}) = 1$. Now use a Taylor expansion to see that, as $n \to \infty$,

$$\begin{aligned}\sqrt{n}\mathbb{E}[\psi(X_1 - \theta_0 - u/\sqrt{n})] &= \sqrt{n}M(\theta_0 + u/\sqrt{n}) \\ &= \sqrt{n}[M(\theta_0) + uM'(\theta_0)/\sqrt{n} + o(|u|/\sqrt{n})] \\ &= uM'(\theta_0) + \sqrt{n}\, o(|u|/\sqrt{n}) \longrightarrow uM'(\theta_0). \qquad (8.4.6)\end{aligned}$$

Also, using (i) and (ii) observe that, as $n \to \infty$,

$$\begin{aligned}&\mathbb{V}\text{ar}[\psi(X_1 - \theta_0 - n^{-1/2}u)] = \sigma^2(\theta_0 + n^{-1/2}u) \\ &= \int \psi^2(x - \theta_0 - n^{-1/2}u)\mathrm{d}F(x) - [M(\theta_0 + n^{-1/2}u)]^2 \to \sigma^2(\theta_0). \qquad (8.4.7)\end{aligned}$$

From (8.4.6) and (8.4.7) it follows that, as $n \to \infty$,

$$a_n = -\frac{\sqrt{n}\mathbb{E}[\psi(X_1 - \theta_0 - u/\sqrt{n})]}{\sqrt{\mathbb{V}\text{ar}[\psi(X_1 - \theta_0 - u/\sqrt{n})]}} \longrightarrow -\frac{uM'(\theta_0)}{\sigma(\theta_0)}. \qquad (8.4.8)$$

To complete the proof, in view of (8.4.5) and (8.4.8), all we have to show is that as $n \to \infty$,

$$P\left(\frac{1}{\sqrt{n}}\sum_{i=1}^{n} Z_{ni} \le -\frac{uM'(\theta_0)}{\sigma(\theta_0)}\right) \longrightarrow \Phi\left(-\frac{uM'(\theta_0)}{\sigma(\theta_0)}\right). \qquad (8.4.9)$$

To see this, suppose that $\{Y_n\}$ is a sequence of random variables and that $\{a_n\}$ is a sequence of real numbers such that $a_n \to a$ as $n \to \infty$, where a is a constant; then note that $Y_n - a_n = (Y_n - a) + (a_n - a)$ so, if $Y_n - a \xrightarrow{D} \mathcal{N}(-a, 1)$, the Slutsky Theorem 7.5.2 implies that $Y_n - a_n \xrightarrow{D} \mathcal{N}(-a, 1)Y$. Thus, we have $P(Y_n - a_n \le 0) = P(Y_n \le a_n) \to \Phi(-a)$ as $n \to \infty$.

Now, to prove (8.4.9), we may relate to the CLT for Triangular Arrays 7.3.5 and it suffices to show that the Lindeberg condition holds, that is, that

$$\mathbb{E}[Z_{n1}^2 I(|Z_{n1}| > \sqrt{n}\varepsilon)] \to 0, \quad \text{as } n \to \infty. \qquad (8.4.10)$$

In this direction, let

$$B_n = [|\psi(x - \theta_0 - u/\sqrt{n}) - M(\theta_0 + u/\sqrt{n})| > \sqrt{n}\varepsilon\sigma(\theta_0 + u/\sqrt{n})]$$

and note that

$$\begin{aligned}\mathbb{E}\left[Z_{n1}^2 I\left(|Z_{n1}| > \sqrt{n}\varepsilon\right)\right] &= \int_{B_n} \frac{\psi^2(x - \theta_0 - u/\sqrt{n})}{\sigma^2(\theta_0 + u/\sqrt{n})}\mathrm{d}F(x) \\ &\quad - \frac{2M(\theta_0 + u/\sqrt{n})}{\sigma^2(\theta_0 + u/\sqrt{n})}\int_{B_n} \psi(x - \theta_0 - n^{-1/2}u)\mathrm{d}F(x) \\ &\quad + \frac{M^2(\theta_0 + u/\sqrt{n})}{\sigma^2(\theta_0 + u/\sqrt{n})}\int_{B_n} \mathrm{d}F(x). \qquad (8.4.11)\end{aligned}$$

Since $M(\theta_0 + u/\sqrt{n})$ and $\sigma^2(\theta_0 + u/\sqrt{n})$ are continuous, it follows that the two last terms of the right-hand side of (8.4.11) converge to zero as $n \to \infty$. Thus, we now have to show that as $n \to \infty$,

$$\int_{\{|\psi(x - \theta - u/\sqrt{n})| > \varepsilon\sqrt{n}\}} \psi^2(x - \theta_0 - n^{-1/2}u)\mathrm{d}F(x) \to 0. \qquad (8.4.12)$$

Since ψ is nonincreasing, given $\varepsilon > 0$, there exists $n_0 = n_0(\varepsilon)$ such that

$$\psi(x - \theta_0 + \varepsilon) \leq \psi\left(x - \theta_0 - t/\sqrt{n}\right) \leq \psi(x - \theta_0 - \varepsilon)$$

for all $n \geq n_0$ and for all $x \in \mathbb{R}$. Therefore, by assumption (ii) and the existence of the integral, (8.4.12) follows readily. ■

***Example* 8.4.2.** Let $X_1, \ldots, X_n$ be i.i.d. random variables with distribution function F and let $\psi(x) = x$; thus, the solution to (8.2.1) defines the LSE, $\widehat{\theta}_n = \overline{X}_n$.

(a) If F corresponds to the $\mathcal{N}(\theta, \sigma^2)$ distribution, then, clearly, $M'(\theta) = -1$ and $\int \psi^2(x-\theta)\mathrm{d}F(x) = \sigma^2$ so that conditions (i) and (ii) of Theorem 8.4.2 are satisfied and, consequently, $\sqrt{n}(\overline{X}_n - \theta) \xrightarrow{D} \mathcal{N}(0, \sigma^2)$.

(b) If F corresponds to a Pearson Type VII distribution with parameters $m = (\nu+1)/2$, $c = \sqrt{\nu}$ and θ, that is, a Student t-distribution with $\nu \geq 3$ degrees of freedom, shifted by θ, we have

$$\begin{aligned} M'(\theta) &= \frac{\partial}{\partial\theta}\int \psi(x-\theta)f(x)\mathrm{d}x \\ &= \frac{\partial}{\partial\theta}\int (x-\theta)\frac{\Gamma[(\nu+1)/2]}{\sqrt{\pi\nu}\Gamma(\nu/2)}\left[1 + \frac{(x-\theta)^2}{\nu}\right]^{-(\nu+1)/2}\mathrm{d}x \\ &= -1 \end{aligned}$$

and $\int \psi^2(x-\theta)f(x)\mathrm{d}x = \nu/(\nu-2)$, so that by Theorem 8.4.2 we may conclude that $\sqrt{n}(\overline{X}_n - \theta) \xrightarrow{D} \mathcal{N}[0, \nu/(\nu-2)]$.

(c) If F is as in (b), but $\nu = 2$, we may not apply Theorem 8.4.2, since condition (ii) does not hold; in such a case, a robust M-estimator for θ may be considered. □

***Example* 8.4.3.** Let $X_1, \ldots, X_n$ be i.i.d. random variables with density function f, symmetric with respect to θ_0, and let ψ be the Huber score function given by (2.4.22). Then, we have

$$\begin{aligned} M(\theta) &= \int_{\theta-k}^{\theta+k}(x-\theta)f(x)\mathrm{d}x + k\int_{\theta+k}^{\infty}f(x)\mathrm{d}x - k\int_{-\infty}^{\theta-k}f(x)\mathrm{d}x \\ &= \int_{-k}^{k}xf(x+\theta)\mathrm{d}x + k\left[P(X \geq \theta+k) - P(X \leq \theta-k)\right]. \end{aligned}$$

Clearly, $M(\theta_0) = 0$ and

$$M'(\theta_0) = \int_{\theta_0-k}^{\theta_0+k}f(x)\mathrm{d}x = P(\theta_0 - k \leq X \leq \theta_0 + k).$$

Also, since ψ is bounded, we have

$$\sigma^2(\theta_0) = \int_{\theta_0-k}^{\theta_0+k}(x-\theta_0)^2 f(x)\mathrm{d}x + 2k^2 P(X \geq \theta_0 + k) < \infty$$

irrespectively of whether $\int_{-\infty}^{+\infty} x^2 f(x)\mathrm{d}x$ is finite or not. Thus, the asymptotic normality of the M-estimator $\widehat{\theta}_n$ obtained as a solution to (8.2.1) with ψ as defined in (2.4.22) follows from Theorem 8.4.2. □

8.5 Asymptotic Efficiency of Estimators

As we have seen, many types of estimators commonly employed in statistical analysis are asymptotically normally distributed around the true value of the parameter. Thus, in order to choose among competing estimators we need further criteria. Consider, for example, sequences $\{\widehat{\theta}_n^{(1)}\}$ and $\{\widehat{\theta}_n^{(2)}\}$ of estimators such that $\sqrt{n}(\widehat{\theta}_n^{(1)} - \theta) \xrightarrow{D} \mathcal{N}(0, \sigma_1^2)$ and $\sqrt{n}(\widehat{\theta}_n^{(2)} - \theta) \xrightarrow{D} \mathcal{N}(0, \sigma_2^2)$. We may use the corresponding asymptotic variances σ_1^2 and σ_2^2 to define the *asymptotic relative efficiency* (ARE) of $\widehat{\theta}_n^{(2)}$ with respect to $\widehat{\theta}_n^{(1)}$ as

$$\mathrm{ARE}(\widehat{\theta}_n^{(2)}|\widehat{\theta}_n^{(1)}) = \sigma_1^2/\sigma_2^2.$$

Obviously, if $\mathrm{ARE}(\widehat{\theta}_n^{(2)}|\widehat{\theta}_n^{(1)}) < 1$, we say that $\widehat{\theta}_n^{(1)}$ is asymptotically more efficient than $\widehat{\theta}_n^{(2)}$.

***Example* 8.5.1.** Let $X_1, \dots, X_n$ be i.i.d. random variables following a Pareto distribution, that is, with density function

$$f(x;\theta) = \theta^{-1}x^{-(1+\theta^{-1})}, \quad x > 1, \quad 0 < \theta < 1/2.$$

Letting $\gamma = \theta^{-1}$, which implies $\gamma > 2$, we have

$$\mu^{(1)} = \mathbb{E}(X_1) = \int_1^\infty x\gamma x^{-1-\gamma}\mathrm{d}x = \frac{\gamma}{\gamma - 1} = \frac{1}{1-\theta}.$$

Therefore, solving for θ and setting $\overline{X}_n = (1-\theta)^{-1}$, we obtain a MME of θ, namely, $\widetilde{\theta}_n = 1 - 1/\overline{X}_n$. Now let $h(\theta) = (1-\theta)^{-1}$ so that $H(\theta) = (\mathrm{d}/\mathrm{d}\theta)h(\theta) = (1-\theta)^{-2}$. Also, observe that

$$\begin{aligned}\mu^{(2)} = \mathbb{E}(X_1^2) &= \int_1^\infty x^2\gamma x^{-1-\gamma}\mathrm{d}x \\ &= \frac{\gamma}{2-\gamma}\int_1^\infty (2-\gamma)x^{1-\gamma}\mathrm{d}x = \frac{\gamma}{\gamma-2} = \frac{1}{1-2\theta}.\end{aligned}$$

Thus,

$$\mathbb{V}\mathrm{ar}(X_1) = \sigma^2 = \frac{1}{1-2\theta} - \frac{1}{(1-\theta)^2} = \frac{\theta^2}{(1-2\theta)(1-\theta)^2},$$

and

$$[H(\theta)]^{-2}\sigma^2 = \frac{\theta^2(1-\theta)^2}{1-2\theta};$$

so from Theorem 8.4.1 it follows that

$$\sqrt{n}(\widetilde{\theta}_n - \theta) \xrightarrow{D} \mathcal{N}[0, \theta^2(1-\theta)^2/(1-2\theta)].$$

Alternatively, the MLE of θ is a solution $\widehat{\theta}_n$ to

$$\frac{\partial}{\partial\theta}\log L_n(\theta) = -\frac{n}{\theta} + \frac{1}{\theta^2}\sum_{i=1}^n \log X_i = 0,$$

which implies $\widehat{\theta}_n = n^{-1}\sum_{i=1}^n \log X_i$. Now observe that

(i)

$$\begin{aligned}\left|\frac{\partial}{\partial\theta}f(x;\theta)\right| &= |\theta^{-2}x^{-(1+\theta^{-1})}+\theta^{-3}x^{-(1+\theta^{-1})}\log x|\\ &\le \theta^{-2}x^{-(1+\theta^{-1})}+\theta^{-3}x^{-(1+\theta^{-1})}\\ &= H_1(x),\end{aligned}$$

which is integrable;

(ii)

$$\begin{aligned}\left|\frac{\partial^2}{\partial\theta^2}f(x;\theta)\right| &= |-2\theta^{-3}x^{-(1+\theta^{-1})}-2\theta^{-4}x^{-(1+\theta^{-1})}\log x+\theta^{-5}x^{-1(1+\theta^{-1})}\log^2 x|\\ &\le 2\theta^{-3}x^{-(1+\theta^{-1})}\left[1+\theta^{-1}\log x+\frac{1}{2}\theta^{-2}\log^2 x\right]\\ &= H_2(x),\end{aligned}$$

which is also integrable;

(iii) $\mathbb{E}(\log X_1)=\theta$ and $[\mathbb{E}(\log X_1)]^2=2\theta^2$, which imply that

$$\begin{aligned}I(\theta) &= \mathbb{E}\left[\frac{\partial}{\partial\theta}\log f(X_1;\theta)\right]^2\\ &= \mathbb{E}\left(\frac{1}{\theta^2}-\frac{2}{\theta^3}\log X_1+\frac{1}{\theta^4}\log^2 X_1\right)=\frac{1}{\theta^2}<\infty;\end{aligned}$$

(iv) $(\partial^2/\partial\theta^2)\log f(x;\theta)=(1/\theta^2)-(2/\theta^3)\log x$, which implies that

$$\begin{aligned}\psi_\delta &= \mathbb{E}\left[\sup_{\{h:|h|\le\delta\}}\left|\frac{1}{(\theta+h)^2}-\frac{1}{\theta^2}-\frac{2}{(\theta+h)^3}\log X_1+\frac{2}{\theta^3}\log X_1\right|\right]\\ &\le \sup_{\{h:|h|\le\delta\}}\left|\frac{1}{(\theta+h)^2}-\frac{1}{\theta^2}\right|+\sup_{\{h:|h|\le\delta\}}\left[\left|\frac{1}{(\theta+h)^3}-\frac{1}{\theta^3}\right|2\mathbb{E}(\log X_1)\right]\\ &= \left|\frac{1}{(\theta+\delta)^2}-\frac{1}{\theta^2}\right|+\left|\frac{1}{(\theta+\delta)^3}-\frac{1}{\theta^3}\right|2\theta\longrightarrow 0\quad\text{as}\quad\delta\to 0.\end{aligned}$$

Therefore, from Theorem 8.3.1 it follows that $\sqrt{n}(\widehat{\theta}_n-\theta)\xrightarrow{D}\mathcal{N}(0,\theta^2)$. Consequently, we have

$$\text{ARE}(\widetilde{\theta}_n|\widehat{\theta}_n)=\theta^2\frac{(1-2\theta)}{\theta^2(1-\theta)^2}=\frac{1-2\theta}{(1-\theta)^2}<1,$$

so that the MME is asymptotically less efficient than the MLE. □

Now let $X_1,\ldots,X_n$ be i.i.d. random variables with density function $f(x;\theta),\theta\in\Theta\subset\mathbb{R}$, such that $\mathbb{E}\{(\partial/\partial\theta)\log f(X_1;\theta)\}^2=I(\theta)<\infty$ and consider a sequence $\{\widehat{\theta}_n\}$ of unbiased estimators of θ, such that $\mathbb{V}\text{ar}(\widehat{\theta}_n)<\infty$. Then note that, by the Cauchy–Schwarz Inequality (1.5.34),

$$\left\{\mathbb{E}\left[\sqrt{n}(\widehat{\theta}_n-\theta)\frac{1}{\sqrt{n}}\frac{\partial}{\partial\theta}\log L_n(\theta)\right]\right\}^2\le\mathbb{E}\left[\sqrt{n}(\widehat{\theta}_n-\theta)\right]^2\mathbb{E}\left[\frac{1}{\sqrt{n}}\frac{\partial}{\partial\theta}\log L_n(\theta)\right]^2. \tag{8.5.1}$$

Expanding the left-hand side of (8.5.1) we obtain

$$\begin{aligned}
&\mathbb{E}\left[\sqrt{n}(\widehat{\theta}_n - \theta)\frac{1}{\sqrt{n}}\frac{\partial}{\partial\theta}\log L_n(\theta)\right] \\
&= \mathbb{E}\left[\widehat{\theta}_n\frac{\partial}{\partial\theta}\log L_n(\theta)\right] - \theta\mathbb{E}\left[\frac{\partial}{\partial\theta}\log L_n(\theta)\right] \\
&= \int \widehat{\theta}_n(x)\left[\frac{\partial}{\partial\theta}\log L_n(\theta;x)\right]L_n(\theta;x)\mathrm{d}x \\
&= \int \widehat{\theta}_n(x)\frac{\partial}{\partial\theta}L_n(\theta;x)\mathrm{d}x \\
&= \frac{\partial}{\partial\theta}\int \widehat{\theta}_n(x)L_n(\theta;x)\mathrm{d}x = \frac{\partial}{\partial\theta}\theta = 1. \qquad (8.5.2)
\end{aligned}$$

Since

$$\mathbb{E}\left[\frac{1}{\sqrt{n}}\frac{\partial}{\partial\theta}\log L_n(\theta)\right]^2 = I(\theta) < \infty,$$

it follows from (8.5.1) and (8.5.2) that:

$$\mathbb{E}[\sqrt{n}(\widehat{\theta}_n - \theta)]^2 = n\mathbb{V}\mathrm{ar}(\widehat{\theta}_n) \geq [I(\theta)]^{-1}, \qquad (8.5.3)$$

which is the *Information Inequality*. This lower bound suggests the definition of the *asymptotic efficiency* of a sequence $\{\widehat{\theta}_n\}$ of estimators for which $\sqrt{n}(\widehat{\theta}_n - \theta) \xrightarrow{D} \mathcal{N}(0, \sigma_\theta^2)$ as the ratio $[I(\theta)]^{-1}/\sigma_\theta^2$. Moreover, if $\sigma_\theta^2 = [I(\theta)]^{-1}$ the corresponding estimator T_n is said to be *asymptotically efficient*. Note that, in general, the asymptotic efficiency of an estimator as defined above may not coincide with the limit of the *Fisher efficiency* given by $[I(\theta)]^{-1}/\lim_{n\to\infty}\mathbb{E}[\sqrt{n}(\widehat{\theta}_n - \theta)^2]$.

In view of Theorem 8.3.1, we may conclude that under quite general regularity conditions, maximum likelihood estimators are asymptotically efficient. More generally, if we recall that equality in (8.5.1) holds if and only if

$$\sqrt{n}(\widehat{\theta}_n - \theta) = Kn^{-1/2}(\partial/\partial\theta)\log L_n(\theta),$$

with $K \neq 0$, it follows that all estimators for which

$$\sqrt{n}(\widehat{\theta}_n - \theta) = Kn^{-1/2}(\partial/\partial\theta)\log L_n(\theta) + o_p(1)$$

are also asymptotically efficient. Estimators in this class are called *best asymptotically normal* (BAN).

***Example* 8.5.2.** Let $X_1, \ldots, X_n$ be i.i.d. random variables with a $\mathcal{N}(\mu, \sigma^2)$ distribution. Since the assumptions of Theorem 8.3.1 are clearly satisfied in this case, it follows that both the (biased) MLE of σ^2, $\widehat{\sigma}_n^2 = n^{-1}\sum_{i=1}^n(X_i - \overline{X}_n)^2$, and the unbiased estimator s_n^2 are such that $\sqrt{n}(\widehat{\sigma}_n^2 - \sigma^2)$ or $\sqrt{n}(s_n^2 - \sigma^2)$ converge in distribution to a random variable following a $\mathcal{N}(0, \mu_4 - \sigma^4)$ distribution, and, consequently, they are BAN estimators. In fact, note that $s_n^2 = \widehat{\sigma}_n^2 - \widehat{\sigma}_n^2/n = \widehat{\sigma}_n^2 + O_p(n^{-1})$. □

The above concepts may be readily extended to the multiparameter case; in this direction, consider i.i.d. random variables, $X_1, \ldots, X_n$ with density function $f(x;\boldsymbol{\theta})$ satisfying the regularity conditions of Theorem 8.2.3 and let $\{\widehat{\boldsymbol{\theta}}_n\}$ be a sequence of unbiased estimators of $\boldsymbol{\theta}$. Here, the Information Inequality (8.5.3) is replaced by the condition that

$$\mathbb{E}[n(\widehat{\boldsymbol{\theta}}_n - \boldsymbol{\theta})(\widehat{\boldsymbol{\theta}}_n - \boldsymbol{\theta})'] - [\boldsymbol{I}(\boldsymbol{\theta})]^{-1} = n\boldsymbol{D}(\widehat{\boldsymbol{\theta}}_n) - [\boldsymbol{I}(\boldsymbol{\theta})]^{-1} \tag{8.5.4}$$

is nonnegative definite. Then, if $\sqrt{n}(\widehat{\boldsymbol{\theta}}_n - \boldsymbol{\theta}) \xrightarrow{D} \mathcal{N}\{\boldsymbol{0}, [\boldsymbol{I}(\boldsymbol{\theta})]^{-1}\}$, the estimator $\widehat{\boldsymbol{\theta}}_n$ is said to be asymptotically efficient. To prove that (8.5.4) is nonnegative definite, first let $\boldsymbol{\lambda}$ and $\boldsymbol{\gamma}$ be arbitrary, but fixed vectors in $\mathbb{R}^q$ take

$$\boldsymbol{U}_n(\boldsymbol{X}, \boldsymbol{\theta}) = (\partial/\partial\boldsymbol{\theta}) \log L_n(\boldsymbol{\theta})$$

and use the Cauchy–Schwarz Inequality (1.5.34) to see that

$$\begin{aligned}
&\mathbb{E}\left[\sqrt{n}\boldsymbol{\lambda}'(\widehat{\boldsymbol{\theta}}_n - \boldsymbol{\theta})\frac{1}{\sqrt{n}}[\boldsymbol{U}_n(\boldsymbol{X}, \boldsymbol{\theta})]'\boldsymbol{\gamma}\right]^2 \\
&\quad \le \mathbb{E}[n\boldsymbol{\lambda}'(\widehat{\boldsymbol{\theta}}_n - \boldsymbol{\theta})(\widehat{\boldsymbol{\theta}}_n - \boldsymbol{\theta})'\boldsymbol{\lambda}]\mathbb{E}\left[\frac{1}{n}\boldsymbol{\gamma}'\boldsymbol{U}_n(\boldsymbol{X};\boldsymbol{\theta})[\boldsymbol{U}_n(\boldsymbol{X}, \boldsymbol{\theta})]'\boldsymbol{\gamma}\right].
\end{aligned} \tag{8.5.5}$$

The same argument employed to show (8.5.2) may be considered to prove that the left-hand side of (8.5.5) is equal to $\boldsymbol{\lambda}'\boldsymbol{\gamma}$; consequently, we may write

$$\mathbb{E}[n\boldsymbol{\lambda}'(\widehat{\boldsymbol{\theta}}_n - \boldsymbol{\theta})(\widehat{\boldsymbol{\theta}}_n - \boldsymbol{\theta})'\boldsymbol{\lambda}] \ge \frac{\boldsymbol{\lambda}'\boldsymbol{\gamma}}{\boldsymbol{\gamma}'\boldsymbol{I}(\boldsymbol{\theta})\boldsymbol{\gamma}}. \tag{8.5.6}$$

Maximizing the right-hand side of (8.5.6) with respect to $\boldsymbol{\gamma}$ we obtain

$$\boldsymbol{\lambda}'[n\mathbb{E}[(\widehat{\boldsymbol{\theta}}_n - \boldsymbol{\theta})(\widehat{\boldsymbol{\theta}}_n - \boldsymbol{\theta})']]\boldsymbol{\lambda} \ge \boldsymbol{\lambda}'[\boldsymbol{I}(\boldsymbol{\theta})]^{-1}\boldsymbol{\lambda}, \tag{8.5.7}$$

and since $\boldsymbol{\lambda}$ is arbitrary, it follows that (8.5.4) is nonnegative definite. Furthermore, it follows that

$$\mathbb{E}[n(\widehat{\boldsymbol{\theta}}_n - \boldsymbol{\theta})(\widehat{\boldsymbol{\theta}}_n - \boldsymbol{\theta})'] = [\boldsymbol{I}(\boldsymbol{\theta})]^{-1}$$

if any only if

$$\sqrt{n}(\widehat{\boldsymbol{\theta}}_n - \boldsymbol{\theta}) = \boldsymbol{K}n^{-1/2}(\partial/\partial\boldsymbol{\theta}) \log L_n(\boldsymbol{\theta})$$

where $\boldsymbol{K}$ is a nonsingular matrix; thus, all estimators $\widehat{\boldsymbol{\theta}}_n$ for which

$$\sqrt{n}(\widehat{\boldsymbol{\theta}}_n - \boldsymbol{\theta}) = \boldsymbol{K}n^{-1/2}(\partial/\partial\boldsymbol{\theta}) \log L_n(\boldsymbol{\theta}) + o_p(\boldsymbol{1})$$

are asymptotically efficient. This is the case with the MLE obtained in situations where the conditions of Theorem 8.3.1 are satisfied.

The asymptotic optimality (or efficiency) of estimators considered here has been tuned to the asymptotic mean squared error in the single parameter case and to the asymptotic covariance matrix in the vector-parameter case. In the literature there are some alternative measures of efficiency of estimators. Among these a noteworthy case is the *Pitman measure of closeness* (PMC). Let T_1 and T_2 be two rival estimators of a real-valued parameter θ. Then, T_1 is said to be closer to θ than T_2 in the Pitman (1937) sense if

$$P_\theta(|T_1 - \theta| \le |T_2 - \theta|) \ge 1/2, \quad \text{for all } \theta, \tag{8.5.8}$$

with strict inequality holding for some θ. This definition is easily generalizable by replacing the simple Euclidean distance by a general loss function $L(a, b)$, and we may say that an estimator T_1 (possibly, vector-valued) is "Pitman-closer" than a rival one, T_2 [with respect to the loss function $L(\cdot)$] if

$$P_\theta\left(L(T_1, \theta) \le L(T_2, \theta)\right) \ge 1/2, \quad \text{for all } \theta, \tag{8.5.9}$$

with strict inequality holding for some θ. In particular, in the vector case, usually, $L(\boldsymbol{a}, \boldsymbol{b})$ is taken as a quadratic loss function, that is, $L(\boldsymbol{a}, \boldsymbol{b}) = (\boldsymbol{a} - \boldsymbol{b})'\boldsymbol{Q}(\boldsymbol{a} - \boldsymbol{b})$ for some given p.d. $\boldsymbol{Q}$. If (8.5.8) [or (8.5.9)] holds for all T_2 belonging to a class $\mathcal{C}$, then T_1 is said to be the "Pitman-closest" estimator within the same class.

In an asymptotic shade, in (8.5.8) or (8.5.9), we may replace the estimators T_1 and T_2 by sequences of estimators $\{T_{n1}\}$ and $\{T_{n2}\}$, respectively, and that

$$\liminf_{n\to\infty} \left\{P_\theta\left[L(T_{n1}, \theta) \le L(T_{n2}, \theta)\right]\right\} \ge 1/2, \quad \text{for all } \theta, \tag{8.5.10}$$

then we say that $\{T_{n1}\}$ is asymptotically closer to θ than $\{T_{n2}\}$ in the Pitman sense.

Whereas a complete treatment of PMC or even the asymptotic PMC is somewhat beyond the scope of this book, we may refer to Keating, Mason, and Sen (1993, chapter 6) for a comprehensive treatise on this topic. They show that within a general class of estimators that are asymptotic normal, a BAN estimator in the sense of having minimum asymptotic variance (as has been stressed earlier) is also the "Pitman-closest" one in an asymptotic setup. Thus, whatever we have discussed about the MLE and other estimators in the light of their BAN characterization pertains to their asymptotic PMC characterization as well. Actually, an asymptotic (partial) ordering of estimators in terms of their asymptotic mean-squared errors is isomorphic to an asymptotic PMC ordering. Thus, viewed from this asymptotic stand, there is no need to prefer the particular definition of asymptotic efficiency/optimality, and it may even be interpreted in a more general setup.

We conclude this section by noting that even if the true underlying distribution function does not match the one utilized in deriving the MLE, it may still have an asymptotic normal distribution, although it may not be asymptotically efficient [see Singer and Sen (1985)]; this illustrates the lack of robustness of MLE and justifies the need of (distributionally) robust estimation methods.

8.6 Asymptotic Behavior of Some Test Statistics

Let $X_1, \ldots, X_n$ be i.i.d. random variables with density function $f(x; \boldsymbol{\theta})$, $\boldsymbol{\theta} \in \boldsymbol{\Theta} \subset \mathbb{R}^q$. We have seen in Section 3.2 that the most powerful test for $H_0 : \boldsymbol{\theta} = \boldsymbol{\theta}_0$ versus $H_1 : \boldsymbol{\theta} = \boldsymbol{\theta}_1$, where $\boldsymbol{\theta}_0$ and $\boldsymbol{\theta}_1$ are two fixed points in $\boldsymbol{\Theta}$, is given by the *Neyman–Pearson criterion*, that is, reject H_0 if $\lambda_n = \lambda_n(X_1, \ldots, X_n) = L_n(\boldsymbol{\theta}_1)/L_n(\boldsymbol{\theta}_0) \ge k_\alpha$ and accept H_0 otherwise, where $k_\alpha = k_\alpha(n, \alpha, \boldsymbol{\theta}_0, \boldsymbol{\theta}_1)$ is determined by $P\{\lambda_n \ge k_\alpha | H_0\} = P_{\boldsymbol{\theta}_0}(\lambda_n \ge k_\alpha) = \alpha$, $0 < \alpha < 1$, and α is the level of significance of the test.

***Example* 8.6.1.** Let $X_1, \ldots, X_n$ be i.i.d. random variables with a $\mathcal{N}(\theta, 1)$ distribution and consider the problem of testing $H_0 : \theta = \theta_0$ versus $H_1 : \theta = \theta_1$. Here, we have

$\log\{L_n(\theta_1)/L_n(\theta_0)\} = n(\theta_1 - \theta_0)\{\overline{X}_n - (\theta_0 + \theta_1)/2\}$, which implies that

$$P_{\theta_0}(\lambda_n \geq k_\alpha) = P_{\theta_0}\left[n\left(\overline{X}_n - \frac{\theta_0 + \theta_1}{2}\right) \geq \frac{\log k_\alpha}{\theta_1 - \theta_0}\right]$$
$$= P_{\theta_0}\left[\sqrt{n}(\overline{X}_n - \theta_0) \geq \frac{\log k_\alpha}{\sqrt{n}(\theta_1 - \theta_0)} + \sqrt{n}\frac{(\theta_1 - \theta_0)}{2}\right]$$

and since $\sqrt{n}(\overline{X}_n - \theta_0) \sim \mathcal{N}(0, 1)$, it follows that

$$k_\alpha = \exp\left[\sqrt{n}(\theta_1 - \theta_0)z_{1-\alpha} - n(\theta_1 - \theta_0)^2\right],$$

where $z_{1-\alpha}$ is the $100(1-\alpha)$th percentile of the $\mathcal{N}(0, 1)$ distribution.

The question is how to determine the constant k_α when the underlying distribution is nonnormal. In this direction, if f denotes the corresponding density function, let

$$Z_i = \log[f(X_i; \theta_1)/f(X_i; \theta_0)]$$

and assume that $\mathbb{V}\mathrm{ar}_{\theta_0}(Z_i) = \sigma_0^2 < \infty$. Then

$$P_{\theta_0}(\lambda_n \geq k_\alpha) = P_{\theta_0}\left[\sum_{i=1}^n \log\frac{f(X_i; \theta_1)}{f(X_i; \theta_0)} \geq \log k_\alpha\right]$$
$$= P_{\theta_0}\left(\sum_{i=1}^n Z_i \geq \log k_\alpha\right)$$
$$= P_{\theta_0}\left\{\frac{1}{\sqrt{n}\sigma_0}\sum_{i=1}^n [Z_i - \mathbb{E}(Z_i)] \geq \frac{\log k_\alpha}{\sqrt{n}\sigma_0} - \frac{\sqrt{n}}{\sigma_0}\mathbb{E}(Z_1)\right\}.$$

By the Central Limit Theorem 7.3.1, it follows that

$$n^{-1/2}\sigma_0^{-1}\sum_{i=1}^n [Z_i - \mathbb{E}(Z_i)] \xrightarrow{D} \mathcal{N}(0, 1)$$

and, then, for sufficiently large n we have

$$\frac{\log k_\alpha}{\sqrt{n}\sigma_0} - \frac{\sqrt{n}}{\sigma_0}\mathbb{E}(Z_1) \simeq z_{1-\alpha},$$

which implies that $k_\alpha \simeq \exp\{\sqrt{n}\sigma_0 z_{1-\alpha} + n\mathbb{E}(Z_1)\}$. □

More generally, for testing $H_0: \boldsymbol{\theta} = \boldsymbol{\theta}_0$ versus $H_1: \boldsymbol{\theta} \neq \boldsymbol{\theta}_0$, most powerful or uniformly most powerful tests do not exist, and we have to rely on other criteria for the choice of an appropriate test statistic. In particular, when the conditions of Theorem 8.3.1 are satisfied, the following three alternatives discussed in Section 3.3 are more frequently considered:

(i) *Wald statistic*:

$$Q_W = n(\widehat{\boldsymbol{\theta}}_n - \boldsymbol{\theta}_0)'\boldsymbol{I}(\widehat{\boldsymbol{\theta}}_n)(\widehat{\boldsymbol{\theta}}_n - \boldsymbol{\theta}_0)$$

where $\widehat{\boldsymbol{\theta}}_n$ is the MLE (or any BAN estimator) of $\boldsymbol{\theta}_0$ and

$$\boldsymbol{I}(\widehat{\boldsymbol{\theta}}_n) = -\frac{1}{n}\sum_{i=1}^n \frac{\partial^2}{\partial\boldsymbol{\theta}\partial\boldsymbol{\theta}'}\log f(X_i; \boldsymbol{\theta})\bigg|_{\widehat{\boldsymbol{\theta}}_n}.$$

If $\boldsymbol{\theta}_0$ specifies the density function completely, then we may use $\boldsymbol{I}(\boldsymbol{\theta}_0)$ instead of $\boldsymbol{I}(\widehat{\boldsymbol{\theta}}_n)$ in the above definition of Q_W.

(ii) *Wilks likelihood ratio statistic:*

$$Q_L = -2\log\lambda_n = 2\{\log L_n(\widehat{\boldsymbol{\theta}}_n) - \log L_n(\boldsymbol{\theta}_0)\}$$

where $\lambda_n = L_n(\boldsymbol{\theta}_0)/\sup_{\boldsymbol{\theta}\in\boldsymbol{\Theta}} L_n(\boldsymbol{\theta})$;

(iii) *Rao efficient score statistic:*

$$Q_R = n^{-1}[\boldsymbol{U}_n(\boldsymbol{\theta}_0)]'[\boldsymbol{I}(\boldsymbol{\theta}_0)]^{-1}\boldsymbol{U}_n(\boldsymbol{\theta}_0), \tag{8.6.1}$$

where $\boldsymbol{U}_n(\boldsymbol{\theta}_0) = \sum_{i=1}^n (\partial/\partial\boldsymbol{\theta})\log f(X_i;\boldsymbol{\theta})|_{\boldsymbol{\theta}_0}$.

In the following theorem we derive the asymptotic null distribution of the three statistics indicated above.

***Theorem* 8.6.1.** *Let $X_1,\ldots,X_n$ be i.i.d. random variables with density function $f(x;\boldsymbol{\theta})$, $\boldsymbol{\theta}\in\boldsymbol{\Theta}\subset\mathbb{R}^q$ satisfying the conditions of Theorem 8.3.1 and consider the problem of testing $H_0:\boldsymbol{\theta}=\boldsymbol{\theta}_0$ versus $H_1:\boldsymbol{\theta}\neq\boldsymbol{\theta}_0$. Then, each of the statistics Q_W, Q_L and Q_R indicated above has an asymptotic χ^2_q distribution under H_0.*

Proof. From Theorem 8.3.1 we know that if $\boldsymbol{\theta}=\boldsymbol{\theta}_0$, the MLE of $\boldsymbol{\theta}$, say $\widehat{\boldsymbol{\theta}}_n$, is such that $\sqrt{n}(\widehat{\boldsymbol{\theta}}_n - \boldsymbol{\theta}_0) \xrightarrow{D} \mathcal{N}\{\boldsymbol{0},[\boldsymbol{I}(\boldsymbol{\theta})]^{-1}\}$, which implies that under H_0,

$$Q_W^* = n(\widehat{\boldsymbol{\theta}}_n - \boldsymbol{\theta}_0)'\boldsymbol{I}(\boldsymbol{\theta}_0)(\widehat{\boldsymbol{\theta}}_n - \boldsymbol{\theta}_0) \xrightarrow{D} \chi^2_q;$$

thus, letting $\boldsymbol{x} = \sqrt{n}(\widehat{\boldsymbol{\theta}}_n - \boldsymbol{\theta}_0)$ we have

$$\frac{Q_W}{Q_W^*} = \frac{\boldsymbol{x}'\boldsymbol{I}(\widehat{\boldsymbol{\theta}}_n)\boldsymbol{x}}{\boldsymbol{x}'\boldsymbol{I}(\boldsymbol{\theta}_0)\boldsymbol{x}}$$

and, using the Courant Theorem 1.5.1, it follows that

$$\mathrm{ch}_q\{\boldsymbol{I}(\widehat{\boldsymbol{\theta}}_n)[\boldsymbol{I}(\boldsymbol{\theta}_0)]^{-1}\} \leq \frac{Q_W}{Q_W^*} \leq \mathrm{ch}_1\{\boldsymbol{I}(\widehat{\boldsymbol{\theta}}_n)[\boldsymbol{I}(\boldsymbol{\theta}_0)]^{-1}\},$$

where $\mathrm{ch}_j(\boldsymbol{A})$ denotes the jth characteristic root of $\boldsymbol{A}$. Now, by the Khintchine WLLN (Theorem 6.3.6) we may conclude that $\boldsymbol{I}(\widehat{\boldsymbol{\theta}}_n)[\boldsymbol{I}(\boldsymbol{\theta}_0)]^{-1} \xrightarrow{P} \boldsymbol{I}_q$ and, consequently, we have

$$\mathrm{ch}_1\{\boldsymbol{I}(\widehat{\boldsymbol{\theta}}_n)[\boldsymbol{I}(\boldsymbol{\theta}_0)]^{-1}\} \xrightarrow{P} 1,$$

and

$$\mathrm{ch}_q\{\boldsymbol{I}(\widehat{\boldsymbol{\theta}}_n)[\boldsymbol{I}(\boldsymbol{\theta}_0)]^{-1}\} \xrightarrow{P} 1.$$

Then it follows that $Q_W/Q_W^* \xrightarrow{P} 1$, and writing $Q_W = Q_W^* \times Q_W/Q_W^*$ we may conclude, via the Slutsky Theorem 7.5.2, that $Q_W \xrightarrow{D} \chi^2_q$ under H_0.

To obtain the asymptotic distribution of Q_L under H_0, first consider the following Taylor expansion:

$$\log L_n(\widehat{\boldsymbol{\theta}}_n) = \log L_n(\boldsymbol{\theta}_0) + \sqrt{n}(\widehat{\boldsymbol{\theta}}_n - \boldsymbol{\theta}_0)' \left[\frac{1}{\sqrt{n}} \frac{\partial}{\partial \boldsymbol{\theta}} \log L_n(\boldsymbol{\theta}) \bigg|_{\boldsymbol{\theta}_0} \right]$$
$$+ \frac{n}{2}(\widehat{\boldsymbol{\theta}} - \boldsymbol{\theta}_0)' \left[\frac{1}{n} \frac{\partial^2}{\partial \boldsymbol{\theta} \partial \boldsymbol{\theta}'} \log L_n(\boldsymbol{\theta}) \bigg|_{\boldsymbol{\theta}_n^*} \right] (\widehat{\boldsymbol{\theta}}_n - \boldsymbol{\theta}_0), \tag{8.6.2}$$

where $\boldsymbol{\theta}_n^*$ belongs to the line segment joining $\boldsymbol{\theta}_0$ and $\widehat{\boldsymbol{\theta}}_n$. Then, consider the following Taylor expansion of the first term within brackets, [ ], on the right-hand side of (8.6.2)

$$\frac{1}{\sqrt{n}} \frac{\partial}{\partial \boldsymbol{\theta}} \log L_n(\boldsymbol{\theta}_0) = \frac{1}{\sqrt{n}} \frac{\partial}{\partial \boldsymbol{\theta}} \log L_n(\boldsymbol{\theta}) \bigg|_{\widehat{\boldsymbol{\theta}}_n}$$
$$+ \left[-\frac{1}{n} \frac{\partial^2}{\partial \boldsymbol{\theta} \partial \boldsymbol{\theta}'} \log L_n(\boldsymbol{\theta}) \bigg|_{\boldsymbol{\theta}_n^{**}} \right] \sqrt{n}(\widehat{\boldsymbol{\theta}}_n - \boldsymbol{\theta}_0)$$
$$= \left[-\frac{1}{n} \frac{\partial^2}{\partial \boldsymbol{\theta} \partial \boldsymbol{\theta}'} \log L_n(\boldsymbol{\theta}) \bigg|_{\boldsymbol{\theta}_n^{**}} \right] \sqrt{n}(\widehat{\boldsymbol{\theta}}_n - \boldsymbol{\theta}_0), \tag{8.6.3}$$

where $\boldsymbol{\theta}_n^{**}$ belongs to the line segment joining $\boldsymbol{\theta}_0$ and $\widehat{\boldsymbol{\theta}}_n$. Substituting (8.6.3) into (8.6.2) we get

$$\log L_n(\widehat{\boldsymbol{\theta}}_n) = \log L_n(\boldsymbol{\theta}_0) - n(\widehat{\boldsymbol{\theta}}_n - \boldsymbol{\theta}_0)' \left[\frac{1}{n} \frac{\partial^2}{\partial \boldsymbol{\theta} \partial \boldsymbol{\theta}'} \log L_n(\boldsymbol{\theta}) \bigg|_{\boldsymbol{\theta}_n^{**}} \right] (\widehat{\boldsymbol{\theta}}_n - \boldsymbol{\theta}_0)$$
$$+ \frac{n}{2}(\widehat{\boldsymbol{\theta}}_n - \boldsymbol{\theta}_0)' \left[\frac{1}{n} \frac{\partial^2}{\partial \boldsymbol{\theta} \partial \boldsymbol{\theta}'} \log L_n(\boldsymbol{\theta}) \bigg|_{\boldsymbol{\theta}_n^*} \right] (\widehat{\boldsymbol{\theta}}_n - \boldsymbol{\theta}_0). \tag{8.6.4}$$

By the Khintchine WLLN (Theorem 6.3.6) under H_0, both terms within brackets, [ ], in (8.6.4) converge in probability to $-\boldsymbol{I}(\boldsymbol{\theta}_0)$; therefore, we have

$$Q_L = 2\left[\log L_n(\widehat{\boldsymbol{\theta}}_n) - \log L_n(\boldsymbol{\theta}_0)\right]$$
$$= n(\widehat{\boldsymbol{\theta}}_n - \boldsymbol{\theta}_0)' \boldsymbol{I}(\boldsymbol{\theta}_0)(\widehat{\boldsymbol{\theta}}_n - \boldsymbol{\theta}_0) + o_p(1) \tag{8.6.5}$$

and by the Slutsky Theorem 7.5.2 it follows that $Q_L \xrightarrow{D} \chi_q^2$, under H_0. Finally, note that, under H_0, the Central Limit Theorem 7.3.1 implies that $n^{-1/2}\boldsymbol{U}_n(\boldsymbol{\theta}_0) \xrightarrow{D} \mathcal{N}[\boldsymbol{0}, \boldsymbol{I}(\boldsymbol{\theta}_0)]$; consequently, it follows that $Q_R \xrightarrow{D} \chi_q^2$ under H_0. ■

Let us now examine the distribution of the three test statistics above under some (fixed) alternative hypothesis of the form $H_{1,\boldsymbol{\delta}} : \boldsymbol{\theta} = \boldsymbol{\theta}_0 + \boldsymbol{\delta}$, where $\boldsymbol{\delta}$ is a fixed vector in $\mathbb{R}^q$ such that $\boldsymbol{\theta} \in \boldsymbol{\Theta}$. Consider, for example, the Wald statistic Q_W and note that

$$\sqrt{n}(\widehat{\boldsymbol{\theta}}_n - \boldsymbol{\theta}) = \sqrt{n}(\widehat{\boldsymbol{\theta}}_n - \boldsymbol{\theta}_0) - \sqrt{n}\boldsymbol{\delta},$$

which implies

$$\sqrt{n}(\widehat{\boldsymbol{\theta}}_n - \boldsymbol{\theta}_0) = \sqrt{n}(\widehat{\boldsymbol{\theta}}_n - \boldsymbol{\theta}) + \sqrt{n}\boldsymbol{\delta}.$$

Now, since, under $H_{1,\boldsymbol{\delta}}$,

$$\sqrt{n}(\widehat{\boldsymbol{\theta}}_n - \boldsymbol{\theta}_0) \xrightarrow{D} \mathcal{N}\{\mathbf{0}, [\boldsymbol{I}(\boldsymbol{\theta}_0 + \boldsymbol{\delta})]^{-1}\}$$

and $\sqrt{n}\|\boldsymbol{\delta}\| \to \infty$ as $n \to \infty$, it follows that $Q_W \to \infty$ as $n \to \infty$. Consequently, for all hypotheses $H_{1,\boldsymbol{\delta}}$, we have $P_{\boldsymbol{\theta}}\{Q_W > K\} \to 1$ as $n \to \infty$, for all $K \in \mathbb{R}$. Therefore, in order to concentrate on meaningful comparisons, we restrict ourselves to *local Pitman-type alternatives* of the form

$$H_{1,n} : \boldsymbol{\theta}_n = \boldsymbol{\theta}_0 + n^{-\frac{1}{2}}\boldsymbol{\delta}. \tag{8.6.6}$$

***Theorem* 8.6.2.** *Assume that the conditions of Theorem 8.6.1 are satisfied. Then, under (8.6.6), each of the statistics Q_W, Q_L and Q_R has an asymptotic $\chi_q^2[\boldsymbol{\delta}'\boldsymbol{I}(\boldsymbol{\theta}_0)\boldsymbol{\delta}]$ distribution.*

Proof. First note that $\sqrt{n}(\widehat{\boldsymbol{\theta}}_n - \boldsymbol{\theta}_0) = \sqrt{n}(\widehat{\boldsymbol{\theta}}_n - \boldsymbol{\theta}_n) + \boldsymbol{\delta}$; now, from Theorem 8.3.1, it follows that under $H_{1,n}$,

$$\sqrt{n}(\widehat{\boldsymbol{\theta}}_n - \boldsymbol{\theta}_0) \xrightarrow{D} \mathcal{N}\{\mathbf{0}, [\boldsymbol{I}(\boldsymbol{\theta}_0)]^{-1}\};$$

thus, using the multivariate version of the Slutsky Theorem 7.5.2 we have $\sqrt{n}(\widehat{\boldsymbol{\theta}}_n - \boldsymbol{\theta}_0) \xrightarrow{D} \mathcal{N}\{\boldsymbol{\delta}, [\boldsymbol{I}(\boldsymbol{\theta}_0)]^{-1}\}$ and

$$Q_W \xrightarrow{D} \chi_q^2[\boldsymbol{\delta}'\boldsymbol{I}(\boldsymbol{\theta}_0)\boldsymbol{\delta}]. \tag{8.6.7}$$

Consider, then, the following Taylor expansion:

$$\begin{aligned}\frac{1}{\sqrt{n}}\sum_{i=1}^{n}\frac{\partial}{\partial\boldsymbol{\theta}}\log f(X_i;\boldsymbol{\theta})\bigg|_{\boldsymbol{\theta}_0+\boldsymbol{\delta}/\sqrt{n}} &= \frac{1}{\sqrt{n}}\sum_{i=1}^{n}\frac{\partial}{\partial\boldsymbol{\theta}}\log f(X_i;\boldsymbol{\theta})\bigg|_{\boldsymbol{\theta}_0} \\ &\quad + \left[\frac{1}{n}\sum_{i=1}^{n}\frac{\partial}{\partial\boldsymbol{\theta}\partial\boldsymbol{\theta}'}\log f(X_i;\boldsymbol{\theta})\bigg|_{\boldsymbol{\theta}_n^*}\right]\boldsymbol{\delta},\end{aligned} \tag{8.6.8}$$

where $\boldsymbol{\theta}_n^*$ belongs to the line segment joining $\boldsymbol{\theta}_0$ and $\boldsymbol{\theta}_0 + (\boldsymbol{\delta}/\sqrt{n})$. Using the Central Limit Theorem 7.3.1, we may show that under $H_{1,n}$, the term on the left-hand side of (8.6.8) converges in distribution to $\mathcal{N}[\mathbf{0}, \boldsymbol{I}(\boldsymbol{\theta}_0)^{-1}]$. Using the Khintchine WLLN (Theorem 6.3.6), if follows that the second term on the right-hand side of (8.6.8) converges in probability to $-\boldsymbol{I}(\boldsymbol{\theta})\boldsymbol{\delta}$. Then, using the Slutsky Theorem 7.5.2 we may conclude that, under $H_{1,n}$,

$$n^{-1/2}\sum_{i=1}^{n}(\partial/\partial\boldsymbol{\theta})\log f(X_i;\boldsymbol{\theta})|_{\boldsymbol{\theta}_0} \xrightarrow{D} \mathcal{N}[\boldsymbol{I}(\boldsymbol{\theta}_0)\boldsymbol{\delta}, \boldsymbol{I}(\theta_0)],$$

which from (8.6.1), implies $Q_R \xrightarrow{D} \chi_q^2[\boldsymbol{\delta}'\boldsymbol{I}(\boldsymbol{\theta}_0)\boldsymbol{\delta}]$.

To complete the proof, note that as in (8.6.5), we have

$$Q_L = n(\widehat{\boldsymbol{\theta}}_n - \boldsymbol{\theta}_0)'\boldsymbol{I}(\boldsymbol{\theta}_0)(\widehat{\boldsymbol{\theta}}_n - \boldsymbol{\theta}_0) + o_p(1).$$

Therefore, using the Slutsky Theorem 7.5.2 and (8.6.7) it follows that under $H_{1,n}$,

$$Q_L \xrightarrow{D} \chi_q^2[\boldsymbol{\delta}'\boldsymbol{I}(\boldsymbol{\theta}_0)\boldsymbol{\delta}]. \qquad \blacksquare$$

Next we consider tests of the (composite) hypothesis

$$H_0 : \boldsymbol{h}(\boldsymbol{\theta}) = \boldsymbol{0} \quad \text{versus} \quad H_1 : \boldsymbol{h}(\boldsymbol{\theta}) \neq \boldsymbol{0}, \tag{8.6.9}$$

where $\boldsymbol{h} : \mathbb{R}^q \to \mathbb{R}^r$ is a vector-valued function such that the $(q \times r)$ matrix $\boldsymbol{H}(\boldsymbol{\theta}) = (\partial/\partial\boldsymbol{\theta})[\boldsymbol{h}(\boldsymbol{\theta})]'$ exists and is continuous in $\boldsymbol{\theta}$ and $\text{rank}[\boldsymbol{H}(\boldsymbol{\theta})] = r$. In this direction, first note that (8.6.9) is equivalent to

$$H_0 : \boldsymbol{\theta} = \boldsymbol{g}(\boldsymbol{\beta}) \quad \text{versus} \quad H_1 : \boldsymbol{\theta} \neq \boldsymbol{g}(\boldsymbol{\beta}), \tag{8.6.10}$$

where $\boldsymbol{g} : \mathbb{R}^{q-r} \to \mathbb{R}^q$ is a vector-valued function, such that the $(q \times q - r)$ matrix $\boldsymbol{G}(\boldsymbol{\beta}) = (\partial/\partial\boldsymbol{\beta}')\boldsymbol{g}(\boldsymbol{\beta})$ exists, $\text{rank}[\boldsymbol{G}(\boldsymbol{\beta})] = q - r$ and $\boldsymbol{\beta} \in \mathbb{R}^{q-r}$. The expression (8.6.9) is known as the *constraint formulation* and (8.6.10) as the *freedom equation formulation* for the specified hypothesis. For example, let $q = 3$, $\boldsymbol{\theta} = (\theta_1, \theta_2, \theta_3)'$ and $\boldsymbol{h}(\boldsymbol{\theta}) = \theta_1 - \theta_2$; then $\boldsymbol{\beta} = (\beta_1, \beta_2)'$ and $\boldsymbol{g}(\boldsymbol{\beta}) = (\beta_1, \beta_1, \beta_2)'$.

The three test statistics discussed above may also be considered in the present context. More specifically, letting $\widehat{\boldsymbol{\theta}}_n$ denote the unrestricted MLE of $\boldsymbol{\theta}$ and $\overline{\boldsymbol{\theta}}_n$ the restricted MLE of $\boldsymbol{\theta}$, that is, subject to $\boldsymbol{h}(\overline{\boldsymbol{\theta}}_n) = \boldsymbol{0}$, we have:

(i) *Wald statistic:*

$$Q_W = n\boldsymbol{h}(\widehat{\boldsymbol{\theta}}_n)'\{[\boldsymbol{H}(\widehat{\boldsymbol{\theta}}_n)]'[\boldsymbol{I}(\widehat{\boldsymbol{\theta}}_n)]^{-1}\boldsymbol{H}(\widehat{\boldsymbol{\theta}}_n)\}^{-1}\boldsymbol{h}(\widehat{\boldsymbol{\theta}}_n)$$

(actually, $\widehat{\boldsymbol{\theta}}_n$ may be replaced by any BAN estimator of $\boldsymbol{\theta}$);

(ii) *Wilk likelihood ratio statistic:*

$$Q_L = -2\log\lambda_n = 2[\log L_n(\widehat{\boldsymbol{\theta}}_n) - \log L_n(\overline{\boldsymbol{\theta}}_n)]$$

where

$$\lambda_n = \sup_{\{\boldsymbol{\theta}\in\boldsymbol{\Theta}:\boldsymbol{h}(\boldsymbol{\theta})=\boldsymbol{0}\}} L_n(\boldsymbol{\theta}) / \sup_{\boldsymbol{\theta}\in\boldsymbol{\Theta}} L_n(\boldsymbol{\theta});$$

note that if we let $\widehat{\boldsymbol{\beta}}_n$ denote the MLE of $\boldsymbol{\beta}$, it follows that

$$\sup_{\boldsymbol{\beta}\in\mathbb{R}^{q-r}} L_n(\boldsymbol{\beta}) = \sup_{\{\boldsymbol{\theta}\in\boldsymbol{\Theta}:\boldsymbol{h}(\boldsymbol{\theta})=\boldsymbol{0}\}} L_n(\boldsymbol{\theta})$$

and the likelihood ratio statistic may be expressed as

$$Q_L = 2[\log L_n(\widehat{\boldsymbol{\theta}}_n) - \log L_n(\widehat{\boldsymbol{\beta}}_n)];$$

(iii) *Rao efficient score statistic*:

$$Q_R = n^{-1}[\boldsymbol{U}_n(\overline{\boldsymbol{\theta}}_n)]'[\boldsymbol{I}(\overline{\boldsymbol{\theta}}_n)]^{-1}\boldsymbol{U}_n(\overline{\boldsymbol{\theta}}_n).$$

The analogue of Theorem 8.6.1 for composite hypotheses is the following:

***Theorem* 8.6.3.** *Let $X_1, \ldots, X_n$ be i.i.d. random variables with density function $f(x;\boldsymbol{\theta})$, $\boldsymbol{\theta} \in \boldsymbol{\theta} \subset \mathbb{R}^q$, satisfying the conditions of Theorem 8.3.1 and consider the problem of testing (8.6.9). Then each of the statistics Q_W, Q_L, and Q_R indicated above has an asymptotic χ_r^2 distribution under H_0.*

Proof. To show that $Q_W \xrightarrow{D} \chi_r^2$ under H_0 we start by using the same argument employed in Theorem 8.4.1 to obtain

$$\sqrt{n}[\boldsymbol{h}(\widehat{\boldsymbol{\theta}}_n) - \boldsymbol{h}(\boldsymbol{\theta})] \xrightarrow{D} \mathcal{N}\{\mathbf{0}, [\boldsymbol{H}(\boldsymbol{\theta})]'[\boldsymbol{I}(\boldsymbol{\theta})]^{-1}\boldsymbol{H}(\boldsymbol{\theta})\}$$

and then repeat the steps of the first part of Theorem 8.6.1.

To deal with Q_L, first recall that

$$n^{-1/2}\boldsymbol{U}_n(\boldsymbol{\theta}) \xrightarrow{D} \mathcal{N}[\mathbf{0}, \boldsymbol{I}(\boldsymbol{\theta})]. \tag{8.6.11}$$

Then, consider the following Taylor expansion:

$$\begin{aligned} 0 &= \frac{1}{\sqrt{n}} \frac{\partial}{\partial \boldsymbol{\theta}} \log L_n(\boldsymbol{\theta})\bigg|_{\widehat{\boldsymbol{\theta}}_n} \\ &= \frac{1}{\sqrt{n}}\boldsymbol{U}_n(\boldsymbol{\theta}) + \left[\frac{1}{n}\frac{\partial^2}{\partial\boldsymbol{\theta}\partial\boldsymbol{\theta}'} \log L_n(\boldsymbol{\theta})\bigg|_{\boldsymbol{\theta}_n^*}\right]\sqrt{n}(\widehat{\boldsymbol{\theta}}_n - \boldsymbol{\theta}), \end{aligned}$$

where $\boldsymbol{\theta}_n^*$ belongs to the line segment joining $\boldsymbol{\theta}$ and $\widehat{\boldsymbol{\theta}}_n$. Now, using the Khintchine WLLN (Theorem 6.3.6) we have

$$n^{-1}(\partial^2/\partial\boldsymbol{\theta}\partial\boldsymbol{\theta}') \log L_n(\boldsymbol{\theta})\big|_{\boldsymbol{\theta}_n^*} \xrightarrow{P} -\boldsymbol{I}(\boldsymbol{\theta}),$$

and, therefore,

$$\sqrt{n}(\widehat{\boldsymbol{\theta}}_n - \boldsymbol{\theta}) = [\boldsymbol{I}(\boldsymbol{\theta})]^{-1}\frac{1}{\sqrt{n}}\boldsymbol{U}_n(\boldsymbol{\theta}) + o_p(\mathbf{1}). \tag{8.6.12}$$

Similarly, we may prove that

$$\sqrt{n}(\widehat{\boldsymbol{\beta}}_n - \boldsymbol{\beta}) = [\boldsymbol{I}^*(\boldsymbol{\beta})]^{-1}\boldsymbol{V}_n(\boldsymbol{\beta}) + o_p(\mathbf{1}), \tag{8.6.13}$$

where

$$\boldsymbol{V}_n(\boldsymbol{\beta}) = n^{-1/2}(\partial/\partial\boldsymbol{\beta}) \log L_n(\boldsymbol{\beta}) \tag{8.6.14}$$

and

$$\boldsymbol{I}^*(\boldsymbol{\beta}) = \mathbb{E}[(\partial^2/\partial\boldsymbol{\beta}\partial\boldsymbol{\beta}') \log L_n(\boldsymbol{\beta})] \tag{8.6.15}$$

is the corresponding Fisher information matrix. Then, using (8.6.5) and (8.6.12), we may write

$$\begin{aligned} Q_L^* &= 2[\log L_n(\widehat{\boldsymbol{\theta}}_n) - \log L_n(\boldsymbol{\theta})] \\ &= n(\widehat{\boldsymbol{\theta}}_n - \boldsymbol{\theta})'\boldsymbol{I}(\boldsymbol{\theta})(\widehat{\boldsymbol{\theta}}_n - \boldsymbol{\theta}) \\ &= n^{-1}[\boldsymbol{U}_n(\boldsymbol{\theta})]'[\boldsymbol{I}(\boldsymbol{\theta})]^{-1}\boldsymbol{U}_n(\boldsymbol{\theta}) + o_p(1). \end{aligned} \tag{8.6.16}$$

Similarly, using (8.6.13), we obtain

$$\begin{aligned} Q_L^{**} &= 2[\log L_n(\widehat{\boldsymbol{\beta}}_n) - \log L_n(\boldsymbol{\beta})] \\ &= n(\widehat{\boldsymbol{\beta}}_n - \boldsymbol{\beta})'[\boldsymbol{I}^*(\boldsymbol{\beta})](\widehat{\boldsymbol{\beta}}_n - \boldsymbol{\beta}) \\ &= n^{-1}[\boldsymbol{V}_n(\boldsymbol{\beta})]'[\boldsymbol{I}^*(\boldsymbol{\beta})]^{-1}\boldsymbol{V}_n(\boldsymbol{\beta}) + o_p(1), \end{aligned} \tag{8.6.17}$$

From (8.6.14), it follows that

$$\boldsymbol{V}_n(\boldsymbol{\beta}) \xrightarrow{D} \mathcal{N}[\boldsymbol{0}, \boldsymbol{I}^*(\boldsymbol{\beta})], \tag{8.6.18}$$

since

$$\begin{aligned}\boldsymbol{V}_n(\boldsymbol{\beta}) &= \frac{1}{\sqrt{n}}\frac{\partial}{\partial \boldsymbol{\beta}} \log L_n(\boldsymbol{\beta}) \\ &= \frac{\partial}{\partial \boldsymbol{\beta}} \boldsymbol{g}(\boldsymbol{\beta}) \frac{1}{\sqrt{n}} \frac{\partial}{\partial \boldsymbol{g}(\boldsymbol{\beta})} \log L_n[\boldsymbol{g}(\boldsymbol{\beta})] \\ &= \frac{1}{\sqrt{n}} [\boldsymbol{G}(\boldsymbol{\beta})]' \boldsymbol{U}_n(\boldsymbol{\beta}),\end{aligned} \tag{8.6.19}$$

and

$$n^{-1/2}[\boldsymbol{G}(\boldsymbol{\beta})]'\boldsymbol{U}_n(\boldsymbol{\theta}) \xrightarrow{D} \mathcal{N}\{\boldsymbol{0}, [\boldsymbol{G}(\boldsymbol{\beta})]'[\boldsymbol{I}(\boldsymbol{\theta})]\boldsymbol{G}(\boldsymbol{\beta})\}.$$

From (8.6.18), we conclude that

$$\boldsymbol{I}^*(\boldsymbol{\beta}) = [\boldsymbol{G}(\boldsymbol{\beta})]'\boldsymbol{I}(\boldsymbol{\theta})[\boldsymbol{G}(\boldsymbol{\beta})]. \tag{8.6.20}$$

From the definition of Q_L and the equivalence between (8.6.9) and (8.6.10), we have $Q_L = Q_L^* - Q_L^{**}$; thus, using (8.6.16)–(8.6.19) it follows that

$$Q_L = n^{-1}[\boldsymbol{U}_n(\boldsymbol{\theta})]' \left\{[\boldsymbol{I}(\boldsymbol{\theta})]^{-1} - \boldsymbol{G}(\boldsymbol{\beta})[\boldsymbol{I}^*(\boldsymbol{\beta})]^{-1}[\boldsymbol{G}(\boldsymbol{\beta})]'\right\} \boldsymbol{U}_n(\boldsymbol{\theta}) + o_p(1). \tag{8.6.21}$$

Using (8.6.20) we have

$$\begin{aligned}&\{[\boldsymbol{I}(\boldsymbol{\theta})]^{-1} - \boldsymbol{G}(\boldsymbol{\beta})\boldsymbol{I}^*(\boldsymbol{\beta})\}\boldsymbol{I}(\boldsymbol{\theta})\{[\boldsymbol{I}(\boldsymbol{\theta})]^{-1} - \boldsymbol{G}(\boldsymbol{\beta})[\boldsymbol{I}^*(\boldsymbol{\beta})]^{-1}[\boldsymbol{G}(\boldsymbol{\beta})]'\} \\ &= [\boldsymbol{I}(\boldsymbol{\theta})]^{-1} - \boldsymbol{G}(\boldsymbol{\beta})[\boldsymbol{I}^*(\boldsymbol{\beta})]^{-1}[\boldsymbol{G}(\boldsymbol{\beta})]'.\end{aligned} \tag{8.6.22}$$

Furthermore,

$$\begin{aligned}&\mathrm{tr}\{[\boldsymbol{I}(\boldsymbol{\theta})]^{-1} - \boldsymbol{G}(\boldsymbol{\beta})[\boldsymbol{I}^*(\boldsymbol{\beta})]^{-1}[\boldsymbol{G}(\boldsymbol{\beta})]'\}\boldsymbol{I}(\boldsymbol{\theta}) \\ &= \mathrm{tr}\{\boldsymbol{I}_q - \boldsymbol{G}(\boldsymbol{\beta})[\boldsymbol{I}^*(\boldsymbol{\beta})]^{-1}[\boldsymbol{G}(\boldsymbol{\beta})]'\boldsymbol{I}(\boldsymbol{\theta})\} \\ &= q - \mathrm{tr}\{[\boldsymbol{I}^*(\boldsymbol{\beta})]^{-1}[\boldsymbol{G}(\boldsymbol{\beta})]'\boldsymbol{I}(\boldsymbol{\theta})\boldsymbol{G}(\boldsymbol{\beta})\} \\ &= q - \mathrm{tr}\{[\boldsymbol{I}^*(\boldsymbol{\beta})]^{-1}\boldsymbol{I}^*(\boldsymbol{\beta})\} = r.\end{aligned} \tag{8.6.23}$$

Then, using expressions (8.6.21)–(8.6.23), the Slutsky Theorem 7.5.2 and Theorem 7.6.3, it follows that, under H_0, $Q_L \xrightarrow{D} \chi^2_r$.

To deal with Q_R, first recall that the restricted likelihood equations are

$$\begin{aligned}\frac{\partial}{\partial \boldsymbol{\theta}} \log L_n(\boldsymbol{\theta}) + \boldsymbol{H}(\boldsymbol{\theta})\boldsymbol{\lambda} &= \boldsymbol{0}, \\ \boldsymbol{h}(\boldsymbol{\theta}) &= \boldsymbol{0},\end{aligned} \tag{8.6.24}$$

where $\boldsymbol{\lambda} \in \mathbb{R}^r$ is a vector of Lagrangian multipliers. Now, letting $\boldsymbol{\theta}_n = \boldsymbol{\theta} + n^{-1/2}\boldsymbol{u}$ where $\|\boldsymbol{u}\| < K, 0 < K < \infty$, we may start by considering the Taylor expansion:

$$\frac{1}{\sqrt{n}}\frac{\partial}{\partial \boldsymbol{\theta}} \log L_n(\boldsymbol{\theta})\bigg|_{\boldsymbol{\theta}_n} = \frac{1}{\sqrt{n}}\boldsymbol{U}_n(\boldsymbol{\theta}) + \left[\frac{1}{n}\frac{\partial^2}{\partial \boldsymbol{\theta}\partial \boldsymbol{\theta}'} \log L_n(\boldsymbol{\theta})\bigg|_{\boldsymbol{\theta}_n^*}\right]\sqrt{n}(\boldsymbol{\theta}_n - \boldsymbol{\theta}),$$

where $\boldsymbol{\theta}_n^*$ belongs to the line segment joining $\boldsymbol{\theta}$ and $\boldsymbol{\theta}_n$. Then, observing that

$$n^{-1}(\partial^2/\partial\boldsymbol{\theta}\partial\boldsymbol{\theta}')\log L_n(\boldsymbol{\theta})\big|_{\boldsymbol{\theta}_n^*} \xrightarrow{P} -\boldsymbol{I}(\boldsymbol{\theta}),$$

we may use the Khintchine WLLN (Theorem 6.3.6) to write

$$\frac{1}{\sqrt{n}}\frac{\partial}{\partial\boldsymbol{\theta}}\log L_n(\boldsymbol{\theta})\bigg|_{\boldsymbol{\theta}_n} = \frac{1}{\sqrt{n}}\boldsymbol{U}_n(\boldsymbol{\theta}) - \boldsymbol{I}(\boldsymbol{\theta})\sqrt{n}(\boldsymbol{\theta}_n - \boldsymbol{\theta}) + o_p(\mathbf{1}). \tag{8.6.25}$$

Also, since $\boldsymbol{H}(\boldsymbol{\theta})$ is continuous in $\boldsymbol{\theta}$, we have

$$\boldsymbol{h}(\boldsymbol{\theta}_n) = [\boldsymbol{H}(\boldsymbol{\theta})]'\sqrt{n}(\boldsymbol{\theta}_n - \boldsymbol{\theta}) + o_p(\mathbf{1}). \tag{8.6.26}$$

Since the restricted MLE, $\overline{\boldsymbol{\theta}}_n$, must satisfy (8.6.24), and in view of (8.6.25) and (8.6.26), it must be a solution to the system

$$\begin{aligned}\frac{1}{\sqrt{n}}\boldsymbol{U}_n(\boldsymbol{\theta}) - \boldsymbol{I}(\boldsymbol{\theta})\sqrt{n}(\overline{\boldsymbol{\theta}}_n - \boldsymbol{\theta}) + \boldsymbol{H}(\boldsymbol{\theta})\frac{1}{\sqrt{n}}\overline{\boldsymbol{\lambda}}_n + o_p(\mathbf{1}) = \mathbf{0},\\ [\boldsymbol{H}(\boldsymbol{\theta})]'\sqrt{n}(\overline{\boldsymbol{\theta}}_n - \boldsymbol{\theta}) + o_p(\mathbf{1}) = \mathbf{0},\end{aligned} \tag{8.6.27}$$

where $\overline{\boldsymbol{\lambda}}_n$ is a vector of Lagrange multipliers. Note that (8.6.27) may be expressed as

$$\begin{pmatrix} \boldsymbol{I}(\boldsymbol{\theta}) & -\boldsymbol{H}(\boldsymbol{\theta}) \\ -[\boldsymbol{H}(\boldsymbol{\theta})]' & \mathbf{0} \end{pmatrix}\begin{pmatrix} \sqrt{n}(\overline{\boldsymbol{\theta}}_n - \boldsymbol{\theta}) \\ n^{-1/2}\overline{\boldsymbol{\lambda}}_n \end{pmatrix} = \begin{pmatrix} \boldsymbol{U}_n(\boldsymbol{\theta}) \\ \mathbf{0} \end{pmatrix} + o_p(\mathbf{1}).$$

Thus,

$$\begin{pmatrix} \sqrt{n}(\overline{\boldsymbol{\theta}}_n - \boldsymbol{\theta}) \\ n^{-1/2}\overline{\boldsymbol{\lambda}}_n \end{pmatrix} = \begin{pmatrix} \boldsymbol{P}(\boldsymbol{\theta}) & \boldsymbol{Q}(\boldsymbol{\theta}) \\ [\boldsymbol{Q}(\boldsymbol{\theta})]' & \boldsymbol{R}(\boldsymbol{\theta}) \end{pmatrix}\begin{pmatrix} n^{-1/2}\boldsymbol{U}_n(\boldsymbol{\theta}) \\ \mathbf{0} \end{pmatrix} + o_p(\mathbf{1}), \tag{8.6.28}$$

where

$$\begin{pmatrix} \boldsymbol{P}(\boldsymbol{\theta}) & \boldsymbol{Q}(\boldsymbol{\theta}) \\ [\boldsymbol{Q}(\boldsymbol{\theta})]' & \boldsymbol{R}(\boldsymbol{\theta}) \end{pmatrix} = \begin{pmatrix} \boldsymbol{I}(\boldsymbol{\theta}) & -\boldsymbol{H}(\boldsymbol{\theta}) \\ -[\boldsymbol{H}(\boldsymbol{\theta})]' & \mathbf{0} \end{pmatrix}^{-1},$$

which implies

$$\begin{aligned}\boldsymbol{P}(\boldsymbol{\theta}) &= [\boldsymbol{I}(\boldsymbol{\theta})]^{-1}\{\boldsymbol{I}_q - \boldsymbol{H}(\boldsymbol{\theta})([\boldsymbol{H}(\boldsymbol{\theta})]'[\boldsymbol{I}(\boldsymbol{\theta})]^{-1}[\boldsymbol{H}(\boldsymbol{\theta})]^{-1}[\boldsymbol{H}(\boldsymbol{\theta})]'[\boldsymbol{I}(\boldsymbol{\theta})]^{-1}\},\\ \boldsymbol{Q}(\boldsymbol{\theta}) &= -[\boldsymbol{I}(\boldsymbol{\theta})]^{-1}\boldsymbol{H}(\boldsymbol{\theta})\{[\boldsymbol{H}(\boldsymbol{\theta})]'[\boldsymbol{I}(\boldsymbol{\theta})]^{-1}\boldsymbol{H}(\boldsymbol{\theta})\}^{-1},\\ \boldsymbol{R}(\boldsymbol{\theta}) &= -\{[\boldsymbol{H}(\boldsymbol{\theta})]'[\boldsymbol{I}(\boldsymbol{\theta})]^{-1}\boldsymbol{H}(\boldsymbol{\theta})\}^{-1}.\end{aligned} \tag{8.6.29}$$

Since by the Central Limit Theorem 7.3.1 and the Cramér–Wold Theorem 7.2.4, we have

$$\{n^{-1/2}[\boldsymbol{U}_n(\boldsymbol{\theta})]', \mathbf{0}'\}' \xrightarrow{D} \mathcal{N}\left\{\mathbf{0}, \begin{pmatrix} \boldsymbol{I}(\boldsymbol{\theta}) & \mathbf{0} \\ \mathbf{0} & \mathbf{0} \end{pmatrix}\right\},$$

it follows from (8.6.28) and the Slutsky Theorem 7.5.2 that

$$\begin{pmatrix} \sqrt{n}(\overline{\boldsymbol{\theta}}_n - \boldsymbol{\theta}) \\ n^{-1/2}\overline{\boldsymbol{\lambda}}_n \end{pmatrix} \xrightarrow{D} \mathcal{N}(\mathbf{0}, \boldsymbol{\Sigma}),$$

where

$$\boldsymbol{\Sigma}(\boldsymbol{\theta}) = \begin{pmatrix} \boldsymbol{\Sigma}_{11}(\boldsymbol{\theta}) & \boldsymbol{\Sigma}_{12}(\boldsymbol{\theta}) \\ \boldsymbol{\Sigma}_{21}(\boldsymbol{\theta}) & \boldsymbol{\Sigma}_{22}(\boldsymbol{\theta}) \end{pmatrix} = \begin{pmatrix} \boldsymbol{P}(\boldsymbol{\theta})\boldsymbol{I}(\boldsymbol{\theta})[\boldsymbol{P}(\boldsymbol{\theta})]' & \boldsymbol{P}(\boldsymbol{\theta})\boldsymbol{I}(\boldsymbol{\theta})\boldsymbol{Q}(\boldsymbol{\theta}) \\ [\boldsymbol{Q}(\boldsymbol{\theta})]'\boldsymbol{I}(\boldsymbol{\theta})[\boldsymbol{P}(\boldsymbol{\theta})]' & [\boldsymbol{Q}(\boldsymbol{\theta})]'\boldsymbol{I}(\boldsymbol{\theta})\boldsymbol{Q}(\boldsymbol{\theta}) \end{pmatrix}.$$

Using (8.6.29) we may show that

$$\boldsymbol{P}(\boldsymbol{\theta})\boldsymbol{I}(\boldsymbol{\theta})[\boldsymbol{P}(\boldsymbol{\theta})]' = \boldsymbol{P}(\boldsymbol{\theta}),$$
$$\boldsymbol{P}(\boldsymbol{\theta})\boldsymbol{I}(\boldsymbol{\theta})\boldsymbol{Q}(\boldsymbol{\theta}) = \boldsymbol{0}$$

and

$$[\boldsymbol{Q}(\boldsymbol{\theta})]'\boldsymbol{I}(\boldsymbol{\theta})\boldsymbol{Q}(\boldsymbol{\theta}) = -\boldsymbol{R}(\boldsymbol{\theta}).$$

Thus, we may write

$$\boldsymbol{\Sigma}_{11}(\boldsymbol{\theta}) = [\boldsymbol{I}(\boldsymbol{\theta})]^{-1}\{\boldsymbol{I}_q - \boldsymbol{H}(\boldsymbol{\theta})([\boldsymbol{H}(\boldsymbol{\theta})]'[\boldsymbol{I}(\boldsymbol{\theta})]^{-1}\boldsymbol{H}(\boldsymbol{\theta}))^{-1}[\boldsymbol{H}(\boldsymbol{\theta})]'\}[\boldsymbol{I}(\boldsymbol{\theta})]^{-1},$$
$$\boldsymbol{\Sigma}_{22}(\boldsymbol{\theta}) = \{[\boldsymbol{H}(\boldsymbol{\theta})]'\boldsymbol{I}(\boldsymbol{\theta})\boldsymbol{H}(\boldsymbol{\theta})\}^{-1},$$

and

$$\boldsymbol{\Sigma}_{12}(\boldsymbol{\theta}) = [\Sigma_{21}(\boldsymbol{\theta})]' = \boldsymbol{0}.$$

Now, using (8.6.24), we have $\boldsymbol{U}_n(\overline{\boldsymbol{\theta}}_n) = -\boldsymbol{H}(\boldsymbol{\theta}_n)\overline{\boldsymbol{\lambda}}_n n^{-1/2}$. Thus,

$$\begin{aligned} Q_R &= n^{-1}[\boldsymbol{U}_n(\overline{\boldsymbol{\theta}}_n)]'[\boldsymbol{I}(\overline{\boldsymbol{\theta}}_n)]^{-1}\boldsymbol{U}_n(\overline{\boldsymbol{\theta}}_n) \\ &= n^{-1}\overline{\boldsymbol{\lambda}}_n'[\boldsymbol{H}(\overline{\boldsymbol{\theta}}_n)]'[\boldsymbol{I}(\overline{\boldsymbol{\theta}}_n]^{-1}\boldsymbol{H}(\overline{\boldsymbol{\theta}}_n)\overline{\boldsymbol{\lambda}}_n \\ &= n^{-1}\overline{\boldsymbol{\lambda}}_n'[\boldsymbol{R}(\overline{\boldsymbol{\theta}}_n)]^{-1}\overline{\boldsymbol{\lambda}}_n. \end{aligned}$$

From (8.6.29) we conclude that

$$Q_R^* = n^{-1}(\overline{\boldsymbol{\lambda}}_n'[\boldsymbol{R}(\boldsymbol{\theta})]^{-1}\overline{\boldsymbol{\lambda}}_n) \xrightarrow{D} \chi_r^2,$$

under H_0, and since $\boldsymbol{R}(\overline{\boldsymbol{\theta}}_n) \xrightarrow{P} \boldsymbol{R}(\boldsymbol{\theta})$, we may use an argument similar to that of Theorem 8.6.1 so that, under H_0,

$$Q_R \xrightarrow{D} \chi_r^2. \qquad \blacksquare$$

See Silvey (1959) for details on the asymptotic distribution of Q_W, Q_L, and Q_R under alternative hypotheses.

8.7 Resampling Methods

Resampling methods pertain to nonparametric estimation (and reduction) of *bias*, variance (or asymptotic variance) as well as sampling distribution of statistics for which analytical tools are often difficult to incorporate. Yet, such resampling methods rest on sound statistical methodology and cover almost all walks of statistical inference. Quenouille (1949) introduced the *jackknife* primarily as a tool for bias reduction while Tukey (1958) incorporated the jackknife for variance estimation, as needed for setting confidence intervals. Hartigan (1971) proposed some resampling methods, which were systematically developed and called *bootstrap* methods by Efron (1979). Both jackknife and bootstrap methods have been extensively studied not only in parametric setups, but also in nonparametric setups and used in almost all areas of application. We provide here only an outline of the basic concepts of such methods.

Consider a statistic $T_n = T(X_1, \ldots, X_n)$ based on a sample of n i.i.d. random variables. Suppose that $\mathbb{E}(T_n) = \theta + \text{bias}$, where the bias can be expressed as $\sum_{j\geq 1} a_j n^{-j}$, with the a_j representing unknown real numbers, not necessarily different from zero. In fact, the a_j may depend on the distribution function F of the underlying random variable X as well as on the form of $T_n(\cdot)$. Thus, we may write

$$\mathbb{E}(T_n) = \theta + a_1 n^{-1} + a_2 n^{-2} + \cdots . \tag{8.7.1}$$

Note that θ may be a parameter or even a functional of F. Then, for $T_{n-1} = T(X_1, \ldots, X_n)$ we have

$$\mathbb{E}(T_{n-1}) = \theta + a_1(n-1)^{-1} + a_2(n-1)^{-2} + \cdots \tag{8.7.2}$$

so that from (8.7.1) and (8.7.2), we have

$$E[nT_n - (n-1)T_{n-1}] = \theta - \frac{a_2}{n(n-1)} + O(n^{-3}).$$

Motivated by this feature, we let

$$T_{n-1}^{(-i)} = T(X_1, \ldots, X_{i-1}, X_{i+1}, \ldots, X_n), \quad 1 \leq i \leq n,$$

and consider *pseudovalues*

$$T_{n,i} = nT_n - (n-1)T_{n-1}^{(-i)}, \quad i = 1, \ldots, n. \tag{8.7.3}$$

The *jackknife estimator* of θ is then defined as

$$T_{n,J} = n^{-1}\sum_{i=1}^{n} T_{n,i} = T_n + \frac{n-1}{n}\sum_{i-1}^{n}(T_n - T_{n-1}^{(-i)}). \tag{8.7.4}$$

It is clear that $\mathbb{E}(T_{nJ}) = \theta - a_2/[n(n-1)] + O(n^{-3})$, so that the order of the bias term is reduced from $O(n^{-1})$ to $O(n^{-2})$.

Next, define the *jackknife variance estimator* as

$$s_{nJ}^2 = \frac{1}{n-1}\sum_{i=1}^{n}(T_{n,i} - T_{n,J})^2. \tag{8.7.5}$$

In the particular case of $T_n = \overline{X}_n$, it follows that $T_{n,i} = X_i$, $1 \leq i \leq n$, $T_{nJ} = \overline{X}_n$, so that s_{nJ}^2 is the usual sample variance. This led Tukey (1958) to conjecture that s_{nJ}^2 should estimate the (asymptotic) variance of $\sqrt{n}(T_n - \theta)$ [or of $\sqrt{n}(T_{nJ} - \theta)$] for other statistics too. For the entire class of U-statistics (see 2.4.28), this conjecture was confirmed by Sen (1977). Exercise 8.7.1 is set to verify that with respect to the subsigmafield $\mathcal{C}_n$ defined in Example 5.2.6,

$$\mathbb{E}(T_{n-1}|\mathcal{C}_n) = \frac{1}{n}\sum_{i=1}^{n} T_{n-1}^{(-i)}, \quad \text{a.e., } n \geq 2. \tag{8.7.6}$$

Furthermore, note that by (8.7.4) and (8.7.6),

$$T_{nJ} = T_n + (n-1)\mathbb{E}[(T_n - T_{n-1})|\mathcal{C}_n].$$

In the same way, if follows that (Exercise 8.7.2)

$$s_{nJ}^2 = n(n-1)\mathbb{V}\text{ar}[(T_n - T_{n-1})|\mathcal{C}_n]. \tag{8.7.7}$$

Thus, basically, jackknifing amounts to first adjusting by $\mathbb{E}[(T_n - T_{n-1})|\mathcal{C}_n]$ to reduce the bias, and then taking the (conditional) distribution of $T_n - T_{n-1}$, given $\mathcal{C}_n$, to generate the pseudovalues leading to the variance estimator s_{nJ}^2. For U-statistics, we may explore the reverse martingale property of $\{U_n, \mathcal{C}_n\}$ and Exercise 8.7.3 is set to show that $s_{nJ}^2 \xrightarrow{a.s.} \mathbb{V}\text{ar}[\sqrt{n}(T_{nJ} - \theta)]$, where $T_{nJ} = T_n$ [Sen (1977)]. For a general family of statistics $\{T_n\}$ such that $\sqrt{n}(T_n - \theta)$ is asymptotically normal, jackknifing preserves the asymptotic normality providing a consistent estimator of the asymptotic variance. The results extend readily to the vector case. Basically, $\mathcal{C}_n$ pertains to the $n!$ equally likely permutations of $(X_1, \ldots, X_n)$, given the collection $\{X_1, \ldots, X_n\}$, and thereby exploits the simple random sampling without replacement scheme. As such, jackknifing applies as well to finite population sampling without replacement (Exercise 8.7.4).

By contrast, bootstrap methods are based on simple random sampling with replacement schemes and Monte Carlo methods. Let $F_n(x)$ be the empirical distribution function based on $X_1, \ldots, X_n$ [see (1.5.82)]; F_n is a $\sqrt{n}$-consistent nonparametric estimator of F, although it is a discrete distribution function whereas F may be continuous. The basic idea is to use simple random sampling without replacement from F_n (i.e., from $X_1, \ldots, X_n$), there being n^n such possible samples. Let $(X_1^*, \ldots, X_n^*)$ be a sample of size n from F_n and let $T_n^* = T(X_1^*, \ldots, X_n^*)$ be the same statistic (T_n) based on the X_i^*. The basic idea is to approximate the sampling distribution of $\sqrt{n}(T_n - \theta)$ by that of $\sqrt{n}(T_n^* - T_n)$ and to use Monte Carlo methods for the latter. Keeping that in mind, we may generate B *bootstrap samples* $\{X_{b_1}^*, \ldots, X_{b_n}^*\}$, $b_j \in \{1, \ldots, n\}$, from F_n by simple random sampling with replacement, and compute

$$T_{nb}^* = T(X_{b_1}^*, \ldots, X_{b_n}^*).$$

Once this is done, we let

$$Z_{nb} = \sqrt{n}(T_{nb}^* - T_n), \quad b = 1, \ldots, B,$$
$$G_n^*(y) = B^{-1} \sum_{b=1}^{B} I(Z_{nb} \le y), \quad y \in \mathbb{R}, \tag{8.7.8}$$

and consider

$$G_n(y) = P[\sqrt{n}(T_n - \theta) \le y], \quad y \in \mathbb{R} \tag{8.7.9}$$

to be the actual sampling distribution of $\sqrt{n}(T_n - \theta)$. The basic idea is to approximate G_n by G_n^*. This works out well if G_n weakly converges to a normal distribution function. Then, G_n^* converges to the same normal distribution function as $n \to \infty$ (see Exercise 8.7.5). In this case, one may also estimate the variance of the limiting distribution function G, that is, the asymptotic mean-squared error of $\sqrt{n}(T_n - \theta)$, by

$$s_{nB}^2 = \frac{1}{B-1} \sum_{b=1}^{B} (Z_{nb} - \overline{Z}_{nB})^2,$$

where

$$\overline{Z}_{nB} = B^{-1} \sum_{b=1}^{B} Z_{nb}.$$

The main advantage of the bootstrap (over the jackknife) is that an estimator of G_n itself is provided by G_n^*, intuitively suggesting that even if $G_n \to G$ (a normal distribution function) does not hold, $G_n^* \to G_n$ if B is chosen large and n is not too small. The jackknife provides distributional approximation through s_{nJ}^2 and, hence, rely on limiting Gaussian distributions. The weak convergence of $G_n^* \to G$ holds for a broad class of distributions (e.g., multinormal, chi-squared, or even convex linear combinations of independent chi-squared variables), which are basically all related to functions of random vectors having multinormal distributions. However, it is to be noted that such results are purely asymptotic, backed by theoretical justification. For small to moderate sample sizes, there may not be adequate theoretical justification for approximating well G_n by G_n^*. Nevertheless, the literature is flooded with numerical and simulation studies justifying bootstrap approximations for small sample sizes and this scenario needs more careful appraisal in the light of methodological perspectives. Furthermore, there are notable cases (not pathological examples – see Exercise 8.7.6) where G_n^* may not converge to G, even if n is large; the situation is worse in small to moderate sample sizes.

There are many variations of the above two resampling methods, justified methodologically as well as by Monte Carlo methods. When n is large, both $n!$ and n^n are large - while $n^n/n! = O[\exp(n)/\sqrt{n}]$, so $n^n >> n!$. This suggests suitable importance sampling for choosing the bootstrap samples [see Efron and Tibshirani (1993)].

8.8 Concluding Remarks

Asymptotic theory of statistical inference is, by itself, a very broad and diverse field of research, much beyond the scope of the present treatise. In earlier chapters, finite-sample (or exact) theory of estimation and hypothesis testing was presented with due emphasis on their scope and limitations. In course of that dissemination, it was observed that under suitable regularity assumptions, the scope of inference could be enhanced in an asymptotic setup allowing the sample size(s) to be large. This is conventionally referred to as asymptotic statistical inference for regular family of distributions (satisfying the Cramér–Rao regularity conditions). Yet, there are other notable cases where such regularity conditions may not hold – the Uniform$[0, \theta]$ density, exponential density over $[\theta, \infty)$ for some real θ, Pareto type distributions are noteworthy examples. In such a case, typically asymptotic normality may not hold – but sample order statistics may have a leading role, and moreover, the rate of convergence may no longer be $n^{1/2}$. Such nonregular cases may be treated case by case with appropriate asymptotic theory, and we pose Exercises 8.8.1–8.8.3 to illustrate this feature of nonregular family.

In actual practice, due to censoring of diverse type, missing observations, incomplete responses and other causes, asymptotic theory of statistical inference for regular distributions may no longer be adaptable. This feature is displayed in Chapter 10, and allied asymptotic inference tools are appraised. In that context, martingale based methodology plays a focal role, and the treatise in this chapter amends well to such abstract setups, often, demand deeper results, form probability theory. This is therefore outlined in Chapter 11 and illustrated how the basic results of Chapter 8 go through with some interpretable and manageable adaptations. In this way, the treatise of Chapter 8 paves the way for more complex setups.

8.9 Exercises

8.2.1 Specify the details of the proof of (8.2.30).

8.3.1 Verify that (8.3.2) holds if the Cramér–Rao–Fréchet conditions are valid.

8.3.2 Verify (8.3.4).

8.3.3 Show that the estimating equation for the Cauchy density function in Example 8.3.1 has $2n-1$ roots.

8.3.4 Work out the details in Example 8.3.2.

8.3.5 Work out the details in Example 8.3.3.

8.3.6 Work out the details in Example 8.3.4.

8.3.7 Work out the details in Example 8.3.5.

8.3.8 Consider the negative binomial law defined by (1.5.50). Show that $(\partial/\partial\pi)\log p(n|m,\pi)$ $[= m/\pi - (n-m)/(1-\pi)]$ and $(\partial^2/\partial\pi^2)\log p(n|m,\pi)$ satisfy all the regularity conditions of Theorem 8.3.1, and that the MLE of π, given by $\widehat{\pi} = m/n$, is asymptotically normal.

8.3.9 For Exercise 8.3.8, show that $n = Y_1 + \cdots + Y_m$ where the Y_i are i.i.d. random variables with mean π^{-1} and a finite variance. Hence, show that as $m \to \infty$, $(n - m/\pi)/\sqrt{m}$ is closely normally distributed. Use the transformation $T_m = m/n = g(n/m)$ to derive the asymptotic normality of $\sqrt{m}(T_m - \pi)$.

8.7.1 Verify expression (8.7.6).

8.7.2 Show (8.7.7)

8.7.3 For U-statistics, show that $s_{nJ}^2 \xrightarrow{a.s.} \mathbb{V}\text{ar}[\sqrt{n}(T_{nJ} - \theta)]$, where $T_{nJ} = T_n$.

8.7.4 Show that jackknife applies as well to finite population sampling under WOR.

8.7.5 Show that if as $n \to \infty$, (8.7.9) weakly converges to a normal distribution function, so does (8.7.8).

8.7.6 Let $X_1, \ldots, X_n$ be i.i.d. random variables having a *stable law distribution* with index α, $1 < \alpha < 2$, for which the characteristic function is given by

$$\phi(x) = \exp\{-|x|^\alpha[1 - i \text{ sign}(x)\tan(\pi\alpha/2)]\}.$$

Show that $\overline{X}$ has the same stable law and $\mathbb{V}\text{ar}(\overline{X}_n)$ does not exist. Hence, or otherwise, compare with the bootstrap distribution of $\overline{X}_n$, which is contended to be normal for large n.

8.8.1 Let $X_1, \ldots, X_n$ be i.i.d. random variables with density $f(x;\theta) = \exp[-(x-\theta)]I(x > \theta)$, $\theta \in \mathbb{R}$. Note that the range $[\theta, \infty)$ depends on θ, violating the Cramér–Rao–Fréchet regularity conditions. Show that the smallest order statistic $X_{n:1}$ is the MLE and further, that $n(X_{n:1} - \theta)$ has asymptotically Exp(1) distribution.

8.8.2 Consider $X_1, \ldots, X_n$ i.i.d. random variables with Pareto distribution such that the survival function is

$$S(x;\theta) = 1 - F(x;\theta) = (\theta_1/x)^{\theta_2}, \quad x \geq \theta_1 > 0,$$

where $\theta_2 > 0$. Obtain the MLE of $\boldsymbol{\theta} = (\theta_1, \theta_2)$ and show that for θ_1, the MLE does not have the asymptotic normality nor the rate of convergence $n^{1/2}$.

8.8.3 Let $X_1, \ldots, X_n$ be i.i.d. random variables with p.d.f.

$$f(x;\theta) = \frac{1}{\theta_2} I\left(\theta_1 - \frac{1}{2}\theta_2 \leq x \leq \theta_1 + \frac{1}{2}\theta_2\right).$$

Obtain the MLE of θ_1, θ_2 and show that neither the asymptotic normality nor the rate of convergence $n^{1/2}$ holds here. What about the asymptotic distribution of the MLE of $\boldsymbol{\theta} = (\theta_1, \theta_2)$?

9

Categorical Data Models

9.1 Introduction

In general, categorical data models relate to *count data* corresponding to the classification of sampling units into *groups* or *categories* either on a qualitative or some quantitative basis. These categories may be defined by the essentially discrete nature of the phenomenon under study (see Example 1.2.11 dealing with the OAB blood classification model) or, often for practical reasons, by the grouping of the values of an essentially continuous underlying distribution (e.g., shoe sizes: 5, $5\frac{1}{2}$, 6, $6\frac{1}{2}$, etc. corresponding to half-open intervals for the actual length of a foot). Even in the qualitative case there is often an implicit ordering in the categories resulting in *ordered categorical data* (i.e., ratings: excellent, very good, good, fair, and poor, for a research proposal under review). Except in some of the most simple cases, exact statistical analysis for categorical data models may not be available in a unified, simple form. Hence, asymptotic methods are important in this context. They not only provide a unified coverage of statistical methodology appropriate for large sample sizes but also suggest suitable modifications, which may often be appropriate for moderate to small sample sizes. This chapter is devoted to the study of this related asymptotic theory.

Although there are a few competing probabilistic models for statistical analysis of categorical data sets, we will find it convenient to concentrate on the *product multinomial model*, which encompasses a broad domain and plays a key role in the development of appropriate statistical analysis tools. Keeping in mind the OAB blood classification model (i.e., Example 1.2.11), we may conceive of r $(\geq k)$ categories (indexed as $1, \ldots, r$, respectively), so that an observation may belong to the jth category with a probability π_j, $j = 1, \ldots, r$, where the π_j are nonnegative numbers adding up to 1. Thus, for a sample of n independent observations, we have the simple multinomial law for the count data $n_1, \ldots, n_r$ (the respective cell frequencies)

$$P\{n_1, \ldots, n_r\} = \left(n! \Big/ \prod_{j=1}^{r} n_j!\right) \prod_{j=1}^{r} \pi_j^{n_j}, \tag{9.1.1}$$

where $\sum_{j=1}^{r} n_j = n$ and $\sum_{j=1}^{r} \pi_j = 1$. Starting with this simple multinomial law, we may conceive of a more general situation corresponding to s (≥ 1) independent samples (of sizes $n_1, \ldots, n_{s\cdot}$, respectively) drawn (with replacement) from s populations whose elements may be classified into r_i (≥ 2) categories, $i = 1, \ldots, s$. Let n_{ij} denote the frequency (count) of sample units from the $n_{i\cdot}$ individuals in the ith sample classified into the jth category

$(1 \le j \le r_i)$ and let

$$\boldsymbol{n}_i = (n_{i1}, \ldots, n_{ir_i})'$$

denote the corresponding vector of observed frequencies for the ith sample, $i = 1, \ldots, s$. Also, let

$$\boldsymbol{\pi}_i = (\pi_{i1}, \ldots, \pi_{ir_i})'$$

denote the corresponding probabilities of classification, $i = 1, \ldots, s$, and let $\boldsymbol{\pi} = (\boldsymbol{\pi}'_1, \ldots, \boldsymbol{\pi}'_s)'$ and $\boldsymbol{n} = (\boldsymbol{n}'_1, \ldots, \boldsymbol{n}'_s)'$. Then, the probability law for $\boldsymbol{n}$ is given by

$$P(\boldsymbol{n}|\boldsymbol{\pi}) = \prod_{i=1}^{s} \left[\left(n_i! \Big/ \prod_{j=1}^{r_i} n_{ij}! \right) \prod_{j=1}^{r_i} \pi_{ij}^{n_{ij}} \right], \tag{9.1.2}$$

where $\sum_{j=1}^{r_i} \pi_{ij} = \boldsymbol{\pi}'_i \mathbf{1}_{r_i} = 1,\ i = 1, \ldots, s$. In the literature this is termed the product multinomial law. For $s = 1$, (9.1.2) reduces to the classical multinomial law in (9.1.1), and, hence, the genesis is easy to comprehend. In this setup, recall that $\boldsymbol{\pi}_i$ belongs to the r_i-dimensional simplex $\mathcal{S}_i = \{\boldsymbol{x} \in \mathbb{R}^{r_i} : \boldsymbol{x} \ge \mathbf{0} \text{ and } \boldsymbol{x}'\mathbf{1}_{r_i} = 1\}$, $i = 1, \ldots, s$, so that $\boldsymbol{\pi} \in \mathcal{S} = \mathcal{S}_1 \times \cdots \times \mathcal{S}_s$. If all the r_i are equal to 2, we have a *product binomial model.* In (9.1.1) [and more generally in (9.1.2)], r, the number of categories may itself be a combination of two or more factors (e.g., grades in English and Mathematics, each grouped into 6 categories so that $r = 36$ possible combinations). This is commonly referred to as *multidimensional contingency tables*, and we shall refer to that later on.

In many problems of practical interest, within the general framework of (9.1.2), the π_{ij} can be expressed as functions of a vector of parameters

$$\boldsymbol{\theta} = (\theta_1, \ldots, \theta_q)',$$

for some $q \le \sum_{j=1}^{s}(r_j - 1)$. For example, in the blood groups model (treated in Example 1.2.11) we have $s = 1, r_1 = r = 4$ and $\boldsymbol{\theta} = (\theta_1, \theta_2)'$ where $0 \le \theta_1 \le 1, 0 \le \theta_2 \le 1$ and $0 \le \theta_1 + \theta_2 \le 1$. In this setup we may write

$$\mathcal{S} = \left\{ (\pi_1, \pi_2, \pi_3, \pi_4) : \pi_j > 0, \quad 1 \le j \le 4 \quad \text{and} \quad \sum_{j=1}^{4} \pi_j = 1 \right\}$$

and

$$\boldsymbol{\theta} = \left\{ (\theta_1, \theta_2, \theta_3) : \theta_i > 0,\ 1 \le i \le 3 \quad \text{and} \quad \sum_{i=1}^{3} \theta_i = 1 \right\},$$

and $\boldsymbol{\pi}$ has the domain $\mathcal{S}$, whereas $\boldsymbol{\theta}$ has the domain $\boldsymbol{\Theta}$. Keeping this in mind, we may write, for the general model in (9.1.2),

$$\boldsymbol{\pi} = \boldsymbol{\pi}(\boldsymbol{\theta}), \quad \boldsymbol{\theta} \in \boldsymbol{\Theta}, \quad \boldsymbol{\pi} \in \mathcal{S}, \tag{9.1.3}$$

and treat the functional forms of the $\pi_{ij}(\boldsymbol{\theta})$ to be given and $\boldsymbol{\theta}$ as an unknown parameter (vector). Based on this formulation, the relevant questions may be expressed in the form of tests of goodness of fit [for the transformation model in (9.1.3)], or more generally of tests of hypotheses about $\boldsymbol{\theta}$ as well as of estimation of the elements of $\boldsymbol{\theta}$.

Section 9.2 is devoted to the study of goodness-of-fit tests for categorical data models. The classical Pearsonian goodness-of-fit test for a simple multinomial distribution is considered as a precursor for general goodness-of-fit tests for the product multinomial model in (9.1.2)–(9.1.3). The relevant parametric (asymptotic) theory (based on BAN estimators of $\boldsymbol{\theta}$) is presented in Section 9.3. The likelihood ratio and Wald tests for general hypotheses on the transformed model (9.1.3) and related large sample theory are presented in a unified, yet simple fashion. In the concluding section, we briefly discuss the asymptotic properties of some other statistics commonly employed in the analysis of categorical data.

9.2 Nonparametric Goodness-of-Fit Tests

To motivate the proposed large sample methodology, first, we consider the simple multinomial law in (9.1.1), so that we have

$$P\{\boldsymbol{n}|\boldsymbol{\pi}\} = \left(n! \Big/ \prod_{j=1}^{r} n_j!\right) \prod_{j=1}^{r} \pi_j^{n_j}, \quad n = \sum_{j=1}^{r} n_j, \quad \sum_{j=1}^{r} \pi_j = 1. \tag{9.2.1}$$

Note that, under (9.2.1),

$$\mathbb{E}(\boldsymbol{n}) = n\boldsymbol{\pi} \quad \text{and} \quad \mathbb{V}\text{ar}(\boldsymbol{n}) = n(\boldsymbol{D}_{\pi} - \boldsymbol{\pi}\boldsymbol{\pi}'), \tag{9.2.2}$$

where $\boldsymbol{\pi} = (\pi_1, \ldots, \pi_r)'$ and $\boldsymbol{D}_{\pi} = \text{diag}\,(\pi_1, \ldots, \pi_r)$. Thus, $\boldsymbol{n} - n\boldsymbol{\pi}$ stands for the deviation of the observed frequencies from their model-based expectations. Hence, to assess the fit of (9.2.1), one may consider the following Pearsonian goodness-of-fit statistic (assuming that $\boldsymbol{\pi}$ is specified):

$$Q_P = Q_P(\boldsymbol{\pi}) = \sum_{j=1}^{r} (n_j - n\pi_j)^2/(n\pi_j); \tag{9.2.3}$$

the case of $\boldsymbol{\pi}$ as in (9.1.3) with unknown $\boldsymbol{\theta}$ will be treated in the next section.

For a given $\boldsymbol{\pi}$, the exact distribution of Q_P can be enumerated by using (9.2.1), when n is not too large. However, this scheme not only depends on the specific $\boldsymbol{\pi}$ but also becomes prohibitively laborious as n becomes large. For this reason, we are interested in obtaining the asymptotic distribution of Q_P in a simpler form. In this context, we set

$$\boldsymbol{Z}_n = (Z_{n1}, \ldots, Z_{nr})'$$

with $Z_{nj} = (n_j - n\pi_j)/\sqrt{n\pi_j}$, $1 \leq j \leq r$, so that $Q_P = \boldsymbol{Z}_n'\boldsymbol{Z}_n = \|\boldsymbol{Z}_n\|^2$. We may also set $\boldsymbol{\pi}^{1/2} = (\pi_1^{1/2}, \ldots, \pi_r^{1/2})'$ and write

$$\boldsymbol{Z}_n = n^{-1/2}\boldsymbol{D}_{\pi^{1/2}}^{-1}(\boldsymbol{n} - n\boldsymbol{\pi}), \tag{9.2.4}$$

where $\boldsymbol{D}_{\pi^{1/2}} = \text{diag}(\pi_1^{1/2}, \ldots, \pi_r^{1/2})$. Then, we have

$$\boldsymbol{Z}_n'\boldsymbol{\pi}^{1/2} = n^{-1/2}(\boldsymbol{n} - n\boldsymbol{\pi})'\boldsymbol{1}_r = n^{-1/2}\sum_{j=1}^{r}(n_j - n\pi_j) = 0$$

and

$$Q_P = \|\boldsymbol{Z}_n\|^2 = \boldsymbol{Z}_n'\boldsymbol{Z}_n = \boldsymbol{Z}_n'[\boldsymbol{I}_r - \boldsymbol{\pi}^{1/2}(\boldsymbol{\pi}^{1/2})']\boldsymbol{Z}_n,$$

that is,

$$Q_P = \boldsymbol{Z}_n'[\boldsymbol{I}_r - \boldsymbol{\pi}^{1/2}(\boldsymbol{\pi}^{1/2})']\boldsymbol{Z}_n, \tag{9.2.5}$$

where

$$[\boldsymbol{I}_r - \boldsymbol{\pi}^{1/2}(\boldsymbol{\pi}^{1/2})'][\boldsymbol{I}_r - \boldsymbol{\pi}^{1/2}(\boldsymbol{\pi}^{1/2})'] = \boldsymbol{I}_r - \boldsymbol{\pi}^{1/2}(\boldsymbol{\pi}^{1/2})' \tag{9.2.6}$$

by virtue of $(\boldsymbol{\pi}^{1/2})'(\boldsymbol{\pi}^{1/2}) = \boldsymbol{\pi}'\boldsymbol{1} = 1$. Let us define

$$\boldsymbol{X}_i = (X_{i1}, \ldots, X_{ir})', \quad i = 1, \ldots, n$$

with

$$X_{ij} = \begin{cases} 1 & \text{if the } i\text{th sample unit is classified in the } j\text{th cell} \\ 0 & \text{otherwise,} \end{cases} \tag{9.2.7}$$

$j = 1, \ldots, r$. This clearly implies that for all $i = 1, \ldots, n$, $j, l = 1, \ldots, r$,

(i) $\mathbb{E}(X_{ij}) = \pi_j$,

(ii) $\mathrm{Cov}(X_{ij}, X_{il}) = \begin{cases} \pi_j(1-\pi_j), & j = l \\ -\pi_j\pi_l, & j \neq l. \end{cases}$

Therefore, $\sum_{i=1}^n \boldsymbol{X}_i = \boldsymbol{n}$, $\mathbb{E}(\boldsymbol{X}_i) = \boldsymbol{\pi}$ and $\mathbb{V}\mathrm{ar}(\boldsymbol{X}_i) = \boldsymbol{D}_{\boldsymbol{\pi}} - \boldsymbol{\pi}\boldsymbol{\pi}'$. Then, write

$$\boldsymbol{Z}_n = n^{-1/2}\sum_{i=1}^n \boldsymbol{D}_{\boldsymbol{\pi}^{1/2}}^{-1}(\boldsymbol{X}_i - \boldsymbol{\pi}) = n^{-1/2}\sum_{i=1}^n \boldsymbol{Y}_i,$$

where

$$\boldsymbol{Y}_i = \boldsymbol{D}_{\boldsymbol{\pi}^{1/2}}^{-1}(\boldsymbol{X}_i - \boldsymbol{\pi}),$$

$$\mathbb{E}(\boldsymbol{Y}_i) = \boldsymbol{0},$$

$$\mathbb{V}\mathrm{ar}(\boldsymbol{Y}_i) = \boldsymbol{D}_{\boldsymbol{\pi}^{1/2}}^{-1}[\boldsymbol{D}_{\boldsymbol{\pi}} - \boldsymbol{\pi}\boldsymbol{\pi}']\boldsymbol{D}_{\boldsymbol{\pi}^{1/2}}^{-1} = \boldsymbol{I}_r - \boldsymbol{\pi}^{1/2}(\boldsymbol{\pi}^{1/2})'$$

and use the (multivariate version of the) Central Limit Theorem 7.3.4 to see that

$$\boldsymbol{Z}_n \xrightarrow{D} N_r[\boldsymbol{0}, \boldsymbol{I}_r - \boldsymbol{\pi}^{1/2}(\boldsymbol{\pi}^{1/2})']. \tag{9.2.8}$$

Finally, given (9.2.5), (9.2.6) and (9.2.8), we may conclude that $Q_P \xrightarrow{D} \chi^2_{(r-1)}$, when (9.2.1) fits by the Cochran Theorem 7.6.1.

Consider now the problem of testing

$$H_0 : \boldsymbol{\pi} = \boldsymbol{\pi}_0 = (\pi_{01}, \ldots, \pi_{0r})'$$

against a fixed alternative

$$H_1 : \boldsymbol{\pi} = \boldsymbol{\pi}_1 = (\pi_{11}, \ldots, \pi_{1r})'$$

using the statistic $Q_P(\boldsymbol{\pi}_0)$. We have seen earlier that when H_0 is true, $Q_P(\boldsymbol{\pi}_0) \xrightarrow{D} \chi^2_{(r-1)}$. Let us examine the power of the test.

When H_1 is true, it follows by the Khintchine Weak Law of Large Numbers (Theorem 6.3.6) that $n^{-1}\sum_{i=1}^{n}\boldsymbol{X}_i \xrightarrow{P} \boldsymbol{\pi}_1$. Then, writing

$$Q_P(\boldsymbol{\pi}_0) = n\sum_{j=1}^{r}(n_j/n - \pi_{0j})^2/\pi_{0j}$$

we may conclude that

$$n^{-1}Q_P(\boldsymbol{\pi}_0) \xrightarrow{P} \sum_{j=1}^{r}(\pi_{1j} - \pi_{0j})^2/\pi_{0j} = K > 0. \tag{9.2.9}$$

Thus,

$$P[Q_P(\boldsymbol{\pi}_0) \geq \chi^2_{(r-1),1-\alpha}|A] = P[n^{-1}Q_P(\boldsymbol{\pi}_0) \geq n^{-1}\chi^2_{(r-1),1-\alpha}|A] \to 1$$

as $n \to \infty$, in view of (9.2.9) and the fact that $n^{-1}\chi^2_{(r-1)}(\alpha) \to 0$ as $n \to \infty$. Hence, for all fixed alternative hypotheses, the power of the test converges to 1, that is, the test is consistent for any departure from the null hypothesis.

Now let us consider what happens to the power of the test when the distance between the alternative and the null hypotheses converges to zero as n increases, that is, for Pitman alternatives of the form

$$H_{1,n} : \boldsymbol{\pi} = \boldsymbol{\pi}_0 + n^{-1/2}\boldsymbol{\delta}$$

where $\boldsymbol{\delta} = (\delta_1, \ldots, \delta_r)'$ is a vector of constants such that $\mathbf{1}_r'\boldsymbol{\delta} = 0$. Note that

$$\mathbb{E}(\boldsymbol{Y}_i|H_{1,n}) = \boldsymbol{D}^{-1}_{\boldsymbol{\pi}_0^{1/2}}\mathbb{E}(\boldsymbol{X}_i - \pi_0|H_{1,n}) = n^{-1/2}\boldsymbol{D}^{-1}_{\boldsymbol{\pi}_0^{1/2}}\boldsymbol{\delta},$$

and

$$\begin{aligned}
\mathbb{V}\text{ar}(\boldsymbol{Y}_i|H_{1,n}) &= \boldsymbol{D}^{-1}_{\boldsymbol{\pi}_0^{1/2}}\mathbb{V}\text{ar}(\boldsymbol{X}_i - \pi_0|H_{1,n})\boldsymbol{D}^{-1}_{\boldsymbol{\pi}_0^{1/2}}\\
&= \boldsymbol{D}^{-1}_{\boldsymbol{\pi}_0^{1/2}}[\boldsymbol{D}_{\boldsymbol{\pi}_0+n^{-1/2}\boldsymbol{\delta}} - (\boldsymbol{\pi}_0 + n^{-1/2}\boldsymbol{\delta})(\boldsymbol{\pi}_0 + n^{-1/2}\boldsymbol{\delta})']\boldsymbol{D}^{-1}_{\boldsymbol{\pi}_0^{1/2}}\\
&= \boldsymbol{D}^{-1}_{\boldsymbol{\pi}_0^{1/2}}[\boldsymbol{D}_{\boldsymbol{\pi}_0+n^{-1/2}\boldsymbol{\delta}} - \boldsymbol{\pi}_0\boldsymbol{\pi}_0' + \mathbf{1}_r O(n^{-1/2})]\boldsymbol{D}^{-1}_{\boldsymbol{\pi}_0^{1/2}}\\
&= \boldsymbol{D}^{-1}_{\boldsymbol{\pi}_0^{1/2}}[\boldsymbol{I}_r - \boldsymbol{\pi}_0\boldsymbol{\pi}_0' + \mathbf{1}_r\mathbf{1}_r' O(n^{-1/2})]\boldsymbol{D}^{-1}_{\boldsymbol{\pi}_0^{1/2}}\\
&= \boldsymbol{I}_r - \boldsymbol{\pi}_0^{1/2}(\boldsymbol{\pi}_0^{1/2})' + \mathbf{1}_r\mathbf{1}_r'[O(n^{-1/2})].
\end{aligned}$$

Therefore, we have

$$\mathbb{E}(\boldsymbol{Z}_n|H_{1,n}) = \boldsymbol{D}^{-1}_{\boldsymbol{\pi}_0^{1/2}}\boldsymbol{\delta},$$

and

$$\mathbb{V}\text{ar}(\boldsymbol{Z}_n|H_{1,n}) = \boldsymbol{I}_r - \boldsymbol{\pi}_0^{1/2}(\boldsymbol{\pi}_0^{1/2})' + \mathbf{1}_r\mathbf{1}_r' O(n^{-1/2})$$

and it follows from the (multivariate version of the) Central Limit Theorem 7.3.4 that, under $H_{1,n}$,

$$\boldsymbol{Z}_n \xrightarrow{D} N_r\big[\boldsymbol{D}^{-1}_{\boldsymbol{\pi}_0^{1/2}}\boldsymbol{\delta},\ \boldsymbol{I}_r - \boldsymbol{\pi}_0^{1/2}(\boldsymbol{\pi}_0^{1/2})'\big],$$

which implies

$$Q_P = \boldsymbol{Z}_n'\big[\boldsymbol{I}_r - \boldsymbol{\pi}_0^{1/2}\big(\boldsymbol{\pi}_0^{1/2}\big)'\big]\boldsymbol{Z}_n \xrightarrow{D} \chi^2_{r-1}(\Delta),$$

where the noncentrality parameter is given by

$$\begin{aligned}\Delta &= \boldsymbol{\delta}'\boldsymbol{D}^{-1}_{\boldsymbol{\pi}_0^{1/2}}\big[\boldsymbol{I}_r - \boldsymbol{\pi}_0^{1/2}\big(\boldsymbol{\pi}_0^{1/2}\big)'\big]\boldsymbol{D}^{-1}_{\boldsymbol{\pi}_0^{1/2}}\boldsymbol{\delta}\\ &= \boldsymbol{\delta}'\boldsymbol{D}^{-1}_{\boldsymbol{\pi}_0^{1/2}}\boldsymbol{D}^{-1}_{\boldsymbol{\pi}_0^{1/2}}\boldsymbol{\delta} - 0\\ &= \sum_{j=1}^{r}\big(\delta_j^2/\pi_{0j}\big).\end{aligned}$$

The extension of the above results to the general case ($s > 1$) is straightforward; using the same arguments as for the case $s = 1$, it is easy to show that under the hypothesis H_0 : $\boldsymbol{\pi} = \boldsymbol{\pi}_0$, where $\boldsymbol{\pi}_0 = (\boldsymbol{\pi}_{01}', \ldots, \boldsymbol{\pi}_{0s}')'$, $\boldsymbol{\pi}_{0i} = (\pi_{0i1}, \ldots, \pi_{0ir})'$, $\boldsymbol{1}_r'\boldsymbol{\pi}_{0i} = 1, i = 1, \ldots, s$, we have

$$Q_P = Q_P(\boldsymbol{\pi}_0) = \sum_{i=1}^{s}\sum_{j=1}^{r}(n_{ij} - n\pi_{0ij})^2/n\pi_{0ij} \xrightarrow{D} \chi^2_{s(r-1)}$$

and that under $H_{1,n} : \boldsymbol{\pi} = \boldsymbol{\pi}_0 + n^{-1/2}\boldsymbol{\delta}$, where $\boldsymbol{\delta} = (\boldsymbol{\delta}_1', \ldots, \boldsymbol{\delta}_s')'$, $\boldsymbol{\delta}_i' = (\delta_{i1}, \ldots, \delta_{ir})'$, $\boldsymbol{1}_r'\boldsymbol{\delta}_i = 0, i = 1, \ldots, s$, we have

$$Q_P = Q_P(\boldsymbol{\pi}_0) \xrightarrow{D} \chi^2_{s(r-1)}(\Delta)$$

where $\Delta = \sum_{i=1}^{s}\boldsymbol{\delta}_i'\boldsymbol{D}^{-1}_{\boldsymbol{\pi}_0^{1/2}}\boldsymbol{\delta}_i = \sum_{i=1}^{s}\sum_{j=1}^{r}\big(\delta_{ij}^2/\pi_{0ij}\big)$.

9.3 Estimation and Goodness-of-Fit Tests: Parametric Case

In general, the "parametric" multinomial model corresponds to (9.1.2) with $\pi_{ij} = \pi_{ij}(\boldsymbol{\theta})$, $i = 1, \ldots, s$, $j = 1, \ldots, r$, where $\boldsymbol{\theta} = (\theta_1, \ldots, \theta_q)'$ is a vector of underlying parameters. A typical example is the Hardy-Weinberg model for the ABO blood group classification system described in Example 1.2.11. There, model (9.1.2) holds with $s = 1, r = 4, \pi_1 = p_O$, $\pi_2 = p_A, \pi_3 = p_B$ and $\pi_4 = p_{AB}$; furthermore, we have $\pi_j = \pi_j(\boldsymbol{\theta})$ where $\boldsymbol{\theta} = (\theta_1, \theta_2, \theta_3)'$ with $\theta_1 = q_O, \theta_2 = q_A$ and $\theta_3 = q_B$.

***Example* 9.3.1.** Consider a two-way contingency table with rows and columns defined by random variables X (with a levels) and Y (with b levels), respectively. The question of interest is to verify whether X and Y are independent. A possible model to describe the corresponding frequency distribution based on n sampling units is (9.1.2) with $s = 1$, $r = ab$ and π_{ij} denoting the probability associated with the (i, j)th cell, $i = 1, \ldots, a$, $j = 1, \ldots, b$. If X and Y are independent, we may write $\pi_{ij} = \pi_{i\cdot}\pi_{\cdot j}$, $i = 1, \ldots, a$, $j = 1, \ldots, b$, where $\pi_{i\cdot}$ and $\pi_{\cdot j}$ represent the marginal probabilities corresponding to X and Y, respectively. Here $\pi_{ij} = \pi_{ij}(\boldsymbol{\theta})$ where $\boldsymbol{\theta} = (\theta_1, \ldots, \theta_{a+b})$ with $\theta_1 = \pi_{1\cdot}, \ldots, \theta_a = \pi_{a\cdot}, \theta_{a+1} = \pi_{\cdot 1}, \ldots, \theta_{a+b} = \pi_{\cdot b}$. □

We now consider the problem of estimating the parameter vector $\boldsymbol{\theta}$ and assessing goodness of fit of such "parametric" models via the three most common approaches, namely, the

ML, *minimum chi-squared* (MCS) and *modified minimum chi-squared* (MMCS) methods. Although this might be considered a special case of the general methodology outlined in Chapter 8, we present the derivations in detail to call the attention to some peculiarities related to the natural restriction that the parameters of the underlying (product) multinomial model must satisfy. We first show the results for the multinomial case ($s = 1$), where the log-likelihood may be expressed as

$$\log L_n(\boldsymbol{\theta}) = \text{constant} + \sum_{j=1}^{r} n_j \log \pi_j(\boldsymbol{\theta}). \tag{9.3.1}$$

We assume that $\pi_j(\boldsymbol{\theta})$, $j = 1, \ldots, r$, possess continuous derivatives up to the order 2.

The MLE of $\boldsymbol{\theta}$ is a solution $\widehat{\boldsymbol{\theta}}$ to

$$\sum_{j=1}^{r} \frac{n_j}{\pi_j(\boldsymbol{\theta})} \frac{\partial}{\partial \boldsymbol{\theta}} \pi_j(\boldsymbol{\theta}) = \mathbf{0} \quad \text{subject to } \textstyle\sum_{j=1}^{r} \pi_j(\boldsymbol{\theta}) = 1. \tag{9.3.2}$$

Using the same technique employed in Theorem 8.3.1 we will derive the asymptotic distribution of $\widehat{\boldsymbol{\theta}}$. First, let $\|\boldsymbol{u}\| \leq K$, $0 < K < \infty$, and consider the Taylor expansion:

$$\begin{aligned}\log L_n(\boldsymbol{\theta} + n^{-1/2}\boldsymbol{u}) &= \log L_n(\boldsymbol{\theta}) + \frac{1}{\sqrt{n}} \boldsymbol{u}' \frac{\partial}{\partial \boldsymbol{\theta}} \log L_n(\boldsymbol{\theta}) \\ &\quad + \frac{1}{2n} \boldsymbol{u}' \frac{\partial^2}{\partial \boldsymbol{\theta} \partial \boldsymbol{\theta}'} \log L_n(\boldsymbol{\theta})\bigg|_{\boldsymbol{\theta}=\boldsymbol{\theta}^*} \boldsymbol{u},\end{aligned}$$

where $\boldsymbol{\theta}^*$ belongs to the line segment joining $\boldsymbol{\theta}$ and $\boldsymbol{\theta} + n^{-1/2}\boldsymbol{u}$. Then, we have

$$\begin{aligned}\lambda_n(\boldsymbol{u}) &= \log L_n(\boldsymbol{\theta} + n^{-1/2}\boldsymbol{u}) - \log L_n(\boldsymbol{\theta}) \\ &= \frac{1}{\sqrt{n}} \boldsymbol{u}' \boldsymbol{U}_n + \frac{1}{2n} \boldsymbol{u}' \boldsymbol{V}_n \boldsymbol{u} + \frac{1}{2n} \boldsymbol{u}' \boldsymbol{W}_n \boldsymbol{u},\end{aligned} \tag{9.3.3}$$

where

$$\begin{aligned}\boldsymbol{U}_n &= \frac{\partial}{\partial \boldsymbol{\theta}} \log L_n(\boldsymbol{\theta}), \\ \boldsymbol{V}_n &= \frac{\partial^2}{\partial \boldsymbol{\theta} \partial \boldsymbol{\theta}'} \log L_n(\boldsymbol{\theta})\end{aligned}$$

and

$$\boldsymbol{W}_n = \frac{\partial^2}{\partial \boldsymbol{\theta} \partial \boldsymbol{\theta}'} \log L_n(\boldsymbol{\theta})\bigg|_{\boldsymbol{\theta}=\boldsymbol{\theta}^*} - \frac{\partial^2}{\partial \boldsymbol{\theta} \partial \boldsymbol{\theta}'} \log L_n(\boldsymbol{\theta}).$$

Now let us examine the terms in the right-hand side of (9.3.3). From the definition of the score statistic,

$$\begin{aligned}n^{-1/2} \boldsymbol{U}_n &= n^{-1/2} \sum_{j=1}^{r} \frac{n_j}{\pi_j(\boldsymbol{\theta})} \frac{\partial}{\partial \boldsymbol{\theta}} \pi_j(\boldsymbol{\theta}) \\ &= n^{-1/2} \sum_{j=1}^{r} \left[\sum_{i=1}^{n} \frac{X_{ij}}{\pi_j(\boldsymbol{\theta})} \frac{\partial}{\partial \boldsymbol{\theta}} \pi_j(\boldsymbol{\theta}) \right] \\ &= \frac{1}{\sqrt{n}} \sum_{i=1}^{n} \boldsymbol{Y}_i,\end{aligned}$$

where

$$\boldsymbol{Y}_i = \sum_{j=1}^{r} \frac{X_{ij}}{\pi_j(\boldsymbol{\theta})} \frac{\partial}{\partial(\boldsymbol{\theta})} \pi_j(\boldsymbol{\theta})$$

and X_{ij} is defined as in (9.2.7). Since

$$\mathbb{E}(\boldsymbol{Y}_i) = \sum_{j=1}^{r} \frac{\mathbb{E}(X_{ij})}{\pi_j(\boldsymbol{\theta})} = \sum_{j=1}^{r} \frac{\partial}{\partial \boldsymbol{\theta}} \pi_j(\boldsymbol{\theta}) = \frac{\partial}{\partial \boldsymbol{\theta}} \sum_{j=1}^{r} \pi_j(\boldsymbol{\theta}) = \mathbf{0}$$

and

$$\begin{aligned}
\mathbb{E}(\boldsymbol{Y}_i \boldsymbol{Y}_i') &= \sum_{j=1}^{r} \frac{\mathbb{E}\left(X_{ij}^2\right)}{[\pi_j(\boldsymbol{\theta})]^2} \frac{\partial}{\partial \boldsymbol{\theta}} \pi_j(\boldsymbol{\theta}) \frac{\partial}{\partial \boldsymbol{\theta}'} \pi_j(\boldsymbol{\theta}) \\
&= \sum_{j=1}^{r} \frac{1}{\pi_j(\boldsymbol{\theta})} \frac{\partial}{\partial \boldsymbol{\theta}} \pi_j(\boldsymbol{\theta}) \frac{\partial}{\partial \boldsymbol{\theta}'} \pi_j(\boldsymbol{\theta}) \\
&= \boldsymbol{I}(\boldsymbol{\theta}),
\end{aligned}$$

it follows from the (multivariate version of the) Central Limit Theorem 7.3.4 that

$$n^{-1/2} \boldsymbol{U}_n \xrightarrow{D} \mathcal{N}[\mathbf{0}, \boldsymbol{I}(\boldsymbol{\theta})]. \tag{9.3.4}$$

Furthermore,

$$\begin{aligned}
n^{-1} \boldsymbol{V}_n &= -\frac{1}{n} \sum_{j=1}^{r} \frac{n_j}{[\pi_j(\boldsymbol{\theta})]^2} \frac{\partial}{\partial \boldsymbol{\theta}} \pi_j(\boldsymbol{\theta}) \frac{\partial}{\partial \boldsymbol{\theta}'} \pi_j(\boldsymbol{\theta}) \\
&\quad + \frac{1}{n} \sum_{j=1}^{r} \frac{n_j}{\pi_j(\boldsymbol{\theta})} \frac{\partial^2}{\partial \boldsymbol{\theta} \partial \boldsymbol{\theta}'} \pi_j(\boldsymbol{\theta}) \\
&= -\sum_{j=1}^{r} \frac{1}{n} \sum_{i=1}^{n} \frac{X_{ij}}{[\pi_j(\boldsymbol{\theta})]^2} \frac{\partial}{\partial \boldsymbol{\theta}} \pi_j(\boldsymbol{\theta}) \frac{\partial}{\partial \boldsymbol{\theta}'} \pi_j(\boldsymbol{\theta}) \\
&\quad + \sum_{j=1}^{r} \frac{1}{n} \sum_{i=1}^{n} \frac{X_{ij}}{\pi_j(\boldsymbol{\theta})} \frac{\partial^2}{\partial \boldsymbol{\theta} \partial \boldsymbol{\theta}'} \pi_j(\boldsymbol{\theta}),
\end{aligned}$$

where X_{ij} is defined as in (9.2.7). Then, using the Khintchine Weak Law of Large Numbers (Theorem 6.3.6), it follows that

$$n^{-1} \boldsymbol{V}_n \xrightarrow{P} -\sum_{j=1}^{r} \frac{1}{\pi_j(\boldsymbol{\theta})} \frac{\partial}{\partial \boldsymbol{\theta}} \pi_j(\boldsymbol{\theta}) \frac{\partial}{\partial \boldsymbol{\theta}'} \pi_j(\boldsymbol{\theta}) + \sum_{j=1}^{r} \frac{\partial^2}{\partial \boldsymbol{\theta} \partial \boldsymbol{\theta}'} \pi_j(\boldsymbol{\theta}) = -\boldsymbol{I}(\boldsymbol{\theta}). \tag{9.3.5}$$

Since $(\partial^2/\partial\boldsymbol{\theta}\partial\boldsymbol{\theta}') \log L_n(\boldsymbol{\theta})$ is a continuous function of $\boldsymbol{\theta}$, we have

$$n^{-1} \boldsymbol{W}_n \to \mathbf{0} \quad \text{as } n \to \infty. \tag{9.3.6}$$

From (9.3.3)–(9.3.6) we may conclude that

$$\lambda_n(\boldsymbol{u}) = n^{-1/2} \boldsymbol{u}' \boldsymbol{U}_n - \frac{1}{2} \boldsymbol{u}' \boldsymbol{I}(\boldsymbol{\theta}) \boldsymbol{u} + o_p(1), \tag{9.3.7}$$

the maximum of which is attained at

$$\widehat{\boldsymbol{u}} = n^{-1/2}[\boldsymbol{I}(\boldsymbol{\theta})]^{-1}\boldsymbol{U}_n + o_p(\mathbf{1}).$$

This is also the maximum of $\log L_n(\boldsymbol{\theta})$, which corresponds to the MLE of $\boldsymbol{\theta}$. Thus,

$$\widehat{\boldsymbol{\theta}}_n = \boldsymbol{\theta} + n^{-1/2}\widehat{\boldsymbol{u}} = \boldsymbol{\theta} + n^{-1}[\boldsymbol{I}(\boldsymbol{\theta})]^{-1}\boldsymbol{U}_n + o_p(\mathbf{1}),$$

which implies that

$$\sqrt{n}(\widehat{\boldsymbol{\theta}} - \boldsymbol{\theta}) = n^{-1/2}[\boldsymbol{I}(\boldsymbol{\theta})]^{-1}\boldsymbol{U}_n + o_p(\mathbf{1}) \tag{9.3.8}$$

and, by Slutsky Theorem 7.5.2, we obtain

$$\sqrt{n}(\widehat{\boldsymbol{\theta}}_n - \boldsymbol{\theta}) \xrightarrow{D} \mathcal{N}\{\mathbf{0}, [\boldsymbol{I}(\boldsymbol{\theta})]^{-1}\}. \tag{9.3.9}$$

Before we proceed with the next approach for the estimation of $\boldsymbol{\theta}$, it is convenient to note that, putting

$$n_j = n\pi_j(\boldsymbol{\theta}) + Z_{nj}(\boldsymbol{\theta})\sqrt{n\pi_j(\boldsymbol{\theta})},$$

where $Z_{nj}(\boldsymbol{\theta})$ is defined as in Section 9.2, the equations (9.3.2) may be re-expressed as

$$\sqrt{n}\sum_{j=1}^{r} Z_{nj}(\boldsymbol{\theta})\frac{1}{\sqrt{\pi_j(\boldsymbol{\theta})}}\frac{\partial}{\partial\boldsymbol{\theta}}\pi_j(\boldsymbol{\theta}) = \mathbf{0}. \tag{9.3.10}$$

As we have mentioned before, the statistic

$$Q_P = Q_P(\boldsymbol{\theta}) = \sum_{j=1}^{r}[n_j - n\pi_j(\boldsymbol{\theta})]^2/n\pi_j(\boldsymbol{\theta})$$

may be used as a measure of the goodness of fit of the model $\pi_j = \pi_j(\boldsymbol{\theta})$, $j = 1, \ldots, r$. The value $\widetilde{\boldsymbol{\theta}}_n$ that minimizes $Q_P(\boldsymbol{\theta})$ is known as the minimum chi-squared (MCS) estimator of $\boldsymbol{\theta}$ and may be obtained as a solution to

$$\begin{aligned}\frac{\partial}{\partial\boldsymbol{\theta}}Q_P(\boldsymbol{\theta}) = &-2\sum_{j=1}^{r}\frac{[n_j - n\pi_j(\boldsymbol{\theta})]}{\pi_j(\boldsymbol{\theta})}\frac{\partial}{\partial\boldsymbol{\theta}}\pi_j(\boldsymbol{\theta})\\ &-\sum_{j=1}^{r}\frac{[n_j - n\pi_j(\boldsymbol{\theta})]^2}{n[\pi_j(\boldsymbol{\theta})]^2}\frac{\partial}{\partial\boldsymbol{\theta}}\pi_j(\boldsymbol{\theta}) = \mathbf{0}.\end{aligned}$$

These equations may be expressed as

$$\begin{aligned}-2\sqrt{n}\Bigg[&\sum_{j=1}^{r} Z_{nj}(\boldsymbol{\theta})\frac{1}{\sqrt{\pi_j(\boldsymbol{\theta})}}\frac{\partial}{\partial\boldsymbol{\theta}}\pi_j(\boldsymbol{\theta})\\ &+\frac{1}{2\sqrt{n}}\sum_{j=1}^{r} Z_{nj}^2(\boldsymbol{\theta})\frac{1}{\pi_j(\boldsymbol{\theta})}\frac{\partial}{\partial\boldsymbol{\theta}}\pi_j(\boldsymbol{\theta})\Bigg] = \mathbf{0},\end{aligned}$$

which, in view of (9.2.8) and Theorem 7.2.5, reduce to

$$\sum_{j=1}^{r} Z_{nj}(\boldsymbol{\theta})\frac{1}{\sqrt{\pi_j(\boldsymbol{\theta})}}\frac{\partial}{\partial\boldsymbol{\theta}}\pi_j(\boldsymbol{\theta}) + O_p(\boldsymbol{n}^{-1/2}) = \mathbf{0}. \tag{9.3.11}$$

Note that the equations in (9.3.11) are equivalent to those in (9.3.10) and consequently to those in (9.3.2) up to the order $n^{-1/2}$. Now consider a Taylor expansion of (9.3.2) around the point $\widetilde{\boldsymbol{\theta}}_n$:

$$\sum_{j=1}^{r}\frac{n_j}{\pi_j(\boldsymbol{\theta})}\frac{\partial}{\partial\boldsymbol{\theta}}\pi_j(\boldsymbol{\theta})\bigg|_{\boldsymbol{\theta}=\widehat{\boldsymbol{\theta}}_n}-\sum_{j=1}^{r}\frac{n_j}{\pi_j(\boldsymbol{\theta})}\frac{\partial}{\partial\boldsymbol{\theta}}\pi_j(\boldsymbol{\theta})\bigg|_{\boldsymbol{\theta}=\widetilde{\boldsymbol{\theta}}_n}$$
$$=\left\{\frac{1}{n}\left[-\sum_{j=1}^{r}\frac{n_j}{\{\pi_j(\boldsymbol{\theta})\}^2}\frac{\partial}{\partial\boldsymbol{\theta}}\pi_j(\boldsymbol{\theta})\bigg|_{\boldsymbol{\theta}=\boldsymbol{\theta}^*}\frac{\partial}{\partial\boldsymbol{\theta}'}\pi_j(\boldsymbol{\theta})\bigg|_{\boldsymbol{\theta}=\boldsymbol{\theta}^*}\right.\right.$$
$$\left.\left.+\sum_{j=1}^{r}\frac{n_j}{\pi_j(\boldsymbol{\theta})}\frac{\partial^2}{\partial\boldsymbol{\theta}\partial\boldsymbol{\theta}'}\pi_j(\boldsymbol{\theta})\bigg|_{\boldsymbol{\theta}=\boldsymbol{\theta}^*}\right]\right\}n(\widehat{\boldsymbol{\theta}}_n-\widetilde{\boldsymbol{\theta}}_n) \tag{9.3.12}$$

where $\boldsymbol{\theta}^*$ belongs to the line segment joining $\widehat{\boldsymbol{\theta}}_n$ and $\widetilde{\boldsymbol{\theta}}_n$. Since by (9.3.5) the term within braces, { }, converges in probability to $-\boldsymbol{I}(\boldsymbol{\theta})$, which, is finite and since the term in the first member is $O_p(\mathbf{1})$, we may write

$$O_p(\mathbf{1})=[-\boldsymbol{I}(\boldsymbol{\theta})+o_p(\mathbf{1})]n(\widehat{\boldsymbol{\theta}}_n-\widetilde{\boldsymbol{\theta}}_n).$$

This implies that $\widetilde{\boldsymbol{\theta}}_n-\widetilde{\boldsymbol{\theta}}_n=O_p(\boldsymbol{n}^{-1})$, and we may conclude that

$$\sqrt{n}(\widehat{\boldsymbol{\theta}}_n-\widetilde{\boldsymbol{\theta}}_n)\xrightarrow{P}\mathbf{0}. \tag{9.3.13}$$

Along the lines of the previous case, we define the MMCS estimator of $\boldsymbol{\theta}$ as the value $\overline{\boldsymbol{\theta}}$ which minimizes

$$Q_N=Q_N(\boldsymbol{\theta})=\sum_{j=1}^{r}[n_j-n\pi_j(\boldsymbol{\theta})]^2/n_j.$$

The corresponding estimating equations are

$$\frac{\partial}{\partial\boldsymbol{\theta}}Q_N(\boldsymbol{\theta})=-2n\sum_{j=1}^{r}\frac{[n_j-n\pi_j(\boldsymbol{\theta})]}{n_j}\frac{\partial}{\partial\boldsymbol{\theta}}\pi_j(\boldsymbol{\theta})=\mathbf{0}$$

or, equivalently,

$$\sum_{j=1}^{r}\frac{n\pi_j(\boldsymbol{\theta})}{n_j}Z_{nj}(\boldsymbol{\theta})\frac{1}{\sqrt{\pi_j(\boldsymbol{\theta})}}\frac{\partial}{\partial\boldsymbol{\theta}}\pi_j(\boldsymbol{\theta})=\mathbf{0}. \tag{9.3.14}$$

Now, since for $x\neq 1$, we have $(1-x)^{-1}=1-x+O(x^2)$ and recalling (9.2.8) we may write

$$\begin{aligned}\frac{n\pi_j(\boldsymbol{\theta})}{n_j}&=\frac{n\pi_j(\boldsymbol{\theta})}{n\pi_j(\boldsymbol{\theta})+Z_{nj}(\boldsymbol{\theta})\sqrt{n\pi_j(\boldsymbol{\theta})}}\\&=\left[1+\frac{Z_{nj}(\boldsymbol{\theta})}{\sqrt{n\pi_j(\boldsymbol{\theta})}}\right]^{-1}\\&=1-\frac{Z_{nj}(\boldsymbol{\theta})}{\sqrt{n\pi_j(\boldsymbol{\theta})}}+O_p\left[\frac{Z_{nj}^2(\boldsymbol{\theta})}{n\pi_j(\boldsymbol{\theta})}\right]\\&=1+O_p(n^{-1/2}).\end{aligned} \tag{9.3.15}$$

Thus, from (9.3.14) and (9.3.15), it follows that the estimating equations corresponding to the MMCS estimator of $\boldsymbol{\theta}$ are equivalent to those in (9.3.2) up to the order $n^{-1/2}$. As in the case of the MCS estimator, it follows that

$$\sqrt{n}(\widetilde{\boldsymbol{\theta}} - \overline{\boldsymbol{\theta}}) \xrightarrow{P} \mathbf{0}. \tag{9.3.16}$$

From (9.3.9) and (9.3.16) we may conclude that all three methods of estimation considered above produce BAN estimators. The choice among the three methods then relies on computational aspects and depends essentially on the intrinsic characteristics of the model $\pi_j = \pi_j(\boldsymbol{\theta})$, $j = 1, \ldots, r$. In general, this involves the solution of a set of nonlinear equations in $\boldsymbol{\theta}$. However, if the π_js are linear functions of $\boldsymbol{\theta}$, that is, $\pi_j = \boldsymbol{x}_j'\boldsymbol{\theta}$, with $\boldsymbol{x}_j$ denoting a vector of constants, $j = 1, \ldots, r$, the MMCS estimator of $\boldsymbol{\theta}$ may be obtained as the solution to linear equations; see Koch, Imrey, Singer, Atkinson, and Stokes (1985) for details.

Goodness-of-fit for such models may be assessed via the statistics $\widehat{Q}_P$, $\widehat{Q}_N$ or $\widehat{Q}_V$, obtained by substituting any BAN estimator $\widehat{\boldsymbol{\pi}}$ for $\boldsymbol{\pi}$ in the expressions of Q_P, Q_N or Q_V respectively. We now show that these statistics follow asymptotic chi-squared distributions. In this direction, first define

$$\boldsymbol{B} = \boldsymbol{B}(\boldsymbol{\theta}) = [\boldsymbol{b}_1, \ldots, \boldsymbol{b}_r] = \frac{\partial[\pi(\boldsymbol{\theta})]'}{\partial\boldsymbol{\theta}}\boldsymbol{D}^{-1}_{\boldsymbol{\pi}^{1/2}(\boldsymbol{\theta})}$$
$$= \begin{bmatrix} \frac{1}{\sqrt{\pi_1(\boldsymbol{\theta})}}(\partial/\partial\theta_1)\pi_1(\boldsymbol{\theta}) & \cdots & \frac{1}{\sqrt{\pi_r(\boldsymbol{\theta})}}(\partial/\partial\theta_1)\pi_r(\boldsymbol{\theta}) \\ \vdots & & \vdots \\ \frac{1}{\sqrt{\pi_1(\boldsymbol{\theta})}}(\partial/\partial\theta_q)\pi_1(\boldsymbol{\theta}) & \cdots & \frac{1}{\sqrt{\pi_r(\boldsymbol{\theta})}}(\partial/\partial\theta_q)\pi_r(\boldsymbol{\theta}) \end{bmatrix}$$

and note that

$$\begin{aligned}\boldsymbol{B}\boldsymbol{B}' &= \frac{\partial}{\partial\boldsymbol{\theta}}[\boldsymbol{\pi}(\boldsymbol{\theta})]'\boldsymbol{D}^{-1}_{\boldsymbol{\pi}(\boldsymbol{\theta})}\frac{\partial}{\partial\boldsymbol{\theta}'}\boldsymbol{\pi}(\boldsymbol{\theta}) \\ &= \sum_{j=1}^{r}\frac{1}{\pi_j(\boldsymbol{\theta})}\frac{\partial}{\partial\boldsymbol{\theta}}\pi_j(\boldsymbol{\theta})\frac{\partial}{\partial\boldsymbol{\theta}'}\pi_j(\boldsymbol{\theta}) = \boldsymbol{I}(\boldsymbol{\theta});\end{aligned}$$

also observe that

$$\begin{aligned}n^{-1/2}\boldsymbol{U}_n &= n^{-1/2}\sum_{j=1}^{r}\frac{n_j}{\pi_j(\boldsymbol{\theta})}\frac{\partial}{\partial\boldsymbol{\theta}}\pi_j(\boldsymbol{\theta}) \\ &= \sum_{j=1}^{r}Z_{nj}(\boldsymbol{\theta})\frac{1}{\sqrt{\pi_j(\boldsymbol{\theta})}}\frac{\partial}{\partial\boldsymbol{\theta}}\pi_j(\boldsymbol{\theta}) = \boldsymbol{B}\boldsymbol{Z}_n,\end{aligned}$$

where $\boldsymbol{Z}_n$ is defined as in Section 9.2. Finally, note that

$$\widehat{\pi}_j = \pi_j(\widehat{\boldsymbol{\theta}}_n) = \pi_j(\boldsymbol{\theta}) + (\widehat{\boldsymbol{\theta}}_n - \boldsymbol{\theta})'\frac{\partial\pi_j(\boldsymbol{\theta})}{\partial\boldsymbol{\theta}}\bigg|_{\boldsymbol{\theta}=\boldsymbol{\theta}^*},$$

where $\boldsymbol{\theta}^*$ lies in the line segment joining $\boldsymbol{\theta}$ and $\widehat{\boldsymbol{\theta}}_n$.

Now, since $(\partial/\partial\boldsymbol{\theta})\pi_j(\boldsymbol{\theta})|_{\boldsymbol{\theta}=\boldsymbol{\theta}^*}$ is finite, we may recall (9.3.9) to conclude that

$$\sqrt{n}[\widehat{\pi}_j - \pi_j(\boldsymbol{\theta})] = O_p(1) \Rightarrow \widehat{\pi}_j - \pi_j(\boldsymbol{\theta}) = O_p(n^{-1/2}). \tag{9.3.17}$$

Then, recalling (9.3.8) write

$$\begin{aligned}\frac{n_j - n\widehat{\pi}_j}{\sqrt{n\pi_j}} &= \frac{n_j - n\pi_j}{\sqrt{n\pi_j}} - \frac{\sqrt{n}(\widehat{\pi}_j - \pi_j)}{\sqrt{\pi_j}} \\ &= Z_{nj} - \sqrt{n}(\widehat{\boldsymbol{\theta}}_n - \boldsymbol{\theta})'\boldsymbol{b}_j + o_p(1) \\ &= Z_{nj} - n^{-1/2}\boldsymbol{U}_n'[\boldsymbol{I}(\boldsymbol{\theta})]^{-1}\boldsymbol{b}_j + o_p(1) \\ &= O_p(1). \end{aligned} \tag{9.3.18}$$

Using (9.3.17) and (9.3.18) and employing an argument similar to that considered in (9.3.15), we have

$$\begin{aligned}\widehat{Q}_P &= \sum_{j=1}^{r} \frac{(n_j - n\widehat{\pi}_j)^2}{n\pi_j}\left[1 + \frac{\widehat{\pi}_j - \pi_j}{\pi_j}\right]^{-1} \\ &= \sum_{j=1}^{r} \frac{(n_j - n\widehat{\pi}_j)^2}{n\pi_j}[1 + O_p(n^{-1/2})] \\ &= \sum_{j=1}^{r} \frac{(n_j - n\widehat{\pi}_j)^2}{n\pi_j} + O_p(n^{-1/2}). \end{aligned} \tag{9.3.19}$$

Now, from (9.3.18) and (9.3.19) it follows that

$$\begin{aligned}\widehat{Q}_P &= \sum_{j=1}^{r}\{Z_{nj} - n^{-1/2}\boldsymbol{U}_n'[\boldsymbol{I}(\boldsymbol{\theta})]^{-1}\boldsymbol{b}_j + o_p(1)\}^2 + O_p(n^{-1/2}) \\ &= \boldsymbol{Z}_n'\boldsymbol{Z}_n + n^{-1}\boldsymbol{U}_n'[\boldsymbol{I}(\boldsymbol{\theta})]^{-1}\boldsymbol{B}\boldsymbol{B}'[\boldsymbol{I}(\boldsymbol{\theta})]^{-1}\boldsymbol{U}_n \\ &\quad - 2n^{-1/2}\boldsymbol{Z}_n'\boldsymbol{B}'[\boldsymbol{I}(\boldsymbol{\theta})]^{-1}\boldsymbol{u}_n + o_p(1) \\ &= \boldsymbol{Z}_n'\boldsymbol{Z}_n + \boldsymbol{Z}_n'\boldsymbol{B}'[\boldsymbol{I}(\boldsymbol{\theta})]^{-1}\boldsymbol{B}\boldsymbol{Z}_n \\ &\quad - 2\boldsymbol{Z}_n'\boldsymbol{B}'[\boldsymbol{I}(\boldsymbol{\theta})]^{-1}\boldsymbol{B}\boldsymbol{Z}_n + o_p(1) \\ &= \boldsymbol{Z}_n'\{\boldsymbol{I}_r - \boldsymbol{B}'[\boldsymbol{I}(\boldsymbol{\theta})]^{-1}\boldsymbol{B}\}\boldsymbol{Z}_n + o_p(1). \end{aligned} \tag{9.3.20}$$

Using the Slutsky Theorem 7.5.2 and observing that both $\boldsymbol{A} = \boldsymbol{I}_r - \boldsymbol{B}'[\boldsymbol{I}(\boldsymbol{\theta})]^{-1}\boldsymbol{B}$ and $\boldsymbol{\Sigma} = \boldsymbol{I}_r - \boldsymbol{\pi}^{1/2}(\boldsymbol{\pi}^{1/2})'$ are idempotent matrices, we may apply Theorem 7.6.3 to conclude that if the model $\pi_j = \pi_j(\boldsymbol{\theta})$, $j = 1, \ldots, r$, holds, $\widehat{Q}_P$ follows asymptotically a chi-squared distribution with degrees of freedom given by

$$\begin{aligned}\operatorname{rank}(\boldsymbol{A}) &= \operatorname{rank}(\boldsymbol{A}\boldsymbol{\Sigma}\boldsymbol{A}) = \operatorname{tr}(\boldsymbol{A}\boldsymbol{A}\boldsymbol{\Sigma}) = \operatorname{tr}(\boldsymbol{A}\boldsymbol{\Sigma}) \\ &= \operatorname{tr}\{\boldsymbol{I}_r - \boldsymbol{\pi}^{1/2}(\boldsymbol{\pi}^{1/2})' - \boldsymbol{B}'[\boldsymbol{I}(\boldsymbol{\theta})]^{-1}\boldsymbol{B} \\ &\quad + \boldsymbol{B}'[\boldsymbol{I}(\boldsymbol{\theta})]^{-1}\boldsymbol{B}\boldsymbol{\pi}^{1/2}(\boldsymbol{\pi}^{1/2})'\} \\ &= \operatorname{tr}(\boldsymbol{I}_r) - \operatorname{tr}\{\boldsymbol{\pi}^{1/2}(\boldsymbol{\pi}^{1/2})'\} - \operatorname{tr}\{[\boldsymbol{I}(\boldsymbol{\theta})]^{-1}\boldsymbol{B}\boldsymbol{B}'\} \\ &\quad + \operatorname{tr}\{\boldsymbol{B}'[\boldsymbol{I}(\boldsymbol{\theta})]^{-1}\boldsymbol{B}\boldsymbol{\pi}^{1/2}(\boldsymbol{\pi}^{1/2})'\} \\ &= r - 1 - q. \end{aligned}$$

The proofs of similar results for $\widehat{Q}_N$ and $\widehat{Q}_V$ are left as exercises.

Also, note that the model $\pi_j = \pi_j(\boldsymbol{\theta})$, $j = 1, \ldots, r$, may be expressed in terms of the $r - q$ restrictions

$$H_0 : \boldsymbol{F}(\boldsymbol{\pi}) = \boldsymbol{0}, \tag{9.3.21}$$

where $\boldsymbol{F}(\boldsymbol{\pi}) = [F_1(\boldsymbol{\pi}), \ldots, F_{r-q}(\boldsymbol{\pi})]'$ are functions obtained by eliminating $\boldsymbol{\theta}$. In particular, if $\boldsymbol{\pi} = \boldsymbol{X\theta}$, (9.3.21) reduces to $H_0 : \boldsymbol{W\pi} = \boldsymbol{0}$ where $\boldsymbol{W}$ is a $(r - q \times r)$ matrix such that $\boldsymbol{WX} = \boldsymbol{0}$ (i.e., the rows of $\boldsymbol{W}$ are orthogonal to the columns of $\boldsymbol{X}$). If the model is nonlinear, let us denote by $\boldsymbol{H}(\boldsymbol{\pi}) = (\partial/\partial\boldsymbol{\pi})\boldsymbol{F}(\boldsymbol{\pi})$ [a $r \times (r - q)$ matrix]. Assume that in an open neighborhood of $\boldsymbol{\pi}$, $\boldsymbol{H}(\boldsymbol{\pi})$ is continuous and (without loss of generality) of rank $r - q$. Then,

$$\boldsymbol{F}(\widehat{\boldsymbol{\pi}}) = \boldsymbol{F}(\boldsymbol{\pi}) + \boldsymbol{H}(\boldsymbol{\pi})(\widehat{\boldsymbol{\pi}} - \boldsymbol{\pi}) + o(||\widehat{\boldsymbol{\pi}} - \boldsymbol{\pi}||),$$

so that

$$\begin{aligned} &n\mathbb{E}[\boldsymbol{F}(\widehat{\boldsymbol{\pi}}) - \boldsymbol{F}(\boldsymbol{\pi})][\boldsymbol{F}(\widehat{\boldsymbol{\pi}}) - \boldsymbol{F}(\boldsymbol{\pi})]' \\ &\quad = \boldsymbol{H}(\boldsymbol{\pi})(\boldsymbol{D}_{\pi} - \boldsymbol{\pi}\boldsymbol{\pi}')[\boldsymbol{H}(\boldsymbol{\pi})]' + o(\boldsymbol{1}\boldsymbol{1}'). \end{aligned}$$

Let

$$\widehat{\boldsymbol{V}}_n = \boldsymbol{H}(\widehat{\boldsymbol{\pi}})(\boldsymbol{D}_{\widehat{\pi}} - \widehat{\boldsymbol{\pi}}\widehat{\boldsymbol{\pi}}')[\boldsymbol{H}(\widehat{\boldsymbol{\pi}})]'$$

and note that $\widehat{\boldsymbol{V}}_n \xrightarrow{P} \boldsymbol{H}(\boldsymbol{\pi})(\boldsymbol{D}_{\pi} - \boldsymbol{\pi}\boldsymbol{\pi}')[\boldsymbol{H}(\boldsymbol{\pi})]'$, as $n \to \infty$. Consequently, under H_0, by Theorem 7.6.1,

$$n[\boldsymbol{F}(\widehat{\boldsymbol{\pi}})]'[\widehat{V}_n]^{-1}\boldsymbol{F}(\widehat{\boldsymbol{\pi}}) \xrightarrow{D} \chi^2_{q-r}.$$

If $\boldsymbol{F}(\boldsymbol{\pi})$ is linear, $\boldsymbol{H}(\boldsymbol{\pi})$ does not depend on $\boldsymbol{\pi}$, and we are back to the conventional MMCS type test. The method of scoring applies to $\widehat{\boldsymbol{\pi}}$ in a more complex setup and, hence, to the Wald-type of test as well. For $\widehat{\boldsymbol{\pi}}$ we do not need it.

For Pitman type local alternatives (see Section 8.6) in either setup we will have appropriate noncentral χ^2 distributions.

It is also worthwhile to mention that for finite sample sizes, the bias associated with such BAN estimators is of the order $o(n^{-1/2})$. To reduce such bias as well as to improve the estimates of the associated asymptotic covariance matrix, one may rely on the classical jackknifing methods, which will be discussed briefly in Chapter 11.

Bhapkar (1966) showed that $\widehat{Q}_W$ is algebraically identical to $\widehat{Q}_N$ up to the first-order approximation and, thus, it shares the same asymptotic properties of the latter. The choice between the competing statistics is generally governed by computational reasons. This interesting equivalence will be discussed under a broader perspective in Section 10.5.

Extensions of the above results for product multinomial data (i.e., $s \geq 2$ in (9.1.2) may be obtained along similar lines with the assumption that n_i/n, $i = 1, \ldots, s$ converge to constants ν_i as $n \to \infty$, with $\sum_{i=1}^{s} \nu_i = 1$. The underling random variables are defined as

$$X_{ijk} = \begin{cases} 1 & \text{if the } i\text{th sample unit is classified in the } j\text{th cell of} \\ & \text{the } k\text{th subpopulation} \\ 0 & \text{otherwise.} \end{cases}$$

9.4 Some Other Important Statistics

First, let us consider a simple 2×2 contingency table relating to two rows (1 and 2) and two columns (1 and 2), so that the cell probability for the ith row, jth column is denoted by π_{ij}, $i, j = 1, 2$. In a sample of size n, the corresponding observed frequencies are n_{ij}, $i, j = 1, 2$. The *odds ratio* or the *cross-product ratio* is defined by

$$\psi = (\pi_{11}\pi_{22})/(\pi_{21}\pi_{12}), \tag{9.4.1}$$

and a departure of ψ from 1 indicates a nonnull association between the random variables defining rows and columns. Based on the sample data, a natural estimator of ψ is

$$\widehat{\psi}_n = (n_{11}n_{22})/(n_{21}n_{12}). \tag{9.4.2}$$

If we denote the sample proportions by

$$p_{ij} = n_{ij}/n, \quad i, j = 1, 2, \tag{9.4.3}$$

then $\widehat{\psi}_n$ is a multiplicative function of p_{ij} to which a multinomial probability law may be associated. However, in this multiplicative form, the (asymptotic or exact) variability of $\widehat{\psi}_n$ about ψ depends on ψ and converges slowly. On the other hand,

$$\theta = \log\psi = \log\pi_{11} + \log\pi_{22} - \log\pi_{21} - \log\pi_{12} \tag{9.4.4}$$

for which the natural estimator is

$$\widehat{\theta}_n = \log\widehat{\psi}_n = \log p_{11} + \log p_{22} - \log p_{21} - \log p_{12}. \tag{9.4.5}$$

Furthermore, as n increases, the asymptotic mean-squared error for $\sqrt{n}(\widehat{\theta}_n - \theta)$ is

$$\sum_{i=1}^{2}\sum_{j=1}^{2}\pi_{ij}^{-1}, \tag{9.4.6}$$

which can, as well, be estimated by

$$n\sum_{i=1}^{2}\sum_{j=1}^{2}n_{ij}^{-1}. \tag{9.4.7}$$

Thus, recalling the (multivariate version of the) Central Limit Theorem 7.3.4 and Theorem 7.5.5, we may conclude that

$$\sqrt{n}(\widehat{\theta}_n - \theta)\Big/\left(\frac{n}{n_{11}} + \frac{n}{n_{12}} + \frac{n}{n_{21}} + \frac{n}{n_{22}}\right)^{1/2} \xrightarrow{D} \mathcal{N}(0, 1), \tag{9.4.8}$$

and this result may be used either to construct confidence intervals for θ (or ψ) or to obtain test statistics for suitable hypotheses on θ (or ψ). Here we note the similarity with the Wald procedure for a specific choice of $F(\boldsymbol{\pi})$ termed the log-linear model. Instead of the logarithmic transformation in (9.4.4), another way to look into the association in a 2×2 table is to use the Yule measure

$$\Upsilon_Y = \frac{\psi - 1}{\psi + 1} = \frac{\pi_{11}\pi_{22} - \pi_{12}\pi_{21}}{\pi_{11}\pi_{22} + \pi_{12}\pi_{21}}, \tag{9.4.9}$$

which satisfies $-1 \leq \Upsilon_Y \leq 1$ and equals 0 when there is no association. Again the Wald method can be used to draw statistical conclusions on φ_Y through $\widehat{\Upsilon}_{Y_n} = (n_{11}n_{22} - n_{12}n_{21})/(n_{11}n_{22} + n_{21}n_{12})$.

For a general $I \times J$ contingency table with probabilities $\{\pi_{ij}\}$, the Goodman–Kruskal *concentration coefficient* is defined as

$$\tau = \left(\sum_{i=1}^{I} \sum_{j=1}^{J} \pi_{ij}^2/\pi_{i.} - \sum_{j=1}^{J} \pi_{.j}^2 \right) \bigg/ \left(1 - \sum_{j=1}^{J} \pi_{.j}^2 \right), \tag{9.4.10}$$

where $\pi_{i.} = \sum_{i=1}^{I} \pi_{ij.}$, and $\pi_{.j} = \sum_{j=1}^{J} \pi_{ij}$, $1 \leq j \leq J$. A related entropy-based measure, called the *uncertainty coefficient*, is

$$\gamma = \left(\sum_{i=1}^{I} \sum_{j=1}^{J} \pi_{ij} \log(\pi_{ij}/\pi_{i.}\pi_{.j}) \right) \bigg/ \left(\sum_{j=1}^{J} \pi_{.j} \log \pi_{.j} \right). \tag{9.4.11}$$

Independence corresponds to $\tau = \gamma = 0$. Large-sample statistical inference for τ and γ can again be drawn by the methods discussed in the earlier sections. Particularly, the Wald method will be very adaptable in such a case of a nonlinear function reducible to a linear one by a first-order Taylor expansion.

Binary response variables constitute an important class of categorical data models. Consider a binary response variable Y (i.e., either $Y = 0$ or 1) and a vector $\boldsymbol{x}$ of design (or regression) variables, such that $\pi(\boldsymbol{x}) = P\{Y = 1|\boldsymbol{x}\} = 1 - P\{Y = 0|\boldsymbol{x}\}$. In a variety of situations, it may be quite appropriate to let

$$\pi(\boldsymbol{x}) = \exp(\alpha + \boldsymbol{x}'\boldsymbol{\beta})/[1 + \exp(\alpha + \boldsymbol{x}'\boldsymbol{\beta})]. \tag{9.4.12}$$

Since (9.4.12) resembles the classical *logistic function*, it is called a *logit model* or a logistic regression model. Note that

$$\log\{\pi(\boldsymbol{x})/[1 - \pi(\boldsymbol{x})]\} = \alpha + \boldsymbol{x}'\boldsymbol{\beta}, \tag{9.4.13}$$

so that we have essentially transformed the original model to a *log-linear model* (or a *generalized linear model*). The likelihood function for the n observations $Y_1, \ldots, Y_n$ is

$$L_n(\alpha, \boldsymbol{\beta}) = \prod_{i=1}^{n} \{\exp[(\alpha + \boldsymbol{x}_i'\boldsymbol{\beta})Y_i][1 + \exp(\alpha + \boldsymbol{x}_i'\boldsymbol{\beta})]^{-1}\}, \tag{9.4.14}$$

so that the ML (or BAN) estimator of α, $\boldsymbol{\beta}$ and tests for suitable hypotheses on $\boldsymbol{\beta}$ can be worked out as in Chapter 8. On the other hand, as in quantal bioassay models, suppose that there are k (≥ 2) design points $\boldsymbol{x}_1, \ldots, \boldsymbol{x}_k$ and that corresponding to the ith point, there are n_i binary responses Y_{ij}, $j = 1, \ldots, n_i$, for which the model (9.4.12) holds for $\boldsymbol{x} = \boldsymbol{x}_i$, $i = 1, \ldots, k$. We let $p_i = n_i^{-1} \sum_{j=1}^{n_i} Y_{ij}$, $i \leq i \leq k$, and

$$Z_i = \log[p_i/(1 - p_i)], \quad i = 1, \ldots, k. \tag{9.4.15}$$

Then, one may consider the measure of dispersion

$$Q_n(\alpha, \boldsymbol{\beta}) = \sum_{i=1}^{k} n_i p_i (1 - p_i)(Z_i - \alpha - \boldsymbol{x}_i'\boldsymbol{\beta})^2 \tag{9.4.16}$$

and minimizing $Q_n(\alpha, \boldsymbol{\beta})$ with respect to α, $\boldsymbol{\beta}$, one gets *minimum logit* estimates of α, $\boldsymbol{\beta}$. Recall that the $n_i p_i$ have binomial distributions, and, hence, by (9.4.15), $\sqrt{n_i}(Z_i - \alpha - \boldsymbol{x}_i'\boldsymbol{\beta})$, $i = 1, \ldots, k$, are asymptotically normally distributed with zero mean and variances $\{\pi(\boldsymbol{x}_i)[1 - \pi(\boldsymbol{x}_i)]\}^{-1}$, $i = 1, \ldots, k$; note also that they are independent. Hence, (9.4.16) is a version of the weighted least-squares setup with the unknown weights $\pi(\boldsymbol{x}_i)[1-\pi(\boldsymbol{x}_i)]$ replaced by their estimates $p_i(1-p_i)$. As such, the logit estimators $\widehat{\alpha}$, $\widehat{\boldsymbol{\beta}}$ are linear in the Z_i with coefficients depending on the $\boldsymbol{x}_i$ and p_i, so that the results of Chapter 8 can be readily used to draw conclusions on the asymptotic multinormality of $\widehat{\alpha}$, $\widehat{\boldsymbol{\beta}}$. A general discussion on the asymptotic properties of models similar to the ones considered here is given in Section 10.5.

We conclude this section with some remarks on the so called *Cochran–Mantel–Haenszel* test for comparing two groups on a binary response, adjusting for control variables. Suppose we have K set of 2×2 contingency tables with the cell probabilities $\{\pi_{ijk}, i, j = 1, 2\}$ and observed frequencies $\{n_{ijk}; i, j = 1, 2\}$, for $k = 1, \dots, K$. Suppose that we want to test the null hypothesis

$$H_0 : \pi_{ij} = \pi_{i \cdot k} \pi_{\cdot jk}, \quad i, j = 1, 2, \quad \text{for every } k = 1, \dots, K. \tag{9.4.17}$$

Under H_0, given the marginals $n_{i \cdot k}, n_{\cdot jk}$, the expected value of n_{11k} is $m_{11k} = n_{1 \cdot k} n_{\cdot 1k} / n_{\cdot \cdot k}$ and the conditional variance of n_{11k} is

$$\nu_k^2 = n_{1 \cdot k} n_{\cdot 1k} n_{2 \cdot k} n_{\cdot 2k} \big/ \left[n_{\cdot \cdot k}^2 (n_{\cdot \cdot k} - 1) \right], \tag{9.4.18}$$

for $k = 1, \dots, K$, and these n_{11k}, $k \geq 1$, are independent too. As such, it is quite intuitive to construct a test statistic

$$Q_T = \sum_{k=1}^{K} (n_{11k} - m_{11k})^2 / \nu_k^2, \tag{9.4.19}$$

which will have asymptotically (under H_0) a chi-squared distribution with K degrees of freedom. Although this test is consistent against any departure from independence, possibly in different models for different strata, in terms of power, it may not be the best (when K is particularly not very small). If, on the other hand, the pattern of dependence is concordant across the K strata, then one may use the test statistic

$$Q_{MH} = \left[\sum_{k=1}^{K} (n_{11k} - m_{11k}) \right]^2 \Big/ \sum_{k=1}^{K} \nu_k^2 \tag{9.4.20}$$

that will have asymptotically (under H_0) a chi-squared distribution with one degree of freedom. This result can be verified directly by showing that under H_0, $n^{-1/2}(n_{11k} - m_{1k})$ asymptotically has a multinormal distribution with null mean vector and dispersion matrix diag $\{\nu_1^2, \dots, \nu_k^2\}$, for $k = 1, \dots, K$, where $n = \sum_{k=1}^{K} n_{\cdot \cdot k}$. Because of the reduction in the degrees of freedom (from K to 1) when the noncentrality of the $n_{11k} - m_{11k}$ are in the same direction, Q_{MH} is likely to have more power than Q_T. On the other hand, if the individual noncentralities are in possibly different directions, the sum $\sum_{k=1}^{n} (n_{11k} - m_{11k})$ may have a very small noncentrality resulting in very little power of Q_{MH} (compared to Q_T). In metaanalyses, the concordance picture is likely to hold, and, hence, Q_{MH} should be preferred to Q_T.

9.5 Concluding Notes

Although product binomial models may be considered as special cases of product multinomial models, they require extra attention because of the specialized nature of some of the related analytical tools. In fact, they are more in tune with those employed in the study of the broader class of (univariate) generalized linear models; these, in turn, are usually analyzed with methods similar to (generalized) least squares, and, thus, we find it convenient to relegate a more detailed asymptotic theory to Chapter 10. In this regard, our brief discussion of some asymptotic results in Section 9.4 deserves some explanation. The major issue may be highlighted by observing that (9.4.16) is a slightly simplified version of a generalized linear model formulation in the sense that the variance function does not depend on the parameters of interest, leading to less complex estimating equations. Even though an asymptotic equivalence between the two approaches may be established, there is a subtle point which merits further discussion. In (9.4.16), usually k (≥ 1) is a fixed positive integer, whereas the sample sizes (n_i) are taken large to justify the asymptotics; in that sense, the p_i are close to the π_i, that is, $|p_i - \pi_i| = O_p(n^{-1/2})$, whereas $(Z_i - \alpha - \boldsymbol{x}_i'\boldsymbol{\beta})^2 = O_p(n^{-1})$. Hence, as in the modified minimum chi-squared method, ignoring the dependence of the $p_i(1-p_i)$ on $(\alpha, \boldsymbol{\beta})$ does not cost much in terms of large sample properties relative to the generalized linear models approach, where such dependence is a part of the scheme. From the computational point of view, however, the picture favors the modified minimum chi-squared method. A similar discussion remains valid for a comparison between the generalized least-squares and the generalized estimating equation approaches, and we will pursue this issue in Section 10.6.

Generally, categorical data models permit easier verification of the regularity conditions of Theorem 8.3.1, which are set up for a broader class of situations. However, in the product multinomial case, if any subset of the cell probabilities approaches zero, the dimension of the parameter space is altered and thereby the whole asymptotic theory may be different. In such a case, the MLE may not be BAN and the likelihood ratio test may not follow an asymptotic chi-squared distribution. We will present some special cases as exercises. In particular, we also include a good example of product Poisson models in the context of contingency tables as related to Example 9.3.1.

9.6 Exercises

9.6.1 (Example 1.2.11) Consider the following table relating to the OAB blood group model:

Blood Group	Observed Frequency	Probability
O	n_O	$p_O = q_O^2$
A	n_A	$p_A = q_A^2 + 2q_Oq_A$
B	n_B	$p_B = q_B^2 + 2q_Oq_B$
AB	n_{AB}	$p_{AB} = 2q_Aq_B$
Total	n	1

The MLE of the gene frequencies $\boldsymbol{q} = (q_O, q_A, q_B)'$ may not be expressible in a closed algebraic form.

(i) Show that $q_O = p_O^{1/2}$, $q_A = 1 - (p_O + p_B)^{1/2}$ and $q_B = 1 - (p_O + p_A)^{1/2}$, and use this to propose some initial estimators:

$$\widehat{q}_O^{(0)} = (n_O/n)^{1/2},$$
$$\widehat{q}_A^{(0)} = 1 - [(n_O + n_B)/n]^{1/2},$$
$$\widehat{q}_B^{(0)} = 1 - [(n_O + n_A)/n]^{1/2}.$$

Hence, or otherwise, use the *method of scoring* to solve for the MLE (or one-step MLE) by an iterative method.

(ii) Show that $\widehat{q}_O^{(0)} + \widehat{q}_A^{(0)} + \widehat{q}_B^{(0)}$ may not be exactly equal to 1. Suggest suitable modifications to satisfy this restriction, and use the method of scoring on this modified version.

(iii) Use Wald statistic, likelihood ratio statistic, and the Pearsonian goodness-of-fit test statistic to test the null hypotheses that the Hardy–Wienberg equilibrium holds. Comment on the relative computational ease of these three methods.

9.6.2 A scientist wanted to study whether some gene frequencies ($\boldsymbol{q}$) varies from one country to another. For this testing for the homogeneity of the $\boldsymbol{q}$, he obtained the following data set:

	Sample				
Blood Group	1	2	...	k	Total
O	n_{1O}	n_{2O}	...	n_{kO}	n_O
A	n_{1A}	n_{2A}	...	n_{kA}	n_A
B	n_{1B}	n_{2B}	...	n_{kB}	n_B
AB	n_{1AB}	n_{2AB}	...	n_{kAB}	n_{AB}
Total	n_1	n_2	...	n_k	n

(i) Treating this as a $4 \times k$ contingency table, construct a completely nonparametric test for homogeneity of the k samples.

(ii) Use the Hardy–Weinberg equilibrium for each country, and in that parametric setup, use the Wald test statistic to test for the homogeneity of the gene frequencies.

(iii) Use the likelihood ratio test under the setup in (ii).

(iv) Comment on the associated degrees of freedom for all the three tests, and also compare them in terms of (a) consistency, (b) power properties and (c) computational ease.

9.6.3 The same scientist has the feeling that the gene frequencies may vary from one race to another (e.g., Mongolian, Negroid, Caucasian) and, hence, from one geographical area to another depending on their relative components. He collected a data set from one of the Caribbean islands where Negroid and Caucasian mixtures are profound (with very little Mongolian presence). As such, he decided to work with the following mixture model:

$$p_X = \pi p_X^{(N)} + (1 - \pi) p_X^{(C)}, \quad \text{for } X = O, A, B, AB,$$

where $0 \leq \pi \leq 1$, and for the $p^{(N)}$ (and $p^{(C)}$), the Hardy–Weinberg equilibrium holds with $\boldsymbol{q} = \boldsymbol{q}^{(N)}$ (and $\boldsymbol{q}^{(C)}$). In this setup, he encountered the following problems:

(a) Under $H_0 : \boldsymbol{q}^{(N)} = \boldsymbol{q}^{(C)}$, there are only two unknown (linearly independent) parameters (π drops out); whereas under the alternative, $\boldsymbol{q}^{(N)} \neq \boldsymbol{q}^{(C)}$, there are five linearly independent ones (including π). Thus, he confidently prescribed a chi-squared test with $5 - 2$ ($= 3$) degrees of freedom. Do you support this prescription? If not, why?

(b) He observed that there are only three (independent) cell probabilities, but five unknown parameters. So he concluded that his hypothesis was not testable in a single sample model. This smart scientist, therefore, decided to choose two different islands (for which the π values are quite different). Using the 4×2 contingency table he obtained, he wanted to estimate $\pi_1, \pi_2, q_O^{(j)}, q_A^{(j)}, j = N, C$. He had six linearly independent cell probabilities and six unknown parameters, so that he was satisfied with the model. Under $H_0 : \boldsymbol{q}^{(N)} = \boldsymbol{q}^{(C)}$, he had two parameters, whereas he had six under the alternative. Hence, he concluded that the degrees of freedom of his goodness-of-fit test would be equal to four. Being so confident this time, he carried out a volume of simulation work to examine the adequacy of the chi-squared approximation. Alas, the fit was very poor! A smarter colleague suggested that the degrees of freedom for the chi-squared approximation should be equal to 2.3 and this showed some improvement. However, he was puzzled why the degrees of freedom was not an integer! Anyway, as there was no theoretical foundation, in frustration, he gave up! Can you eliminate the impasse?

(c) Verify that for this mixture model there is a basic *identifiability issue*: if $\boldsymbol{q}^{(N)} = \boldsymbol{q}^{(C)}$ regardless of whether the π_j $(j \leq k)$ are on the boundary (i.e., $\{0\}, \{1\}$) or not, the number of unknown parameters is equal to 2, whereas this number jumps to $4 + k$ when H_0 does not hold. Thus, the parameter point belongs to a boundary of the parameter space under H_0. Examine the impact of this irregularity on the asymptotics underlying the usual goodness-of-fit tests.

(d) Consider the $4 \times k$ contingency table (for $k \geq 2$) and discuss how the nonparametric test overcomes this problem?

(e) Can you justify the Hardy–Weinberg equilibrium for the mixture model from the random mating point of view?

9.6.4 Consider the multinomial model in (9.1.1). Let $n_1, \ldots, n_r$ be r independent Poisson variables with parameters $m_1, \ldots, m_r$, respectively. Show that $n = n_1 + \cdots + n_r$ has the Poisson distribution with the parameter $m = m_1 + \cdots + m_r$. Hence, or otherwise, show that the conditional probability law for $n_1, \ldots, n_r$, given n, is multinomial with the cell probabilities $\pi_j = m_j/m$, $1 \leq j \leq r$.

9.6.5 Using the Poisson to Normal convergence, show that $(n_j - m_j)^2/m_j$ are independent and asymptotically distributed as χ_1^2, $1 \leq j \leq r$. Hence, using the Cochran Theorem (7.6.1), verify that given n, $\sum_{j=1}^{r}(n_j - m_j)^2/m_j$ has asymptotically chi-squared law with $r - 1$ degrees of freedom.

9.6.6 For the product multinomial model (9.1.2), show that treating the n_{ij} as independent Poisson variables with parameters m_{ij}, $1 \leq j \leq r_j$, $1 \leq i \leq s$, and conditioning on the n_i $(= \sum_{j=1}^{r_i} n_{ij})$, $1 \leq i \leq s$, one has the same multinomial law, where $\pi_{ij} = m_{ij}/m_i$, $1 \leq j \leq r_i$, $m_i = \sum_{j=1}^{r_i} m_{ij}$, $1 \leq i \leq s$.

9.6.7 Let n_{ij} $(1 \leq i \leq r, 1 \leq j \leq c)$ be the observed cell frequencies of a $r \times c$ contingency table, let $n_{i\cdot} = \sum_{j=1}^{c} n_{ij}$ and $n_{\cdot j} = \sum_{i=1}^{r} n_{ij}$ be the marginal totals and $n = \sum_{i=1}^{r}\sum_{j=1}^{c} n_{ij}$. Treat the n_{ij} as independent Poisson variables with parameters m_{ij}, and hence, or otherwise, show that conditional on the $n_{i\cdot}$ and $n_{\cdot j}$ being given, the n_{ij} have the same multinomial law as arising in the classical $r \times c$ table with marginals fixed. Use the Poisson to Normal convergence along with the Cochran Theorem (7.6.1) to derive the appropriate asymptotic chi-squared distribution [with $(r - 1) \times (c - 1)$ degrees of freedom] for the classical contingency table test.

9.6.8 Consider the log-linear model

$$\log m_{ij} = \mu + \alpha_i + \beta_j + \gamma_{ij}, \quad 1 \leq i \leq r, \quad 1 \leq j \leq c,$$

where $\sum_{i=1}^{r} \alpha_i = 0$, $\sum_{j=1}^{c} \beta_j = 0$ and $\sum_{i=1}^{r} \gamma_{ij} = \sum_{j=1}^{c} \gamma_{ij} = 0$, for all i, j. Show that the classical independence model relates to $\gamma_{ij} = 0$, for all i, j. Critically examine the computational aspects of the estimation of αs, βs, and γ's by the ML, MCS, and MMCS methods, and show that the test considered in Exercise 9.6.7 is appropriate in this setup too.

9.6.9 The lifetime X of an electric lamp has the simple exponential survival function $\overline{F}(x) = \exp(-x/\theta)$, $\theta > 0$, $x \geq 0$. A factory has n lamps which are lighted day and night, and every week the burnt out ones are replaced by new ones. At a particular inspection time, it was observed that out of n, n_1 were dead, and $n - n_1 = n_2$ were still alive. Find out a BAN estimator of θ based on this data set. Obtain the Fisher information on θ from this binary model as well as from the ideal situation when $X_1, \ldots, X_n$ would have been all observed. Compare the two and comment on the loss of information due to this grouping.

9.6.10 Consider a 2×2 contingency table with the cell frequencies n_{ij} and probabilities π_{ij}, for $i, j = 1, 2$. Under the independence model, $\pi_{ij} = \pi_{i.}\pi_{.j}$, and its estimator is $n_{i.}n_{.j}/n_{..}^2 = \widehat{\pi}_{ij}^*$, say

(i) Obtain the expressions for $\mathbb{E}(\widehat{\pi}_{ij}^*)$ and $\mathbb{V}\text{ar}(\widehat{\pi}_{ij}^*)$ when the independence model may not hold.

(ii) Hence, or otherwise, comment on the rationality of the test for independence based on the odds ratio in (9.4.1).

9.6.11 Consider the genetical data

Type	AA	Aa	aA	aa	Total
Probability	$(1+\theta)/4$	$(1-\theta)/4$	$(1-\theta)/4$	$(1+\theta)/4$	1
Obs. frequency	n_1	n_2	n_3	n_4	n

Find the MLE of the linkage factor θ. Compute the mean and variance of the estimator, and find out the appropriate variance stabilizing transformation.

9.6.12 Use the results in the previous exercise to provide an asymptotic $100\gamma\%$ confidence interval for θ.

9.6.13 Suppose that there are k (≥ 2) independent data sets each pertaining to the model in Exercise 9.6.11 (with respective θ-values being $\theta_1, \ldots, \theta_k$). Use a variance stabilizing transformation to test for the homogeneity of $\theta_1, \ldots, \theta_k$. Comment on the desirability of the transformation in this setup.

10

Regression Models

10.1 Introduction

The general denomination of *regression models* is used to identify statistical models for the relationship between one or more *explanatory* (independent) *variables* and one or more *response* (dependent) *variables*. Typical examples include the investigation of the influence of:

(i) the amount of fertilizer on the yield of a certain crop;
(ii) the type of treatment and age on the serum cholesterol levels of patients;
(iii) Driving habits and fuel type on the gas mileage of a certain make of automobile; and
(iv) the type of polymer, extrusion rate, and extrusion temperature on the tensile strength and number of defects/unit length of synthetic fibers.

Within the class of regression models, the *linear models* play an important role for statistical applications; they are easy to interpret, mathematically tractable and may be successfully employed for a variety of practical situations as in (i)–(iv) above. They include models usually considered in *Linear Regression Analysis*, *Analysis of Variance* (ANOVA), and *Analysis of Covariance* (ANCOVA) and may easily be extended to include *Logistic Regression Analysis*, *Generalized Linear Models*, *Multivariate Regression*, *Multivariate Analysis of Variance* (MANOVA) or *Multivariate Analysis of Covariance* (MANCOVA).

In matrix notation, the univariate linear model may be expressed as

$$\boldsymbol{Y}_n = \boldsymbol{X}_n\boldsymbol{\beta} + \boldsymbol{\varepsilon}_n, \tag{10.1.1}$$

where $\boldsymbol{Y}_n = (Y_1, \ldots, Y_n)'$ is an $(n \times 1)$ vector of observable response variables, $\boldsymbol{X}_n = ((x_{ij}^{(n)}))$ is an $(n \times q)$ matrix of known constants $x_{ij}^{(n)}$ (which for ease of notation we denote x_{ij}, omitting the superscript) representing the value of the jth explanatory variable for the ith sample unit, $\boldsymbol{\beta} = (\beta_1, \ldots, \beta_q)'$ denotes the $(q \times 1)$ vector of unknown parameters, and $\boldsymbol{\varepsilon}_n = (\varepsilon_1, \ldots, \varepsilon_n)'$ is an $(n \times 1)$ vector of unobservable errors assumed to follow a distribution with not necessarily fully specified distribution function F such that $\mathbb{E}(\boldsymbol{\varepsilon}_n) = \boldsymbol{0}$ and $\mathbb{V}\text{ar}(\boldsymbol{\varepsilon}_n) = \sigma^2\boldsymbol{I}_n, \sigma^2 < \infty$, that is, the elements of $\boldsymbol{\varepsilon}_n$ are uncorrelated, have zero means and constant (but unknown) variance. This is known in the literature as the *Gauss–Markov setup*. It is usual to take $x_{i1} = 1$, $i = 1, \ldots, n$, and in the case of a single explanatory variable, (10.1.1) reduces to the *simple linear regression model*:

$$Y_i = \alpha + \beta x_i + \varepsilon_i, \quad i = 1, \ldots, n, \tag{10.1.2}$$

where α and β are the (unknown) intercept and slope, respectively, and x_i denotes the value of the explanatory variable for the ith sample unit, $i = 1, \ldots, n$.

Statistical analysis under (10.1.1) or (10.1.2) usually involves estimation of the parameter vector $\boldsymbol{\beta}$ [α, β in (10.1.2)] and tests of hypotheses about its components. The well-known *least-squares method* (LSM) for estimating $\boldsymbol{\beta}$ consists of minimizing the residual sum of squares

$$\boldsymbol{\varepsilon}_n'\boldsymbol{\varepsilon}_n = (\boldsymbol{Y}_n - \boldsymbol{X}_n\boldsymbol{\beta})'(\boldsymbol{Y}_n - \boldsymbol{X}_n\boldsymbol{\beta}), \tag{10.1.3}$$

which, for (10.1.2) reduces to minimizing

$$\sum_{i=1}^n \varepsilon_i^2 = \sum_{i=1}^n (Y_i - \alpha - \beta x_i)^2. \tag{10.1.4}$$

Since (10.1.3) is a quadratic function of $\boldsymbol{\beta}$ it is easy to show that its minimum is attained at the point $\widehat{\boldsymbol{\beta}}_n$, identifiable as a solution to the *estimating equations*

$$\boldsymbol{X}_n'\boldsymbol{X}_n\widehat{\boldsymbol{\beta}}_n = \boldsymbol{X}_n'\boldsymbol{Y}_n. \tag{10.1.5}$$

The solution $\widehat{\boldsymbol{\beta}}_n$ is called the *least-squares estimator* (LSE) of $\boldsymbol{\beta}$. In the special case of (10.1.2), we may show that the LSE of (α, β) in (10.1.2) is a solution to

$$\begin{aligned}\sum_{i=1}^n Y_i &= n\widehat{\alpha}_n + \widehat{\beta}_n \sum_{i=1}^n x_i,\\ \sum_{i=1}^n x_i Y_i &= \widehat{\alpha}_n \sum_{i=1}^n x_i + \widehat{\beta}_n \sum_{i=1}^n x_i^2.\end{aligned} \tag{10.1.6}$$

If $\text{rank}(\boldsymbol{X}_n) = q \leq n$, (10.1.5) has a single solution, given by

$$\widehat{\boldsymbol{\beta}}_n = (\boldsymbol{X}_n'\boldsymbol{X}_n)^{-1}\boldsymbol{X}_n'\boldsymbol{Y}_n; \tag{10.1.7}$$

otherwise, there are infinite solutions, which may be obtained by replacing $(\boldsymbol{X}_n'\boldsymbol{X}_n)^{-1}$ with a generalized inverse $(\boldsymbol{X}_n'\boldsymbol{X}_n)^-$ in (10.1.7). In fact, if the x_i are not all equal in (10.1.2), the unique solution to (10.1.6) is

$$\begin{aligned}\widehat{\beta}_n &= \sum_{i=1}^n Y_i(x_i - \overline{x}_n) \Big/ \sum_{i=1}^n (x_i - \overline{x}_n)^2,\\ \widehat{\alpha}_n &= \overline{Y}_n - \widehat{\beta}_n \overline{x}_n.\end{aligned} \tag{10.1.8}$$

Since, for $\boldsymbol{X}_n$ not of full rank, $\boldsymbol{X}_n'\boldsymbol{\beta}$ can be written as $\boldsymbol{X}_n^{*\prime}\boldsymbol{\beta}^*$, where $\boldsymbol{X}_n^*$ is of full rank and $\dim(\boldsymbol{\beta}^*) \leq \dim(\boldsymbol{\beta})$, for all practical purposes the *model specification matrix* $\boldsymbol{X}_n$ can be chosen of full rank q. We will restrict our attention to this case, unless otherwise stated.

Given the assumptions underlying (10.1.1), it follows immediately from (10.1.7) that $\mathbb{E}(\widehat{\boldsymbol{\beta}}_n) = \boldsymbol{\beta}$ and $\mathbb{V}\text{ar}(\widehat{\boldsymbol{\beta}}_n) = \sigma^2(\boldsymbol{X}_n'\boldsymbol{X}_n)^{-1}$. In particular, for (10.1.2), we obtain

$$\mathbb{V}\text{ar}\begin{pmatrix}\widehat{\alpha}_n\\ \widehat{\beta}_n\end{pmatrix} = \frac{\sigma^2}{\sum_{i=1}^n (x_i - \overline{x}_n)^2}\begin{bmatrix}\sum_{i=1}^n x_i^2/n & -\overline{x}_n\\ -\overline{x}_n & 1\end{bmatrix}.$$

Moreover, under the same assumptions, the well-known *Gauss–Markov Theorem* [see Searle (1971)] states that for every fixed q-vector $\boldsymbol{c} \in \mathbb{R}^q$, $\boldsymbol{c}'\widehat{\boldsymbol{\beta}}_n$ is the *best linear unbiased estimator* (BLUE) of $\boldsymbol{c}'\boldsymbol{\beta}$, in the sense that it has the smallest variance in the class of linear unbiased estimators of $\boldsymbol{c}'\boldsymbol{\beta}$. Although this constitutes an important optimality property of the LSE, it is of little practical application unless we have some idea of the corresponding distribution. If we consider the additional assumption that the random errors in (10.1.1) [or (10.1.2)] are normally distributed, it follows that the LSE coincides with the MLE and that

$$(\boldsymbol{X}_n'\boldsymbol{X}_n)^{1/2}\widehat{\boldsymbol{\beta}}_n \sim \mathcal{N}_q(\boldsymbol{\beta}, \sigma^2\boldsymbol{I}_q).$$

This assumption, however, is too restrictive and rather difficult to verify in practice, indicating that some approximate results are desirable. This chapter is devoted to the study of the large sample properties of the LSE and some other alternatives for the estimation of $\boldsymbol{\beta}$ in (10.1.1).

In Section 10.2 the asymptotic properties of the LSE are laid down. In Section 10.3 we motivate the use of alternative (robust) estimation methods for the linear model; essentially we expand the brief discussion presented in Section 10.2 to this more interesting case and outline the proofs of some asymptotic results. Section 10.5 deals with extensions of the results to Generalized Linear Models (GLS). The relation between Generalized Estimating Equations and Generalized Least Squares is outlined in Section 10.6 and, finally, in Section 10.7, asymptotic results for Nonparametric Regression Models are briefly described.

10.2 Generalized Least-Squares Procedures

We begin this section with a result on the asymptotic distribution of the LSE for the simple linear regression model. A detailed proof is presented since similar steps may be successfully employed in other situations.

***Theorem* 10.2.1.** *Consider the simple regression model 10.1.2 under the Gauss–Markov setup discussed in the previous section, with uncorrelated and identically distributed error terms ε_i, $i = 1, \ldots, n$, and let $\widehat{\alpha}_n$ and $\widehat{\beta}_n$ be the LSE of α and β, respectively. Write*

$$t_n = \sum_{i=1}^n (x_i - \overline{x}_n)^2,$$
$$c_{ni} = (x_i - \overline{x}_n)/\sqrt{t_n}$$

and assume further that:

$$\max_{1\le i\le n} c_{ni}^2 \to 0 \quad \text{as } n \to \infty \text{ (the Noether condition)}; \tag{10.2.1}$$

$$\lim_{n\to\infty} \overline{x}_n = \overline{x} < \infty, \qquad \text{where } \overline{x} \text{ is some constant}; \tag{10.2.2}$$

$$\lim_{n\to\infty} n^{-1}t_n = \lim_{n\to\infty} n^{-1}\sum_{i=1}^n (x_i - \overline{x}_n)^2 = t < \infty, \qquad \text{for some constant } t. \tag{10.2.3}$$

Then,

$$\sqrt{n}\begin{pmatrix}\widehat{\alpha}_n - \alpha \\ \widehat{\beta}_n - \beta\end{pmatrix} \xrightarrow{D} \mathcal{N}_2\left(\boldsymbol{0}, \begin{bmatrix} 1+\overline{x}^2/t & -\overline{x}/t \\ -\overline{x}/t & 1/t\end{bmatrix}\sigma^2\right). \tag{10.2.4}$$

Proof. From (10.1.8) and (10.1.2) we may write

$$\widehat{\beta}_n = t_n^{-1} \sum_{i=1}^{n} (x_i - \overline{x}_n)[\alpha + \beta(x_i - \overline{x}_n) + \varepsilon_i]$$

$$= \beta + t_n^{-1} \sum_{i=1}^{n} (x_i - \overline{x}_n)\varepsilon_i, \tag{10.2.5}$$

which implies that

$$\sqrt{t_n}(\widehat{\beta}_n - \beta) = \sum_{i=1}^{n} \frac{(x_i - \overline{x}_n)}{\sqrt{t_n}} \varepsilon_i = \sum_{i=1}^{n} c_{ni}\varepsilon_i. \tag{10.2.6}$$

Observe that $\sum_{i=1}^{n} c_{ni} = 0$ and $\sum_{i=1}^{n} c_{ni}^2 = 1$; then, in view of (10.2.1), it follows from the Hájek–Šidak Central Limit Theorem 7.3.6 that:

$$\sqrt{t_n}(\widehat{\beta}_n - \beta) \xrightarrow{D} \mathcal{N}(0, \sigma^2).$$

Using (10.2.3) and the Slutsky Theorem 7.5.2 we obtain

$$\sqrt{n}(\widehat{\beta}_n - \beta) \xrightarrow{D} \mathcal{N}(0, \sigma^2/t).$$

Again, using (10.1.8) and (10.1.2), we may write

$$\begin{aligned}\widehat{\alpha}_n &= \overline{Y}_n - \widehat{\beta}_n\overline{x}_n \\ &= \alpha + \beta\overline{x}_n + \overline{\varepsilon}_n - \widehat{\beta}_n\overline{x}_n \\ &= \alpha - (\widehat{\beta}_n - \beta)\overline{x}_n + \overline{\varepsilon}_n,\end{aligned} \tag{10.2.7}$$

which implies that

$$\begin{aligned}\sqrt{n}(\widehat{\alpha}_n - \alpha) &= \sqrt{n}\overline{\varepsilon}_n - \sqrt{n}(\widehat{\beta}_n - \beta)\overline{x}_n \\ &= \sqrt{n}\overline{\varepsilon}_n - \sqrt{n/t_n}\overline{x}_n\sqrt{t_n}(\widehat{\beta}_n - \beta)\end{aligned}$$

and, which in view of (10.2.6), may be re-expressed as

$$\sqrt{n}(\widehat{\alpha}_n - \alpha) = \frac{1}{\sqrt{n}} \sum_{i=1}^{n} \varepsilon_i - \sqrt{n/t_n}\overline{x}_n \sum_{i=1}^{n} c_{ni}\varepsilon_i = \sum_{i=1}^{n} d_{ni}\varepsilon_i, \tag{10.2.8}$$

where

$$d_{ni} = n^{-1/2} - \sqrt{n/t_n}\overline{x}_n c_{ni}.$$

Observing that $\sum_{i=1}^{n} d_{ni} = \sqrt{n}$ and $\sum_{i=1}^{n} d_{ni}^2 = 1 + n\overline{x}_n^2/t_n$ and applying the Hájek–Šidak Central Limit Theorem 7.3.6, we obtain

$$\sqrt{n}\left(1 + \frac{n\overline{x}_n^2}{t_n}\right)^{-1/2} (\widehat{\alpha}_n - \alpha) \xrightarrow{D} \mathcal{N}(0, \sigma^2).$$

Using (10.2.2), (10.2.3), and the Slutsky Theorem 7.5.2 it follows that

$$\sqrt{n}(\widehat{\alpha}_n - \alpha) \xrightarrow{D} N\left[0, \sigma^2\left(1 + \frac{\overline{x}^2}{t}\right)\right].$$

Now let $\boldsymbol{\lambda} = (\lambda_1, \lambda_2)' \in \mathbb{R}^2$, $\boldsymbol{\lambda} \neq \mathbf{0}$, be an arbitrary but fixed vector and use (10.2.6) and (10.2.8) to see that

$$\lambda_1\sqrt{n}(\widehat{\alpha}_n - \alpha) + \lambda_2\sqrt{n}(\widehat{\beta}_n - \beta) = \sum_{i=1}^{n} f_{ni}\varepsilon_i,$$

where

$$f_{ni} = \lambda_1 d_{ni} + \lambda_2\sqrt{n/t_n}c_{ni}.$$

Then, we have $\sum_{i=1}^{n} f_{ni} = \lambda_1\sqrt{n}$, and

$$\sum_{i=1}^{n} f_{ni}^2 = \lambda_1^2\left(1 + \overline{x}_n^2 n/t_n\right) + \lambda_2^2 n/t_n - 2\lambda_1\lambda_2\overline{x}_n n/t_n.$$

Letting

$$\boldsymbol{V}_n = \begin{bmatrix} 1 + n\overline{x}_n^2/t_n & -n\overline{x}_n/t_n \\ -n\overline{x}_n/t_n & n/t_n \end{bmatrix}$$

and applying the Hájek–Šidak Central Limit Theorem 7.3.6 it follows that

$$\left[(\lambda_1, \lambda_2)\boldsymbol{V}_n\begin{pmatrix}\lambda_1\\ \lambda_2\end{pmatrix}\right]^{-1/2}[\lambda_1\sqrt{n}(\widehat{\alpha}_n - \alpha) + \lambda_2\sqrt{n}(\widehat{\beta}_n - \beta)] \xrightarrow{D} \mathcal{N}(0, \sigma^2)$$

and using the Slutsky Theorem 7.5.2 we have, in view of (10.2.2) and (10.2.3),

$$\lambda_1\sqrt{n}(\widehat{\alpha}_n - \alpha) + \lambda_2\sqrt{n}(\widehat{\beta}_n - \beta) \xrightarrow{D} \mathcal{N}(0, \sigma^2\boldsymbol{\lambda}'\boldsymbol{V}\boldsymbol{\lambda}),$$

where

$$\boldsymbol{V} = \begin{bmatrix} 1 + \overline{x}^2/t & -\overline{x}/t \\ -\overline{x}/t & 1/t \end{bmatrix}.$$

Since $\boldsymbol{\lambda}$ is arbitrary, the result follows directly from the Cramér–Wold Theorem 7.2.4. ■

It is possible to relax (10.2.3) in Theorem 10.2.1 to a certain extent. For example, we have $t_n^{1/2}(\widehat{\beta}_n - \beta) \xrightarrow{D} \mathcal{N}(0, 1)$ whenever (10.2.1) holds in conjunction with

$$\liminf_{n\to\infty} n^{-1}t_n > 0. \tag{10.2.9}$$

On the other hand, if $n^{-1}t_n \to \infty$ as $n \to \infty$, $t_n^{1/2}(\widehat{\alpha}_n - \alpha)$ does not have a limiting law, and, hence, the bivariate result in (10.2.4) may not hold.

As a direct consequence of the above theorem and a multivariate generalization of Theorem 7.2.5 it follows that $\sqrt{n}(\widehat{\alpha}_n - \alpha, \widehat{\beta}_n - \beta)' = O_P(\mathbf{1})$, which implies that $\widehat{\alpha}_n$ and $\widehat{\beta}_n$ are weakly consistent. In fact, they are strongly consistent too. Also, for this consistency property, asymptotic normality is not needed, and, hence, some of the assumed regularity conditions for Theorem 10.2.1 may also be relaxed. To see this note that by (10.2.6), it follows that

$$t_n(\widehat{\beta}_n - \beta) = \sum_{i=1}^{n}(x_i - \overline{x}_n)\varepsilon_i$$
$$= \sum_{i=1}^{n}(x_i - \overline{x})\varepsilon_i + n(\overline{x} - \overline{x}_n)\overline{\varepsilon}_n. \qquad (10.2.10)$$

Let $A_n = \sum_{i=1}^{n}(x_i - \overline{x})\varepsilon_i$, $B_n = n(\overline{x} - \overline{x}_n)\overline{\varepsilon}_n$ and observe that $\overline{\varepsilon}_n = n^{-1}\sum_{i=1}^{n}\varepsilon_i \xrightarrow{a.s.} 0$ by the Khintchine SLLN (Theorem 6.3.13). Then, using (10.2.2) and (10.2.3) as in Theorem 10.2.1 we have

$$\frac{B_n}{t_n} = \frac{n}{t_n}(\overline{x} - \overline{x}_n)\overline{\varepsilon}_n = O(1)o(1)o(1) = o(1) \text{ a.e.} \qquad (10.2.11)$$

Also,

$$A_{n+1} = A_n + \varepsilon_{n+1}(x_{n+1} - \overline{x}), \quad n \geq 1$$

so that, for all $n \geq 1$,

$$\mathbb{E}(A_{n+1}|\varepsilon_1, \ldots, \varepsilon_n) = A_n,$$

implying that $\{A_n\}$ is a martingale. Then, in view of (10.2.2) and (10.2.3) of Theorem 10.2.1 together with $\mathbb{E}(\varepsilon_i^2) = \sigma^2 < \infty$, it follows from the Kolmogorov SLLN for martingales (Theorem 6.4.2) that

$$A_n/t_n \xrightarrow{a.s.} 0. \qquad (10.2.12)$$

Substituting (10.2.11) and (10.2.12) in (10.2.10) we may conclude that $\widehat{\beta}_n - \beta \xrightarrow{a.s.} 0$ and via (10.2.7) also that $\widehat{\alpha}_n - \alpha \xrightarrow{a.s.} 0$. In this context, we may even relax the second moment condition on the ε_i to a lower-order moment condition, but the proof will be somewhat more complicated.

In order to make full use of Theorem 10.2.1 to draw statistical conclusions on (α, β), we need to provide a consistent estimator of the unknown variance σ^2. In this direction, an unbiased estimator of σ^2 is

$$s_n^2 = (n-2)^{-1}\sum_{i=1}^{n}(Y_i - \widehat{\alpha}_n - \widehat{\beta}_n x_i)^2. \qquad (10.2.13)$$

Using (10.1.2), (10.2.5) and (10.2.7), we obtain, from (10.2.13), that

$$s_n^2 = (n-2)^{-1}\sum_{i=1}^{n}[\varepsilon_i - \overline{\varepsilon}_n - (\widehat{\beta}_n - \beta)(x_i - \overline{x}_n)]^2$$
$$= (n-2)^{-1}\sum_{i=1}^{n}(\varepsilon_i - \overline{\varepsilon}_n)^2 - (\widehat{\beta}_n - \beta)^2(n-2)^{-1}\sum_{i=1}^{n}(x_i - \overline{x}_n)^2$$
$$= \frac{(n-1)}{(n-2)}\frac{1}{(n-1)}\sum_{i=1}^{n}(\varepsilon_i - \overline{\varepsilon}_n)^2 - (\widehat{\beta}_n - \beta)^2(n-2)^{-1}t_n. \qquad (10.2.14)$$

Now, using (6.3.84) it follows that

$$(n-1)^{-1}\sum_{i=1}^{n}(\varepsilon_i-\overline{\varepsilon}_n)^2 \xrightarrow{a.s.} \sigma^2.$$

Recalling (10.2.3), we may write $n^{-1}t_n = O(1)$ and, finally, using $\widehat{\beta}_n - \beta \xrightarrow{a.s.} 0$, we may conclude from (10.2.14) that $s_n^2 \xrightarrow{a.s.} \sigma^2$. This, in turn, implies that $s_n^2 \xrightarrow{P} \sigma^2$.

Let us consider now an extension of Theorem 10.2.1 to the multiple regression model. In this context, note that (10.1.1) and (10.1.7) imply

$$\widehat{\boldsymbol{\beta}}_n - \boldsymbol{\beta} = (\boldsymbol{X}_n'\boldsymbol{X}_n)^{-1}\boldsymbol{X}_n'\boldsymbol{\varepsilon}_n \tag{10.2.15}$$

so that $\mathbb{E}(\widehat{\boldsymbol{\beta}}_n) = \boldsymbol{\beta}$ and $\mathbb{V}\mathrm{ar}(\widehat{\boldsymbol{\beta}}_n) = \sigma^2(\boldsymbol{X}_n'\boldsymbol{X}_n)^{-1}$. If we consider an arbitrary linear combination $Z_n = \widehat{\boldsymbol{\lambda}}'(\widehat{\boldsymbol{\beta}}_n - \boldsymbol{\beta})$, $\boldsymbol{\lambda} \in \mathbb{R}^q$, we get $Z_n = \boldsymbol{c}_n'\boldsymbol{\varepsilon}_n$ with $\boldsymbol{c}_n = \boldsymbol{X}_n(\boldsymbol{X}_n'\boldsymbol{X}_n)^{-1}\boldsymbol{\lambda}$. Then, to obtain the asymptotic distribution of Z_n (conveniently standardized) all we need is to verify that $\boldsymbol{c}_n$ satisfies the regularity condition of the Hájek–Šidak Central Limit Theorem 7.3.6. Denoting the kth row of $\boldsymbol{X}_n$ by $\boldsymbol{x}_{nk}'$ and taking into account that the result must hold for every (fixed) but arbitrary $\boldsymbol{\lambda} \in \mathbb{R}^q$ such regularity condition may be reformulated by requiring that as $n \to \infty$,

$$\sup_{\boldsymbol{\lambda}\in\mathbb{R}^q}\left\{\max_{1\le i\le n}[\boldsymbol{\lambda}'(\boldsymbol{X}_n'\boldsymbol{X}_n)^{-1}\boldsymbol{x}_{ni}\boldsymbol{x}_{ni}'(\boldsymbol{X}_n'\boldsymbol{X}_n)^{-1}\boldsymbol{\lambda}/\boldsymbol{\lambda}'(\boldsymbol{X}_n'\boldsymbol{X}_n)^{-1}\boldsymbol{\lambda}]\right\} \to 0. \tag{10.2.16}$$

Now in view of Courant Theorem 1.5.1, we have

$$\begin{aligned}
&\sup_{\boldsymbol{\lambda}\in\mathbb{R}^q}[\boldsymbol{\lambda}'(\boldsymbol{X}_n'\boldsymbol{X}_n)^{-1}\boldsymbol{x}_{ni}\boldsymbol{x}_{ni}'(\boldsymbol{X}_n'\boldsymbol{X}_n)^{-1}\boldsymbol{\lambda}/\boldsymbol{\lambda}'(\boldsymbol{X}_n'\boldsymbol{X}_n)^{-1}\boldsymbol{\lambda}]\\
&\quad = \mathrm{ch}_1[(\boldsymbol{X}_n'\boldsymbol{X}_n)^{-1}\boldsymbol{x}_{ni}\boldsymbol{x}_{ni}']\\
&\quad = \boldsymbol{x}_{ni}'(\boldsymbol{X}_n'\boldsymbol{X}_n)^{-1}\boldsymbol{x}_{ni},
\end{aligned}$$

implying that (10.2.16) reduces to the generalized Noether condition

$$\max_{1\le k\le n}\boldsymbol{x}_{nk}'(\boldsymbol{X}_n'\boldsymbol{X}_n)^{-1}\boldsymbol{x}_{nk} \to 0 \quad \text{as } n \to \infty. \tag{10.2.17}$$

Likewise, (ii) and (iii) in Theorem 10.2.1 should be replaced by

$$\lim_{n\to\infty} n^{-1}(\boldsymbol{X}_n'\boldsymbol{X}_n) = \boldsymbol{V}, \quad \text{finite and p.d.} \tag{10.2.18}$$

A direct application of Hájek–Šidak Theorem 7.3.6 to the Z_ns in conjunction with the Cramér–Wold Theorem 7.2.4 may then be used to prove the following theorem.

***Theorem* 10.2.2.** *Consider the model (10.1.1) and assume that the ε_is are uncorrelated and identically distributed random variables with zero mean and finite positive variance σ^2. Then, under (10.2.17) and (10.2.18),*

$$\sqrt{n}(\widehat{\boldsymbol{\beta}}_n - \boldsymbol{\beta}) \xrightarrow{D} \mathcal{N}_q(\boldsymbol{0}, \sigma^2\boldsymbol{V}^{-1}) \tag{10.2.19}$$

or, equivalently,

$$(\boldsymbol{X}_n'\boldsymbol{X}_n)^{1/2}(\widehat{\boldsymbol{\beta}}_n - \boldsymbol{\beta}) \xrightarrow{D} \mathcal{N}_q(\boldsymbol{0}, \sigma^2\boldsymbol{I}_q). \tag{10.2.20}$$

Note that if the individual coordinates of $\widehat{\boldsymbol{\beta}}_n - \boldsymbol{\beta}$ converge to zero (in probability or almost surely), then $\|\widehat{\boldsymbol{\beta}}_n - \boldsymbol{\beta}\|$ does too. For such coordinate elements, the proof sketched for the simple regression model goes through neatly. Hence, omitting the details, we claim that $\widehat{\boldsymbol{\beta}}_n - \boldsymbol{\beta} \xrightarrow{a.s.} 0$ (and, hence in probability, also). Moreover, for the multiple regression model, an unbiased estimator of σ^2 is

$$
\begin{aligned}
s_n^2 &= (n-q)^{-1} \sum_{i=1}^{n} (Y_i - \boldsymbol{x}'_{ni} \widehat{\boldsymbol{\beta}}_n)^2 \\
&= (n-q)^{-1} \left[\sum_{i=1}^{n} \varepsilon_i^2 - (\widehat{\boldsymbol{\beta}}_n - \boldsymbol{\beta})'(\boldsymbol{X}'_n \boldsymbol{X}_n)(\widehat{\boldsymbol{\beta}}_n - \boldsymbol{\beta}) \right]
\end{aligned} \tag{10.2.21}
$$

so that using the Khintchine SLLN (Theorem 6.3.13) on $n^{-1}\sum_{i=1}^{n} \varepsilon_i^2$, using (10.2.18) on $n^{-1}(\boldsymbol{X}'_n\boldsymbol{X}_n)$, and appealing to the fact that $\widehat{\boldsymbol{\beta}}_n - \boldsymbol{\beta} \xrightarrow{a.s.} \mathbf{0}$, we have that $s_n^2 \xrightarrow{a.s.} \sigma^2$.

Let us now examine the performance of the LSE when the errors ε_i are uncorrelated but not necessarily identically distributed. For simplicity we go back to the simple regression model and let $\mathbb{E}(\varepsilon_i^2) = \sigma_i^2$, $i \geq 1$. Since $\mathbb{E}(\varepsilon_i) = 0$, $i \geq 1$, it follows from (10.2.6) that $\mathbb{E}(\widehat{\beta}_n) = \beta$ and thus,

$$
\mathbb{V}\mathrm{ar}(\widehat{\beta}_n) = t_n^{-1} \sum_{i=1}^{n} (x_i - \overline{x}_n)^2 \sigma_i^2. \tag{10.2.22}
$$

The Noether condition in Theorem 10.2.1 can be modified to extend the asymptotic normality result to this heteroskedastic model, but there is a subtle point that we would like to make clear at this stage. If the ε_is are normally distributed, then ε_i/σ_i are all standard normal variables, and, hence, they are identically distributed. On the other hand, if the distribution of ε_i, $i \geq 1$, does not have a specified form, the ε_i/σ_is may not all have the same distribution (unless we assume that the distribution function F_i of ε_i, $i \geq 1$, is of the scale form $F_0(e/\sigma_i)$, where F_0 has a functional form independent of the scale factors). Under such a scale family model, we may write

$$
(x_i - \overline{x}_n)\varepsilon_i = [(x_i - \overline{x}_n)\sigma_i]\varepsilon_i/\sigma_i, \quad i \geq 1, \tag{10.2.23}
$$

so that the Noether condition, as well as (10.2.2) and (10.2.3) in Theorem 10.2.1 can be formulated in terms of the $(x_i - \overline{x}_n)\sigma_i$, $i \geq 1$, and the conclusion in (10.2.4) can be validated when the dispersion matrix is adjusted accordingly. However, these elements are then functions of the given x_i as well as of the unknown σ_i^2, $i \geq 1$, and, hence, to draw statistical conclusions from this asymptotic normality result, one needs to estimate the dispersion matrix from the observed sample. This may turn out to be much more involved. To illustrate this point, we consider the marginal law for $\widehat{\beta}_n - \beta$, so that we need to estimate $\mathbb{V}\mathrm{ar}(\widehat{\beta}_n)$ in (10.2.22). If the true ε_i, $i \geq 1$, were observable, we might have taken this estimator as

$$
V_n^0 = t_n^{-1} \sum_{i=1}^{n} (x_i - \overline{x}_n)^2 \varepsilon_i^2. \tag{10.2.24}
$$

Thus, if we assume that as $n \to \infty$,

$$
\frac{\sum_{i=1}^{n} |(x_i - \overline{x}_n)|^{2+\delta} \mathbb{E}[(|\varepsilon_i^2 - \sigma_i^2|)^{1+\delta}]}{\left[\sum_{i=1}^{n} (x_i - \overline{x}_n)^2 \sigma_i^2\right]^{1+\delta}} \to 0 \tag{10.2.25}
$$

for some $\delta > 0$, we may invoke the Markov LLN (Theorem 6.3.7) to conclude that

$$\frac{V_n^0}{\mathbb{V}\mathrm{ar}(\widehat{\beta}_n)} = \left[\sum_{i=1}^{n}(x_i - \overline{x}_n)^2\varepsilon_i^2\right]\left[\sum_{i=1}^{n}(x_i - \overline{x}_n)^2\sigma_i^2\right]^{-1} \xrightarrow{P} 1. \tag{10.2.26}$$

Note that (10.2.25) holds whenever the ε_is have finite moments of order $r > 2$ and the x_is are bounded, as in many practical situations.

Since the $\varepsilon_i, i \geq 1$, are unobservable, we may consider the residuals $\boldsymbol{e}_n = (e_{n1}, \ldots, e_{nn})'$ given by

$$\begin{aligned}
\boldsymbol{e}_n &= \boldsymbol{Y}_n - \widehat{\alpha}_n\mathbf{1}_n - \widehat{\beta}_n\boldsymbol{x}_n \\
&= \boldsymbol{\varepsilon}_n + (\alpha - \widehat{\alpha}_n)\mathbf{1}_n + (\beta - \widehat{\beta}_n)\boldsymbol{x}_n \\
&= \boldsymbol{\varepsilon}_n - \boldsymbol{A}_n\boldsymbol{\varepsilon}_n = (\boldsymbol{I}_n - \boldsymbol{A}_n)\boldsymbol{\varepsilon}_n,
\end{aligned} \tag{10.2.27}$$

where $\boldsymbol{A}_n$ is an $(n \times n)$ matrix. Thus, we may define an estimator of $\mathbb{V}\mathrm{ar}(\widehat{\beta}_n)$ as

$$V_n = t_n^{-1}\sum_{i=1}^{n}(x_i - \overline{x}_n)^2 e_{ni}^2 = t_n^{-1}\boldsymbol{e}_n'\boldsymbol{L}_n'\boldsymbol{L}_n\boldsymbol{e}_n \tag{10.2.28}$$

where

$$\boldsymbol{L}_n = \mathrm{diag}\{x_1 - \overline{x}_n, \ldots, x_n - \overline{x}_n\}.$$

Also, letting $\boldsymbol{\sigma}_n = (\sigma_1^2, \ldots, \sigma_n^2)'$, we have,

$$\mathbb{V}\mathrm{ar}(\widehat{\beta}_n) = t_n^{-1}\boldsymbol{\sigma}_n'\boldsymbol{L}_n'\boldsymbol{L}_n\boldsymbol{\sigma}_n$$

and

$$V_n^0 = t_n^{-1}\boldsymbol{\varepsilon}_n'\boldsymbol{L}_n'\boldsymbol{L}_n\boldsymbol{\varepsilon}_n,$$

so that taking (10.2.27) and (10.2.28) into account we get

$$\begin{aligned}
\frac{V_n - V_n^0}{\mathbb{V}\mathrm{ar}(\widehat{\beta}_n)} &= \frac{\boldsymbol{\varepsilon}_n'(\boldsymbol{I}_n - \boldsymbol{A}_n)'\boldsymbol{L}_n'\boldsymbol{L}_n(\boldsymbol{I}_n - \boldsymbol{A}_n)\boldsymbol{\varepsilon}_n - \boldsymbol{\varepsilon}_n'\boldsymbol{L}_n'\boldsymbol{L}_n\boldsymbol{\varepsilon}_n}{\boldsymbol{\sigma}_n'\boldsymbol{L}_n'\boldsymbol{L}_n\boldsymbol{\sigma}_n} \\
&= \left[\frac{\boldsymbol{\varepsilon}_n'\boldsymbol{L}_n'\boldsymbol{L}_n\boldsymbol{\varepsilon}_n}{\boldsymbol{\sigma}_n'\boldsymbol{L}_n'\boldsymbol{L}_n\boldsymbol{\sigma}_n}\right]\left[\frac{\boldsymbol{\varepsilon}_n'\boldsymbol{A}_n'\boldsymbol{L}_n'\boldsymbol{L}_n\boldsymbol{A}_n\boldsymbol{\varepsilon}_n}{\boldsymbol{\varepsilon}_n'\boldsymbol{L}_n'\boldsymbol{L}_n\boldsymbol{\varepsilon}_n} - 2\frac{\boldsymbol{\varepsilon}_n'\boldsymbol{A}_n'\boldsymbol{L}_n'\boldsymbol{L}_n\boldsymbol{\varepsilon}_n}{\boldsymbol{\varepsilon}_n'\boldsymbol{L}_n'\boldsymbol{L}_n\boldsymbol{\varepsilon}_n}\right] \\
&= \left[\frac{\boldsymbol{\varepsilon}_n'\boldsymbol{L}_n'\boldsymbol{L}_n\boldsymbol{\varepsilon}_n}{\boldsymbol{\sigma}_n'\boldsymbol{L}_n'\boldsymbol{L}_n\boldsymbol{\sigma}_n}\right](a_n - b_n),
\end{aligned} \tag{10.2.29}$$

where

$$a_n = \frac{\boldsymbol{\varepsilon}_n'\boldsymbol{A}_n'\boldsymbol{L}_n'\boldsymbol{L}_n\boldsymbol{A}_n}{\boldsymbol{\varepsilon}_n'\boldsymbol{L}_n'\boldsymbol{L}_n\boldsymbol{\varepsilon}_n}$$

and

$$b_n = 2\frac{\boldsymbol{\varepsilon}_n'\boldsymbol{A}_n'\boldsymbol{L}_n'\boldsymbol{L}_n\boldsymbol{\varepsilon}_n}{\boldsymbol{\varepsilon}_n'\boldsymbol{L}_n'\boldsymbol{L}_n\boldsymbol{\varepsilon}_n}.$$

Since, by (10.2.26), the first term on the right-hand side of (10.2.29) converges in probability to 1, we may apply Courant Theorem 1.5.1 to a_n and b_n and conclude that sufficient conditions for $(V_n - V_n^0)/\mathbb{V}\mathrm{ar}(\widehat{\beta}_n) \xrightarrow{P} 0$ and, consequently, for $V_n/\mathbb{V}\mathrm{ar}(\widehat{\beta}_n) \xrightarrow{P} 1$ are

$$\mathrm{ch}_1(\boldsymbol{A}_n'\boldsymbol{L}_n'\boldsymbol{L}_n\boldsymbol{A}_n(\boldsymbol{L}_n'\boldsymbol{L}_n)^{-1}) \to 0 \quad \text{and} \quad \mathrm{ch}_1(\boldsymbol{A}_n + \boldsymbol{A}_n') \to 0, \tag{10.2.30}$$

as $n \to \infty$. The regularity conditions needed for this stochastic convergence result depend all on the set of (x_i, σ_i^2), $1 \le i \le n$, and can be verified for a given design $\{\boldsymbol{x}_n\}$ by imposing suitable variational inequality conditions on the σ_i as in the following example.

Example* 10.2.1 *(Two-sample problem). Consider the simple regression model (10.1.2) with $n = 2m$, $x_i = 1$, for $1 \le i \le m$, $x_i = -1$, for $m+1 \le i \le 2m$, and assume that $\mathbb{V}\text{ar}(\varepsilon_i) = \sigma_i^2$, $1 \le i \le n$. Then $\overline{x}_n = 0$ and $t_n = 2m$. Also, using (10.2.6) and (10.2.7) we may write

$$\widehat{\beta}_n - \beta = t_n^{-1}\sum_{i=1}^{n}(x_i - \overline{x}_n)\varepsilon_i = \frac{1}{2m}\left(-\mathbf{1}_m', \mathbf{1}_m'\right)\boldsymbol{\varepsilon}_n$$

and

$$\widehat{\alpha}_n - \alpha = (\widehat{\beta}_n - \beta)\overline{x}_n + \overline{\varepsilon}_n = \frac{1}{2m}\left(\mathbf{1}_m', \mathbf{1}_m'\right)\boldsymbol{\varepsilon}_n,$$

which, in connection with (10.2.27), yields

$$\begin{aligned}
\boldsymbol{e}_n &= \boldsymbol{\varepsilon}_n - \frac{1}{2m}\left(\mathbf{1}_m', \mathbf{1}_m'\right)\boldsymbol{\varepsilon}_n\begin{pmatrix}\mathbf{1}_m\\ \mathbf{1}_m\end{pmatrix} - \frac{1}{2m}\left(-\mathbf{1}_m', \mathbf{1}_m'\right)\boldsymbol{\varepsilon}_n\begin{pmatrix}-\mathbf{1}_m\\ \mathbf{1}_m\end{pmatrix}\\
&= \boldsymbol{\varepsilon}_n - \frac{1}{2m}\begin{bmatrix}\mathbf{1}_m\mathbf{1}_m' & \mathbf{1}_m\mathbf{1}_m'\\ \mathbf{1}_m\mathbf{1}_m' & \mathbf{1}_m\mathbf{1}_m'\end{bmatrix}\boldsymbol{\varepsilon}_n - \frac{1}{2m}\begin{bmatrix}\mathbf{1}_m\mathbf{1}_m' & -\mathbf{1}_m\mathbf{1}_m'\\ -\mathbf{1}_m\mathbf{1}_m' & \mathbf{1}_m\mathbf{1}_m'\end{bmatrix}\boldsymbol{\varepsilon}_n\\
&= \boldsymbol{\varepsilon}_n - \boldsymbol{A}_n\boldsymbol{\varepsilon}_n,
\end{aligned}$$

where

$$\boldsymbol{A}_n = \frac{1}{m}\begin{bmatrix}\mathbf{1}_m\mathbf{1}_m' & \mathbf{0}\\ \mathbf{0} & \mathbf{1}_m\mathbf{1}_m'\end{bmatrix}.$$

Now, since $\boldsymbol{L}_n = \text{diag}\{\boldsymbol{I}_m, -\boldsymbol{I}_m\}$, it follows that $\boldsymbol{L}_n'\boldsymbol{L}_n = \boldsymbol{I}_{2m}$ so that $\text{ch}_1[\boldsymbol{A}_n'\boldsymbol{L}_n'\boldsymbol{L}_n\boldsymbol{A}_n(\boldsymbol{L}_n'\boldsymbol{L}_n)^{-1}] = m^{-2}$ and $\text{ch}_1(\boldsymbol{A}_n + \boldsymbol{A}_n') = 2m^{-1}$, both converge to zero as $m \to \infty$, that is, (10.2.30) holds. Now, if we assume further that $\mathbb{E}(|\varepsilon_i|^{2+\delta}) < \infty$, the boundedness of $|x_i|$ implies (10.2.25) and, therefore, (10.2.28) is a strongly consistent estimate of $\mathbb{V}\text{ar}(\widehat{\beta}_n)$. □

The above discussion casts some light on the robustness of the estimator V_n with respect to a possible heteroskedasticity of the error components. A similar, but admittedly more complex analysis can be made for the general linear model (10.1.1) when the ε_is have possibly different variances. For some of these details, we refer to Eicker (1967).

Let us go back to model (10.1.1) and consider another generalization wherein we assume that $\mathbb{V}\text{ar}(\boldsymbol{\varepsilon}_n) = \sigma^2\boldsymbol{G}_n$ with $\boldsymbol{G}_n$ being a known p.d. matrix. The *generalized least-squares estimator* (GLSE) of $\boldsymbol{\beta}$ is obtained by minimizing the quadratic form

$$(\boldsymbol{Y}_n - \boldsymbol{X}_n'\boldsymbol{\beta})'\boldsymbol{G}_n^{-1}(\boldsymbol{Y}_n - \boldsymbol{X}_n'\boldsymbol{\beta}) = \|\boldsymbol{Y}_n - \boldsymbol{X}_n'\boldsymbol{\beta}\|_{\boldsymbol{G}_n}^2 \tag{10.2.31}$$

with respect to $\boldsymbol{\beta}$ and is given by:

$$\widetilde{\boldsymbol{\beta}}_n = (\boldsymbol{X}_n'\boldsymbol{G}_n^{-1}\boldsymbol{X}_n)^{-1}\boldsymbol{X}_n'\boldsymbol{G}_n^{-1}\boldsymbol{Y}_n \tag{10.2.32}$$

whenever $(\boldsymbol{X}_n'\boldsymbol{G}_n^{-1}\boldsymbol{X}_n)$ is of full rank [otherwise, we may replace the inverse by a generalized inverse in (10.2.32)]. In the cases where $\boldsymbol{G}_n$ is a diagonal matrix, $\widetilde{\boldsymbol{\beta}}_n$ in (10.2.32) is known as the *weighted least-squares estimators* (WLSE) of $\boldsymbol{\beta}$. Before we proceed to consider the asymptotic properties of the GLSE (WLSE) of $\boldsymbol{\beta}$, we need some preliminary considerations.

If in (10.1.1) $\boldsymbol{\varepsilon}_n$ has an n-variate multinormal distribution with null mean vector and dispersion matrix $\sigma^2\boldsymbol{G}_n$, then $\widetilde{\boldsymbol{\beta}}_n$ in (10.2.32), being linear in $\boldsymbol{\varepsilon}_n$, has also a multinormal law with mean vector $\boldsymbol{\beta}$ and dispersion matrix $\sigma^2(\boldsymbol{X}_n'\boldsymbol{G}_n^{-1}\boldsymbol{X}_n)^{-1}$. But this simple conclusion may not generally follow when $\boldsymbol{\varepsilon}_n$ does not have a multinormal law. We may notice that, by assumption, $\boldsymbol{\varepsilon}_n^* = \boldsymbol{G}_n^{-1/2}\boldsymbol{\varepsilon}_n$ has mean vector $\boldsymbol{0}$ and dispersion matrix $\sigma^2\boldsymbol{I}_n$. Thus, the elements of $\boldsymbol{\varepsilon}_n^*$ are mutually uncorrelated. However, in general, uncorrelation may not imply independence, and homogeneity of variances may not imply identity of distributions. Thus, to adopt the GLSE(WLSE) methods in a given context, we may need to spell out the model more precisely, so that the components of $\boldsymbol{\varepsilon}_n^*$ are independent and identically distributed. We do this by rewriting (10.1.1) as

$$\boldsymbol{Y}_n = \boldsymbol{X}_n\boldsymbol{\beta} + \boldsymbol{G}_n^{1/2}\boldsymbol{\varepsilon}_n^*, \tag{10.2.33}$$

where $\boldsymbol{\varepsilon}_n^* = (\varepsilon_1^*, \ldots, \varepsilon_n^*)'$ has independent and identically distributed components ε_i^*, such that $\mathbb{E}(\varepsilon_i^*) = 0$ and $\mathbb{E}[(\varepsilon_i^*)^2] = \sigma^2$. For the WLSE, $\boldsymbol{G}_n^{1/2}$ is a diagonal matrix, so that putting $\boldsymbol{G}_n^{1/2} = \boldsymbol{D}_n^{1/2} = \text{diag}(\sigma_1, \ldots, \sigma_n)$ where the σ_is are nonnegative scalar constants, we may also write

$$\boldsymbol{G}_n^{1/2}\boldsymbol{\varepsilon}_n^* = \boldsymbol{D}_n^{1/2}\boldsymbol{\varepsilon}_n^* \tag{10.2.34}$$

and conclude that $\mathbb{E}[(\varepsilon_i^*)^2] = 1$, $i \geq 1$. This is called the *heteroskedastic error model.* In the current context, we assume that $\sigma_i^2 = \sigma^2 h_i$, $i = 1, \ldots, n$, where the h_i, $i \geq 1$, are given and σ^2 is unknown. Then we may write

$$\boldsymbol{Y}_n^* = \boldsymbol{G}_n^{-1/2}\boldsymbol{Y}_n \quad \text{and} \quad \boldsymbol{X}_n^* = \boldsymbol{G}_n^{-1/2}\boldsymbol{X}_n, \tag{10.2.35}$$

which, in conjunction with (10.2.33), leads to

$$\boldsymbol{Y}_n^* = \boldsymbol{X}_n^*\boldsymbol{\beta} + \boldsymbol{\varepsilon}_n^*. \tag{10.2.36}$$

Now, from (10.2.32) and (10.2.35) we have

$$\widetilde{\boldsymbol{\beta}}_n = (\boldsymbol{X}_n^{*t}\boldsymbol{X}_n^*)^{-1}\boldsymbol{X}_n^{*t}\boldsymbol{Y}_n^*, \tag{10.2.37}$$

which resembles (10.1.7) with $\boldsymbol{X}_n$, $\boldsymbol{Y}_n$ (and $\boldsymbol{\varepsilon}_n$) being replaced by $\boldsymbol{X}_n^*$, $\boldsymbol{Y}_n^*$ (and $\boldsymbol{\varepsilon}_n^*$), respectively. Thus, along the same lines as in (10.2.17) and (10.2.18), we assume that the following two conditions hold:

$$\max_{1\leq i\leq n} [\boldsymbol{x}_{ni}^{*t}(\boldsymbol{X}_n^{*t}\boldsymbol{X}_n^*)^{-1}\boldsymbol{x}_{ni}^*] \to 0 \quad \text{as } n \to \infty \tag{10.2.38}$$

and

$$\lim_{n\to\infty} n^{-1}(\boldsymbol{X}_n^{*t}\boldsymbol{X}_n^*) = \boldsymbol{V}^*, \quad \text{finite and p.d.} \tag{10.2.39}$$

Then, Theorem 10.2.2 extends directly to the following:

***Theorem* 10.2.3.** *For the model (10.2.33) with the* ε_i^* *being independent and identically distributed random variables with 0 mean and finite positive variance* σ^2*, it follows that under (10.2.38) and (10.2.39)*

$$\sqrt{n}(\widetilde{\boldsymbol{\beta}}_n - \boldsymbol{\beta}) \xrightarrow{D} \mathcal{N}_q(\mathbf{0}, \sigma^2 \boldsymbol{V}^{*^{-1}}). \tag{10.2.40}$$

The transformation model in (10.2.33) is a very convenient one for asymptotic theory, and in most of the practical problems (particularly for $\boldsymbol{G}_n$ diagonal), the assumed regularity conditions on the ε_n^* are easy to justify (without imposing normality on them). Also (10.2.38) and (10.2.39) are easy to verify for the specification matrix $\boldsymbol{X}_n$ and the matrix $\boldsymbol{G}_n$ usually considered in practice. In this context, a very common problem relates to the *replicated model*, where each y_i is an average of a number of observations (say, m_i) which relate to a common $\boldsymbol{x}_i$, so that $\boldsymbol{G}_n = \mathrm{diag}\{m_1^{-1}, \ldots, m_n^{-1}\}$, with the m_i, $i \geq 1$, known. This allows us to handle unbalanced designs in a general setup.

As may be the case with many practical problems, the observations $y_i, i \geq 1$, may be due to different investigators and/or made on different measuring instruments (whose variability may not be homogeneous) so that we may have a model similar to (10.2.33), but with $\boldsymbol{G}_n$ unknown. Thus, it may be more realistic to consider the following general model:

$$\boldsymbol{Y}_n = \boldsymbol{X}_n \boldsymbol{\beta} + \boldsymbol{\Sigma}_n^{1/2} \boldsymbol{\varepsilon}_n^* \tag{10.2.41}$$

where the components of $\boldsymbol{\varepsilon}_n^*$ are independent and identically distributed with mean zero and unit variance, whereas $\boldsymbol{\Sigma}_n$ is a $(n \times n)$ unknown positive definite matrix. The fact that the dimension of $\boldsymbol{\sigma}_n$ increases with n may cause some technical problems, which we illustrate with the following example.

Example* 10.2.2 *(Neyman–Scott problem). Let $\boldsymbol{X}_n = \mathbf{1}_n$, and consider the model

$$\boldsymbol{Y}_n = \theta \mathbf{1}_n + \boldsymbol{\varepsilon}_n,$$

where $\theta \in \Theta \subset \mathbb{R}$ and $\boldsymbol{\varepsilon}_n = (\varepsilon_1, \ldots, \varepsilon_n)'$ is a vector of independent random errors with zero means and unknown variances $\sigma_1^2, \ldots, \sigma_n^2$, respectively. Thus, here $\boldsymbol{\sigma}_n = \mathrm{diag}\{\sigma_1^2, \ldots, \sigma_n^2\}$. As in Neyman and Scott (1948), we assume that the ε_i, $i \geq 1$, are normally distributed, but, as we will see, even this strong assumption does not help much in the study of the asymptotic properties of the estimator of the location parameter θ. If the σ_i^2, $i \geq 1$, were known, the WLSE of θ would have been

$$\widehat{\theta}_n^* = \sum_{i=1}^n \sigma_i^{-2} Y_i \Big/ \sum_{i=1}^n \sigma_i^{-2}. \tag{10.2.42}$$

Here, $\widehat{\theta}_n^*$ is also the MLE of θ when the ε_i, $i \geq 1$, are normally distributed with known variances. When the σ_i, $i \geq 1$ are unknown, to obtain the MLE of θ (as well as of σ_i^2, $i \geq 1$), one needs to solve for $(n+1)$ simultaneous equations, which for the normal model reduce to

$$\sum_{i=1}^n (Y_i - \theta)/\sigma_i^2 = 0, \quad \sigma_i^{-1} = (Y_i - \theta)^2/\sigma_i^3, \quad 1 \leq i \leq n, \tag{10.2.43}$$

and the solution $\widetilde{\theta}_n^*$, obtainable from

$$\sum_{i=1}^{n}(Y_i - \widetilde{\theta}_n^*)/(Y_i - \widetilde{\theta}_n^*)^2 = 0$$

or, equivalently, from

$$\sum_{i=1}^{n}|Y_i - \widetilde{\theta}_n^*|^{-1}\text{sign}(Y_i - \widetilde{\theta}_n^*) = 0, \tag{10.2.44}$$

may not even be consistent; this is particularly due to the fact that the $Y_i, i \geq 1$, arbitrarily close (but not identical to) $\widetilde{\theta}_n^*$ may have unbounded influence on the estimating equation (10.2.44). Thus, it may be better to consider some alternative estimators $\widehat{\sigma}_1^2, \ldots, \widehat{\sigma}_n^2$ (of $\sigma_1^2, \ldots, \sigma_n^2$, respectively) and formulate an estimator of θ as

$$\breve{\theta}_n^* = \sum_{i=1}^{n}\widehat{\sigma}_i^{-2}Y_i \Big/ \sum_{i=1}^{n}\widehat{\sigma}_i^{-2}. \tag{10.2.45}$$

The question, however, remains open: is $\breve{\theta}_n^*$ consistent for θ? If so, is it asymptotically normally distributed? Recall that by (10.2.42)

$$\begin{aligned}\mathbb{E}[(\widehat{\theta}_n^* - \theta)^2] &= \sum_{i=1}^{n}\sigma_i^{-4}\sigma_i^2 \Big/ \left(\sum_{i=1}^{n}\sigma_i^{-2}\right)^2 = \left(\sum_{i=1}^{n}\sigma_i^{-2}\right)^{-1} \\ &= n^{-1}\left(n^{-1}\sum_{i=1}^{n}\sigma_i^{-2}\right)^{-1} \leq n^{-1}\left(n^{-1}\sum_{i=1}^{n}\sigma_i^2\right)\end{aligned} \tag{10.2.46}$$

where the last step is due to the Arithmetic/Geometric/Harmonic Mean Inequality (1.5.40) for nonnegative numbers. Thus, whenever

$$\overline{\sigma}_n^2 = n^{-1}\sum_{i=1}^{n}\sigma_i^2 = o(n), \tag{10.2.47}$$

(10.2.46) converges to zero as $n \to \infty$ and, by the Chebyshev Inequality (6.2.27), $\widehat{\theta}_n^* \xrightarrow{P} \theta$. In fact, (10.2.47) is only a sufficient condition; an even weaker condition is that the harmonic mean $\overline{\sigma}_{nH}^2 = (n^{-1}\sum_{i=1}^{n}\sigma_i^2)^{-1}$ is $o(n)$. To see this, consider the particular case where $\sigma_i^2 \propto i, i \geq 1$; then $\overline{\sigma}_n^2 \propto (n+1)/2 = O(n)$, whereas $\overline{\sigma}_{nH}^2 \propto (n^{-1}\log n)^{-1} = n/\log n = o(n)$. On the other hand, in (10.2.45), the $\widehat{\sigma}_i^{-2}$, $i \geq 1$, are themselves random variables and, moreover, they may not be stochastically independent of the Y_is. Thus, the computation of the mean square error of $\breve{\theta}_n^*$ may be much more involved; it may not even be $o(1)$ under (10.2.47) or some similar condition. To illustrate this point, we assume further that the model corresponds to a replicated one, where $n = 2m$ and $\sigma_{2i}^2 = \sigma_{2i-1}^2 = \gamma_i, i = 1, \ldots, m$. Then $(Y_{2i} - Y_{2i-1})^2/2 = \widehat{\gamma}_i, i = 1, \ldots, m$, are unbiased estimators of the $\gamma_i, i = 1, \ldots, m$, and, moreover, $(Y_{2i} + Y_{2i-1})/2$ and $(Y_{2i} - Y_{2i-1})/2$ are mutually independent so that $\widehat{\gamma}_i$ is independent of $(Y_{2i} + Y_{2i-1})/2, i = 1, \ldots, m$. Thus, we may rewrite (10.2.45) as

$$\breve{\theta}_n^* = \left[\sum_{i=1}^{n}\widehat{\gamma}_i^{-1}(Y_{2i} + Y_{2i-1})/2\right]\left[\sum_{i=1}^{m}\widehat{\gamma}_i^{-1}\right]^{-1}. \tag{10.2.48}$$

Using the aforementioned independence [between the $\widehat{\gamma}_i$ and $(Y_{2i} + Y_{2i-1}]$, the conditional mean square error of $\check{\theta}_n^*$ given $\widehat{\gamma}_1, \ldots, \widehat{\gamma}_m$ is equal to

$$\frac{1}{2}\left[\sum_{i=1}^{m} \widehat{\gamma_i}^{-2}\gamma_i\right]\left[\sum_{i=1}^{m} \widehat{\gamma_i}^{-1}\right]^{-2} \tag{10.2.49}$$

so that the unconditional mean square error is given by

$$\frac{1}{2}\mathbb{E}\left[\left(\sum_{i=1}^{m} \gamma_i/\widehat{\gamma_i^2}\right)\left(\sum_{i=1}^{m} \widehat{\gamma_i}^{-1}\right)^{-2}\right]. \tag{10.2.50}$$

Using again the Arithmetic/Geometric/Harmonic Mean Inequality (1.5.40), we may bound (10.2.49) from above by

$$\left[1/2m^{-2}\sum_{i=1}^{m} \widehat{\gamma_i}^{-2}\gamma_i\right]\left[m^{-1}\sum_{i=1}^{m} \widehat{\gamma_i}\right]^2,$$

which is, in turn, bounded from above by

$$\left[\frac{1}{2m}\frac{1}{m}\sum_{i=1}^{m} \widehat{\gamma_i}^{-2}\gamma_i\right]\left[\frac{1}{m}\sum_{i=1}^{m} \widehat{\gamma_i^2}\right]. \tag{10.2.51}$$

Now we could think of appealing to the Cauchy–Schwarz Inequality (1.5.34) to obtain an upper bound for (10.2.50), but even so, we may not have the required luck, as $\mathbb{E}(\widehat{\gamma_i}^{-2}) = +\infty$ for every $i \geq 1$, whereas $\mathbb{E}(\widehat{\gamma_i^2}) = 2\gamma_i^2$. Thus, inherent independence of the $\widehat{\gamma}_i$ and $(Y_{2i} + Y_{2i-1})$ as well as the normality of errors may not contribute much toward the use of the Chebyshev Inequality (6.2.27) as a simple way to prove the consistency of $\check{\theta}_n^*$. Also note that (10.2.49) is bounded from below by

$$\frac{1}{2}\left[\left(\sum_{i=1}^{m} \widehat{\gamma_i}^{-1}\gamma_i^{1/2}\right)\left(\sum_{i=1}^{m} \widehat{\gamma_i}^{-1}\right)^{-1}\right]^2 \tag{10.2.52}$$

so that the fact that $\mathbb{E}(\widehat{\gamma_i}^{-1}) = +\infty$, $i \geq 1$, continues to pose difficulties in the computation of the mean squared error. This suggests that more structure on the σ_i^2, either in terms of a fixed pattern or possibly some Bayesian model, may be necessary to eliminate this drawback and to render some tractable solutions to the corresponding asymptotic study. □

Let us return to the general linear model (10.2.41). If $\boldsymbol{\Sigma}_n$ were known, the GLSE of $\boldsymbol{\beta}$ would be

$$\widehat{\boldsymbol{\beta}}_n = (\boldsymbol{X}_n'\boldsymbol{\Sigma}_n^{-1}\boldsymbol{X}_n)^{-1}\boldsymbol{X}_n'\boldsymbol{\Sigma}_n^{-1}\boldsymbol{Y}_n. \tag{10.2.53}$$

Assuming that (10.2.38) and (10.2.39) with $\boldsymbol{G}_n$ replaced by $\boldsymbol{\Sigma}_n$ both hold, one may claim via Theorem 10.2.3 that

$$\sqrt{n}(\widehat{\boldsymbol{\beta}}_n - \boldsymbol{\beta}) \xrightarrow{D} \mathcal{N}_q(\boldsymbol{0}, \boldsymbol{V}^{-1}) \tag{10.2.54}$$

where $\boldsymbol{V} = \lim_{n\to\infty} n^{-1}\boldsymbol{V}_n$ and $\boldsymbol{V}_n = \boldsymbol{X}_n'\boldsymbol{\Sigma}_n^{-1}\boldsymbol{X}_n$. Note that the consistency of $\widehat{\boldsymbol{\beta}}_n$ follows directly from (10.2.54). Since, in general, $\boldsymbol{\Sigma}_n$ is unknown, we may attempt to replace it by an estimate $\widehat{\boldsymbol{\Sigma}}_n$ in (10.2.53) obtaining the *two-step (Aitken) estimator* :

$$\widehat{\widehat{\boldsymbol{\beta}}}_n = (\boldsymbol{X}_n'\widehat{\boldsymbol{\Sigma}}_n^{-1}\boldsymbol{X}_n)^{-1}\boldsymbol{X}_n'\widehat{\boldsymbol{\Sigma}}_n^{-1}\boldsymbol{Y}_n. \tag{10.2.55}$$

Letting $\widehat{\boldsymbol{V}}_n = \boldsymbol{X}_n'\widehat{\boldsymbol{\Sigma}}_n^{-1}\boldsymbol{X}_n$ it follows from (10.2.41) and (10.2.55) that

$$\widehat{\widehat{\boldsymbol{\beta}}}_n = \boldsymbol{\beta} + \widehat{\boldsymbol{V}}_n^{-1}\boldsymbol{X}_n'\widehat{\boldsymbol{\Sigma}}_n^{-1}\boldsymbol{\Sigma}_n^{1/2}\boldsymbol{\varepsilon}_n^*, \tag{10.2.56}$$

whereas by (10.2.41) and (10.2.53) we have

$$\widehat{\boldsymbol{\beta}}_n = \boldsymbol{\beta} + \boldsymbol{V}_n^{-1}\boldsymbol{X}_n'\boldsymbol{\Sigma}_n^{-1/2}\boldsymbol{\varepsilon}_n^*. \tag{10.2.57}$$

Therefore, from (10.2.56) and (10.2.57), we may write:

$$\begin{aligned}
\widehat{\widehat{\boldsymbol{\beta}}}_n - \widehat{\boldsymbol{\beta}}_n &= (\widehat{\boldsymbol{V}}_n^{-1} - \boldsymbol{V}_n^{-1})\boldsymbol{X}_n'\widehat{\boldsymbol{\Sigma}}_n^{-1}\boldsymbol{\Sigma}_n^{1/2}\boldsymbol{\varepsilon}_n^* \\
&\quad + \boldsymbol{V}_n^{-1}\boldsymbol{X}_n'\big[\widehat{\boldsymbol{\Sigma}}_n^{-1} - \boldsymbol{\Sigma}_n^{-1}\big]\boldsymbol{\Sigma}_n^{1/2}\boldsymbol{\varepsilon}_n^* \\
&= (\widehat{\boldsymbol{V}}_n^{-1} - \boldsymbol{V}_n^{-1})\boldsymbol{X}_n'\boldsymbol{\Sigma}_n^{-1/2}\boldsymbol{\varepsilon}_n^* \\
&\quad + (\widehat{\boldsymbol{V}}_n^{-1} - \boldsymbol{V}_n^{-1})\boldsymbol{X}_n'\big[\widehat{\boldsymbol{\Sigma}}_n^{-1} - \boldsymbol{\Sigma}_n^{-1}\big]\boldsymbol{\Sigma}_n^{1/2}\boldsymbol{\varepsilon}_n^* \\
&\quad + \boldsymbol{V}_n^{-1}\boldsymbol{X}_n'\big[\widehat{\boldsymbol{\Sigma}}_n^{-1} - \boldsymbol{\Sigma}_n^{-1}\big]\boldsymbol{\Sigma}_n^{1/2}\boldsymbol{\varepsilon}_n^*.
\end{aligned} \tag{10.2.58}$$

The first term on the right-hand side of (10.2.58) may be expressed as

$$(\widehat{\boldsymbol{V}}_n^{-1} - \boldsymbol{V}_n^{-1})\boldsymbol{V}_n(\widehat{\boldsymbol{\beta}}_n - \boldsymbol{\beta}) = (\widehat{\boldsymbol{V}}_n^{-1}\boldsymbol{V}_n - \boldsymbol{I}_n)(\widehat{\boldsymbol{\beta}}_n - \boldsymbol{\beta}). \tag{10.2.59}$$

Invoking (10.2.54) one may claim that $\widehat{\boldsymbol{\beta}}_n - \boldsymbol{\beta} \xrightarrow{P} \boldsymbol{0}$. Thus, (10.2.59) is $o_p(\boldsymbol{1})$ whenever $\widehat{\boldsymbol{V}}_n^{-1}\boldsymbol{V}_n - \boldsymbol{I}_n = O_p(\boldsymbol{1})$, that is, whenever the characteristic roots of $\widehat{\boldsymbol{V}}_n^{-1}\boldsymbol{V}_n$ are bounded in probability. Similarly, the second term on the right-hand side of (10.2.58) may be re-expressed as

$$\begin{aligned}
&(\widehat{\boldsymbol{V}}_n^{-1}\boldsymbol{V}_n - I_n)\boldsymbol{V}_n^{-1}\boldsymbol{X}_n'(\widehat{\boldsymbol{\Sigma}}_n^{-1} - \boldsymbol{\Sigma}_n^{-1})\boldsymbol{\Sigma}_n^{1/2}\boldsymbol{\varepsilon}_n^* \\
&\quad = (\widehat{\boldsymbol{V}}_n^{-1}\boldsymbol{V}_n - \boldsymbol{I}_n)\boldsymbol{V}_n^{-1}\boldsymbol{X}_n'\boldsymbol{\Sigma}_n^{-1}(\boldsymbol{\Sigma}_n\widehat{\boldsymbol{\Sigma}}_n^{-1} - \boldsymbol{I}_n)\boldsymbol{\Sigma}_n^{1/2}\boldsymbol{\varepsilon}_n^*.
\end{aligned} \tag{10.2.60}$$

Thus, if all the characteristic roots of $\boldsymbol{\Sigma}_n\widehat{\boldsymbol{\Sigma}}_n^{-1} - \boldsymbol{I}_n$ are $o_p(1)$, whereas the characteristic roots of $\widehat{\boldsymbol{V}}_n^{-1}\boldsymbol{V}_n$ are $O_p(1)$, (10.2.60) converges stochastically to $\boldsymbol{0}$. A very similar treatment holds for the last term on the right-hand side of (10.2.58). Thus, we may conclude that $\widehat{\widehat{\boldsymbol{\beta}}}_n - \widehat{\boldsymbol{\beta}} \xrightarrow{P} \boldsymbol{0}$ whenever (10.2.38) and (10.2.39) hold in conjugation with

$$\mathrm{ch}_1(\boldsymbol{V}_n\widehat{\boldsymbol{V}}_n^{-1}) = O_p(1) \tag{10.2.61}$$

and

$$\mathrm{ch}_1(\boldsymbol{\Sigma}_n\widehat{\boldsymbol{\Sigma}}_n^{-1}) - 1 = o_p(1) = \mathrm{ch}_n(\boldsymbol{\Sigma}_n\widehat{\boldsymbol{\Sigma}}_n^{-1}) - 1. \tag{10.2.62}$$

On the other hand, if we replace (10.2.61) by

$$\mathrm{ch}_1(\boldsymbol{V}_n\widehat{\boldsymbol{V}}_n^{-1}) - 1 = o_p(1) = \mathrm{ch}_n(\boldsymbol{V}_n\widehat{\boldsymbol{V}}_n^{-1}) - 1, \tag{10.2.63}$$

we may conclude that

$$\sqrt{n}(\widehat{\widehat{\boldsymbol{\beta}}}_n - \widehat{\boldsymbol{\beta}}_n) \xrightarrow{P} \boldsymbol{0} \tag{10.2.64}$$

so that via (10.2.54) and (10.2.64), we obtain

$$\sqrt{n}(\widehat{\widehat{\boldsymbol{\beta}}}_n - \boldsymbol{\beta}) \xrightarrow{D} \mathcal{N}_q(\mathbf{0}, \boldsymbol{V}^{-1}). \tag{10.2.65}$$

These regularity conditions framed for the consistency and asymptotic normality of $\widehat{\widehat{\boldsymbol{\beta}}}_n$ are sufficient ones and are easily verifiable in terms of the specification matrix $\boldsymbol{X}_n$, the covariance matrix $\boldsymbol{\Sigma}_n$ and its consistent estimator $\widehat{\boldsymbol{\Sigma}}_n$. Here we interpret the consistency of $\widehat{\boldsymbol{\Sigma}}_n$ (with respect to $\boldsymbol{\Sigma}_n$) in the sense of (10.2.62) or, more explicitly, as $\boldsymbol{\Sigma}_n\widehat{\boldsymbol{\Sigma}}_n^{-1} \xrightarrow{P} \boldsymbol{I}_n$.

In passing, we remark that both $\boldsymbol{\Sigma}_n$ and $\widehat{\boldsymbol{\Sigma}}_n$ are $(n \times n)$ matrices and, hence, in general, they may not satisfy this consistency property. However, in many problems of practical interest, both $\boldsymbol{\Sigma}_n$ and $\widehat{\boldsymbol{\Sigma}}_n$ are generated by a finite number of finite dimensional matrices for which it may be possible to verify the above conditions. This point is illustrated in the exercises.

We finalize this section by noting that a direct application of the Cramér–Wold Theorem 7.2.4 may be used to extend the above results to the standard *multivariate linear model*:

$$\mathbf{Y}_n = \boldsymbol{X}_n\boldsymbol{B} + \boldsymbol{E}_n \tag{10.2.66}$$

where $\mathbf{Y}_n = (\boldsymbol{Y}_1, \ldots, \boldsymbol{Y}_n)'$ is a $(n \times p)$ matrix of observable response variables with rows $\boldsymbol{Y}_i' = (Y_{i1}, \ldots, Y_{ip})$, $\boldsymbol{X}_n$ is defined as in (10.1.1), $\boldsymbol{B} = (\boldsymbol{\beta}_1, \ldots, \boldsymbol{\beta}_p)$ is a $(q \times p)$ matrix of unknown parameters [with the jth column, $j = 1, \ldots, p$, having the same interpretation as $\boldsymbol{\beta}$ in (10.1.1) with respect to the jth response variate] and $\boldsymbol{E}_n = (\boldsymbol{\varepsilon}_1, \ldots, \boldsymbol{\varepsilon}_n)'$ is an $(n \times p)$ matrix of unobservable random errors, the rows of which are assumed independent and following a (generally) unknown distribution with distribution function F such that $\mathbb{E}(\boldsymbol{\varepsilon}_i) = \mathbf{0}$ and $\mathbb{V}\text{ar}(\boldsymbol{\varepsilon}_i) = \boldsymbol{\Sigma}$, finite and p.d.

10.3 Robust Estimators

The method of least squares and its ramifications are particularly appealing in the linear model setup because, generally, the corresponding estimators are computationally simple (i.e., the allied "estimating equations" have explicit solutions) and are asymptotically normal under quite general regularity conditions (wherein the underlying error distribution function F belongs to a general class of which the normal law is an important member). From the general results presented in Section 10.2 we know that the LSE of $\boldsymbol{\beta}$ in model (10.1.1) is such that $\mathbb{V}\text{ar}(\widehat{\boldsymbol{\beta}}_n) = \sigma^2(\boldsymbol{X}_n'\boldsymbol{X}_n)^{-1}$ where $\sigma^2 = \mathbb{V}\text{ar}(\varepsilon)$, whereas $(\boldsymbol{X}_n'\boldsymbol{X}_n)^{-1}$ depends solely on the specification matrix $\boldsymbol{X}_n$ and is independent of the distribution function F. This may well be a vintage point for looking at the picture from a wider perspective, when in (10.1.1) we do not necessarily assume that F is normal. In fact, all we have assumed is that $\mathbb{E}_F(\varepsilon) = \int x\mathrm{d}F(x) = 0$ and that $\sigma_F^2 = \mathbb{E}(\varepsilon^2) = \int x^2\mathrm{d}F(x)$ is finite and positive. Also note that if we assume that F has a differentiable density function f having its support the entire $\mathbb{R}$,

$$\begin{aligned} 1 = \int f(x)\mathrm{d}x &= [xf(x)] - \int xf'(x)\mathrm{d}x \\ &= \int x[-f'(x)/f(x)]\mathrm{d}F(x) \end{aligned} \tag{10.3.1}$$

so that using the Cauchy–Schwarz Inequality (1.5.34), we have

$$1 \le \left[\int x^2\mathrm{d}F(x)\right]\left[\int\{-f'(x)/f(x)\}^2\mathrm{d}F(x)\right] \quad \text{or} \quad \sigma_F^2 \ge 1/I(f), \tag{10.3.2}$$

where $I(f) = \mathbb{E}_F\{[-f'(\varepsilon)/f(\varepsilon)]^2\}$ is the Fisher information. The equality in (10.3.2) holds when, for some $k \neq 0$,

$$-f'(\varepsilon)/f(\varepsilon) = k\varepsilon, \quad \forall\varepsilon \quad \text{a.e.}, \tag{10.3.3}$$

this is, when f is a normal density function. Thus, in the conventional linear model, where we assume that the errors are normally distributed, $\sigma_F^2 = 1/I(f)$ is a minimum, and, hence, the LSE is optimal. This rosy picture may disappear very fast if F departs from a normal form, even locally. To illustrate this point, we consider the following example.

***Example* 10.3.1.** Suppose that f corresponds to a double exponential distribution, *i.e.*, $f(x) = (1/2)\exp(-|x|)$, $-\infty < x < \infty$. In this case, $-f'(x)/f(x) = \text{sign}(x)$, $x \in \mathbb{R}$, and $I(f) = 1$. On the other hand,

$$\sigma_F^2 = \frac{1}{2}\int_{-\infty}^{+\infty} x^2 e^{-|x|}\mathrm{d}x = \int_0^{\infty} x^2 e^{-x}\mathrm{d}x = 2! = 2.$$

Thus, σ_F^2 is twice as large as $1/I(f)$ and, therefore, the LSE of $\boldsymbol{\beta}$ in model (10.1.1) entails about 50 percent loss of efficiency in this case. □

Example* 10.3.2 *(Error contamination model). Suppose that there exist an α $(0 < \alpha < 1)$ and a positive $k < 1$ such that

$$f(\varepsilon) = (1-\alpha)\varphi(\varepsilon) + \alpha k^{-1}\varphi(\varepsilon/k), \quad \varepsilon \in \mathbb{R},$$

where φ denotes the standard normal density function. According to this model, $100(1-\alpha)\%$ of the observations come from a normal distribution (with zero mean and unit variance), whereas the remaining $100\alpha\%$ correspond to a normal distribution with zero mean but a larger variance k^2; this accounts for the error contamination or presence of gross errors in a plausible manner. Here, we have [on using the fact that $\varphi'(x) = -x\varphi(x)$],

$$\frac{-f'(\varepsilon)}{f(\varepsilon)} = \frac{(1-\alpha)\varepsilon\varphi(\varepsilon) + \alpha k^{-2}\varepsilon\varphi(\varepsilon/k)}{(1-\alpha)\varphi(\varepsilon) + \alpha k^{-1}\varphi(\varepsilon/k)} = \varepsilon\left\{\frac{1-\alpha[\varphi(\varepsilon) - k^{-2}\varphi(\varepsilon/k)]}{1-\alpha[\varphi(\varepsilon) - k^{-1}\varphi(\varepsilon/k)]}\right\}$$

where the term within brackets, [ ], depends very much on ε (when $k \neq 1$), so that (10.3.3) does not hold, and, hence, $\sigma_F^2 > 1/I(f)$ for all $\alpha \in (0,1)$ when $k \neq 1$. The expression for $I(f)$ is not that simple to obtain. □

***Example* 10.3.3.** Suppose that F corresponds to the logistic distribution, that is, $F(x) = [1+\exp(-x)]^{-1}$, $-\infty < x < \infty$ so that $f(x) = \exp(-x)[1+\exp(-x)]^{-2} = F(x)[1-F(x)]$, $x \in \mathbb{R}$. Thus, $-f'(x) = f(x)[2F(x)-1]$, $x \in \mathbb{R}$, so that $I(f) = \int_0^1 (2u-1)^2 \mathrm{d}u = 1/3$. Since the density function f is symmetric about zero, it follows that

$$\sigma_F^2 = \int_{-\infty}^{+\infty} x^2 e^{-x}(1+e^{-x})^{-2}\mathrm{d}x = \pi^2/3 > 3 = 1/I(f).$$

□

If the underlying error distribution in model (10.1.1) corresponds to one of the three in the examples discussed above, the LSE of $\boldsymbol{\beta}$ fails to achieve the lower bound for the variance, even asymptotically. This holds for almost all nonnormal underlying error distribution function F even when they are quite close to a normal distribution function, say Φ_σ, in the sense that $\|F - \Phi_\sigma\|$ is small. Because of this discouraging factor, we say that the LSE is not *robust*.

We may appeal to the results in Chapter 8 and construct MLE/BAN estimators of $\boldsymbol{\beta}$ when the form of F is assumed to be given. For example, under the double-exponential distribution discussed in Example 10.3.1, the MLE of $\boldsymbol{\beta}$ is given by

$$\widehat{\boldsymbol{\beta}}_n^* = \arg\min\left(\sum_{i=1}^n |Y_i - \boldsymbol{x}_{ni}'\boldsymbol{\beta}| : \boldsymbol{\beta} \in \mathbb{R}^q\right). \tag{10.3.4}$$

Although an explicit solution for (10.3.4) is not available (unlike the case of the LSE), we may note that $\sum_{i=1}^n |Y_i - \boldsymbol{x}_{ni}'\boldsymbol{\beta}|$ is a convex function of $\boldsymbol{\beta}$ (for given $\boldsymbol{Y}_n$ and $\boldsymbol{X}_n$), so that, in principle, we may agree to abandon the role of the squared distance (i.e., $\sum_{i=1}^n |Y_i - \boldsymbol{x}_{ni}'\boldsymbol{\beta}|^2$) in favor of the absolute distance (i.e., $\sum_{i=1}^n |Y_i - \boldsymbol{x}_{ni}'\boldsymbol{\beta}|$). The situation is somewhat more complicated for the other two examples, although it may be possible to use the distance function $\sum_{i=1}^n[-\log f(Y_i - \boldsymbol{x}_{ni}'\boldsymbol{\beta})]$. This feature motivates us to consider an arbitrary nonnegative function $\rho : \mathbb{R} \to \mathbb{R}^+$ and define

$$Q_n(\boldsymbol{\beta}) = \sum_{i=1}^n \rho(Y_i - \boldsymbol{x}_{ni}'\boldsymbol{\beta}) \tag{10.3.5}$$

so that minimization of $Q_n(\boldsymbol{\beta})$ with respect to $\boldsymbol{\beta}$ would lead to the desired estimator. The choice of ρ can, of course, be based on the assumed density function f; this ρ, however, may not be robust in the sense that it may generate inefficient estimators of $\boldsymbol{\beta}$ whenever the true underlying distribution does not exactly match the assumed one. The picture is very much similar to that of the LSE discussed above. In addition to this situation, there is the basic fact that there exists no single ρ that would lead to (at least asymptotically) optimal estimators for all F belonging to a certain class. An exception to this feature is the case where ρ is data dependent, that is, chosen *adaptively* so as to yield asymptotically optimal estimators within a broad class. But this *adaptative estimation theory* requires some mathematical analysis that stands at a far higher level than that we wish to contemplate, and, hence, we shall not pursue this topic further.

From a practical point of view, the *robustness* picture can be judged from either a *local* or *global* perspective, depending on the level of confidence one would have on the assumed form of the density function f. If we refer to Example 10.3.2, choosing α close to zero implies that we have high confidence in f in the sense of it being close to a normal density, although we would like to have some protection against the presence of error contamination/gross errors/outliers. This would fit under the denomination of *local robustness*. On the other hand, we may suppose that the density f belongs to a class (e.g., the class of all symmetric, unimodal, absolutely continuous density functions with finite second moments), so that we desire to have an estimator of $\boldsymbol{\beta}$ which performs well for all underlying error distributions in this class. In this setup, we may have a larger class of estimators that are robust in a global sense, and within this class, we may want to choose specific ones which achieve high (asymptotic) efficiency for specific (important) members of the class of distributions under consideration.

Within the class of M-estimators (as defined in Section 2.4) one may easily construct robust estimators in a local sense, whereas rank based (or *R-estimators*) retain robustness to a higher degree of global perspective. We illustrate this for the parameter β in the simple regression model (10.1.2). The LSE of β is given by

$$\widehat{\beta}_{LS} = \sum_{i=1}^{n}(x_i - \overline{x}_n)Y_i \Big/ \sum_{i=1}^{n}(x_i - \overline{x}_n)^2,$$

and the corresponding ρ function is $\rho_{LS}(y) = y^2,\ y \in \mathbb{R}$. As we have seen, $\widehat{\beta}_{LS}$ is not so robust against departures from the assumed normality of the error distribution function F. To clarify this picture, we first rewrite

$$\widehat{\beta}_{LS} = \sum_{1 \le i < j \le n} (x_i - x_j)(Y_i - Y_j) \Big/ \sum_{1 \le i < j \le n} (x_i - x_j)^2.$$

Note that by assumption, $x_1, \ldots, x_n$ are not all equal, but we are not ignoring the possibility of ties among them; thus, without loss of generality, we may set $x_1 \le x_2 \le \cdots \le x_n$ with at least one strict inequality sign. Also, consider the set

$$S_n = \{(i, j) : 1 \le i < j \le n \quad \text{and} \quad x_i \ne x_j\} \tag{10.3.6}$$

and let N be the cardinality of S_n. Then, we may write

$$\begin{aligned}
\widehat{\beta}_{LS} &= \frac{\sum_{S_n}(x_j - x_i)(Y_j - Y_i)}{\sum_{S_n}(x_j - x_i)^2} \\
&= \frac{\sum_{S_n}(x_j - x_i)^2[Y_j - Y_i)/(x_j - x_i)]}{\sum_{S_n}(x_j - x_i)^2} \\
&= \sum_{S_n} w_{nij} Z_{ij},
\end{aligned}$$

where the summation $\sum_{S_n}$ extends over all possible pairs (i, j) belonging to the set S_n, $w_{nij} = (x_j - x_i)^2 / \sum_{S_n}(x_j - x_i)^2$ are nonnegative weights adding up to 1 and the $Z_{ij} = (Y_j - Y_i)/(x_j - x_i)$, $(i, j) \in S_n$, are the "divided differences". Recall that by (10.1.2), for every $(i, j) \in S_n$,

$$Z_{ij} = \beta + (\varepsilon_j - \varepsilon_i)/(x_j - x_i) = \beta + \varepsilon_{ij}^*,$$

where the $\varepsilon_{ij}^* = (\varepsilon_j - \varepsilon_i)/(x_j - x_i)$ have distribution functions symmetric about zero but with variances that may not all be the same. Thus, $\widehat{\beta}_{LS}$ is a weighted average of the basic estimators Z_{ij}, $(i, j) \in S_n$, with weights proportional to the distances $(x_j - x_i)^2$. This explains why outliers/gross errors have considerable influence on $\widehat{\beta}_{LS}$. To enhance robustness, we look at the set $\{Z_{ij}, (i, j) \in S_n\}$, order the Z_{ij}s in ascending order and denote these ordered variables by $Z_{(1)} \le Z_{(2)} \le \cdots \le Z_{(N)}$. Note that the Z_{ij} are not all mutually independent, so that the $Z_{(k)}$ do not relate to the conventional set of independent and identically distributed random variables). Now consider the alternative estimator

$$\begin{aligned}
\widehat{\beta}_{TS} &= \text{median}[Z_{ij}, (i, j) \in S_n] \\
&= \begin{cases} Z_{(M+1)} & \text{if } N = 2M + 1 \\ \frac{1}{2}[Z_{(M)} + Z_{(M+1)}] & \text{if } N = 2M. \end{cases}
\end{aligned} \tag{10.3.7}$$

In the literature, $\widehat{\beta}_{TS}$ is known as the *Theil–Sen estimator* of β [Sen (1968a)]. This is indeed a very robust estimator in a global sense and it is a member of the class of R-estimators of β. To motivate $\widehat{\beta}_{TS}$, let us start with the usual *Kendall tau statistic*

$$T_n = \sum_{1 \le i < j \le n} \text{sign}(Y_j - Y_i)\, \text{sign}(x_j - x_i) = T_n(\boldsymbol{Y}_n, \boldsymbol{x}_n).$$

If we replace $\boldsymbol{Y}_n$ by $\boldsymbol{Y}_n - \boldsymbol{x}_n b$, $b \in \mathbb{R}$, then

$$\begin{aligned} T_n = T_n(b) &= \sum_{1 \le i < j \le n} \text{sign}(Y_j - x_j b - Y_i + x_i b)\, \text{sign}(x_j - x_i) \\ &= \sum_{1 \le i < j \le n} \text{sign}[Y_j - Y_i - b(x_j - x_i)]\, \text{sign}(x_j - x_i) \\ &= \sum_{S_n} \text{sign}\left(\frac{Y_j - Y_i}{x_j - x_i} - b\right) [\text{sign}(x_j - x_i)]^2 \\ &= \sum_{S_n} \text{sign}(Z_{ij} - b), \end{aligned}$$

so that $T_n(b)$ is nonincreasing in $b \in \mathbb{R}$. Moreover, $T_n(\beta)$ has a distribution symmetric about zero. Thus, "equating $T_n(b)$ to zero" yields the estimator $\widehat{\beta}_{TS}$ (as the median of the Z_{ij}). Furthermore, for any $\varepsilon > 0$,

$$\begin{aligned} P_\beta(\widehat{\beta}_{TS} > \beta + \varepsilon) &\le P_\beta[T_n(\beta + \varepsilon) \ge 0] = P_0[T_n(\varepsilon) \ge 0], \\ P_\beta(\widehat{\beta}_{TS} < \beta - \varepsilon) &\le P_\beta[T_n(\beta - \varepsilon) \le 0] = P_0[T_n(-\varepsilon) \le 0]. \end{aligned} \tag{10.3.8}$$

Noting that $\binom{n}{2}^{-1} T_n(\pm\varepsilon)$ are U-statistics, we may use the results in Chapters 6 and 7 to conclude that both of the above two terms converge to zero as $n \to \infty$, so that $\widehat{\beta}_{TS}$ is consistent. Similarly, we may approximate $P_\beta[\sqrt{n}(\widehat{\beta}_{TS} - \beta) \le a]$ by

$$\begin{aligned} &P_0\left[T_n\left(-\frac{a}{\sqrt{n}}\right) \le 0\right] \\ &\quad = P_0\left\{T_n\left(-\frac{a}{\sqrt{n}}\right) - E_0\left[T_n\left(-\frac{a}{\sqrt{n}}\right)\right] \le -E_0\left[T_n\left(-\frac{a}{\sqrt{n}}\right)\right]\right\} \end{aligned} \tag{10.3.9}$$

and derive the asymptotic normality of $\sqrt{n}(\widehat{\beta}_{TS} - \beta)$. We pose most of these results in the form of exercises (see Exercises 10.3.3–10.3.4); see Sen (1968a) for details.

In the same vein, we may consider the Wilcoxon signed rank statistic (see Example 3.4.2)

$$W_n = W(Y_1, \ldots, Y_n) = \sum_{i=1}^{n} \text{sign}(Y_i) R_{ni}^{+} \tag{10.3.10}$$

where $R_{ni}^{+} = R_{ni}^{+}(\boldsymbol{Y}_n)$ is the rank of $|Y_i|$ among $|Y_1|, \ldots, |Y_n|$, $i = 1, \ldots, n$. Now, using the identity

$$\sum_{i=1}^{n} I(Y_i > 0) R_{ni}^{+} = \sum_{1 \le i \le j \le n} I(Y_i + Y_j > 0)$$

we may conclude that

$$\sum_{1\le i\le j\le n} I(Y_i + Y_j > 0) = \sum_{1\le i\le j\le n} I\left(\frac{1}{2}(Y_i + Y_j) > 0\right)$$
$$= \frac{n(n+1)}{4} + \frac{1}{2}W_n. \tag{10.3.11}$$

Then, letting $W_n(a) = W(Y_1 - a, \dots, Y_n - a), a \in \mathbb{R}$, it follows from (10.3.11) that $W_n(a)$ is nonincreasing in a (see Exercise 10.3.5). Also, taking $\beta = 0$ in model (10.1.2) it is clear that $W_n(\alpha)$ has a distribution function symmetric about zero. Thus, for the simple location model, we may equate $W_n(\alpha)$ to zero to obtain a robust estimator of the intercept α. If we write $M_{ij} = \frac{1}{2}(Y_i + Y_j)$, $1 \le i \le j \le n$, and order these M_{ij} as

$$M_{(1)} \le \cdots \le M_{\binom{n+1}{2}},$$

then the estimator turns out to be the median $M_{(n^*)}$, where $n^* = \frac{1}{2}\binom{n+1}{2}$ if $\binom{n+1}{2}$ is odd and $\frac{1}{2}[M_{n^*} + M_{(n^*+1)}]$ if $\binom{n+1}{2}$ is even ($= 2n^*$). Here also W_n is a linear function of two U-statistics (see Exercise 10.3.6), so that the consistency and asymptotic normality of $W_n(a)$ may be used to derive the parallel results for the estimator (see Exercise 10.3.7). When β is unknown, we consider the estimator $\widehat{\beta}_{TS}$ and write $\widehat{Y}_i = Y_i - x_i\beta_{TS}$, $i = 1, \dots, n$; the Wilcoxon signed rank statistic is based on the $\widehat{Y}_i - a$.

Motivated by this simple case, we now consider a general class of rank statistics. Let

$$L_n = L_n(Y_i, \dots, Y_n; \boldsymbol{x}_n) = \sum_{i=1}^{n}(x_i - \overline{x}_n)a_n(R_{ni}),$$

where $a_n(1) \le \cdots \le a_n(n)$ are suitable scores and R_{ni} denotes the rank of Y_i among $Y_1, \dots, Y_n$, $1 \le i \le n$. Then let

$$L_n(b) = L_n(Y_1 - bx_1, \dots, Y_n - bx_n, \boldsymbol{x}_n), \quad b \in \mathbb{R}.$$

Then, it is easy to verify that $L_n(b)$ is nonincreasing in b and $L_n(\beta)$ has mean 0. Thus, we may set $\widehat{\beta}_{n1} = \sup\{b : L_n(b) > 0\}$, $\widehat{\beta}_{n2} = \inf\{b : L_n(b) < 0\}$ and then $\widehat{\beta}_{nR} = (\widehat{\beta}_{n1} + \widehat{\beta}_{n2})/2$, the R-estimator of β, is translation invariant and $\boldsymbol{x}_n$ regression equivariant. The consistency of $\widehat{\beta}_{nR}$ can be established (as in the case of $\widehat{\beta}_{TS}$) through that of $L_n(\beta \pm \varepsilon)$, $\varepsilon > 0$, and, similarly, the asymptotic normality of $\sqrt{n}(\widehat{\beta}_{nR} - \beta)$ can be studied via the asymptotic normality of $n^{-1/2}L_n(\beta + n^{-1/2}a)$. This calls for a study of the asymptotic properties of linear rank statistics, which is somewhat beyond the scope of this treatise. A fair amount of details is provided, however, in the related material of Section 11.6 [see (11.6.1)–(11.6.12)].

We may also extend the definition of L_n to the vector case by taking $\boldsymbol{L}_n = \sum_{i=1}^{n}(\boldsymbol{x}_i - \overline{\boldsymbol{x}}_n)a_n(R_{ni})$ and defining $\boldsymbol{L}_n(b)$ with $\boldsymbol{b} \in \mathbb{R}^q$. Here we may take

$$\widehat{\boldsymbol{\beta}}_{nR} = \arg\min\left(\sum_{j=1}^{q}|L_{nj}(\boldsymbol{b})| : \boldsymbol{b} \in \mathbb{R}^q\right).$$

The asymptotic properties of such estimator requires a more sophisticated treatment; see Puri and Sen (1985, Chapter 6) and Jurečková and Sen (1996, Chapters 4–6) for details.

Finally, we may consider a general signed rank statistic as

$$S_n = \sum_{i=1}^{n} \text{sign}(Y_i) a_n(R_{ni}^+) = S_n(\boldsymbol{Y}_n),$$

where R_{ni}^+ is the rank of $|Y_i|$ among $|Y_1|, \ldots, |Y_n|$, $1 \leq i \leq n$, and as in the case of W_n, we may work with $S_n(\boldsymbol{Y}_n - \boldsymbol{X}_n' \widehat{\boldsymbol{\beta}}_{nR} - a\boldsymbol{1}_n)$, which, when equated to zero, yields an (aligned) R-estimator of α.

Up to this point, the main focus of our discussion in this section was on some asymptotic properties of regression parameters estimators that are robust in some global sense. In many practical situations, however, there is evidence in favor of some particular model and in such cases we would like to use this information to the extent possible, while providing some protection against mild departures from the proposed model, that is, we want to deal with locally robust estimators. In this context, the M-methods introduced in Section 2.4 for the location model may be extended to the more general models under discussion here. We now proceed to examine some of their asymptotic properties.

An M-estimator of the parameter $\boldsymbol{\beta}$ in model (10.1.1) is defined as

$$\overline{\boldsymbol{\beta}}_n = \arg\min \left[\sum_{i=1}^{n} \rho(Y_i - \boldsymbol{x}_{ni}' \boldsymbol{\beta}) : \boldsymbol{\beta} \in \mathbb{R}^q \right], \tag{10.3.12}$$

where ρ is some arbitrary function chosen in accordance with the discussion in Chapters 4 and 6. When $\psi(z) = (\partial/\partial z)\rho(z)$ exists (10.3.12) corresponds to a solution to the estimating equations

$$\boldsymbol{M}_n(\overline{\boldsymbol{\beta}}_n) = \sum_{i=1}^{n} \psi(Y_i - \boldsymbol{x}_{ni}' \overline{\boldsymbol{\beta}}_n) \boldsymbol{x}_{ni} = \boldsymbol{0}, \tag{10.3.13}$$

which is generally obtained via some iterative procedure. We are interested in obtaining the asymptotic distribution of $\sqrt{n}(\overline{\boldsymbol{\beta}}_n - \boldsymbol{\beta})$. Note that even if ψ is monotone, the elements of

$$\boldsymbol{M}_n(\boldsymbol{u}) = \sum_{i=1}^{n} \psi(Y_i - \boldsymbol{x}_{ni}' \boldsymbol{u}) \boldsymbol{x}_{ni}$$

are not monotone with respect to the elements of $\boldsymbol{u}$ and, therefore, we may not apply the same technique used to prove Theorem 8.4.2. Here we have to rely on an important linearity result due to Jurečková (1977) in order to establish the required asymptotic normality of the estimators. A complete rigorous treatment of this topic is beyond the scope of our text, but we present a general outline that serve as a guideline. Consider the linear model (10.1.1) with the assumption that $\sigma^2 < \infty$ replaced by:

(A1) The distribution function F of the error random variables is absolutely continuous with density function f such that f' exists a.e.

(A2) The distribution function F has a finite Fisher information with respect to location, that is,

$$I(\mu) = \int \left[f'(\varepsilon)/f(\varepsilon) \right]^2 f(\varepsilon) \mathrm{d}\varepsilon < \infty.$$

Note that (A2) ensures that $\int |f'(x)| \mathrm{d}x \leq I^{1/2}(\mu) < \infty$.

Assume further that:

(B1) $\max_{1\le k\le n} \boldsymbol{x}'_{nk}(\boldsymbol{X}'_n\boldsymbol{X}_n)^{-1}\boldsymbol{x}_{nk} \to 0$ as $n \to \infty$;
(B2) $\lim_{n\to\infty} n^{-1}(\boldsymbol{X}'_n\boldsymbol{X}_n) = \boldsymbol{V}$, finite and p.d.

Consider the M-estimator defined by (10.3.12) and assume that the score function ψ is a nonconstant function expressible in the form

$$\psi(x) = \sum_{\ell=1}^{s} \psi_\ell(x),$$

where $s \ge 1$ and each ψ_ℓ is monotone and is either an absolutely continuous function on any bounded interval in $\mathbb{R}$ with derivative ψ'_ℓ almost everywhere or is a step function. Assume that it satisfies

(C1) $\int \psi(\varepsilon)f(\varepsilon)\mathrm{d}\varepsilon = 0$,
(C2) $\sigma_\psi^2 = \int \psi^2(\varepsilon)f(\varepsilon)\mathrm{d}\varepsilon < \infty$,
(C3) $\gamma = -\int \psi(\varepsilon)f'(\varepsilon)\mathrm{d}\varepsilon < \infty$.

Note that whenever ψ is absolutely continuous, by partial integration, we have $\gamma = \int \psi'(\varepsilon)f(\varepsilon)\mathrm{d}\varepsilon$.

***Theorem* 10.3.1.** *For the linear model (10.1.1), under (A1), (A2), (B1), (B2), (C1), (C2), and (C3) it follows that*

$$\sqrt{n}(\overline{\boldsymbol{\beta}}_n - \boldsymbol{\beta}) \xrightarrow{D} \mathcal{N}_q\left(\boldsymbol{0}, \gamma^{-2}\sigma_\psi^2\boldsymbol{V}^{-1}\right). \tag{10.3.14}$$

Proof. (Outline) First we have to show that, given $K > 0$ and $\varepsilon > 0$, as $n \to \infty$

$$P\left[\sup_{\sqrt{n}\|\overline{\boldsymbol{\beta}}_n-\boldsymbol{\beta}\|\le K} n^{-1/2}\|\boldsymbol{M}_n(\overline{\boldsymbol{\beta}}_n) - \boldsymbol{M}_n(\boldsymbol{\beta}) + n\gamma\boldsymbol{V}(\overline{\boldsymbol{\beta}}_n - \boldsymbol{\beta})\| > \varepsilon\right] \to 0. \tag{10.3.15}$$

The proof of this linearity result for the standard multivariate linear model is given in Singer and Sen (1985) and is based on a similar result proved by Jurečková (1977) under more restrictive conditions. Next we have to show that given $\eta > 0$ and $K > 0$,

$$P\left(\min_{\sqrt{n}\|\boldsymbol{u}-\boldsymbol{\beta}\|\le K} \frac{1}{\sqrt{n}}\|\boldsymbol{M}_n(\boldsymbol{u})\| > \eta\right) \to 0 \quad \text{as } n \to \infty, \tag{10.3.16}$$

or, in other words, that for every $K > 0$, $\frac{1}{\sqrt{n}}\boldsymbol{M}_n(\boldsymbol{u}) \xrightarrow{P} \boldsymbol{0}$ uniformly in the compact set $\sqrt{n}\|\boldsymbol{u} - \boldsymbol{\beta}\| \le K$, which is equivalent to showing that $\sqrt{n}\|\overline{\boldsymbol{\beta}}_n - \boldsymbol{\beta}\|$ is bounded in probability. See Jurečková (1977) for a proof of (10.3.16).

Now, in view of (10.3.15) and (10.3.16), it follows that $\sqrt{n}\boldsymbol{V}\gamma(\overline{\boldsymbol{\beta}}_n - \boldsymbol{\beta})$ has the same asymptotic distribution as $n^{-1/2}\boldsymbol{M}_n(\boldsymbol{\beta}) = n^{-1/2}\sum_{i=1}^n \boldsymbol{Z}_{ni}$, where $\boldsymbol{Z}_{ni} = \psi(Y_i - \boldsymbol{x}'_{ni}\boldsymbol{\beta})\boldsymbol{x}_{ni}$, $i = 1, \ldots, n$, are independent random variables such that $\mathbb{E}(\boldsymbol{Z}_{ni}) = \boldsymbol{0}$ and $\mathbb{V}\mathrm{ar}(\boldsymbol{Z}_{ni}) = \sigma_\psi^2\boldsymbol{x}_{ni}\boldsymbol{x}'_{ni}$. Using assumptions (B1) and (B2) in connection with Cramér–Wold Theorem 7.2.4 and Hájek–Šidák Theorem 7.3.6 we may conclude that $n^{-1/2}\boldsymbol{M}_n(\boldsymbol{\beta}) \xrightarrow{D} \mathcal{N}_q(\boldsymbol{0}, \boldsymbol{V}\sigma_\psi^2)$, and (10.3.14) follows. ■

Since M-estimators are not scale invariant, the above result is of little practical utility; however, it is easily extended to cover situations where (10.3.12) is replaced by

$$\boldsymbol{M}_n(\overline{\boldsymbol{\beta}}_n) = \sum_{i=1}^{n} \psi\left(\frac{Y_i - \boldsymbol{x}'_{ni}\overline{\boldsymbol{\beta}}_n}{\widehat{\sigma}_n}\right)\boldsymbol{x}_{ni} = \mathbf{0}, \tag{10.3.17}$$

where $\widehat{\sigma}_n$ is a $\sqrt{n}$-consistent estimate of the scale parameter σ associated to the distribution of the error random variables provided that F has a finite Fisher information with respect to scale, that is,

$$I(\sigma) = -1 + \int \varepsilon^2 \left[\frac{f'(\varepsilon)}{f(\varepsilon)}\right]^2 f(\varepsilon)\mathrm{d}\varepsilon < \infty.$$

The reader is referred to Singer and Sen (1985) for a proof as well as for extensions of the result to multivariate linear models.

10.4 Nonlinear Regression Models

Before proceeding to introduce *generalized linear models* (GLM), let us present an outline of nonlinear regression models and exhibit how the generalized least squares methodology and robust estimation procedures extend themselves in this setup. We conceive of response variables $Y_1, \ldots, Y_n$ (independent) with associated explanatory or covariates $\boldsymbol{x}_1, \ldots, \boldsymbol{x}_n$ (possibly vectors). Consider then the regression model

$$Y_i = g(\boldsymbol{x}_i, \boldsymbol{\beta}) + e_i, \quad 1 \le i \le n, \tag{10.4.1}$$

where the e_i are independent and identically distributed random variables with 0 mean and a finite positive variance σ^2. Then, we have the usual norm

$$Q_n(\boldsymbol{\beta}) = \sum_{i=1}^{n} [Y_i - g(\boldsymbol{x}, \boldsymbol{\beta})]^2, \tag{10.4.2}$$

and we proceed to minimize $Q_n(\boldsymbol{\beta})$ with respect to $\boldsymbol{\beta}$, yielding the estimating equation

$$\sum_{i=1}^{n} [Y_i - g(\boldsymbol{x}_i, \boldsymbol{\beta})] \frac{\partial}{\partial \boldsymbol{\beta}} g(\boldsymbol{x}_i, \boldsymbol{\beta}) = \mathbf{0}. \tag{10.4.3}$$

In general, (10.4.3) need to be solved by some iterative procedure. A preliminary estimator $\widetilde{\boldsymbol{\beta}}_n$, possible based on the method of moments, provides an iterative procedure similar to (8.2.1).

Instead of $Q_n(\boldsymbol{\beta})$, we may use some other objective functions, such as $\sum_{i=1}^{n} |Y_i - g(\boldsymbol{x}_i, \boldsymbol{\beta})|$ or Huber-type functions. The results of the previous section go over this situation (the asymptotic linearity result being the main tool). See Gallant (1987) for details. We pose Exercises 10.4.1–10.4.3.

A more general case may evolve when the e_i are independent but not necessarily identically distributed, or the variance of e_i may depend on $(\boldsymbol{x}_i, \boldsymbol{\beta})$. This will introduce additional complications, and will be outlined in the context of generalized estimating equation in the next section.

10.5 Generalized Linear Models

The terminology *generalized linear models*, due to Nelder and Wedderburn (1972), has become a household word in the statistical community. Their systematic study showed how linearity in common statistical models could be exploited to extend classical linear models (as treated in earlier sections) to more general statistical models, cropping up in diverse applications, with a view to unifying the apparently diverse statistical inference techniques commonly employed in such situations. Generalized linear models include, as special cases, linear regression and analysis of variance models, log-linear models for categorical data, product multinomial response models, logit and probit models in biological assays (quantal responses), and some simple statistical models arising in survival analysis. An excellent treatise of this subject matter is due McCullagh and Nelder (1989), where a variety of useful numerical illustrations has also been considered along with the relevant methodology. Since most of the inferential results based on such models are valid for large samples, there is a need to look into the allied asymptotic theory so that applications can be governed by a sound methodological basis. At a level consistent with the rest of this book, we plan to provide here the desired asymptotics.

We start by introducing some notation and assumptions underlying the definition of generalized linear models. Consider a vector of observations $\boldsymbol{y}_N = (y_1, \ldots, y_N)'$ corresponding to N independent random variables $Y_i, i = 1, \ldots, N$, where Y_i has a distribution in the *exponential family*, that is, with density function

$$f(y_i, \theta_i, \phi) = c(y_i, \phi)\exp\{[y_i\theta_i - b(\theta_i)]/a(\phi)\} \tag{10.5.1}$$

for $i = 1, \ldots, N$, so that the corresponding joint density is

$$\begin{aligned}\prod_{i=1}^{N} & c(y_i, \phi)\exp\{[y_i\theta_i - b(\theta_i)]/a(\phi)\} \\ &= c_N(\boldsymbol{y}_N, \phi)\exp\left\{\sum_{i=1}^{N}[y_i\theta_i - b(\theta_i)]/a(\phi)\right\}. \end{aligned} \tag{10.5.2}$$

Here $\theta_i, i = 1, \ldots, N$, are the parameters of interest, ϕ (> 0) is a scale (nuisance) parameter and a, b, and c are functions of known form. Note that by (10.5.1), for each i,

$$\int c(y, \phi)\exp[y\theta_i/a(\phi)]\mathrm{d}\nu(y) = \exp[b(\theta_i)/a(\phi)], \tag{10.5.3}$$

so that differentiating both sides with respect to θ, we have

$$\int c(y, \phi)\left[\frac{y}{a(\phi)}\right]\exp\left[\frac{y\theta_i}{a(\phi)}\right]\mathrm{d}\nu(y) = \frac{b'(\theta_i)}{a(\phi)}\exp\left[\frac{b(\theta_i)}{a(\phi)}\right]$$

or, equivalently

$$\mathbb{E}(Y_i)/a(\phi) = b'(\theta_i)/a(\phi), \quad i = 1, \ldots, N, \tag{10.5.4}$$

where $b'(\theta) = (\partial/\partial\theta)b(\theta)$. Differentiating one more time with respect to θ_i we have

$$\int c(y,\phi)\left[\frac{y}{a(\phi)}\right]^2 \exp\left[\frac{y\theta_i}{a(\phi)}\right] d\nu(y) = \left\{\frac{b''(\theta_i)}{a(\phi)} + \left[\frac{b'(\theta_i)}{a(\phi)}\right]^2\right\} \exp\left[\frac{b(\theta_i)}{a(\phi)}\right],$$

which yields $\mathbb{E}(Y_i^2) = b''(\theta_i)a(\phi) + [\mathbb{E}(Y_i)]^2$, so that

$$\mathbb{E}(Y_i) = \mu_i = \mu_i(\theta_i) = b'(\theta_i), \quad \mathbb{V}\text{ar}(Y_i) = b''(\theta_i)a(\phi). \tag{10.5.5}$$

We may also write

$$\mathbb{V}\text{ar}(Y_i) = a(\phi)\frac{\partial}{\partial\theta_i}\mu_i(\theta_i) = a(\phi)v_i[\mu_i(\theta_i)], \tag{10.5.6}$$

where $v_i[\mu_i(\theta_i)]$ is known as the *variance function* and it depends solely on $\mu_i(\theta_i)$, $1 \le i \le N$. In the particular case of a normal density, $\mu_i = \theta_i$ so that $b'(\theta_i) \equiv \theta_i$ and, hence, $b''(\theta_i) = 1$, that is, $v_i[\mu_i(\theta_i)] = 1$ implying $\mathbb{V}\text{ar}(Y_i) = a(\phi)$. This explains why $a(\phi)$ is referred to as the *scale factor*. Further, since the form of $b(\theta_i)$ is assumed to be the same for every i (they differ only in the θ_i), $v_i(y) = v(y)$, for all i, so that we have a common variance function $v[\mu(\theta_i)]$, $i = 1, \ldots, N$. The $\mu_i(\theta_i)$ play a basic role in this formulation; in the usual case of linear models, $\mu_i(\theta_i) = \theta_i$, $i = 1, \ldots, N$, are expressible linearly in terms of a finite parameter vector $\boldsymbol{\beta} = (\beta_1, \ldots, \beta_q)'$, say, and of a given $(N \times q)$ specification matrix $\boldsymbol{X}_N$, so that $\boldsymbol{\theta}_N = (\theta_1, \ldots, \theta_N)' = \boldsymbol{X}_N\boldsymbol{\beta}$. Moreover, by (10.5.5), $\mu_i(\theta_i) = b'(\theta_i)$. This suggests that it may be possible to conceive of a transformation

$$g[\mu(\theta_i)] = g(\mu_i), \quad i = 1, \ldots, N, \tag{10.5.7}$$

where $g(t)$ is a monotone and differentiable function of t, such that

$$\boldsymbol{G}_N(\boldsymbol{\mu}_N) = [g(\mu_1), \ldots, g(\mu_N)]' = \boldsymbol{X}_N\boldsymbol{\beta} \tag{10.5.8}$$

with $\boldsymbol{\mu} = (\mu_1, \ldots, \mu_N)'$, $\boldsymbol{X}_N = (\boldsymbol{x}_{N1}, \ldots, \boldsymbol{x}_{NN})'$ denoting a known $(N \times q)$ matrix of known constants and $\boldsymbol{\beta} = (\beta_1, \ldots, \beta_q)'$ is a q-vector of unknown parameters. Thus,

$$g[\mu(\theta_i)] = \boldsymbol{x}'_{Ni}\boldsymbol{\beta}, \quad i = 1, \ldots, N, \tag{10.5.9}$$

and, hence, g provides the link between the mean μ_i and the linear predictor parameter $\boldsymbol{x}'_{Ni}\boldsymbol{\beta}$, $i = 1, \ldots, N$. That is why g is called a *link function* [e.g., Nelder and Wedderburn (1972)]. In view of (10.5.9), we have

$$\theta_i = (g \circ \mu)^{-1}(\boldsymbol{x}'_{Ni}\boldsymbol{\beta}), \quad i = 1, \ldots, N, \tag{10.5.10}$$

so that when $g = \mu^{-1}$, $g \circ \mu$ is the identity function, and, hence, $\theta_i = \boldsymbol{x}'_{Ni}\boldsymbol{\beta}$, in such a case, g is called a *canonical link function*.

***Example* 10.5.1.** Let Z_i, $i = 1, 2$, be independent random variables with the $\text{Bin}(n, \pi_i)$ distribution and take $Y_i = Z_i/n$, $i = 1, 2$. Then Y_i, $i = 1, 2$, has probability function

$$f(y; n, \pi_i) = \binom{n}{ny}\pi_i^{ny}(1 - \pi_i)^{n-ny}, \quad y = 0, 1/n, 2/n, \ldots, 1.$$

This may be re-expressed as

$$\begin{aligned} f(y; n, \pi_i) &= \binom{n}{ny} \exp[ny \log \pi_i + (n - ny)\log(1 - \pi_i)] \\ &= \binom{n}{ny} \exp\left\{\left[y \log \frac{\pi_i}{1 - \pi_i} - \log(1 - \pi_i)\right] / n^{-1}\right\} \\ &= c(y, \phi) \exp\{[y\theta_i - b(\theta_i)]/a(\phi)\}, \end{aligned}$$

where $\theta_i = \log[\pi_i/(1 - \pi_i)]$, $b(\theta_i) = \log[1 + \exp(\theta_i)]$, $\phi = n^{-1}$, $c(y, \phi) = \binom{n}{ny}$ and $a(\phi) = n^{-1}$. Two generalized linear models for this setup are given by

(i) $\theta_i = \log[\pi_i/(1 - \pi_i)] = \alpha + \beta x_i, i = 1, 2$, with $x_1 = 1, x_2 = -1$ and α and β denoting two unknown scalars, if we choose the canonical link;

(ii) $\pi_i = \exp(\theta_i)/[1 + \exp(\theta_i)] = \alpha + \beta x_i, i = 1, 2$, with x_1, x_2, α and β as above, if we consider the link function $g(z) = z$.

Either model would serve the purpose of verifying the homogeneity of the two distributions, which is implied by $\beta = 0$. □

***Example* 10.5.2.** Let Y_i, $i = 1, \ldots, N$, be independent random variables with the Poisson(λ_i) distribution, that is, with probability function

$$f(y; \lambda_i) = e^{-\lambda_i} \lambda_i^y / y!, \quad y = 0, 1, 2 \ldots,$$

which may be rewritten as

$$f(y; \lambda_i) = (y!)^{-1} \exp(y \log \lambda_i - \lambda_i) = c(y, \phi) \exp[y\theta_i - b(\theta_i)],$$

where $\theta_i = \log(\lambda_i)$, $b(\theta_i) = \exp \theta_i$, $\phi = 1$, $c(y, \phi) = (y!)^{-1}$ and $a(\phi) = 1$. Assume further that to each Y_i we associate a vector $\boldsymbol{x}_i$ corresponding to the observed values of q explanatory variables. Two generalized linear models for the association between the response and explanatory variables may be expressed as

(i) $\theta_i = \log \lambda_i = \boldsymbol{x}_i'\boldsymbol{\beta}$, $i = 1, \ldots, N$, where $\boldsymbol{\beta}$ is a vector of unknown constants, if we choose the canonical link; or

(ii) $\lambda_i = \exp \theta_i = \boldsymbol{x}_i'\boldsymbol{\beta}$, $i = 1, \ldots, N$, with $\boldsymbol{\beta}$ as above, if we consider the link function $g(z) = z$.

The model in (i) is a log-linear model for Poisson counts whereas that in (ii) is a linear model similar in spirit to the usual (normal theory) regression models. □

***Example* 10.5.3.** Let Y_i, $i = 1, \ldots, N$, be independent random variables following Bernoulli(π_i) distributions, which may be expressed as

$$f(y; \pi_i) = c(y, \phi) \exp[y\theta_i - b(\theta_i)],$$

where $\theta_i = \log[\pi_i/(1 - \pi_i)]$, $b(\theta_i) = \log[1 + \exp(\theta_i)]$, $\phi = 1$, $c(y, \phi) = 1$ and $a(\phi) = 1$. Also, let $\boldsymbol{x}_i$ be as in Example 10.5.2. A generalized linear model commonly used to describe the relation between the Y_is and $\boldsymbol{x}_i$s is the logistic regression model

$$\theta_i = \log\left(\frac{\pi_i}{1 - \pi_i}\right) = \boldsymbol{x}_i'\boldsymbol{\beta}, \quad i = 1, \ldots, N, \tag{10.5.11}$$

which may also be expressed as

$$\pi_i = \pi(\boldsymbol{x}_i) = [1 + \exp(-\boldsymbol{x}_i'\boldsymbol{\beta})]^{-1}, \quad i = 1, \ldots, N.$$

□

Although we have defined generalized linear models in terms of the two-parameter exponential family of distributions (10.5.1), it is possible to consider extensions to the multiparameter case. Typical examples are the log-linear models for product multinomial data discussed in Chapter 9 and Cox's proportional hazards models for survival data as discussed in Cox and Oakes (1984), for example.

In what follows, we define $h = (g \circ \mu)^{-1}$, so that (10.5.10) reduces to

$$\theta_i = h(\boldsymbol{x}'_{Ni}\boldsymbol{\beta}), \quad i = 1, \ldots, N, \tag{10.5.12}$$

where h is monotone and differentiable. In view of (10.5.9) and (10.5.12), given the explanatory vectors $\boldsymbol{x}_{Ni}, i = 1, \ldots, N$, on the same spirit as in classical linear models, here, also, our main interest centers around the parameter vector $\boldsymbol{\beta}$. Looking back at (10.5.1) and (10.5.2), we may note that the nuisance parameter ϕ does not affect the estimating equations leading to the MLE of $\boldsymbol{\beta}$ and that it influences the Fisher information matrix, say, $\boldsymbol{I}(\boldsymbol{\beta})$ by a multiplicative factor, $[a(\phi)]^{-2}$, which may be estimated consistently; thus, for the sake of simplicity and without loss of generality, we take $a(\phi) \equiv 1$, so that the log-likelihood function (in terms of $\boldsymbol{\beta}$) corresponding to (10.5.2) and (10.5.12) is given by

$$\log L_N(\boldsymbol{\beta}) = \sum_{i=1}^{N} \left\{ Y_i h(\boldsymbol{x}'_{Ni}\boldsymbol{\beta}) - b\left[h(\boldsymbol{x}'_{Ni}\boldsymbol{\beta})\right]\right\} - K, \tag{10.5.13}$$

where K is a constant not depending on $\boldsymbol{\beta}$. Also let

$$\mu_i(\boldsymbol{\beta}) = \mu[h(\boldsymbol{x}'_{Ni}\boldsymbol{\beta})] = b'[h(\boldsymbol{x}'_{Ni}\boldsymbol{\beta})] \tag{10.5.14}$$

and

$$v_i(\boldsymbol{\beta}) = v_i[h(\boldsymbol{x}'_{Ni}\boldsymbol{\beta})] = b''[h(\boldsymbol{x}'_{Ni}\boldsymbol{\beta})] \tag{10.5.15}$$

and note that, on letting $y = h(x)$, that is, $x = h^{-1}(y) = (g \circ \mu)(y)$, it follows that

$$\begin{aligned}(\mathrm{d}/\mathrm{d}x)h(x) &= (\mathrm{d}/\mathrm{d}x)(g \circ \mu)^{-1}(x) = [(\mathrm{d}/\mathrm{d}y)(g \circ \mu)(y)]^{-1} \\ &= \{g'[\mu(y)\}\mu'(y)]\}^{-1} = [g'\{\mu[h(x)]\}\mu'(y)]^{-1} \\ &= [g'\{\mu[h(x)]\}b''\{h(x)\}]^{-1}\end{aligned} \tag{10.5.16}$$

so that

$$\begin{aligned}(\partial/\partial\boldsymbol{\beta})h(\boldsymbol{x}'_{Ni}\boldsymbol{\beta}) &= [g'\{b'[h(\boldsymbol{x}'_{Ni}\boldsymbol{\beta})]\}b''\{h(\boldsymbol{x}'_{Ni}\boldsymbol{\beta})\}]^{-1}\boldsymbol{x}_{Ni} \\ &= [g'\{\mu_i(\boldsymbol{\beta})\}v_i(\boldsymbol{\beta})]^{-1}\boldsymbol{x}_{Ni}\end{aligned} \tag{10.5.17}$$

and

$$\begin{aligned}(\partial/\partial\boldsymbol{\beta})b[h(\boldsymbol{x}'_{Ni}\boldsymbol{\beta})] &= b'[h(\boldsymbol{x}'_{Ni}\boldsymbol{\beta})](\partial/\partial\boldsymbol{\beta})h(\boldsymbol{x}'_{Ni}\boldsymbol{\beta}) \\ &= \mu_i(\boldsymbol{\beta})\{g'[\mu_i(\boldsymbol{\beta})]v_i(\boldsymbol{\beta})\}^{-1}\boldsymbol{x}_{Ni}.\end{aligned} \tag{10.5.18}$$

Therefore, from (10.5.13)–(10.5.18), the score statistic is given by:

$$\begin{aligned}\boldsymbol{U}_N(\boldsymbol{\beta}) &= (\partial/\partial\boldsymbol{\beta})\log L_N(\boldsymbol{\beta}) \\ &= \sum_{i=1}^{N}[Y_i\{g'[\mu_i(\boldsymbol{\beta})]v_i(\boldsymbol{\beta})\}^{-1}\boldsymbol{x}_{Ni} - \mu_i(\boldsymbol{\beta})[g'\{\mu_i(\boldsymbol{\beta})\}v_i(\boldsymbol{\beta})]^{-1}\boldsymbol{x}_{Ni}] \\ &= \sum_{i=1}^{N}\frac{\{Y_i - \mu_i(\boldsymbol{\beta})\}}{[g'\{\mu_i(\boldsymbol{\beta})\}v_i(\boldsymbol{\beta})]}\boldsymbol{x}_{Ni}. \end{aligned} \qquad (10.5.19)$$

Note that if g is a canonical link function, $h(x) = x$, so that, in (10.5.16), $(\mathrm{d}/\mathrm{d}x)h(x) = 1$, in (10.5.17), $(\partial/\partial\boldsymbol{\beta})h(\boldsymbol{x}'_{Ni}\boldsymbol{\beta}) = \boldsymbol{x}_{Ni}$ and also in (10.5.18), $(\partial/\partial\boldsymbol{\beta})b[h(\boldsymbol{x}'_{Ni}\boldsymbol{\beta})] = b'(\boldsymbol{x}'_{Ni}\boldsymbol{\beta})\boldsymbol{x}_{Ni}$ implying that the denominators of the individual terms on the right hand side of (10.5.19) are all equal to 1. Thus, the *estimating equations* (for the MLE $\widehat{\boldsymbol{\beta}}_N$) reduce to

$$\sum_{i=1}^{N}[Y_i - b'(\boldsymbol{x}'_{Ni}\boldsymbol{\beta})]\boldsymbol{x}_{Ni} = \mathbf{0}. \qquad (10.5.20)$$

As a check, the reader is invited to verify that for the classical linear model in (10.1.1) with the ε_i normally distributed, the link function is canonical and, further, that $b'(y) = y$, implying that (10.5.20) leads to the MLE (or LSE)

$$\widehat{\boldsymbol{\beta}}_N = \left(\sum_{i=1}^{N}\boldsymbol{x}_{Ni}\boldsymbol{x}'_{Ni}\right)^{-1}\left(\sum_{i=1}^{N}\boldsymbol{x}_{Ni}Y_i\right).$$

In the general case of h being not necessarily an identity function, by virtue of the assumed monotonicity and differentiability of g, $g'[\mu_i(\boldsymbol{\beta})]v_i(\boldsymbol{\beta})$ is a nonnegative function of the unknown $\boldsymbol{\beta}$ and the given $\boldsymbol{x}_{Ni}$, so that (10.5.19) may be interpreted as an extension of the corresponding estimating equations in the weighted least-squares method [see (10.1.5)], where the weights, besides being nonnegative, may, in general, depend on the unknown parameter $\boldsymbol{\beta}$. For this reason, we call

$$\sum_{i=1}^{N}[g'\{\mu_i(\widehat{\boldsymbol{\beta}}_N)\}v_i(\widehat{\boldsymbol{\beta}}_N)]^{-1}[Y_i - \mu_i(\widehat{\boldsymbol{\beta}}_N)]\boldsymbol{x}_{Ni} = \mathbf{0} \qquad (10.5.21)$$

a *generalized estimating equation* (GEE). If we let

$$\boldsymbol{r}_N(\boldsymbol{\beta}) = [Y_1 - \mu_1(\boldsymbol{\beta}), \ldots, Y_N - \mu_N(\boldsymbol{\beta})]' \qquad (10.5.22)$$

and

$$\boldsymbol{D}_N(\boldsymbol{\beta}) = \mathrm{diag}\big[g'\{\mu_1(\boldsymbol{\beta})\}v_1(\boldsymbol{\beta}), \ldots, g'\{\mu_N(\boldsymbol{\beta})\}v_N(\boldsymbol{\beta})\big],$$

then the GEE in (10.5.21) may also be written as

$$\boldsymbol{X}'_N\boldsymbol{D}_N^{-1}(\boldsymbol{\beta})\boldsymbol{r}_N(\boldsymbol{\beta})|_{\boldsymbol{\beta}=\widehat{\boldsymbol{\beta}}_N} = \mathbf{0}. \qquad (10.5.23)$$

Drawing a parallel between the general asymptotic results on the MLE in Section 8.3 and the GLSE in Section 10.2, we study here the desired large sample properties of $\widehat{\boldsymbol{\beta}}_N$, the MLE for the generalized linear model in (10.5.8). In this context, we like to clarify the role of N in the asymptotics to follow. Compared to the linear model (10.1.1), usually here also we assume that in the asymptotic case, $N \to \infty$. However, there may be a situation where for each i, the Y_i may be a statistic based on a subsample of size n_i, $i \geq 1$ (see Example 10.5.1). In such a case, we might consider a second type of asymptotics where it is not crucial to have N large, provided the n_is are themselves large. For simplicity of presentation we take $N = 1$, and recalling (10.5.1), we consider a sequence $\{f_{(n)}\}$ of densities of the form

$$\begin{aligned} f_{(n)}(y;\theta,\phi) &= f(y;\theta_{(n)},\phi_{(n)}) \\ &= c(y,\phi_{(n)})\exp\{[y\theta_{(n)} - b_{(n)}(\theta_{(n)})]/a_{(n)}(\phi_{(n)})\}, \end{aligned} \tag{10.5.24}$$

for $n \geq 1$ and observe that in many situations, as $n \to \infty$

$$\frac{Y - b'_{(n)}(\theta_{(n)})}{[b''_{(n)}(\theta_{(n)})a_{(n)}(\phi_{(n)})]^{1/2}} \xrightarrow{D} \mathcal{N}(0, 1). \tag{10.5.25}$$

This is the case with each Y_i in Example 10.5.1, where $b'_{(n)}(\theta_{(n)}) = \pi_i$, $b''_{(n)}(\theta_{(n)}) = \pi_i(1 - \pi_i)$ and $a_{(n)}(\phi_{(n)}) = n^{-1}$. In Example 10.5.2 we may have a similar picture for each Y_i, if we observe that $b'_{(n)}(\theta_{(n)}) = b''_{(n)}(\theta_{(n)}) = \lambda_i$, $a_{(n)}(\phi_{(n)}) = 1$ and interpret $n \to \infty$ as $\lambda_i = \lambda_i(n) \to \infty$ as in the well-known approximation of the binomial distribution by the Poisson distribution. Since this second type of asymptotics may be examined directly by the methods described in Chapter 7, we will not elaborate on them further and intend to concentrate on the regular case, where $N \to \infty$.

Note that the "weight matrix" $\boldsymbol{D}_N(\boldsymbol{\beta})$ in (10.5.23) may depend on the parameter vector $\boldsymbol{\beta}$, as opposed to the GLSE case where $\boldsymbol{G}_n$ was independent of $\boldsymbol{\beta}$. This introduces additional complications, and, hence, further regularity conditions may be needed to formulate the general asymptotics. It follows from (10.5.19) that whenever g' and $v \equiv b'$ are both differentiable,

$$\begin{aligned} -\frac{\partial^2}{\partial\boldsymbol{\beta}\partial\boldsymbol{\beta}'}\log L_N(\boldsymbol{\beta}) &= \sum_{i=1}^N \frac{b''[h(\boldsymbol{x}'_{Ni}\boldsymbol{\beta})]}{[g'\{\mu_i(\boldsymbol{\beta})\}v_i(\boldsymbol{\beta})]^2}\boldsymbol{x}_{Ni}\boldsymbol{x}'_{Ni} + \sum_{i=1}^N \frac{[Y_i - \mu_i(\boldsymbol{\beta})]}{\{g'[\mu_i(\boldsymbol{\beta})]v_i(\boldsymbol{\beta})\}^2} \\ &\quad \times \left\{[v_i(\boldsymbol{\beta})]^2 g''[\mu_i(\boldsymbol{\beta})] + \frac{g'[\mu_i(\boldsymbol{\beta})]b'''[h(\boldsymbol{x}'_{Ni}\boldsymbol{\beta})]}{g'[\mu_i(\boldsymbol{\beta})]v_i(\boldsymbol{\beta})}\right\}\boldsymbol{x}_{Ni}\boldsymbol{x}'_{Ni} \\ &= \sum_{i=1}^N \{g'[\mu_i(\boldsymbol{\beta})]\}^{-2}[v_i(\boldsymbol{\beta})]^{-1}\boldsymbol{x}_{Ni}\boldsymbol{x}'_{Ni} + \sum_{i=1}^N [Y_i - \mu_i(\boldsymbol{\beta})] \\ &\quad \times \left\{\frac{g''[\mu_i(\boldsymbol{\beta})]}{[g'\{\mu_i(\boldsymbol{\beta})\}]^2} + \frac{b'''[h(\boldsymbol{x}'_{Ni}\boldsymbol{\beta})]}{[g'\{\mu_i(\boldsymbol{\beta})\}]^2[v_i(\boldsymbol{\beta})]^3}\right\}\boldsymbol{x}_{Ni}\boldsymbol{x}'_{Ni}. \end{aligned} \tag{10.5.26}$$

Since $\mathbb{E}(Y_i) = \mu_i(\boldsymbol{\beta})$ for all $i \geq 1$, we obtain from (10.5.26) that

$$\begin{aligned} \boldsymbol{I}_N(\boldsymbol{\beta}) &= \mathbb{E}_{\boldsymbol{\beta}}\left[-\frac{\partial^2}{\partial\boldsymbol{\beta}\partial\boldsymbol{\beta}'}\log L_N(\boldsymbol{\beta})\right] \\ &= \sum_{i=1}^N \left\{[g'\{\mu_i(\boldsymbol{\beta})\}]^{-2}v_i(\boldsymbol{\beta})^{-1}\right\}\boldsymbol{x}_{Ni}\boldsymbol{x}'_{Ni}. \end{aligned} \tag{10.5.27}$$

Thus, we may re-express (10.5.26) as

$$-\frac{\partial^2}{\partial\boldsymbol{\beta}\partial\boldsymbol{\beta}'}\log L_N(\boldsymbol{\beta}) = \boldsymbol{I}_N(\boldsymbol{\beta}) + \boldsymbol{r}_N(\boldsymbol{\beta}), \tag{10.5.28}$$

where

$$\boldsymbol{r}_N(\boldsymbol{\beta}) = \sum_{i=1}^{N}[Y_i - \mu_i(\boldsymbol{\beta})]\left\{\frac{g''\{\mu_i(\boldsymbol{\beta})\}}{[g'\{\mu_i(\boldsymbol{\beta})\}]^2} + \frac{b'''\{h(\boldsymbol{x}'_{Ni}\boldsymbol{\beta})\}}{[g'\{\mu_i(\boldsymbol{\beta})\}]^2[v_i(\boldsymbol{\beta})]^3}\right\}\boldsymbol{x}_{Ni}\boldsymbol{x}'_{Ni}. \qquad (10.5.29)$$

Note that for canonical link functions, $\boldsymbol{r}_N(\boldsymbol{\beta}) = \mathbf{0}$.

Now, in order to obtain the asymptotic distribution of the estimator of the parameter vector $\boldsymbol{\beta}$ under the setup outlined above, we first discuss the required assumptions, comparing them to those employed previously in similar contexts. First, along the same lines considered in the case of the GLSE in Section 10.2 we require that, when divided by N, the left-hand side of (10.5.28) satisfies at least a weak consistency property. In this direction we must first assume that

$$\lim_{N\to\infty}\frac{1}{N}\boldsymbol{I}_N(\boldsymbol{\beta}) = \boldsymbol{I}(\boldsymbol{\beta}), \quad \text{finite and p.d.} \qquad (10.5.30)$$

Also, from (10.5.29), we see that $\boldsymbol{r}_N(\boldsymbol{\beta})$ is a sum of independent centered random variables with finite variances $v_i[\mu_i(\boldsymbol{\beta})]$ and nonstochastic matrix coefficients,

$$\boldsymbol{G}_i = \left\{\frac{g''[\mu_i(\boldsymbol{\beta})]}{(g'[\mu_i(\boldsymbol{\beta})])^2} + \frac{b'''[h(\boldsymbol{x}'_{Ni}(\boldsymbol{\beta})]}{(g'[\mu_i(\boldsymbol{\beta})])^2[v_i(\boldsymbol{\beta})]^3}\right\}\boldsymbol{x}_{Ni}\boldsymbol{x}'_{Ni}.$$

Thus, if we consider the assumption

$$\lim_{N\to\infty} N^{-2}\sum_{i=1}^{N} v_i[\mu_i(\boldsymbol{\beta})]\mathrm{tr}(\boldsymbol{G}_i\boldsymbol{G}'_i) = 0, \qquad (10.5.31)$$

a direct application of Chebyshev Inequality (6.2.27) to (10.5.29) will enable us to show that $N^{-1}\boldsymbol{r}_N(\boldsymbol{\beta}) \xrightarrow{P} \mathbf{0}$ and, thus, that the required consistency holds when the model is true.

Recalling (10.5.27), observe that $\boldsymbol{I}_N(\boldsymbol{\beta})$ depends on g and v as well as on the $\boldsymbol{x}_{Ni}$; thus, it is natural to expect some further assumptions related to the link function and to the variance function in addition to those on the explanatory vectors in order to establish the regularity conditions under which the asymptotic properties of the estimators will follow. In this direction we recall that in Theorem 8.3.1, dealing with the large sample properties of MLEs, we have used a compactness condition on the second derivative of the log-likelihood function, which we extend to the generalized linear model setup under consideration. This condition essentially rests on the uniform continuity of $\{g'[\mu_i(\boldsymbol{\beta}^*)]\}^{-2}$, $[v_i(\boldsymbol{\beta}^*)]^{-3}$, $g''[\mu_i(\boldsymbol{\beta}^*)]$ and $b'''[h(\boldsymbol{x}'_{Ni}\boldsymbol{\beta}^*)]$ in an infinitesimal neighborhood of the true $\boldsymbol{\beta}$, that is, in the set

$$B(\delta) = \{\boldsymbol{\beta}^* \in \mathbb{R}^q : \|\boldsymbol{\beta}^* - \boldsymbol{\beta}\| < \delta\}, \quad \delta \downarrow 0. \qquad (10.5.32)$$

Essentially, if we let

$$w_{1i}(\boldsymbol{\beta}) = \{g'[\mu_i(\boldsymbol{\beta})]\}^{-2}[v_i(\boldsymbol{\beta})]^{-1}$$

and

$$w_{2i}(\boldsymbol{\beta}) = \mu_i(\boldsymbol{\beta})\left\{g''[\mu_i(\boldsymbol{\beta})]\{g'[\mu_i(\boldsymbol{\beta})]\}^{-2} + b'''[h(\boldsymbol{x}'_{Ni}\boldsymbol{\beta})]\{g'[\mu_i(\boldsymbol{\beta})]\}[v_i(\boldsymbol{\beta})]^{-3}\right\},$$

the required compactness condition may be expressed as

$$\sup_{\boldsymbol{\beta}^* \in B(\delta)} N^{-1} \sum_{i=1}^{N} \|[w_{ki}(\boldsymbol{\beta}^*) - w_{ki}(\boldsymbol{\beta})]\boldsymbol{x}_{Ni}\boldsymbol{x}'_{Ni}\| \to 0, \tag{10.5.33}$$

for $k = 1, 2$, as $\delta \downarrow 0$, and

$$\mathbb{E}_{\boldsymbol{\beta}} \left\{ \sup_{\boldsymbol{\beta}^* \in B(\delta)} N^{-1} \sum_{i=1}^{N} |Y_i| \|[w_{2i}(\boldsymbol{\beta}^*) - w_{2i}(\boldsymbol{\beta})]\boldsymbol{x}_{Ni}\boldsymbol{x}'_{Ni}\| \right\} \to 0, \tag{10.5.34}$$

as $\delta \downarrow 0$. Then we may consider the following theorem, the proof of which is based on the same technique as that used to demonstrate Theorem 8.3.1.

***Theorem* 10.5.1.** *Consider the generalized linear model (10.5.8) and let $\widehat{\boldsymbol{\beta}}_N$ denote the MLE of $\boldsymbol{\beta}$ under the likelihood corresponding to (10.5.13). Then, if (10.5.30)–(10.5.34) hold, it follows that*

$$\sqrt{N}(\widehat{\boldsymbol{\beta}}_N - \boldsymbol{\beta}) \xrightarrow{D} \mathcal{N}_q[\boldsymbol{0}, \boldsymbol{I}^{-1}(\boldsymbol{\beta})]. \tag{10.5.35}$$

Proof. First let $\|\boldsymbol{u}\| < K$, $0 < K < \infty$ and consider a Taylor expansion of $\log L_N(\boldsymbol{\beta} + N^{-1/2}\boldsymbol{u})$ around $\log L_N(\boldsymbol{\beta})$ to define

$$\begin{aligned} \lambda_N(\boldsymbol{u}) &= \log L_N(\boldsymbol{\beta} + N^{-1/2}\boldsymbol{u}) - \log L_N(\boldsymbol{\beta}) \\ &= \frac{1}{\sqrt{N}}\boldsymbol{u}'\boldsymbol{U}_N(\boldsymbol{\beta}) + \frac{1}{N}\boldsymbol{u}'\frac{\partial^2}{\partial\boldsymbol{\beta}\partial\boldsymbol{\beta}'}\log L_N(\boldsymbol{\beta})\bigg|_{\boldsymbol{\beta}^*}\boldsymbol{u}, \end{aligned} \tag{10.5.36}$$

where $\boldsymbol{U}_N(\boldsymbol{\beta}) = (\partial/\partial\boldsymbol{\beta})\log L_N(\boldsymbol{\beta})$ and $\boldsymbol{\beta}^*$ belongs to the line segment joining $\boldsymbol{\beta}$ and $\boldsymbol{\beta}^*$. Then, let

$$\begin{aligned} Z_n(\boldsymbol{u}) = \frac{1}{N}\bigg[&\boldsymbol{u}'\frac{\partial^2}{\partial\boldsymbol{\beta}\partial\boldsymbol{\beta}'}\log L_N(\boldsymbol{\beta})\bigg|_{\boldsymbol{\beta}^*}\boldsymbol{u} - \boldsymbol{u}'\frac{\partial^2}{\partial\boldsymbol{\beta}\partial\boldsymbol{\beta}'}\log L_N(\boldsymbol{\beta})\bigg|_{\boldsymbol{\beta}}\boldsymbol{u} \\ &+ \boldsymbol{u}'\frac{\partial^2}{\partial\boldsymbol{\beta}\partial\boldsymbol{\beta}'}\log L_N(\boldsymbol{\beta})\bigg|_{\boldsymbol{\beta}}\boldsymbol{u} + \boldsymbol{u}'\boldsymbol{I}_N(\boldsymbol{\beta})\boldsymbol{u}\bigg] \end{aligned}$$

and note that (10.5.36) may be re-expressed as

$$\lambda_N(\boldsymbol{u}) = \frac{1}{\sqrt{N}}\boldsymbol{u}'\boldsymbol{U}_N(\boldsymbol{\beta}) - \frac{1}{2N}\boldsymbol{u}'\boldsymbol{I}_N(\boldsymbol{\beta})\boldsymbol{u} + Z_N(\boldsymbol{u}). \tag{10.5.37}$$

Now observe that

$$\begin{aligned} &\sup_{\|\boldsymbol{u}\|<K} \|Z_N(\boldsymbol{u})\| \\ &\leq \frac{1}{2}\sup_{\boldsymbol{\beta}^* \in B(K/\sqrt{n})}\left\|\frac{1}{N}\frac{\partial^2}{\partial\boldsymbol{\beta}\partial\boldsymbol{\beta}'}\log L_N(\boldsymbol{\beta})\bigg|_{\boldsymbol{\beta}^*} - \frac{1}{N}\frac{\partial^2}{\partial\boldsymbol{\beta}\partial\boldsymbol{\beta}'}\log L_N(\boldsymbol{\beta})\bigg|_{\boldsymbol{\beta}}\right\| \\ &\quad + \frac{K^2}{2}\left\|\frac{1}{N}\frac{\partial^2}{\partial\boldsymbol{\beta}\partial\boldsymbol{\beta}'}\log L_N(\boldsymbol{\beta}) + \frac{1}{N}\boldsymbol{I}_N(\boldsymbol{\beta})\right\|. \end{aligned} \tag{10.5.38}$$

Using assumptions (10.5.30) and (10.5.31) in connection with (10.5.28) we may conclude that the last term on the right-hand side of (10.5.38) converges to zero as $N \to \infty$. On the other hand, assumptions (10.5.33) and (10.5.34) may be employed to show that the first term on the right hand side of (10.5.38) also converges to zero as $N \to \infty$. Thus, it follows that $Z_N(\boldsymbol{u}) = o_p(1)$ uniformly in $\boldsymbol{u}$, $\|\boldsymbol{u}\| < K$, so that (10.5.37) may be written as

$$\lambda_n(\boldsymbol{u}) = \frac{1}{\sqrt{N}}\boldsymbol{u}'\boldsymbol{U}_N(\boldsymbol{\beta}) - \frac{1}{2N}\boldsymbol{u}'\boldsymbol{I}_N(\boldsymbol{\beta})\boldsymbol{u} + o_p(1)$$

so that an argument similar to that used in Theorem 8.3.1 completes the proof. In this regard, note that, by (10.5.19), the score statistic $\boldsymbol{U}_N(\boldsymbol{\beta})$ involves independent centered summands, and, hence, we may directly appeal to the Lindeberg–Feller Central Limit Theorem 7.3.3 to establish its asymptotic normality. [If the $\boldsymbol{x}_{Ni}$ are dependent on N as well, we may need to use the Central Limit Theorem 7.3.5 for Triangular Schemes.] In this context, it may be easier to verify the Liapunov condition as in Theorem 7.3.2. Also note that in a generalized linear model setup, the $\boldsymbol{x}_{Ni}$ need not be all the same, and, hence, the $\mu_i(\boldsymbol{\beta})$ may be different. Thus, the Y_i may not be identically distributed, so that the Hájek–Šidák Theorem 7.3.6 may not be appropriate here. ■

Suppose we assume that $\mathbb{E}(Y_i) = \mu_i$ and $\mathbb{V}\text{ar}(Y_i) = \sigma^2 v_i(\mu_i)$, where σ^2 is a positive scalar, possibly unknown, and v_i is a completely known variance function, but make no specific assumptions about the functional form of the density. Then there is no (finite dimensional) likelihood function and so we cannot obtain maximum-likelihood estimates in the usual way. However, the equation

$$\sum_{i=1}^{N} \frac{d\mu_i}{d\boldsymbol{\beta}}[v_i(\mu_i)]^{-1}(Y_i - \mu_i) = \boldsymbol{0} \tag{10.5.39}$$

can be solved to obtain estimates of $\boldsymbol{\beta}$. Under this setup (10.5.39) is referred to as the *quasiscore* or *quasilikelihood* estimating equation Wedderburn (1976). In the above setup the response $\boldsymbol{Y}_i$ can be an $n_i \times 1$ vector, and the variance function v_i would be a mapping into an $n_i \times n_i$ matrix.

Note that in generalized linear models the roots of the estimating equations may not be unique or not exist. See Wedderburn (1976), Pratt (1981), Silvapulle (1981), and Silvapulle and Burridge (1986) for details. The conditions needed for existence and uniqueness are essentially similar to those for the MLE. So for large-sample properties the roots are assumed to exist, at least with high probability, in addition to further regularity conditions. See Fahrmeir and Kaufmann (1985) for details.

The large sample consistency and asymptotic normality of the quasilikelihood estimators are obtained with assumptions about the quasiscore rather than the likelihood function. Godambe and Heyde (1987) provide further insight into this problem.

It is important to point out that in the above section the variance function involves no unknown parameters. The "generalized estimating equations (GEE)" extension of Liang and Zeger (1986) allows for $v_i(.)$ to include a finite dimensional vector $\boldsymbol{\alpha}$ of unknown parameters to be estimated. Specifically, the variance matrix is decomposed into the correlation matrix and a diagonal matrix of variances. The diagonal matrix is a completely known function of μ_i. The correlation matrix is parametrized as a function of $\boldsymbol{\alpha}$. Usually specific patterns of correlation, such as exchangeable and autoregressive ones, are assumed. Liang and Zeger (1986) showed consistency and asymptotic normality of the GEE estimators. Their arguments require a $\sqrt{N}$-consistent estimator of $\boldsymbol{\alpha}$ which may be obtained by using simple moment estimators and this is in line with the one-step procedure mentioned earlier.

***Example* 10.5.4.** Consider the setup described in Example 10.5.2, where for the sake of simplicity we assume that there is only one explanatory variable, that is, $q = 1$ and to avoid a degenerate Poisson distribution, we take $x_i \neq 0$. For the canonical link function $g(z) = \log z$, we get $g'(z) = z^{-1}$ and $v_i(\beta) = \exp(\beta x_i)$, so that $w_{1i}(\beta) = [\exp(-\beta x_i)]^{-2}[\exp \beta x_i)]^{-1} = \exp(\beta x_i)$ and the assumptions required for Theorem 10.5.1 reduce to

$$\lim_{N\to\infty} \frac{1}{N} \sum_{i=1}^{N} e^{\beta x_i} x_i^2 = I(\beta) < \infty \tag{10.5.40}$$

and for some $K: 0 < K < \infty$, as $N \to \infty$

$$\sup_{|h|\le K/\sqrt{N}} N^{-1} \sum_{i=1}^{N} |[e^{(\beta+h)x_i} - e^{\beta x_i}]x_i^2| = N^{-1} \sum_{i=1}^{N} [e^{(\beta+K/\sqrt{N})x_i} - e^{\beta x_i}]x_i^2 \to 0. \tag{10.5.41}$$

If we suppose that the range of the explanatory variable is bounded, then clearly both (10.5.40) and (10.5.41) hold, and it follows that the MLE $\widehat{\beta}_N$ of β is such that

$$\sqrt{N}(\widehat{\beta}_N - \beta) \xrightarrow{D} \mathcal{N}[0, I^{-1}(\beta)].$$

Take, for example, the two-sample location problem, where $x_i = 1$ if the ith sample unit is obtained from the first population and $x_i = -1$, otherwise, $i = 1, \ldots, N$. Let $N_1 = \sum_{i=1}^{N} I(x_i = 1)$ and assume that as $N \to \infty$, $N_1/N \to \pi, 0 < \pi < 1$. Then,

$$I(\beta) = \lim_{N\to\infty} \left[\frac{N_1}{N} e^{\beta} + \frac{(N - N_1)}{N} e^{-\beta}\right] = \pi e^{\beta} + (1 - \pi)e^{-\beta} < \infty$$

and

$$N^{-1} \sum_{i=1}^{N} [e^{(\beta+K/\sqrt{N})x_i} - e^{\beta x_i}]x_i^2$$
$$= \frac{N_1}{N}[e^{\beta+K/\sqrt{N}} - e^{\beta}] + \frac{N - N_1}{N}[e^{-(\beta+K/\sqrt{N})} - e^{-\beta}],$$

which clearly converges to zero as $N \to \infty$, implying that

$$\sqrt{N}(\widehat{\beta}_N - \beta) \xrightarrow{D} \mathcal{N}\{0, [\pi e^{\beta} + (1 - \pi)e^{-\beta}]^{-1}\}.$$

Extension to the multisample location problem is immediate (see Exercise 10.5.3).

If we consider the noncanonical link function $g(z) = z$, a convenient model is given by $\mu_i = \exp(\theta_i) = \alpha + \beta x_i$. In such a case we have $g'(z) = 1$, $g''(z) = 0$, $v_i(\boldsymbol{\beta}) = \alpha + \beta x_i$, $h(x) = \log x$, and $b'''(x) = \exp(x)$, implying that $w_{1i}(\boldsymbol{\beta}) = (\alpha + \beta x_i)^{-1}$; also,

$$\boldsymbol{x}_i \boldsymbol{x}_i' = \begin{pmatrix} 1 & x_i \\ x_i & x_i^2 \end{pmatrix},$$

so that defining

$$A(N) = \sup_{\boldsymbol{h}\in H} \frac{1}{N} \sum_{i=1}^{N} \left\| \left[\frac{1}{(\alpha + h_1) + (\beta + h_2)x_i} - \frac{1}{\alpha + \beta x_i}\right] \boldsymbol{x}_i \boldsymbol{x}_i' \right\|$$

where $H = \{\boldsymbol{h} \in \mathbb{R}^2 : \|\boldsymbol{h}\| \le K/\sqrt{N}\}$ and

$$B(N) = \mathbb{E}_{\boldsymbol{\beta}} \left\{ \sup_{\boldsymbol{h}\in H} \frac{1}{N} \sum_{i=1}^{N} |Y_i| \left\| \left[\frac{1}{(\alpha + h_1) + (\beta + h_2)x_i} - \frac{1}{\alpha + \beta x_i} \right] \boldsymbol{x}_i \boldsymbol{x}_i' \right\| \right\}$$

the assumptions required for Theorem 10.5.1 reduce to

$$\lim_{N\to\infty} \frac{1}{N} \sum_{i=1}^{N} \frac{1}{\alpha + \beta x_i} \begin{pmatrix} 1 & x_i \\ x_i & x_i^2 \end{pmatrix} = \boldsymbol{I}(\boldsymbol{\beta}) \quad \text{is finite and p.d.,} \tag{10.5.42}$$

$$\lim_{N\to\infty} \frac{1}{N^2} \sum_{i=1}^{N} \frac{1 + x_i^2}{\alpha + \beta x_i} = 0, \tag{10.5.43}$$

$$A(N) \to 0 \quad \text{as } N \to \infty, \tag{10.5.44}$$

and

$$B(N) \to 0 \quad \text{as } N \to \infty. \tag{10.5.45}$$

Again let us return to the two-sample location problem where, for convenience, we let $x_i = 0$ if the ith sample unit is obtained from the first population and $x_i = 1$, otherwise, $i = 1, \ldots, N$. Let N_1 and π have the same interpretation as before. Then, we have

$$\begin{aligned} \boldsymbol{I}(\boldsymbol{\beta}) &= \lim_{N\to\infty} \frac{N_1}{N} \frac{1}{\alpha} \begin{pmatrix} 1 & 0 \\ 0 & 0 \end{pmatrix} + \lim_{n\to\infty} \frac{N - N_1}{N} \frac{1}{\alpha + \beta} \begin{pmatrix} 1 & 1 \\ 1 & 1 \end{pmatrix} \\ &= \begin{bmatrix} \pi/\alpha + (1-\pi)/(\alpha+\beta) & (1-\pi)/(\alpha+\beta) \\ (1-\pi)/(\alpha+\beta) & (1-\pi)/(\alpha+\beta) \end{bmatrix} \end{aligned} \tag{10.5.46}$$

is finite and p.d.,

$$\begin{aligned} &\lim_{N\to\infty} \frac{1}{N^2} \sum_{i=1}^{N} \frac{1 + x_i^2}{\alpha + \beta x_i} \\ &= \lim_{N\to\infty} \left[\frac{N_1}{N} \left\{ \frac{1}{N\alpha} \right\} + \frac{N - N_1}{N} \left\{ \frac{2}{N(\alpha+\beta)} \right\} \right] = 0, \end{aligned} \tag{10.5.47}$$

$$\begin{aligned} A(N) \le{}& \frac{N_1}{N} \left| \frac{1}{\alpha + K/\sqrt{N}} - \frac{1}{\alpha} \right| \\ &+ \frac{N - N_1}{N} \left| \frac{1}{\alpha + \beta + 2K/\sqrt{N}} - \frac{1}{\alpha + \beta} \right| \to 0 \quad \text{as } N \to \infty, \end{aligned} \tag{10.5.48}$$

$$\begin{aligned} B(N) \le{}& \frac{N_1}{N} \alpha \left| \frac{1}{\alpha + K/\sqrt{N}} - \frac{1}{\alpha} \right| \\ &+ \frac{N - N_1}{N} \left| \frac{1}{\alpha + \beta + 2K/\sqrt{N}} - \frac{1}{\alpha + \beta} \right| \to 0 \quad \text{as } N \to \infty. \end{aligned} \tag{10.5.49}$$

Since (10.5.42)–(10.5.45) hold, the asymptotic normality of $\sqrt{N}(\widehat{\boldsymbol{\beta}}_N - \boldsymbol{\beta})$ follows from Theorem 10.5.1. It is important to note that the linear regression model imposes a restriction on the parameter space, induced by the fact that $\alpha + \beta x_i$ must be positive. Thus, the assumptions of Theorem 10.5.1 remain valid when $\boldsymbol{\beta}$ lies in the interior of the (restricted) parameter space. If we allow $\boldsymbol{\beta}$ to lie on the boundary of this parameter space, the computations in (10.5.17), (10.5.18), and (10.5.26) are to be modified according to the boundary restraints on $\boldsymbol{\beta}$, and the resulting equations may be highly complex. In fact, for such boundary points, we have a different solution which may be regarded as analogous to that of the classical restricted MLE case. This is outside the scope of our book; see Barlow, Bartholomew, Bremmer, and Brunk (1972) for a very useful account on the subject. □

As a final remark, we point out that even though the generalized linear models approach permits a great flexibility, the associated asymptotic results rely heavily on the underlying exponential family distribution as well as on the choice of the link function. Therefore, we may question the robustness of such procedures with respect to departures from the highly structured models on which they are based and look for alternative approaches. This topic will be discussed briefly in the next section.

10.6 Generalized Least-Squares Versus Generalized Estimating Equations

As we have seen in Section 10.5, the MLE of $\boldsymbol{\beta}$ under the generalized linear model $\boldsymbol{G}_N(\boldsymbol{\mu}_N) = \boldsymbol{X}_N\boldsymbol{\beta}$ is a solution to the GEE (10.5.21). A convenient way to solve each equations is via the Fisher scoring algorithm briefly introduced in Section 8.2. In particular, for the case under investigation here, the $(\nu+1)$th step of the algorithm corresponds to

$$\boldsymbol{\beta}^{(\nu+1)} = \boldsymbol{\beta}^{(\nu)} + \boldsymbol{I}_N^{-1}(\boldsymbol{\beta}^{(\nu)})\boldsymbol{U}_N(\boldsymbol{\beta}^{(\nu)}), \tag{10.6.1}$$

where $\boldsymbol{I}_N(\boldsymbol{\beta}^{(\nu)})$ and $\boldsymbol{U}_N(\boldsymbol{\beta}^{(\nu)})$ are, respectively, the Fisher information matrix (10.5.27) and the score statistic (10.5.9) evaluated at the estimate $\boldsymbol{\beta}^{(\nu)}$ obtained at the νth step. From (10.6.1) we may write

$$\boldsymbol{I}_N(\boldsymbol{\beta}^{(\nu)})\boldsymbol{\beta}^{(\nu+1)} = \boldsymbol{I}_N(\boldsymbol{\beta}^{(\nu)})\boldsymbol{\beta}^{(\nu)} + \boldsymbol{U}_N(\boldsymbol{\beta}^{(\nu)}). \tag{10.6.2}$$

Now, letting $\omega_i(\boldsymbol{\beta}^{(\nu)}) = \{g'[\mu_i(\boldsymbol{\beta}^{(\nu)})]\}^{-2}[v_i(\boldsymbol{\beta}^{(\nu)})]^{-1}$, $i = 1, \ldots, N$, we have

$$\begin{aligned} &\boldsymbol{I}_N(\boldsymbol{\beta}^{(\nu)})\boldsymbol{\beta}^{(\nu)} + \boldsymbol{U}_N(\boldsymbol{\beta}^{(\nu)}) \\ &\quad = \left[\sum_{i=1}^N \omega_i(\boldsymbol{\beta}^{(\nu)})\boldsymbol{x}_{Ni}\boldsymbol{x}'_{Ni}\right]\boldsymbol{\beta}^{(\nu)} + \sum_{i=1}^N \omega_i(\boldsymbol{\beta}^{(\nu)})\boldsymbol{x}_{Ni}[Y_i - \mu(\boldsymbol{\beta}^{(\nu)})]g'[\mu_i(\boldsymbol{\beta}^{(\nu)})], \end{aligned} \tag{10.6.3}$$

which, upon considering the first-order Taylor expansion:

$$\boldsymbol{G}_N(\boldsymbol{Y}_N) \cong \boldsymbol{G}_N[\boldsymbol{\mu}_N(\boldsymbol{\beta}^{(\nu)})] + \dot{\boldsymbol{G}}_N[\boldsymbol{\mu}_N(\boldsymbol{\beta}^{(\nu)})][\boldsymbol{Y}_N - \boldsymbol{\mu}_N(\boldsymbol{\beta}^{(\nu)})] \tag{10.6.4}$$

where $\boldsymbol{Y}_N = (Y_1, \ldots, Y_N)'$ and

$$\dot{\boldsymbol{G}}_N[\boldsymbol{\mu}_N(\boldsymbol{\beta}^{(\nu)})] = \text{diag}\{g'[\mu_1(\boldsymbol{\beta}^{(\nu)})], \ldots, g'[\mu_N(\boldsymbol{\beta}^{(\nu)})]\}.$$

Hence, we may re-express the left-hand side of (10.6.3) as

$$\begin{aligned} &\boldsymbol{I}_N(\boldsymbol{\beta}^{(\nu)})\boldsymbol{\beta}^{(\nu)} + \boldsymbol{U}_N(\boldsymbol{\beta}^{(\nu)}) \\ &= \sum_{i=1}^{N} \omega_i(\boldsymbol{\beta}^{(\nu)})\boldsymbol{x}_{Ni}\{\boldsymbol{x}'_{Ni}\boldsymbol{\beta}^{(\nu)} - [Y_i - \mu_i(\boldsymbol{\beta}^{(\nu)})]g'[\mu_i(\boldsymbol{\beta}^{(\nu)})]\} \\ &= \boldsymbol{X}'_N \boldsymbol{W}_N(\boldsymbol{\beta}^{(\nu)})\boldsymbol{G}_N(\boldsymbol{Y}_N), \end{aligned} \tag{10.6.5}$$

where $\boldsymbol{W}_N(\boldsymbol{\beta}^{(\nu)}) = \text{diag}[\omega_1(\boldsymbol{\beta}^{(\nu)}), \ldots, \omega_N(\boldsymbol{\beta}^{(\nu)})]$. Thus, (10.6.2) reduces to the estimating equations

$$\boldsymbol{X}'_N \boldsymbol{W}_N(\boldsymbol{\beta}^{(\nu)})\boldsymbol{X}_N\boldsymbol{\beta}^{(\nu+1)} = \boldsymbol{X}'_N \boldsymbol{W}_N(\boldsymbol{\beta}^{(\nu)})\boldsymbol{G}_N(\boldsymbol{Y}_N)$$

which lead to

$$\boldsymbol{\beta}^{(\nu+1)} = \{\boldsymbol{X}'_N \boldsymbol{W}_N(\boldsymbol{\beta}^{(\nu)})\boldsymbol{X}_N\}^{-1}\boldsymbol{X}'_N \boldsymbol{W}_N(\boldsymbol{\beta}^{(\nu)})\boldsymbol{G}_N(\boldsymbol{Y}_N). \tag{10.6.6}$$

Since (10.6.6) has the same form of the GLSE (10.2.55), with weights that are recomputed at each step, the algorithm is known as *iterative reweighted least squares*. The MLE $\widehat{\boldsymbol{\beta}}_N$ is obtained by letting $\nu \to \infty$.

Suppose that consistent [but not necessarily $O_p(n^{-1/2})$] estimators $\widehat{\mu}_i$ of μ_i such that $\mathbb{V}\text{ar}(\widehat{\mu}_i) = v_i(\mu_i)$, $i = 1, \ldots, N$, are available. Then, writing

$$\begin{aligned} \boldsymbol{G}_N(\widehat{\boldsymbol{\mu}}_N) &= \widehat{\boldsymbol{G}}_N = [g(\widehat{\mu}_1), \ldots, g(\widehat{\mu}_N)]', \\ \boldsymbol{V}_N(\widehat{\boldsymbol{\mu}}_N) &= \text{diag}\{v_1(\mu_1), \ldots, v_N(\mu_N)\} \end{aligned}$$

and using a first-order Taylor expansion similar to (10.6.4) we may consider the model

$$\widehat{\boldsymbol{G}}_N = \boldsymbol{X}_N\boldsymbol{\beta} + \boldsymbol{\varepsilon}_N, \tag{10.6.7}$$

where $\boldsymbol{\varepsilon}_N = \boldsymbol{G}_N(\boldsymbol{\mu}_N)(\widehat{\boldsymbol{\mu}}_N - \boldsymbol{\mu}_N) + o_p(\boldsymbol{1})$ is such that $\mathbb{E}(\boldsymbol{\varepsilon}_N) = \boldsymbol{0} + o(\boldsymbol{1})$ and $\mathbb{V}\text{ar}(\boldsymbol{\varepsilon}_N) = \boldsymbol{W}_N^{-1}(\boldsymbol{\mu}_N) + o(\boldsymbol{1})$ [here we remind the reader to recall that $\boldsymbol{W}_N^{-1}(\boldsymbol{\mu}_N) = \dot{\boldsymbol{G}}_N(\boldsymbol{\mu}_N) \boldsymbol{V}_N(\boldsymbol{\mu}_N)\dot{\boldsymbol{G}}'_N(\boldsymbol{\mu}_N)$]. Neglecting the $o(\boldsymbol{1})$ terms, we may relate (10.6.7) to the model (10.2.41) with $\boldsymbol{\Sigma}_N = \boldsymbol{W}_N^{-1}(\boldsymbol{\mu}_N)$ and, therefore, consider the GLSE

$$\widehat{\widehat{\boldsymbol{\beta}}}_N = [\boldsymbol{X}'_N \boldsymbol{W}_N(\widehat{\boldsymbol{\mu}}_N)\boldsymbol{X}_N]^{-1}\boldsymbol{X}'_N \boldsymbol{W}_N(\widehat{\boldsymbol{\mu}}_N)\widehat{\boldsymbol{G}}_N \tag{10.6.8}$$

as an alternative to the MLE. Then, in view of (10.2.61) and (10.2.62) we may conclude that $\sqrt{N}(\widehat{\widehat{\boldsymbol{\beta}}}_N - \boldsymbol{\beta})$ has the same asymptotic distribution as $\sqrt{N}(\widehat{\boldsymbol{\beta}}_N - \boldsymbol{\beta})$, namely, $\mathcal{N}[\boldsymbol{0}, \boldsymbol{I}^{-1}(\boldsymbol{\beta})]$. But, (10.2.61)–(10.2.62) may be needed to justify the last result.

This approach is particularly appealing when $\boldsymbol{Y}_N$ itself is a consistent estimate of $\boldsymbol{\mu}_N$. In such cases, (10.6.8) constitutes the first step of the Fisher–Rao scoring algorithm corresponding to (10.6.6). Some further relations between the two approaches will be discussed in the following examples.

***Example* 10.6.1.** Let $Y_i, i = 1, \ldots, s$, be independent random variables with a Poisson(λ_i) distribution and consider the generalized linear model $\lambda_i = N_i h(\boldsymbol{x}'_i\boldsymbol{\beta}), i = 1, \ldots, s$, where h is a convenient function, $\boldsymbol{x}_i$ is a $(q \times 1)$ vector of explanatory variables and N_i is a measure of exposure (i.e., geographic area, volume) In this context, the parameters $\nu_i = \lambda_i/N_i$, $i = 1, \ldots, s$, may be interpreted as rate parameters at which the events are counted in Y_i per unit of exposure N_i. Under this Poisson regression model, the log-likelihood may be expressed as

$$\log L_N(\boldsymbol{\beta}) = \sum_{i=1}^{s} [Y_i \log \lambda_i(\boldsymbol{\beta}) - \lambda_i(\boldsymbol{\beta})] + \text{constant} \tag{10.6.9}$$

so that, here, the MLE $\widehat{\boldsymbol{\beta}}_N$ may be obtained as a solution to the GEE (10.5.21), which reduces to

$$\sum_{i=1}^{s} \left[\frac{Y_i - \lambda_i(\boldsymbol{\beta})}{\lambda_i(\boldsymbol{\beta})}\right] \frac{\partial \lambda_i}{\partial \boldsymbol{\beta}}(\boldsymbol{\beta}) = \mathbf{0}. \tag{10.6.10}$$

On the other hand, we may rewrite (10.6.9) as

$$\log L_N(\boldsymbol{\beta}) = \sum_{i=1}^{s} \left\{ Y_i \log \left[1 + \frac{\lambda_i(\boldsymbol{\beta}) - Y_i}{Y_i}\right] - \lambda_i(\boldsymbol{\beta}) \right\} + \text{constant}$$

and using the expansion

$$\log(1 + x) = x - \frac{1}{2}x^2 + \frac{1}{3}x^3 - \cdots, \quad |x| < 1,$$

to write

$$\log L_N(\boldsymbol{\beta}) = -\sum_{i=1}^{s} \frac{[Y_i - \lambda_i(\boldsymbol{\beta})]^2}{Y_i} - \sum_{i=1}^{s} \frac{[Y_i - \lambda_i(\boldsymbol{\beta})]^3}{Y_i^2} - \cdots + K,$$

with K a constant. Taking $N = \max(N_1, \ldots, N_s)$ and using (10.5.25), we may see that

$$\sum_{i=1}^{s} \frac{[Y_i - \lambda_i(\boldsymbol{\beta})]^3}{Y_i^2} = \sum_{i=1}^{s} \frac{[Y_i - N_i h(\boldsymbol{x}_i'\boldsymbol{\beta})]^3}{Y_i^2} = O_P(N^{-3/2}).$$

Thus, by neglecting the $O_P(N^{-3/2})$ terms, maximization of (10.6.9) will be equivalent to the minimization of $\sum_{i=1}^{s}[Y_i - \lambda_i(\boldsymbol{\beta})]^2/Y_i$, which, in turn, corresponds to solving

$$\sum_{i=1}^{s} \left[\frac{Y_i - \lambda_i(\boldsymbol{\beta})}{Y_i}\right] \frac{\partial \lambda_i(\boldsymbol{\beta})}{\partial \boldsymbol{\beta}} = \mathbf{0}. \tag{10.6.11}$$

If we consider the linear model $h(\boldsymbol{x}_i'\boldsymbol{\beta}) = \boldsymbol{x}_i'\boldsymbol{\beta}$, (10.6.11) reduces to

$$\sum_{i=1}^{s} N_i \boldsymbol{x}_i = \sum_{i=1}^{s} \frac{N_i^2}{Y_i} \boldsymbol{x}_i' \widehat{\widehat{\boldsymbol{\beta}}}_N \boldsymbol{x}_i,$$

a solution of which is given by the GLSE

$$\widehat{\widehat{\boldsymbol{\beta}}}_N = (\boldsymbol{X}'\boldsymbol{D}_N\boldsymbol{D}_Y^{-1}\boldsymbol{D}_N\boldsymbol{X})^{-1}\boldsymbol{X}'\boldsymbol{D}_N\mathbf{1}_s,$$

where $\boldsymbol{D}_N = \text{diag}\{N_1, \ldots, N_s\}$ and $\boldsymbol{D}_Y = \text{diag}\{Y_1, \ldots, Y_s\}$. For nonlinear models, such as $h(\boldsymbol{x}_i'\boldsymbol{\beta}) = \exp(\boldsymbol{x}_i'\boldsymbol{\beta})$, (10.6.11) are also nonlinear, but may be approximated by the linear equations which define the corresponding GLSE. See Koch et al. (1985) for details. □

***Example* 10.6.2.** Consider the situation described in Example 10.5.1 in a more general setup where we allow for s (≥ 2) groups and for a ($q \times 1$) vector of covariates $\boldsymbol{x}_i$ associated with each Z_i. Writing $Y_i = \sum_{i=1}^{n} Y_{ij}$ where $Y_{ij} = n^{-1}$ with probability π_i and $Y_{ij} = 0$ with probability $(1 - \pi)$, the GEE (10.5.21) may be expressed as

$$\sum_{i=1}^{s}\sum_{j=1}^{n}\left[\frac{Y_{ij} - \pi_i(\boldsymbol{\beta})}{\omega_i(\boldsymbol{\beta})}\right]\boldsymbol{x}_{N_i} = \boldsymbol{0}, \tag{10.6.12}$$

where $\omega_i(\boldsymbol{\beta}) = g'[\pi_i(\boldsymbol{\beta})]\pi_i(\boldsymbol{\beta})[1 - \pi_i(\boldsymbol{\beta})]$, $i = 1, \ldots, s$. Whenever a consistent preliminary estimate $\widetilde{\boldsymbol{\beta}}$ of $\boldsymbol{\beta}$ is available, we may consider the GLSE, which is a solution to

$$\sum_{i=1}^{s}\sum_{j=1}^{n}\left[\frac{Y_{ij} - \pi_i(\boldsymbol{\beta})}{\omega_i(\widetilde{\boldsymbol{\beta}})}\right]\boldsymbol{x}_{N_i} = \boldsymbol{0}. \tag{10.6.13}$$

Now, since

$$\begin{aligned}\sum_{i=1}^{s}\sum_{j=1}^{n}\left[\frac{Y_{ij} - \pi_i(\boldsymbol{\beta})}{\omega_i(\widetilde{\boldsymbol{\beta}})}\right]\boldsymbol{x}_{N_i} &= \sum_{i=1}^{s}\sum_{j=1}^{n}\left[\frac{Y_{ij} - \pi_i(\boldsymbol{\beta})}{\omega_i(\boldsymbol{\beta})}\right]\frac{\omega_i(\boldsymbol{\beta})}{\omega_i(\widetilde{\boldsymbol{\beta}})}\boldsymbol{x}_{N_i}\\ &= \sum_{i=1}^{s}\sum_{j=1}^{n}\left[\frac{Y_{ij} - \pi_i(\boldsymbol{\beta})}{\omega_i(\boldsymbol{\beta})}\right]\boldsymbol{x}_{N_i} + \sum_{i=1}^{s}\frac{\omega_i(\boldsymbol{\beta}) - \omega_i(\widetilde{\boldsymbol{\beta}})}{\omega_i(\widetilde{\boldsymbol{\beta}})}\sum_{j=1}^{n}\left[\frac{Y_{ij} - \pi_i(\boldsymbol{\beta})}{\omega_i(\boldsymbol{\beta})}\right]\boldsymbol{x}_{N_i},\end{aligned}$$

it follows that (10.6.12) and (10.6.13) will be asymptotically equivalent whenever

$$\max_{1\le i\le s}\left|\frac{\omega_i(\boldsymbol{\beta}) - \omega_i(\widetilde{\boldsymbol{\beta}})}{\omega_i(\widetilde{\boldsymbol{\beta}})}\right| \xrightarrow{P} 0.$$

Note that this is true even if $n = 1$, although in such a case it may be difficult to obtain the preliminary estimate $\widetilde{\boldsymbol{\beta}}$. □

10.7 Nonparametric Regression

It seems clear from the previous sections that in parametric statistical modeling (be it in a classical normal setup or via the generalized linear model approach), the form of the underlying density of the error component plays a vital role in establishing the asymptotic properties of conventional statistical inference procedures. To a certain extent, the M-methods discussed in Sections 8.4 and 10.3 can accommodate mild violations of the proposed model, but in some practical problems it may be quite unreasonable to make even the less restrictive assumptions required by these procedures. As a simple illustration, consider a set of n independent and identically distributed random vectors (X_i, Y_i), $i = 1, \ldots, n$, and suppose that our primary interest centers on

$$\mu(x) = \mathbb{E}(Y|X = x), \quad x \in \mathbb{R}, \tag{10.7.1}$$

where $\mu(x)$ is referred to as the regression function of Y on X. In the spirit of Section 10.2, we may set

$$\mu(x) = \alpha + \beta x, \quad x \in \mathbb{R}, \tag{10.7.2}$$

where α and β are unknown parameters, so that (10.7.1) is completely specified as a function of (α, β). Such a model is, of course, justified when linearity of regression and homoskedasticity, that is, $\mathbb{V}\text{ar}(Y|x) = \sigma^2$ for all x hold. This amounts to adopting the bivariate normal model as the underlying distribution. A departure from the assumed normality may not only distort (10.7.2) but may also affect considerably the homogeneity of the conditional variance. To enhance more generality, we may as well replace the linear model (10.7.2) by a generalized linear model, as described in Section 10.5; this allows us to cover a much broader scope of applications, not only by permitting a nonlinear link between the mean response μ and the explanatory variable X but also by relaxing the homoskedasticity assumption. Even though, such models are still not robust, in the sense that small departures from the assumed form of the conditional density of Y given $X = x$, $f(y|x)$, may cause considerable damage to the efficiency of the related statistical inference procedures. Alternative models that are not as dependent on a specific form for the underlying random error distribution have deserved a fair amount of research in the recent past. They can be classified into the following two categories:

(i) *Semiparametric models*, where one assumes that the conditional density $f(y|x)$ has a parametric structure, yet allowing a nonparametric formulation for the underlying error distribution. More specifically, such models may be expressed as

$$f(y|x) = f_0[y - \beta(x)], \quad x \in \mathbb{R}, \quad y \in \mathbb{R}, \tag{10.7.3}$$

where $\beta(x)$ is a function of x of known form, involving a finite number of parameters and f_0 has an arbitrary form. A typical example is the Cox (1972) proportional hazards model frequently employed in the analysis of survival data. Here the hazard at time t for a sample unit with covariate vector $\boldsymbol{x}$ is given by $h(t, \boldsymbol{x}) = \lambda(t) \exp(\boldsymbol{x}'\boldsymbol{\beta})$, where λ is an arbitrary function defined only at the points where the events of interest occur and which does not have any relevance with respect to the estimation of the parameter vector $\boldsymbol{\beta}$. These semiparametric models may work out better for nonstochastic explanatory variables, as in the case of R-estimators, or the above mentioned Cox proportional hazards model.

(ii) *Nonparametric regression models*, where no parametric structures are imposed on $f(y|x)$. Thus, we may express

$$\mu(x) = \int y f(y|x) \mathrm{d}y, \quad x \in \mathbb{R}, \tag{10.7.4}$$

and then try to draw statistical conclusions about $\mu(x)$ through the estimation of the density $f(y|x)$ in a completely nonparametric fashion. In this formulation we may even generalize further, by replacing the conditional mean $\mu(x)$ by comparable functionals of $f(y|x)$ [or the corresponding conditional distribution function $F(y|x)$], such as median, trimmed mean, etc. In this setup we allow the explanatory variables to be stochastic as well.

For the semiparametric linear model in the simplest case, where $\beta(x) = \alpha + \beta(x)$ in (10.7.3), the R-estimators briefly introduced in Section 10.3 may work out quite well; see, for example, Sen (1968b). More general methodology, developed recently, involves some novel counting processes approaches resting on a much higher mathematical level, and will not be discussed in detail.

In the nonparametric regression problem, the crucial step is the estimation of the density $f(y|x)$, $x \in \mathbb{R}$. Note that even in the case of marginals, the empirical distribution function of the Y_i, say, $F_n(y)$, $y \in \mathbb{R}$, is a step function, so that the corresponding empirical density function $f_n(y)$ assumes the value zero at all y for which $y \neq Y_i$ for some $i = 1, \dots, n$. On the other hand, at each of these Y_i, $f_n(y) = (\mathrm{d}/\mathrm{d}y)F_n(y)$ is not defined. The situation is more complicated for the conditional density $f(y|x)$. Because of this fundamental problem, the first requirement of a statistical approach to this problem is to formulate suitable *smooth* estimators for the density function. In this direction we describe some of such *smoothing techniques* for the problem of estimating the density function of a random variable at a given point. Let $Y_1, \ldots, Y_n$ be i.i.d. random variables with density $f(y)$, $y \in \mathbb{R}$, and suppose that we are interested in estimating $f(a)$, where $a \in \mathbb{R}$ is a continuity point of f. Let $F_n(y) = n^{-1}\sum_{i=1}^{n} I(Y_i \le y)$ and for a small $h > 0$, consider the estimator

$$\widehat{f}_{n,h}(a) = \frac{1}{2h}[F_n(a+h) - F_n(a-h)]. \tag{10.7.5}$$

Recall that

$$(2nh)\widehat{f}_{n,h}(a) = \sum_{i=1}^{n} I(a - h \le Y_i \le a + h) \sim \mathrm{Bin}[n, \pi_h(a)], \tag{10.7.6}$$

where

$$\pi_n(a) = F(a+h) - F(a-h) = 2hf(a) + o(h). \tag{10.7.7}$$

We may improve the order of the remainder term from $o(h)$ to $o(h^2)$ by assuming further that $f'(x)$ exists and is finite at $x = a$. Using the Central Limit Theorem 7.3.1 for the binomial law, whenever

$$n\pi_h(a)[1 - \pi_h(a)] \to \infty \text{ with } n \to \infty,$$

that is, $nh \to \infty$, we obtain

$$n^{-1/2}\{\pi_h(a)[1 - \pi_h(a)]\}^{-1/2}[2nh\widehat{f}_{n,h}(a) - n\pi_h(a)] \xrightarrow{D} \mathcal{N}(0,1), \tag{10.7.8}$$

which can be rewritten as

$$\frac{2nh[\widehat{f}_{n,h}(a) - f(a) - o(1)]}{[2nhf(a) + o(nh)]^{1/2}} \xrightarrow{D} \mathcal{N}(0,1). \tag{10.7.9}$$

This, in turn, is equivalent to saying that

$$(2nh)^{1/2}[\widehat{f}_{n,h}(a)/f^{1/2}(a) - f^{1/2}(a) - o(1)]$$

converges in law to a standard normal variate as $n \to \infty$. On the other hand, in this setup $(nh)^{1/2}o(1)$ is not generally $o(1)$, so that the bias term in (10.7.7) may have a significant role. Here, also, the picture is slightly better when it is possible to assume that $f'(x)$ exists and is continuous at $x = a$, so that $o(h)$ can be replaced by $o(h^2)$. In such a case, to control the bias term we need to choose $h\ (= h_n)$, such that, as $n \to \infty$, $(nh_n)^{1/2}h_n \not\to \infty$ or $nh_n^3 \not\to \infty$, i.e., $h_n = O(n^{-1/3})$. Further improvements depend on additional smoothness assumptions on the density function. If we assume that $f''(x)$ exists and is finite in a neighborhood of a, then, in (10.7.7), we may replace $o(h)$ by $O(h^3)$, so that we may even take $h\ (= h_n)$ such that as $n \to \infty$, $(nh_n)^{1/2}h_n \not\to \infty$, that is, $nh_n^5 = O(1)$ or $h_n = O(n^{-1/5})$, a rate that is generally recommended in practice. In this case, $nh_n \sim n^{4/5}$, so that, on writing

$$\pi_{h_n}(a) = 2h_n f(a) + \frac{1}{3}h_n^3 f''(a) + o(h_n^3) \tag{10.7.10}$$

we have

$$\sqrt{2}n^{2/5}\left[\widehat{f}_{n,h_n}(a) - f(a) - \frac{1}{6}h_n^2 f''(a)\right] - o(n^{2/5}h_n^{2/5}) \xrightarrow{D} \mathcal{N}[0, f(a)] \quad (10.7.11)$$

so that, for $h_n \sim n^{-1/5}$,

$$n^{2/5}[\widehat{f}_{n,h_n}(a) - f(a)] \xrightarrow{D} \mathcal{N}[f''(a)/6, f(a)/2]. \quad (10.7.12)$$

Although for any practical purposes, both parameters of the limiting distribution must be estimated, the crucial problem here relates to the estimation of the bias term $f''(a)/6$; this may be accomplished either by estimating $f''(a)$ or by choosing $h_n = o(n^{1/5})$, so that $n^{2/5}h_n^2 \to 0$ as $n \to \infty$. In either case, the rate of convergence (i.e., $n^{2/5}$ or less) is slower than the conventional rate of $n^{1/2}$.

A more popular method, known as the *kernel method*, can be described as follows. Choose some known density $K(x)$ possessing some smoothness properties (usually one takes a unimodal symmetric density such as the normal) and let $\{h_n\}$ be a sequence of nonnegative real numbers such that $h_n \downarrow 0$ as $n \to \infty$. Then, the required estimator is

$$\begin{aligned}\widehat{f}_{n,K}(a) &= \int_{\mathbb{R}} \frac{1}{h_n} K\left(\frac{a-y}{h_n}\right) \mathrm{d}F_n(y) \\ &= \frac{1}{nh_n}\sum_{i=1}^n K\left(\frac{a-Y_i}{h_n}\right) \\ &= \frac{1}{n}\sum_{i=1}^n Z_{ni}(Y_i; a, h_n),\end{aligned} \quad (10.7.13)$$

where

$$Z_{ni}(Y_i; a, h_n) = \frac{1}{h_n} K\left(\frac{a-Y_i}{h_n}\right), i = 1, \ldots, n,$$

are i.i.d. random variables (in a triangular scheme). Thus, we may use the Central Limit Theorem for triangular arrays (7.3.5), carry out a similar analysis as in the case of (10.7.5) and, thus, obtain the asymptotic normality of the standardized form of $\widehat{f}_{n,K}(a)$. The resulting picture is very much comparable to (10.7.12).

We now return to the original nonparametric regression model (10.7.4). Here the situation is somewhat more complex, since to estimate the conditional density $f(a|x)$ for given a and x, we need to choose appropriate neighborhoods of both points. Before commenting on the use of the nonparametric methods outlined above, we introduce a third one, known as the *k nearest neighborhood (k-NN) method*, which is specially appealing in this bivariate context. Corresponding to a given x, we consider the distances $|X_i - x|$, $i = 1, \ldots, n$, and denote by $Z_1 \leq \cdots \leq Z_n$ their ordered values. Choose k $(= k_n)$ such that $k_n \sim n^{\lambda}$ for some $\lambda \in (1/2, 4/5]$. Also, let $O_1, \ldots, O_{k_n}$ be the subscripts of the X_i's corresponding to the smallest k_n Z_is, so that $Z_j = |X_{O_j} - x|$, $j = 1, \ldots, k_n$. To estimate $\mu(x)$ in (10.7.4) we take the empirical distribution function $F^*_{k_n}$ of the Y_{O_j}, $j = 1, \ldots, k_n$, and consider the estimator

$$\widehat{\mu}_n(x) = \int_{\mathbb{R}} y \mathrm{d}F^*_{k_n}(y) = \frac{1}{k_n}\sum_{j=1}^{k_n} Y_{O_j}. \quad (10.7.14)$$

As we have noted earlier, under this approach, it is not necessary to choose $\mu(x)$ as the regression function. Based on robustness considerations, it may as well be appropriate to choose a conditional quantile function, say the median, that is, $\mu(x) = F_{Y|x}^{-1}(1/2) = \inf\{y : F(y|x) \geq 1/2\}$ so that we may select the corresponding sample quantile of Y_{O_j}, $j = 1, \ldots, n$, as our estimator. Recall that the Y_{O_j}, $j = 1, \ldots, k_n$, given the Z_j, $j = 1, \ldots, k_n$, are conditionally independent, but not necessarily identically distributed random variables, so that the results of Chapters 7 and 8 can be used to study the asymptotic normality of $k_n^{1/2}[\widehat{\mu}_n(x) - \mu(x)]$. Here also, the rate of convergence is $k_n^{1/2}(\sim n^{2/5})$ and not $n^{1/2}$. Moreover, we may note that $F_{k_n}^*(y)$, even in a conditional setup, does not estimate $F_{Y|x}(y)$ unbiasedly; the bias is of the order $n^{-2/5}$ and, as such, for $k_n \sim n^{4/5}$, $k_n^{1/2}$ is of the same order as $n^{2/5}$. This makes it necessary to incorporate a small bias correction, or to choose $k_n \sim o(n^{4/5})$. Again, from a robustness point of view, we may argue that $\widehat{\mu}_n(x)$ in (10.7.14) may not be preferred to $\widetilde{\mu}_n(x) = \inf\{y : F_{k_n}^*(y) \geq 1/2\}$. We may also remark that instead of the k − NN method, we could have used the kernel method with $h_n \sim n^{-1/5}$, and both estimators would have the same asymptotic behavior. For details to Gangopadhyay and Sen (1990, 1992) and Eubank (1999), where a list of other pertinent references is also given.

Finally, we note that the above discussion was centered on the case of stochastic explanatory variables; in the nonstochastic case, the appropriate treatment reduces to grouping the observations according to the values of the explanatory variables and applying the density estimation techniques presented above to each of these groups.

10.8 Concluding Notes

The asymptotic theory for linear models may be focused in an unified manner to cover the cases of weighted least squares and generalized least squares as well as of some robust procedures. Along with the general asymptotic results developed in Chapter 8, this methodology provides the foundations for the large sample theory of generalized linear models. In this setup, the generalized (weighted) least-squares method plays a central role, especially for categorical data, given their popularization by Grizzle, Starmer, and Koch (1969). These authors applied generalized least-squares methods to many of the models usually considered by the generalized linear models approach. Since, in both, cases, we must rely on their asymptotic properties for inferential purposes, the analysis of the relationship between the two methodologies requires some attention. In this context, it is proper for us to clarify a subtle point of difference between the two approaches when viewed from an asymptotic perspective. In the generalized least-squares theory for categorical data models, we have an asymptotic situation comparable to that of (10.5.25) where the sample size N is usually fixed (and not large) but the observations themselves correspond to large subsamples. On the other hand, the general thrust on the asymptotics of generalized linear models is mainly on N being large. We hope that the examples and exercises set in this context will contribute to the clarification of this issue.

Although nonlinear statistical models have deserved a good deal of attention in the past few years, the developments so far have not yet crossed the fence of pure academics. In order to adopt such methodology in applied research, one may need sample sizes so large that they do not lie within limits of practical consideration. The generalized linear models or the robust procedures discussed in this chapter are generally applicable for moderately large sample sizes and have good efficiency properties, too. The nonparametric regression methods examined in Section 10.7 require somewhat larger sample sizes, but these methods still are far better than completely arbitrary nonlinear models. *Spline methodology* is an important omission on our part, but, again, its inclusion would require a much higher level of mathematical sophistication than that for which we aimed. For more information on this topic, see the excellent expository monographs by Thompson and Tapia (1990) and Stone, Hansen, Koopenberg, and Truong (1997). We are also somewhat skeptical about the genuine prospects of the semiparametric models in applied research. As with parametric models, in semiparametrics robustness although currently in use without any reservation, may be a main issue, in the sense that even a small departure from an assumed semiparametric model may lead to a serious bias in the large sample context.

10.9 Exercises

10.2.1 Under the general linear model (10.1.1) show that $\widehat{\boldsymbol{\beta}}_n - \boldsymbol{\beta} \xrightarrow{a.s.} \mathbf{0}$.

10.2.2 Consider the general linear model (10.1.1) and s_n^2 defined by (10.2.21). Show that $s_n^2 \xrightarrow{a.s.} \sigma^2$.

10.2.3 Show that (10.2.25) implies (10.2.26).

10.2.4 Consider the simple regression model (10.1.2) and assume further that $\overline{x}_n \to 0$ as $n \to \infty$ and that $\sigma_i^2 \propto |x_i - \overline{x}_n|^a$ for some $a \geq 0$, $1 \leq i \leq n$. Verify (10.2.25) under appropriate conditions on the x_i, $1 \leq i \leq n$. (For $a = 0$, 1 or 2, we may end up with the LSE, ratio estimator and regression estimator formulations.)

10.2.5 Consider the estimator $\breve{\theta}_n^*$ in (10.2.48). Let $\boldsymbol{\Sigma}_n = \boldsymbol{I}_2 \otimes \operatorname{diag}\{\sigma_1^2, \ldots, \sigma_m^2\}$ and $\widehat{\boldsymbol{\Sigma}}_n = \boldsymbol{I}_2 \otimes \operatorname{diag}\{\widehat{\gamma}_1, \ldots, \widehat{\gamma}_m\}$; show that (10.2.62) does not hold.

10.2.6 Under a setup similar to that of Example 10.2.2, consider a model where for some k (≥ 1), $Y_{(m-1)k+1}, \ldots, Y_{mk}$ are i.i.d. random variables with a $\mathcal{N}(\theta, \sigma_m^2)$ distribution for $m = 1, \ldots, n^*$ and $n = mn^*$. Verify that (10.2.62) and (10.2.63) hold whenever k is large and n^* is fixed.

10.3.1 For the location model, that is, $\boldsymbol{x}_{ni} = 1, i = 1, \ldots, n$, show that $\widehat{\boldsymbol{\beta}}_n^*$ in (10.3.4) reduces to the sample median.

10.3.2 For the Theil–Sen estimator $\widehat{\beta}_{TS}$, show that $n - 1 \leq \text{cardinality}(S_n) \leq \binom{n}{2}$. When are the lower and upper bounds attained?

10.3.3 Show that the right-hand side of both inequalities in (10.3.8) converge to zero as $n \to \infty$.

10.3.4 Use (10.3.9) to show that the Theil–Sen estimator $\widehat{\beta}_{TS}$ (conveniently standardized) is asymptotically normal.

10.3.5 Use (10.3.11) to show that the statistic $W_n(a)$ is nonincreasing in a.

10.3.6 Show that W_n defined in (10.3.10) is such that

$$W_n = \frac{n-1}{n+1} U_n + \frac{2}{n+1} U_n^*,$$

where

$$U_n = \binom{n}{2}^{-1} \sum_{1 \leq i < j \leq n} \operatorname{sign}(Y_i + Y_j)$$

and

$$U_n^* = n^{-1} \sum_{i=1}^{n} \operatorname{sign}(Y_i).$$

10.3.7 Use the result of Exercise 10.3.6 to verify the asymptotic normality of W_n.

10.3.8 Use the result of Exercise 10.3.7 to derive the asymptotic normality of the estimator of the location model parameter α, obtained by equating $W_n(\alpha)$ to zero.

10.4.1 Consider the model $g(x_i, \boldsymbol{\beta}) = \exp(\beta_0 + \beta_1 x_i)$ and work out the estimating equation in (10.4.3). Verify the asymptotic linearity results in (10.3.15) for this specific case and, hence, comment on the asymptotic normality of the estimates of β_0 and β_1 in (10.4.3).

10.4.2 Consider the *Hill model*, $g(x, \boldsymbol{\theta}) = \theta_1 x^{\theta_2}/(x^{\theta_2} + \theta_3^{\theta_2})$, where $\theta_1 > 0$, $\theta_2 > 0$, and $\theta_3 > 0$. Verify the asymptotic linearity result (in $\boldsymbol{\theta}$) in this case.

10.4.3 Consider the logit model $g(x, \boldsymbol{\theta}) = \{1 + \exp[-(x - \theta_1)/\theta_2]\}^{-1}$. Verify the asymptotic linearity (in θ_1 and θ_2) result and comment on the asymptotic normality of the estimators. Compare the results with those outlined in Section 9.4.

10.5.1 For normally distributed response variables, show that (10.5.20) reduce to the classical normal equations.

10.5.2 For the logistic regression model (10.5.11) obtain the GEE as given by (10.5.21) and compare it with the minimum logit equations obtained from (6.4.16).

10.5.3 Consider an extension of Example 10.5.2 to the case where the number of populations is greater than 2. Obtain the asymptotic distribution of the corresponding parameter vector.

10.5.1 Let Z_i, $i = 1, \ldots, s$, be independent random variables following the $\text{Bin}(n, \pi_i)$ distribution and assume that associated to each Z_i there is a q-vector of explanatory random variables $\boldsymbol{x}_i$. Consider the logistic regression model $\log \pi_i/(1 - \pi_i) = \boldsymbol{x}_i'\boldsymbol{\beta}$, $i = 1, \ldots, s$. Obtain the expressions for the iterative reweighted least-squares procedure and for the generalized least-squares estimator of $\boldsymbol{\beta}$, as well as for the variances of the corresponding asymptotic distributions.

10.6.1 For (10.7.3), show that if x is a discrete random variable having finitely many masspoints, $\beta(x)$ can be treated rather arbitrarily, but the estimators may have slower rate of convergence.

10.6.2 Consider the Cox (1972) model with a single explanatory variable x assuming only two values, 0 and 1. Show that $h(t, x)$ relates equivalently to the Lehmann (1953) model $\overline{G} \equiv \overline{F}^{\lambda}$.

10.6.3 In (10.7.13), let K be the standard normal density. Derive an expression for the asymptotic bias of $\widehat{f_{n,K}}(a)$.

11

Weak Convergence and Gaussian Processes

11.1 Introduction

The general asymptotics presented in Chapter 6 and 7 play an important role in large sample methods in statistical inference. Yet, there are some asymptotic methods in statistical inference that rest on more sophisticated probability tools developed mostly in the past forty years, albeit most of these sophisticated tools are beyond the reach of the contemplated level of this text. We therefore take an intermediate route, introduce the basic concepts in these weak convergence approaches in less abstract ways, and emphasize on their fruitful statistical use. Among the stochastic processes entering our contemplated coverage, there are a few ones that are most commonly used in statistical inference:

(i) *Partial sum processes*;
(ii) *empirical distributional processes*; and
(iii) *statistical functionals.*

In order to treat them in a unified way, there is a need to explain how the weak convergence results in Chapter 7 have led to more abstract but impactful *weak invariance principles* (Section 11.2). Section 11.3 deals with partial sum processes, and Section 11.4 with empirical processes. Section 11.5 is devoted to statistical functionals. Some applications of weak invariance principles are considered in Section 11.6. In particular, some resampling plans (e.g., jackknifing and bootstraping) are appraised in the light of these weak invariance principles the embedding of the Wiener process is outlined in Section 11.7. Some general remarks are presented in the last section.

11.2 Weak Invariance Principles

In Chapter 5 we have introduced some stochastic processes, including the random walk model, which are discrete time-parameter processes. Consider the simpler case of $X_n = Y_1 + \cdots + Y_n$, $n \geq 1$, $X_0 = 0$, where the Y_j are i.i.d. Bernoulli variables, such that

$$P(Y_j = 1) = P(Y_j = -1) = 1/2, \quad \text{for all } j \geq 1. \tag{11.2.1}$$

The passage from a random walk to a *diffusion process* hinges on some asymptotics concepts where the timepoint j is replaced by hj, $h > 0$, $j \geq 1$, the displacements ± 1 are replaced by $\pm\delta$, both h and δ are made to converge to 0 in a manner in which the discrete time parameter process converges in law to a Gaussian process. This process is called *Brownian motion* or *Wiener process* and has continuous sample paths. This transition from a discrete time parameter process to a continuous time parameter process relies on a *compactness* or *tightness* concept that we will explain.

Let us consider a second example. Let $U_1, \ldots, U_n$ be i.i.d. random variables having the Uniform[0, 1] distribution. Thus, $G(t) = P(U_i \leq t) = t$, for $0 \leq t \leq 1$. Also, the empirical distribution function is defined as $G_n(t) = n^{-1}\sum_{i=1}^{n} I(U_i \leq t)$, $0 \leq t \leq 1$. Define an empirical (distributional) process Z_n by

$$Z_n(t) = \sqrt{n}[G_n(t) - G(t)], \quad 0 \leq t \leq 1. \tag{11.2.2}$$

Then, $Z_n = \{Z_n(t),\ 0 \leq t \leq 1\}$ has continuous time parameter but, as G_n is a step function, it has jump discontinuities of magnitude $n^{-1/2}$. Here also Z_n converges in law to a Gaussian process called a *Brownian bridge* or a *tied-down Wiener process*, which has continuous sample paths.

In both the cases, the probability law, say P_n, is generated by the n-fold distribution of the n i.i.d. random variables (Y_j or U_j, $j = 1, \ldots, n$). On the other hand, the probability law, say P, governing the Gaussian process is defined on a different probability space. Hence, in order to study the weak convergence of $\{P_n\}$ to P, as $n \to \infty$, we need to introduce some basic criteria that provide the leading steps. These are (a) *convergence of finite-dimensional laws* and (b) *compactness* or *tightness*.

For any $m \geq 1$ and arbitrary $t_1 < \cdots < t_m$, let $P_n(t_1, \ldots, t_n)$ be the m-variate probability distribution of the m-vector $[Z_n(t_1), \ldots, Z_n(t_m)]'$. Similarly, let $P(t_1, \ldots, t_m)$ be the joint distribution of $[Z(t_1), \ldots, Z(t_m)]'$ for a suitable process $\mathbf{Z}$. If, for every (fixed) m and $t_1, \ldots, t_m$,

$$P_n(t_1, \ldots, t_m) \quad \text{converges weakly to } P(t_1, \ldots, t_m), \tag{11.2.3}$$

in the sense of Theorem 7.3.1, then we say that the finite-dimensional distributions of $\mathbf{Z}_n$ converge to those of $\mathbf{Z}$, as $n \to \infty$. The formulation of the tightness concept needs a bit deeper appraisal.

Let $C[0, 1]$ be the space of all continuous functions $f = \{f(t),\ t \in [0, 1]\}$ on $[0, 1]$. We may define the *modulus of continuity* $\omega_f(\delta)$ as

$$\omega_f(\delta) = \sup_{0 \leq s \leq t \leq s+\delta \leq 1} [|f(t) - f(s)|], \quad \delta > 0, \tag{11.2.4}$$

so that $f \in C[0, 1]$ if and only if

$$\omega_f(\delta) \to 0 \quad \text{as } \delta \downarrow 0. \tag{11.2.5}$$

Since $[0, 1]$ is a compact interval and $f \in C[0, 1]$ is uniformly continuous [by (11.2.5)], it follows that

$$\|f\| = \sup_{0 \leq t \leq 1} [|f(t)|] < \infty. \tag{11.2.6}$$

Perhaps, it would be convenient to introduce at this stage the *uniform metric* ρ as

$$\rho(f_1, f_2) = \|f_1 - f_2\| = \sup_{0 \leq t \leq 1} [|f_1(t) - f_2(t)|] \tag{11.2.7}$$

(the topology specified by ρ is termed the *uniform topology*).

In our context, $f = [f(t),\ 0 \leq t \leq 1]$ is generally a stochastic function. If P stands for the probability law (measure) (on $C[0, 1]$) associated with f, we say that P is *tight* or *relatively compact*, if for every $\varepsilon > 0$, there exists a compact K_ε (depending on ε), such that

$$P(f \in K_\varepsilon) > 1 - \varepsilon. \tag{11.2.8}$$

In view of (11.2.5), (11.2.6), and (11.2.8), we may formalize the tightness concept.

Tightness. A sequence $\{X_n = [X_n(t),\ t \in [0, 1]]\}$ of stochastic processes with probability measures $\{P_n\}$ (on $C[0, 1]$) satisfies the tightness axiom if and only if for every $\varepsilon > 0$, there exists an $M_\varepsilon < \infty$, such that

$$P_n(\|X_n\| > M_\varepsilon) \leq \varepsilon, \quad \forall n \geq n_o, \tag{11.2.9}$$

and

$$\lim_{\delta \downarrow 0} \left\{ \limsup_{n\to\infty} P_n[w_{X_n}(\delta) > \varepsilon] \right\} = 0. \tag{11.2.10}$$

Let us now go back to (11.2.1), and in the same fashion, we consider a sequence $\{Z_n = [Z_n(t),\ t \in [0, 1]]\}$ of stochastic processes (on $C[0, 1]$). Let $\{P_n\}$ be a sequence of probability laws (measures) on $C[0, 1]$ (induced by $\{Z_n\}$). Similarly, let Z be another random element of $C[0, 1]$ with a probability law (measure) P. (As in Chapter 7, $\{P_n\}$ and P are not necessarily defined on the same measure space.) Then, we have the following theorem.

***Theorem* 11.2.1.** *If $\{P_n\}$ is tight and the convergence of finite dimensional distributions in (11.2.3) holds for all $t_1, \ldots, t_m$ belonging to $[0, 1]$ and $m \geq 1$, then $\{P_n\}$ converges weakly to P, as $n \to \infty$ (i.e., $\{P_n\} \Rightarrow P$).*

See Billingsley (1968) for an excellent treatise of this weak convergence result (along with others). With this, we may formalize the definition of weak convergence in a general metric space S as follows.

Weak convergence in metric space. A sequence $\{P_n, n \geq 1\}$ of probability measures (on S) is said to converge weakly to a measure P (denoted by $P_n \Rightarrow P$) if

$$\int g dP_n \to \int g dP \quad \text{as} \quad n \to \infty, \tag{11.2.11}$$

for all g belonging to the space $C(S)$, the class of all bounded, continuous, real functions on S.

This definition implies that for every continuous functional $h(Z_n)$ of Z_n on S,

$$P_n \Rightarrow P \quad \Longrightarrow \quad h(Z_n) \xrightarrow{D} h(Z). \tag{11.2.12}$$

This last result is of immense statistical importance, and we will call it the *weak invariance principle* for $\{P_n\}$ (or $\{Z_n\}$). This definition is not confined only to the $C[0, 1]$ space, and more general metric spaces can readily be accommodated by formalizing the concept of tightness in the respective context and then rephrasing Theorem 11.2.1 accordingly. This was the main theme in Parthasarathy (1967) and Billingsley (1968), the two most notable texts in this novel area, and we like to advocate these sources for supplementary reading.

The space $C[0, 1]$ can be trivially enlarged to the space $C[0, T]$ for any finite T; even the space $C[0, \infty)$ can be included in this setup by a simple modification of the uniform metric

ρ in (11.2.7); see, for example, Sen (1981, p. 23). Let us consider now another important metric space, namely, $D[0, 1]$, which includes $C[0, 1]$ as a subspace and is rich enough to cover most of the important statistical applications we have in mind.

The metric space $D[0, 1]$. The space denoted by $D[0, 1]$ is the space of all $f = \{f(t),\ t \in [0, 1]\}$ such that $f(t)$ is right-continuous and has a left-hand limit that is, $f(t) = \lim_{\delta\downarrow 0} f(t + \delta)$ exists for every $0 \leq t < 1$, $f(t - 0) = \lim_{\delta\downarrow 0} f(t - \delta)$ exists for every $0 < t \leq 1$, but $f(t)$ and $f(t - 0)$ may not agree everywhere. Thus, f has only discontinuities of the first kind. A function $f \in D[0, 1]$ has a jump at $t \in [0, 1]$ if $|f(t) - f(t - 0)| > 0$.

Recall that for the $C[0, 1]$ space, $f(t) = f(t - 0)$ for every $t \in [0, 1]$, so that $C[0, 1]$ is a subspace of $D[0, 1]$. If is easy to verify that $\{Z_n\}$ in (11.2.2) belongs to $D[0, 1]$ but not to $C[0, 1]$. In view of possible jumps (and jump points) in the $D[0, 1]$ space, we need to modify the definitions in (11.2.4) and (11.2.7). Compared to (11.2.7), we introduce the *Skorokhod metric*

$$d_S(f_1, f_2) = \inf\{\varepsilon > 0 : \|\lambda(t) - t\| < \varepsilon \text{ and } \|f_1(t) - f_2(\lambda(t))\| < \varepsilon\}, \qquad (11.2.13)$$

where the uniform metric $\|\ \|$ is defined in (11.2.7) and $\lambda = \{\lambda(t),\ t \in [0, 1]\} \in \Lambda$, the class of all strictly increasing, continuous mappings $\lambda\colon [0, 1] \to [0, 1]$. (This metric generates the *Skorokhod (J_1-) topology* on $D[0, 1]$.) For a function $f \in D[0, 1]$, the definition of tightness in (11.2.8) is valid, but, in view of the Skorokhod metric in (11.2.13) [replacing the uniform metric in (11.2.7)], in the formulation of a compact K_ε, we may need to replace the modulus of continuity $\omega_f(\delta)$ in (11.2.4) by

$$\omega'_f(\delta) = \sup_{0\leq s\leq u\leq t\leq s+\delta\leq 1} \{\min[|f(u) - f(s)|, |f(t) - f(u)|]\}, \qquad (11.2.14)$$

which takes into account the possible jump discontinuities of $f(\in D[0, 1])$. It is easy to verify that for every $0 < \delta < 1/2$ and $f \in D[0, 1]$,

$$\omega'_f(\delta) \leq \omega_f(2\delta), \qquad (11.2.15)$$

so that the definition of compactness of $\{X_n\}$ $(\in C[0, 1])$ in (11.2.9)–(11.2.10) goes through for $D[0, 1]$-valued functions as well; we may replace $\omega_{X_n}(\delta)$ by $\omega'_{X_n}(\delta)$ in (11.2.10). In practice, often this may not be that crucial. As such, Theorem 11.2.1 and (11.2.11) hold for the $D[0, 1]$ space too, so that we may use (11.2.12) with advantage for plausible statistical applications. The inequality in (11.2.15) and Theorem 11.2.1 ensure that if $\{Z_n\}$ does not belong to the $C[0, 1]$ space (belongs to $D[0, 1]$), we may still use (11.2.3), (11.2.9) and (11.2.10) to ensure the weak convergence of $\{P_n\}$ to P. The crux of the problem is, therefore, to verify (11.2.9)–(11.2.10) along with (11.2.3).

As in the case with $C[0, 1]$, the space $D[0, 1]$ can also be readily extended to $D[0, T]$, for an arbitrary T $(< \infty)$. Even $D[0, \infty)$ can be covered with a modification of the Skorokhod metric in (11.2.13).

11.3 Weak Convergence of Partial Sum Processes

Let us consider first the following simple model, which includes the random walk model in (11.2.1) as a particular case. Let $\{X_i, i \geq 1\}$ be a sequence of i.i.d. random variables with finite first and second moments, and without any loss of generality, we let $\mu = \mathbb{E}(X) = 0$

and $\sigma^2 = \mathbb{E}(X^2) = 1$. Also, conventionally, we let $X_0 = 0$. Then, let

$$S_k = \sum_{j \le k} X_j, \quad k = 0, 1, 2, \dots. \tag{11.3.1}$$

Define $\lfloor a \rfloor = \max\{k \colon k \le a\}$ and for each $n \ge 1$,

$$W_n(t) = n^{-1/2} S_{\lfloor nt \rfloor}, \quad 0 \le t \le 1. \tag{11.3.2}$$

Note that

$$W_n(k/n) = n^{-1/2} S_k, \quad k = 0, 1, 2, \dots, n, \tag{11.3.3}$$

and

$$W_n(t) = W_n(k/n), \quad \frac{k}{n} \le t < \frac{k+1}{n}, \quad k \ge 0. \tag{11.3.4}$$

Thus, for each n, $\boldsymbol{W}_n = \{W_n(t), t \in [0, 1]\}$ belongs to the $D[0, 1]$ space (with the jump points k/n, $1 \le k \le n$). By virtue of (11.3.1) and (11.3.3), we note that

$$\mathbb{E}[W_n(k/n)] = 0, \quad \mathbb{E}[W_n^2(k/n)] = k/n \tag{11.3.5}$$

and that, for $q \ge k$,

$$\begin{aligned}\mathbb{E}[W_n(k/n)W_n(q/n)] &= n^{-1}\mathbb{E}(S_k S_q) \\ &= n^{-1}\mathbb{E}\{S_k[S_k + (S_q - S_k)]\} \\ &= n^{-1}[\mathbb{E}(S_k^2) + 0] = k/n.\end{aligned} \tag{11.3.6}$$

Therefore, by (11.3.4), (11.3.5), and (11.3.6), we have

$$\mathbb{E}[W_n(t)] = 0 \quad \text{and} \quad \mathbb{E}[W_n(s)W_n(t)] = n^{-1}\lfloor n(s \wedge t) \rfloor, \tag{11.3.7}$$

So, we obtain that for every $s, t \in [0, 1]$

$$\lim_{n \to \infty} \mathbb{E}[W_n(t)W_n(s)] = \lim_{n \to \infty} \mathbb{C}\text{ov}[W_n(t), W_n(s)] = s \wedge t \tag{11.3.8}$$

Now, for every (fixed) t, $0 < t \le 1$, $W_n(t) = n^{-1/2} S_{\lfloor nt \rfloor}$ involves a sum of $\lfloor nt \rfloor$ i.i.d. random variables where, as $n \to \infty$, $\lfloor nt \rfloor \to \infty$ and $n^{-1}\lfloor nt \rfloor \to t$. Hence, by the Central Limit Theorem 7.3.1, as $n \to \infty$,

$$W_n(t) \xrightarrow{D} \mathcal{N}(0, t), \quad t \in (0, 1]. \tag{11.3.9}$$

Let us next consider two points t_1, t_2, $0 < t_1 < t_2 \le 1$, and take $k_{n_1} = \lfloor nt_1 \rfloor$ and $k_{n_2} = \lfloor nt_2 \rfloor$. Then, note that

$$W_n(t_2) - W_n(t_1) = n^{-1/2} \sum_{i = k_{n_1}+1}^{k_{n_2}} X_i \tag{11.3.10}$$

is independent of $W_n(t_1)$, and further, by (11.3.6), that as $n \to \infty$

$$\mathbb{E}\{[W_n(t_2) - W_n(t_1)]^2\} = (k_{n_2} - k_{n_1})/n \to (t_2 - t_1). \tag{11.3.11}$$

In general, for any (fixed) $m \ge 1$ and $0 < t_1 < \cdots < t_m \le 1$, by the same decomposition as in (11.3.10), it follows that

$$W_n(t_j) - W_n(t_{j-1}), \quad 1 \le j \le m, \tag{11.3.12}$$

with $t_0 = 1$, are mutually stochastically independent, and as in (11.3.11), for each $j = 1, \ldots, m$, as $n \to \infty$

$$\mathbb{E}\{[W_n(t_j) - W_n(t_{j-1})]^2\} = n^{-1}\{\lfloor nt_j \rfloor - \lfloor nt_{j-1} \rfloor\} \to (t_j - t_{j-1}). \quad (11.3.13)$$

As such, we may consider the m-vector

$$[W_n(t_1), \ldots, W_n(t_m)]' \quad (11.3.14)$$

and writing $U_{nj} = W_n(t_j) - W_n(t_{j-1})$, $1 \leq j \leq m$, we rewrite (11.3.14) as

$$(U_{n1}, U_{n1} + U_{n2}, \ldots, U_{n1} + \cdots + U_{nm})', \quad (11.3.15)$$

where the U_{nj} are mutually independent and asymptotically normal with 0 mean and variance $t_j - t_{j-1}$, for $j = 1, \ldots, m$. Thus, (11.3.14) has asymptotically a multinormal law with null mean vector and dispersion matrix $((t_j \wedge t_{j'}))$. Let us now consider a Gaussian random function $\boldsymbol{W} = \{W(t), t \geq 0\}$ having the following properties:

(i) The process has independent and homogeneous increments.
(ii) For every $s < t$, $W(t) - W(s)$ is Gaussian with 0 mean and variance $t - s$.

Then, $\boldsymbol{W}$ is called a standard *Brownian motion* or *Wiener process* on $\mathbb{R}^+ = [0, \infty)$. Note that by virtue of (i) and (ii), we have

$$\begin{aligned}\mathbb{E}[W(t)W(s)] &= \mathbb{E}\{W^2(s) + W(s)[W(t) - W(s)]\}\\ &= \mathbb{E}[W^2(s)] = s, \quad s \leq t.\end{aligned}$$

Also, by virtue of (i) and (ii), for every $m \geq 1$ and $t_1 < \cdots < t_m$, the vector $[W(t_1), \ldots, W(t_m)]'$ has the m-variate normal distribution with null mean vector and dispersion matrix $((t_j \wedge t_{j'}))$. Also, since for every $t > s$, $W(t) - W(s)$ is $\mathcal{N}(0, t - s)$, we can argue that the sample paths of W are continuous, with probability one. By an abuse of notation, we denote by $\boldsymbol{W} = \{W(t),\ t \in [0, 1]\}$, the standard Brownian motion process on $[0, 1]$. Then $\boldsymbol{W}$ belongs to the $C[0, 1]$ space, with probability one. Moreover, for every positive integer k and $t > s$, by virtue of (ii)

$$\mathbb{E}[(|W(t) - W(s)|)^{2k}] = \frac{(2k)!}{2^k k!}(t - s)^k, \quad (11.3.16)$$

so that by using (11.3.16) with $k = 2$, in a version of the Billingsley Inequality [Billingsley (1968)] with continuous time parameter, it follows that the Gaussian probability measure (on $C[0, 1]$) induced by $\boldsymbol{W}$ is tight.

By the discussion following (11.3.15) and the above characterization of $\boldsymbol{W}$, it follows that for every (fixed) $m \geq 1$ and $0 < t_1 < \cdots < t_m \leq 1$, as $n \to \infty$,

$$[W_n(t_1), \ldots, W_n(t_m)] \xrightarrow{D} [W(t_1), \ldots, W(t_m)], \quad (11.3.17)$$

and $W_n(0) = W(0) = 0$, with probability one. Thus, (11.2.3) holds. Therefore, to verify Theorem 11.2.1, in this case, we need to verify the two conditions in (11.2.9) and (11.2.10) for $\boldsymbol{W}_n$. Recall that

$$\|\boldsymbol{W}_n\| = n^{-1/2} \max_{1 \leq k \leq n} |S_k|, \quad (11.3.18)$$

where the S_k, $k > 0$, form a zero-mean martingale sequence. Hence, by the Kolmogorov Maximal Inequality (Theorem 6.4.1), for every $K > 0$

$$P(\|\boldsymbol{W}_n\| > K) \leq K^{-2}n^{-1}\mathbb{E}(S_n^2) = K^{-2} < \varepsilon, \tag{11.3.19}$$

by choosing K adequately large. Thus, (11.2.9) holds. Next, we may note that, if for every $0 < \delta \leq \delta_0$, $n \geq n_0$,

$$P\left[\sup_{0\leq u\leq\delta} |W_n(s+u) - W_n(s)| > \varepsilon\right] < \eta\delta, \tag{11.3.20}$$

for every $s \in [0, 1)$, where $\varepsilon > 0$ and $\eta > 0$ are arbitrary, then (11.2.10) holds for $\boldsymbol{W}_n$. Thus, if the X_i in (11.3.1) have a finite fourth order moment, we may again use the Kolmogorov Maximal Inequality for the (submartingale) $[W_n(s+u) - W_n(s)]^2$, $0 \leq u \leq \delta$, and verify that (11.3.20) holds. So that, by Theorem 11.2.1, we may conclude that as $n \to \infty$

$$\boldsymbol{W}_n \xrightarrow{D} \boldsymbol{W}, \quad \text{on } D[0, 1]; \tag{11.3.21}$$

or, equivalently, writing $\boldsymbol{P}_n = P(\boldsymbol{W}_n)$, $n \geq n_0$,

$$\boldsymbol{P}_n \Rightarrow \boldsymbol{P} = P(\boldsymbol{W}). \tag{11.3.22}$$

The fourth moment condition, as assumed above, is a sufficient but not necessary condition; (11.3.20) holds even under finiteness of the second moment. To verify this, we may need the following inequality [due to Brown (1971b)], which is presented without proof.

***Theorem* 11.3.1.** *Let $\{X_n,\ n \geq 1\}$ be a submartingale. Then, for every $t > 0$,*

$$P\left(\max_{1\leq k\leq n} |X_k| > 2t\right) \leq t^{-1}\mathbb{E}[|X_n|I(|X_n| \geq t)]. \tag{11.3.23}$$

For any $s \geq 0$, let $a = \lfloor ns \rfloor$ and $a + b = \lfloor n(s+\delta) \rfloor$. Then

$$\begin{aligned}
&\sup_{0\leq u\leq\delta} [|W_n(s+u) - W_n(s)|] \\
&\quad = n^{-1/2} \max_{1\leq k\leq b} |X_{a+1} + \cdots + X_{a+k}| \\
&\quad \stackrel{D}{=} n^{-1/2} \max_{1\leq k\leq b} |S_k|,
\end{aligned} \tag{11.3.24}$$

where $\stackrel{D}{=}$ stands for the equality of distributions, and where the $|S_k|$, $k \geq 0$, form a submartingale sequence. Thus, by (11.3.23) and (11.3.24), we have for every $\varepsilon > 0$,

$$\begin{aligned}
&P\left[\sup_{0\leq u\leq\delta} |W_n(s+u) - W_n(s)| > \varepsilon\right] \\
&\quad = P\left(\max_{1\leq k\leq b} |S_k| > \varepsilon\sqrt{n}\right) \\
&\quad \leq \frac{2}{\varepsilon\sqrt{n}}\mathbb{E}[|S_b|I(|S_b| \geq \varepsilon\sqrt{n}/2)] \\
&\quad \leq \frac{2}{\varepsilon\sqrt{n}}\sqrt{\mathbb{E}(S_b^2)}[P(|S_b| \geq \varepsilon\sqrt{n}/2)]^{1/2}.
\end{aligned} \tag{11.3.25}$$

Note that $n^{-1}\mathbb{E}(S_b^2) \le \delta$, whereas for any $\delta > 0$, as $n \to \infty$,

$$n^{-1/2} S_b \xrightarrow{D} \mathcal{N}(0, \delta), \tag{11.3.26}$$

so that, as $n \to \infty$, we have

$$P\left(|S_b| \ge \frac{1}{2}\varepsilon\sqrt{n}\right) = P\left(n^{-1/2}\frac{|S_b|}{\sqrt{\delta}} \ge \frac{\varepsilon}{2\sqrt{\delta}}\right) \longrightarrow 2\left[1 - \Phi\left(\frac{\varepsilon}{2\sqrt{\delta}}\right)\right]. \tag{11.3.27}$$

Recall that as $x \to \infty$, $1 - \Phi(x) \cong (2\pi)^{-1/2}x^{-1}\exp(-1/2x^2)$, so that for every (fixed) $\varepsilon > 0$, choosing $\delta > 0$ sufficiently small (so that $\varepsilon/2\sqrt{\delta}$ is large), we may take the right-hand side of (11.3.25), for large n, as

$$\frac{4\delta^{3/4}\varepsilon^{-3/2}}{(2\pi)^{1/4}} e^{-\varepsilon^2/16\delta} < \eta\delta, \quad \delta \le \delta_0, \tag{11.3.28}$$

as $4\delta^{-1/4}\varepsilon^{-3/2}\exp(-\varepsilon^2/16\delta) \to 0$ as $\delta \downarrow 0$. Therefore, (11.3.20) holds under the classical second moment condition (needed for the Central Limit Theorem), and, hence, (11.3.21) holds under the same condition as in Theorem 7.3.1.

The weak convergence result in (11.3.21) is not confined only to the simplest case treated in (11.3.1)–(11.3.4). This holds under the full generality of the Lindeberg–Feller Central Limit Theorem 7.3.3. The formulation is quite simple and is presented below.

Let $\{X_i, i \ge 1\}$ be a sequence of independent random variables with $\mu_i = \mathbb{E}(X_i)$ and $\sigma_i^2 = \mathbb{V}\text{ar}(X_i) < \infty$, $i \ge 1$. Let $X_0 = \mu_0 = 0$, and for every $k \ge 1$, let

$$S_k^0 = \sum_{j \le k}(X_j - \mu_j), \quad \tau_k^2 = \sum_{j \le k}\sigma_j^2. \tag{11.3.29}$$

For every $n \ge 1$, define $\boldsymbol{W}_n = \{W_n(t), t \in [0, 1]\}$ by letting

$$W_n(t) = \tau_n^{-1} S_{k_n(t)}^0, \quad 0 \le t \le 1, \tag{11.3.30}$$
$$k_n(t) = \max\{k : \tau_k^2 \le t\tau_n^2\}, \quad 0 \le t \le 1. \tag{11.3.31}$$

Then $\boldsymbol{W}_n$ belongs to the $D[0, 1]$ space. Assume that the classical Lindeberg condition holds, *i.e.*, for every positive ε, as $n \to \infty$,

$$\frac{1}{\tau_n^2}\sum_{i=1}^{n}\mathbb{E}[(X_i - \mu_i)^2 I(|X_i - \mu_i| > \varepsilon\tau_n)] \to 0. \tag{11.3.32}$$

The computations leading to (11.3.8) and (11.3.13) are routine in this setup too, whereas the Lindeberg–Feller Central Limit Theorem 7.3.3 ensures (11.3.9) as well as its extension to the finite-dimensional distributions in (11.3.14). The tightness part of $\boldsymbol{W}_n$ follows from the martingale property of the S_k^0 and, hence, we arrive at the following theorem.

***Theorem* 11.3.2.** *Under the same hypotheses of the Lindeberg–Feller Theorem 7.3.3, for $\boldsymbol{W}_n$ defined by (11.3.30) and (11.3.31), the weak convergence result in (11.3.21) holds.*

Because of the implications in (11.2.12) [i.e., for arbitrary continuous f], Theorem 11.3.2 may also be termed a functional central limit theorem. Let us illustrate the scope of such a functional central limit theorem by some examples.

Example* 11.3.1 *(CLT for random sample size). As in (11.3.1) consider a sequence $\{X_i, i \geq 1\}$ of i.i.d. random variables whose mean μ and variance $\sigma^2 < \infty$ are unknown. Based on a sample of size n (i.e., $X_1, \ldots, X_n$), conventional estimators of μ and σ^2 are, respectively,

$$\overline{X}_n = n^{-1}\sum_{i=1}^{n} X_i \quad \text{and} \quad s_n^2 = \frac{1}{n-1}\sum_{i=1}^{n}(X_i - \overline{X}_n)^2. \tag{11.3.33}$$

In a variety of situations, the sample size n is not fixed a priori but is based on a *stopping rule*, that is, $n\ (= N)$ is a positive integer value random variable, such that for every $n \geq 1$, the event $[N = n]$ depends on $(X_1, \ldots, X_n)$. For example, suppose that one wants to find a confidence interval for μ (based on $\overline{X}_n, s_n^2$) such that (a) the coverage probability is equal to some fixed $1 - \alpha$ and (b) the width of the interval is bounded from above by some (fixed) $2d, d > 0$. For this problem, no fixed sample-size solution exists [Dantzig (1940)] and one may have to consider two- (or multi-)stage procedures [Stein (1945)] or some sequential procedures [Chow and Robbins (1965)]. In the Stein (1945) two-stage procedure, an initial sample of size n_0 (≥ 2) is incorporated in the definition of the stopping number N as

$$N_d = \max\{n_0, \lfloor t_{1,1-\alpha/2}^2 d^{-2} s_{n_0}^2 \rfloor + 1\}, \tag{11.3.34}$$

whereas in the Chow and Robbins (1965) sequential case, we have

$$N_d = \min\{k : k \geq n_0 \quad \text{and} \quad s_k^2 \leq d^2 k / a_k^2\}, \tag{11.3.35}$$

where $n_0 \geq 2$ is the starting sample size and $a_k \to z_{1-\alpha/2}$, (as $k \to \infty$); it may as well be taken equal to $z_{\alpha/2}$.

If σ^2 were known and d (> 0) is small, by virtue of the Central Limit Theorem 7.3.1, a prefixed sample size (n_d) would have led to the desired solution, where

$$n_d = d^{-2}\sigma^2 z_{1-\alpha/2}^2. \tag{11.3.36}$$

By (6.3.84), $s_n^2 \xrightarrow{a.s.} \sigma^2$, whenever $\sigma^2 < \infty$. Hence, comparing (11.3.35) and (11.3.36), we note that

$$N_d / n_d \xrightarrow{a.s.} 1 \quad \text{as } d \downarrow 0. \tag{11.3.37}$$

On the other hand, N_d is stochastic, whereas n_d is not. Moreover, as $d \downarrow 0$, by Theorem 7.3.1,

$$n_d^{1/2}(\overline{X}_{n_d} - \mu)/\sigma \xrightarrow{D} \mathcal{N}(0, 1).$$

Thus, it may be of interest to see whether by virtue of (11.3.37), n_d can be replaced by N_d, when $d \to 0$. The answer is in the affirmative. Note that

$$\begin{aligned} N_d^{1/2}\frac{\overline{X}_{N_d} - \mu}{s_{N_d}} &= \left(\frac{\sigma}{s_{N_d}}\right) N_d^{-1/2}\left(\frac{S_{N_d}^0}{\sigma}\right) \\ &= \left(\frac{\sigma}{s_{N_d}}\right)\left(\frac{n_d}{N_d}\right)^{1/2} W_{n_d}(N_d/n_d), \end{aligned} \tag{11.3.38}$$

where the S_k^0 and $W_n(k/n)$ are defined as in (11.3.29) and (11.3.30). Recall that by the tightness property of the probability measure $\boldsymbol{P}_{n_d}$ induced by $\boldsymbol{W}_{n_d}$, for every $\varepsilon > 0$ and $\eta > 0$, there exist a $\delta_0 < 1$ and a sample size n_0, such that for $n \geq n_0$

$$P[|W_n(t) - W_n(1)| > \varepsilon, \quad \text{for some } t : |t - 1| < \delta] < \eta. \tag{11.3.39}$$

Since $N_d/n_d \xrightarrow{P} 1$ and $\sigma/s_{N_d} \xrightarrow{P} 1$, as $d \downarrow 0$, by (11.3.38), (11.3.39), and Slutsky Theorem 7.5.2, we have for $d \downarrow 0$,

$$\left|N_d^{1/2}(\overline{X}_{N_d} - \mu)/s_{N_d} - W_{n_d}(1)\right| \xrightarrow{P} 0, \tag{11.3.40}$$

where $W_{n_d} \xrightarrow{D} \mathcal{N}(0, 1)$. A similar result holds for (11.3.34) whenever n_0 $[= n_0(d)]$ increases with $d \downarrow 0$. Let us now formalize the asymptotic normality result in (11.3.40) in a general form. Define $\{\overline{X}_n, n \geq 1\}$ as in (11.3.33) and let N_n be a positive integer-valued random variable indexed by n, a nonstochastic positive integer such that

$$N_n/n \xrightarrow{P} C\ (> 0). \tag{11.3.41}$$

Then, whenever the Central Limit Theorem holds for $\overline{X}_n$, it holds for $\overline{X}_{N_n}$ as well. The key to this solution is given by (11.3.39), which is a by-product of the tightness part of the weak convergence of $\boldsymbol{W}_n$ to $\boldsymbol{W}$ (which holds under no extra conditions on the Central Limit Theorem). In this way, we do not need to verify the Anscombe (1952) condition

$$P\left(\max_{m:|m-n|<\delta n} \sqrt{n}|\overline{X}_m - \overline{X}_n| > \varepsilon\right) < \eta \tag{11.3.42}$$

for $n \geq n_0(\varepsilon, \eta)$ and $\delta \leq \delta_0$. In fact, the Anscombe condition is itself a by-product of the tightness part of $\boldsymbol{W}_n$. Actually, there are even stronger results relating to the weak convergence of $\{W_{N_n}\}$ to W, for these results, see Billingsley (1968). □

Example* 11.3.2 *[Repeated significance test (RST)]. Consider the same model as in Example 11.3.1, and suppose that one may want to test for

$$H_0 : \mu = \mu_0 = 0 \quad \text{versus} \quad H_1 : \mu \neq 0. \tag{11.3.43}$$

In order that the test has a good power (when μ is not too far away from the null value), one generally requires that the sample size n be large. In clinical trials/medical studies and other areas, often, the sample units are not all available at the same time point, and, hence, their collection may demand a considerable *waiting time*. In this context, *interim analyses* are sometimes made on the accumulating data set whereby either periodically over time a test is made for H_0 versus H_1 in (11.3.43) or after each observation is available, the test statistic is updated. See Armitage (1975) for an excellent statistical account of RST in medical studies. In either way, the target sample size n provides an upper bound for the actual sample size, and we have a sequence $\{n_1, \ldots, n_k = n\}$ of increasing sample sizes, such that, at each n_j, the test statistic is used to draw statistical conclusions; the number k may be fixed in advance or may even be equal to n (i.e., $n_j = j$, $1 \leq j \leq n$). Also, if periodically over time is the prescription, then the n_j may also be stochastic. For the model (11.3.43), if σ is assumed to be given, based on a sample of size m, an appropriate test statistic is

$$T_m^* = m^{1/2}\overline{X}_m/\sigma. \tag{11.3.44}$$

Recall that $\{T_m, m \leq n\}$ contains a set of test statistics that are not independent. Even if they were independent, making multiple tests, each one at a level of significance, say, α $(0 < \alpha < 1)$, can easily make the actual overall level of significance quite large (compared to α). Thus, the basic issue is how to control the overall significance level in this RST scheme. Such a scheme may also be regarded as a version of a *truncated sequential probability*

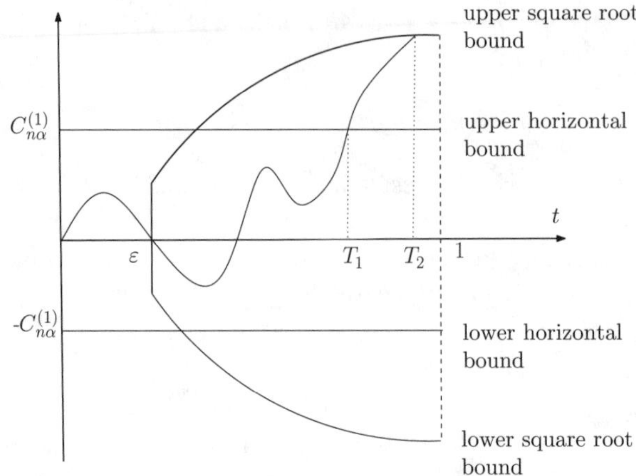

Figure 11.3.1. Stopping times for repeated significance tests. T_i = stopping time for $\boldsymbol{W}_n$ according to the ith type bound, $i = 1, 2$ $\varepsilon(> 0)$ = lower truncation point for square root bound.

ratio test (TSPRT) with truncation at $N = n$. However, there are generally some subtle differences in the boundaries relating to the RST and TSPRT schemes. Nevertheless, we may motivate the large-sample theory in a unified manner as follows.

Define $\boldsymbol{W}_n$ as in (11.3.30) and (11.3.31) with $\mu_j = 0, \sigma_j = \sigma$, for all $j \geq 1$. Plot $\boldsymbol{W}_n(t)$ against $t \in [0, 1]$ as indicated in Figure 11.3.1. Then we may describe the RST in terms of $\boldsymbol{W}_n$ as follows. First, consider the RST in the form of a TSPRT, resulting in horizontal boundaries. For a two-sided alternative [as in (11.3.43)], we take the test statistic as

$$Z_n^{(1)} = \max_{0 \leq k \leq n} [|W_n(k/n)|] = \sup_{0 \leq t \leq 1} [|W_n(t)|], \tag{11.3.45}$$

so that the stopping time T_1 is given by

$$T_1 = \min\{\min[k \geq 1 : \ |W_n(k/n)| \geq C_{n\alpha}^{(1)}], \ n\}, \tag{11.3.46}$$

where $C_{n\alpha}$ specifies the boundary, such that

$$P(Z_n^{(1)} \geq C_{n\alpha}^{(1)} | H_0) = \alpha. \tag{11.3.47}$$

At this stage, we appeal to the weak convergence result in (11.3.21), so that for large n, we may approximate $C_{n\alpha}^{(1)}$ by $C_{\alpha}^{(1)}$, where

$$P\left(\sup_{0 \leq t \leq 1} |W(t)| \geq C_{\alpha}^{(1)} \right) = \alpha. \tag{11.3.48}$$

There are some well-known results on the fluctuations of Wiener processes, which may be used in this context. For example, we know that for every $\lambda > 0$ [Anderson (1960)]

$$P\left(\sup_{0 \leq t \leq T} |W(t)| < \lambda \right) = \sum_{k=-\infty}^{+\infty} (-1)^k \{\Phi[(2k+1)\lambda/\sqrt{T}] - \Phi[(2k-1)\lambda/\sqrt{T}]\}, \tag{11.3.49}$$

where Φ is the standard normal distribution function; for the one-sided case, we have a much simpler expression. For all $\lambda \geq 0$,

$$P\left[\sup_{0 \leq t \leq T} W(t) < \lambda\right] = 1 - 2[1 - \Phi(\lambda/\sqrt{T})]. \qquad (11.3.50)$$

Thus, for the one-sided case, $C_\alpha^{(1)} = z_{1-\alpha/2}$, whereas for the two-sided case, $C_\alpha^{(1)} \leq z_{1-\alpha/4}$, and the approximate equality sign holds for α not too large. There are some related results for drifted Wiener processes [i.e., for $W(t) + \mu(t)$, with linear $\mu(t)$] [e.g., Anderson (1960)], which may even be used to study local asymptotic power of the test based on $Z_n^{(1)}$.

Let us next consider the case of RST based on the sequence in (11.3.44). Recall that by definition, for $m \leq n$,

$$T_m^* = W_n(m/n)[(m/n)^{-1/2}]. \qquad (11.3.51)$$

So, we need to use a square root boundary as has been suggested in Figure 11.3.1. There is a technical point that merits some consideration at this stage. Note that (11.3.21) does not ensure that

$$[W_n(t)/q(t),\ 0 \leq t \leq 1] \xrightarrow{D} [W(t)/q(t),\ 0 \leq t \leq 1] \qquad (11.3.52)$$

for every $\{q(t), t \in [0, 1]\}$. If $q(t)$ is bounded away from 0 (from below), then (11.3.52) holds. But, in (11.3.51), we have $q(t) = t^{1/2}$, so that $q(t) \to 0$ as $t \to \infty$. Also,

$$\int_0^1 q^{-2}(t)\mathrm{d}(t) = \int_0^1 t^{-1}\mathrm{d}t = \infty.$$

Hence, we are not in a position to use (11.3.21) to conclude on (11.3.52) for this specific $q(t)$. In fact, by the law of iterated logarithm for the Wiener process,

$$\lim_{t \downarrow 0} |W(t)|/(-2t \log t)^{1/2} = 1 \quad \text{a.s.} \qquad (11.3.53)$$

so that $|W(t)/\sqrt{t}| \xrightarrow{a.s.} +\infty$ as $t \to 0$. To avoid this technical problem, often a small truncation (at 0) is made, so that one considers the statistic

$$Z_n^{(2)} = \max_{k_0 \leq k \leq n} (|T_k^*|), \qquad (11.3.54)$$

where

$$k_0/n \to \varepsilon > 0. \qquad (11.3.55)$$

Then, we have, under H_0,

$$Z_n^{(2)} \xrightarrow{D} \sup_{\varepsilon \leq t \leq 1} [|W(t)|/\sqrt{t}] \qquad (11.3.56)$$

so that the cutoff point $C_{\alpha,\varepsilon}^{(2)}$ is defined by

$$P[|W(t)/\sqrt{t}| \leq C_{\alpha,\varepsilon}^{(2)},\quad \varepsilon \leq t \leq 1] = 1 - \alpha. \qquad (11.3.57)$$

It is clear that the choice of ε has an important role in this context. The larger the value of ε, the smaller $C_{\alpha,\varepsilon}^{(2)}$ is. Then, for $k^* = \min\{k \geq k_0 : T_k^* \geq C_{\alpha,\varepsilon}^{(2)}\}$, the stopping time T_2 is defined as

$$T_2 = \begin{cases} k^* & \text{if } k^* \leq n, \\ n & \text{otherwise.} \end{cases} \qquad (11.3.58)$$

For both T_1 and T_2 in (11.3.46) and (11.3.58), the corresponding stopping numbers are given by nT_1 and nT_2, respectively. Although both the tests have significance level $0 < \alpha < 1$, their power properties (and the behavior of the stopping numbers) depend on the kind of alternatives and the choice of $\varepsilon > 0$.

In some cases, only a (fixed) finite number of tests are made on the accumulating data. For example, for some prefixed $k \geq 2$, at sample sizes $n_1, \ldots, n_k$, the statistics $T^*_{n_1}, \ldots, T^*_{n_k}$ are computed, so that the test statistic, similar to in (11.3.54) is given by

$$Z_n^{(3)} = \max_{j=1,\ldots,k} (|T_j^*|). \tag{11.3.59}$$

This involves the joint distribution of the vector

$$[W_n(n_1/n)/\sqrt{n_1/n}, \ldots, W_n(n_k/n)/\sqrt{n_k/n}]. \tag{11.3.60}$$

If the n_j/n are prefixed, then the critical level $C_\alpha^{(3)}$ can be obtained by considering the multinormal distribution with null mean vector and dispersion matrix

$$(([n_i/n_{j'}]^{-1/2})). \tag{11.3.61}$$

However, for $k \geq 3$, this task can become quite involved. If k is ≥ 8 and the n_i/n are scattered over $(\varepsilon, 1)$ evenly, $C_{\alpha,\varepsilon}^{(2)}$ may provide a reasonably close upper bound for $C_\alpha^{(3)}$. Again, the larger the value of ε, the better this approximation is. □

We conclude this section with some additional remarks on the weak convergence of $\{\boldsymbol{W}_n\}$ (to $\boldsymbol{W}$). In Chapter 7, we have briefly discussed the generalizations of the CLT to both triangular schemes of random variables and some dependent sequences too. Extensions of the weak convergence result in (11.3.21) to such triangular schemes of random variables or to dependent sequences have been worked under very general conditions. The conditions are essentially the same as in the case of the respective CLTs, but instead of requiring them to be true at the sample size n only, we may need that for each $t \in (0, 1]$, they hold for the respective partial sample size $n(t)$. For some details, see Sen (1981, chapter 2) and Sen (1985) where various nonparametric statistics have been incorporated in the formulation of suitable weak invariance principles.

11.4 Weak Convergence of Empirical Processes

We start with the (reduced) empirical (distribution) process

$$Z_n = [Z_n(t), \quad 0 \leq t \leq 1],$$

defined by (11.2.2). Recall that $G_n(t)$, $0 \leq t \leq 1$, is a step function having n jumps of magnitude $1/n$ at each of the order statistics $0 < Y_{n:1} < \cdots < Y_{n:n} < 1$. Therefore, Z_n has also n points of discontinuity, namely, the $Y_{n:i}$, $1 \leq i \leq n$, and at each of these points, the jump is of the magnitude $n^{-1/2}$ (which can be made small when $n \to \infty$). Thus, for every finite n, Z_n belongs to the $D[0, 1]$ space. Also, note that by (11.2.2), we have

$$\mathbb{E}[Z_n(t)] = 0, \quad 0 \leq t \leq 1, \tag{11.4.1}$$

$$\begin{aligned}\mathbb{E}[Z_n(s)Z_n(t)] &= \mathbb{E}[I(Y_1 \le s) - s][I(Y_1 \le t) - t] \\ &= \mathbb{E}[I(Y_1 \le s)I(Y_1 \le t)] - st \\ &= \min(s, t) - st, \quad 0 \le s, t \le 1. \end{aligned} \tag{11.4.2}$$

Moreover, for arbitrary $m \ge 1$ and $0 \le t_1 < \cdots < t_m \le 1$, consider the vector $[Z_n(t_1), \ldots, Z_n(t_m)]' = \boldsymbol{Z}_{nm}$, say. Then, for an arbitrary $\boldsymbol{\lambda} = (\lambda_1, \ldots, \lambda_m)'$, we have

$$\boldsymbol{\lambda}'\boldsymbol{Z}_{nm} = n^{-1/2}\sum_{i=1}^{n}\left\{\sum_{j=1}^{m}\lambda_j[I(Y_i \le t_j) - t_j]\right\} = n^{-1/2}\sum_{i=1}^{n}U_i, \tag{11.4.3}$$

where the U_i are i.i.d. random variables, with $\mathbb{E}(U_i) = 0$ and

$$\mathbb{E}(U_i^2) = \boldsymbol{\lambda}'\boldsymbol{\Gamma}_m\boldsymbol{\lambda}, \tag{11.4.4}$$

where $\boldsymbol{\Gamma}_m = ((t_i \wedge t_j - t_it_j))$ is p.d. Hence, by the multivariate Central Limit Theorem 7.3.9, we claim that as $n \to \infty$,

$$\boldsymbol{Z}_{nm} \xrightarrow{D} \boldsymbol{Z}_m = (Z_1, \ldots, Z_m)', \tag{11.4.5}$$

where

$$\boldsymbol{Z}_m \sim \mathcal{N}_m(\boldsymbol{0}, \boldsymbol{\Gamma}_m). \tag{11.4.6}$$

In Section 11.3, we have introduced a special Gaussian function W, called the Brownian motion or the Wiener process. Let us define a stochastic process $\boldsymbol{Z} = \{Z(t),\ t \in [0, 1]\}$ by letting

$$Z(t) = W(t) - tW(1), \quad 0 \le t \le 1, \tag{11.4.7}$$

where $\boldsymbol{W} = \{W(t),\ 0 \le t \le 1\}$ is a standard Wiener process on [0,1]. Then $\boldsymbol{Z}$ is a Gaussian process. Also,

$$\mathbb{E}[Z(t)] = t - t = 0, \quad 0 \le t \le 1; \tag{11.4.8}$$

and

$$\begin{aligned}\mathbb{E}[Z(s)Z(t)] &= \mathbb{E}[W(s)W(t)] + st\mathbb{E}[W^2(1)] \\ &\quad - t\mathbb{E}[W(s)W(1)] - s\mathbb{E}[W(t)W(1)] \\ &= \min(s, t) + st - st - st = \min(s, t) - st. \end{aligned} \tag{11.4.9}$$

As such, for arbitrary $m \ge 1$ and $0 \le t_1 < \cdots < t_m \le 1$, the vector $[Z(t_1), \ldots, Z(t_m)]'$ follows a $\mathcal{N}_m(\boldsymbol{0}, \boldsymbol{\Gamma}_m)$ distribution with $\boldsymbol{\Gamma}_m = ((\min\{t_j, t_{j'}\} - t_jt_{j'}))$. Thus, by (11.4.5) and (11.4.7)–(11.4.9), the finite-dimensional distribution of $\boldsymbol{Z}_n$ converges to that of Z, as $n \to \infty$. Moreover, since $\boldsymbol{W}$ is tight and t belongs to a compact interval [0, 1], $\boldsymbol{Z}$ is also tight. The process $\boldsymbol{Z} = \{Z(t), 0 \le t \le 1\}$ is called a *Brownian bridge* or a *tied-down Wiener process*. We may remark that $Z(0) = Z(1) = 0$, with probability 1, so that $\boldsymbol{Z}$ is tied down (to 0) at both ends of [0, 1], whereas its perceptible fluctuations are in the open interval $0 < t < 1$, mostly in the central part. This might have been the reason why the terms bridge or tied-down part became associated with $\boldsymbol{Z}$. We may also reverse the role of $\boldsymbol{Z}$ and $\boldsymbol{W}$ in (11.4.7) in a different representation. Consider the stochastic process

$\boldsymbol{\xi} = \{\xi(t),\ 0 \le t < \infty\}$, where

$$\xi(t) = (t+1)Z(t/(t+1)), \quad 0 \le t < \infty. \tag{11.4.10}$$

Then using (11.4.8), (11.4.9), and (11.4.10), we obtain that

$$\mathbb{E}[\xi(t)] = 0, \quad t \in [0,\infty), \tag{11.4.11}$$

$$\mathbb{E}[\xi(s)\xi(t)] = (s+1)(t+1)\mathbb{E}\left[Z\left(\frac{s}{s+1}\right)Z\left(\frac{t}{t+1}\right)\right] = \min(s,t), \tag{11.4.12}$$

for $s, t \in \mathbb{R}^+ = [0, \infty)$. Hence, ξ is a standard Wiener process on $\mathbb{R}^+$. We will find this representation very useful in some applications.

Combining (11.4.5) with the above discussion, we notice that the finite-dimensional distributions of $\{\mathbf{Z}_n\}$ converge to those of $\mathbf{Z}$. So, to establish the weak convergence of $\mathbf{Z}_n$ to $\mathbf{Z}$, all we need is to verify that $\{\mathbf{Z}_n\}$ is tight. Toward this verification, we may note that $G_n(t) = n^{-1}\sum_{i=1}^{n} I(Y_i \le t), 0 \le t \le 1$, so that writing for $0 \le s \le t \le 1$,

$$I(Y_i \le t) = I(Y_i \le s) + I(s \le Y_i \le t), \quad 1 \le i \le n,$$

and noting that

$$\mathbb{E}[I(s \le Y_i \le t)|I(Y_i \le s)] = \begin{cases} 0, & I(Y_i \le s) = 1 \\ (t-s)/(1-s), & I(Y_i \le s) = 0, \end{cases} \tag{11.4.13}$$

we obtain by some routine steps that for every $n \ge 1$,

$$[Z_n(t)/(1-t), \quad 0 \le t < 1] \quad \text{is a martingale.} \tag{11.4.14}$$

As such, using the Kolmogorov Inequality for martingales, we obtain that for every $\lambda > 0$,

$$\begin{aligned} P\left(\sup_{0\le t\le 1/2} |Z_n(t)| > \lambda\right) &\le P\left(\sup_{0\le t\le 1/2} \frac{|Z_n(t)|}{1-t} > \lambda\right) \\ &\le 4\lambda^{-2}\mathbb{E}[Z_n^2(1/2)] = \lambda^{-2}. \end{aligned} \tag{11.4.15}$$

Thus, completing the other half by symmetry, we have

$$P\left(\sup_{0\le t\le 1} |Z_n(t)| > \lambda\right) \le 2\lambda^{-2}, \quad \forall \lambda > 0. \tag{11.4.16}$$

Then choosing λ adequately large, we obtain that (11.2.9) holds for Z_n. To verify (11.2.10), we proceed as in (11.3.20). Recall that for every $s \in (0,1)$ and $\delta > 0$, $s + \delta \le 1$,

$$\begin{aligned} |Z_n(s+u) - Z_n(s)| &= \left|(1-s-u)\frac{Z_n(s+u)}{(1-s-u)} - (1-s)\frac{Z_n(s)}{1-s}\right| \\ &\le \left|\frac{Z_n(s+u)}{1-s-u} - \frac{Z_n(s)}{1-s}\right| + \delta\left|\frac{Z_n(s)}{1-s}\right|, \quad \forall u \le \delta. \end{aligned} \tag{11.4.17}$$

Therefore,

$$\begin{aligned} P\Big(\sup_{0\le u\le\delta}|Z_n(s+u)-Z_n(s)|>\varepsilon\Big) &\le P\left[\sup_{0\le u\le\delta}\left|\frac{Z_n(s+u)}{1-s-u}-\frac{Z_n(s)}{1-s}\right|>\varepsilon/2\right] \\ &\quad + P[|Z_n(s)|>(1-s)\varepsilon/2\delta]. \end{aligned} \tag{11.4.18}$$

By (11.4.16), the second term on the right-hand side of (11.4.18) converges to 0 as $\delta \downarrow 0$. For the first term, we use (11.4.14) along with Theorem 11.3.1, and using the same technique as in (11.3.25)–(11.3.28), we obtain that it converges to 0 as $\delta \downarrow 0$ ($n \to \infty$). Hence, the tightness of $\{\mathbf{Z}_n\}$ holds. Therefore, we conclude that as $n \to \infty$

$$\mathbf{Z}_n \xrightarrow{D} \mathbf{Z}. \tag{11.4.19}$$

Let us exploit this basic weak convergence result in various statistical applications. Before that, we state some of the probability laws for same common functionals of $\mathbf{Z}$, which we may use for similar functionals of $\{\mathbf{Z}_n\}$. First, we consider the case of

$$D^+ = \sup_{0\le t\le 1}\{Z(t)\}. \tag{11.4.20}$$

Note that by virtue of (11.4.10), we have for every $\lambda \ge 0$,

$$\begin{aligned} P(D^+\ge\lambda) &= P[Z(t)\ge\lambda, \text{ for some } t\in[0,1]] \\ &= P\left[Z\left(\frac{u}{u+1}\right)\ge\lambda, \text{ for some } u\in[0,\infty)\right] \\ &= P\left[(u+1)Z\left(\frac{u}{u+1}\right)\ge\lambda(u+1), \text{ for some } u\in\mathbb{R}^+\right] \\ &= P[\xi(t)\ge\lambda+\lambda t, \text{ for some } t\in\mathbb{R}^+] \\ &= e^{-2\lambda^2}, \end{aligned} \tag{11.4.21}$$

where the final step follows from a well-known result on the Brownian motion ξ (on $\mathbb{R}^+$) [e.g., Anderson (1960)]. In a similar manner, it follows on letting

$$D = \sup_{0\le t\le 1}|Z(t)| \tag{11.4.22}$$

that for every $\lambda > 0$,

$$P(D>\lambda) = 2(e^{-2\lambda^2}-e^{-8\lambda^2}+e^{-18\lambda^2}-\cdots). \tag{11.4.23}$$

If we consider a simple linear functional

$$L = \int_0^1 g(t)Z(t)\mathrm{d}t, \tag{11.4.24}$$

we immediately obtain that L is normal with 0 mean and

$$\mathbb{V}\mathrm{ar}(L) = \int_0^1 G^2(t)\mathrm{d}t - \left[\int_0^1 G(t)\mathrm{d}t\right]^2, \tag{11.4.25}$$

where

$$G(t) = \int_0^t g(u)\mathrm{d}u, \quad 0 \le t \le 1, \tag{11.4.26}$$

and the square integrability of $G(t)$ is a part of the regularity assumptions. Further, by (11.4.7), we have

$$L = \int_0^1 g(t)W(t)\mathrm{d}t - W(1)\int_0^1 tg(t)\mathrm{d}t, \tag{11.4.27}$$

so that, by some simple steps, we have

$$\int_0^1 g(t)W(t)\mathrm{d}t \sim \mathcal{N}(0, \sigma_g^2), \tag{11.4.28}$$

with

$$\sigma_g^2 = \int_0^1 G^2(t)\mathrm{d}t + G^2(1) - 2G(1)\int_0^1 G(t)\mathrm{d}t. \tag{11.4.29}$$

Let us also consider the statistic

$$D_g^* = \int_0^1 g(t)Z^2(t)\mathrm{d}t, \tag{11.4.30}$$

where the weight function g, satisfies the condition that

$$\int_0^1 t(1-t)g(t)\mathrm{d}t < \infty. \tag{11.4.31}$$

In particular, for $g \equiv 1$, D^* has the same distribution as

$$\sum_{j \ge 1}(j^2\pi^2)^{-1}X_j^2, \tag{11.4.32}$$

where the X_j are i.i.d. random variables such that

$$X_j \sim \mathcal{N}(0, 1). \tag{11.4.33}$$

In a general setup, we may write

$$D_g^* \stackrel{D}{=} \sum_{j \ge 1}\lambda_j X_j^2, \tag{11.4.34}$$

where the real numbers $\{\lambda_j, j \ge 1\}$ depend on g, the X_j are defined as in (11.4.33) and, under (11.4.31),

$$\sum_{j \ge 1}\lambda_j < \infty. \tag{11.4.35}$$

Example* 11.4.1 *(One-sample goodness-of-fit tests). Suppose that $X_1, \ldots, X_n$ are n i.i.d. random variables from a continuous distribution function F, defined on $\mathbb{R}$. We want to test for a null hypothesis $H_0 : F \equiv F_0$, where F_0 has a specified form. Then, by making use of the probability integral transformation $Y = F_0(X)$, we have $Y_1, \ldots, Y_n$, n i.i.d. random variables with a distribution function $G(= F \circ F_0^{-1})$, and the null hypothesis reduces to $H_0 : G =$ Uniform(0,1) distribution function. Let G_n be the empirical distribution function of the Y_i. Then, the classical Kolmogorov–Smirnov statistics for testing the goodness of fit of the distribution function F_0 are given by

$$D_n^+ = \sup_{0 \le t \le 1} \{\sqrt{n}[G_n(t) - t]\} = \sup_{x \in \mathbb{R}} \{\sqrt{n}[F_n(x) - F_0(x)]\}, \tag{11.4.36}$$

$$D_n^- = \sup_{0 \le t \le 1} \{\sqrt{n}[t - G_n(t)]\} = \sup_{x \in \mathbb{R}} \{\sqrt{n}[F_0(x) - F_n(x)]\} \tag{11.4.37}$$

and

$$D_n = \sup_{0 \le t \le 1} \{|\sqrt{n}(G_n(t) - t)|\} = \sup_{x \in \mathbb{R}} \{|\sqrt{n}[F_n(x) - F_0(x)]|\}. \tag{11.4.38}$$

Under the null hypothesis, the distribution of D_n^+, D_n^- or D_n is generated by the uniform order statistics, and, hence, it does not depend on F_0. Thus, these statistics are all exactly distribution-free (EDF). For small values of n, one can obtain this distribution by direct enumeration, but it becomes too laborious as $n \to \infty$. On the other hand, noting that

$$D_n^+ = \sup_{0 \le t \le 1} [Z_n(t)] \quad \text{and} \quad D_n = \sup_{0 \le t \le 1} [|Z_n(t)|], \tag{11.4.39}$$

we may as well use (11.4.19) along with (11.4.20) and (11.4.21) to conclude that D_n^+ (or D_n^-) has asymptotically the same distribution as D^+, and $D_n \xrightarrow{D} D$. Thus, for large n, one may use (11.4.21) and (11.4.23) to provide a good approximation to the critical values of D_n^+ (D_n^-) and D_n, which may be used for making one-sided and two-sided tests.

An alternative test for this goodness-of-fit problem may rest on the Cramér–von Mises statistic

$$\text{CvM}_n = \int_{\infty}^{\infty} n[F_n(x) - F_0(x)]^2 \mathrm{d}F_0(x) = \int_0^1 Z_n^2(t)\mathrm{d}t, \tag{11.4.40}$$

so that by virtue of (11.4.19), we may again refer to D_g^* in (11.4.30) with $g \equiv 1$, and conclude that the asymptotic distribution of CvM_n is given by the distribution of the statistic

$$V = \sum_{j \ge 1} \left(\frac{1}{j^2\pi^2}\right) U_j, \tag{11.4.41}$$

where the U_j are i.i.d. random variables, each U_j having the central chi-squared distribution function with 1 degree of freedom. Some tabulations of the percentile points of U are available in the literature. □

Example* 11.4.2 *(Two-sample problem). Consider now two independent samples $X_1, \ldots, X_{n_1}$ and $Y_1, \ldots, Y_{n_2}$ drawn randomly from two populations with distribution functions F and G, respectively, both defined on $\mathbb{R}$. We assume that both F and G are continuous a.e. and frame the null hypothesis ($H_0 : F \equiv G$) as the equality of F and G, against alternatives that F and G are not identical. This is the classical two-sample nonparametric model; more restrictive models relate to $G(x) = F[(x - \theta)/\lambda]$, where θ and λ are the location and scale factors and F is generally treated as unknown. Then, the null hypothesis $F \equiv G$ reduces to $\theta = 0$, $\lambda = 1$, and in this setup, treating F as nuisance, one may frame alternatives in terms of θ and/or λ. There are various types of nonparametric tests for such semiparametric models. Although these tests may perform quite well for such shift/scale variation models, they may not be consistent against all alternatives $F \not\equiv G$. For this reason, often one prescribes Kolmogorov–Smirnov–type of tests, which remain consistent globally.

To set the things in the proper perspectives, let

$$F_{n_1}(x) = n_1^{-1} \sum_{i=1}^{n_1} I(X_i \le x), \quad x \in \mathbb{R}, \tag{11.4.42}$$

$$G_{n_2}(x) = n_2^{-1} \sum_{i=1}^{n_2} I(Y_i \le x), \quad x \in \mathbb{R}. \tag{11.4.43}$$

Then the two-sample (one-sided and two-sided) Kolmogorov–Smirnov test statistics are

$$D^+_{n_1 n_2} = \left(\frac{n_1 n_2}{n_1 + n_2}\right)^{1/2} \sup_x [F_{n_1}(x) - G_{n_2}(x)], \tag{11.4.44}$$

$$D_{n_1 n_2} = \left(\frac{n_1 n_2}{n_1 + n_2}\right)^{1/2} \sup_x |F_{n_1}(x) - G_{n_2}(x)|. \tag{11.4.45}$$

The normalizing factor $[n_1 n_2/(n_1 + n_2)]^{1/2}$ is included to produce proper nondegenerate distributions (under H_0) for large sample sizes. Note that if the null hypothesis $H_0 : F \equiv G$ holds, then all the $n = n_1 + n_2$ observations $X_1, \ldots, X_{n_1}, Y_1, \ldots, Y_{n_2}$ are from a common distribution function (F), and, hence, their joint distribution remains invariant under any permutation of the coordinates. This leads to a totality of $\binom{n}{n_1}$ $[= \binom{n}{n_2}]$ possible partitioning of the n observations into two subsets of n_1 and n_2 elements, each of which has the same (conditional) probability $\binom{n}{n_1}^{-1}$. On the other hand, $D^+_{n_1 n_2}$ or $D_{n_1 n_2}$ remains invariant under any (strictly) monotone transformation on the Xs and Ys. Thus, if we use the (common) probability integral transformation $[Z = F(X)$ or $F(Y)]$, we may reduce these n observations to i.i.d. random variables coming from the Uniform(0, 1) distribution. Hence, the permutation law stated above can be readily adapted to the n-fold Uniform(0, 1) distribution function, so that these test statistics are EDF (under H_0). This feature may not be true if F and G are not continuous a.e., and also in the multivariate case, they are only permutationally (or conditionally) distribution-free (as the joint distribution of the coordinate-wise probability integral transformed variables depends generally on the unknown joint distribution of the vector). It is clear that as n_1, n_2 both increase, the totality of such partitions [i.e., $\binom{n}{n_1}$] increases at an alarming rate, and, hence, the exact enumeration of the permutation distribution of $[D^+_{n_1 n_2}]$ or $[D_{n_1 n_2}]$ becomes prohibitively laborious. Fortunately, (11.4.19) can be very effectively used to simplify the large sample properties.

We assume that there exists a number ρ, $0 < \rho < 1$, such that

$$n_1/n \to \rho \quad \text{as } n \to \infty. \tag{11.4.46}$$

Also, without any loss of generality, we assume that both F and G are uniform distribution function over (0,1), that is, $F(x) \equiv G(x) \equiv x$, $0 \le x \le 1$. As in (11.2.2), we define the two-sample reduced empirical processes by

$$\mathbf{Z}^{(1)}_{n_1} = \{Z^{(1)}_{n_1}(t) = n_1^{1/2}[F_{n_1}(t) - t], \quad 0 \le t \le 1\}$$

and

$$\mathbf{Z}^{(2)}_{n_2} = \{Z^{(2)}_{n_2}(t) = n_2^{1/2}[G_{n_2}(t) - t], \quad 0 \le t \le 1\},$$

respectively. Also, let

$$\mathbf{Z}^*_n = [Z^*_n(t), \quad 0 \le t \le 1] \tag{11.4.47}$$

where

$$Z_n^*(t) = \left(\frac{n_2}{n}\right)^{1/2} Z_{n_1}^{(1)}(t) - \left(\frac{n_1}{n}\right)^{1/2} Z_{n_2}^{(2)}(t), \quad 0 \le t \le 1.$$

Then, by (11.4.44), (11.4.45), and (11.4.47), we obtain that

$$D_{n_1 n_2}^{+} = \sup_{0 \le t \le 1} [Z_n^*(t)]; \; D_{n_1 n_2} = \sup_{0 \le t \le 1} [|Z_n^*(t)|]. \tag{11.4.48}$$

Thus, if we are able to show that as $n \to \infty$,

$$Z_n^* \xrightarrow{D} Z \quad \text{(Brownian Bridge)}, \tag{11.4.49}$$

then (11.4.21) and (11.4.23) become adaptable here. For this, note that each of the $Z_{n_i}^{(i)}$ converges weakly to a Brownian bridge, say, $\mathbf{Z}_i, i = 1, 2$, where $\mathbf{Z}_{n_1}^{(1)}$ and $\mathbf{Z}_{n_2}^{(2)}$ are independent and, hence, $\mathbf{Z}_1$ and $\mathbf{Z}_2$ are independent too. Moreover, for each $\mathbf{Z}_{n_i}^{(i)}$, the tightness property follows as in the discussion before (11.4.19). Therefore, $\mathbf{Z}_n^*$, being a linear function of $\mathbf{Z}_{n_1}^{(1)}$ and $\mathbf{Z}_{n_2}^{(2)}$, has also the tightness property. Thus, it suffices to show that the finite-dimensional distributions of $\{\mathbf{Z}_n^*\}$ converge to those of a Brownian motion $\mathbf{Z}$. Note that for an arbitrary $m \ge 1$ and $0 \le t_1 < \cdots < t_m \le 1$,

$$\begin{aligned}[Z_n^*(t_1), \ldots, Z_n^*(t_m)] = {} & \sqrt{\frac{n_2}{n}}[Z_{n_1}^{(1)}(t_1), \ldots, Z_{n_1}^{(1)}(t_m)] \\ & - \sqrt{\frac{n_1}{n}}[Z_{n_2}^{(2)}(t_1), \ldots, Z_{n_2}^{(2)}(t_m)],\end{aligned} \tag{11.4.50}$$

so that using (11.4.5) for each of the two vectors on the right-hand side (along with their independence), we obtain that (11.4.5) holds for $\mathbf{Z}_n^*$ as well. Thus, (11.4.50) holds, and our task is accomplished. In this connection, note that in (11.4.46), we assumed that as $n \to \infty$, $n_1/n \to \rho$, $0 < \rho < 1$. If n_1/n converges to 0 or 1 as $n \to \infty$, then in (11.4.47) one of the terms drops out and the weak convergence result in (11.4.49) still holds if $\min(n_1, n_2) \to \infty$. However, $n_1 n_2/(n_1 + n_2)$ becomes $o(n_1 + n_2) = o(n)$, when $\rho \to 0$ or 1, and, hence, the convergence rate becomes slower. For this reason, it is desirable to limit ρ to the central part of (0,1), i.e., to make n_1 and n_2 of comparable magnitudes. We may also consider the two-sample Cramér–von Mises statistic

$$\int_{-\infty}^{\infty} \frac{n_1 n_2}{n_1 + n_2}[F_{n_1}(x) - G_{n_2}(x)]^2 \mathrm{d}H_n(x), \tag{11.4.51}$$

where $H_n(x) = (n_1/n)F_{n_1}(x) + (n_2/n)G_{n_2}(x)$, $x \in \mathbb{R}$. Again noting that under $H_0 : F \equiv G$, H_n converges a.s. to F and proceeding as before, we may conclude that the Cramér–von Mises statistic in (11.4.51) converges in law to

$$\int_0^1 Z^2(t)\mathrm{d}t, \tag{11.4.52}$$

where $Z(t)$, $t \in [0, 1]$, is a Brownian bridge, so that as in (11.4.49) and (11.4.51), we may use the same distribution as that of V in (11.4.41).

Some other nonparametric tests for $H_0 : F \equiv G$ are based on some specific functionals for which simpler asymptotic theory is available. For example, consider the two-sample

Wilcoxon–Mann–Whitney statistic

$$W_{n_1n_2} = n_1^{-1}\sum_{i=1}^{n_1} R_i - \frac{n+1}{2}, \tag{11.4.53}$$

where

$$R_i = \text{rank of } X_i \text{ among the } n \text{ observations, } 1 \leq i \leq n_1. \tag{11.4.54}$$

Note that by definition of the ranks,

$$n^{-1}R_i = H_n(X_i), \quad i = 1, \ldots, n_1, \tag{11.4.55}$$

so that we have

$$\begin{aligned}
(n+1)^{-1}W_{n_1n_2} &= \frac{n}{n+1}\frac{1}{n_1}\sum_{i=1}^{n_1} H_n(X_i) - \frac{1}{2} \\
&= \frac{n}{n+1}\int H_n(x)\mathrm{d}F_{n_1}(x) - \frac{1}{2} \\
&= \frac{n}{n+1}\left[\frac{n_1}{n}\int F_{n_1}(x)\mathrm{d}F_{n_1}(x) + \frac{n_2}{n}\int G_{n_2}(x)\mathrm{d}F_{n_1}(x)\right] - \frac{1}{2} \\
&= \frac{n}{n+1}\left[\frac{n_1}{n}\frac{1}{n_1}\frac{n_1(n_1+1)}{2n_1} + \frac{n_2}{n}\int G_{n_2}(x)\mathrm{d}F_{n_1}(x)\right] - \frac{1}{2} \\
&= \frac{n_1+1}{2(n+1)} + \frac{n_2}{n+1}\int G_{n_2}(x)\mathrm{d}F_{n_1}(x) - \frac{1}{2} \\
&= \frac{n_2}{n+1}\int G_{n_2}(x)\mathrm{d}F_{n_1}(x) - \frac{n_2}{2(n+1)} \\
&= \frac{n_2}{n+1}\left[\int G_{n_2}(x)\mathrm{d}F_{n_1}(x) - \int F(x)\mathrm{d}F(x)\right].
\end{aligned} \tag{11.4.56}$$

Thus, writing $G_{n_2} = F + (G_{n_2} - F)$ and $F_{n_1} = F + (F_{n_1} - F)$, we have from (11.4.56)

$$\begin{aligned}
\frac{n^{1/2}}{n+1}W_{n_1n_2} = \frac{n_2}{n+1}\bigg[&\int F(x)\mathrm{d}\{\sqrt{n}[F_{n_1}(x) - F(x)]\} \\
&+ \int \sqrt{n}\{G_{n_2}(x) - F(x)\}\mathrm{d}F(x) \\
&+ \int \sqrt{n}[G_{n_2}(x) - F(x)]\mathrm{d}[F_{n_1}(x) - F(x)]\bigg],
\end{aligned} \tag{11.4.57}$$

where $\sqrt{n}\|G_{n_2} - F\| = O_p(1)$ and $\|F_{n_1} - F\| = o_p(1)$. Hence, the last term on the right-hand side of (11.4.57) is $o_p(1)$, whereas by partial integration of the first term, we have

$$\begin{aligned}
\frac{n^{1/2}}{n+1}W_{n_1n_2} = \frac{n_2}{(n+1)}\bigg\{&\int \sqrt{n}[G_{n_2}(x) - F(x)]\mathrm{d}F(x) \\
&- \int \sqrt{n}[F_{n_1}(x) - F(x)]\mathrm{d}F(x)\bigg\} + o_p(1)
\end{aligned} \tag{11.4.58}$$

and, as such, we may appeal to the definition of $\mathbf{Z}_{n_1}^{(1)}$ and $\mathbf{Z}_{n_2}^{(2)}$ presented before (11.4.47), so that $n^{-1/2}W_{n_1n_2}$ is distributed as a linear functional of a difference between two independent

Brownian bridges, and then

$$n^{-1/2} W_{n_1 n_2} \xrightarrow{D} \mathcal{N}\{0, [12\rho(1-\rho)]^{-1}\}, \tag{11.4.59}$$

where ρ is defined by (11.4.46). This method of attack based on the weak convergence of $\mathbf{Z}_{n_1}^{(1)}$ and $\mathbf{Z}_{n_2}^{(2)}$ works also neatly when $F \not\equiv G$ and also for general linear rank statistics involving some score function $a_n(k)$, $1 \le k \le n$, where

$$a_n(k) = \psi(k/(n+1), \quad 1 \le k \le n, \tag{11.4.60}$$

and ψ is a smooth function having finitely many points of discontinuity. In (11.4.56), we may need to replace $H_n(x)$ by $\psi\{[n/(n+1)]H_n(x)\}$, and a Taylor expansion takes care of the situation. In this way, the weak invariance principles play a basic role in the asymptotic theory of nonparametric statistics. We will discuss more about this in Section 11.6. □

Example* 11.4.3 *(Life-testing model). Consider the same two-sample model as in Example 11.4.2 but suppose now that the X_i and Y_j are the failure times (nonnegative random variables), so that F and G are both defined on $\mathbb{R}^+$. Consider a very simple case where the X_i are the lifetimes of electric lamps of a particular brand (say, A) and the Y_i, for a second brand (say, B). If we put all these n $(= n_1 + n_2)$ lamps to life testing at a common point of time (say, 0), then the successive failures occur sequentially over time, so that at any time point $t > 0$, the observable random elements relate to

(i) actual *failure times* of the lamps if the failure times are $\le t$, and
(ii) operating status for other lamps which have not failed before t (i.e., the *censoring* event).

Thus, to collect the entire set of failure times, one may need to wait until all the failures have taken place. Often, based on time and cost considerations, it may not be possible to continue the experimentation all the way to the end, and one may need to curtail the study at an intermediate point (say, t^*). This relates to a *truncated* scheme. Note that even if t^* is fixed in advance, r_{t^*}, the number of failures occurring before t^* is a random variable. Alternatively, one may curtail the study after a certain number (say r_n) of failures have occurred. In this way, r_n may be fixed in advance, but T^*, the time period, becomes stochastic; such a scheme is termed a *censoring* one. In either case, the Kolmogorov–Smirnov and the Cramér–von Mises tests discussed in (11.4.48), (11.4.51) and (11.4.52) can be adapted by truncating the range $\mathbb{R}^+$ to $(0, t^*]$ or $(0, T^*]$. In the censoring scheme, r_n is nonstochastic, and, hence, such censored tests are all EDF under H_0. On the other hand, in the truncation scheme, these truncated tests are only conditionally EDF (given r_{t^*}) under H_0. However, in the large sample case, under H_0, $n^{-1} r_{t^*} \xrightarrow{a.s.} F(t^*)$, so that letting $r_n = nF(t^*)$, both the schemes can be studied in a common manner. Basically, they rest on the weak convergence of

$$Z_{n_1}^{(1)0} = \left[Z_{n_1}^{(1)}(t), \quad 0 \le t \le F(t^*) = p^*\right]$$

and

$$Z_{n_2}^{(2)0} = \left[Z_{n_2}^{(2)}(t), \quad 0 \le t \le p^*\right]$$

to $Z^0 = [Z(t),\ 0 \le t \le p^*]$, which is a direct corollary to the general result in the earlier example. □

In view of the accumulating nature of the data set, it is not uncommon to adapt a *time-sequential* scheme in such a life-testing model. In this scheme, one may want to monitor the study from the beginning with the objective that if at any early point of time there is a significant difference between the two-sample responses, the study may be curtailed at that time along with the rejection of the null hypothesis. This formulation is quite important in clinical trials/medical studies, where patients may be switched on to the better treatment if a significant treatment difference is detected. One of the nice properties of the Kolmogorov–Smirnov and the Cramér–von Mises tests is that once the critical values are determined by reference to a single truncation/censoring point, the tests are automatically adaptable to a time-sequential setup. For example, $|F_{n_1}(x) - G_{n_2}(x)|$ may be sequentially (in $x \geq 0$) monitored and if at any x, for the first time, this difference exceeds the critical level of $D^o_{n_1 n_2}$ $[= \sup_{x \leq t^*}\{|F_{n_1}(x) - G_{n_2}(x)|\}]$, we stop the study along with the rejection of H_0. The distribution of the stopping time, in an asymptotic setup, may thus be related to the distribution of exit times for a Brownian bridge over an appropriate interval $J \subset [0, 1]$. This explains the importance of the weak invariance principles in asymptotic theory of statistical inference. With rank statistics, the picture is somewhat more complicated, because the form of the functional may itself depend on the truncation point. Nevertheless, the key solution is provided by the weak convergence results studied in Section 11.3 and 11.4. We provide a brief discussion of this important topic at the end of the next section.

11.5 Weak Convergence and Statistical Functionals

In Chapters 7 and 8, we observed that in a nonparametric setup, often a parameter θ is expressed as a functional $\theta(F)$ of the underlying distribution function F. For example, *estimable parameters* or *regular functionals* in the Hoeffding (1948) sense are of the form

$$\int \cdots \int \psi(x_1, \ldots, x_m) \mathrm{d}F(x_1) \cdots \mathrm{d}F(x_m), \quad F \in \mathcal{F}, \tag{11.5.1}$$

where $m \geq 1$ is a finite positive integer and ψ is a *kernel* (of specified form). In such a case, it may be quite natural to replace the unknown distribution function F by the sample distribution function F_n and estimate θ by the *von Mises* functional $\theta(F_n)$. Whereas the U-statistics considered in Chapter 8 are unbiased, these von Mise's type functionals are generally not unbiased but they share other optimal properties with the U-statistics. There are some other functionals $\theta(F)$, which may not be expressible in the form of (11.5.1) with a fixed $m \geq 0$. A very simple example is the quantile functional

$$\theta(F) = F^{-1}(p), \quad p \in (0, 1), F \in \mathcal{F}, \tag{11.5.2}$$

where $\mathcal{F}$ is the class of all distribution functions that admit a unique quantile. In such a case, we may as well consider an estimator $\theta(F_n)(= T_n$, say), although to define it uniquely we may need to go by some convention (e.g., Section 1.5). If we use the estimator $T_n = \theta(F_n)$ corresponding to (11.5.1), that is,

$$T_n = \int \cdots \int \psi(x_1, \ldots, x_m) \mathrm{d}F_n(x_1) \cdots \mathrm{d}F_n(x_m), \tag{11.5.3}$$

we may write

$$\mathrm{d}F_n(x_j) = \mathrm{d}F(x_j) + \mathrm{d}[F_n(x_j) - F(x_j)], \quad 1 \leq j \leq m,$$

so that T_n can be decomposed into 2^m terms that can be recollected as

$$\begin{aligned} T_n &= \theta(F) + \binom{m}{1}\int \psi_1(x_1)\mathrm{d}[F_n(x_1) - F(x_1)] \\ &+ \binom{m}{2}\int\int \psi_2(x_1, x_2)\mathrm{d}[F_n(x_1) - F(x_1)]\mathrm{d}[F_n(x_2) - F(x_2)] \\ &+ \cdots + \binom{m}{m}\int\cdots\int \psi(x_1, \ldots, x_m)\prod_{j=1}^{m}\mathrm{d}[F_n(x_j) - F(x_j)], \end{aligned} \qquad (11.5.4)$$

where

$$\psi_j(x_1, \ldots, x_j) = \mathbb{E}[\psi(X_1, \ldots, X_j)|X_i = x_i, i \le j], \quad 1 \le j \le m. \qquad (11.5.5)$$

In the literature, this is known as the *Hoeffding decomposition* of a symmetric function. For a general *statistical function*, $T_n = T(F_n)$, a similar expansion can be worked out whenever $\mathbb{E}_F(T_n^2) < \infty$; however, it need not have $(m + 1)$ orthonormal terms for some fixed $m \ge 1$, and, second, the components, especially, the higher order ones, may not be that simple to be adaptable for further analysis. For this reason, often, we may look at a representation of T_n involving only one or two terms, which would suffice for the desired asymptotic studies. For example, if we need only to study the (weak or strong) consistency of T_n [as an estimator of $T(F)$], we may express

$$T_n = T(F_n) = T[F + (F_n - F)], \qquad (11.5.6)$$

where (by Theorem 6.3.16) $\|F_n - F\| \xrightarrow{a.s.} 0$, so that the desired result follows if the functional $T(\cdot)$ is continuous in an appropriate norm. Let us assume that F is continuous, so that $Y = F(X)$ has the Uniform (0,1) distribution function, and the reduced empirical distribution function G_n, defined by (6.3.98), belongs to the $D[0, 1]$ space, for every $n \ge 1$. Then, we may write

$$T(H) = T[F^{-1}(F \circ H)] = \tau(F \circ H), \qquad (11.5.7)$$

where τ is the reduced functional. Then, writing $T_n = \tau(G_n)$, denoting by U the Uniform(0, 1) distribution function, and noting that $\theta = T(F) = \tau(U)$, we have

$$T_n - \theta = \tau(G_n) - \tau(U). \qquad (11.5.8)$$

Therefore, if τ, defined on the $D[0, 1]$ space, is continuous relative to the Skorokhod metric in (11.2.13), then $\|G_n - U\| \xrightarrow{a.s.} 0$ ensures that $T_n - \theta(F) \xrightarrow{a.s.} 0$. Similarly, if we have in mind the study of the asymptotic normality of $n^{1/2}[T_n - \theta(F)]$, we may write

$$\mathbf{Z}_n = \{Z_n(t) = \sqrt{n}[G_n(t) - t], \quad 0 \le t \le 1\}$$

[as in (11.2.2)], and note that

$$n^{1/2}[T_n - \theta(F)] = n^{1/2}[\tau(U + n^{-1/2}Z_n) - \tau(U)]. \qquad (11.5.9)$$

Thus, if the functional τ is differentiable in a suitable manner, so that the right-hand side of (11.5.9) can be expanded by a first-order Taylor expansion (albeit in a functional space), then, we may write

$$n^{1/2}[T_n - \theta(F)] = \int \mathring{\tau}(U; t)\mathrm{d}Z_n(t) + R_n, \qquad (11.5.10)$$

where

$$|R_n| = o(\|Z_n\|), \tag{11.5.11}$$

and $\overset{\circ}{\tau}(U;\cdot)$ is the derivative of τ at U. Note that $\|\mathbf{Z}_n\| = O_p(1)$ [by (11.4.16)], whereas $\overset{\circ}{\tau}(U;t)$ is a linear functional, so that

$$\sigma^2_{\overset{\circ}{\tau}} = \int_0^1 \overset{\circ}{\tau}{}^2(U;t)\mathrm{d}t < \infty$$

ensures that

$$\int \overset{\circ}{\tau}(U;t)\mathrm{d}Z_n(t) \xrightarrow{D} \mathcal{N}\big(0, \sigma^2_{\overset{\circ}{\tau}}\big). \tag{11.5.12}$$

As such, by (11.5.10)–(11.5.12), one obtains that as $n \to \infty$,

$$n^{1/2}[T_n - \theta(F)]/\sigma_{\overset{\circ}{\tau}} \xrightarrow{D} \mathcal{N}(0, 1). \tag{11.5.13}$$

$\overset{\circ}{\tau}(U;t)$ [$\equiv \overset{\circ}{\tau}{}^*(F;x)$] is called the *influence function* of $T(F)$ at x. The mode of differentiation in (11.5.9)–(11.5.11) needs some further clarification. This can, however, be made more precisely if we are permitted to use some basic results in functional analysis. However, consistent with our level of presentation, we state the following results in a bit less generality [e.g., Fernholz (1983)].

Suppose that V and W are topological vector spaces, and let $L(V, W)$ be the set of continuous linear transformations form V to W. Let $\mathcal{S}$ be a class of compact subsets of V such that every subset consisting of a single point is in $\mathcal{S}$, and let A be an open subset of V. A function $T : A \to W$ is called the *Hadamard differentiable* at $F \in A$ if there exists a $T'_F \in L(U, V)$, such that for any $K \in \mathcal{S}$, uniformly for $H \subset K$,

$$\begin{aligned}
&\lim_{t\downarrow 0} t^{-1}[T(F + tH) - T(F) - T'_F(tH)] \\
&\quad = \lim_{t\downarrow 0} t^{-1}[R(T, F; tH)] \mathrel{\overline{\overline{\cdot}}} 0.
\end{aligned} \tag{11.5.14}$$

The *Hadamard* (or *compact*) derivative T'_F reduces to the linear functional on the right-hand side of (11.5.10), whereas identifying t as $n^{-1/2}$, R_n in (11.5.10)–(11.5.11) corresponds to the remainder term $R(\cdot)$ in (11.5.14). There are other modes of differentiability (such as the Gateaux and Fréchet ones), which will not be discussed here.

However attractive such a differentiable approach may appear to be, there are certain limitations:

(i) The functional $\tau = T \circ F^{-1}$ may depend on the unknown F in a rather involved manner, and verification of its continuity or Hadamand differentiability condition may require rather complicated analysis.
(ii) As is the case with MLE, M- and R- estimators, the functional T [or τ] is defined implicitly as the root of some other (estimating) functionals. Refined treatment of such implicit functionals is even more delicate.
(iii) In many cases τ turns out to be a bounded functional in order that Hadamard differentiability holds, and that may exclude many practically important cases.
(iv) This approach is not quite in line with those in earlier chapters.

In view of these points, we will not pursue this approach in further details.

A more convenient approach relates to a *first-order representation* for a general statistical functional T_n, in the sense that

$$T_n - \theta(F) = n^{-1} \sum_{i=1}^{n} \Psi_F(X_i) + R_n, \tag{11.5.15}$$

where $\Psi_F(x)$ generally depends on θ (through F),

$$\mathbb{E}_F[\Psi_F(X)] = 0, \quad \mathbb{E}_F[\Psi_F^2(X)] = \sigma_\Psi^2 < \infty, \tag{11.5.16}$$

$$R_n = o_p(n^{-1/2}). \tag{11.5.17}$$

We observed in Chapter 7 that the projection method (e.g. Theorem 7.5.1) actually yields such a representation in many cases (including U-statistics, von Mises' functionals and many nonparametric statistics). In Chapter 8, we observed that for the MLE, (11.5.15) holds under the usual regularity conditions, where $\Psi_F(x) = -(\partial/\partial\theta)\log f(x,\theta)$. For R-estimators, (11.5.15) holds under more general regularity conditions than those pertaining to (11.5.14). Also, in general, (11.5.15) requires a less complicated analysis than the analysis required for the Hoeffding decomposition, which goes further to decompose R_n into multiple orthogonal components of stochastically decreasing order of magnitudes. Also the first-order case can be strengthened to the second order case by appealing to some relatively more stringent regularity conditions. We will, therefore, be more inclined to the adaptation of (11.5.15) in various situations at hand. In this respect, the treatment for functionals of the form (11.5.2) needs a somewhat different approach, which we will examine briefly.

Consider the reduced empirical process $\mathbf{Z}_n = \{Z_n(t), t \in [0,1]\}$, defined by (11.2.2) and studied thoroughly in (11.4.1)–(11.4.19). Note that by virtue of (11.4.18) and (11.4.19), for every (fixed) p, $0 < p < 1$, and every $\varepsilon > 0$, $\eta > 0$, there exists a $\delta_0 > 0$, such that for every $\delta \leq \delta_0$

$$P\bigg(\sup_{|t-p|<\delta} |Z_n(t) - Z_n(p)| > \varepsilon \bigg) < \eta, \quad \forall n \geq n_0. \tag{11.5.18}$$

On the other hand, for the Uniform(0,1) distribution function, if $\tilde{Y}_{n,p}$ stands for the sample p-quantile, then for every $\delta > 0$, we have, on noting that $G_n(\tilde{Y}_{n,p}) = p + o(n^{-1/2})$,

$$\lim_{n\to\infty} P(|\tilde{Y}_{n,p} - p| > \delta) = 0. \tag{11.5.19}$$

Thus, combining (11.5.18) and (11.5.19), we have for $n \to \infty$

$$n^{1/2}[G_n(\tilde{Y}_{n,p}) - G(\tilde{Y}_{n,p}) - G_n(p) + p] \xrightarrow{P} 0, \tag{11.5.20}$$

so that as n increases,

$$[n^{1/2}(\tilde{Y}_{n,p} - p)] - \{-n^{1/2}[G_n(p) - p]\} \xrightarrow{P} 0, \tag{11.5.21}$$

as $G(y) = y, 0 \leq y \leq 1$. On the other hand, if the distribution function F has a continuous and positive density function f [at $\xi_p : F(\xi_p) = p$], then, for the sample p-quantile $\tilde{X}_{n,p}$, noting that $\tilde{Y}_{n,p} = F(\tilde{X}_{n,p})$, we have from (11.5.21),

$$\begin{aligned} n^{1/2}[\tilde{X}_{n,p} - \xi_p] &= n^{1/2}[\tilde{X}_{n,p} - \xi_p](\tilde{Y}_{n,p} - p)/(\tilde{Y}_{n,p} - p) \\ &= n^{1/2}(\tilde{Y}_{n,p} - p)\{[F(\tilde{X}_{n,p}) - F(\xi_p)]/(\tilde{X}_{n,p} - \xi_p)\}^{-1} \\ &= n^{1/2}(\tilde{Y}_{n,p} - p)\{f(\xi_p) + o_p(1)\}^{-1} \\ &= -[f(\xi_p)]^{-1} n^{1/2}[G_n(p) - p] + o_p(1), \end{aligned} \tag{11.5.22}$$

so that

$$\tilde{X}_{n,p} - \xi_p = n^{-1}\sum_{i=1}^{n} \Psi_F(X_i) + o_p\left(\frac{1}{\sqrt{n}}\right), \tag{11.5.23}$$

where

$$\Psi_F(X_i) = -\frac{1}{f(\xi_p)}[I(X_i \leq \xi_p) - p], \quad 1 \leq i \leq n. \tag{11.5.24}$$

Thus, the first-order representation in (11.5.15) holds for sample quantiles under the same regularity conditions, as in Theorem 7.7.2 pertaining to its asymptotic normality. We may, of course, get a stronger result under additional regularity conditions. Note that by virtue of (11.4.14), for every $n \geq 1$,

$$\{\exp[(1-t)^{-1}Z_n(t)], \quad t \in [0, 1]\} \quad \text{is a submartingale.} \tag{11.5.25}$$

Hence, using the Kolmogorov–Hájek–Rényi–Chow Inequality and proceeding as in (11.4.17)–(11.4.18), we obtain that as $n \to \infty$,

$$\sup\left[\sqrt{n}|G_n(t) - t - G_n(p) + p| : |t - p| \leq \frac{1}{\sqrt{n}}\log n\right] = O(n^{-1/4}\log n) \quad \text{a.s.} \tag{11.5.26}$$

Note that for the uniform distribution function

$$\sup_{0<p<1} [|\tilde{Y}_{n,p} - p|] = \|G_n(t) - t\| = n^{-1/2}\|(1-t)W_n(t)\| \leq n^{-1/2}\|W_n(t)\|, \tag{11.5.27}$$

where $\{W_n(t) = Z_n(t)/(1-t), 0 \leq t \leq 1\}$ is a martingale. Hence, using the Kolmogorov Inequality we can verify that

$$n^{-1/2}\|W_n(t)\|/\log n \xrightarrow{a.s.} 0$$

which by virtue of (11.5.27) implies that

$$|\tilde{Y}_{n,p} - p| < n^{-1/2}\log n \quad \text{a.s.}$$

Therefore, from the above two formulas, as $n \to \infty$,

$$\tilde{Y}_{n,p} - p = -[G_n(p) - p] + O(n^{-3/4}\log n) \quad \text{a.s.} \tag{11.5.28}$$

Thus, if we assume that the distribution function F admits an absolutely continuous density function f, such that $f'(x)$ is finite in a neighborhood of ξ_p, then from (11.5.28), we have for $n \to \infty$,

$$\tilde{X}_{n,p} - \xi_p = n^{-1}\sum_{i=1}^{n} \psi_F(X_i) + R_n, \tag{11.5.29}$$

where $\psi_F(x)$ is defined as in (11.5.24) and as $n \to \infty$,

$$R_n = O(n^{-3/4} \log n) \quad \text{a.s.} \tag{11.5.30}$$

In the literature, this is known as the *Bahadur representation of sample quantiles* [Bahadur (1966)]. Equation (11.5.24) is a weaker version of (11.5.29), without requiring the finiteness (and existence) of $f'(x)$ at ξ_p. Also note that in (11.5.29), $R_n = O(n^{-3/4} \log n)$ and not $O_p(n^{-1})$ or $O(n^{-1} \log n)$ a.s., which could have been the case with some other smooth estimators. For discontinuous score functions, typically R_n is $O_p(n^{-3/4})$ or $O(n^{-3/4} \log n)$ a.s. [not $O_p(n^{-1})$]. See Sen (1981, chapter 7) for a detailed account of this topic.

11.6 Weak Convergence and Nonparametrics

In nonparametric methods, rank-based procedures are generally adapted so as to induce invariance under appropriate groups of transformations and, thereby, to achieve more robustness. By definition, the ranks are closely related to the order statistics and the empirical distribution function, and, therefore, weak convergence and related invariance principles for order statistics and empirical processes play a basic role in the asymptotic theory of nonparametric methods.

We may start with a remark that whether be it a single sample or a multisample model, the ranks are not independent random variables, so that the laws of large numbers, probability inequalities and central limit theorems developed for independent summands (in Chapters 6 and 7) may fail to be directly applicable in nonparametrics. There have been several alternative tracks to bridge this gap, some of which are the following:

(i) Express a nonparametric statistic in the form of a (generalized) U-statistic, and then use the methodology discussed in Chapter 8. This may not workout generally (e.g., linear rank statistics/signed rank statistics when the score function is not a polynomial function).

(ii) Consider an integral representation as in (11.4.56)–(11.4.57), and then use invariance principles for empirical processes to draw the desired conclusions. This method works out well for a larger class of nonparametric statistics, although it may require appropriate regularity conditions and elaborate analysis.

(iii) Under appropriate hypotheses of invariance (with respect to suitable groups of transformations that map the sample space onto itself), there are generally some (sub)martingale or reversed martingale properties shared by various nonparametric statistics. As such, exploiting weak (or strong) invariance principles for such dependent sequences, asymptotic theory for a general class of nonparametric statistics can be obtained (under the hypothesis of invariance) under essentially minimal conditions. Using then the concept of *contiguity of probability measures*, the asymptotic theory can then be extended to cover alternative models.

Although in (i) or (ii) one need not be confined only to such local alternatives, in terms of applicability, they often dominate the asymptotics, so that the third approach may generally be the most simple one and may entail essentially the minimal regularity assumptions.

In this section, we intend to provide a very brief introduction to this third approach along with some applications. [For details, see Sen (1981).] Consider the two-sample problem treated in (11.4.53) through (11.4.59). Put it in the following general framework:

$X_1, \ldots, X_n$ are independent random variable with distribution functions $F_1, \ldots, F_n$ and the null hypothesis H_0 states that $F_1 \equiv \cdots \equiv F_n \equiv F$ (unknown). A general linear rank statistic is given by

$$T_n = \sum_{i=1}^{n} (c_i - \overline{c}_n) a_n(R_{ni}), \tag{11.6.1}$$

where $\{c_i, i \geq 1\}$ is a sequence of (regression) constants, $\overline{c}_n = n^{-1} \sum_{i=1}^{n} c_i$, R_{ni} is the rank of X_i among $X_i, \ldots, X_n$, $i = 1, \ldots, n$, and

$$a_n(k) = \mathbb{E}[\psi(Y_{n:k})], \quad k = 1, \ldots, n, \tag{11.6.2}$$

where $Y_{n:1} < \cdots < Y_{n:n}$ are the ordered random variables of a sample of size n from the Uniform(0, 1) distribution function, and ψ is a square integrable score function; $\psi(u) = u - 1/2$, leads to the Wilcoxon statistic in (11.4.53). Recall that (even under H_0) T_n does not have independent summands. But (Exercise 11.6.1) note that by (11.6.2), for every $n \geq 1$,

$$\frac{k}{n+1} a_{n+1}(k+1) + \frac{n+1-k}{n+1} a_{n+1}(k) = a_n(k), \quad 1 \leq k \leq n. \tag{11.6.3}$$

As such (Exercise 11.6.2), it is easy to show that

$$\mathbb{E}(T_{n+1} | \boldsymbol{R}_n, H_0) = T_n \quad \text{a.e. for all } n \geq 1, \tag{11.6.4}$$

where $\boldsymbol{R}_n = (R_{n1}, \ldots, R_{nn})'$. Given this martingale property of the T_n, $n \geq 1$, we can readily use the appropriate results for martingales (e.g., inequalities, laws of large numbers, central limit theorems) to draw conclusions on the asymptotic behavior of T_n (under H_0) (Exercise 11.6.3). Thus, by invoking the asymptotic normality (under H_0) of T_n, that is,

$$T_n/(C_n A_n) \xrightarrow{D} \mathcal{N}(0, 1), \tag{11.6.5}$$

where

$$C_n^2 = \sum_{i=1}^{n} (c_i - \overline{c}_n)^2 \quad \text{and} \quad A_n^2 = \frac{1}{n-1} \sum_{i=1}^{n} [a_n(i) - \bar{a}_n]^2, \tag{11.6.6}$$

we are able to obtain an asymptotic coverage probability by

$$P(-C_n A_n z_{1-\alpha/2} \leq T_n \leq C_n A_n z_{1-\alpha/2} | H_0) \cong 1 - \alpha. \tag{11.6.7}$$

Consider now a simple regression model

$$F_i(x) = F(x - \theta - \beta c_i), \quad 1 \leq i \leq n, \tag{11.6.8}$$

so that under $\beta = 0$, the F_i are all the same. If each X_i is replaced by $X_i - bc_i$, the resulting ranks are denoted by $R_{ni}(b)$ and replacing the R_{ni} in (11.6.1) by the corresponding $R_{ni}(b)$, we denote the (aligned) rank statistic by $T_n(b)$, $b \in \mathbb{R}$. If $\psi(n)$ is nondecreasing in u, $0 \leq u \leq 1$, so that the $a_n(i)$ are also increasing in i $(1 \leq i \leq n)$, we have

$$T_n(b) \quad \text{is nonincreasing in } b \in \mathbb{R}. \tag{11.6.9}$$

Therefore, equating $T_n(b)$ to $-C_n A_n z_{1-\alpha/2}$ and $C_n A_n z_{1-\alpha/2}$, we get an upper and lower bound $\widehat{\beta}_{U,n}$ and $\widehat{\beta}_{L,n}$ respectively, where, by (11.6.7),

$$P(\widehat{\beta}_{L,n} \leq \beta \leq \widehat{\beta}_{U,n}) \cong 1 - \alpha. \tag{11.6.10}$$

Thus, we have a distribution-free confidence interval for β, and asymptotically it takes on a simple form. In fact, equating $T_n(b)$ to 0, we get the point estimate $\widehat{\beta}_n$, termed the R-estimator of β. $\widehat{\beta}_n$, is a translation-equivariant, robust, consistent estimator of β with the property that as $n \to \infty$,

$$C_n(\widehat{\beta}_n - \beta)/A_n \xrightarrow{D} \mathcal{N}(0, \gamma^{-2}), \quad \text{with } \gamma = \int_{-\infty}^{\infty} \psi[F(x)]f^2(x)\mathrm{d}x, \tag{11.6.11}$$

The study of the asymptotic properties of (11.6.10) and (11.6.11) is facilitated by a *uniform asymptotic linearity result* (in b) of $T_n(b)$ (near β) – but this is outside the scope of the current work. See Chapters 4 and 5 of Sen (1981) and Jurečková and Sen (1996) for some of these details. In passing, we may, however, comment that by virtue of (11.6.9), proceeding as in the proof of the Glivenko–Cantelli Theorem 6.3.16, we may show that for every finite K,

$$\begin{aligned} &\sup_{|b|\leq K} \{C_n^{-1}A_n^{-1}|T_n(\beta + C_n^{-1}b) - T_n(\beta) + b\gamma C_n|\} \\ &\leq \max_{j\leq a}\{C_n^{-1}A_n^{-1}|T_n(\beta + C_n^{-1}b_j) - T_n(\beta) + b_j\gamma C_n|\} + \varepsilon/2, \end{aligned} \tag{11.6.12}$$

where a is a finite positive number, depending on ε and K. Then, for the pointwise convergence one may consider standard weak convergence results in nonparametrics.

For M-estimators of location and regression a very similar linearity theorem holds, and that provides the easy access to the study of asymptotic properties of the estimators [e.g., Sen (1981, chapter 8)]. There is another important aspect of nonparametrics that we may like to discuss briefly here. Typically, in an estimation problem (parametric or nonparametric), we have an estimator T_n of a parameter θ, such that

$$n^{1/2}(T_n - \theta)/\nu_\theta \xrightarrow{D} \mathcal{N}(0, 1) \tag{11.6.13}$$

where ν_θ^2, the asymptotic variance of T_n, generally depends on the unknown parameter θ or on the unspecified distribution function F. In a parametric setup, F is of known form, and, hence ν_θ can be estimated by various methods (e.g., MLE). In a nonparametric setup, $\nu_\theta = \nu(F)$ is itself a functional of the unknown distribution function F. If the form of this functional ν is known, a very simple and convenient method of estimating $\nu(F)$ would be to use the sample distribution function F_n for F and take the estimator as $v_n = \nu(F_n)$. This is generally related to the *Delta-method* where expanding $\nu(F_n)$ around $\nu(F)$ (under appropriate smoothness conditions), consistency, and other properties can be studied. The estimator is, in general, not unbiased, and, often, the bias can be of serious considerations. A more complicated situation may arise where the form of ν may lead to very high level of bias by the Delta-method or ν may not be at all of a simple form.

In such a problem, *resampling plans* are often used to estimate $\nu(F)$ in a nonparametric fashion. Among these resampling methods, the two most popular ones are the *jackknife* and *bootstrap* methods, discussed in Section 8.7. We now illustrate them with statistical functionals, so that, considering the notation introduced in Section 8.7 we have from (8.7.3) that, for each i, $1 \leq i \leq n$,

$$T_{ni} = \theta(F) + \psi_F(X_i) + R_{ni}, \tag{11.6.14}$$

so that

$$T_{nJ} = \theta(F) + \bar{\psi}_n + R_{nJ}, \quad \text{with } \bar{\psi}_n = \frac{1}{n}\sum_{n-1}^{n} \psi_F(X_i). \tag{11.6.15}$$

Therefore, by (8.7.5) and (11.6.15), we have

$$s_{nJ}^2 = \frac{1}{n-1}\sum_{i=1}^{n}[\psi_F(X_i) - \bar{\psi}_n]^2 + \frac{1}{n-1}\sum_{i=1}^{n}[R_{ni} - R_{nJ}]^2$$
$$+ \left[\frac{2}{n-1}\sum_{i=1}^{n}[\psi_F(X_i) - \bar{\psi}_n][R_{ni} - R_{nJ}]\right]. \tag{11.6.16}$$

The first term on the right-hand side of (11.6.16) converges a.s. to $\nu^2(F) = \sigma_\psi^2 = \mathbb{E}[\psi_F^2(X_i)]$ as $n \to \infty$. Thus, a sufficient condition for the convergence of s_{nJ}^2 to $\nu^2(F)$ is that as $n \to \infty$

$$\frac{1}{n-1}\sum_{i=1}^{n}[R_{ni} - R_{nJ}]^2 \xrightarrow{a.s.} 0. \tag{11.6.17}$$

Although (11.6.17) is stronger than (11.5.17), it can be verified under quite general condition. We shall not, however, enter into these details. But we prescribe (11.6.17) as a key solution to (11.6.23).

In Section 8.7 we remarked, initially, that jackknifing was motivated by the *bias reduction* objective. The first order representation in (11.5.15) entails that the asymptotic bias of T_{nJ} is $o(n^{-1/2})$ but may not be $O(n^{-1})$. If the bias of order $O(n^{-\lambda})$ for some $1/2 < \lambda \leq 1$, then for $\lambda < 1$, the bias reduction of T_{nJ} is not that effective (Exercise 11.6.4). If, however, we have a *second-order representation*

$$T_n - \theta(F) = \frac{1}{n}\sum_{i=1}^{n}\psi_{1F}(X_i) + \frac{1}{n^2}\sum_{i=1}^{n}\sum_{j=1}^{n}\psi_{2F}(X_i, X_j) + R_{2n},$$

where ψ_{1F} and ψ_{2F} are centered at zero and are orthogonal to each other, then it follows that (see Exercise 11.6.5)

$$T_n - \theta(F) = \frac{1}{n}\sum_{i=1}^{n}\psi_{1F}(X_i) + \frac{1}{n}\left[\frac{1}{n}\sum_{i=1}^{n}\psi_{2F}(X_i, X_i)\right] + \frac{n-1}{n}U_n^{(2)} + R_{nJ}, \tag{11.6.18}$$

where

$$U_n(2) = \binom{n}{2}^{-1}\sum_{1\leq i<j\leq n}\psi_{2F}(X_i, X_j)$$

and $\mathbb{E}[\psi_{1f}(X_i)] = \mathbb{E}[\psi(X_i, X_j)] = 0$, while $\psi_{2F}(X_i, X_j)$ may have an expectation different from 0. As such, the centering of $T_{nJ} - \theta(F)$ is $O(n^{-1})$. Therefore, it follows that

$$T_{ni} - \theta(F) = \psi_{1F}(X_i) - \frac{1}{n-1}\left[\psi_{2F}(X_i, X_i) - \frac{1}{n}\sum_{r=1}^{n}\psi_{2F}(X_r, X_r)\right]$$
$$+ \frac{n-1}{n}U_n^{(2)} - \frac{n-2}{n-1}U_{n-1,i}^{(2)} + O_p(n^{-3/2}),$$

where $U_{n-1,i}^{(2)}$ corresponds to $U_n^{(2)}$ deleting the ith observation. It follows that the asymptotic bias of T_{nJ} is $O(n^{-3/2}) = o(n^{-1})$.

Let us recall that $T_n = T(X_1, \ldots, X_n)$ and F_n is the empirical distribution function of the X_i. Let $X_1^*, \ldots, X_n^*$ be n independent observations drawn with replacement in an equal probability sampling (SRSWR) scheme from $X_1, \ldots, X_n$; we say that $X_1^*, \ldots, X_n^*$ are (conditionally) *i.i.d.* random variables drawn from F_n. Then, let

$$T_n^* = T(X_1^*, \ldots, X_n^*). \tag{11.6.19}$$

Also let us draw M such independent sets of n observations, denoted by $\boldsymbol{X}_i^* = (X_{i1}^*, \ldots, X_{in}^*)'$, and let the corresponding T_n^* be denoted by T_{ni}^*, $i = 1, \ldots, M$. Let then

$$G_{nM}^*(y) = M^{-1} \sum_{i=1}^{M} I[n^{1/2}(T_{ni}^* - T_n) \leq y], \quad y \in \mathbb{R}, \tag{11.6.20}$$

and

$$G_n(y) = P[n^{1/2}(T_n - \theta) \leq y], \quad y \in \mathbb{R}. \tag{11.6.21}$$

Also, let

$$v_{nB}^2 = \frac{1}{M} \sum_{i=1}^{M} n(T_{ni}^* - T_n)^2. \tag{11.6.22}$$

Then, under appropriate regularity conditions, for large M (and n),

$$\|G_{nM}^* - G_n\| \stackrel{P}{\longrightarrow} 0, \tag{11.6.23}$$

and the *bootstrap* variance estimator $v_{MB}^{*2} \stackrel{P}{\longrightarrow} v^2(F)$. Perhaps, it will be convenient for us to motivate this result through the functional approach is Section 11.5. We denote the empirical distribution function for $X_1^*, \ldots, X_n^*$ by F_n^*, so that corresponding to the M copies of the bootstrap samples, we have the empirical distribution functions $F_{n1}^*, \ldots, F_{nM}^*$. As in (11.5.6), we take $T_n = T(F_n)$, so that

$$T_{ni}^* = T(F_{ni}^*), \quad i = 1, \ldots, M, \tag{11.6.24}$$

which are (conditionally on $X_1, \ldots, X_n$ being given) i.i.d. random variables. Thus, by (11.6.22) and (11.6.24), we have

$$\begin{aligned}
v_{nB}^{*2} &= \frac{1}{M} \sum_{i=1}^{M} n[T(F_{ni}^*) - T(F_n)]^2 \\
&= \int_{-\infty}^{\infty} x^2 \mathrm{d}G_{nM}^*(y) \\
&= \int_{-\infty}^{\infty} x^2 \mathrm{d}G_n(y) + \int_{-\infty}^{\infty} x^2 \mathrm{d}[G_{nM}^*(y) - G_n(y)] \\
&= \mathbb{E}[n(T_n - \theta)^2] + \frac{1}{\sqrt{M}} \int_{-\infty}^{\infty} x^2 \mathrm{d}\{\sqrt{M}[G_{nM}^*(y) - G_n(y)]\},
\end{aligned} \tag{11.6.25}$$

where the first term converges to $\nu^2(F)$, and under quite general regularity conditions, conditionally on F_n, when n and M are large,

$$M^{1/2}(G^*_{nM} - G_n) \xrightarrow{D} W, \tag{11.6.26}$$

where W is a tied down Gaussian process. Thus, by (11.6.25) and (11.6.26), we have for large M (and n),

$$v^{*2}_{nB} = \mathbb{E}[n(T_n - \theta)^2] + O_p\left(\frac{1}{\sqrt{M}}\right) \xrightarrow{P} \nu^2(F). \tag{11.6.27}$$

Actually, (11.6.23) is a direct consequence of (11.6.26), so that for both (11.6.23) and (11.6.27), the weak convergence result in (11.6.26) provides the basic key. On the other hand, noting that $n^{1/2}(F_n - F)$ converges weakly to a Gaussian process, so that $n^{1/2}\|F_n - F\| = O_p(1)$, by imposing appropriate smoothness conditions of F, (11.6.26) can be studied as in Section 11.4 under the conditional setup (given F_n). We omit these details. There are some situations (e.g., sample quantiles) where bootstrapping may work better than jackknifing. But, in general, they perform equally well in a majority of regular cases.

The resampling plans become more involved for regression and other models where the observable random elements are not identically distributed. For a regression model $Y_i = \boldsymbol{\beta}'\boldsymbol{x}_i + e_i$, $1 \le i \le n$, the errors e_i are i.i.d. but for $\boldsymbol{\beta} \neq \boldsymbol{0}$, the Y_i are not identically distributed. Thus, if we consider the *weighted empirical process*

$$\boldsymbol{F}_{nw}(y) = \boldsymbol{Q}_n^{-1/2}\sum_{i=1}^{n}(\boldsymbol{x}_i - \overline{\boldsymbol{x}}_n)I(Y_i \le y), \quad y \in \mathbb{R},$$

where $\boldsymbol{Q}_n = (\boldsymbol{x}_i - \overline{\boldsymbol{x}}_n)(\boldsymbol{x}_i - \overline{\boldsymbol{x}}_n)'$, weak convergence properties of $\boldsymbol{F}_{nw}$ will depend on the unknown $\boldsymbol{\beta}$. If $\boldsymbol{\beta} = \boldsymbol{0}$, then $\boldsymbol{F}_{nw}$ will have a similar weak convergence pattern (as the empirical process in the i.i.d. case) and jackknifing will be readily applicable. This prompted researchers to formulate some weighted jackknife methods that are not free from arbitrariness. There is a modified procedure: Use the LSE of $\boldsymbol{\beta}$, say $\widehat{\boldsymbol{\beta}}_n$, and define the residuals $\widehat{e}_i = Y_i - \widehat{\boldsymbol{\beta}}_n'\boldsymbol{x}_i$, $1 \le i \le n$. In the definition of $\boldsymbol{F}_{nw}$, replace Y_j by $\widehat{e}_j$ and denote it by $\widehat{\boldsymbol{F}_{nw}}$. Use this jackknife methods, albeit a bit heuristically, on $\widehat{\boldsymbol{F}_{nw}}$ and then construct the jackknife estimators and their dispersion matrix estimator. This idea also lend itself to bootstrapping with resampling of the residuals. Theoretical adjustments are necessary, and they, in turn, demand some extra regularity conditions on these regression residuals. Compactness and weak invariance principles are of good help in this context.

We conclude this section with some remarks on the EM-algorithms that are viewed as viable alternatives to the method of scoring. There are essentially two aspects of convergence to validate and justify in this context: (i) convergence of the iterative process to the desired solution (which is itself stochastic) and (ii) stochastic convergence of the iteratively obtained solution to its population counterparts. The second aspect is largely asymptotic (and belongs to our treatise) while the first aspect is mainly computational and merits a much more critical appraisal. The EM-algorithm has two steps: *E step* – computation of the expectation and *M step* – maximization. These two steps have been well motivated; however, the theory surrounding the algorithm still awaits for a fully generalized and unified treatment. Neither (i) nor (ii) is guaranteed and the complexities of the models pertain more toward the lack of optimality or desirability of EM algorithms. In Example 11.4.3, through life-testing model we have motivated some aspects of survival analysis and outlined some simple truncation and censoring schemes. More general counting process–based models were introduced in Section 5.4. In the rest of that section we outlined how (local) martingale methods apply to such counting processes. Similar models arise in *competing risks* where two or more risk factos, not necessarily independent, simultaneously introduce some incompleteness in the observable random elements. Missing data models are also prevalent in the literature. To a much lesser extent, grouped data models involve some loss of statistical information due to grouping. In all these cases, the classical likelihood principle, such as the *partial-*, *pseudo-*, *incomplete-*, *conditional-*, *profile-* and *quasi-*likelihood methods, have been advocated. Some of these methods invoke some degrees of subjective intervention and more notably loss of information, even the asymptotic optimality of the classical likelihood methods. In genetics models too, *genotype–phenotype* complexity entails similar incompleteness. From computation point of view, the method of scoring described in Section 8.3, may encounter impasses. An alternative approach, called the EM-algorithm [Dempster, Laird, and Rubin (1977)], has been extensively used in such incomplete data models. A complete theoretical and methodological justification of EM algorithms has yet to emerge in full generality – the developments are piecemeal and tailored to specific areas of application. This being outside the level of our book will not be covered here. We refer to McLachlan and Krishnan (1997) and Lange (2002) for some useful accounts.

11.7 Strong Invariance Principles

The invariance principle studied in Section 11.2 [see (11.2.12)] is called the *weak invariance principle*. There are some important applications in Statistics (especially, in sequential analysis) where a stronger mode is more convenient. Following Skorokhod (1956) we may consider this as follows.

Let $\{X_i, i \geq 1\}$ be a sequence of i.i.d. random variables with a distribution function F, defined on $\mathbb{R}$, and for simplicity of presentation, we take $\mathbb{E}(X) = 0$ and $\mathbb{V}\mathrm{ar}(X) = 1$. Consider then the partial sum sequence

$$S_n = X_1 + \cdots + X_n, \quad n \geq 1; \quad S_0 = 0. \tag{11.7.1}$$

As in Section 8.3, let us consider a Wiener process $\boldsymbol{W} = \{W(t), t \in [0, \infty)\}$. Then the *Skorokhod embedding of Wiener process* asserts that there exists a sequence $\{T_i, i \geq 1\}$ of nonnegative and independent random variables, such that for every $n \geq 1$,

$$\begin{pmatrix} X_1 \\ X_2 \\ \vdots \\ X_n \end{pmatrix} \stackrel{D}{=} \begin{pmatrix} W(T_1) \\ W(T_1 + T_2) - W(T_1) \\ \vdots \\ W(\sum_{i=1}^{n} T_i) - W(\sum_{i=1}^{n-1} T_i) \end{pmatrix} \tag{11.7.2}$$

and

$$\mathbb{E}(T_i) = \mathbb{V}\mathrm{ar}(T_i) = 1, \quad i \geq 1. \tag{11.7.3}$$

This embedding has been extended not only to nonidentically distributed random variables, but also to martingales and some other dependent sequences. Komlós, Major, and Tusnády (1975) have shown that whenever the X_i are independent with mean 0 and variances equal to 1 and have finite-moment generating functions (in a neighborhood of 0), as $n \to \infty$,

$$S_n = W(n) + O(\log n) \quad \text{a.s.,} \tag{11.7.4}$$

while Strassen (1967), under a fourth moment condition on the X_i, showed that as $n \to \infty$,

$$S_n = W(n) + O[(n \log\log n)^{1/4}(\log n)^{1/2}] \quad \text{a.s.} \tag{11.7.5}$$

Strassen's results also pertain to martingales under a very mild regularity condition on their variance function.

Almost sure invariance principles have also been established for empirical processes. Consider a two-parameter Gaussian process $\boldsymbol{Y} = \{Y(s,t) : 0 \le s < \infty, 0 \le t < \infty\}$, such that $\mathbb{E}(\boldsymbol{Y}) = 0$ and

$$\mathbb{E}[Y(s,t)Y(s',t')] = \min(s,s')\min(t,t'), \quad \forall (s,t),(s',t'). \tag{11.7.6}$$

$\boldsymbol{Y}$ is called a *Brownian sheet*. Then, let

$$\boldsymbol{Y}^* = \{Y^*(s,t), \quad 0 \le s \le 1,\ 0 \le t < \infty\}$$

be defined by

$$Y^*(s,t) = Y(s,t) - sY(1,t), \quad 0 \le s \le 1, 0 < t < \infty. \tag{11.7.7}$$

$\boldsymbol{Y}^*$ is called a *Kiefer process*. Consider now the reduced empirical process $\boldsymbol{Z}^* = \{Z^*(s,t);\ 0 \le s \le 1, t \in [0,\infty)\}$ defined by

$$Z^*(s,t) = Z_{[t]}(s), \quad s \in [0,1], \quad t \in \mathbb{R}^+, \tag{11.7.8}$$

where $\boldsymbol{Z}_{[t]} \equiv 0$ for $0 < t < 1$. Then, as $t \to \infty$

$$\sup_{0 \le s \le 1} |Z^*(s,t) - Y^*(s,t)| = O((\log t)^2), \quad \text{a.s.} \tag{11.7.9}$$

Note that in (11.7.4)–(11.7.5), $\{S_n\}$ and $\boldsymbol{W}$ are not necessarily defined on a common probability space, and similarly in (11.7.9), $\boldsymbol{Z}^*$ and $\boldsymbol{Y}^*$ may not be defined on a common probability space. However, these a.s. invariance principles allow us to replace the S_n (or $\boldsymbol{Z}^*$) by $\boldsymbol{W}$ (or $\boldsymbol{Y}^*$) for diverse probabilistic analyses, and, hence, they have great scope in various asymptotic analyses. For some of these applications, see Sen (1981).

11.8 Concluding Notes

The topics covered in this chapter may deserve a more sophisticated rigorous treatment, but that would have been a violation of the uniformity of the level of presentation of the current treatise. We hope our readers could understand our motivation in providing, mostly, a basic survey of these specialized (and advanced) topics at an intermediate level. The reader may skip a greater part of this chapter and yet may have a comprehensive account of the major undercurrents. The examples cited throughout this chapter are, of course, geared more toward potential applications. Therefore, we suggest that one should look into them, even if one may not be interested in the theoretical aspects. Nevertheless these topics, treated in more generality in Parthasarathy (1967) and Billingsley (1968), help in acquiring a better understanding of the modern methodology in asymptotic methods in statistics, and we sincerely hope that the survey outlined in this chapter makes it convenient to look into the directions for further reading with a view to incorporating more of these basic tools into the main stream of large-sample methods in statistics. However, we like to make it a point of distinction: mathematical and probabilistic tools are indispensable in the study of large sample methods, although the emphasis should be primarily on the applicable methodology, not merely on abstract concepts that may not be appealing to statisticians at large.

11.9 Exercises

11.3.1 Let W_n be defined as in (11.3.2)–(11.3.4). Then show that $\{W_n(t), t \geq 0\}$ is a martingale. Hence, for every (fixed) s (≥ 0), $\{W_n(s+t) - W_n(s), t \geq 0\}$ is a martingale, so that $\{[W_n(s+t) - W_n(s)]^2, t \geq 0\}$ is a submartingale.

11.3.2 Show that (11.3.24) holds, that is,

$$\sup_{0 \leq u \leq \delta} [|W_n(s+u) - W_n(s)|] \overset{D}{=} \max_{0 \leq j \leq m} [n^{-1/2}|S_j|],$$

where $m = [n\delta]$. Hence, use Theorem 6.4.1 to verify that (11.3.20) holds when the X_j [in (11.3.1)] have finite moments up to the order r, for some $r > 2$.

11.3.3 Verify (11.3.28) for the right-hand side of (11.3.25).

11.3.4 Use Exercises 11.3.1 and 11.3.2 to verify the tightness part of Theorem 11.3.2.

11.4.1 Verify (11.4.14).

11.4.2 Verify (11.4.29). (You may use partial integration.)

11.4.3 For Z_n^* defined by (11.4.47), verify that the tightness of $Z_{n_1}^{(1)}$ and $Z_{n_2}^{(2)}$ ensure the same for Z_n^*.

11.4.4 Verify (11.4.59) from (11.4.58).

11.5.1 Verify (11.5.26).

11.6.1 Verify (11.6.3).

11.6.2 Verify (11.6.4).

11.6.3 Verify (11.6.5).

11.6.4 Let $\mathbb{E}(T_n) = \theta + n^{-\lambda}a + n^{-\lambda-1}b + \cdots$, for some $\lambda > 1/2$. Assume the same pattern for T_{n-1}. Then, show that

$$\mathbb{E}[nT_n - (n-1)T_{n-1}] = \theta + a[n^{1-\lambda} - (n-1)^{1-\lambda}] + b[n^{-\lambda} - (n-1)^{-\lambda}] + \cdots.$$

Also, use the expansion

$$(n-1)^\gamma = n^\gamma\left(1 - \frac{1}{n}\right)^\gamma = n^\gamma - \gamma n^{\gamma-1} + \frac{\gamma(\gamma-1)}{2}n^{\gamma-2} + \cdots$$

and show that

$$\mathbb{E}[nT_n - (n-1)T_{n-1}] = \theta + a(1-\lambda)n^{-\lambda} + O(n^{-\lambda-1}).$$

Thereby, conclude that the bias of T_{nj} is $O(n^{-\lambda})$ if $\lambda < 1$ and $O(n^{-2})$ if $\lambda = 1$.

11.6.5 Verify (11.6.18).

Bibliography

Aalen, O. O. (1978). Nonparametric inference for a family of counting processes, *Annals of Statistics* **6**(4): 701–726.

Andersen, P. K., Borgan, Ø., Gill, R., and Keiding, N. (1996). *Statistical Models Based on Counting Processes*, Springer-Verlag, New York.

Anderson, T. W. (1960). A modification of the sequential probability ratio test to reduce the sample size, *Annals of Mathematical Statistics* **31**: 165–197.

Anscombe, F. J. (1948). The transformation of Poisson, binomial and negative binomial data, *Biometrika* **35**: 246–254.

Anscombe, F. J. (1952). Large sample theory of sequential estimation, *Proceedings of the Cambridge Philosophical Society*, Vol. 48, pp. 600–607.

Armitage, P. (1975). *Sequential Medical Trials*, John Wiley, New York.

Bahadur, R. R. (1966). A note on quantiles in large samples, *Annals of Mathematical Statistics* **37**: 577–580.

Barlow, R., Bartholomew, D., Bremmer, J. M., and Brunk, H. D. (1972). *Statistical Inference under Order Restrictions*, John Wiley, New York.

Berger, J. O. (1993). *Statistical Decision Theory and Bayesian Analysis*, Springer-Verlag, New York.

Berry, A. C. (1941). The accuracy of the Gaussian approximation to the sum of independent variates, *Transactions of the American Mathematical Society* **49**: 122–136.

Bhapkar, V. P. (1966). A note on the equivalence of two test criteria for hypotheses in categorical data, *Journal of the American Statistical Association* **61**: 228–235.

Bhattacharya, A. (1946). On some analogues of the information and their use in statistical estimation, *Sankhyā* **8**: 1–14.

Billingsley, P. (1968). *Convergence of Probability Measures*, John Wiley, New York.

Billingsley, P. (1995). *Probability and Measure*, John Wiley & Sons, New York.

Brown, B. M. (1971a). Martingale central limit theorems, *Annals of Mathematical Statistics* **42**: 59–66.

Brown, B. M. (1971b). A note on convergence of moments, *Annals of Mathematical Statistics* **42**: 777–779.

Chapman, D. G. and Robbins, H. (1951). Minimum variance estimation without regularity assumptions, *Annals of Mathematical Statistics* **22**: 581–586.

Chiang, C. L. (1980). *An Introduction to Stochastic Processes and Their Applications*, Krieger, New York.

Chow, Y. S. and Robbins, H. (1965). On the asymptotic theory of fixed-width sequential confidence intervals for the mean, *Annals of Mathematical Statistics* **36**: 457–452.

Chow, Y. S. and Teicher, H. (1978). *Probability Theory: Independence, Interchangeability, Martingales*, Springer-Verlag, New York.

Cox, D. R. (1972). Regression models and life tables (with discussion), *Journal of the Royal Statistical Society B* **17**: 187–220.

Cox, D. R. and Hinkley, D. V. (1974). *Theoretical Statistics*, Chapman and Hall, London.

Cox, D. R. and Oakes, D. (1984). *Analysis of Survival Data*, Chapman and Hall, London.

Cramér, H. (1946). *Mathematical Methods of Statistics*, Princeton University Press, New Jersey.

Daniels, H. A. (1945). The statistical theory of the strength of bundles of threads, *Proceedings of the Royal Society of London A*, vol. 183, pp. 405–435.

Dantzig, G. B. (1940). On the non-existence of tests of student's hypothesis having power functions independent of σ, *Annals of Mathematical Statistics* **11**: 186–192.

David, H. A. (1970). *Order Statistics*, John Wiley & Sons, New York.

Dempster, A. P., Laird, N. M., and Rubin, D. B. (1977). Maximum likelihood from incomplete data via the em algorithm, *Journal of the Royal Statistical Society B* **39**: 185–197.

Diggle, P., Heagerty, P., Liang, K.-Y., and Zeger, S. (2002). *Analysis of Longitudinal Data*, Oxford University Press, Oxford.

Dvoretzky, A. (1971). Asymptotic normality of sums of dependent random variables, *Proceedings of the Sixth Berkeley Symposium on Mathematical Statistics and Probability, Vol. 2*, University of California Press, Berkeley, pp. 513–536.

Eaton, M. L. (2007). *Multivariate Statistics: A vector space approach*, Vol. 53 of *Lecture Notes – Monograph Series*, Institute of Mathematical Statistics.

Efron, B. (1979). Bootstrap methods: another look at the jackknife, *Annals of Statistics* **7**: 1–26.

Efron, B. and Tibshirani, R. J. (1993). *An Introduction to the Bootstrap*, Chapman & Hall, New York.

Eicker, F. (1967). Limit theorems for regression with unequal and dependent errors, *Proceedings of the Fifth Berkeley Symposium on Mathematical Statistics and Probability, Vol. 1*, University of California Press, Berkeley, pp. 59–82.

Eubank, R. (1999). *Nonparametric regression and spline smoothing*, Marcel Dekker, New York.

Fahrmeir, L. and Kaufmann, H. (1985). Consistency and asymptotic normality of the maximum likelihood estimator in generalized linear models, *Annals of Statistics* **13**: 342–368.

Feller, W. (1971). *An Introduction to Probability Theory and its Applications*, vol. 2, 2nd ed., John Wiley & Sons, New York.

Fernholz, L. T. (1983). *Von Mises Calculus for Statistical Functionals*, Vol. 19 of *Lecture Notes in Statistics*, Springer-Verlag, New York.

Fleming, T. R. and Harrington, D. P. (2005). *Counting Processes and Survival Analysis*, John Wiley & Sons, New York.

Foutz, R. V. (1977). On the unique consistent solution to the likelihood equations, *Journal of the American Statistical Association* **72**: 147–148.

Fuller, W. A. (1986). *Measurement Error Models*, John Wiley & Sons, New York.

Gallant, A. R. (1987). *Nonlinear Statistical Models*, John Wiley & Sons, New York.

Gangopadhyay, A. K. and Sen, P. K. (1990). Bootstrap confidence intervals for conditional quantile functions, *Sankhya A* **52**: 346–363.

Gangopadhyay, A. K. and Sen, P. K. (1992). Contiguity in nonparametric estimation of a conditional functional. In *Nonparametric Statistics and Related Topics*, A. E. Saleh (ed.), North-Holland, Amsterdam, pp. 141–162.

Gayen, A. K. (1951). The frequency distribution of the product moment correlation coefficient in random samples of any size drawn from non-normal universes, *Biometrika* **38**: 219–247.

Gnedenko, B. V. (1943). Sur la distribution limite du terme maximum d'une série aléatoire, *Annals of Mathematics* **44**: 423–453.

Godambe, V. P. and Heyde, G. C. (1987). Quasi-likelihood and optimal estimation, *International Statistical Review* **55**: 231–244.

Grizzle, J. E., Starmer, C. F., and Koch, G. G. (1969). The analysis of categorical data by linear models, *Biometrics* **25**: 489–504.

Gumbel, E. J. (1958). *Statistics of Extremes*, Columbia University Press, New York.

Hájek, J. (1968). Asymptotic normality of simple linear rank statistics under alternatives, *Annals of Mathematical Statistics* **39**: 325–346.

Halmos, P. R. (1946). The theory of unbiased estimation, *Annals of Mathematical Statistics* **17**: 34–43.

Hartigan, J. A. (1971). Error analysis by replaced samples, *Journal of the Royal Statistical Society B* **33**: 98–110.

Hedges, L. V. and Olkin, I. (1985). *Statistical Methods for Meta-Analysis*, Academic Press, San Diego, CA.

Hoeffding, W. (1948). A class of statistics with asymptotically normal distribution, *Annals of Mathematical Statistics* **19**: 293–325.

Hoeffding, W. (1963). Probability inequalities for sums of bounded random variables, *Journal of the American Statistical Association* **58**: 13–30.

Huber, P. J. (1964). Robust estimation of a location parameter, *Annals of Mathematical Statistics* **35**: 73–101.

Huber, P. J. (1981). *Robust Statistics*, John Wiley & Sons, New York.

James, W. and Stein, C. (1961). Estimation with quadratic loss, *Proceedings of the Fourth Berkeley Symposium on Mathematical Statistics and Probability, Vol. 1*, University of California Press, Los Angeles, pp. 311–319.

Jurečková, J. (1977). Asymptotic relations of M-estimates and R-estimates in linear regression model, *Annals of Statistics* **5**: 464–472.

Jurečková, J. and Sen, P. K. (1996). *Robust Statistical Procedures: Asymptotics and Interrelations*, John Wiley & Sons, New York.

Kass, R. and Raftery, A. (1995). Bayes factors, *Journal of the American Statistical Association* **90**: 777–795.

Keating, J. P., Mason, R. L., and Sen, P. K. (1993). *Pitman's Measure of Closeness: A Comparison of Statistical Estimators*, SIAM, Philadelphia.

Kiefer, J. (1952). On minimum variance estimators, *Annals of Mathematical Statistics* **23**: 627–629.

Koch, G. G., Imrey, P. B., Singer, J. M., Atkinson, S. S., and Stokes, M. E. (1985). *Analysis of Categorical Data*, Les Presses de l'Uni-versité de Montréal, Montréal.

Komlós, J., Major, P., and Tusnády, G. (1975). An approximation of partial sums of independent RV's and the sample DF.I, *Zeitschrift für Wahrscheinlichkeitstheorie und Verwandte Gebiete* **32**: 111–131.

Lange, K. (2002). *Mathematical and Statistical Methods for Genetic Analysis*, 2nd ed., Springer Verlag, New York.

Lehmann, E. L. (1953). The power of rank tests, *Annals of Mathematical Statistics* **24**: 23–43.

Lehmann, E. L. (1983). *Theory of Point Estimation*, John Wiley & Sons, New York.

Lehmann, E. L. (1986). *Testing Statistical Hypotheses*, John Wiley & Sons, New York.

Liang, K.-Y. and Zeger, S. L. (1986). Longitudinal data analysis using generalized linear models, *Biometrika* **73**: 13–22.

Loynes, R. M. (1970). An invariance principle for reversed martingales, *Proceedings of the American Mathematical Society*, vol. 25, American Mathematical Society, pp. 56–64.

McCullagh, P. and Nelder, J. A. (1989). *Generalized Linear Models*, 2nd ed., Chapman and Hall, London.

McLachlan, G. J. and Krishnan, T. (1997). *The EM Algorithm and Extensions*, John Wiley & Sons, New York.

McLeish, D. L. (1974). Dependent central limit theorems and invariance principles, *Annals of Probability* **2**: 620–628.

Mood, A. M. (1941). On the joint distribution of the median in samples from a multivariate population, *Annals of Mathematical Statistics* **12**: 268–278.

Mosteller, F. (1946). On some useful "inefficient" statistics, *Annals of Mathematical Statistics* **17**: 377–408.

Nelder, J. A. and Wedderburn, R. W. M. (1972). Generalized linear models, *Journal of the Royal Statistical Society A* **135**: 370–384.

Nelson, W. (1969). Hazard plotting for incomplete failure data, *Journal of Quality Technology* **1**: 25–52.

Neyman, J. and Scott, E. L. (1948). Consistent estimates based on partially consistent observations, *Econometrica* **16**: 1–32.

Parthasarathy, K. R. (1967). *Probability Measures on Metric Spaces*, Academic Press, New York.

Pearson, K. (1900). On the criterion that a given system of deviations from the probable in the case of a correlated system of variables is such that it can be reasonably supposed to have arisen from random sampling, *Philosophy Magazine* **50**: 157–172.

Pitman, E. J. G. (1937). The closest estimate of statistical parameters, *Proceedings of the Cambridge Philosophical Society*, vol. 33, pp. 212–222.

Pratt, J. W. (1981). Concavity of the log likelihood, *Journal of the American Statistical Association* **76**: 103–106.

Puri, M. L. and Sen, P. K. (1985). *Nonparametric Methods in General Linear Models*, John Wiley, New York.

Pyke, R. and Root, D. (1968). On convergence in r-mean normalized partial sums, *Annals of Mathematical Statistics* **39**: 379–381.

Quenouille, M. (1949). Approximate tests of correlation in time series, *Journal of the Royal Statistical Society B* **11**: 18–44.

Robert, C. P. and Casella, G. (2004). *Monte Carlo Statistical Method*, 2nd ed., Springer-Verlag, New York.

Roy, S. N. (1953). On a heuristic method of test construction and its use in multivariate analysis, *Annals of Mathematical Statistics* **24**: 220–238.

Sarhan, A. E. and Greenberg, B. (1962). *Contributions to Order Statistics*, John Wiley & Sons, New York.

Searle, S. R. (1971). *Linear Models*, John Wiley & Sons, New York.

Sen, P. K. (1960). On some convergence properties of U-statistics, *Calcutta Statistical Association Bulletin* **10**: 1–18.

Sen, P. K. (1964). On some properties of the rank weighted means, *Journal of the Indian Society of Agricultural Statistics* **16**: 51–61.

Sen, P. K. (1968a). Estimation of regression coefficients based on Kendall's tau, *Journal of the American Statistical Association* **63**: 1379–1389.

Sen, P. K. (1968b). Robustness of some nonparametric procedures in linear models, *Annals of Mathematical Statistics* **39**: 1913–1922.

Sen, P. K. (1977). Some invariance principles relating to jackknifing and their role in sequential analysis, *Annals of Statistics* **5**: 316–329.

Sen, P. K. (1981). *Sequential Nonparametrics: Invariance Principles and Statistical Inference*, John Wiley & Sons, New York.

Sen, P. K. (1985). *Theory and Applications of Sequential Nonparametrics*, SIAM, Philadelphia.

Sen, P. K. (2007). Union-intersection principle and constrained statistical inference, *Journal of Statistical Planning and Inference* **137**: 3741–3752.

Sen, P. K., Bhattacharyya, B. B., and Suh, M. W. (1973). Limiting behavior of the extrema of certain sample functions, *Annals of Statistics* **1**: 297–311.

Sen, P. K. and Ghosh, M. (1976). Comparison of some bounds in estimation theory, *Annals of Statistics* **4**: 1247–1257.

Sen, P. K. and Singer, J. M. (1993). *Large Sample Methods in Statistics: An Introduction with Applications*, Chapman and Hall, New York.

Serfling, R. J. (1980). *Approximation Theorems of Mathematical Statistics*, John Wiley & Sons, New York.

Silvapulle, M. J. (1981). On the existence of maximum likelihood estimators for the binomial response models, *Journal of the Royal Statistical Society B* **43**: 310–313.

Silvapulle, M. J. and Burridge, J. (1986). Existence of maximum likelihood estimates in regression models for grouped and ungrouped data, *Journal of the Royal Statistical Society B* **48**: 100–106.

Silvapulle, M. J. and Sen, P. K. (2004). *Constrained Statistical Inference: Inequality, Order, and Shape Restrictions*, John Wiley & Sons, New York.

Silvey, S. D. (1959). The Lagrangian multiplier test, *Annals of Mathematical Statistics* **30**: 389–407.

Singer, J. M. and Andrade, D. F. (1997). Regression models for the analysis of pretest/posttest data, *Biometrics* **53**: 729–735.

Singer, J. M., Pedroso de Lima, A. C., Tanaka, N. I., and González-López, V. A. (2007). To triplicate or not to triplicate? *Chemometrics and Intelligent Laboratory Systems* **866**: 82–85.

Singer, J. M. and Sen, P. K. (1985). M-methods in multivariate linear models, *Journal of Multivariate Analysis* **17**: 168–184.

Skorokhod, A. V. (1956). Limit theorems for stochastic processes, *Theory of Probability and Applications* **1**: 261–290.

Stein, C. (1945). A two sample test for a linear hypothesis whose power is independent of the variance, *Annals of Mathematical Statistics* **16**: 243–258.

Stein, C. (1956). Inadmissibility of the usual estimator for the mean of a multivariate normal distribution. In *Proceedings of the Third Berkeley Symposium on Mathematical Statistics and Probability, Vol. 1*, J. Neyman (ed.), University of California Press, Los Angeles, pp. 187–195.

Stone, C. J., Hansen, M. H., Koopenberg, G., and Truong, Y. K. (1997). Polynomial splines and their tensor products in extended linear modeling, *Annals of Statistics* **25**: 1371–1425.

Strassen, V. (1967). Almost sure behavior of sums of independent random variables and martingales, *Proceedings of the Fifth Berkeley Symposium in Mathematical Statistics and Probability, Vol. 2*, University of California Press, Berkeley, pp. 315–343.

Thompson, J. R. and Tapia, R. A. (1990). *Nonparametric Function Estimation, Modelling and Simulation*, SIAM, Philadelphia.

Tucker, H. G. (1967). *A Graduate Course in Probability*, Academic Press, New York.

Tukey, J. W. (1958). Bias and confidence in not quite large samples, *Annals of Mathematical Statistics* **29**: 614. Abstract.

van Beeck, P. (1972). An application of Fourier methods to the problem of sharpening the Berry-Eeséen inequality, *Zeitschrift für Wahrscheinlichkeitstheorie und Verwandte Gebiete* **23**: 187–197.

von Bahr, B. (1965). On the convergence of moments in the central limit theorem, *Annals of Mathematical Statistics* **36**: 808–818.

von Mises, R. (1947). On the asymptotic distribution of differentiable statistical functionals, *Annals of Mathematical Statistics* **18**: 309–348.

Wald, A. (1943). Test of statistical hypothesis concerning several parameters when the number of observations is large, *Transactions of the American Mathematical Society* **54**: 426–482.

Wedderburn, R. W. (1976). On the existence an uniqueness of the maximum likelihood estimates for certain generalized linear models, *Biometrika* **63**: 27–32.

Zolotarev, M. R. (1967). A sharpening of the inequality of Berry-Esséen, *Zeitschrift für Wahrscheinlichkeitstheorie und Verwandte Gebiete* **8**: 332–342.

Index